6th International Conference on Energy and City of the Future (EVF'2019)

E3S Web of Conferences Volume 170 (2020)

Pune City, India
18 – 20 December 2019

Editors:

Sudarshan B. Sanap
Virendra K. Bhojwani
Mathew V K

ISBN: 978-1-7138-1292-0

Printed from e-media with permission by:

Curran Associates, Inc.
57 Morehouse Lane
Red Hook, NY 12571

Some format issues inherent in the e-media version may also appear in this print version.

For additional information, please contact EDP Sciences – Web of Conferences
at the address below.

EDP Sciences – Web of Conferences
17, Avenue du Hoggar
Parc d'Activité de Courtabœuf
BP 112
F-91944 Les Ulis Cedex A
France

Phone: +33 (0) 1 69 18 75 75
Fax: +33 (0) 1 69 28 84 91

contact-edps@webofconferences.org

Additional copies of this publication are available from:

Curran Associates, Inc.
57 Morehouse Lane
Red Hook, NY 12571 USA
Phone: 845-758-0400
Fax: 845-758-2633
Email: curran@proceedings.com
Web: www.proceedings.com

TABLE OF CONTENTS

Preface

The 6th Edition of the International Conference on **Energy and City of the Future" EVF2019"** took place from Dec 18th to Dec 20th, 2019, at MIT-ADT University, Pune City in India. **The EVF Series "Energy and City of the Future" EVF' is created by ECAM-EPMI, Cergy-Pontoise (France), University of Lorraine (France), and Queens University (Canada).** EVF Series Chairpersons are Prof. Moumen DARCHERIF, Director-General ECAM-EPMI, (France), Prof. Mohammed EL GANAOUI, University de Loraine (France), and Prof. Jean-Michel NUNZI, Queen's University (Canada).

The 6th International Conference on Energy and City of the Future, EVF2019 was co-organized under the leadership of Dr. Vishwananth KARAD, Founder of MIT group of Institution, Dr. Mangesh KARAD, Executive President, MITADT University, Prof. Sunita KARAD (Director, MITCOM), Dr. Sunil RAI, Vice-Chancellor of MITADT University, Dr. Kishore RAVANDE, Principal, MITSOE, and Dr. Sudarshan SANAP, HOD, Mechanical, MITSOE the Convenor for the conference. The conference had seven different themes/tracks in various domains ranging from, 1. Factories of the future, 2. Energy and Management, 3. Waste management, 4. Cultural infrastructure, 5. Transportation, 6. E-Health and 7. City and Building of the Future.

The EVF aims to bring together academic researchers, industry, scientists, and city political actors to exchange their experiences and research results about all aspects (historical, societal, environmental, conceptual, methodological, practices.) dealing with the Energy and the City of the Future. The previous editions were organized in Cergy-Pontoise, France (2014), Tetouan Morrocco (2015), Queens, Canada (2016), Longwy, France, (2017) and Fès, Morrocco (2018)"

It was a great honor and privilege for MIT-ADT University to have the Nobel Prize Winner and Emeritus Professor of the Built Environment, University of Manchester, Hon'ble. Dr. Geoffrey LEVERMORE as Chief Guest, who graced the occasion with this august presence. He is amongst the prominent keynote speakers for the conference who talked about climate change and the urban heat island effect.

Mr. Rajendra SHENDE, Former Director United Nations Environment Programme (UNEP) and Chairman, TERRE Policy Centre, enlighten the audience with 17 different Sustainable development goals (SDGs) and the Road to The Future; Making University SDGs Ready.". Dr. K. G. NARAYANKHEDKAR who is Chancellor MGMCET, Panvel, Navi Mumbai, and Dr. D. N. MALKHEDE who is Adviser to AICTE. Other prominent persons from Industry who chaired various sessions were Dr. Dadasaheb SHENDAGE, Senior General Manager H2e Power, and Dr. HITENDRA Patel, Mechanical Architect, Philip's Health care.
The theme of the conference was to debate and present ideas required for cities of the future in line with the 17 Sustainable Development Goals set by the United Nations for the whole world.

The major goal and feature of the conference are to bring academic scientists, engineers, and industry researchers together to exchange and share their experiences and research results, and discuss the practical challenges encountered and the solutions adopted.

We are honored by EVF Series Chairperson for opting MIT ADT University as the organizer for the 6th Edition of the International Conference on Energy and City of the Future" EVF2019.

Desk of EVF Series

Cites emerge from the organization of life and communications, which constructs their unique signature. This also applies to any form of science, like mechanics, energy, and knowledge in general that builds up to understand, formalize, unify, master, use, predict and anticipate the notion of tomorrow.

Indeed, prediction of the future responds to the implicit obsession that humans develop as an alternative to the impossibility of reversing the axis of time, which forces them to predict the future quickly and far away.

In this move, points of view have been at the origin of ruptures in intellectual practices that have enabled humanity to take giant steps forward in its march towards progress and therefore towards risk.

Thermodynamics, through the notions of work and energy, has made it possible to evidence a common currency (the joule) that has always been implicitly used by humanity in its exchanges since the beginning. Progress towards abstraction led to the capitalization of the written text towards variables relevant to the economy of mind that they have fostered (enthalpy, entropy, training, etc.). Institution that would not have delivered its certainties without mathematical progress (number theory, group theory, statistics, geometry, infinitesimal calculation, differentials, implicit functions, etc.).

In fluid mechanics, Euler and Lagrange are associated in their vision of the evolution of the particle in its movement between an immediate future and a recomposed past. Two fertile points of view for a progress in terrestrial mobility achieved today without comparison and without the use of advanced theories of physics. This largely suggests what the future could look like if the potential of physics was more fully exploited.

It is therefore not strange that we find a common vocabulary between science and the city through words like Energy, Entropy, Economy, Turbulence, Light, Life, etc. this is the real link if we look at it.

Between energy, work, and their manifestations, the contours of tomorrow's city will continue to be sketched out in the movement of prediction and correction on the path of the advancement of knowledge. Boundary conditions shape the solution and confront it with reality. The challenges of tomorrow are therefore facing these borders require trade-offs and choices.

The choice includes in its definition an elimination and therefore a responsibility in the writing of the city's energy history. The role of our community will remain through the exchange and confrontation of points of view that will generate new orientations allowing a gain on the future and an acceleration of the trajectory towards the unknown.

In its present, the city contains possible insights into its future, since here and elsewhere metropolises do not take shape in the same way, citizens do not react in the same way, constraints are managed differently, sometimes bypassed, sometimes reversed in favor of the hoped-for development. The apparent contradiction invites innovation; innovation shapes lifestyles and relationships to space and time. The speed of change cannot remain a random parameter in the design of the city of the future, the human being is forced to understand and anticipate, and learned encounters are the ideal place for such an exchange.

A conference dedicated to the city of the future with the advantage of opening up a participatory reflection to all communities who, through their scientific, technical and artistic approaches, will propose points of view and participate in choices that will inspire the continuous mutation of the city of tomorrow between the freedom of dreams and the pragmatism of the laws of physics.

The city has also to ensure strong and quick adaptation to new threats affecting the continuity and the quality of life.

The singular health situation experienced by the world in 2020 has shown more than ever the eyes of a city been turned towards scientists to provide explanations, visions, recommendations and then immediate solutions, for short time followed by permanent solutions to ensure the survival of our species and then the environment for a better life. This belief in science, reason, and proof is more than ever the hope that every city of tomorrow cultivates guaranteeing life and continuity.

By
Pr. R. BENNACER,
Pr. M. EL GANAOUI,
Pr. A-M. DARCHERIF,
Pr. J-M. NUNZI
Pr. G. LEVERMORE.

Opening Address

It is with great pleasure for MIT-ADT University to organize the **6th International Conference on Energy and City of the future, EVF2019.** The quest for knowledge has been from the beginning of time but knowledge only becomes valuable when it is disseminated and applied to benefit humankind. It's the first time in India that the 6th Edition of EVF Series with ECAM-EPMI was held at MIT ADT University Campus, Pune, India. The seven themes of the conference are Energy & Management, City and Building of the future, Waste Management, Factories of the Future, e-Health, Transport & Mobility, Cultural Infrastructure. We have received around 178 full-length papers across the globe. International Authors from France, Canada, Algeria, Russia, and Bulgaria have participated. National authors from Andhra Pradesh, Punjab, Uttar Pradesh, Tamil Nadu & Maharashtra from the Institute like IITs, NIT, and COEP have participants. After a stringent and double-blind review process by International and National reviewers' panel, we had selected only 125 papers for presenting the paper at the conference.

EVF2019 consists mainly of four sessions, the keynote speeches from specially invited speakers which would inspire the participants in their future work, the oral presentations from the participants, the thematic session especially for the researchers across the globe, and the panel discussion led by the Industry experts and Academic experts. With these four sessions, people will learn and get benefited from the researches of others and the free exchange of ideas is ensured.

The aftermath of this event is this volume devoted to issues discussed during the conference.

Editors of the Conference Proceedings

Dr. A.-Moumen DARCHERIF
Dr. M. EL GANAOUI
Dr. J.-M. NUNZI

Dr. Sudarshan BSanap
Dr. VirendraK Bhojwani
Prof. Mathew V K

Panel Discussion on "Sustainable Development Goals"

The panel discussion was organized during the International conference EVF2019 to create awareness on Sustainable Development Goals from the Academic and Industry perspective. The panel discussion was led by Dr. Mohit Dubey CEO, AIC-MIT ADT Incubator Forum, Mr. Manojit Bose, Chief Knowledge Officer, Pune Smart City Development Corporation Limited, Mr. Nikhil Malhotra, Global Head of Innovation at Tech-Mahindra, Mr. Manish Mehta, Vice President & Global Delivery Executive, Capgemini, Pune, Dr. Moumen Darcheriff, Director, ECAM-EPMI, France and Dr. M. EL GANAOUI, University de Loraine (France). Each panel Member had a debate , discussion, and put forth their contributions. The perspective of a few panel members is discussed so that the audience referring to the proceeding of EVF2029 are benefited at a larger scale.

Innovation evolving between knowledge accumulation and knowledge creation

When a society loses the memory of things, it loses its historical background and compromises its present and future. When a society loses the education of new generations, it loses its present and compromises its future. When it loses the ability to produce abstraction, It loses immediately any hope of the future.

The community of education anywhere in the world has the privilege to work with abstraction as an artist is handling the primitive matter, this privilege has to be merited after a longer period of listening, learning, making, doing and correcting errors. In any technology, the prediction - correction process is a part of the way from research to usual use.

The period when artificial intelligence has to assist human intelligence has passed. Today, a real menace of losing our ability to create an abstraction to facilitate our direction towards the good sense of things is permanently present and this increases the gap between the notion of technical smartness and humanist intelligence.

It is urgent to wait and to make a new skin allowing us better communication with earth and universe. This updated layer will be our better sensor to guide us in a short way to what is really needed by a future Village, City to guarantee, health, food, security, education, and happiness to their habitats. Hope and more realistic dreams will be the immediate consequence of such quality of life.

All the gadgets making our lives easier today come from hard work done in the last centuries on mathematics, physics, chemistry ..etc. and the guidance has been working hard without waiting for any compensation and understanding 'how it worked' before 'why I'm working'.

It is evident that academic time and industrial time are not similar, and a dialogue has to be installed to create a common understanding. This can come from the elementary society's needs locally coupled to a global vision of the global system.

Pragmatic approaches, methods, philosophy can be discussed together to make a model of spirit economy taking advantage of the difference between the approaches in different success schools developing internationally.

Anywhere in the world, the time is a necessary remedy for any medication to work. While waiting for its effect, we should still encourage the exchange of methods and practices to make the dialog place in a city further and stronger.

This round table made in Pune was an open space for a French-Indian and International dialogue with a convergence on our manner of seeing the future in such described spirit, this first meeting have to be reconducted periodically and illustrated by illustrative approaches made in our respective Universities and knowledge centers.

Pr. M. El Ganaoui,

University of Lorraine, France

Head of Research Group and in Charge of International cooperation, innovation and territory valorization in Institute Henri Poincaré of Longwy/Univ. Of Lorraine.

EVF 2019 - Panel discussion - Intervention of M. DARCHERIF

Sustainable development is a major concern for all economic, political and academic actors around the world. It is a complex human concept with multiple ramifications and impacts. It is a transgenerational problem that affects our present and the future of our children. Its complexity is also due to its multi-variable character. It can be approached in different ways and admit various solutions.

From my point of view, as Professor and Academic Manager, facing the challenges and challenges of SD requires acting on at least four levers:

- *The first one, is to set a new global fighting disparities between countries. The sustainability is impossible if it's not Fair, Equitable and Responsible. We need to work together for building a better future. We share the planet, There is no south no north. More than a billion earthlings do not have access to food, while our planet has enough to feed 12 billion people.*

- *The second lever, is to Ensure perfect control of human resources, especially : energy, food and water. Technological advances are obviously needed, but it will be useless if political commitments are not made at the international level.*

- *The third one, is to Promote Innovation, Research and Entrepreneurship. We need to encourage our researchers to be part of an entrepreneurial approach, promoting the transfer of technologies and the incubation of innovative ideas.*

- *The last one is to raise awareness, to sensitize young people to the issues of tomorrow and above all to ensure them a high level education.*

This last lever is fundamental. Why ? Because the decision's makers of tomorrow are our current students. So, we must prepare them now. We have no time to loose. Our future engineers and managers must " think globally & Act locally ".

To do it, we need to adapt the content and the philosophy of our training, to connect our universities, and to promote students mobility. We must break the boundaries of knowledge. That's a big job easier said than done.

Our responsibility as trainers is complete. We need to act everywhere to better prepare our young people.

Before ending my intervention, I want to give you some data concerning French education experience.

In France we have more almost 3 million students (undergraduate, graduate and post-graduate levels), almost 70 Universities, around 300 Engineering and Business Graduates schools, and more than 4000 high schools with three main sections (General one, technological one, and Professional one).

More than 20% of the French graduate schools prepare their students to deal with the main problematics of ecology and energy transition. That's good but not enough.

French high schools are more committed into Sustainable development. The most important technological section in France is called STI 2D (STI as "Sciences and Technologies for Industry". And 2D as "Développement Durable", which means Sustainable Development in French)

At ECAM-EPMI, Graduate school, more than 30 % of the Master's courses are directly oriented towards sustainable development issues: "Energy transition", "Smart industry", and "Green Technologies". We have also a full degree dedicated to Energy and Materials of tomorrow.

Prof. Moumen DARCHERIF,

CEO of ECAM-EPMI, French Graduate School of Engineering, since 2001.
Vice-President of Paris-Seine University in charge of student live.
Senior member of IEEE.

EVF 2019 - Panel discussion - Intervention of Mohit Dubey (CEO, Atal Incubation Centre @MIT ADT University)

In September 2015, the General Assembly adopted the 2030 Agenda for Sustainable Development that includes 17 Sustainable Development Goals (SDGs). Building on the principle of "leaving no one behind", the new Agenda emphasizes a holistic approach to achieving sustainable development for all.

Given the urgency and need for Actions to achieve SDGs by 2030, a new beginning is required. A new world order should be built by our partnerships to address some of the key challenges faced by billions of people. As a key take away from the EVF discussions and putting them against the SDGs backdrop, we should focus on building programs for:

- ✓ ***SDG 4: Quality Education***
- ✓ ***SDG 8: Decent Work and Economic Growth***
- ✓ ***SDG 11: Sustainable Cities and Communities***
- ✓ ***SDG 17: Partnerships for the Goals***

As MIT ADT University, Pune (India) provided a platform to deliberate upon the contours of the action item; I have below suggestions to work for the above SDGs:

1. Set up a Multi Skill Development Centre (MSDC) with Centre of Excellence (CoE) approach for various domains for eg; climate change, renewable energy (Solar, Wind), cultural exchange

2. Mentoring programs for students to orient them for career pathways across two countries based on the skills acquired

3. Research partnership leading to Innovation and Venture Development opportunities at Incubator

4. Annual Dialogue between Academia, Industry and Government partners of both countries

While we all agreed the need and pivotal role of ecosystem with both local and Global players, India presents a huge opportunity given the volume of young graduates (8 million every year). We should put together the systems where the energy of these youngsters can be channelized in the right direction. With a huge network of 900+ Universities and 40,000 colleges, India has a potential to transform these youths into 'Skill Capital' for the World.

Govt. of India has recently launched several programs like Skill India, Smart Cities, Digital India, Atal Innovation Mission (AIM), Make in India, Clean India Mission etc with huge potential. Hence over the next decade if the French partners can bring the Technological wisdom and Financial support to India, I'm sure the Indian Academia, Corporate and Government partners are ready to play their part in creating World class Solutions for the World.

Dr. Mohit Dubey CEO, AIC-MIT ADT Incubator Forum

Statement of Peer review

In submitting conference proceedings to E3S Web of Conferences, I certify to the Publisher that I adhere to the **Policy on Publishing Integrity** of the journal in order to safeguard good scientific practice in publishing.

1. All articles have been subjected to peer review administered by the proceedings editors.
2. Reviews have been conducted by expert referees, who have been requested to provide unbiased and constructive comments aimed, whenever possible, at improving the work.
3. Proceedings editors have taken all reasonable steps to ensure the quality of the materials they publish and their decision to accept or reject a paper for publication has been based only on the merits of the work and the relevance to the journal.

Title, date and place of the conference

6th International Conference on Energy and City of the Future, " EVF2019" Dec 18th to Dec 20th, 2019, at MIT-ADT University, Pune City ,India

Proceedings editor(s):

Dr. Sudarshan B Sanap
Dr. Virendra K Bhojwani
Prof. Mathew V K

Date and editor's signature

18th May 2020

MIT SCHOOL OF ENGINEERING MIT ADT UNIVERSITY PUNE

Dr. Sudarshan B Sanap

Dr. Virendra K Bhojwani

Prof. Mathew V K

A comparative study and graphical analysis in designing and operation of Solar Thermal circular concentrator for enhancing efficiency of solar concentrating system

Mohan Kulkarni[1,*], *Sunil Dingre* [1] *and Chandrakant Kulkarni*[2]
[1]MIT School of Engineering, MIT ADT University, Pune, India
[2]Zeal College of Engineering, Pune, India

Abstract. The present line concentrator system with constant concentration ratio exhibits rise in temperature of working media, however if the difference between outlet and inlet temperature of working media is large then they exhibit lower efficiency. Also the rate of fall of efficiency with increase in its temperature difference is high. To overcome this problem it is proposed to have a variable concentration ratio concentrator system. The variable concentration ratio is achieved by employing receiver consisting of the pipes having different diameters; with the larger diameter pipe at start followed by small diameter receiver. Thus, the concentrator system will have different diameter receivers offering variable concentration ratio system. This concept is confirmed with the help of G.O. Lof, Fester and Duffie Beck paper. The present paper describes above concept by graphical analysis carried out for the newly proposed circular line concentrator with variable concentration ratio. The results of superimposition of graphs leads to confirmation for the promisingly use of variable concentration ratio receivers for enhancing efficiency of solar concentrating system.

1 Introduction

Energy is the fundamental and moreover strategic tool to have a better quality life. The conventional energy resources are limited; also their excess utilization has led towards major environmental issues like global warming.[7] The solar energy can be harvested either by the technology referred as solar thermal or by photovoltaic. In case of photovoltaic technology the solar radiations are converted directly into electricity where in the solar thermal technology the heat from solar rays is extracted and is used for down line applications such as power generation, or material processing. In the present study attempts are made to design and develop effective solar concentrator system for midrange temperature yielding high efficiency with minimum heat loss.[8]Also, it is proposed to have the system with minimum change in a existing system, so that the existing system can be improved upon within less time and less expenses. In the present paper concentrator with variable diameter is proposed after the experimental study and graphical analysis of previous research work done by authors G.O. Lof, Fester and Duffie J.A., William A. Beckman [1] as a experimental output of these authors which concludes "Energy balance on a parabolic cylindrical solar collector" where they uses size of receiver 0.060meter and 0.027 meter and 1.89 is the reflector aperture where they illustrate that for largest diameter receiver factor of intercept and heat losses are in excess amount.[1] B.S. Singh et al [2] studied and illustrates the relation between increment in the thermal losses with the developing area of aperture. Houtan Moaveni [3] has performed the experimental analyses, mathematical simulation and explained change of temperature in receiver is more sensible to the intercept factor of collector and also for the mirror reflectivity.

2 Theoretical Considerations

The figure.1 illustrates that as size of receiver tube decreases concentration ratio increases. Concentration ratio is represented by using lower limit curve where absorbed energy and thermal losses are equal [4]. Concentration ratio is nothing but the ratio of area of the collector aperture to the area of the surface of receiver. [5] A flux concentration ratio is defined as the ratio of the average energy flux on the receiver to that on the aperture, but generally there are substantial variations in energy flux over the surface of a receiver. Concentration ratio (CR) is defined as the ratio of the collector aperture area to the surface area of the receiver. The area concentration ratio is

$$C = Aa/ Ar$$

* Corresponding author: **mohan.kulkarni@mituniversity.edu.in**

 https://doi.org/10.1051/e3sconf/202017001001

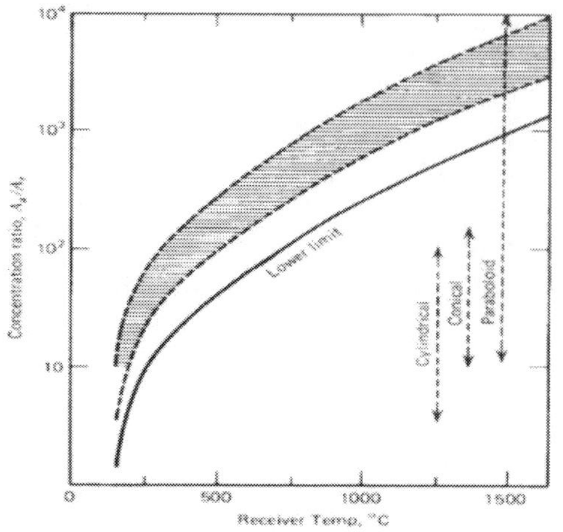

Figure1: Graph plotted between receiver temperature and concentration ratio [4]

Relation between Intercept factor and receiver diameter have evaluated by author G.O.G.LOF et al. In figure 2 graph plotted illustrates that 58% reflected radiations has been intercepted by the receiver having 0.027m diameter. If there is a application of normal distribution which is shifted, interception would have been 57% and if symmetrical normal distribution is applied interception is 70%.Observed values for 0.060m receiver are 89%, 95% and 90%. Hence it illustrates that for a receivers having small diameters, actual distribution is transferred to theoretical focus while analysing the performance of collector which may lead to considerable error from symmetrical normal distribution. The experimental values are used to plot the graph by using theoretical tool named Sol trace software. This is a new optical modelling tool used to plot graph between receiver diameter and intercept factor. So during analysis of these experimental studies very minute observations of actual intercept factor are received related with shifted normal distribution of system. So for the referred system reflection of solar rays are in more amount for the receiver having large diameter in comparison with receiver having small diameter. However major losses are there for receiver having larger diameter [3].

The energy distribution of the incident rays for small and large sized receivers for full sized reflectors is shown in figure 3 and figure 4 which is a function of conditions compared as solar radiations, velocity of wind and ambient air temperature [3]. These illustrations show evident distribution of losses and direction logically to proceed further for performance improvement. In case of tubes having small diameter optical losses are more while at higher temperature for tubes having larger diameter optical and thermal losses are equally important. For reducing radiation loss and convection loss one can use transparent envelope but it may increases optical loss. By using desired coatings thermal loss could be reduced; however some of the imperfect absorbtivity could be generated for less than 095 solar radiations.

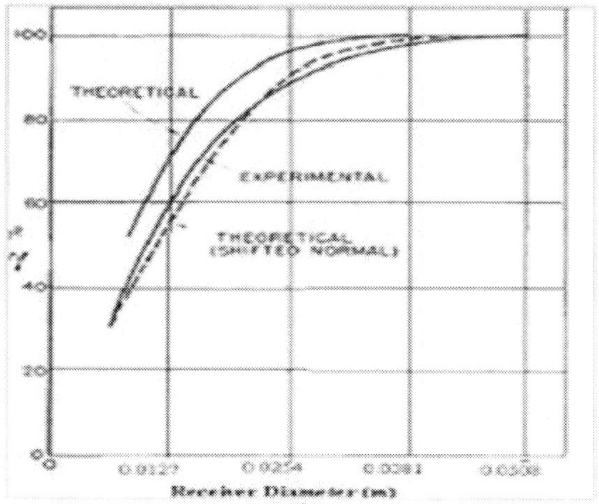

Figure 2: Graph plotted between Receiver diameter and intercept factor [1]

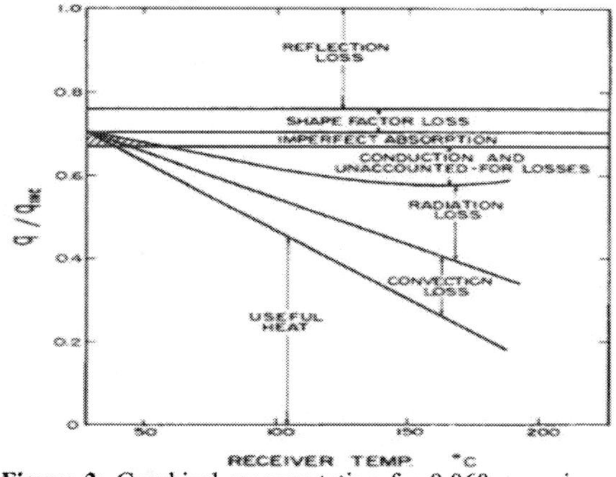

Figure 3: Graphical representation for 0.060m receiver diameter and 1.89 m reflector aperture [3]

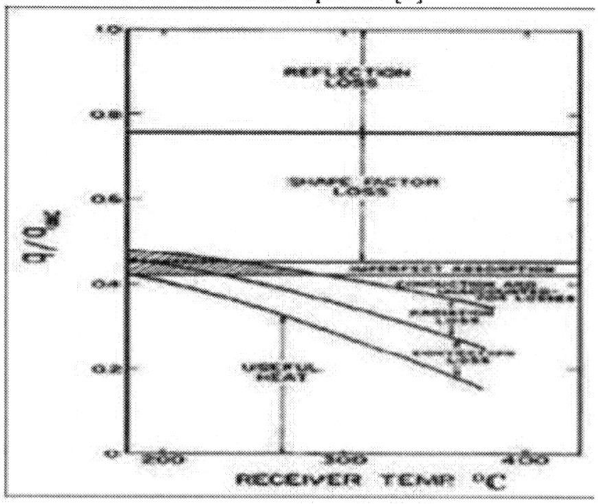

Figure4: Graphical representation for energy distribution incident for 1.89m aperture reflector and 0.27m receiver diameter.[3]

The outcome of the above theoretical consideration illustrates that after employing a receiver with constant diameter convection and radiation heat losses are more and due to uniform diameter of receiver, the efficiency obtained is also low.

Figure 3 and figure 4 illustrates [1] the distribution of solar energy beam which are incident on collector and useful for effective heat gain and for finding losses for receiver having 0.060m diameter and 0.027m diameter. It shows evident magnitude for the losses. This graphical analysis shows that due to size variation in receiver diameter collector efficiency is affected.

Intercept factor is a receiver size function and leading to the increment in the energy intercepted by receiver. Where based on diameter of receiver there is a dependency of heat losing areas. For lower temperature, there is a major impact on thermal performance having large diameter. However as there is a major heat loss at high temperature, efficiency decreases. In another case where receiver having small diameter and concentration ratio is high and having less intercepted energy, there is low heat loss at higher temperature because of receiver is having less surface area. Hence for obtaining the good performance it is needed to optimize concentration ratio for varying temperature of fluid.

At the time of performance evaluation of solar concentrating system references are ready for receiver temperature. Where temperature of receiver (Tr) and temperature of water (Tw) works parallel. It is illustrated in the (figure 5). In the proposed solar concentrating system, having design specifications as aperture of reflector is 0.6 m and its length 6 m,with variable diameters without any glass cover receiver tubes with black coating. Experimental performance of this set up is shown in figure 5. For a given length difference between temperature of water (Tw) and temperature of receiver (Tr) is evident. The temperature difference between these Tr and Tw is less at starting and increases as length of receiver increases.[6] As in the proposed research work by author,for the preparation of experimental set up of solar thermal circular concentrator having dimension of reflector aperture 0.6m and 6m long in length,different diameter receiver without transparent cover is used in this set up. The difference between Tr and Tw is less at initial stage while as length increases difference increases due to variation in concentration ratio. Peak temperature difference between these Tw and Tr is obtained at the end was 142^{0}c

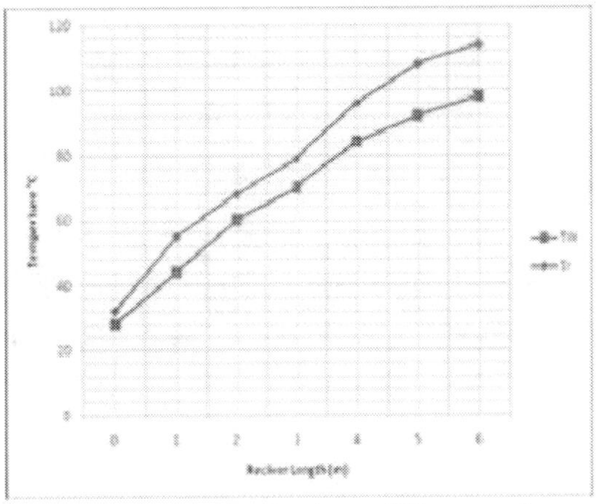

Figure 5: Graph showing difference between temperature of water and temperature of receiver

3. Graphical collation of figure 3 and figure 4

After the analysis of collation of figure 3 and figure 4, presented by Lof et al [3] author have drawn superimposed resultant and shown in figure 6.Receiver having 0.060 m diameter shows low performance up to temperature 170^{0}c and receiver having 0.027m diameter having higher temperature. Here these superimposed graphs are used to put forth new development of solar concentrating system with variable concentration ratio for enhancing the efficiency of concentrator system. In the present trends concentration ratio with constant receiver diameter is used which leads to low performance. Hence it proves that whenever there is a need of substantial difference between the exit and inlet, a solar concentrating system with variable concentrating ratio offers and may enhance the efficiency of concentrator system.

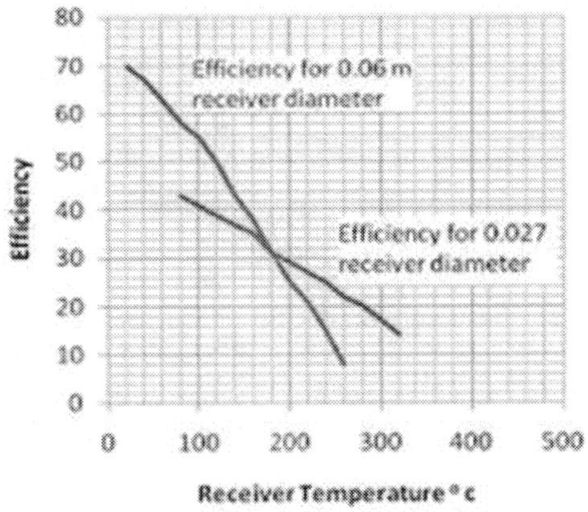

Figure 6: Graph showing superimposition of figure 3 and figure 4 [3]

https://doi.org/10.1051/e3sconf/202017001001

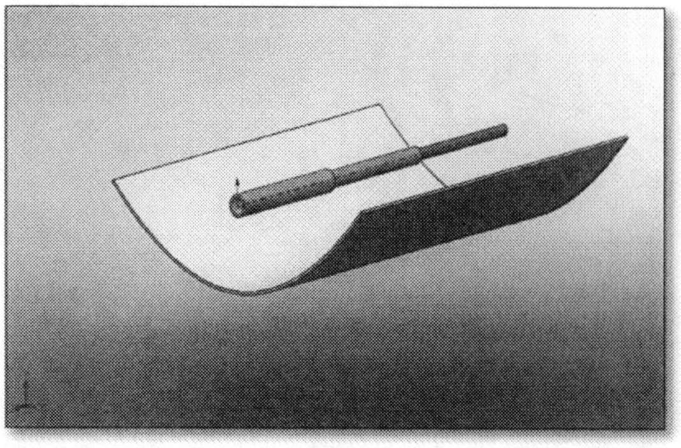

Figure 7: Variable concentration ratio receiver for constant aperture

Figure 8: Showing flow and dimensions of panel for Variable concentration ratio receiver

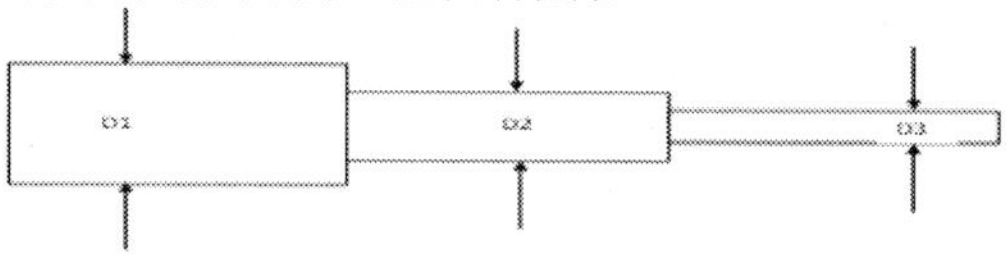

Figure 9: Variable concentration ratio receiver for constant aperture

4. Conclusion

As shown in Figure 6, intersection of the efficiency for the receiver having diameter 0.06m and 0.027m meets at one point. It illustrates that for large diameter, larger is the energy intercepted by receiver. While, as receiver diameter increases, heat loss also increases for that temperature of receiver. Hence for the improvement in efficiency there is a need to balance the energy intercepted and heat loss. It can be achieved by using larger diameter receiver at initial position where concentration ratio is lower and temperature of water is also low. Further portion of the concentrator where fluid temperature is higher, a high concentrator ratio can be used to reduce the loss of heat through reduction of area of the receiver tube (shown in Figure 7).As shown in Figure 8, authors have contributed for the preparation of experimental set up of solar thermal circular concentrator having dimension of reflector aperture 0.6m and 6m long in length. Different diameter receiver without transparent cover is used in this set up. Hence this experimental set up which is having receiver with variable concentration ratio is appropriate for the improvement in efficiency of solar thermal circular concentrator. In the experimental set up used by researchers, 0.027m receiver diameter receiver offers higher efficiency as compared with 0.06m diameter. Hence authors have drawn the conclusion to use a receiver having stepped receiver with diameter 0.06m upto 170^{0}c and then 0.027m can be used to adopt higher efficiency. Keeping the same theme authors have proposed to use a multistepped receiver with variable concentration ratio based on the range of temperature and fluid for the system in reference.

Appellation

A_a	Area of aperature (m2)	q_u	Heat gain rate (w)
A_r	Receiver area (m2)	T_r	Temp.of receiver surface (^{0}C)
Υ	Factor of intercept	T_w	Water Temperature (^{0}C)

References

[1] G.O.G.LOF "Energy balance on a parabolic cylindrical solar collector" ASME PP.24-32, Jan1962

[2] Balbir Singh Mahinder Singh, Fauziah Sulaiman, "Designing A Solar Thermal Cylindrical Parabolic trough Concentrator By Simulation "International Rio3 Congress, World Climate and Energy Event, pp. 1-5,December 2003

[3] Houtan Moaveni, "Numerical Analysis and Experimental Study in Design and Operation of Three Solar Parabolic Thermal Power Plants" ,powering, pp.12-14, April 2007

[4] John A. Duffie, William A. Beckman, "Solar Engineering of Thermal Processes", John Wiley and Sons INC, pp. 324-373, 2006.

[5] Sukhatme S., "Solar Energy, Principles of thermal collection and storage," Tata McGraw Hill-Publication, pp.71-257, 2010.

[6] Enhancing Performance of Solar Line Concentrator System Using Variable Concentration Ratio, Science Direct, Materials Today: Proceedings 4 (2017) 7533–7545

[7] William Stine and Raymond W Harrington,(1985) A Book on "Solar Energy Fundamentals and Design", John Wiley and sons.

[8] Brain Norton,(2014)A Book on "Harnessing Solar Heat", 2nd Edition, Springer

Implementation of BIM Modelling and Simulation Tools in Reducing Annual Energy Consumption of Multifamily Dwellings

Subbarao Yarramsetty[1], Siva Kumar MVN[1], and Anand Raj P[1]*
[1]Department of Civil Engineering, NIT Warangal, Telangana State-506 004, India.

Abstract. In current research, building modelling and energy simulation tools were used to analyse and estimate the energy use of dwellings in order to reduce the annual energy use in multifamily dwellings. A three-story residential building located in Kabul city was modelled in Revit and all required parameters for running energy simulation were set. A Total of 126 experiments were conducted to estimate annual energy loads of the building. Different combinations from various components such as walls, roofs, floors, doors, and windows were created and simulated. Ultimately, the most energy efficient option in the context of Afghan dwellings was figured out. The building components consist of different locally available construction materials currently used in buildings in Afghanistan. Furthermore, the best energy efficient option was simulated by varying, building orientation in 15-degree increments and glazing area from 10% to 60% to find the most energy efficient combination. It was found that combination No. 48 was best option from energy conservation point of view and 120-degree rotational angle from north to east, of the existing building was the most energy-efficient option. Also, it was observed that 60% glazing area model consumed 24549 kWh more electricity compared to the one with 10% glazing area.

1 Introduction

The energy efficiency of buildings is a much debated and research area in Architecture, Engineering, and Construction (AEC) sector nowadays. It plays an important role in impacting environment, determining the life-cycle cost of projects, helping carbon reduction and global warming and ensuring sustainability issues. Minimizing energy demand of buildings by increasing building energy efficiency is the need of the hour due to the growth of population and high demand for fuel and electricity. Thus, clean energy solutions and energy conservation measures are absolutely essential for the future.

40% of total primary energy production of the world is consumed by buildings and they are responsible for producing 30% carbon dioxide emissions [1]. Applying of insulation materials to building envelopes significantly reduces energy demand, but most of the builders in Afghanistan don't use insulation materials in buildings, as energy performance of residential areas is comparatively low. The building industry in Afghanistan consumes about 71% of total domestic electricity [2].

Energy consumption analysis of dwelling is the most important factor in AEC industry because of its direct impact on lifecycle cost of buildings and environmental issues. Energy simulation needs to be conducted in early design phases of the project in order to take appropriate decision on design alternatives. Traditional energy analysis techniques are not adequate to deliver accurate simulation results due to their drawbacks and shortages. Utilizing proper software tools and systems for the anticipation of energy demand in buildings is very essential nowadays. Therefore, building information modeling (BIM) is used as a dominant contemporary program for this purpose.

Climate variation in Afghanistan is complicated. It depends on different zones and different seasons. Some regions of the country are very hot and some regions are very cold. Generally, Afghanistan is known as a dry continental climate country with cold and relatively rainy weather in winter and hot and sunny weather in summer. Even though numerous researches have been conducted on the building energy behavior on energy consumption based on climate variation worldwide, not enough literature has been published on this topic in the context of Afghanistan.

Energy consumption will increase in the future all over the world because of increase in population, enhancement of social and economic conditions, and high demand for services which need energy [3]. According to International Energy Agency (IEA), the global total primary energy consumption and CO_2 emissions grew by 49% and 43% respectively, with an average annual rise of 2% and 1.8% from 1984 to 2004. While in 2012, energy use and CO_2 discharges increased by 85% and 75% respectively. Total world energy use is anticipated to increase by almost 32% and CO_2 emission is projected to grow by approximately 16% until 2035 [4,5]. Table 1 represents global energy demand projection of different organizations including British multinational oil & gas Company (BP), European Commission (EC), Energy Information Administration (EIA), International Atomic Energy Agency (IAEA), International Energy Agency, International Institute for Applied Systems Analysis (IIASA), Shell International, World Energy Council

*Corresponding author: **sraoystp@student.nitw.ac.in**

(WEC) and Tellus Institute for the future. According to IIASA's perception, global energy consumption is anticipated to increase up to 1740 EJ (exajoules) in 2100 [6].

Table 1. World Energy Anticipation (from 2020 To 2100), in EJ (EJ = 1018J), Adopted from [6]

Institutes	Year	2020	2030	2050	2100
BP	(2011)	565–635	600–760	NA	NA
EC	(2006)	570–610	650–705	820–935	NA
EIA	(2010)	600–645	675–780	NA	NA
IAEA	(2009)	585–650	670–815	NA	NA
IEA	(2010)	NA	605–705	NA	NA
IIASA	(2007)	555–630	NA	800–1175	985–1740
Shell International	(2008)	630–650	690–735	770–880	NA
WEC	(2008)	615–675	700–845	845–1150	NA
Tellus Institute	(2010)	504–644	489–793	425–1003	243–1200

1.1 BIM & BEP

For better assessment of energy demand in the buildings, the following factors must be taken into consideration which includes equipment use, geographic location, number of occupancies, construction materials, size and space of buildings, etc. Much of the information must be provided early in the design phases by the designer [7]. Most of the countries believe that construction sector is the most complicated and risky industry [8]. Today BIM is used in USA, Europe, and some other countries for energy simulation and validation purposes [9]. Compared to BIM, other energy simulation tools are very expensive, time-consuming, difficult to run and unable to evaluate all characteristics of energy in building simulation trend [10]. BIM allows designers to select the most energy effective construction materials for the buildings by investigating various alternatives and systems early in the conceptual phase of the project. By utilizing BIM, users don't need to reinter geometry and other data required for energy simulation which leads comprehensive saving of time [11].

Building energy performance (BEP) is a vast computer-based building analysis software which is used for different purposes, such as assessment of loads, complete energy analysis of buildings, inside air quality and its flow analysis, lighting system design, code compliance, retrofit, analysis of renewable energy and economic assessment [12]. BEP tools are commonly divided into two parts: design tools and simulation tools. Design tools are used to study the worst-case scenarios for determining optimum size of HVAC systems. Energy analysis tools take dynamic calculations into consideration for whole year studies [13]. Despite numerous building analysis software tools, the real implementation of the tools is mostly confined to computation of energy load for complete designing of heat, ventilation as well as air-condition (HVAC) systems[14]. Energy Plus, eQuest, and Ecotect are examples of Building Performance Simulation (BPS) software tools. Engineers, scholars, and architects in general use BEP tools for a detailed evaluation of energy consumption such as heating and cooling loads as well as other required parameters of both new and existing buildings by preparing energy models to increase the thermal efficiency of the buildings [15].

1.2 Buildings in Afghanistan:

The selection of construction materials and techniques depends on the local situations and conditions such as affordability, climate changes, material availability, culture, and social living schema, etc.

People in Afghanistan build different types of houses. the main types of construction material used in buildings are Sun-dried (Adobe) brick or mud walls with wooden roofs and floors, fired brick walls (350mm) with barrel-arch (Taqe-Zarbi) roofs and floors, fired or adobe brick with dome roofs. Contemporary buildings generally use fired brick (220mm) and concrete masonry unit block (200mm) walls with Reinforced Cement Concrete (RCC) roofs and floors. The majority of existing houses are constructed with adobe brick walls and wooden slabs as well as mud walls and dome roofs all over the country. The primary construction material locally available is Mud, which contributed to 95% of construction material in Afghanistan. Contemporary materials are used only in urban areas[16].

A typical family in Afghanistan is relatively big and varies between 8-14 persons. Traditionally married and unmarried brothers used to live with parents together in the same building. For this reason, people prefer building multifamily houses. The area of the buildings starts from $100m^2$ onwards. In urban areas, people extend their houses vertically while in rural areas extension of buildings is done horizontally.

Today energy is one of the most important factors which helps countries to enhance their economy and bring numerous facilities and services in societies. Afghanistan is a rich country from the point of energy resources. According to the Ministry of Energy and Water 2016, Afghanistan has the capacity of generating 2300MW of electricity from hydropower plant, 158500MW from wind power, 4000MW from biomass, and 3500MW from solar power as a clean and environmentally friendly energy source. But due to political and security issues, no significant investment has been made in this sector yet. Thus, the major energy needs of the country come from abroad. On the consumption side, houses in Afghanistan contributed to consumption of 71% of overall electricity use in the country, industrial buildings 11%, public buildings – 11% and commercial buildings – 7% [2]. Nowadays most of the houses consume wood, coal, charcoal, gas, diesel, kerosene, and even animal dung, which leads to various environmental problems. Houses in Afghanistan, need the highest amount of energy in winter season compared to summer [2,17].

2 Methodology

The main purpose of this research was to minimize annual energy consumptions of multifamily residential dwellings without compromising comfort, health or productivity levels of the buildings, utilizing BIM-based modelling and Energy simulation software tools. The research objectives are presented below:

- ✓ Gathering information regarding available materials used in buildings in Afghanistan
- ✓ Creating 2D Plans of the benchmark buildings based on site measurement data in AutoCAD
- ✓ Creating a 3D BIM model by importing 2D plans of the model to Autodesk Revit Architecture
- ✓ Utilizing Revit energy plug-in and GBS for energy simulation.
- ✓ Evaluating energy analysis results of different combinations of building's components to achieve the best combination from energy use point of view.
- ✓ Investigate energy performance by varying the orientation of the building.
- ✓ Check impact of increasing glazing area on energy consumption of the building.

3 Case Study Building

The case study building is an existing three-story residential typical dwelling, located in Kabul city as shown in Figure 1. For research purpose, we are assuming that the building is still in design phase. The building plan area is $125m^2$.

(a) (b)

Fig. 1. (a) Location – satellite image, (b) 3D Front view of the Building.

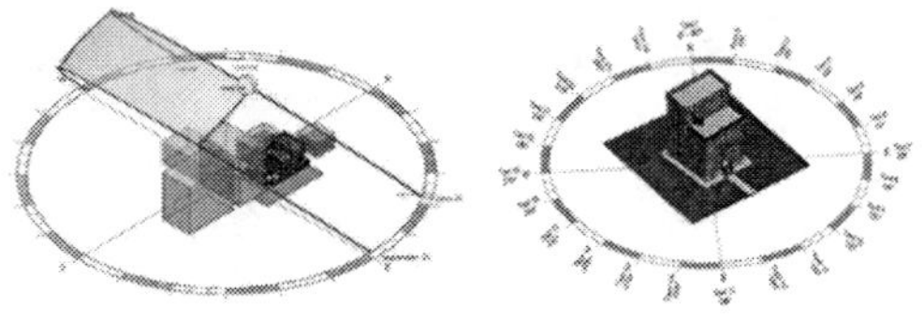

(a) (b)

Fig. 2. (a) existing surrounding buildings, (b) Test scenarios for orientation analysis.

Basement walls of the building consist of 600mm thick stone masonry, where conventional stones are used in basement walls and foundations in construction of buildings in Afghanistan. This wall remains unchanged with all combinations across the energy performance assessment process. The nearest weather station for the building is station No. 699799 which was assigned to software for energy simulation. The building has a box shape and its roof is flat. South-East façade of the building has its entrance on the Ground floor. The North-West and South-West facades of the building have small windows while North-East and South-East façades have larger windows.

4 BIM modelling and Analysis

BIM modelling is the most essential step involved in thermal analysis process. Modelling for energy performance analysis is slightly different than that of architecture designs. Any likely mistakes in this process hugely impact the result of the analysis. For this purpose, sufficient and correct information must be set into the model. Energy analysis is one of the important features of the rivet which performs whole building energy analysis. There are two ways in Revit for energy analysis simulation: 1) Using conceptual masses where actual physical parts of the building components are not required to be included in the model. 2) Using building elements when actual model with all materials and elements including their properties need to be included in the model. In the present study, building element method was used as the base to perform simulation analysis. Fig. 3. Shows the model generated using Revit and the process involved in extracting energy analysis. As soon as BIM model was created, a few modifications were made to obtain accurate result from the energy analysis process.

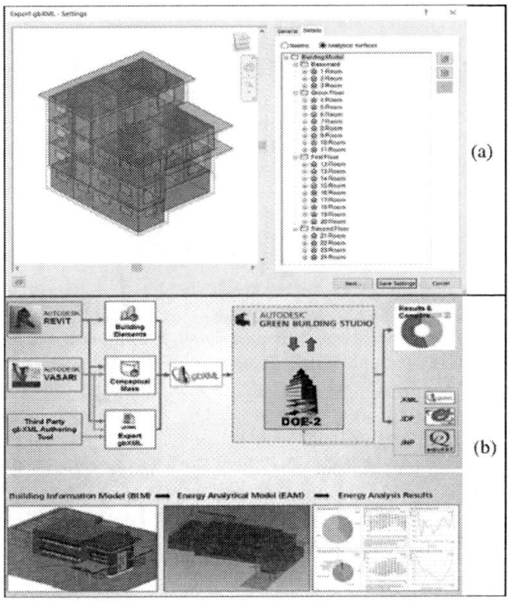

Fig. 3. (a) gbXML file of the model in Revit, (b) Energy Analysis Work Flow in Autodesk tools

4.1 Building envelope Variation Impacts

In the current study, gbXML files of all combinations were imported from Revit to GBS cloud. Table 2 shows the default simulation parameters used in GBS studio. Locally available materials which are typically used in the envelop of residential buildings in Afghanistan were used. A total of 127 models (96 building envelope + 24 orientation analysis + 7 analysis for glazing impacts) were analyzed for energy estimates. For this purpose, 6 types of walls, 4 types of floors, and 4 types of roofs were taken into consideration. 96 (C1 to C96) different combinations of Afghan local construction materials for the envelope of the residential buildings were studied. As the result of energy analysis experiments, it was found that a combination-C48 of wall type (W3)- Adobe brick wall, roof type (R4)- Mud Roof with timber as core structure and Floor type (F4)-Mud floor with timber as structure, was the best option that consumed least amount of energy leading to reduction in annual electricity demand. The work carried in identifying the best combination of building material is not presented here, and only its outcome is used for orientation and glazing analysis. The minimum energy consumption combination is tabulated in Table 3.

Table 2. Default values for multifamily building in GBS

Building Type	EPD (W/ft²)	LP (W/ft²)	Outside Air Flow/Person (cfm/Person)	Outside Air Flow/Area (cfm/ft²)	Infiltration Flow (ACH)
Muli Family	1	0.7	N/A	0.06	0.25

Table 3. Minimum energy demand - combination C48

Wall Type	Roof Type	Floor Type	Electricity Demand (KWh)	Electricity Annual Cost (AF)
W3	R4	F4	29,496.00	2,13,846.00

4.2 Orientation Variation Impacts

The building was rotated clockwise from the North side of the building with 15° rotational angle interval. A total of 24 tests were performed; test details are marked in Fig. 2 (b) along with actual orientation. The result shows that test 7 with 90° rotation angle was the most efficient option from an energy conservation point of view. It indicates 8.22% energy efficiency improvement compared to actual oriented option (test 23 with 330° rotational angle) which is the worst case from the energy conservation point of view. Fig. 4 show complete results of the tests for all rotated angles.

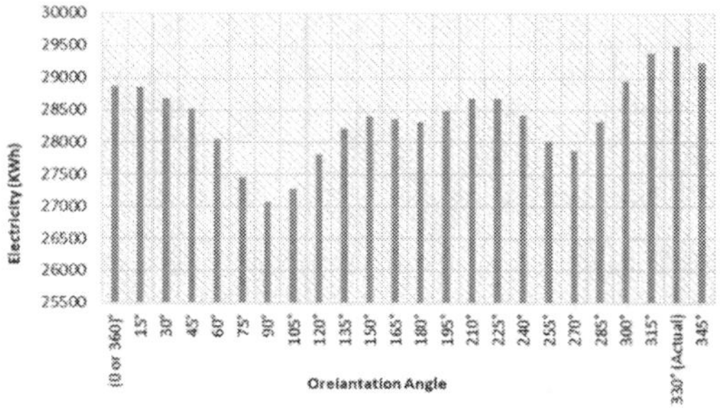

Fig. 4. Electricity demand of all orientation change increment

4.3 Glazing area variation results

Glazing area of the building envelope of the most energy efficient case (C48) was increased from 10% to 60% with 10% WWR increment intervals. For this purpose, the percentage of glazing area of models for 10% to 60% as well as the actual model were calculated in Microsoft Excel and new windows with new dimensions were assigned to the models accordingly. Then energy simulation was conducted for all 6 new models. Finally, electricity demand results in all models were compared. Detailed result of Energy demand is shown in Table 4 and Fig. 5.

Table 4. Electricity demand for different glazing area

Com. No.	Windows to Wall Ratio (WWR)	Electricity Demand (KWh)	Electricity Cost (AF)
C48_1	10%	24356	176584
C48_2	20%	28442	206204
C48_3 (Actual)	20.3%	29496	213846
C48_4	30%	33831	245276
C48_5	40%	39728	288025
C48_6	50%	44126	319911
C48_7	60%	48885	354417

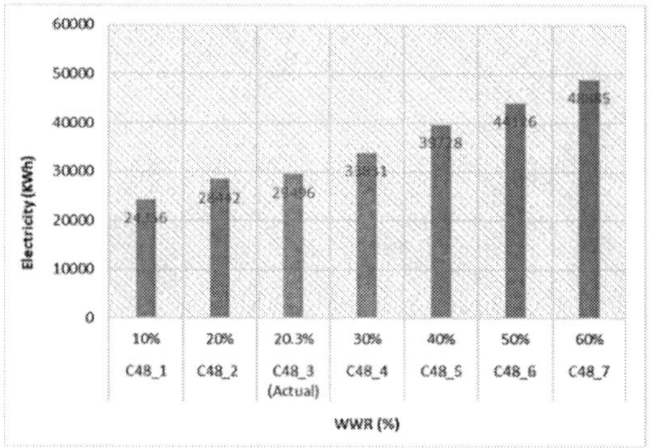

Fig. 5. Variation of Electricity demand for variation in glazing area

5 Conclusion

Due to technological developments, numerous software tools have been developed in energy assessment in the construction industry. Building energy evaluation allows the users to make proper decisions at different stages of design to enhance sustainability by applying specific solutions.

The aim of this research was to identify proper measures for improving living standards by reducing annual energy consumption in Afghan multifamily dwellings with the implementation of BIM technology. The main reason for using BIM based software tools in this project was the potential they have to achieve the project goals quickly and accurately without imposing any charges for software acquisition. From the other side, learning BIM software tools for 3D modelling and energy analysis is very easy and many online resources are available to learn the software quickly.

Revit Architecture and Green Building Studio two BIM based software tools were used to create and simulate a three-story building located in Kabul city. AutoCAD 2D plans were used as a baseline to build the 3D model quickly, and Microsoft Excel was used for cost estimation and creating various tables and graphs in the research project.

The main objective of the research was to investigate local construction material behaviour on building energy consumption. For this purpose, different local construction materials which are conventionally used in buildings were assigned to main components of the 3D model in Revit. In all 7 types of walls, 4 types of roofs, and 4 types of floors were created from various local materials, and all the aforementioned building elements were combined to gather and 96 models were built. gbXML format of all 96 models was imported to GBS cloud for energy simulation. Finally, it was observed that combination No. 48 which was made of adobe brick walls and mud roofs and floors was the best one from energy conservation point of view as it consumes 29496 KWh electricity annually which is the minimum electricity consumption among all combinations.

Furthermore, building orientation variation and glazing area enlargement impact were investigated to find out the best rotation angle and favourable percentage of WWR from energy conservation point of view. For this purpose, the most energy efficient combination model was rotated clockwise with 15^0 increment angle and 24 new models were built for orientation changing tests and 6 new models were prepared for 10% to 60% WWR experiments. It was observed that 120^0 rotation angle for existing building orientation state was the best choice. By increasing WWR percentages, electricity demand also increased accordingly. So, to build energy efficient buildings, smaller WWR is recommended.

References

1. A. Costa, M. M. Keane, J. I. Torrens, and E. Corry, Appl. Energy **101**, 310 (2013).
2. O. T. Mohmmad, *Energy Efficiency Guidbook for Buildings*. (2017).
3. N. Nebojia and J. M. Jefferson, *Global Energy Perspectives to 2050 and Beyond* (1995).
4. X. Cao, X. Dai, and J. Liu, Energy Build. **128**, 198 (2016).
5. L. Pérez-Lombard, J. Ortiz, and C. Pout, Energy Build. **40**, 394 (2008).
6. P. Moriarty and D. Honnery, Renew. Sustain. Energy Rev. **16**, 244 (2012).
7. H. Kim and K. Anderson, J. Comput. Civ. Eng. **27**, 203 (2013).
8. C. M. Clevenger, M. E. Ozbek, H. Mahmoud, and B. Fanning, *Impacts and Benefits of Implementing Building Information Modeling on Bridge Infrastructure Projects* (2014).
9. T. Laine and A. Karola, in *Clima 2007 WellBeing Indoors* (2007).
10. A. Venkataraman and K. M. Ramesh, in *Proc. Int. Conf. Adv. Civ. Eng. AETACE* (2013).
11. F. Jalaei and A. Jrade, J. Inf. Technol. Constr. **19**, 494 (2014).
12. V. Ugursal, Buildings **4**, 762 (2014).
13. M. Senave and S. Boeykens, in *Build. Inf. Model. Des. Constr. Oper.* (2015).
14. C. J. Hopfe, *Uncertainty and Sensitivity Analysis in Building Performance Simulation for Decision Support and Design Optimization* (2009).
15. Y. Ham and M. Golparvar-Fard, Energy Build. **63**, (2013).
16. A. S. Arya, *Guidelines for Earthquake Resistant Design, Construction, and Retrofitting of Buildings in Afghanistan* (2003).
17. A. M. Ershad, R. J. Brecha, and K. Hallinan, Renew. Energy **85**, (2016).

The influence of *Typha Australis* as an insulation panel on the thermal performance of building

Labouda Ba[1,2&3*], *Ikram EL ABBASSI*[2], *Cheikh S.E Kane*[3], *A-Moumen Darcherif*[1], *Mamoudou Ndongo*[3]
[1]Laboratoire de Quartz, Université de Cergy-Pontoise (France)
[2] Laboratoire de Recherche en Eco-innovation Industrielle et Energétique, ECAM-EPMI,13 bd de l'hautil 95092,Cergy-Pontoise (France)
[3]Unité de recherche appliquée sur les énergies renouvelables, Université de Nouakchott Al Asriya, Nouakchott (Mauritanie)
Laboratoire de Mécanique et Matériaux du Génie Civil, Université de Cergy-Pontoise (France)

Abstract. *Typha Australis* is a plant that grows abundantly in fresh water. The proliferation of this plant causes health problems, so several measures have been taken to eradicate this problem such as: cutting the plant, coal production. So this article is about the valorisation of this plant as a bio-based material in order to solve the energy problem in the building. In this study, clay was used as a binder with a given percentage. The mixture of clay and Typha was used as an insulation panel and a comparison was made with a conventional habitat without an insulation panel. A dynamic thermal simulation was performed on TRNSYS to evaluate the influence of the use of this insulation panel on the energy consumption in the building. The results of the comparison showed that the use of this insulation panel, which is a mixture of clay and *Typha Australis*, reduced the energy requirement by 23%, which is a satisfactory result.

1 Introduction

In recent years, we have witnessed environmental scourges that impact human well-being and health. As a result, the use of high-performance building materials and new technologies in the building sector has been adopted with the development of materials based on renewable plant resources in order to address these environmental problems [1]. These renewable vegetable fibres have low environmental impacts because they are less energy consuming (six times less energy consuming than glass fibres and twelve times less than carbon fibres) and generally have low densities [2]. The use of these fibres as a building material is a solution to overcome the problem of thermal comfort in the building and also to optimize the energy requirement. This optimization could be carried out through the thermal performance of these bio-based building materials. Indeed, several studies have been conducted to evaluate the thermal performance of these bio-based building materials [3, 4, 5]. For example, materials have been studied such as mixing peanut shells and plaster, peanut shells allow materials to have good thermal stability compared to plaster alone for wall and roof coverings [3], there are also studies that have shown thermomechanical characteristics of cork or lightweight concrete based on olive residue, providing ideas that prove that cork is a suitable insulating material for a wall and the heating plate is used to determine the thermal properties of mixtures. [4, 5].
These materials being spread in several countries of the world. In our study we chose to study the valorisation of *Typha Australis* which is a plant of West Africa which grows in profusion in the Senegal River valley. Indeed, recent studies have shown that this fiber could be used in construction [6-7]. In this article we will study the impact of the use of *Typha Australis* on the energy requirement of a building. This study will be done using Trnsys software, a dynamic thermal simulation software that has been used by several researchers around the world to evaluate the thermal effects of buildings [8]. The binder used in this article is clay, because it is a local material and with almost no transformation, resulting in a lower carbon impact. The next part of this article will show the methodology and modelling performed for the simulation.

2 Method and modelling

Two habitats were studied, the first is a conventional habitat without insulation panel which will be named in the following article by building ($H_{no\text{-}ip}$) and the second is a conventional habitat but with an insulation panel which will be named by Building (H_{ip}). the insulation panel used is the mixture of clay and Typha Australis which will be inserted into the walls of this conventional habitat as well as among its compositions.

2.1 Building description

The two habitats have the same geometric shape which is 4m*5m*3m similar to offices with a wooden door and double glazed windows and their dimensions are given in Table 1. The location site is Nouakchott (the climate of the site is given by Meteonorm).

*Corresponding author: **laboudaa@gmail.com**

Table 1. Building size

	Length (m)	Width (m)
Walls and roof	5	4
Door	2.1	0.8
Window	1	0.8

2.2 Thermo-physical characteristics of the insulation panel

The thermo-physical characterizations of the insulation panel have been collected from the literature [6], the mixture is composed of 22.9% *Typha Australis* and the data are presented in Table 2:

Table 2. characteristics of the clay mixture and Typha Australis

Effusivity (E) $J.m^2.K^{-1}.s^{-1/2}$	242
Conductivity (λ) $W.m^{-1}.K^{-1}$	0.127
Mass heat (Cp) KJ/K	461
Density (ρ) Kg/m^3	1000

As the thermo-physical conductivity is relatively low, this mixture could be considered as an insulating.

2.3 Composition of building (H_{no-ip})

Building (Hno-ip) is the so-called conventional habitat, it does not contain an insulation panel, the composition of the walls and roof of this habitat are shown in Table 3 and Table 4 respectively.

Table 3. Composition of wall (Hno-ip)

Wall	Compositions	Thickness (m)
Int wall	Cement mortar	0,03
	Parpaing	0,15
	Plaster	0.,02
Out wall	Mortar cement	0,03
	Plaster	0,02

Table 4. Composition of roof (Hno-ip)

Composition of roof for	Thickness (m)
Cement Mortar	0,1
Parpaing	0,15
Cement Mortar	0,1
Plaster	0,05

2.4 Composition of building (H_{ip})

Building H_{ip} is the building consisting of an insulation panel. Cement mortar in conventional building is replaced by the mixture of *Typha Australis* and clay (insulation board). The table 5 and Table 6 represent the composition of this building.

Table 5. Composition of wall (H_{ip})

Wall	Compositions	Thickness (m)
Int wall	Mixture of *Typha Australis* and clay	0,03
	Parpaing	0,15
	Plaster	0.,02
Out wall	Mixture of *Typha Australis* and clay	0,03
	Plaster	0,02

Table 6. Composition of roof (H_{ip})

Composition of roof for	Thickness (m)
Mixture of *Typha Australis* and clay	0,1
Parpaing	0,15
Mixture of *Typha Australis* and clay	0,1
Plaster	0,05

2.5 Modelling

The simulation was carried out on TRNSYS which uses the Fortran programming language, the habitat characteristics were implemented in the software. The simulation was carried out over a full year, i.e. over 8760 hours. The set temperature has been set at 25°C. This is the difference between the maximum and minimum ambient temperature (max 40°C and min 15°C). The energy system (air conditioning) is used when the office is busy, i.e. Monday to Friday from 8am to 6pm.

2.6 Mathematical Model

The Trnsys "type 56" module was used to simulate the transfer phenomena in the building [9]. The walls are governed by the three modes of transfer: conduction (in the wall), convection (between the wall and the fluid 'air') and short and long wavelength radiation.
The exchange by conduction on TRNSYS is the Z-transform method which is more efficient than the finite difference method [10].

Exchange by conduction on the walls [11]:

$$\varphi_{ce}(t)=-\sum_{i=0}^{Na} a_i T_{se}(t-i\Delta t)+\sum_{i=0}^{Nb} b_i T_{si}(t-i\Delta t)-\sum_{i=0}^{Nd} d_i \varphi_c \quad (1)$$

$$\varphi_{ci}(t)=-\sum_{i=0}^{Na} b_i T_{se}(t-i\Delta t)+\sum_{i=0}^{Nb} c_i T_{si}(t-i\Delta t)-\sum_{i=0}^{Nd} d_i \varphi_c \quad (2)$$

φ_c: conductive flux density (W/m²)
a_i, b_i, c_i: Coefficients of the Z-transform of functions transfer rates for surface temperatures (W/m²K)
d_i: Coefficients of the Z-transform of transfer functions related to flow densities
Na, Nb, Nd: number of coefficients a_i, b_i, c_i and d_i
Δt: simulation step
$T_{se}(t-i\Delta t)$: Temperature of the outer surface of the wall at $(t-i\Delta t)$ (K)
$T_{si}(t-i\Delta t)$: Temperature of the inner surface of the wall at $(t-i\Delta t)$ (K)

Exchange by conduction on the glazing [11]:
Since the glazing is considered to be inertial, the equation that models the conduction exchange is in steady state.

$$\varphi_c=K(T_{si}-T_{se}) \quad (3)$$

φ_c: conductive flux density for glazing
K: Conductivity 'surface - surface' of the glass wall (W/m²K)
Convection exchange

$$\varphi_{conv,e}=h_{ce}(T_{se}-T_e) \quad (4)$$
$$\varphi_{conv,i}=h_{ci}(T_{si}-T_i) \quad (5)$$

φ_{conv}:: convective flux density
$T_e$and T_i: Temperature of the outdoor and indoor environment
h_{ce}, h_{ci}: Coefficients of convection on the outer and inner sides
Radiation exchange

$$\varphi_{rde}=\varphi_{CLOe}+\varphi_{GLOe} \quad (6)$$

φ_{rde}: radiative flux density
φ_{CLO}: Short wavelength radiative flux density
φ_{GLO}:Long wavelength radiative flux density

$$\varphi_{GLOe}=h_{r,c}(T_{se}-T_c)+h_{r,e}(T_{se}-T_e) \quad (7)$$

$h_{r,c}$: Coefficient of GLO radiative exchange with the environment
$h_{r,e}$: GLO radiation exchange coefficient with the celestial vault
Short wavelength radiative exchanges, the data are derived from meteorological data.
The heat transfer balance in a room (the building) [11]:

$$\sum_{i=0}^{Na} S_i h_{ci}(T_{si}-T_i)+\acute{m}_a c_a(T_a-T_i)+A_p=\rho_a c_a V\frac{dT_i}{dt} \quad (8)$$

$\acute{m}_a$:Air flow rate into the zone (kg/s)
ρ_a:Air density (assumed to be constant) (kg/m³)
V: volume (m³)
A_p:Internal convective gain (W)
T_a: Air temperature (K)

2.7 Numerical simulation model

The following figure represents the numerical model of the simulation. The inputs and outputs are shown.

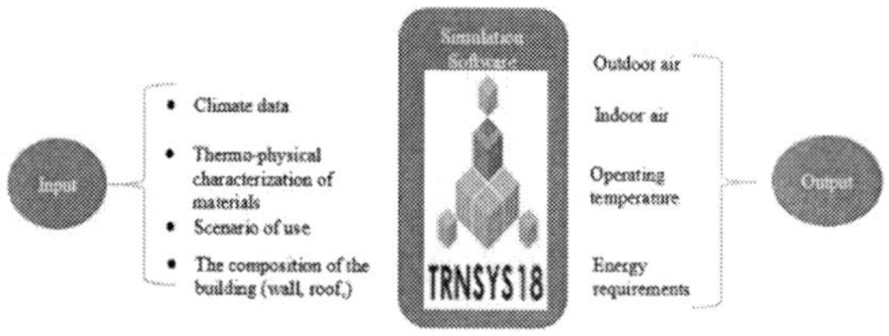

Fig.1.Numerical model

3 Results

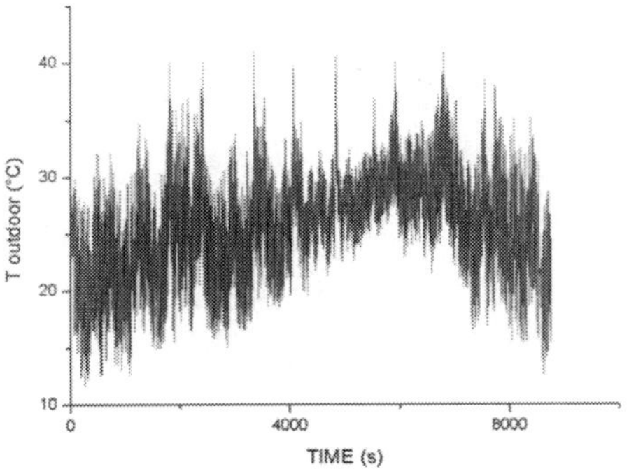

Fig.2.Outdoor temperature

In this section we discuss the results of the simulations in order to evaluate the rate of reduction in energy demand by substituting conventional building by a building with an insulation panel. Indeed, this rate is assessed by comparing these two buildings.

Figure 2, Figure 3 and Figure 4 represent the outdoor, Air and operating temperature of the building, we can notice that the operating temperature is close to the indoor temperature and that the latter is equal to the indoor temperature if the air conditioning is not used otherwise it is equal to the set temperature and that the outdoor temperature fluctuates between 15°C and 40°C.

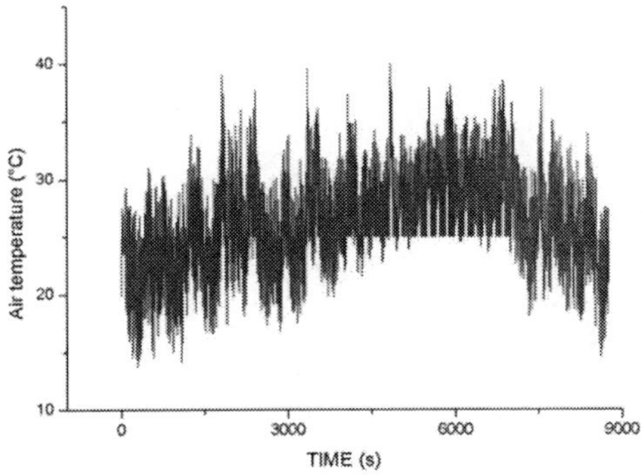

Fig. 3. Air temperature

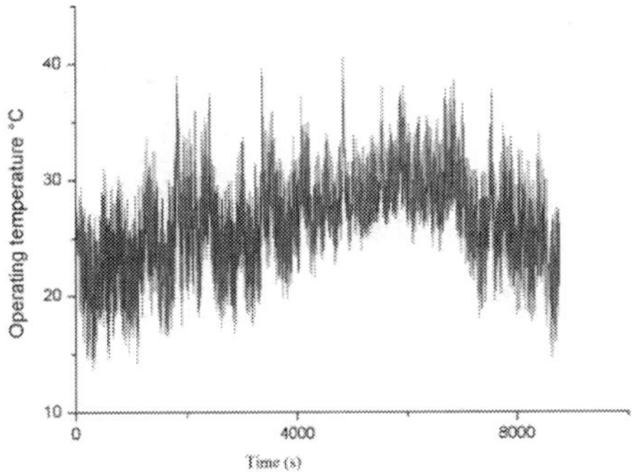

Fig. 4. Operating temperature

Figure 5 represents the energy consumption in order to keep the set temperature when it is imposed, this consumption is zero when the air conditioning is not used.

The curve in Figure 6 shows the energy requirement for air conditioning when the building is occupied and the temperature is set at 25°C.

The numerical simulation on Trnsys, provided us with the annual air-conditioning requirement which is 5903.2 kwh for the building without insulation panel and 4532.06 kwh for the building with insulation panel, hence a 23% reduction in energy consumption.

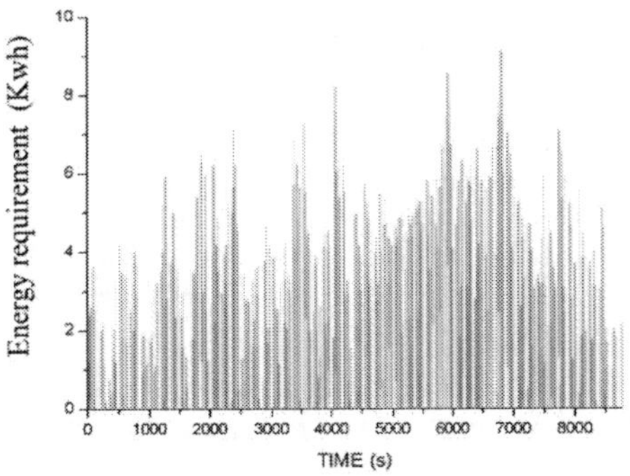

Fig.5. Energy requirement

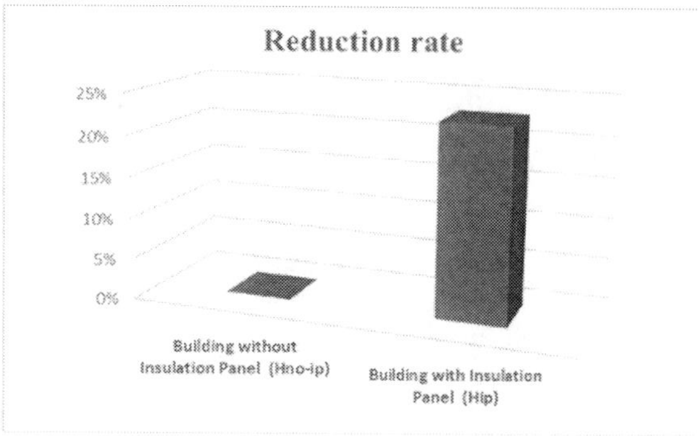

Fig.6.Reduction rate

This result allows us to state that, given the low thermal conductivity of the clay/Typha Australis mixture, the building's energy needs could be reduced by 23% by using this mixture as an insulation panel. this will allow us to valorise this plant as a bio-sourced material.

4 Conclusion

In this article, we have studied the impact of using Typha Australis in a habitat. This study was carried out using clay as a binder and taking conventional buildings with the same characteristics to compare the rate of air conditioning use.

The result allowed us to evaluate the impact of this mixture on a building, because by substituting the cement mortar of the conventional building with the mixture of clay and Typha Australis, the rate of energy requirement decreased by 23% year-round, which is a satisfactory rate

In perspective, we can consider changing the position of this insulation panel in order to better evaluate this optimization.

References

1. Y.BROUARD. Thèse de Doctorat, Université Tours, (2018)

N.Dujardin, D.Thevenot. (25ème Journées Scientifques de l'Environnement - L'économie verte en question, 2014)

M. Lamrani, N. Laaroussi, A. Khabbazi, M. Khalfaoui, M. Garoum, A. Feiz, cas stu in con mat **7,** (2017)

A. El Bakkouri, Thèse de Doctorat en Sciences, Faculté des sciences Tétouan Maroc (2004).

A. Khabbazi, M. Garoum, O. Terahmina, AMSE, Jou.Adv.Of.mode.and.Sim 74,(2015)

Y. Dieye, V.Sambou, M.Faye, A.Thiam, M.Adj,D.Azilinon, Jou of Buil Eng **9**,(2017)

M.T.Diatta,S.Gaye,A.Thiam,D.Azilinon, Cong. SFT, Perpignan (2011)

S. Klei, *A Transient System Simulation Program, Solar Energy Laboratory*, (University of Wisconsin-Madison, Madison,WI, USA ,2000)

TRNSYS 17, Upgrade Description

B. Berges,, (Thèse de Doctorat, Université de Nice, 1986)

M. Annabi, A Mokhtari, TA Hafrad Rev des Ene Ren, **9** (2006)

STRATEGY OF CONTAINER ASSEMBLY FOR A POSITIVE ENERGY IMPACT IN POONA INDIA

F. Mustapha[1,2*], *I. El Abbassi*[1,2,], *A. kaci*[2] ,*E. Kadri*[2]. *M. Darcherif*[1,3]
[1]Laboratoire de Recherche en Eco-innovation Industrielle et Energétique, ECAM-EPMI, Cergy-Pontoise
[2]Laboratoire de Mécanique et Matériaux du Génie Civil, Université de Cergy-Pontoise.
[3]Laboratoire QUARTZ, (EA 7393), Université Paris-Seine.

Abstract. Every year the Indian government is involved in the construction of nearly two million new buildings excluding the tertiary sector. These homes are not only intended primarily to house Indian populations, but the authorities concerned have also placed emphasis on sustainable development, since as all over the world the residential sector is the one that consumes the most energy. Indeed, active and passive strategies to reduce the energy consumption of dwellings have been established and even applied. Among the passive strategies, the global shape of the building, initially specific to modular architecture impacts the aesthetic part of the building but also its energetic bills. The concept of modular architecture is mainly the assembly of identical cells and, in different ways, the overall energy consumption depends on the initial assembly of the cells and therefore on the resulting shape of this assembly. The purpose of this article is to determine the impact of the shape on the energy balance. For this study, the container is the archetype used, in fact the metal box being a standardized, economical and easily transformable product in living space. The complex system resulting from the assembly of several containers offers a new way of thinking about housing from an architectural, economic, aesthetic and energetic point of view. In this article, we will focus on studying the impact of the overall shape of container-type housing on annual energy bill, with the aim of optimizing energy efficiency and achieving an eco-container model that meets international eco-sustainability criteria. The results demonstrate the undeniable correlation between geometric and energy aspects, the latter being mainly related to the internal thermal comfort of the container housing cells.

1.INTRODUCTION

The energy transition is a global issue, precisely for the residential sector, one of the most energy consuming in terms of heating, cooling or lighting, the success of this transition must not be at odds on the user who remains the first consumer, human and housing interaction is very present because the environment has a direct impact on the mental and physical well-being of the individual, rethinking user comfort is a multi-faceted, aesthetic, economic and thermal task, the latter is on the centrepiece of all government and local initiatives.

The reduction of the energy bill is done by optimizing the overall consumption of the building [1], to do so, existing solutions are applied by all the project managers, including passive architecture, which consists in intervening on several levels, the first is the envelope using natural and local materials, then the definition of the thicknesses of the materials and their physical and thermal characterization, the proportion of the size of the glazing compared to the facade, WWR (Windows to Wall Ratio), this last impact considerably the quantity of energy acquired or lost by the whole house. Other passive strategies exist such as geometry, in which interest is growing, this criterion impacts both the aesthetic aspect of the building and its energy efficiency, expressed in heating or air conditioning needs [2].

Many studies have been conducted on the subject of energy transition, and especially on its ecological flight, these surveys now encourage architects to extend their creativity towards the reuse of materials and have a recycling approach, among these concepts, container-type architecture, which was recommended for its economic benefits and speed of execution [3].

In this article, we will focus on this archetype, used in the residential sector, in order to evaluate the impact of geometries on the overall energy bill, and since the Indian climate and more particularly the city of Poona is considered to be hot or even semi-arid, with an annual average of temperature varying between 20° and 28°C [4], this is why the primary interest will focus on the treatment of air conditioning or cooling demand. This comparative study will focus on dwellings made in containers with different morphologies but with the same inner surface area and therefore the same number of container cells. Indeed, having a different assembly necessarily implies a different energy efficiency, aware that the container, initially designed for the logistical sector, represents a cell standardized by the international standardization

https://doi.org/10.1051/e3sconf/202017001004

organization [5]. As such the 20' container has the following dimensions: W: 6058 mm; H: 2591 mm; I: 2438 mm; taking into account these information's and the thermal performance of the envelope and the materials that compose it in order to proceed on numerical study on TRNSYS by applying all the passive strategy principals.

2. Methodology

2.1 Model definition:

The first step consisted in the definition of the envelope and its materials, for this purpose, Table 1 & figure 2 highlights the types of walls used as well as the U-value relating to each one. Then the choice was made for the total number of containers, which is 8 cells in layout. For a total surface area of 110.88m². before moving on to simulations on Trnsys, it is necessary to define the input data for each cell, first the use scenario that is typical for a house (see **Figure 1**) and which is equivalent to the occupancy schedule, then a natural ventilation was imposed. Given the local climate of Poona, the heating will never be active, so the cooling system is the only one imposed with a set temperature of 25°C, other parameters are also taken into account by the software which are as follows:

- Meteorological data: the weather data of Poona India (meteonorme of TRNSYS17)
- Atmospheric pressure: 101325 [Pa].
- Heat of vaporization of water: 2454 [kJ/ kg].
- Stefan Boltzmann Constant: $2.041e^{-007}$ [kJ/hm²K].
- Approximation of average surface temperature: 293.15 [K].

Table 1. Wall type of the container

Wall type	Materials	Thickness [m]	U-Value [W/m²K]
Out wall	Wall board	0.013	0.093
	Wood Structure and Air gap	0.05	
	Hamp wool	0.12	
	Corten steel structure	0.08	
Windows	Glass	0.004	1.4
	Air gap	0.016	
	Glass	0.004	
Roof	Wall board	0.013	0.470
	Wood Structure and Air gap	0.05	
	Hamp wool	0.12	
	Corten steel structure	0.08	
Ground	Keruing wood	0.15	0.852

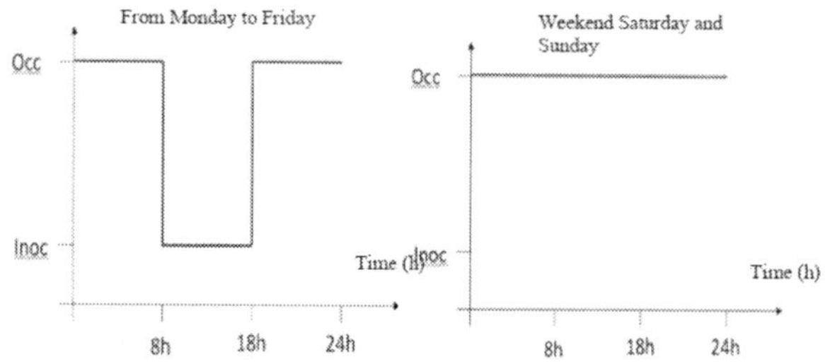

Fig. 1. Users scenario during the week (unoccupied "inc" occupied "occ")

2.2 Simulation Process:

The next phase of the study consist on the definition of the geometries specific to each configuration, we want to demonstrate the impact of morphology on the primary demand for cooling energy, the strategy chosen and to vary the designs according to a criterion which is the number of exposed surfaces, and therefore naturally vary the compacity of these container assembly. The first objective is not only to reduce the energy consumption of the building only with the overall shape but also to increase the indoor comfort for the user. These containers layouts are described on the table below (See **Table 2 & Figures 3-6)**

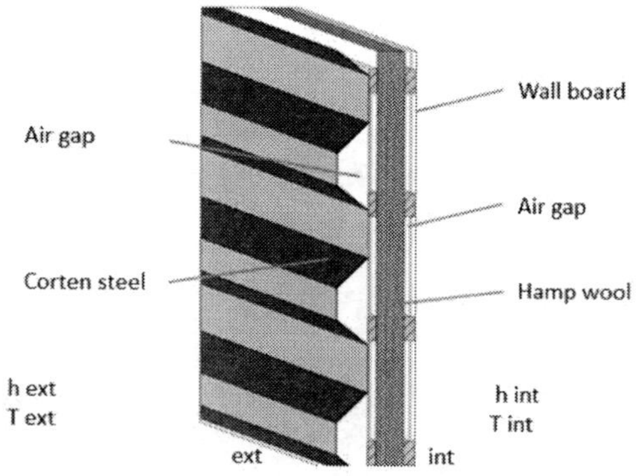

Fig. 2. Axonometric wall section

Table 2. Configurations descriptions

Configurations	Perimeters (number of exposed facades)	Surfaces [m²]
Configuration a	12	216.10
Configuration b	14	244.30
Configuration c	16	243.82
Configuration d	18	272.02

Ours models present the same solar exposition, in fact all the doors are oriented west, and all the windows are oriented south, and a WWR around 20%.
When the simulation is launched, results representing an hour-by-hour assessment of the studied area for the whole year, this numerical calculation is based on the following equation (see equation (1)):

$$Qbal = DQ_{air} + Q_{heat} - Q_{cool} + Q_{vent} + Q_{inf} + Q_{trans} + Q_{gainint} + Q_{sol} \quad [KJ/H] \quad (1)$$

Where : Qbal represents the energy bal of the set geometry.
$dQair/dt$: Indicates the variation in enthalpy of the air in the zone, it reflects either a decrease or an increase in temperature. [kJ]

Q_{HEAT}: Indicates the heating requirements by the system to reach the set temperature. [kJ]

Q_{COOL}: Indicates the cooling requirements by the system to reach the cooling setpoint temperature. [kJ]

Q_{INF}: Indicates losses through infiltration. [kJ]

Q_{VENT}: Indicates losses through ventilation. [kJ]

Q_{TRANS}: Indicates losses related to wall transmissions. (Surface and linear losses). [kJ]

$Q_{GAININT}$: Indicates internal inputs. [kJ]

Q_{SOL}: Indicates internal gains related to solar gains (short waves). [kJ]
The thermal energy transfer rates of ventilation air infiltration and flow are calculated by the following equations, respectively :

$$Q_{inf} = m_{inf} C_{air} (T_{air} - T_{ext}) \quad (2)$$

$$Q_{vent} = m_{vent} C_p (T_{vent,ext} - T_{vent,int}) \quad (3)$$

$$Q_{trans} = S\, h_{conv}\, (T_{surf} - T_{air}) \quad (4)$$

Where :
m_{inf}: Air flow rate due to infiltration [kg / s]
Cair: The specific heat of the air is assumed to be constant and estimated at 1008 [m2 s- 2 K-1, d kg-1 K-1]
Cp : specific heat [J kg-1 K-1]
m_{vent}: Air circulation due to ventilation [kg / s]
T_{air} : Air temperature [°C]
h_{conv} : Coefficient of heat flux exchanged by convection [w m-2 K-1]
S: the solar factor, represents the total amount of energy that is allowed to pass through glazing in relation to incident solar energy.

The gross requirements representing the total losses of the building are the addition of $Q_{TRANS} + Q_{INF} + Q_{VENT}$ and the free contributions are the sum of $Q_{GAININT} + Q_{SOL}$ that we subtract from the total losses in order to know our net requirement which is the cooling requirement.

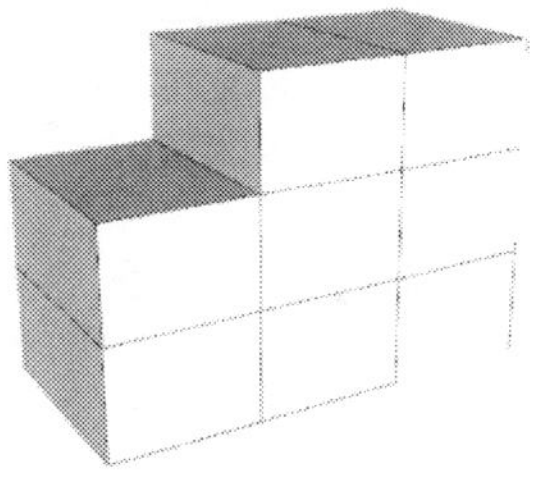

Fig. 3. Configuration a

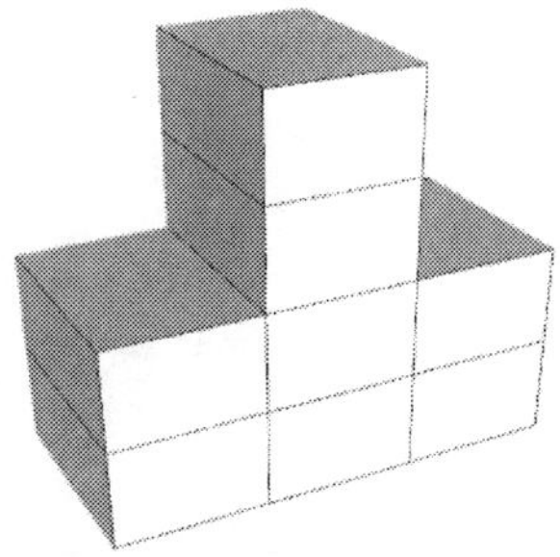

Fig. 4. Configuration b

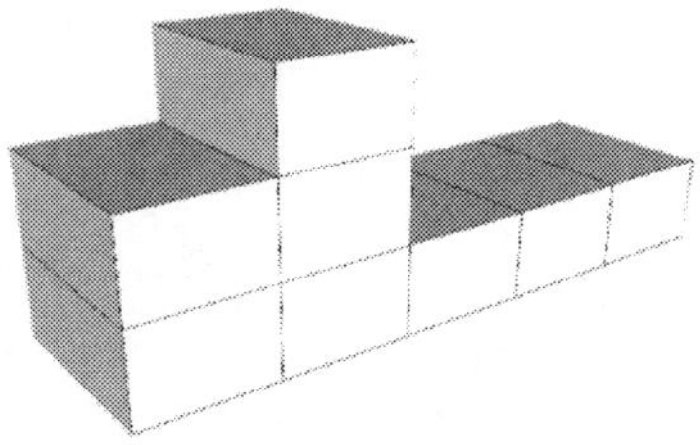

Fig.5. Configuration c

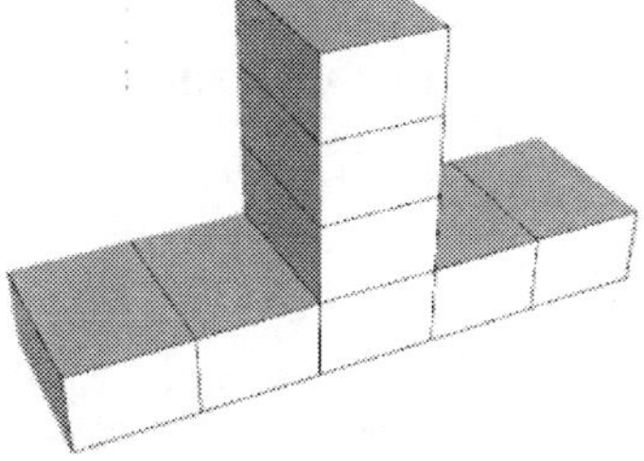

Fig. 6. Configuration d

In the next step of the study is to study the compactness ratio relative to each geometry, we have chosen to define it by the ratio of the spreading of geometry on the ground and exposed surfaces (see equation 2). The interest of this part is demonstrated by the extent of its impact on the energy balance.

$$Rc = \frac{Aext}{Af} \quad (5)$$

Where:

Rc is the ratio of compacity.

A*ext*: represents the external area [m²].

A*f*: represents area of the ground[m²].

Table 3. Compacity ratio by configuration

Rc configuration a	Rc configuration b	Rc configuration c	Rc configuration d
5.20	5.88	3.52	3.93

The energy balance produced by TRNSYS on the proposed models has been simulated for a year with a time step of one hour. The hourly results were integrated to obtain the monthly and annual summaries in the results section.

3. Results

The results show an improvement in terms of cooling energy demand, especially configuration c, which gives a better profitability (see **Table**) , this geometry presents not only an economy but also a considerable exposed roof area 69.3m². This morphology is interesting from an architectural point of view, the latter has an irregular reading and Skyline with a difference in level in steps, which in turn provides a specific shading that improves indoor comfort, and thus directly impacts the energy efficiency of the design.
As our model is a container with a metallic outer shell, this highly conductive material makes it possible to benefit from natural heat.

The configurations a & b show a similarity both in terms of compacity ratio and cooling demand, but at the formal level the two designs are indeed different, configuration a has only one lateral layer and therefore one possibility of exploitation, while configuration b has two lateral surfaces that can be used later on, during the space distribution and interior design phase.

Table 4. Energy balance by configurations during a year

Configurations	Configuration a	Configuration b	Configuration c	Configuration d
QCOOL [KWh /Year]	34361.11	34444.44	33583.33	35055.56
QINF [KWh /Year]	-79.78	-77.67	-78.56	-71.75
QTRANS [KWh /Year]	-3230.56	-3097.22	-2289.17	-2474.17
QGAININT [KWh /Year]	22902.78	22902.78	22902.78	22902.78
QSOLGAIN [KWh /Year]	14769.44	14730.56	13055.56	14711.11
QCOOL [KWh /Year]	309.89	310.65	302.88	316.16

All these developed models allow not only an energy saving, but also provide a shade that is specific to the shape of the building, so this passive design strategy permits an economic saving on the overall cost of the eco-container building.

Configuration c shows an energy inertia that is explained by its compacity ratio (see **Table 3**) which is the lowest of all the designs, this ratio is defined, for the Poona climate.

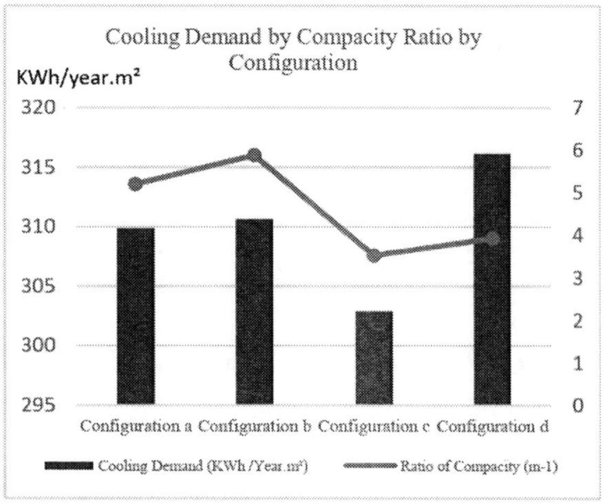

Fig. 7. Cooling Demand by Compacity Ratio and by Configuration.

This type of architecture is recommended especially for its ecological approach; however, the energy approach must be done from the design phase and would be appropriate to India.
Figure 7 illustrates the correlation between the compacity ratio as defined by us and the air conditioning demand expressed by each of the configurations, and we can observe that the higher the compacity ratio, the greater the demand expressed.

Since configuration c is the one that has demonstrated the lowest demand in terms of air conditioning, and for this reason, the study is extended in more detail on this model, in order to better target the most critical period.

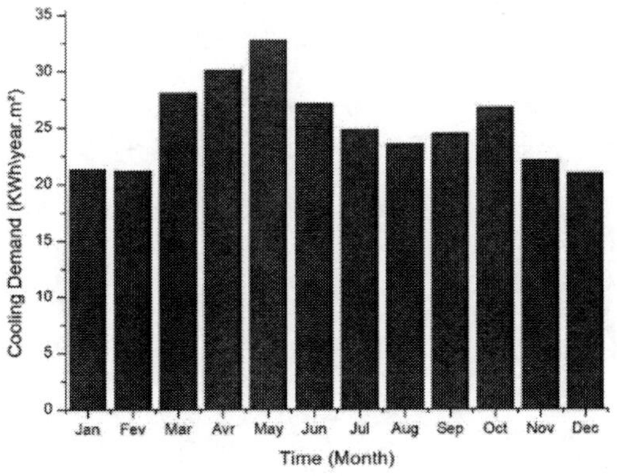

Fig. 8. Cooling demand per months of configuration c

The figure above (see **Figure8**) shows the cooling demand per month and over a whole year for configuration c. We observe that the most energy-intensive period is from March to July. Nevertheless, for the rest of the year, consumption is more or less close.

4.CONCLUSION

Bioclimatic design is based on several parameters such as materials, orientation, wall composition, compacity, but also the climatic factor. The simulations carried out made it possible to evaluate the annual energy consumption of an eco-container used as elementary cells in our study, and shows the correlation between shape and energy in building sector.
With the resulting shape of the containers, the result is interesting not only from an architectural point of view but also in terms of energy. The dynamics of the stacked configurations has allowed the creation of new spaces that can take on another function, such as a terrace or even a garden, from a structural point of view the containers are designed to be stacked and juxtaposed.
This study represents the advantage of the construction of the container in semi-arid or even arid climates, the economic opportunity is such that today we can design, build and live in a container without fear of discomfort.

Given the results obtained, we will work on more complex shapes and thus enlarge the scale, move on to collective housing and the eco-neighbourhood, with taking into consideration the comfort of the user and compliance with the thermal regulations in force on the site concerned.

References

1. G. O. J. a. N. T. Roshan, *Ene. Pol.* **49**. 731-739, (2012).
2. C.Hachem-Vermette, *Sol Ene.* ***159***. 710-721, (2018).
3. J. Vijayalaxmi, *Loc Env.* **15**. 13. 245-259, (2010).
4. YASHADA, «Nati Symp on Nema Mana: *Nematological Society of India and Indian Council of Agricultural Research (*2015).
5. «International Organization for Standardization,» [En ligne]. Available: https://www.iso.org/about-us.html.

Numerical study of the effect of DSF walls geometrical shape on heat transfer

Mohamed Elamine GHEDHAB[1,2*], *Ikram El ABBASSI*[1], *Rafik ABSI*[1] *and Yannick MÉLINGE*[2]
[1]ECAM-EPMI, LR2E, 13 Boulevard de l'Hautil, 95092 Cergy-Pontoise, France
[2]Université de Cergy-Pontoise, L2MGC, 5 Mail Gay-Lussac 95031 Cergy-Pontoise cedex, France.

Abstract. The building envelope is an important element that influences energy performance, both in terms of gains and losses. Indeed, much research has focused on improving and optimizing this element. The façade represents the main component of the envelope; it plays a crucial role lying in the protection of the internal environment of the building from external climate variations, by providing thermal comfort to its occupants. In order to gain in energy efficiency, a numerical investigation (CFD) on the influence of the geometric shape of façades on heat exchange in the building is carried out in this work. More particularly, the study will be established on double skin façades (DSF) representing the most commonly used facade typology on high-rise buildings. The simulations will be carried out taking into account the climate in Paris region. The first part will be dedicated to the definition of the basic model, which will be considered as a reference case with plan surfaces on the outside and inside of the building. In the second part, the influence of the shape of the external façade on heat exchange will be discussed.

Keywords:

Heat transfer, double skin façade, CFD simulation

1) Introduction

Conjugate heat transfer inside enclosures is a classical model used to study numerous engineering and physical phenomena, such as combustion engine, nuclear energy, food industry, solar systems and civil engineering [1]. To use this model in the building application, several experiences and numerical investigations were conducted in this last years, mixed natural convection-conduction were studied for different configurations, like double skin facades.

Many researches on the conjugate natural convection, conduction and radiation heat transfer has been done. Li et *al*. [2], studied the influence of the optical constants of semi-transparent wall on laminar natural convection, their study shown that the convective transfer decreases in the semi-transparent configuration compared to the opaque one. Xaman et al. [3] studied free convection combined with thermal radiation inside two-dimensional square cavity with a simple glass wall, they concluded that the flow is not symmetrical, this behaviour was explained by the fact that there is a temperature variation over the glass and the radiation interchange. González-Julián et al. [4] investigated four cases of double panel window for a hot climate, taking into account the three heat transfer methods, results show that the configuration with a reflective treatment glass can reduce the heat flux passing this windows about 72% compared to the their reference configuration with à simple glass window. Khatamifar et al. [5] examined the effects of partition position and thickness on the unsteady-state conjugate natural convection heat transfer in a differentially heated partitioned enclosure over a wide range of Rayleigh numbers, and found that when the thickness of the partition is increasing the average Nusselt number decreases. They also found that the position of the middle wall has a negligible effect on the Nusselt number. Liqun et al. [6] presented a two-dimensional analysis of turbulent combined convection radiation inside a divided enclosure with double semi-transparent walls, it was found that the transmittance has a great impact on the heat and flow characteristics inside the enclosure. Also it was concluded that the semi-transparent walls reduce the heat losses.

This study is focused on the effect of geometrical shape of the external wall, for the double skin façade on rectangular enclosure. The aim of this configuration is to evaluate the temperature distribution in the internal environment when we modify the external semi-transparent form. The present study is organised as follow: introduction is presented in section 1. Section 2 introduces the problem formulation. Numerical approach is described in the section 3. In section 4, the results, analysis and discussion are reported. Finally, conclusion and future research areas are summarized in the section 5.

 https://doi.org/10.1051/e3sconf/202017001005

2) Problem formulation

The geometrical model configurations are shown in the figure 1, they represent two different shapes (case a and case b) of the external façades with two cavities in each case. The dimensions of the first and the second case are mentioned in the Table 1, the equivalent area of the cavity 1 is the same as of the case 1. The considered fluid and solid materials are respectively the air and the glass, their thermo-physical properties are presented in Table 2.

The ambient temperature is T_a=5°C with an external convective coefficient of 6.8 W/m².K, which corresponds to a value of air velocity of 3 m/s [3], the mean solar radiation[a] is 73 W/m², the inner temperature is fixed to 19°C.

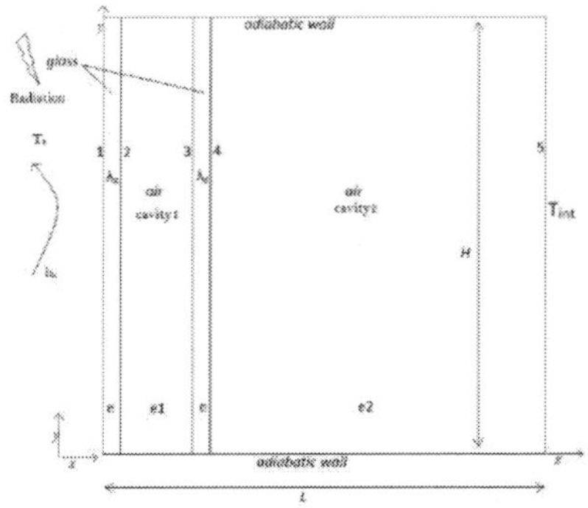

Case (a)

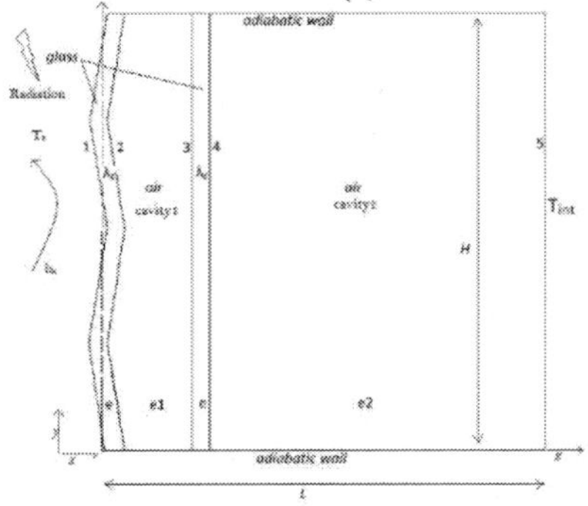

Case (b)

Fig 1. Schematic representations of the two studied cases

Tab 1. Geometrical dimensions

H =L[mm]	100
e[mm]	4
e1[mm]	16
e2[mm]	76mm

[a] and www.meteorange.fr

Tab 2. Thermo-physical properties

	symbol	Air (12°C)	Glass
Density	**ρ[kg/m³]**	1.238	2500
Specific heat	**Cp[J/kg.K]**	1006	720
Thermal conductivity	**λ[W/m.K]**	2.454x10^{-2}	1
Thermal diffusivity	**α[m²/s]**	1.944x10^{-05}	-
Viscosity	**μ[kg/m.s]**	1.788x10^{-05}	-
Emissivity	**ε**	-	0.85
Absorption coef.	**a**	-	0.14
Thermal expansion coef	**β[1/K]**	3.5069x10^{-03}	
Gravity	**g[m/s²]**	9.81	

Assuming that the fluid is incompressible, Newtonian and the flow is in the steady-state and laminar, the fluid flow in both cavities is represented by the following 2D: continuity, momentum (Naviers-Stokes) and energy equations, taking into account the Boussinesq approximation for buoyancy forces, with the temperature equation in Cartesian coordinates presented as follow:

$$\frac{\partial u}{\partial x}+\frac{\partial v}{\partial y}=0 \quad (1)$$

$$u\frac{\partial u}{\partial x}+v\frac{\partial u}{\partial y}=\frac{1}{\rho}\left[-\frac{\partial p}{\partial x}+\mu\left(\frac{\partial^2 u}{\partial x^2}+\frac{\partial^2 u}{\partial y^2}\right)\right] \quad (2)$$

$$u\frac{\partial v}{\partial x}+v\frac{\partial v}{\partial y}=\frac{1}{\rho}\left[-\frac{\partial p}{\partial y}+\mu\left(\frac{\partial^2 v}{\partial x^2}+\frac{\partial^2 v}{\partial y^2}\right)+\rho g\beta\Delta T\right] \quad (3)$$

$$u\frac{\partial T}{\partial x}+v\frac{\partial T}{\partial y}=\alpha\left(\frac{\partial^2 T}{\partial x^2}+\frac{\partial^2 T}{\partial y^2}\right) \quad (4)$$

Additionally, the boundary conditions are considered as:
- No slip condition on walls surfaces (u=v=0);
- Top and bottom walls are set as adiabatic (q=0);
- Wall 1: $(T_1-T_a)h_a+\sigma\varepsilon(T_1^4-T_a^4)=-\lambda_g\frac{\partial T_{g,1-2}}{\partial x}$;
- Wall 2: $-\lambda_g\frac{\partial T_{g,1-2}}{\partial x}=-\lambda\frac{\partial T}{\partial x}+q_{2,rad}+q_{2,conv}$;
- Wall 3: $-\lambda_g\frac{\partial T_{g,3-4}}{\partial x}=-\lambda\frac{\partial T}{\partial x}+q_{3,rad}+q_{3,conv}$;
- Wall 4: $-\lambda_g\frac{\partial T_{g,3-4}}{\partial x}=-\lambda\frac{\partial T}{\partial x}+q_{4,rad}+q_{4,conv}$;
- Wall 5: $T_5=T_{int}$.

3) Numerical approach

In this investigation, the governing equations are discretised by the finite volume method by using the solver Fluent V16.0. The SIMPLEC algorithm is chosen for the velocity-pressure coupling [7] [8]. The other discretization schemes are maintained in the second order and pressure-based solver is done by PRESTO. The Discrete Ordinates (DO) scheme is fixed in the first order [9].
In order to validate the proposed model, simple semi-transparent wall's cavity is suggested to compare the obtained results with the numerical study of Xaman [3].

To this end, The Rayleigh number is estimated to 10^6 inside the enclosure. Figure 2 illustrates the dimensionless temperature distribution at mid-height of the cavity compared to Xamán et al. simulations for $Ra=10^6$ [3].

$$X^* = \frac{x - e_1}{L} \qquad T^* = \frac{T - T_2}{T_{int} - T_2}$$

Where T_2 is the temperature in interface 2.

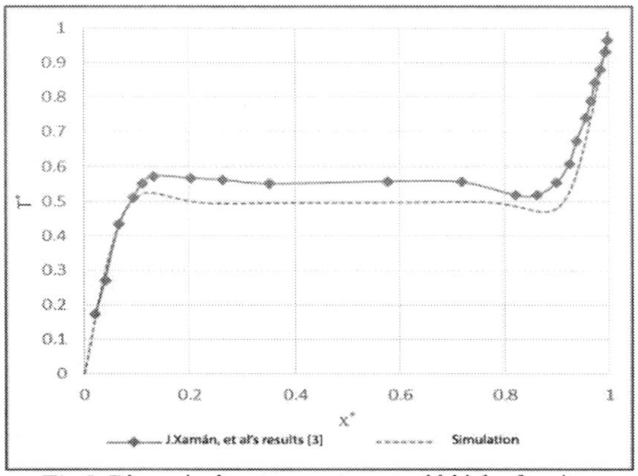

Fig 2. Dimensionless temperature at mid-high of cavity

4) Results and discussion

The simulations are performed taking into consideration the data climate of Paris region[b]. The mesh used for the first configuration has a rectangular form with equal sizes in the x and y-axis ($\Delta x=\Delta y=8x10^{-4}m$). For both cases, previously presented in Figure 1, the value of the Rayleigh number is Ra $\approx 3x10^5$ for each case of the second cavity.

Figure 3 represents the temperature profile in the internal surface 4 of the internal wall for the two cases.

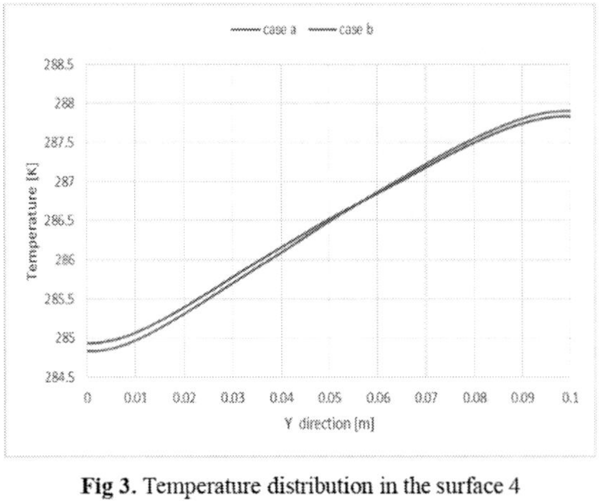

Fig 3. Temperature distribution in the surface 4

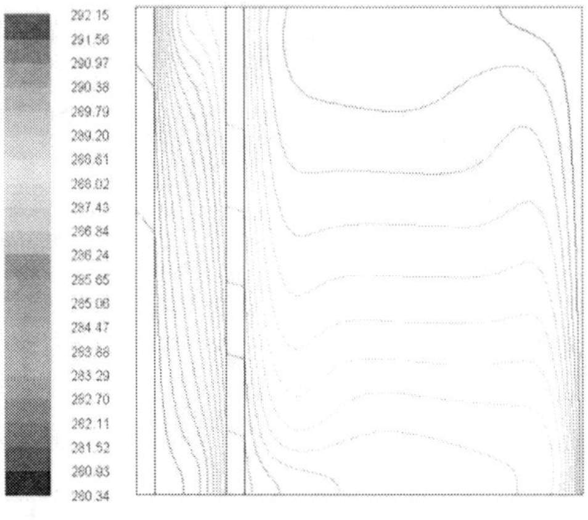

Case (a)

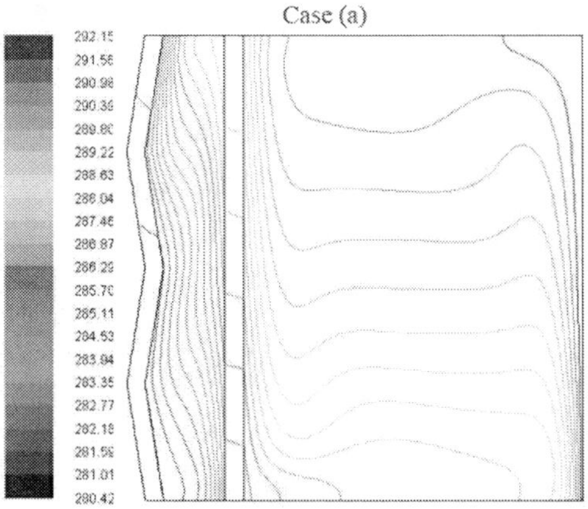

Case (b)

Fig 4. Isothermal contours

The results show that the temperature profiles are symmetrical with the mean value, but there is a small difference between the two cases. In case (b) the temperature is higher in the bottom and lower on the top compared to case (a). This involves a lower mean temperature in case (b) with the proposed modified geometry (figure 3).

According to figure 4, it can be clearly seen that the isothermal contours have not a considerable variation. However, refereeing to the colormaps, the value of the temperature in case (b) is lower than the value in case (a) in term of the mean temperature. More analysis is needed since external surfaces and external heat convective coefficients are not the same in the two cases.

Knowing that a common colormaps scale is used to evaluate the slight difference in the stream function contours, it can be observed through figure 5 that the values of the stream function of case (a) are greater than in case (b) in both cavities. This means that the flow is weak in the enclosure with the modified external wall.

[b] www.climate-data.org

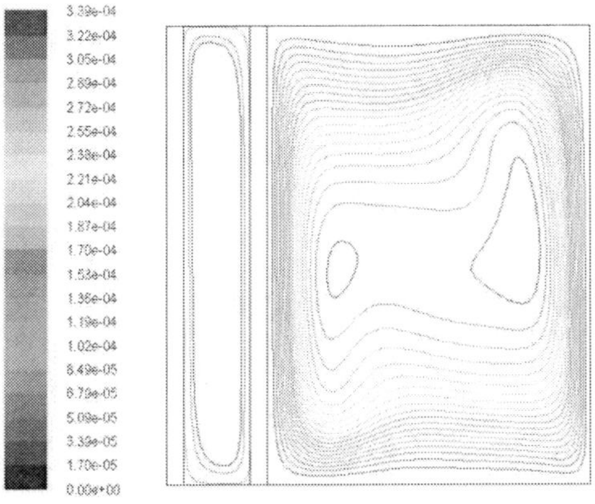

Case (a)

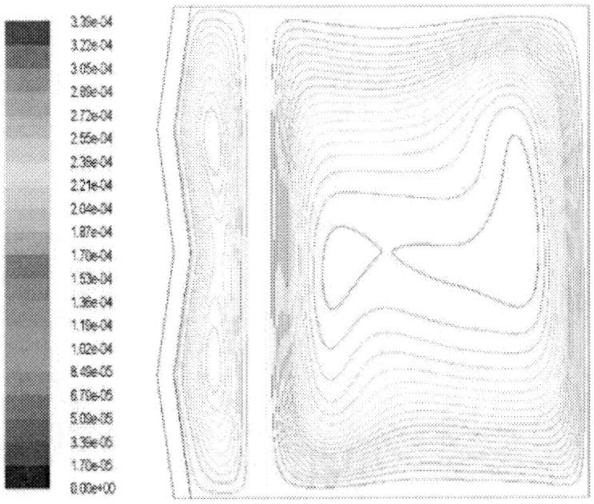

Case (b)

Fig 5. Stream-function contours

5) Conclusion and perspectives

In this work, a numerical study has been conducted for a double skin façade configuration so as to evaluate the effect of external shape on the thermal transfer and inside environment flow pattern. The obtained results show that the bottom and top temperatures on the indoor surface of the internal wall are impacted (when the geometry is different), we observed that the temperature difference values are small between the convectional and proposed geometry. This difference could be related to the considered small scale of our cases compared to the real building dimensions on the one hand and to the flow regime on the other hand. Additionally, we took the same value for the convective coefficient in both cases. However, this value should be different when the external surface is modified (case b). This will be considered in our future simulations. Simulations are currently being carried out with realistic dimensions corresponding to real building configurations in order to provide practical solutions for architects and civil engineers in term of thermal comfort.

References

[1] A.Baïri, E.Zarco-Pernia and J.-M. d. María, *Applied Thermal Engineering,* **63**, 304-322, 2014.

[2] D. Li, Q. Ai, X. Xia and G. Wu, *International Journal of Heat and Mass Transfer,* **67**, 724-733, 2013.

[3] J.Xamán, J.Arce, G.Álvarez and Y.Chávez, " *International Journal of Thermal Sciences,* **47**(12), 1630-1638, 2008.

[4] E.González-Julián, J.Xamán, N. O.Moraga, Y.Chávez, I.Zavala-Guillén and E.Simá, *Applied Thermal Engineering,* **143**, 100-111, 2018.

[5] M. Khatamifar, . W. Lin, S. Armfield, D. Holmes and M. Kirkpatrick, *International Communications in Heat and Mass Transfer,* **81**, 92-103, 2017.

[6] Z. Liqun, L. Jiaying, H. Qunwu and W. Yiping, *Transactions of Tianjin University,* 25(5), 472-487, 2019.

[7] R. Absi, in *Proc. 27th International Conference on Coastal Engineering (ICCE), July 16-21, 2000,* Sydney, Australia, 2000.

[8] N. El-Gharbi, R. Absi and A. Benzaoui, *J. Appl. Fluid Mech.,* **5**, 63-70, 2012.

[9] "ANSYS Fluent User's Guide, Release 15.0. 2620 p.," ANSYS INC, 2013.

Scheduling problems with energy consideration: State of the art and opportunities

Ghita Lebbar[1], *Ikram El Abbassi*[1], *Moumen Darcherif*[1], *Abdellah El Barkany*[2]
[1]ECAM-EPMI, 13 Boulevard de l'Hautil, 95092 Cergy-Pontoise
[2]Mechanical Engineering Laboratory, Faculty of Sciences and Techniques, Sidi Mohammed Ben Abdellah University, Fez, Morocco

Abstract. The target of this study is to find an appropriate research framework on the basis of the main published works designed to resolve the scheduling issues with energy aspect consideration. Relatively, the major part of work is placed in the context when modeling scheduling is subject only to one constraint and criteria, which in reality is not constantly the case. This study is handled in the framework where job transactions are interrupted at the similar time by set of technological, environmental and especially energetic constraints that has been previously little discussed in the literature. As soon as the objective functions, constraints, models, approaches and the scheduling environments that are defined as the fundamental elements of scheduling problems resolution are determined, a literature review of the present state of the art is introduced. Then, an analysis is carried out to highlight the fundamental ideas leading to fix a future research area that still require a huge deal of attention.

1 Introduction

The actuality of climate change and the magnitude of this total challenge are well established. The need and urgency for action is now recognized by mutually the scientific researchers and political leaders. Since Copenhagen, a global ambition level has been defined, with the purpose of preventing a warming of more than 2°C compared to the pre-industrial era.

Climate change is intrinsically linked to our dependence on fossil resources: it results from the fact that humanity, by burning at high speed fossil organic fuels accumulated over millennia, releases big quantities of carbon into the atmosphere. Climate change and energy scarcity are two front of the same coin; these crises cannot be dealt with separately. To cope with this, it is obviously necessary to spotlight on energy demand in the various activities sectors and mainly, the manufacturing one.

This augment in energy consumption has become a considerable interest in particular for energy-intensive industries, counting mussels, chemicals and petroleum. For example, the energy use of the ferrous metal casting and pressing industry (mainly injection moulds and stamping dies) represents about 28.8% of the overall energy consumption of the industrial area in 2014.

Under those conditions, minimizing energy consumption represents a major challenge in these environments, leading not only to an efficient industrial system in terms of cost, quality and time. But also towards a global system that effectively responds to the enormous environmental, economic and technological hazards at the same time. This optimization can be attained d by using several keys representing problems made complex by the application of the real constraints of the environments in question. In particular, the problem of scheduling tasks which is considered as one of the significant tools that have an impact on machine consumption by affecting the quality, efficiency and cost of the production implementation.

In this context, this research work revolves around a literature review and a detailed bibliography analysis of scheduling problems aimed at optimizing mainly energy criteria with energy considerations in order to fix the critical issues that still necessitate a great interest and determine future research area.

The rest of this paper is structured as follows. After explaining the research approach adopted to select interesting documents, in Section 2, fundamentals elements, which are considered essential for the definition of a scheduling problem with energy aspect are introduced in section 3. Section 4 reports a global literature review for the production scheduling problems under investigation. Finally, conclusions are drawn together with perceptions for future research in Section 5.

2 RESEARCH APPROACH

Given that, the interesting papers are dispersed on different sources, it is extremely arduous to establish an exhaustive bibliographic study of the studied problem. Consequently, a documentary research tool is used in this research work. It is divided on publication paper, conference papers, student reports and theses which have been deeply examined so as to attain a comprehensive bibliography on the scheduling problem with energetic considerations like: Scopus-Elsevier, Taylor and Francis Group, IEEEXplore, Springer Link...

Firstly, the following keyword was used: 'scheduling' 'energy constraint' and 'energy consumption'. This procedure originally produced interesting papers; afterwards, each paper has been attentively investigated to erase those that are not linked to the requested search criteria. Finally, the pertinent papers are chosen and arranged by year of publication as mentioned in Figure 1.

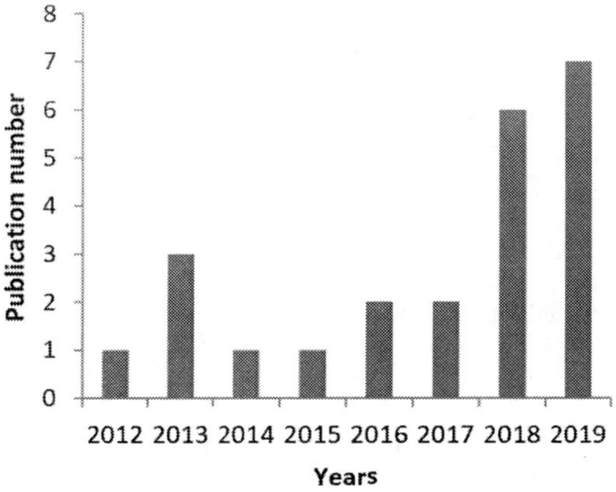

Fig.1: Evolution of publications of scheduling problems with energy constraints

It can be noted that, the growth of publications for the studied problem has increased rapidly in recent years, although during a few years the quantity of publications has been almost invariable, there is a trend of progress in the research for flow shop type of scheduling problems with energetic aspect, which further demonstrates the importance of looking more closely at these types of problems.

3 Fundamental elements of production scheduling with energetic aspects

According to [1], Scheduling the operation of an industrial production system consists in managing the allocation of resources over time, while optimizing a set of criteria as much as possible. From an industrial point of view, it consists of a set of N jobs J=(J1,J2,....,Jn) that have to be handled on a set of m machines M=(M1,M2,...,Mm) in accordance with the order which mainly depends on the type of scheduling workshop studied.

Basing on the references that had handled the scheduling problems with energetic consideration, we selected the most used configurations in the literature.

In our research work, we are mainly concerned with the scheduling problem with energetic aspect, whose main target is to find the best possible scheduling sequence that not only minimizes processing time criteria such as delay, advance, makespan and flow time. But also energy efficiency measures such as minimizing energy consumption, energy costs, peak power and total carbon emissions.

Consequently, figure 2 includes different the scheduling environments adopted, the developed models, the adopted approaches, the targeted criteria and the constraints considered in the literature.

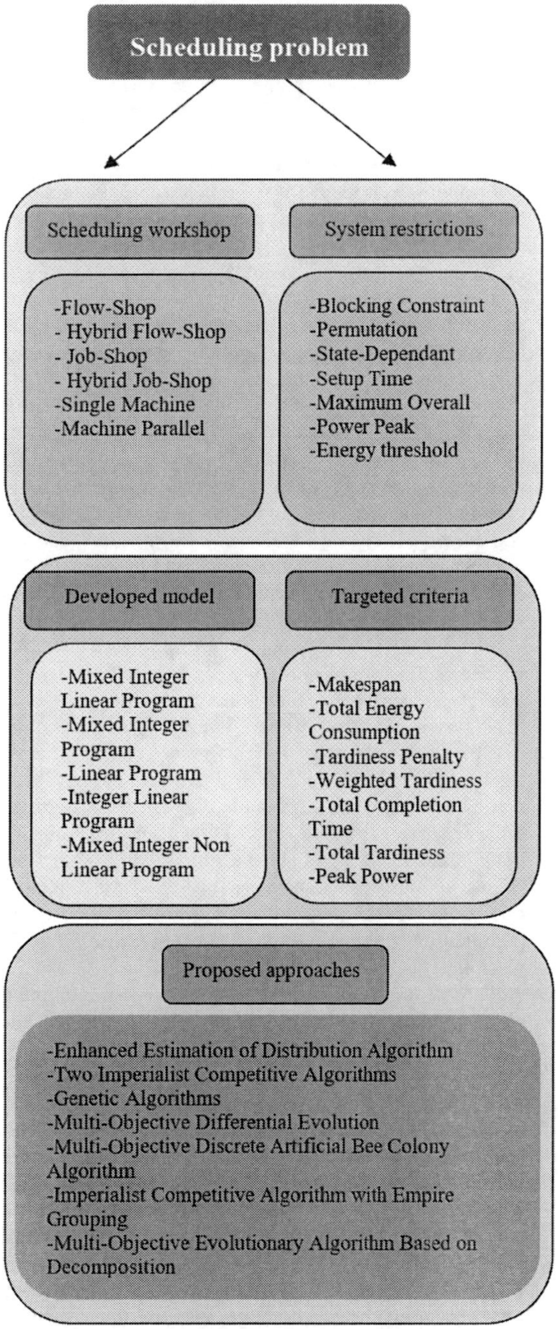

Fig. 2: The different configurations handled in the literature

4 Production scheduling with environmental aspects-literature review

Among the studies that have been published in this direction, Yan et al. [2] handled the hybrid flow shop scheduling problem (HFSP) aiming at minimizing makespan and total energy consumption with a multi-stage optimization approach. Luo et al. [3] adressed the hybrid flow shop scheduling problem (HFS) where the electric power cost under time-of-use (TOU) electricity prices and production efficiency are considered, to handle this problem, a multi-objective ant colony optimization algorithm was inroduced. An original technique to modify the scheduling of the jobs to account for peak power is suggested by Bruzzone et al. [4]. To optimize maximum of completion time and total energy consumption in flexible flow shop scheduling, a genetic-simulated annealing algorithm have been developed by [5]. Luo et al. [6] introduced a novel ant colony optimization aiming at minimizing the electricity consumption cost for the hybrid flow shop scheduling problem. An improved particle swarm optimization (PSO) have been suggested by [7] for solving the flexible flow shop scheduling problem. Lei et al. [8] developed a lexicographical tool to optimize total tardiness and total energy consumption considered as key target and as non-key one, respectively. A multi-objective optimization scheduling model where cost-weighted processing quality, makespan, and energy consumption, have been optimized, is suggested by Jiang et al. [9], they have applied an improved NSGA-II approach regarding crossover and mutation strategy aiming to surmount the prematurity of population. Mokhtari and Hassni [10] have handled the HFS problem (FJSP) by suggesting a multi-objective optimization model with three objective functions namely, total completion time, the total availability of the system, and total energy cost of production and maintenance operations. Liu et al. [11] have addressed the fuzzy flow shop scheduling problem where the target was to transmit operations to the ressources and to fix the optimal permutation and state shift of each machine so as to minimize tardiness and energy consumption. To handle the problem, they have proposed two common model corresponding implementation and heuristics and then they have combined them with the famous genetic algorithm. A multi-objective optimization algorithm (EA-MOA) is proposed by [12] to optimize simoultanously the makespan and the energy consumptions for a HFS problem with the consideration of the setup energy consumptions. A Multi-criteria MILP formulation for solving the HFS scheduling is developed by [13]. In another study [14], a MILP formulation is suggested to look for the optimal solutions for the HFS Scheduling Problem. The purpose of this study was to optimize the total production costs counting transportation and energy costs, basing on a (GA) with a detailed matrix encoding technique. Salido et al. [15] have handled a jobshop scheduling problem designed to optimize energy efficiency, robustness, and makespan. Wang et al [16] have investigated the flow shop scheduling with the constraint of the limited buffer between consecutive resources and the aim of minimizing maximum completion time and energy consumption. To resolve this problem, they have proposed a multi-objective parallel variable neighborhood search (MPVNS) algorithm. Chengzhi Guo and Deming Lei [17] suggested a novel imperialist competitive algorithm (ICA) for solving the multi-objectif FJSS problem . The main target of their work was to optimize total tardiness and makespan with a threshold of the total energy consumption imposed. Zhong et al [18] have solved the multi-objective based green flow shop scheduling problem (MOPFS) using a hybrid shuffle frog leaping algorithm based on the cuckoo search algorithm (HFLCS), the performance measures was the makespan and energy consumption. Recently, Lu et al. [19] have developed a MIP model for solving the HFS scheduling problem considering energy consumption and noise pollution . Then , to handle this problem, they suggested a MOCGWO approach with the aim to balance exploration and exploitation by integrating the qualities of variable neighborhood search (VNS) for intensification and cellular automata (CA) for diversification. Min et al. [20] addressed the FJSS shop scheduling problem with transportation constraints, the aim of their study is to minimize energy consumption and makespan. Li et al [21] investigated the Energy-efficient hybrid flow shop scheduling problem with makespan ,total tardiness, and total energy consumption criteria. The problem resolution was based on a new Pareto dominance and a two-level imperialist competitive algorithm (TICA). Zhang et al. [22] proposed a scheduling approach for the MOFJSP and the resultant mathematical model, considering both the total production energy consumption and makespan. Besides, transportation with its influance on the energy consumption was considered in their model for realistic application. An approached tool based on the NSGA-II was applied to solve the considered problem. Zhou et al. [23] have solved the HFS scheduling problem with fuzzy processing time by proposing a multi-objective formulation that minimizes simultaneously total energy consumption andtotal weighted delivery penalty. In their study they have considered Setup times as sequence-dependent and in-stage parallel, resources are unrelated. Jiang et al. A MILP model have been suggested by [24] for solving the HFS scheduling problem with restricted buffers. The aim of their study was to minimize non-processing energy (NPE) and the total weighted tardiness (TWT).

Analysis and discussion

The greater part of the research that has been published has paying attention on the optimization of energy consumption in conjunction with makespan in order to realize a production environment that uses maximum resources but with minimal energetic consumption. For manufacturing environments, the HFS is the most studied. With consider to resolution approaches the majority of research teams have leaned towards approximate resolution methods. Finally, it can be noted that a lot of energetic and technological restrictions related to the short planning energetic systems has been handeled in the literature.

5 Conclusion

In the recent years, with the massive increase in energy consumption and its dangerous consequences essentially affecting the international problem of climate change. industries are seeking to improve their production processes by paying particular attention to the energy aspect. In this context, the research on the integration of the problems encountered at the level of the industrial environment by considering the energy constraints with a very important weight has been of great interest in the last decades. Among those problems, short-term production planning – or scheduling is considered as one of the key instruments that make it possible to achieve an economically and energetic optimal production system. As a result, studies resolving the scheduling problem environmental aspect consider actually the energetic restriction as a perturbation to the production process. In this context, our research work focuses on a comprehensive literature review of the works that have addressed, a and analyzed workshop scheduling issues with energy considerations. The analysis of the bibliography can led to fix research perspectives that still necessitate attention.

In perspective, we intend to develop a mathematical formulation for the scheduling environment having an important historical place in the literature by considering mainly the energetic criteria and the energetic and technological constraints simultaneously. Regarding solution approach, a hybrid evolutionary algorithm will be considered since this combination will be adjusted by a strong theoretical complexity.

References

[1]. F.A. Rodammer et K. Preston White, A recent survey of production scheduling, IEEE Transaction on Systems, Man and Cybernetics, 6-18, (1999).

[2]. J. Yan, L. Li, F. Zhao, F. Zhang, Q. Zhao, A multi-level optimization approach for energy-efficient flexible flow shop scheduling, J. Cleaner Prod. 137, 1543–1552, (2016).

[3]. H. Luo, B. Du, G.Q. Huang, H. Chen, X. Li, Hybrid flow shop scheduling considering machine electricity consumption cost, Int. J. Prod. Econ. 146, 423–439, (2013).

[4]. A.A.G. Bruzzone, D. Anghinolfi, M. Paolucci, F. Tonelli, Energy-aware scheduling for improving manufacturing process sustainability: a mathematical model for flexible flow shops, CIRP Ann. - Manuf. Technol. 61, 459–462, (2012).

[5]. M. Dai, D.B. Tang, A. Giret, M.A. Salido, W.D. Li, Energy-efficient scheduling for a flexible flow shop using an improved genetic-simulated annealing algorithm, Robo. Comput.-Int. Manuf. 29 (5),418–429, (2013).

[6]. H. Luo, D. Du, G.Q. Huang, H.P. Chen, X.L. Li, Hybrid flow shop scheduling considering machine electricity consumption cost, Int. J. Prod. Econ. 146 (2), 423–439, (2013).

[7]. D.B. Tang, M. Dai, M.A. Salido, A. Giret, Energy-efficient dynamic scheduling for a flexible flow shop using an improved particle swarm optimization, Comput. Ind. (30) 223–232, (2015).

[8]. D.M. Lei, L. Gao, Y.L. Zheng, A novel teaching-learning-based optimization algorithm for energy-efficient scheduling in hybrid flow shop, IEEE Trans. Eng. Manag. 65 (2) 330–340, (2018).

[9]. Jiang, Z., & Le, Z. Study on multi-objective flexible job-shop scheduling problem considering energy consumption. Journal of Industrial Engineering and Management (JIEM), 7(3), 589-604, (2014).

[10]. Mokhtari, H., & Hasani, A. An energy-efficient multi-objective optimization for flexible job-shop scheduling problem. Computers & Chemical Engineering, 104, 339-352, (2017).

[11]. Liu, G. S., Zhou, Y., & Yang, H. D. Minimizing energy consumption and tardiness penalty for fuzzy flow shop scheduling with state-dependent setup time. Journal of cleaner production, 147, 470-484, (2017).

[12]. Li, J. Q., Sang, H. Y., Han, Y. Y., Wang, C. G., & Gao, K. Z. Efficient multi-objective optimization algorithm for hybrid flow shop scheduling problems with setup energy consumptions. Journal of Cleaner Production, 181, 584-598, (2018).

[13]. Schulz, S. A multi-criteria MILP formulation for energy aware hybrid flow shop scheduling. In Operations Research Proceedings 2016 (pp. 543-549). Springer, Cham, (2018).

[14]. Schulz, S., Neufeld, J. S., & Buscher, U. A multi-objective iterated local search algorithm for comprehensive energy-aware hybrid flow shop scheduling. Journal of Cleaner Production, 224, 421-434, (2019).

[15]. Salido, M. A., Escamilla, J., Barber, F., Giret, A., Tang, D., & Dai, M. Energy efficiency, robustness, and makespan optimality in job-shop scheduling problems. AI EDAM, 30(3), 300-312, (2016).

[16]. Wang, F., Deng, G., Jiang, T., & Zhang, S. Multi-objective parallel variable neighborhood search for energy consumption scheduling in blocking flow shops. IEEE Access, 6, 68686-68700, (2018).

[17]. Guo, C., & Lei, D. Multi-objective Flexible Job Shop Scheduling Problem with Energy Consumption Constraint Using Imperialist Competitive Algorithm. In International Conference on Intelligent Computing (pp. 659-669). Springer, Cham, (2018).

[18]. Zhong, L. C., Qian, B., Hu, R., & Zhang, C. S. The Hybrid Shuffle Frog Leaping Algorithm Based on Cuckoo Search for Flow shop

Scheduling with the Consideration of Energy Consumption. In International Conference on Intelligent Computing (pp. 649-658). Springer, Cham, (2018).

[19]. Lu, C., Gao, L., Pan, Q., Li, X., & Zheng, J. A multi-objective cellular grey wolf optimizer for hybrid flow shop scheduling problem considering noise pollution. Applied Soft Computing, 75, 728-749, (2019).

[20]. Dai, M., Tang, D., Giret, A., & Salido, M. A. Multi-objective optimization for energy-efficient flexible job shop scheduling problem with transportation constraints. Robotics and Computer-Integrated Manufacturing, 59, 143-157, (2019).

[21]. Li, M., Lei, D., & Cai, J. Two-level imperialist competitive algorithm for energy-efficient hybrid flow shop scheduling problem with relative importance of objectives. Swarm and Evolutionary Computation, (2019).

[22]. Zhang, B., Pan, Q. K., Gao, L., Li, X. Y., Meng, L. L., & Peng, K. K. A multiobjective evolutionary algorithm based on decomposition for hybrid flow shop green scheduling problem. Computers & Industrial Engineering, 136, 325-344, (2019)..

[23]. Zhou, B., & Liu, W. Energy-efficient multi-objective scheduling algorithm for hybrid flow shop with fuzzy processing time. Proceedings of the Institution of Mechanical Engineers, Part I: Journal of Systems and Control Engineering, 0959651819827705, (2019).

[24]. Jiang, S. L., & Zhang, L. Energy-Oriented Scheduling for Hybrid Flow shop With Limited Buffers Through Efficient Multi-Objective Optimization. IEEE Access, 7, 34477-34487, (2019).

EFFECT OF PHASE CHANGE MATERIAL USING IN BUILDING THERMAL COMFORT APPLICATIONS THROUGH SEVERAL CLIMATE CONDITIONS.

Marwa El Yassi[1, 2*], *Ikram El Abbassi*[1, 2], *Alexandre Pierre*[2] *and Yannick Melinge*[2]
[1]Laboratoire de Recherche en Eco-innovation Industrielle et Energétique (LR2E), ECAM-EPMI, Cergy, France.
[2]Laboratoire de Mécanique et Matériaux du Génie Civil (L2MGC-EA 4114), Université de Cergy-Pontoise, France.

Abstract. Nowadays, buildings sector contributes to climate change by consuming a considerable amount of energy to afford thermal comfort for occupants. Passive cooling techniques are a promising solution to increase the thermal inertia of building envelopes, and reduce temperature fluctuations. The phase change materials, known as PCM, can be efficiently employed to this purpose, because of their high energy storage density. Among the various existing solutions, the present study is dedicated to solid-liquid phase change materials. Temperature evolution (according to their defined temperature range) induces the chemical change of the material and its state. For building applications, the chemical transition can be accomplished from liquid to solid (solidification) and from solid to liquid (melting). In fact, this paper presents a comparative thermal analysis of several test rooms with and without phase change materials embedded in a composite wallboard in different climates. The used PCM consist in a flexible sheet of 5 mm thickness (Energain, manufactured by the company DuPont de Nemours). The main properties of such a commercial solution have been delivered by the manufacturer and from analyses. The room model was validated using laboratory instrumentations and measurements of a test room in four cities: Lyon; Reading and Casablanca. Results indicate that this phase change material board can absorb heat gains and also reduce the indoor air temperature fluctuations during daytime. The aim of the study is to show the benefits of this layer with phase change material and compare it in different climatic zones.

1 Introduction

In recent years, interest in energy consumption is increasing steadily. The building sector is responsible for causing one part of global CO_2 emissions. That is why it's necessary to reduce this consumption and try to improve the energy efficiency.

Recently, many researchers focused their studies on thermal energy storage (TES) methods, which it can be classified as sensible, thermochemical and latent heat energy storage. These systems can provide better energy efficiency and low carbon footprint.

Phase change materials, known as PCM, represent one of these methods that use latent heat energy storage (LHTES). These materials can be directly integrated in the envelope of a building. And it is a way to improve thermal comfort and control temperatures within a specific range.
These materials can be immersed in the wall or in combination with a mixture of other wall panels, for example in concrete [1, 2, 3] or gypsum [4, 5, 6].

The main purpose of this article is to present a comparative analysis of one phase change material integrated into the wall of a building, by proposing three experimental studies in order to evaluate the energy performance and the influence of PCM boards on the temperature of the interior wall.

2 Literature review

This section clearly describes the main evolution in the integration of PCMs into the building sector. **Fig. 1** shows the development of publications per year, characterized by an increasing number of scientific works from the beginning of the years 2005 until 2018.

Fig. 1 Evolution of research in passive use of PCM.

Several researchers installed PCM in the wall and the ceiling of the building because of their various advantages. In fact, the impact of these materials is the enhancement of the thermal comfort inside buildings. In other words, they have the ability to absorb the energy coming from outside during the phase change (solid to liquid) and to release it in the cavity during the state change (liquid to solid).

* Corresponding author: m.elyassi@ecam-epmi.com

https://doi.org/10.1051/e3sconf/202017001007

Recently, some manufacturers have started to commercialize different panels with phase change materials such as Rubitherm GmbH [7] and BASF [8] based in Germany, Phase change energy solutions [9] based in United States. Many researches integrated these products in different forms: For instance, new gypsum composite containing RT27 from Rubitherm GmbH [7] was studied [10], and shows a higher heat storage capacity by reducing the energy saving than ordinary gypsum. Or, microencapsulated paraffin manufactured from BASF [8] into a plaster mortar. This one is able to decrease the ambient temperature contrary to a classic mortar [11]

3 Methods and methodology

In this paper, the evaluation is focused on the analysis of the PCM applied in building in order to evaluate the total energy savings and temperature fluctuations in different climatic zones.

3.1. PCM characterization

The PCM analysed, ENERGAIN has been manufactured by the company DuPont de Nemours (Luxembourg) and is a mixture of paraffin wax (60%) and a mixture of ethylene based polymer (40%). The product is a rectangular panel of 5 mm thickness (see Fig. 2)

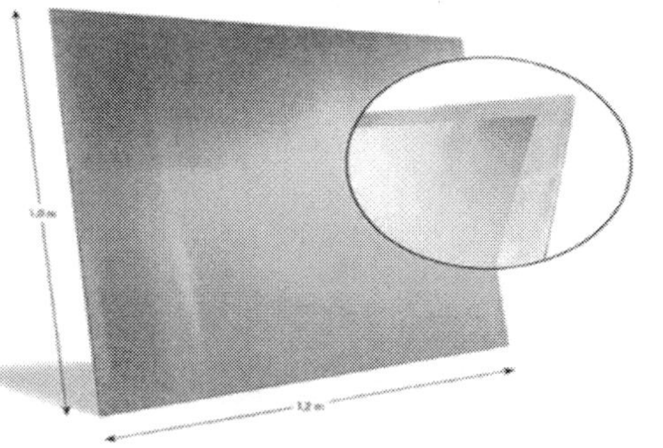

Fig. 2 Phase change material by DuPont de Nemours.

The PCM used has been measured [12]. The thermophysical properties of this product are listed in Table 1.

Table 1 Thermophysical properties of PCM [12].

Thermophysical property	Value
Thermal conductivity in solid phase (W/mK)	0,18
Thermal conductivity in liquid phase (W/mK)	0,22
Melting point (°C)	21
Freezing point (°C)	13,6
Latent heat (J/g)	106

This article presents the results of works done by [12, 13, 14] with the same PCM installed under several climatic conditions in three cities in order to compare a potential solution for cooling applications during heating season.

3.2. Experimental setup

This section describes the experimental set up of the different studies. The first one is located in the Department of Civil Engineering of Applied Sciences in Lyon. Fig. 3 represents the test room known as MINIBAT. The volume of the room is: 3.10×3.10×2.5 m^3. In this study, the PCM was integrated on the north, east and west walls with a night ventilation.

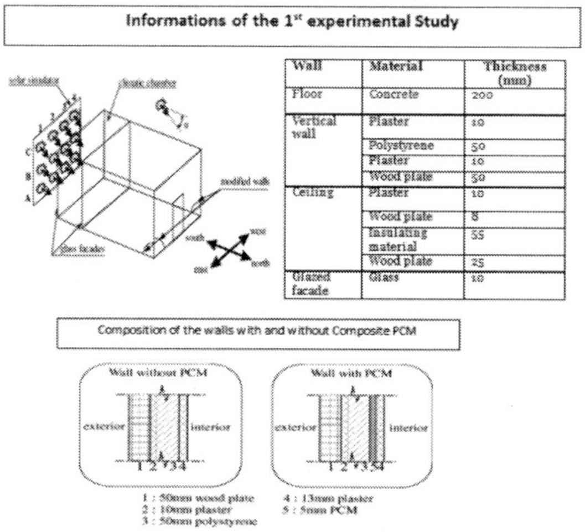

Fig. 3 Information of the first experimental study in Lyon [12]

The second study is located in Reading. Fig. 4 shows the experimental chamber with the following dimensions: 4×3×2.5 m^3. The PCM was installed on the Wall 5 (see Fig. 4).

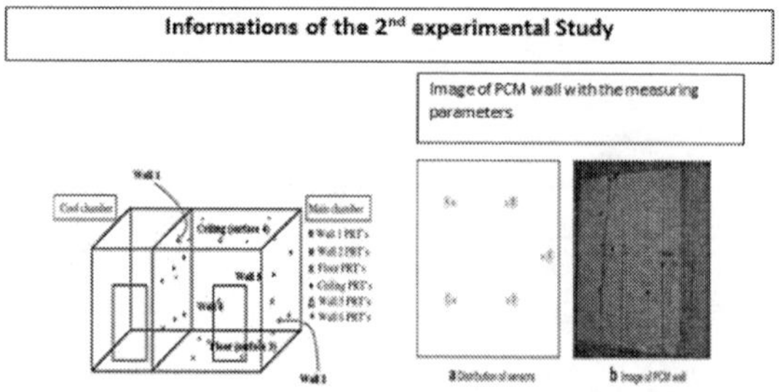

Fig. 4 Information of the second experimental study in Reading [13]

The last study is located in the Faculty of Sciences Aîn Chock, Casablanca. The experience was conducted in two same full-scale cubicles (see Fig. 5). The PCM was integrated on the ceiling and vertical walls in one of the cubicles. The volume of the PCM-Cavity is: 3×3×3 m^3.

https://doi.org/10.1051/e3sconf/202017001007

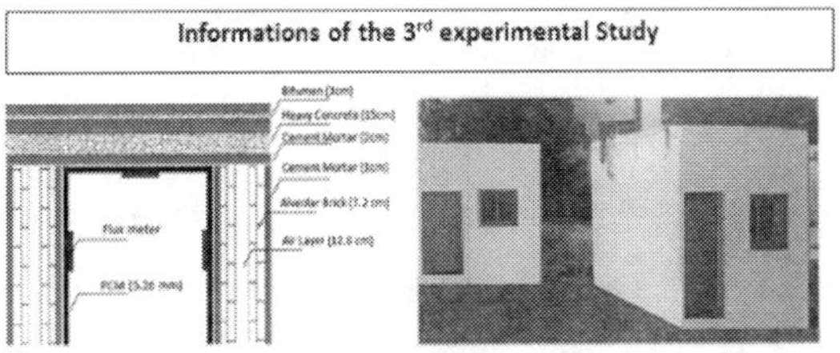

Wall	Composition	Thickness (cm)	Thermal conductivity (W/m.K)	Density (kg/m^3)	Heat capacity (kJ/kg.K)	U-Heat transfer coefficient (W/m^2.K)
Vertical walls	Cement mortar	1	1.15	1700	1	0.531
	Alveolar brick	7.2	1.15	1700	0.794	
	Air layer	12.6	0 0.08	1	1.227	
	Alveolar brick	7.2	1.15	1700	0.794	
	Cement mortar	1	1.15	1700	1	
Ceiling	Cement mortar	2	1.15	1700	1	2.722
	Bitumen	3	0.5	1700	1	
	Cement mortar	2	1.15	1700	1	
	heavy concrete	15	1.75	2300	0.92	
	Cement mortar	2	1.15	1700	1	
Floor	Cement mortar	5	1.15	1700	1	0.694
	Clay	3	0.6	2000	1.5	
	Polystyrene	2	0.039	25	1.38	
	Heavy Concrete	4	1.75	2300	0.92	
	Limestone	10	2.25	2400	0.8	

Fig. 5 Information of the third experimental study in Casablanca [14]

3.2 Instrumentation and measurements

For each study, different instrumentations are used to measure ambient temperatures of the test rooms, such as resistance thermometers (PRT) sensors and thermocouples (See Fig. 4, Fig. 5). These one are placed on the wall of the test room exposed to the local weather or a climatic chamber where the temperature can be controlled (See Fig. 3, Fig. 4). In this case, there are three weathers: Lyon, Reading and Casablanca.

4 Cooling energy performance

This part describes the different experimental results in order to compare the cooling energy performance of this PCM.

Fig. 6 shows the room air temperatures of different walls (North, East and West) of the test room located in Lyon. It's interesting to see the difference between the temperature with and without PCM. The wall with PCM presents a temperature fluctuation lower than the wall without. For the west and east walls, the PCM have the same effect on the temperature evolution and can reduce it of 3.5°C.

Fig. 7 shows the ambient temperature with two probes at heights 0.85 m and 1.70 m in order to evaluate the thermal stratification of the ambient air. With this material, the temperature variation can be decreased of about 3.9 °C. It can be noticed an interesting observation: a difference of about 1.3 °C between the two probes.

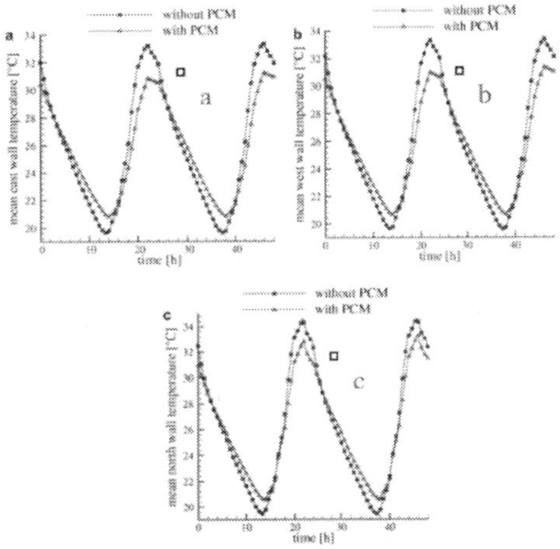

Fig. 6 Temperatures of the east wall (a), the west wall (b) and the north wall (c) for the sunny season [12]

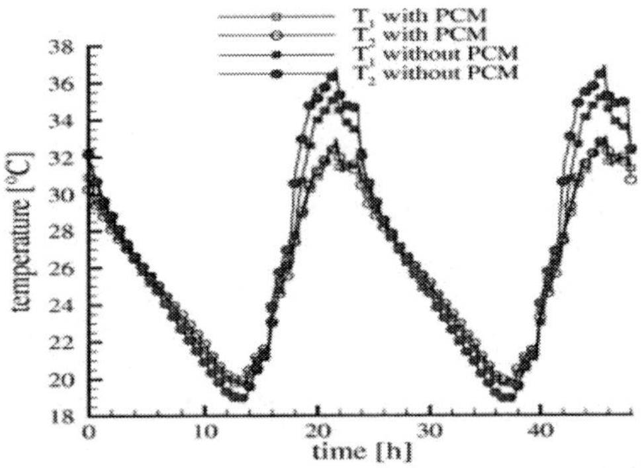

Fig. 7 Test room temperatures – T1 and T2 at height 0.85m and 1.70m. [12]

Concerning the second study located in Reading, the efficiency of the phase change material wall can be shown in Fig. 8, which represents a comparison between the wall with and without PCM. This one illustrates that the wall with PCM can cause a decrease of about 1.3 °C.

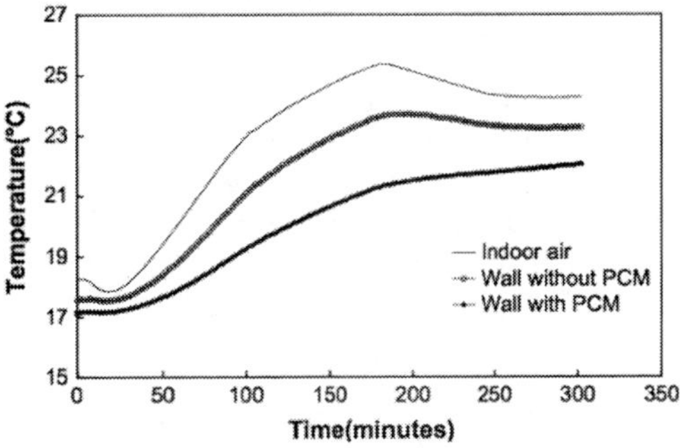

Fig. 8 Evolution of indoor temperature with phase change material in a complete-cycle [13]

https://doi.org/10.1051/e3sconf/202017001007

Conversely, it was found that during spring season in Casablanca, the application of this PCM board have the ability to increase the ambient temperature inside the cavities of about 2.2 °C as shown in Fig. 9.

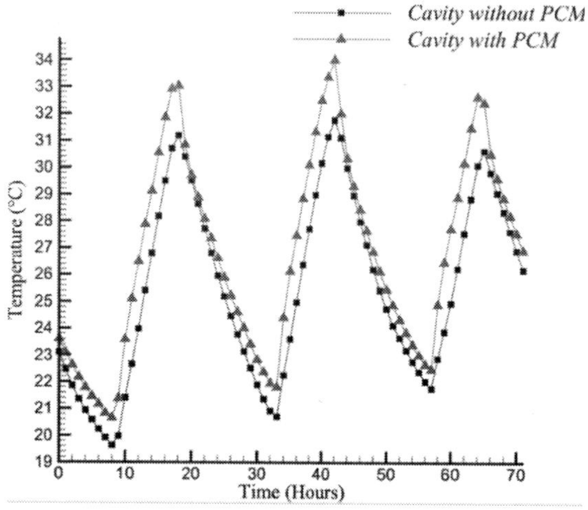

Fig. 9 Evolution of the temperature of the wall with and without PCM [14]

5 Analysis and discussion

This paper discusses three experimental studies in order to analyse the impacts of a PCM integrated in a building. These experiments can show that the cooling energy and peak temperature are reduced during the hottest periods with different climate conditions and experimental techniques. Their results are local and can be compared with each other. The experiments have shown that the phase change materials tested have the ability effectively maintain the temperature variation in the test room within the comfort zone by reducing the air temperature. However, these materials can reduce the maximum air temperature depending on several conditions such as the climate, the building envelops and the ventilation. Taking into consideration all of these conditions of the experiments, the results were different of each other. Contrary to Lyon and Reading, in Casablanca, the PCM wallboard was able to keep the temperature of the cavity higher than the reference-cell and destroy the thermal stratification. In Lyon, a night ventilation was switched on to help the PCM to complete the fully cycle (melting/solidification), to improve the thermal inertia of the building envelop and contribute to air renewal. Moreover, thermal stratification doesn't exist for the PCM's case with a temperature reduction. In Reading, the results were lower than Lyon.

To conclude, it can be noted that the results of these experiences were different mainly because of the local climate: In the Oceanic climate, the more the outdoor temperature is higher, the more the PCM wallboard is effective: In comparison to Lyon, Reading has the coolest climate with the lower results in terms of temperature variation (Table 2). For the Mediterranean climate, the impact of the PCM wallboard was different: The inverse effect on the ambient variation has been noted.

Thus, according to these studies, the geographical location is significantly important for better results. Despite the noted difference, there is an interesting impact of a PCM on the variation of air temperature in the Oceanic climate even if in the caseof Reading. For Casablanca, it can be a great solution for a passive heating application during cold season. In fact, all of the results of this product can represent a potential solution to improve the thermal comfort of building.

City	Köppen-Geiger Climate classification	Average temperature in the year
Lyon (France)	Oceanic climate (Cfb)	11.6 °C
Reading (United Kingdom)	Oceanic climate (Cfb)	9.9 °C
Casablanca (Morocco)	Mediterranean climate (Csa)	17.7 °C

Table 2 Climate characteristics according to Köppen-Geiger classification of selected cities. [15]

6 Conclusions

In this article, a comparative study was made for the same PCM installed under several climates. The use of this phase change material wallboard can decrease the fluctuation ambient temperature in the test cavity, enhance the thermal inertia and try to minimize the thermal stratification. This paper has analysed results of the three experiments and has confirmed that this solution is particularly interesting to reduce energy consumption of buildings.

In perspectives, supplementary investigations are needed in order to validate the application of this phase change material wallboard for architecture envelope. Other numerical simulations must be realized too: It will be interesting to evaluate this PCM under Parisian climate.

References

1. Cabeza, Luisa F., C. Castellón, M. Nogués, M. Medrano, R. Leppers, O. Zubillaga. *Ene and Buil* 39, n° 2 (2007): 113-19.
2. Memon, Shazim Ali, H. Z. Cui, Hang Zhang, et Feng Xing. *App Ene* 139 (2015): 43-55.
3. Stritih, U., V. V. Tyagi, R. Stropnik, H. Paksoy, F. Haghighat, et M. Mastani Joybari. *Sus Cit and Soc* 41 (2018): 286-95.
4. Athienitis, A. K., C. Liu, D. Hawes, D. Banu, et D. Feldman. *Buil and Env* 32, n° 5 (1997): 405-10.
5. Borreguero, Ana M., A. Serrano, I. Garrido, Juan F. Rodríguez, Manuel Carmona. *Ene Conv and Man* 87 (2014): 138-44.
6. Voelker, Conrad, O. Kornadt, Milan Ostry. *Ene and Buil* 40, n° 5 (2008): 937-44.
7. « Rubitherm GmbH »
8. « BASF – France »
9. Phase Change. « Phase Change Energy Solutions: Leader in Efficient Building Materials ».
10. Sá, Ana Vaz, M. Azenha, H. de Sousa, et A. Samagaio. *Ene and Buil* 49 (2012): 16-27.
11. Borreguero, Ana M., A. Serrano, I. Garrido, Juan F. Rodríguez, M. Carmona. *Ene Con and Mant* 87 (2014): 138-44.
12. Kuznik, Frédéric, J. Virgone. *App Ene* 86, n° 10 (2009): 2038-46.
13. Liu, Hongim, Hazim B. Awbi. *Buil and Env* 44, n° 9 (2009): 1788-93.
14. A. Mourid, M. El Alami, F. Kuznik. *Sus City and Soc* 41 (2018): 35-43.
15. Climate-Data.org ». 2019.

E3S Web of Conferences 170, 01008 (2020)
EVF'2019
https://doi.org/10.1051/e3sconf/202017001008

Energy from Waste

Rupankar Saha[1*], and *Binay* Kumar Singh[1]

[1]Maharashtra Academy of Naval Education & Training, Pune-412201, India

Abstract. The objective of this paper is to discuss the various technology and methods for producing energy from waste & its advantages. These technologies are incineration, gasification, plasma arc gasification, pyrolysis, anaerobic digestion. These technologies reduce volume of waste, environmental influence threat to public health and the dependency of the fossil fuel for generating power. The efficiency of this technology is up to 20-40%. As per rough idea, a typical 100000 tonnes per annum waste to energy plant will produce around 7MW of electricity, which is sufficient to power approximately 10,000 homes and cost of the energy is around 0.03$ -0.05$ per Kilo-watt Hours depend upon the technologies.

1. Introduction:

Now-a-days all developing countries are facing challenges regarding collection of waste, transport and deposing of waste, while advanced country are using innovative technology to produce energy from different by-product of waste. The advancement in waste to energy technology provides economical and optical solution for energy and support the energy crisis. The main reason behind for massive waste generation are population boom, high standard of living, better economic growth and increase in global trades [1]. For example in 2018, 62 million metric tons of waste was produced in India and it is prospect that it will become 165MT by 2031 & by year 2050, it will become 436 million tons. The land required to dump 62 million MTs of untreated waste is nearly 1240 hectors per year & it will expected that the requirement may be expand to 66000 hectors of land required for dumping the waste and also required man power for treating waste with proper technology [2]. This technology help to reduce the volume of waste and assist in energy production. Waste to energy conservation technique support in minimizing the production of waste and keeps environment clean. The world is facing short supply of energy due to resource constraints. Hence, the idea of producing energy from waste is suitable option.

2. Waste Hierarchy:

It is a tool used for protecting the environment by managing the waste in terms of what is best for the environment. The aim behind the waste hierarchy is that it helps to reduce the generation waste, reduce the amount of waste goes for landfill and increase in recovery and re-use of resources. It helps in reduction in green houses gases and pollutants which harm environment.

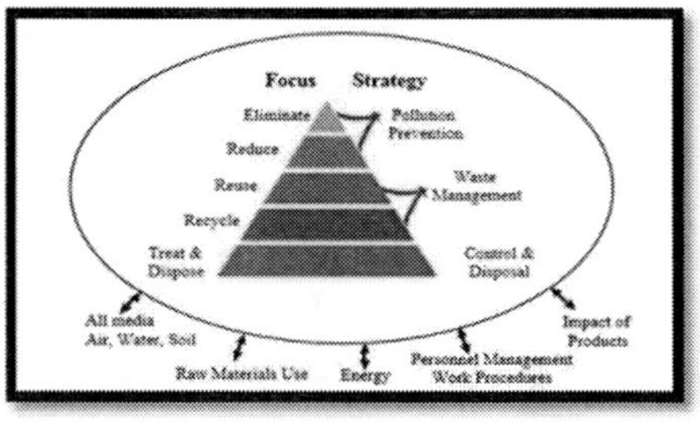

Fig. 1. Waste management hierarchy

3. 5 R Technique for Waste Management:

Fig. 2. 5 R Technique for waste management

*Corresponding author: rupankar444@gmail.com

4. Types of waste:

Waste from urban	**Municipal Solid Waste**: Paper, Glass, Metals, Man-made polymers such as epoxy, Teflon, polyethylene, polyester, nylon, cords, coils, Gadgets, Hides and leather discard pharmaceuticals wastes ,kitchen waste **Sewage**: 1) Bulk excretory matters (Feces and urine), 2.) Wastes from the body (necrosis, spit, tears 3.) Baths, 4).Laundry wastes (Detergent and soap precipitates).
Waste from industry	**Solid:** 1) Paper and pulp waste 2) Bagasse (sugarcane, bamboo, etc.), 3) Press mud, 4) Brine mud, 5) Metallurgical slags, 6) Gypsum polyesters, etc. , **Liquid**: 1) Non-soluble of miscible oil, Paint with emulsion, waste from oil, 2) Tarry waste and Residues/High density metals 3) Industrial effluents (textile dyes) 4) A mixture of liquid with manure, a mixture of a liquid with cement 5).By-product after cheese/casein manufacture from milk
Waste from biomass	1) Crop debris (stalks, straws, cobs, husks, oil cakes, shells, pods, etc.), 2) Diary forms from where we get milk 3) Abattoir wastes (Flesh, Bones, Organs, and Blood etc.) 4) Bulk forest wastes (wood, stubbles and Humus, 5) Wastes from forests wood wastes, leaves, grains, spores in trees
Waste from biomedical	1) Anatomical wastes and pathological waste (muscles, glands, body portion) 2) Veterinary wastes from animals 3) Laboratory reagents, 4) Discarded medicines and cytotoxic drugs, 5) Fluids from the body such as plasma, urine etc.

5. Incineration:

It is a thermochemical technology. It is the most common and oldest method for producing waste to energy. In this process waste is fed into the combustion chamber for burning at a temperature of 900°c -950°c with the presence of oxygen & release heat energy. This heat is utilize to generate steam, which is used to drive the turbine & produce electricity.

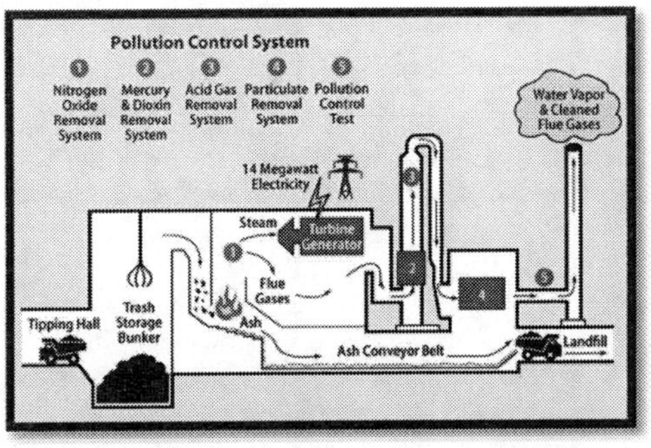

Fig. 3. Schematic diagram of the incineration process

By-products of this process are ash and air pollution control residue.in this process, waste reduction potential is around 80% and efficiency is up to 18-28% [3].Depending upon the types of furnace, there are 3 types-

- Moving Grate
- Fluidised Bed
- Rotary Kiln

5.1. Moving Grate: This is the oldest & widely used waste to energy technology process. In this process, the waste is fed into a grate by the help of reciprocating, rocking, travelling or rolling movement to convey the waste through combustion chamber. The power efficiency in this process is up to 30% and combined heat & power more than 70% [4]. Types of waste burn in this systems are Municipal solid waste (MSW), Commercial and Industrial (C&I) waste, Refuse Derived Fuels (RDF), Woods, Hazardous Waste, Clinical waste. There are over 900 plants is present across the world. The indicative cost of this technology $450-$700 per tonnent. Treatment capacity of this technology is up to 120 MW [5].

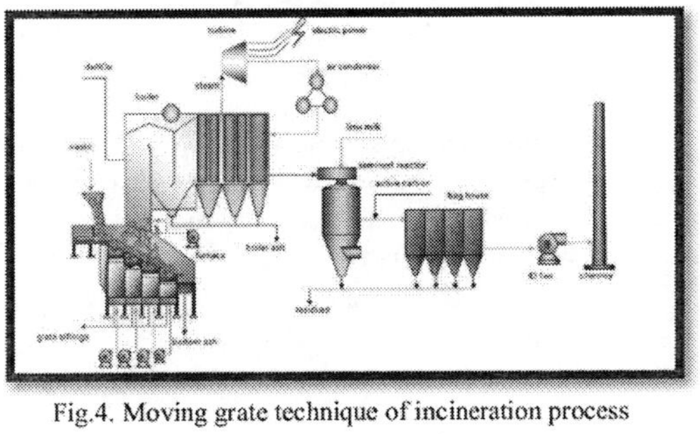

Fig.4. Moving grate technique of incineration process

5.2. Fluidised bed*:* Fluidised bed having a vertical combustion chamber. The process is mainly used for coal, raw sewage, sludge and bio-mass. A hot bed is present in the bottom part of combustion chamber. Waste is burned in the hot bed. The material of the bed consisting of sand, ash or other inert material.

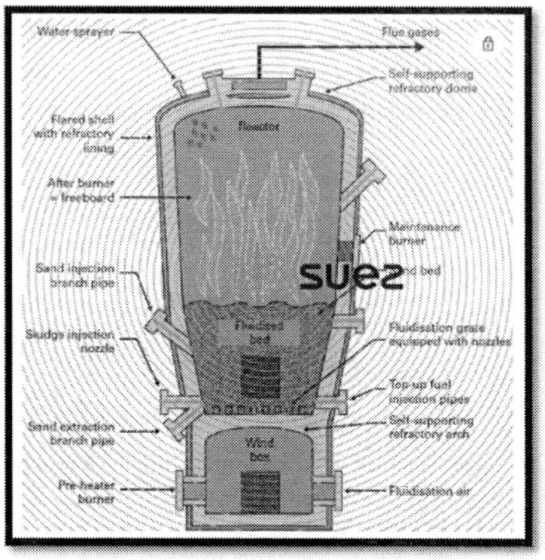

Fig. 5. Fluidised bed technique of incineration

The waste is fed into hot bed from the top or side. Holes are provided for supplying the pre-heated air into the combustion chamber. Power efficiency for this technology is up to 25% and combined heat and power can reach in this process greater than 70%. Treatment capacity of this technology is up to 90MW. This technology is intensively used in Japan.

If the waste is heterogeneous for this technology requires a preparatory process so it conforms to size specification. For this technology the acceptable limit of the waste particle size less than 200mm.

This technology is categorized on the basis of gas speed and nozzle design gas speeds and design of the nozzle plate:

- Stationary / Bubbling) fluidized bed
- Rotating fluidized bed
- Circulating fluidized bed

i. **Stationary / Bubbling fluidized bed:** This technology can be used for generating energy by treating sewage and industrial sludge.

ii. **Rotating fluidized bed:** This type of fluidized bed is revolved in the combustion chamber, so it's achieve longer residence time in the incineration chamber. This are used for mixed municipal waste.

iii. **Circulating fluidized bed:** This circulating fluidized bed is used for calcination of dried sewage sludge having high heating value.

- **Case study** (3grate incinerator lines & fluidized bed incinerator): Indaver Doel [6].

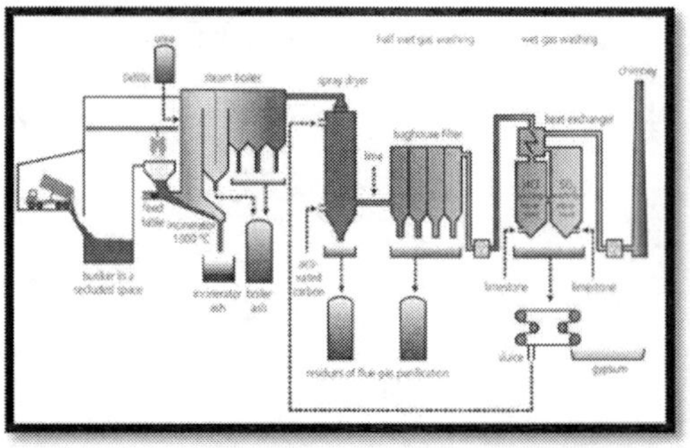

Fig. 6. Grate process of incineration

Types of waste: Non-recyclable MSW and Commercial waste @400000tpa for 3 grate incineration lines.

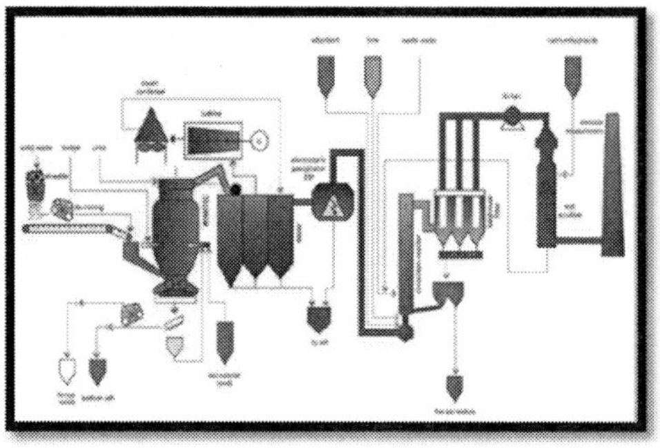

Fig. 7. Fluidized bed incineration

Types of waste: Non-recyclable MSW and commercial waste @600000tpa for Fluidized bed incineration.

Result: It produces electricity to 170000 household & it supply 30tph steam to a chemical producing company "Ineos".

5.3. Rotary kiln: It consist of a cylindrical vessel which is inclined at the horizontal axis. This vessel is usually lying on the roller, so the kilns spin or oscillated around its axis normally reciprocating motion.

The waste is introduce to the kiln at higher end and conveyed through the kiln by the gravity as it rotates & resulting ash discharge through at the lower end.

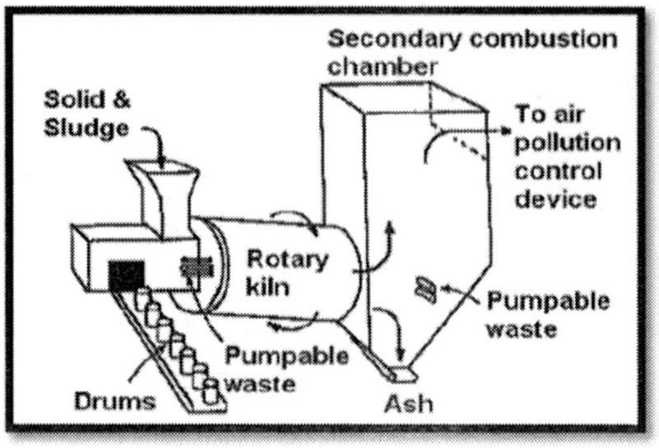

Fig. 8. Rotary kiln process of incineration

The operating temperature of this kiln from 500°c to 1450°c. If the hazardous waste is used for this technology, the temperature range 900°- 1200°c [7] hazardous waste and also used for clinical waste, sewage sludge and waste water sludge. Power efficiency of this technology is up to 25% and combined power and heat recovery more than 70% [8]

6. Pyrolysis:

In this technology organic material (i.e. cellulose, hemicellulose and lignin) is exposed at high temperatures with in inert atmosphere (absence of oxygen). In this treatment, material is exposed to elevated temperature and passes through chemical and physical separation into different molecules. It is mostly used for treating organic material and plastic. Depending upon the configuration of pyrolysis equipment there are 3 types such as fluidised bed, Moving bed and Rotating cone. Power efficiency of this technology is up to 23% & regenerative efficiency i.e. power and heat up to 81% through the steam turbine [9].

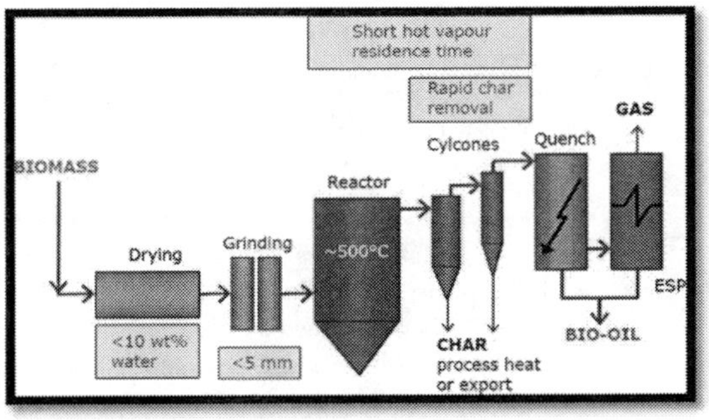

Fig. 9. Schematic diagram of pyrolysis technique.

Types of pyrolysis:-

1) **Slow pyrolysis**:- In this process the heating temperature is in between 400°c-500°c and the residence time is 5 -30 min.
2) **Fast pyrolysis**:- In this process the heating temperature is in between 400°c-650°c and the residence time is 0.5-2s. The major products are bio-oil, gases and char.
3) **Flash pyrolysis**:- In Flash pyrolysis, heating temperature is in between 700°c - 1000°c i.e. very high and the residence time is < 0.5sec.

❖ **Categorization of pyrolysis process on the basis of heating rate, major products and residence time** [10].

Method	Temperature (°c)	Residence time	Heating rate (°c/s)	Major products
Slow Pyrolysis	400°c -500°c	5-30 min	10	Gases Char Bio-oil (tar)
Fast pyrolysis	400°c -650°c	0.5-2 s	100	Bio-oil (thinner) Gases Char
flash pyrolysis	700°c -1000°c	<0.5s	>500	Gases Bio-oil

❖ **Output product from Pyrolysis and its use :**

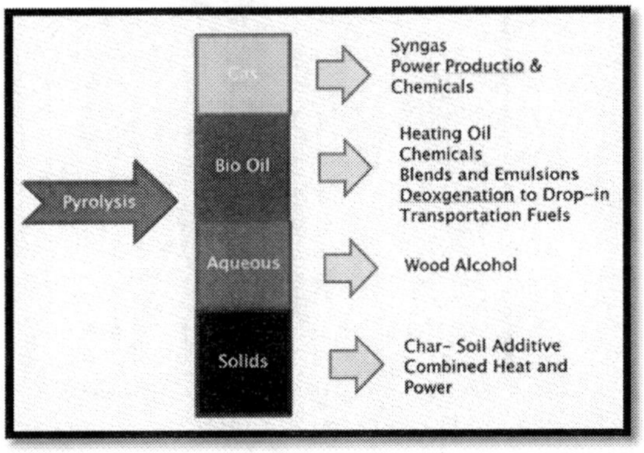

Fig. 10. Products of pyrolysis

7. Gasification:

It is a partial oxidation process where the 80% of the chemical energy in the waste specially carbon & hydrogen being converted into syngas or synthesis gas. This gas can be used to provide energy to steam boiler and gas engines.

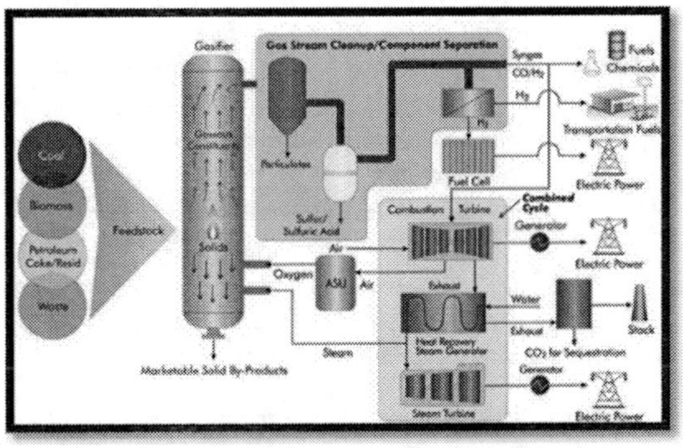

Fig. 11. Schematic diagram of Gasification process

The temperature range of this process 500°c -1800°c and power efficiency in this process is up to 23% and regenerative efficiency i.e. power and heat is up to 81%

using steam turbine [11]. This process helps to recover the energy present in biomass and municipal solid waste by converting those material into valuable products and eliminating the need for incinerator landfilling. Products obtain from syngas includes chemical (45%), liquid fuel (28%), gaseous fuels (8%) and electric power (19%). This process helps in reducing the volume of solid waste up to 85% to 92% and reduce the amount of SOx, NOx and dioxins.

8. Plasma Gasification:

This technology uses an electric arc to produce high temperature within the reactor to convert solid waste or hazardous waste (asbestos and radioactive) to syn gas and melt arising from inorganic species of feedstock.

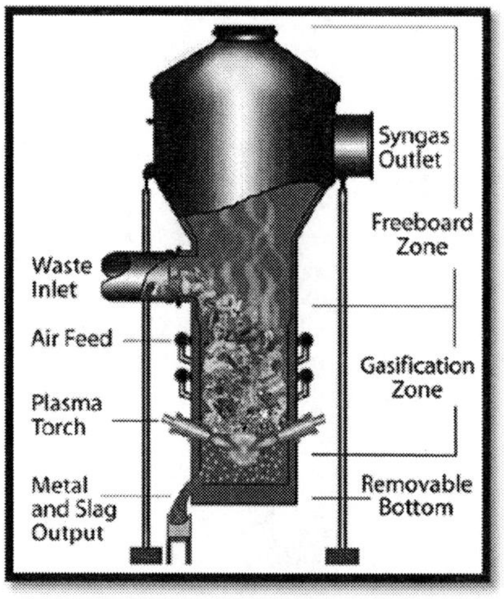

Fig. 12. Plasma arc Gasification system

This technology reduce the tar content by 1000 times less than of conventional Gasification [12].Temperature range of this process as high as 7000°c. Syngas is produced in this process more cleaner than conventional combustion. Power efficiency of this process is up to 23% [13]. Plasma is an electrically conductive gas (nitrogen & argon) which are heated by an electrical current to very high temperature. Requirement of Oxygen is very less in this process compared to conventional Gasification. Example of this type of plant is Alter NRG (Westinghouse). There are 3 mains types of plasma gasification available such as-

- Direct exposure to plasma torch (only for high level hazardous waste).
- Plasma assisted gasification
- Plasma for syngas polishing.

9. Anaerobic Digestion:

In this process biological treatment of organic materials is done with the help of micro-organism which break down organic material into smaller particle in oxygen poor environment with the help of different types of bacteria & release combustible bio-gas methane & carbon di-oxide which are used to produce heat electricity & bio-fuel.

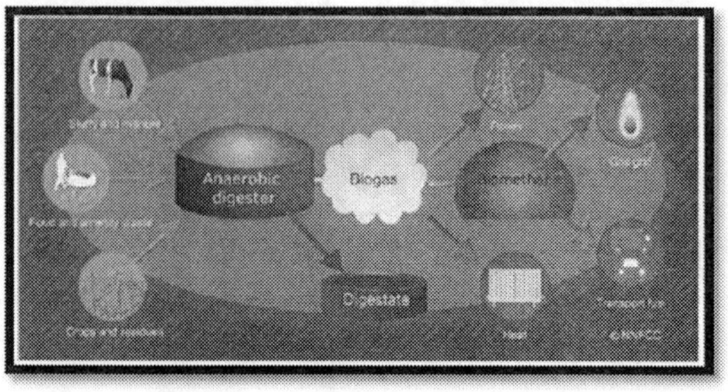

Fig. 13. Anaerobic Digestion Process

For this technology we can use various feedstock such as Bio-solids, Food waste, green waste, crop residue, sewage sludge & other higher moisture content waste. There are main four steps of Anaerobic Digestion process such as Hydrolysis, Acidogenesis, Acetogenesis, Methanogenesis.

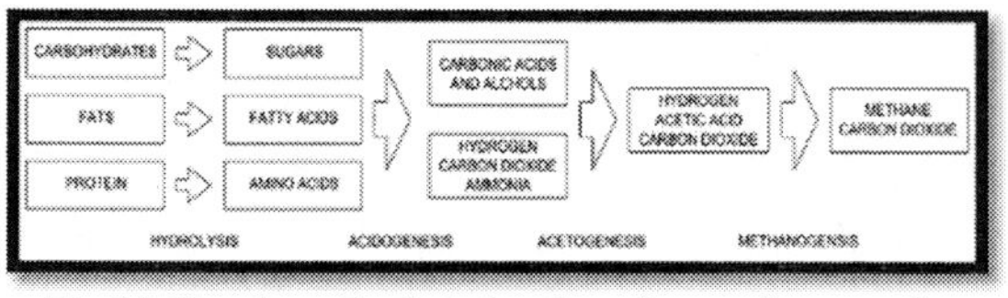

Fig. 14. Steps to production of methane from various organic compound.

- **Hydrolysis:** In this process large protein molecules, complex of fats, carbohydrates are brake through with the help of hydrolysis and produce amino acids, long chain fatty acids & sugar.
- **Acidogenesis:** In this stage sugar, fatty acid, amino acid converted into three, four & five carbon volatile fatty acid such as lactic, butyric acid and valeric acid.
- **Acetogenesis:** Here, volatile fatty acids are converted into acetic acid, carbon dioxide and hydrogen with the help of Acetogenesis bacteria.
- **Methanogenesis:** In this stage, methane gas is produce with the help of methane-forming bacteria which digest acetate, carbon dioxide and hydrogen.

The end product of this process is bio-gas, which consist of methane (CH_4) (50-70%), Co_2 (30-45%), & H_2S (200-4000ppm) [14].

❖ Example for this type of plant is **Tilburg** plant which uses Valogra process. It has an annual waste capacity 52,000 tons of garden and vegetable waste per year. This plant having two digester, each of 3300 m^3 capacity & produces 2.8 million m^3 of methane per year. Each ton of waste produce approximately 106m^3 bio-gas. The retention duration time is 20 days at mesophilic temperature of 38°c [15].

10. Conclusion:

India is an energy deficient country with the dense populations in the urban areas. The energy from every source is an important and essential necessitate. With increase in population, the production of waste is also increased and thus we can use this waste for producing energy. Life cycle assessment studies suggest that waste to energy production is better way to tackle the problem arise from waste management. Power produced from the WTE production activity can reduce the load on natural resources. Most of the electricity in India produce from fossil fuels, so fossil fuel are reducing day by day whereas waste is rising progressively. Electricity production from waste helps in reducing the load on fossil fuels and also helps in reduction on CO_2 emission by 20-25million kg/year. It also helps in reduction of SO_2, NO_X etc. WTE technology counter the problem regarding waste management and energy generation for example waste from Kanpur city produced 33MW electricity which satisfy 5% electricity demand of Kanpur city. So, it reduces the problem regarding waste management by decrementing volume of waste by 90%.

11. REFERENCE:

1. M. Rehan, J. Gardy, A. Demirbas, U. Rashid, W. M. Budzianowski, D. Pant, and A. S. Nizami, Bioresour. Technol. 250:17-25 (2018)
2. J. Charles Rajesh Kumar, B. Mary Arunsi, R. Jenova, and M. A. Majid, Int. J. Renew. Energy Res. **Vol.9**, No.1, (March 2019).
3. EU-Indonesia Trade Cooperation Facility, 82, (2015).
4. Ricardo-AEA/R/ED58135/Issue Number 5 – Final Report, 11, (02/07/2013).
5. S. Boshoff, *Overview of thermal technologies for waste to energy application,* 15, (7 April, 2016).
6. S. Boshoff, *Overview of thermal technologies for waste to energy application,* 19-20, (7 April, 2016).
7. A. Bosmans and L. Helsen, in *Third Int. Symp. Energy from Biomass Waste* (November, 2010)
8. Ricardo-AEA/R/ED58135/Issue Number 5 – Final Report, 14, (02/07/2013).
9. Ricardo-AEA/R/ED58135/Issue Number 5 – Final Report, 26, (02/07/2013).
10. S. I. Ahmed, A. Al Fatta, M. J. Hossain, and S. Tanvir, *Int. Conf. Energy Environ.* 3 (2018)
11. Ricardo-AEA/R/ED58135/Issue Number 5 – Final Report, 26, (02/07/2013).
12. M. Saghir, M. Rehan, and A.-S. Nizami, in *Gasif. Low-Grade Feed,* 106, (11th July 2018).
13. Ricardo-AEA/R/ED58135/Issue Number 5 – Final Report, 24, (02/07/2013).
14. S. Verma, Anaerobic digestion of biodegradable organics in municipal solid wastes. Fu Foundation School of Engineering and Applied Science, Columbia University, (2002).
15. M. McCrea, T.K. Tan, H. H. Ting, X. Zuo, A Cost Benefit Analysis of Different Waste to Energy Technologies for the Management of Municipal Solid Waste in Singapore, 7.

Experimental Study on Effect of Water Spray Characteristics on Performance of Cooling Tower

Akshay S. Dhurandhar[1,*], *Amarsingh B. Kanase-Patil*[1]
Department of Mechanical Engineering, Sinhgad College of Engineering, Pune 411041, Maharashtra, India

Abstract. Cooling tower is an indispensable part, used as a direct contact type heat exchanger mainly for evaporative cooling. Cooling tower generally dissipates, remove heat from thermal power plants. In an induced draft cooling tower of counter flow, used for a mini-steam power plant, hot water enters at the top, while the air is introduced at the bottom and exits at the top, air is allowed to come in contact with falling water droplets, causing evaporative cooling. A possibility of desired change with different spray angle, patterns, is tried and analysed. On findings, best suited spray nozzle angle resulted is 90°, and amongst three spray patterns, full cone, hollow cone and spiral type nozzle; full cone nozzle of 90° spray angle helps achieving efficiency up to 82%. The range increases successively from 9.8°C to 15.5°C for FC nozzle, in approach to WBT; the desirable fall of 3.56°C is attained with effectiveness of 81.63%.

1. INTRODUCTION

Cooling towers are used to remove temperature of hot recirculating fluid in the process plant, which otherwise subject to lesser efficiency of process without of it, power generation in case of steam power plant for an instance and keeping the temperature under stipulated limit ensures safety. Counter-flow cooling towers are very common amongst the various types of cooling towers used, with direct contact between the water and the air by evaporating some part of the water. Cooled water falls into the tower basin while the remove heat leaves the device as warm air, at the cost of some mass of water evaporates as it encounters outside air. Present study emphasizes on the thermal performance of counter flow.

Fig. 1 shows the packaged type cooling tower and water spraying system inside on which the study carried out. Mechanical draft cooling towers are available in the following airflow arrangements:

a. Counter flow forced draft.
b. Counter flow induced draft.
c. Cross flow induced draft

In an induced draft cooling tower of counter flow, hot water enters at the top falling through spraying nozzle while the air is introduced at the bottom and exits at the top.

Fig. 1. Packaged type cooling tower and water spraying system inside

As the inverted blades (7°-8°) of an induced draft fan settled a top, on rotating at high speed, creates vacuum so allowing the air particles outside to enter and entrapped upward, through inside the cooling tower where the hot fluid from condenser is falling down through water distribution system; that's where the evaporative cooling needs to be happened effectively. In the existing setup used for evaporative cooling they fitted a spray nozzle of 60° hollow cone spray pattern and an induce draft fan not working properly, leading to spray water and extracting air-water droplet stream mass to top from bottom incomplete.

Fig. 2 shows the one of in context cooling tower, counter flow induced draft cooling tower used for a mini-steam power plant is analyzed in this study.

*Corresponding author: **dhurandharakshay@yahoo.com**

Castro M. et al. [1], developed an optimization model for a cooling water system which supplies a heat exchanger network. The objective was to minimize the total operating cost of the system, which includes cooling water make-up and energy costs. From different climatic conditions point to the fact that the most important influence on the cooling system performance is not the ambient temperature itself, but its humidity, forced withdrawal of water upstream of the tower is an important resource for fulfilling the cooling duty requirements. Sun et al. [3] studied and evaluated natural draft dry cooling tower (NDDCT), presented it with the concentrating solar power plant. The high ambient temperature during summer days leads to a 20% net power reduction for power plants using a dry cooling system, in fact, for a low solar rays concentration it worsens to even 50% of net power reduction. The inlet air pre-cooling would improve the efficiency of NDDCT and thus reduce power generation loss under high-ambient conditions.

Fisenko S. et al. [4], studied, proposed a new mathematical model of a mechanical draft cooling tower; performance has been developed, concluded points, for mechanical draft cooling towers, this dependence is weaker than for other types of cooling towers. For small values Qw=Qa, when they change from 0.01 to 1, the thermal efficiency of the cooling tower practically does not change. The flow rate of evaporated water can reach 3% of the mass flow rate of water, the evaporation is stronger at lower atmospheric pressure as diffusion coefficient of the water vapor increases. Satish Kumar M. presented study, the factors affecting the performance like environmental conditions, cooling water quality have been studied on Induced draft cooling tower of 32 Mw thermal power plant. The concluding remarks given as: It was observed that in part load condition liquid to gas ratio is decreased even for same heat removal and same temperature differential as in full load condition. The climatic conditions like air dry bulb and wet bulb temperature, relative humidity will affect the performance of the cooling tower.

Lucas M., et al. [8], studied with experimentation with a gravity type water distribution system (GWDS) for six drift eliminators and when no drift eliminator was fitted. Slope up to 6° shown increase in performance after 6° the performance decreases. The study brought out findings as; the pressure water distribution systems type (PWDS) achieves better performances than the GWDS. Maximum averaged of 38.66% in terms of cooling tower performance have been obtained between the two water systems.

2. EXPERIMENTAL SETUP

2.1 Experimental Setup:

A mini-steam power plant which is situated in the college building workshop, runs on a fuel intake (usually high-speed diesel oil), operates on the working pressure of 15 bar, evaporation rate meets at 800 kg/hr (wet steam) in boiler of water tube packaged (Non-IBR). Generates 10 bar, 250 °C and 3000 rpm of inlet pressure, temperature and speed respectively in a steam turbine of a single stage type, gives an output of 6 KVA for a generator set. In Fig. 2, an actual photograph of a plant mentioned above is shown,

Fig. 2. Mini-steam power plant

At last, it is a cooling tower outside near window on the cold-water tank (basin). A counter flow, packaged type cooling tower is used in conjunction with the plant to reduce the heat of the incoming fluid from condenser outlet, with all the entrained heat of whole plant say. An induced draft fan is provided at the top, above the drift eliminator on the frame. It absorbs the air through small inlet doors of cooling tower (louvers), enhancing evaporative cooling by facilitating more air with contact with falling literally hot water droplets, the enriched heat exchange happens there.

The experimental setup consists of a boiler, a condenser, turbine, pump, pipelines to flow fluid along, and the portion of interest of this text; a cooling tower, water tank (basin), at last other miscellaneous but important elements cum instruments.

Initially boiler heats cold water, converts water into steam. Pressure of it rises certainly, and this steam then goes to turbine. High velocity, high pressure steam makes the turbine rotate at speed, electricity to be produced from that. Afterwards is allowed to be passed through the condenser, exchanging heat with the cold fluid (from cold water tank) steam gets converted into hot condensate, so give up its latent heat of vaporization. Finally, this hot fluid is fed to a cooling tower with the duty to proceed evaporative cooling, where it is sprayed from top over fills and upcoming outside air currents surrendering conditioned fluid to water tank and again to recirculate. Fig. 3 shows a packaged type cooling tower, setup:

The cooling tower possesses the elements lead it towards an efficient operation, so acquiring good cooling. It is having:

1. water distribution system (water spray nozzles),
2. an induced draft fan,
3. fills, drift eliminator,
4. Resin structured cooling tower body,
5. Water tank below it.

The water distribution system also called spray nozzles fitted inside the tower, the first important part of cooling tower, sprays water over the fills, in minute droplets uniformly distributes enhancing evaporative cooling.

Fig. 3. A packaged type cooling tower; setup

An induced draft fan entraps outside air in tower, extracts air up to top and so making it smoother for cooling, as air meets falling hot fluid droplets. Fig. 4 shows a fiber induced draft fan.

Fig. 4. Induced draft fiber fan

Fills are placed below the water spray nozzle, over which spray of hot water spreads a good large area so enabling much heat transfer in between, Fig. 5 shows fills used in cooling tower, nozzles spraying water. Drift eliminator restricts, traps the moisture droplets which air-stream particles entrained within when it is expelled from top with the help of an induced draft fan. Fig. 5 also shows drift eliminator.

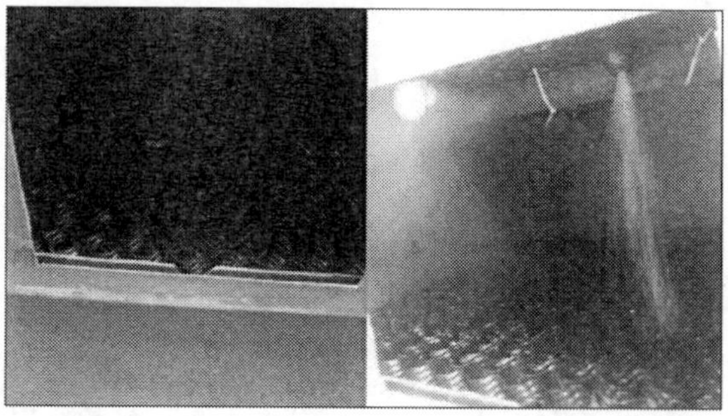

Fig. 5. Fills and drift eliminator

2.2 Spray Characteristics (angle, spray pattern) of a Water Spray Nozzle

Water spray nozzles are used to distribute water onto the cooling fills in a cooling tower. It is one of the important parameters for the selection of nozzles. Spray nozzles made up of brass, stainless steel or PVC plastic as per requirement and convenience. Nozzle spray characteristic describes the distribution of water from the axis of the nozzle along a fill.

Spray angle varies from the distance where it is going to spray. It is converge-diverge conical angle between. They are semi sphere inside and conical while going outward. Spray angle converges and diverges with respect to vertical axis. It may collapse or broaden with increase in distance from the orifice. Spray angle and its coverage is shown in the schematic Fig. 6 below,

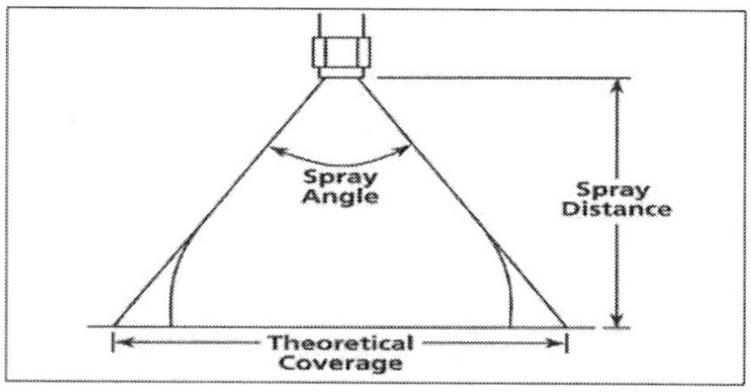

Fig. 6. Nozzle Spray angle, coverage, distance

from the standard table the best suited angle for the application is decide by finding the theoretical coverage of spray pattern as calculate from the included spray and the distance from the nozzle orifice up to fills. In this it doesn't make any sense to set an experiment for all standard spray angles.

The spray pattern also has an influence on the evaporative cooling, so on the heat exchange, thereby the performance. Spray patterns are namely: -

1. Hollow cone spray nozzle
2. Full cone (solid spray) spray nozzle
3. Spiral spray nozzle

Full cone and spiral cone of stainless steel is shown in a photograph, in Fig. 7 below,

Fig. 7. Full cone and spiral nozzle (spray angle: - 90°)

For the packaged type cooling tower used in this case, the best suited angle calculated as 90°, and distance from the nozzle orifice is around 200mm. though one previously available PVC nozzle of 60°, given rise for more results. For all the spray patterns, 90° is the most suitable angle, it means the spray coverage is about 60cm for a height from nozzle orifice. As desirable it the spray of nozzle slightly get mixed with each other and at the wall, it splashes a little.

2.3 Testing Procedure for a Cooling Tower

For any cooling tower used in a steam power plant or other, it becomes necessary to probe whether it is performing reasonably or not. To get all needed quantitative plus qualitative attributes one must learn about the testing procedures which are proposed to prepare for, put on several standard conditions based on previous experimental research, which helps in getting all answers of required aspects important in the initial and final process.

It involves conditions of test, operating conditions, duration of test, frequency of test and constancy of test conditions. Scheduling a regular, thorough inspection is an essential step in safeguarding the efficiency and lifespan of your cooling tower. The cooling tower tests are conducted in accordance with the Cooling Technology Institute (CTI) Acceptance Test Code for Water-Cooling Towers ATC-105. This test code provides details as to how a cooling tower is to be tested, what test instrumentation is to be a used, acceptable test condition, and how the results are to be interpreted.

3. RESULTS

Cooling tower of counter flow (an induced draft, small packaged type) held with an experimental trial, all performance parameters have been recorded in quantitative form, and qualitative conclusions acquired. An observation table is prepared as per the previous study of testing procedure for cooling tower. Readings have been taken for the table, for related properties as mentioned in the table. All performance calculations carried out as per the parameters enclosed in the annexure, considered for performance of cooling tower.

Range, approach and effectiveness are the most important parameter of cooling tower to know how good it is operating. Other than these three mentioned before, there are more to be acknowledge i.e. cooling capacity, evaporative loss, liquid-gas ratio, etc. A high CT Range means that the cooling tower has been able to reduce the water temperature effectively for a large span of temperature of hot water inlet to cooling tower to cold water outlet of cooling tower and thus, performing well and the lower the approach the better the cooling tower performance.

The inlet cooling tower temperature in above case is nearly 50°C, it is prior to note that the load was increased successively from initial load to 3000W and remained at that load for around 2-5 minutes in 200W of interval. In further trials, it made to run on full load so inlet CW temperature T_1 may increase up to 55°C than first case (as direct increase of extreme load on plant put some extra burden at start). But all other i.e. outlet CW temperature, WBT, air stream velocity, humidity etc. of cooling tower governs same importance, affecting the performance.

3.1 Range and Approach

These are the two basic performance indicating parameters of cooling tower, puts initial picture of span of maximum hot water temperature it has chance to reduce to lowest cold-water temperature; whereas approach deals with the possible fall of outlet CW temperature to ambient wet bulb temperature, means the difference in between. It is desirable to have larger range, however possible lesser approach. Fig. 8 shows the graph for range-approach for different nozzle types with 60°, 90° angles.

In the graph below blue line indicates variation of range, orange line indicates approach for different nozzle types of cooling tower. It is observed that the range increases successively from hollow cone nozzle to full cone nozzle, shown onto graph. It is 9.5°C at initial state for HC-60°, starts increasing from hollow cone 60° to 90° by 11.8°C, 12.23°C up to spiral type 90°-12.72°C and incredibly increases for full cone nozzle of 90°, followed by a rise in blue line after SC-90°, attains range of 15.5°C.

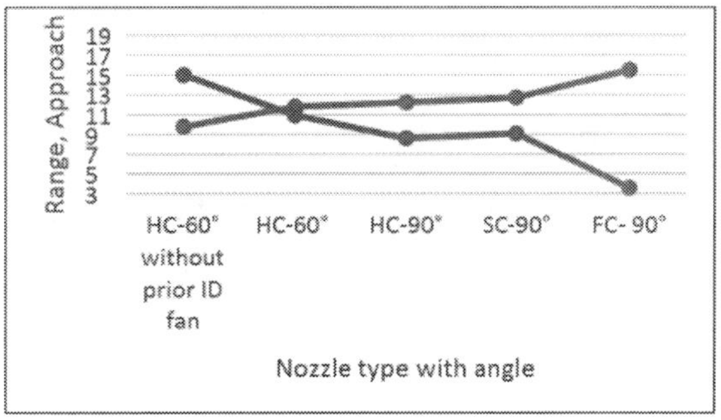

Fig. 8. Range-Approach for different nozzle types with 60°, 90° angles

Approach to WBT is seen obeyed, as desired to gets lesser difference between outlet CW temperature to WBT. At first for hollow cone approach of 15°C is observed, later the desirable fall in the orange curve is 10.93°C for HC and 8.6°C, 9.1°C for HC-90° and

SC-90° respectively. Finally, 3.56°C difference is attained followed by sudden fall after spiral nozzle.

3.2 Overall Performance Average Results

Table no. 1 shows overall average results of all performance parameters, including range, approach, effectiveness, Efficiency, etc. for the load of 3000W.

Table no. 1 Overall performance results

Sr. no.	Nozzle type	Range (°C)	Approach (°C)	Effectiveness (%)	Efficiency (%)	Load (watt)
1.	HC-60° without fan	9.8	15	38.27	38	0-3000
2.	HC-60°	11.8	10.93	51.91	52	3000
	HC-90°	12.23	8.6	58.05	58	3000
4.	Spiral type-90°	12.72	9.1	58.65	59	3000
5.	FC- 90°	15.5	3.56	81.63	82	3000

Range of 15.5°C is attained, approach of 3.56°C is reached and the effectiveness of 81.63 % is achieved for full cone spray nozzle of 90° angle, as mentioned in the table above with all other nozzle types. Water spray nozzles are therefore influence the evaporative cooling significantly. Characteristics like nozzle cone angle spray coverage and spray pattern affect incredibly on the performance of cooling tower.

3.3 Effectiveness Vs Spray Nozzle Type

The obtained values of Efficiency and Effectiveness are almost same. Therefore, the bar graph in Fig. 9, below shows the effectiveness, efficiency for all the spray patterns and two of spray angle. Effectiveness is the ratio between the range and the ideal range, the greater this ratio, the greater is the cooling tower effectiveness. Whereas the efficiency of the cooling tower is calculated as, inlet water temp. – outlet water temp. / inlet water temp. - Wbt.

Existing hollow cone nozzle with 60°, only able to attain 38%, on fitting prior induce fan, it reaches to 52% followed by bar lift of 58% for HC-90°, showing importance of spray angle on the efficiency. It achieves very little lift later for spiral nozzle type, and finally for the full cone of 90°, it attains sudden increment in percentage, 82% followed by greater rise of bar from SC-90°.

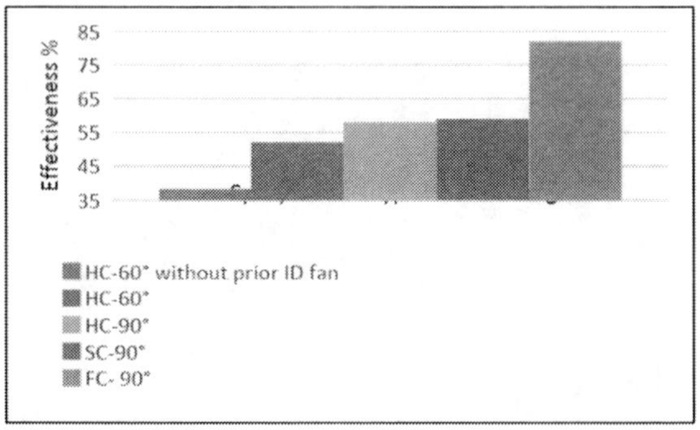

Fig. 9. Effectiveness Vs Spray nozzle types

3.4 Wet bulb temperature, relative humidity and cooling capacity

It is observed that WBT, Relative humidity and cooling capacity tend to increase successively with the change in spray pattern, angle and so that shares credit in All these three parameters are shown in the Table no. 2, given below.

The WBT is the lowest temperature that reaches under current ambient condition by evaporation of water only. At 100% RH, the wet bulb temperature is equal to air temperature (DBT). Cooling capacity is heat rejected in kCal/hr.

Fig. 10 shows the graph for properties mentioned above, three lines showing successive increment in the values leading to increased efficiency. First blue line near x- axis indicating WBT, middle gray line is for RH and line at top indicates increment of cooling capacity with all the types of spray nozzles.

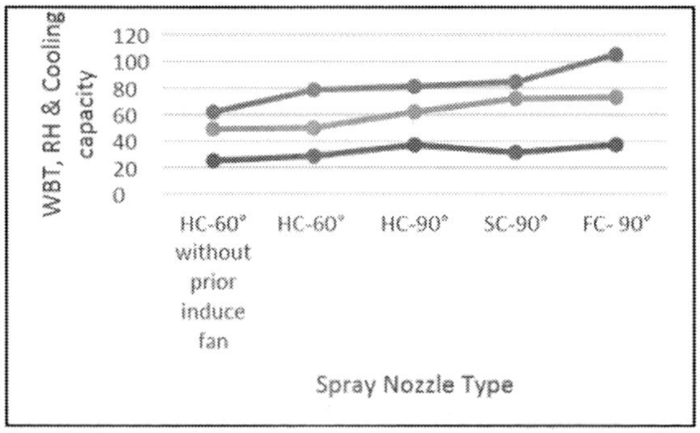

Fig. 10. WBT, RH & Cooling capacity Vs Spray nozzle type

Large variation in WBT, RH influences performance of cooing tower, and the increased cooling capacity and above two for the HC, SC and FC are indicated on the graph by the lines; starting from HC at 25.2, 49 and 69 followed by the rise of all three lines lead to 37.3, 73 and 105.20 for FC-90, WBT, RH and cooling capacity respectively. The other intermediate values mentioned in the Table no. 2 below:

Table no. 2. WBT, RH & Cooling Capacity

Sr. no.	Nozzle type	WBT	Relative Humidity (%)	Cooling capacity
1.	HC-60° without fan modification	25.2	49	61.84
2.	HC-60. with fan modification	28.67	50	78.47
3.	HC-90°	37.07	62	81.13
4.	Spiral type-90°	31.45	72	84.59
5.	FC- 90°	37.2	73	105.20

4. CONCLUSIONS

A desirable possibility for change with different spray nozzle angle, patterns is checked, tried and analyzed. An angle of 90° is considered as the best suited angle, then for three spray patterns: full cone, hollow cone and spiral spray nozzle, as per the trials done, readings are recorded.

1. The range of cooling tower is observed increased by 58.16% and the approach has been improved (decreased as desired to less difference with WBT) by 76.26%.
2. The cooling tower effectiveness increases with increase in inlet water temperature, it is found with an impressive increase of 113.30% (38.27% to 81.63%). The efficiency of cooling tower increases with increase in the inlet water temperature.
3. Amongst all spray pattern and angle full cone (solid spray) found to be more efficient for evaporative cooling. An efficiency of 82% is achieved for Full cone nozzle of 90°.
4. Cooling capacity is observed with an increase of 70.16%

REFERENCES

1. M. Castro, T Song, Pinto J., Trans IChemE, **78,** 192-201, (2000)
2. J Khan, M. Yaqub, S. Zubair., Energy Convers. and Manag., **44,** 2073–2091 (2003)
3. Y. Sun, Z. Guan,K. Hooman., Renew. and Sust. Energy Revs., **79**, 618–637 (2017)
4. S. Fisenko, A. Brin, A. Petruchik., Int.Journ. of HMTrans., **47,** 165–177 (2004)
5. M. Maiya, *Heat Rec. Sys.& CHP.*, **15**, 293-303 (1995)
6. T. Muangnoi, W. Asvapoositkul,S. Wongwises., Energy Conv. and Man., **49**, 2795–2800 (2008)
7. S.M. Kumar., Int. Journ. On Emerg. Trends in Tech., **38**, 442-448 (2016),
8. M. Lucas, J. Ruiz, P. Martínez, A. S. Kaiser. A. Viedma, B. Zamora B., *Applied Thermal Engineering.*, **50**, 282-292 (2013)
9. N. Tao, C. Daotong, J. Mingxiao, W. Jinshi, Y. Junjie., *Applied Thermal Engineering*., **78,** 228-235 (2015)
10. A. Tomás, S. Araujo, M.Paes, A. Primo, J. Da Costa, A. Ochoa., *Applied Thermal Engineering*, **144**, 310-319 (2018)
11. Y. Sun, Z. Guan, H. Gurgenci,J. Wang, P. Dong,K. Hooman., *Energy*, **168,** 273-284 (2018)
12. Gammon Cooling Tower India (GTPL) Ltd., *Cooling Tower Performance Testing Procedure*, **1,** 1-4 (2013)

E3S Web of Conferences **170**, 01010 (2020)
EVF'2019
https://doi.org/10.1051/e3sconf/202017001010

Assessing the Evolution of Human Settlements with 'Energy' as a Determinant

Manas Vijayan[1,*], *Akshay* Patil[1], *and Vijay* Kapse[1]
[1]Department of Architecture and Planning, VNIT Nagpur, Maharashtra, 440010, India

Abstract: Human settlements have evolved from caves in the *Paleolithic Age* to high rise buildings and cities in the modern era. Energy is one of the major driving forces in shaping the settlements of today. It is a fundamental of our everyday life and will continue to influence the future generations. It is also responsible for the many major looming threats faced by the world today, like climate change, ozone layer depletion, acid rains and global warming. Hence it is essential to investigate the influence of energy in shaping the settlements of the past, to understand the present, and to develop a vision for the future settlements. This paper is an attempt to study the evolution of human settlements based on the '*urban form determinants*' framework developed by A.E.J. Morris with 'energy' as an additional determinant. The investigation proposes how energy has influenced in shaping the settlements of the past, and the correlation between energy and other urban form determinants. This study will help various stakeholders in developing an understanding on how energy can play a role in shaping a sustainable future, and also in identifying the parameters which influence them.

1. Introduction

We live in a period of time which is currently facing several megatrends that threat a sustainable development of the planet earth. These threats include depleting resources, deteriorating environment, climate change, pollution of air, water & soil; most of which can be directly associated with the outcomes of various human activities. It goes without saying that the threats faced today cannot be seen in isolation but as a result of a series of events which happened in the past. It is hence essential to investigate the events in the past to understand the present, and to plan for the future.

Interestingly most of these threats can be directly related with energy consumed, and their associated impacts. The impacts of energy on the nature is a research area that has emerged from the 1980s [1]. The direct and indirect impacts of energy consumed include adverse problems like global climate change, stratospheric ozone depletion and acid rains [2]. Other impacts include health issues to the people and other living organisms [1], global warming, emission of radioactive substances, forest destruction [2], greenhouse gas (GHG) emission[3], and pollution of air, water and land. The by-products from the extraction, production, networking and consumption of energy define the largest single environmental loading of society when taken together [3].

This paper is an attempt to investigate the impact of energy as an *urban form determinant* (UFD) [4] in shaping the human settlements. The use of energy by human beings has changed from the time when the evolutionary predecessors of *homo-sapiens* were foragers in the *Stone Age*, to people of modern history. During *Stone Age*, the energy used by human beings was primarily for hunting and gathering. This accounted for about 2500 calories per day, and the average amount of calories consumed by one person living in USA (in 1968) was found to be 200,000 [5]. When human beings discovered different forms of resources from which they can extract energy; the lifestyle, the kinds of shelter, the food habits, the types and locations of settlements changed. They began to use energy for providing, building indoor comfort, transporting goods & people, and for other activities. The average energy used in different ages of human history is shown in **Fig. 1.**

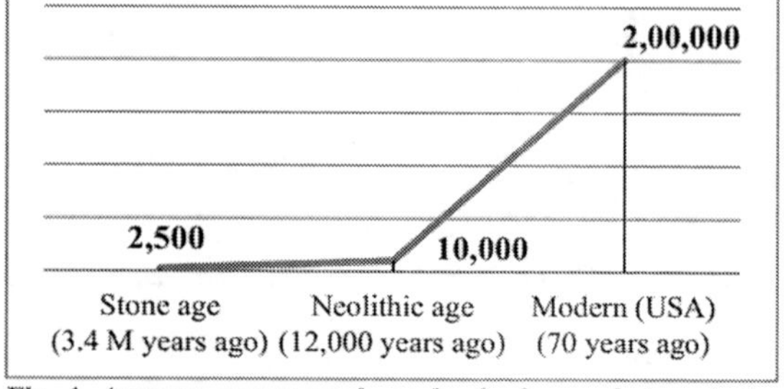

Fig. 1. Average energy used per day by human beings during different ages of history. [5]

*Corresponding author: **manas@students.vnit.ac.in**

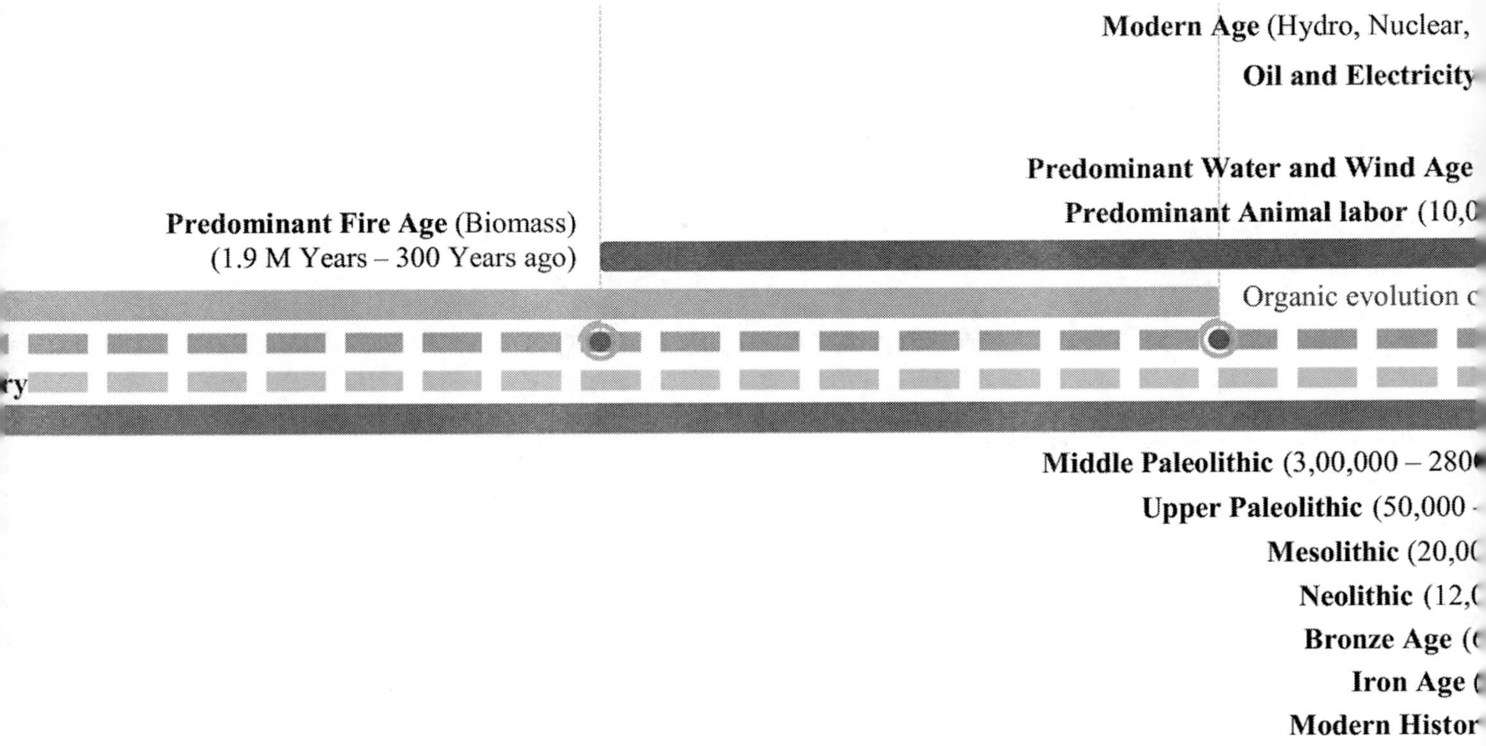

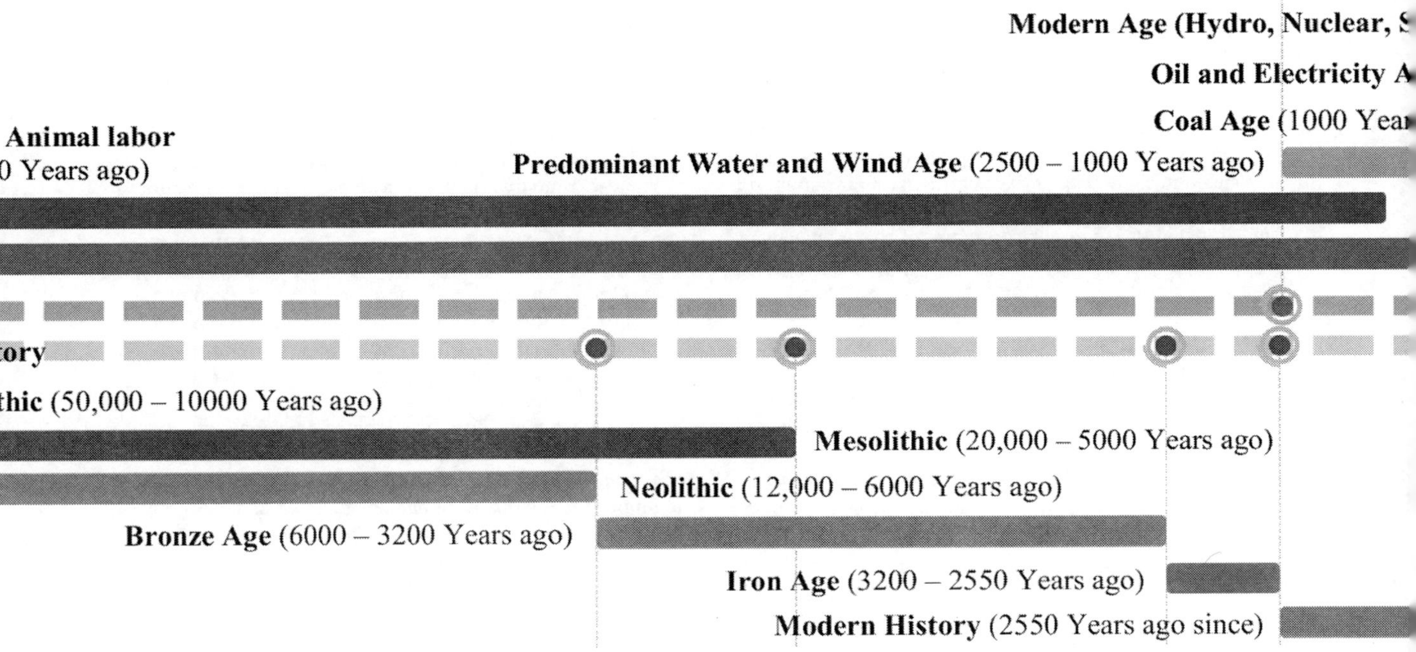

ine and energy use timeline from the Stone Age (3.4 million years ago). B) Human history timeline and energy *years before present*

This can be better understood by comparing the human history with the history of energy used; which is shown in **Fig. 2**. The figure explains the beginning and ending of different human prehistoric ages along with a timeline of energy. Since the exact ending of all energy eras mentioned in the energy timeline will not be a specific year, the diagram shows the predominant energy used at a specific time period. It is clearly visible that the discovery of new types of energy forms has had a massive impact in the development of settlements. The discovery of new sources of energy meant that humans started extracting more energy and storing them, instead of using them at once.

The '*energy balance*' as quoted by Smil [6], between energy extracted and used was at equilibrium at some point during the *foragers* era. This balance was lost in the human settlements that followed, which lead to unsustainable development, with humans consuming more resources than what is required for their survival. Between 1500 AD to 2014 AD, the human population has increased 14-fold (500 million to 7 billion) and the energy use by humanity has increased 115-fold (13 trillion calories/day to 1500 trillion calories/day) [7]. The modern history (2550 years ago since) of human settlements have witnessed a great number of developments in the energy sector and the 'threats' mentioned above can be associated with these events. To understand the changes in the human settlements in this timeline, one settlement from each of the energy eras in **Fig. 2** were selected. Framework for assessing these settlements was selected from literature.

2. Methodology & analysis

2.1 Structure of the study

Different frameworks have been used to study the evolution of human settlements, to understand its patterns and principles, the cultural base and socio-economic infrastructure. The prominent ones among these include the one developed by Patrick Geddes based on key units of *work, folk* and *place* [8]; UFD framework by Morris [4], *Central Place Theory* by Walter Christaller [9], *General Systems Theory* by Ludwig von Bertalanffy [10], *Regional Input-Output Analysis* by Isard [11], *Growth Centers and Human Settlements* by Hansen [12], *Business Organisation and Spatial Organisation* by Lasuen [13]. The framework used for this study is the UFD framework by Morris, for the depth in analysis provided in the framework using sixteen UFDs (three natural and thirteen man-made), and one additional UFD (energy), which are listed in **Table 1**. Using these determinants will also enable a researcher to understand the interdependence of various determinants.

Table 1. UFDs used in the framework for the study.

Natural determinants		Code
1	Topography	N_1
2	Climate	N_2
3	Materials and construction technology	N_3
Man-made determinants		
1	Economic	M_1
2	Political	M_2
3	Religious	M_3
4	Pre-urban cadastre	M_4
5	Defence	M_5
6	Aggrandizement	M_6
7	Grid-iron	M_7
8	Urban mobility	M_8
9	Aesthetic	M_9
10	Legislation	M_{10}
11	Urban infrastructure	M_{11}
12	Social, religious and ethnic groups	M_{12}
13	Leisure	M_{13}
Additional determinant considered		
1	Energy	E_1

2.2 Selection of settlements

Permanent physical units for living did not exist before 14000 BC and the first civilizations were established during the late stages of *Bronze Age*. The decisive step towards the formation of urban civilization was due to the end of the last of *ice ages* around 7000 BC, which made the environment, friendly for the humans, to exploit the organic world [4].

The settlements are selected as representative case studies which fall in the energy eras shown in **Fig. 2**. To understand the influence of energy in shaping the settlements, case studies are selected such that they had minimal external influences in their development; and they fall into similar climatic conditions and geographic regions. Since the earliest known human civilizations *(Sumerian civilization)* are from 5000 BC [4], settlements before 5000 BC are not selected for the study. The settlements selected, and their characteristic features are shown in **Table 2**; and their layouts are shown in **Fig. 4**, **Fig. 5** and **Fig. 6**. **Table 2** explains the energy era and human history era to which the selected settlements belong to; the predominant energy source and the important elements of the settlements.

2.3 Data presentation

To understand the UFDs which influence the energy consumed in the case settlements, it is essential to know the various activities which consumed energy in a specific human era. The average per capita annual energy consumption for various activities over the history of humans stated by Smil [6] is shown in Fig. 3.

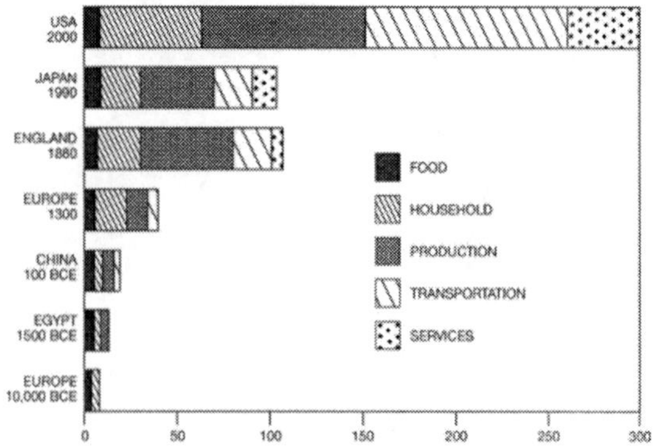

Fig. 3. Per capita energy consumption (GJ/year) (Source: [6]).

Table 2. Settlements selected for the study and their characteristics.

Sl. No	Settlement	Reason for emergence	Energy era*	Energy influencing UFDs**	Predominant energy source	Human history era	Important elements of settlement	Ref
1	'Ur of the Chaldees'	Unplanned development grown out of long evolutionary process	(b,c)	(N_{1-2}, $M_{1,2,8}$)	i.Human & animal labour ii.Biomass iii.Wind	Neolithic: Mesopotamia n *(Sumerian)*	i.North harbour ii.West harbour *iii.'Temanos'* *iv.*Housing district	[4]
2	Athens	Organic growth with special attention to spatial relationships and layouts determined by inherited constraints.	(d,b)	(N_{1-3}, $M_{1-3,6-9,13}$)	i.Human & animal labour ii.Biomass iii.Water & wind	*Bronze Age*	*i.Acropolis* *ii.Agora* *iii.Pnyx* iv.Long walls v.*Olympieion* temple	[4]
3	Winchester	Medieval town influenced by the original gridiron structure of Roman Venta Belgarum.	(d,b)	($M_{1-2,4-7,9-10}$)	i.Human & animal labour ii.Biomass iii.Water & wind	Medieval	i.Castle ii.Market place iii.Street iv.Defence wall v.Square	[4, 14]
4	Turin	Roman colony in the 15th century which was extended during Renaissance period.	(e,b)	(N_3, $M_{1-2,4-9,11-12}$)	i.Human & animal labour ii.Biomass iii.Water & wind iv.Coal	Renaissance	i.Fortification ii.Primary straight street/stations iii.Gridiron districts iv.Enclosed spaces	[4]
5	Paris	Reorganization and remodelling of the city in 1850s, which took place after the industrial revolution began.	(e)	(N_2, $M_{1-2,4-11,13}$)	i.Coal ii.Human & animal labour iii.Water & wind	Modern era-Industrial revolution	i.Civic centres ii.Traffic iii.Gridiron iv.Symmetry & balance	[15]
6	Vancouver	Post-industrial city with emphasis on offices, middle-class residential land-use, recreation, and entertainment facilities	(f,e)	(N_3, $M_{1-4,6-13}$)	i.Fossil fuels ii.Electricity iii.Coal iv.Human & animal labour	Modern era-Post-industrialism	i.Gentrified neighbourhoods ii.Leisure, luxury & entertainment hubs iii.Gridiron iv.Slums outside city	[16-17]
7	Neighbourhood unit	Revitalizing existing neighbourhoods to reduce urban expansion, provide social equity, and enhance sustainability with smart technologies.	(f,g, e)	(N_3, $M_{1-2,4,8-13}$)	i.Electricity ii.Fossil fuels iii.Coal iv.Renewable energy v.Solar powered cities	Post-modernism	i.Smart & sustainable neighbourhoods ii.Transit oriented development iii.Pedestrian networks	[17-18]

Note:
(*)- Stone age (a), Fire age (b), Animal labour (c), Water &wind age (d), Coal age (e), Oil & electricity age (f), Modern age (g)
(**)- Codes as shown in **Table 1**.

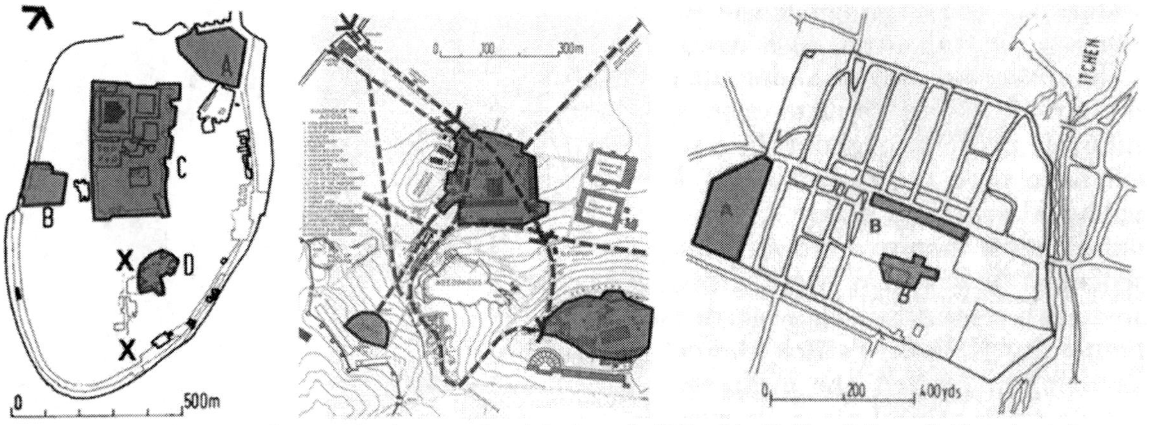

Fig. 4. Layouts and important elements of settlements selected for the study. 1) Ur of the Chaldees (left), 2) Athens (centre), 3) Winchester (right) (Source: [4])

 https://doi.org/10.1051/e3sconf/202017001010

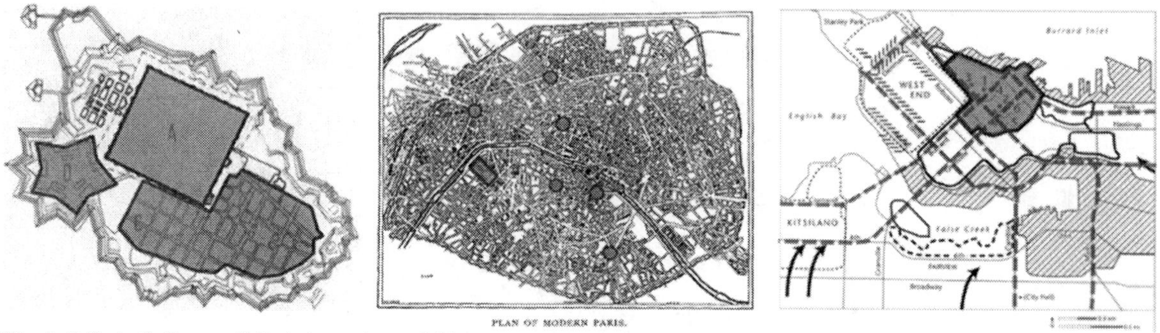

Fig. 5. 4) Turin (left), 5) Paris (centre), 6) Vancouver (right) (Source: [4,16,18])

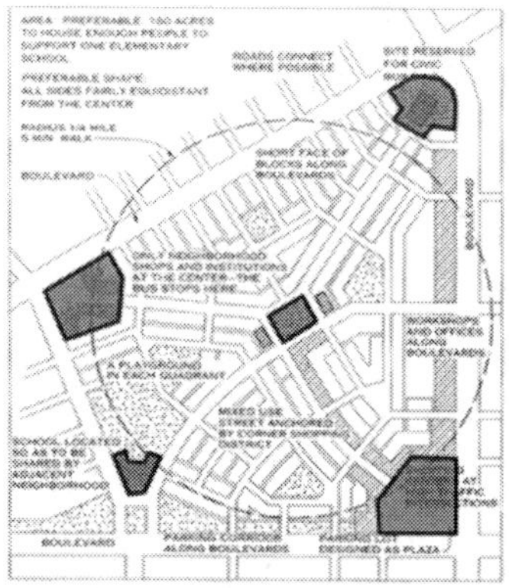

Fig. 6. 7) Duany Plater-Zyberk's Neighborhood Unit based on compact cities and transit-oriented development. (Source:[19])

The city of Ur was located on the banks of the Euphrates River in such a way that it was secure from flooding, and can benefit from trading [4]. The buildings were designed to respond to the local climate. The main energy consuming activities at the time being food, production and household [6], the UFDs which had a major influence in the energy consumption are *topography* (N_1), *climate* (N_2), *economic* (M_1) and *urban mobility* (M_8). Since *Ur* had leaders in power, *political* (M_2) is also considered as a UFD influencing the amount of energy consumed. The same method was used for identifying the UFDs which influence energy consumption for other six case settlements.

3. Observations and discussion

The distribution of natural and man-made UFDs influencing energy consumed is shown in **Fig. 7**. The diagram shows that the early settlements had minimal number of UFDs influencing energy, and the number increasing till Vancouver, and a dip in curve at the end. To investigate natural and man-made UFDs separately, **Fig. 8** was generated.

The diagram indicates settlements being driven by man-made determinants progressively with time, and a decreasing influence of natural determinants. This observation can be validated by the trend of prime movers stated by Smil [6] which is shown in **Fig. 9**.
The diagram shows the way in which humans have started relying progressively on man-made energy sources, and inevitably moving away from nature.

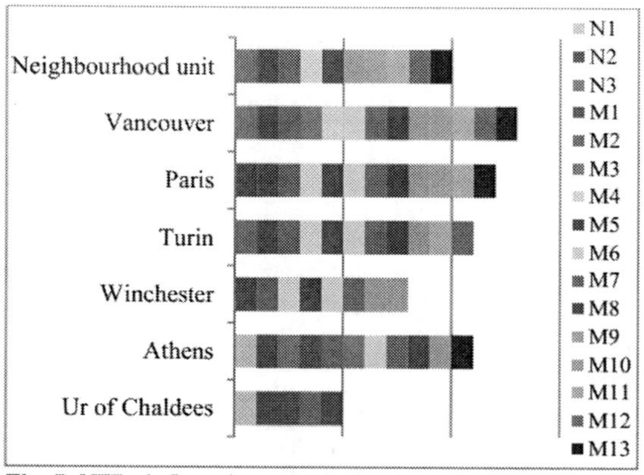

Fig. 7. UFDs influencing energy consumed in the selected settlements. (UFD codes as given in **Table 1**)

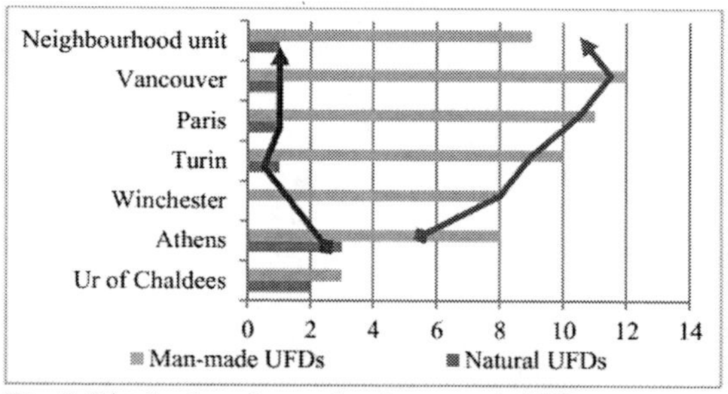

Fig. 8. Distribution of natural and man-made UFDs influencing energy consumed in the selected settlements.

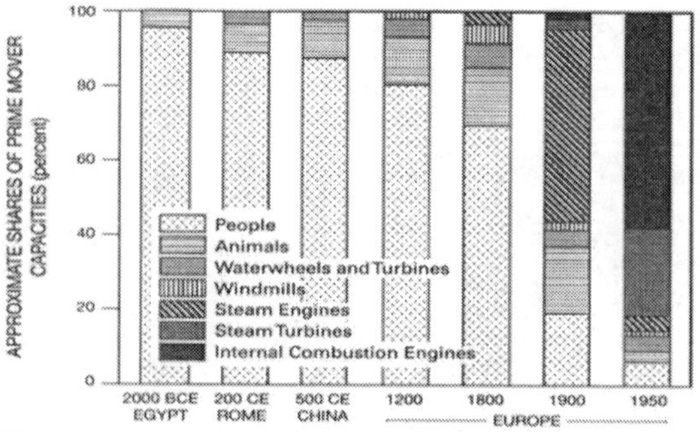

Fig. 9. Share of prime movers (in percentage)(Source: [6])

The earliest of human settlements *(barbarism)* were completely based on human muscular energy and the settlements thrived in areas located closer to the sources of energy, which ensured highest *energy returns*. The highest population density during the *Palaeolithic Age* were in coastal settlements, where hunting of much larger mammals was possible and the population density was ten times higher than forest settlements [20]. The assessment also indicates the use of more advanced energy sources in the prehistoric settlements. For instance, in *Ur*, the existence of harbour and market places beside them indicates the use of wind energy for transportation of goods, even though the settlement belongs to the *fire* and *animal labour* energy era. Such overlaps exist depending on the materials and technology available with a specific settlement.

The development of methods of transmitting energy meant that settlements no longer needed to be located close to the source of energy, which led to settlements moving away from the coastal areas and river valleys, into the inlands. It is safe to say that the settlements that followed had one of its objectives to make this transmission of energy uncomplicated. This statement can be validated by the fact that five of the six case settlements, after humans started using energy source other than *human labour* and *animal labour*, were influenced by *grid-iron* and *urban mobility,* which will make this transmission unsophisticated. The peak of this change was during the industrial revolution and post-industrial period. The post-modern cities, which are trying to create low-carbon sustainable settlements, are planning layouts to exploit renewable energy sources. The influence of energy in shaping the settlements is evident from the study. It can be questioned if we can understand and learn from our early settlements better, when energy is included as one of the determinants in the frameworks used for assessing them.

4. Conclusions

Energy as a determinant had a major role in shaping the early settlements and is a major driving force which shapes the current settlements. The use of energy in human settlements has changed with time. When humans started relying progressively on man-made energy sources, they have inevitably started moving away from nature. This change is what triggered the unsustainable development patterns, as energy was no longer the deciding factor for the survival of human beings but was a driving force in changing their lifestyle from *foragers*, to *farmers* with permanent settlements, and to the cities of modern era.

Political & economic UFDs had the most influence in energy consumption of past settlements, which shows that the settlements were not driven by energy efficiency awareness but social & political awareness.

To make the future development sustainable, we will have to gain insight from the earliest settlements, where energy was derived directly from the source, the *energy returns* were at equilibrium, and ensured minimal impact to the surrounding ecosystem. The future sustainable settlements will have to derive energy from direct sources like kinetic energy of the oceans, and heat from the sun rather than converting them to unsustainable forms [5]. It is the need of the hour that we question if compact cities are actually the solution to make the future development sustainable; to harness renewable energy from sunlight, heat, wind and rainwater; to employ the landscape of the city to purify our air, water and land; and to exploit the human resources for nature, against our past trends.

5. References

1. W. H. Qu, L. Xu, G. H. Qu, Z. J. Yan, and J. X. Wang, Nat. Hazards **87**, 675 (2017)
2. I. Dincer, Energy Policy **27**, 845 (1999)
3. S. Bilgen, Renew. Sustain. Energy Rev. **38**, 890 (2014)
4. A. E. J. Morris, *History of Urban Form before the Industrial Revolutions* (Routledge, New York, 1972)
5. C. A. Doxiadis, Athens Cent. Ekistics **26**, 455 (1968)
6. V. Smil, *Energy and Civilisation- A History* (2017)
7. Y. N. Harari, *Sapiens: A Brief History of Humankind.* (Random House, 2014)
8. P. Geddes, *Cities in Evolution* (London, 1949)
9. C. W. B. & W. C. B.J.I. Berry, Econ. Geogr. **43**, 275 (1967)
10. L. Von Bertalanffy, Gen. Syst. Theory Found. Dev. Appl. 296 (1973)
11. W. Isard, Rev. Econ. Stat. **33**, 318 (1951)
12. N. Hansen, Reg. Sci. Assoc. **38**, (1977)
13. J. R. Lasuen, Urban Stud. **6**, (1969)
14. T. B. James and E. Roberts, Mediev. Archaeol. **44**, 181 (2000)
15. R. Schermerhorn Jr, Brooklyn Eng. Club, Proc. 102 (1912)
16. T. A. Hutton, Urban Stud. **41**, 1953 (2004)
17. A. Mair, Polit. Geogr. Q. **5**, 351 (1986)
18. A. Kylili and P. A. Fokaides, Sustain. Cities Soc. **15**, 86 (2015)
19. A. Sharifi, Sustain. Cities Soc. **20**, 1 (2016)
20. V. Smil, Encycl. Energy **6**, 549 (2004)

Design and prototyping of dual axis solar tracking system for performance enhancement of solar photo-voltaic power plant

Ashish Patil[1], Mangesh Dhavalikar[1], Sunil Dingare[1] and Virendra Bhojwani[1]*
[1]MIT School of Engineering, MIT ADT University, Pune, India.

Abstract. Increase in the world energy demand has made renewable energy source as one of the best option. It is finding wide utility in comparison with the other sources of energy. Solar energy is one of the major source of renewable energy. The abundance of solar energy in India makes it highly desirable to exploit but due to the limitations of technology, only a fraction of the energy can be extracted and used. This paper describes a concept for solar detection sensor implementation in Photo-Voltaic (PV) dual-axis solar tracking systems. The system uses only three units of identical light-detecting- resistors (LDR) arranged on the three sides of tetrahedron geometry and is able to track the light source position at the strongest intensity of visible light. On the basis of comparative study of outputs of all the three LDRs, the panel will tilt so as to get maximum output from the system. This system results in a wide field of view with minimum error and also has advantages in terms of sensor quantity, size of the system, accuracy and effectiveness.

1 Introduction

In India around 470 million tons coal is used in 2017by various thermal power stations which produces lots of pollution.[8] Also, in year 2017-18 the total energy requirement in India was around 915123 million units (MU) but total energy available was around 908650 million units (MU). So, in 2017-18 there was energy shortage of around 6473 million units which is of 0.71% of total energy requirement in India. Similarly, peak demand in India in year 2017-18 was around 164066 Megawatt (MW) and peak met was around 160752 megawatt (MW). So, peak shortage in India in year 2017-18 was around 3314 megawatt (MW) which was of 2% of total peak demand. [8]
In today's era of modern science and technology, the socio-economic development depends on electric energy. However, in developing countries, this electric energy is in demand and very poorly mismanaged. Also now a days there is a shortage of fossil fuels and pollution incurred upon their use. So to reduce this energy crisis the renewable energy can play the very important role. As the solar energy is a clean renewable energy source, it gets more and more attention, especially in the field of electricity generation. There are different tracking systems with several tracking strategies designed and tested by various scientists all over the world.

2 Literature Review

[1] In 2016, Yuwaldi Away and M. Ikhsan from Indonesia newly designed a dual axis sun tracker sensor based on tetrahedron geometry. Three LDRs were arranged on three sides of tetrahedron. Depending upon comparison of outputs of three LDRs, the system rotates to get the equal outputs from all three LDRs. This system is further tested in single and multiple light sources arrangements. It has an advantage of less complexity with less number of LDRs and has wide range of Factor of View (FOV).
[2] In 2017, Meita Rumbayan, Muhamad Dwisnanto Putro from Indonesia proposed asingle axis tracking system to improve the panel efficiency. There were eight LDRs as sensors - four at four corners and four at four sides of a solar panel, a servo motor as actuator and an Arduino as the controller. The proposed system results consists of system design, hardware design andalgorithm design.
[3] In 2014, Shahriar Bazyari, Reza Keypour, Shahrokh Farhangi, Amir Ghaedi, Khashayar Bazyari, compared different photo-voltaic solarpanel tracking systems in Qushm Island. This paper also include the received solar radiation density associated in summer, autumn and in winter. Further it includes the graphical comparison in fixed panel system, single axis tracking system and the dual axis tracking system. Accordingly concluded that the panel with single axis tracker receives 1.35 times more energy than the fixed panel, while the dual axis tracker receives 1.04 times more energy than the single axis tracker.
[4] In 2017, Jerin Kuria kose Tharamuttam, Andrew Keong Ng from Singapore designed an active type solar tracking system which can precisely track the sun positions. The developed system contains the hybrid algorithm strategy which contains both sensor (four LDRs) as well as the mathematical models of the sun position to detect the current position of sun and accordingly Arduino Controller will rotate the panel with the help of two motors. Further the real time monitoring of the solar data is done on the webpage also the graphical

https://doi.org/10.1051/e3sconf/202017001011

comparison between the active, chronological, fixed mounted and the proposed hybrid system on sunny as well as on cloudy day is done. This proposed hybrid system has advantage in all-weather condition as compare to the other systems.

[5] In 2014, Yingxue Yao, Yeguang Hu, ShengdongGao, Gang Yang, Jinguang Du designed and prototyped a dual axis solar tracking system which works on both solar time base strategy as well as on sensor based sun positioning strategy. The proposed system rotates 15°/hr. Due to combination of both the strategies, the proposed system requires less energy input with tracking error which is below 0.15°.

[6] In 2014, Jeng-Nan Juang and R. Radharamanan designed & developed a dual axis solar tracking system for standard sized homes as a supplementary power source or as an independent power source. The proposed system contains four LDRs as sensors, two linear actuators and an Arduino Uno as micro-controller. This system has low cost, low power consumption, user friendly light weight structure with wheels mounting.

[7] In 2014, Mayank Kumar Lokhande, designed & developed an automatic single axis solar tracking system for low power and residential usage applications. The proposed system contains the two LDRs as sensors, anew design of controller parts and the geared DC motor as actuator. This system can successfully track the current positions of the sun and can be used for non-critical and low power applications.

The existing tracking systems have complicated sensor strategies so has high cost and maintenance issues. Many researchers from all over the world have developed so many solar radiations sensing devices and tracking strategies but they contain large number of LDRs which makes them complicated. So, use of minimum number of LDRs for the development of Perfect Tracking System is the objective of this work.

3 Problem identification

The common problem in solar Tracking is the complex movement of sun. For a specific longitude, the sun moves from east to west along a fixed solar path everyday. But the sun moves through 46° north and south during 21st June to 21st December which is as shown in below figure 1. A dual axis solar tracker can account for both the daily and seasonal motion of the sun.

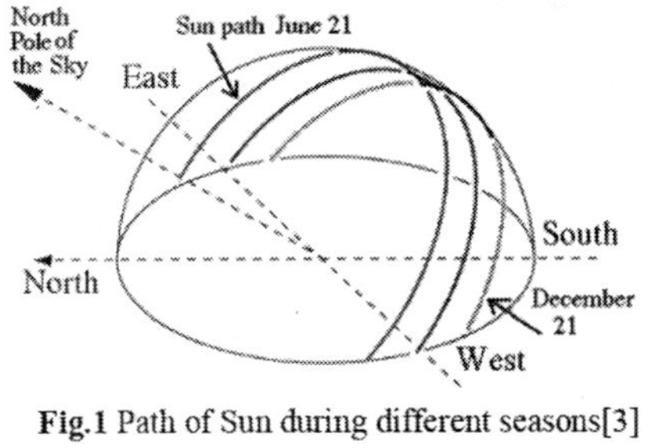

Fig.1 Path of Sun during different seasons[3]

4 Principle of Solar Tracking

In a day, sun moves from east to west and seasonally it also moves in north and south direction. But the PV panel system gives the maximum ammount of power absorption when the panel is perpendicular to the sun rays. So, it is advantageous to have the solar panels track the location of the sun, to get the maximum power output.

There are two types of Solar tracking systems as active & passive. Active solar tracking system uses electronic sensors, actuators to track the current positions of the sun. Whereas the passive solar tracking system uses the liquid od low boiling point so that it gets evaporated due to sun's heat and accordingly the solar tracking is done.

5 Methodology

To achieve the perfect tracking, active type solar trackers are more precise as they work on real time data coming from the sensors. Here, the light sensing sensors such as photodiodes OR Light Dependent Resisters (LDR's) detect the intensity of light. This light intensity is converted into the equivalent resistance by the LDR.

This output of LDR is given to microcontroller which detects the maximum intensity light direction according to the Program fed to it. Microcontroller with the help of program instructions will actuate the motor/s, enabling the system to move to a particular direction.

The block diagram of the system is shown in figure 2. As per the signals coming from microcontroller the motor (in case of single axis tracking system) or motors (in case of dual axis solar tracking) will turn the system to the direction of maximum intensity light.

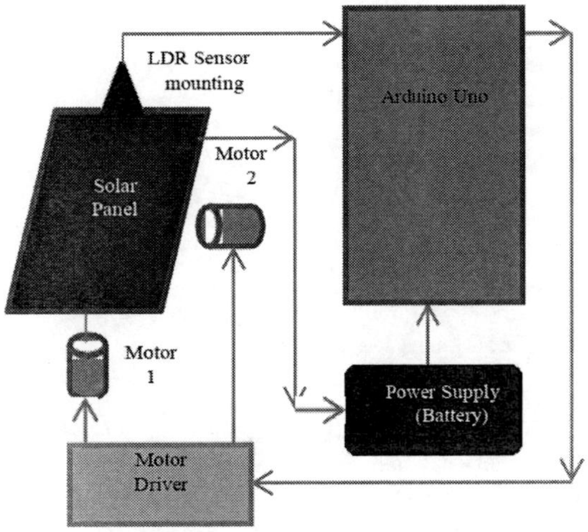

Fig.2 Building Block diagram of the system

6 Design of System

6.1 Modelling in Catia V5R20 Software

The basic model of the proposed system is drawn in Catia V5R20 software as shown in below figure 3. This basic model contains the tetrahedron sensor geometry, Solar

 https://doi.org/10.1051/e3sconf/202017001011

panel platform, Two Motors, base plates, etc.

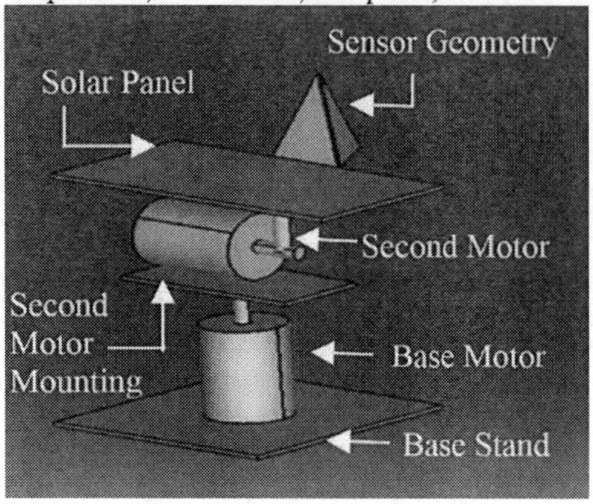

Fig.3. Catia Model of the System

6.2 Energy consumption by system

As system is of active type, it will require energy to operate. The load calculations as well as the system consumption are calculated as given below:

6.2.1 *Panel Calculations*

Energy generation by a panel
= Solar Irradiance * Solar Cell Area * Cell Conversion Efficiency (1)
Energy generation by a panel
= 5440 Whr/m^2/day * (0.17 m * 0.32 m) * 0.16
Energy generation by a panel = 47.4 Whr/day

6.2.2 *Battery Calculations*

Battery Power = Voltage * Current (2)
Battery Power = 12 V * 14 Ah
Battery Power = 168 Whr

Battery can be dischargeable up to 80 % Therefore,
Actual Power = 168 * 0.8
Actual Power = 134.4 Whr

6.2.3 *Motor Torque*

Weight of Solar Panel = 1.100 kg

Therefore,
Torque required to rotate the panel
= Weight of panel * perpendicular distance between motor shaft and the panel (3)
Torque required to rotate the panel = 1.100 kg * 5 cm
Torque required to rotate the panel = 5.5 kg-cm

6.2.4 *Load considerations*

LED bulbs – 3 (5 Watt each) Backup – 8 hours
So,
Load Energy Consumption = 120 Whr

System consumption

Motor Power = P = V*I (4)
Motor Power = P = 12 * 0.3
Motor Power = P = 3.6 Watt

Considering tracking of sun for 12 hours from East to West, total 180° (i.e. -90° to +90°)
Therefore,
Angle of rotation per Hour for motor1 = $\frac{\text{Total angle}}{\text{Time}}$ (5)
Angle of rotation per Hour for motor 1 = 180°/12
Angle of rotation per Hour for motor 1 = 15°/hour

Motor speed is 3.5 rpm
So, for one revolution motor takes 60 seconds /3.5 rpm =17 seconds
So, to rotate half revolution required time is 17/2
= 8.5 sec. = 0.0025 hr

Therefore,
Energy consumption by motor1=Motor Power*Time (6)
Energy consumption by motor1= 3.6 * 0.0025
Energy consumption by motor1= 0.00875 Whr
Energy consumption by motor1= 8.75 mWhr

On consideration of seasonal tracking motor will rotate 47° * 2 = 94°/year

So,
Angle of rotation per Hour for motor 2 = $\frac{\text{Total angle}}{\text{Time}}$ (7)

Angle of rotation per Hour for motor 2 = $\frac{94°}{24hr*365days}$

Angle of rotation per Hour for motor 2 = 0.01°/hr

As 0.01° is so small we can neglect it. Therefore,
Total Energy consumption
= Load Energy Consumption + System Consumption (8)
Total Energy consumption = 120 + 0.00875
Total Energy consumption = 120.00875 Whr

7 Selection of Components

The components used to achieve the dual axis solar tracking are mentioned as follows along with their specifications.

7.1 Light Dependent Resister (LDR)

It is a light sensor whose resistance changes with the change in light intensity. It has wide range of performance, it can sense the small change in light intensity.

Fig.4 Light Dependent Resister (LDR)

 https://doi.org/10.1051/e3sconf/202017001011

7.2 Arduino Uno Development Board

Microcontroller : Atmega328P
Operating Voltage : 5V
Input Voltage (limits) : 6-20V
Digital I/O Pins : 14
Analog Input Pins : 6
Clock Speed : 16 MHz

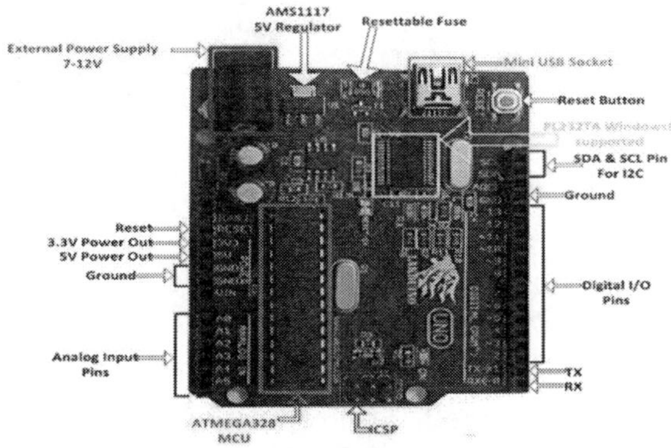

Fig.5 Arduino Uno

7.3 DC Motor

We require a high torque (to withstand the weight of the system) and a low speed (to achieve a precise angle of rotation) motor. So here we use 3.5 rpm geared DC motor.

Specifications and Features:
RPM : 3.5
Operating Voltage : 12 V
DC Gearbox : Attached Spur gearbox
Shaft Diameter : 6mm with internal hole
Torque : 7 kg-cm
No-Load Current : 60 mA
Load Current : 300 mA

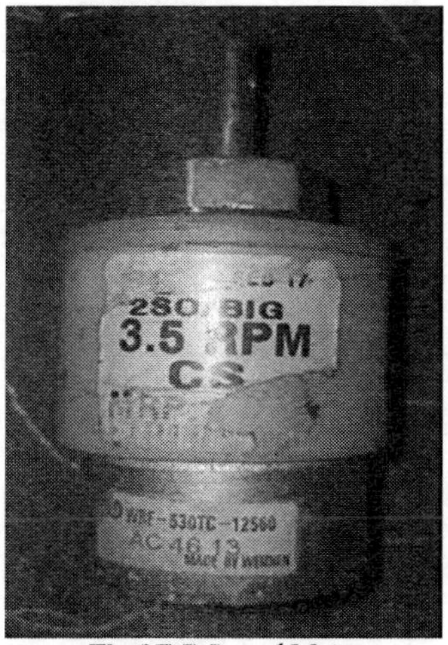

Fig.6 DC Geared Motor

7.4 Solar Panel

Panel Technology : Mono crystalline
Output Power : 10 Watt
Operating Voltage : 12 Volt
Cell Conversion Efficiency : 16%
Number of Cells : 36
Weight: 1.100 kg

Fig.7 Solar Panel

7.5 Battery

Brand : Exide
Rated Capacity : 7 Ah
Battery Voltage : 12 V
Items in Pack : 2 (Parallelly Connected)

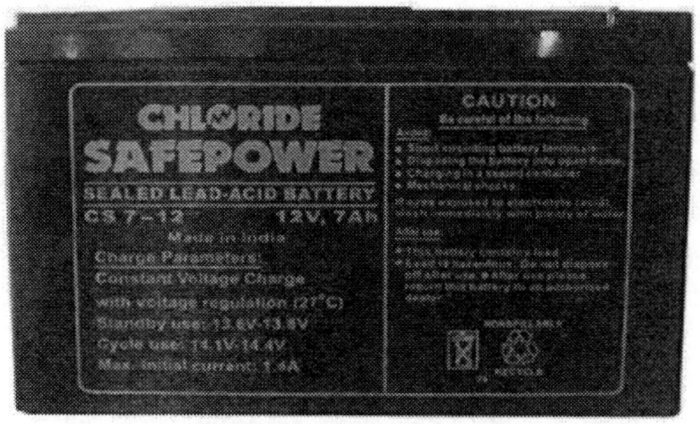

Fig.8 Battery

7.6 H Bridge Motor Driver

L298N Dual Motor Controller Module allows us to control the speed and direction of two DC motors. This can be used with motors that have a voltage between 5 and 35 V DC.
Double H Bridge Driver Chip : L298N
Logical Voltage : 5V
Drive Voltage : 5V-35V
Logical Current : 0-36 mA
Drive Current : 2A

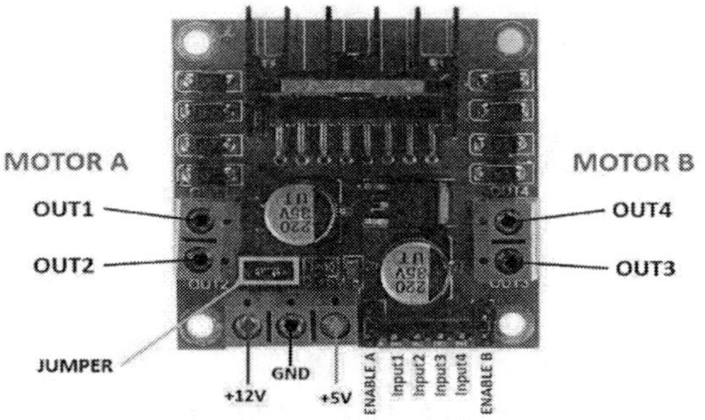

Fig.9 L298N H-Bridge Motor Driver

The all components required for execution of proposed dual axis solar tracking system are listed in the bellow given Bill of Material Table.

Table 1. Bill of Material

Sr. No.	Component Name	Component Description	Quantity
1	Arduino Uno kit	Generic Uno R3 ATmega328P	1
2	H bridge motor controller	L298N Dual Motor Controller Module	1
3	LDRs	10mm Diameter	3
4	Motors	3.5 RPM 12V Geared DC Motor	2
5	Solar Panel	10W Mono-Crystalline Panel	1
6	Battery	Exide 12 V, 7 Ah Lead-Acid Battery	2
7	Wires	3 meter	3
8	LED Bulb	5 Watt DC LED Bulb	1
9	Base Stand	CI material	1
10	Sensor Geometry	Aluminum Sheet	1
11	Black color	Metal Spray Color	1

8 Arduino Programing

First the analog data coming from the three LDR sensors are read and saved to the variable memory locations a, b and c. Then according to the equation given in the Arduino program flow chart the values of X and Y are calculated. According to the values of X and Y variables and the If-Else statements used in the program the further digital outputs are given to the pins 9, 10, 11 and 12 on the Arduino board to ON and OFF the two motors.

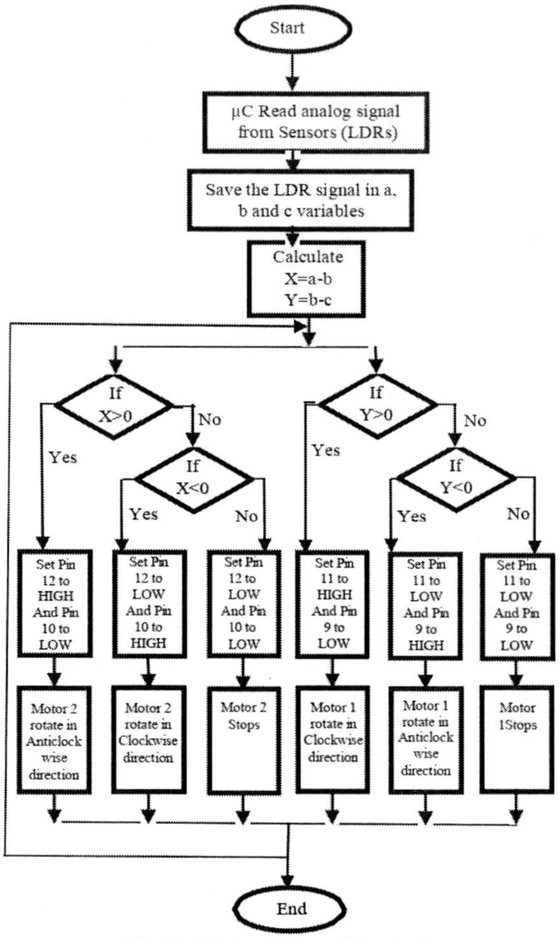

Fig.10 Arduino program Flowchart

9 Fabrication & Assembly

The various components such as sensor geometry, base motor and second motor mounting are done separately and then assembled with each other as well as with the electronic system also.

9.1 Sensor Geometry

The tetrahedron geometry is used to mount the three LDRs on the three sides of tetrahedron and the connecting wires are taken from the base side of the same.[1]

The whole structure of the tetrahedron is made in CR steel material. The one connection of each LDR is connected to the 100 KOhm resister whereas the other connection of LDR is connected to the +5 Volt DC supply. The other connection of 100 KOhm resister is connected to the Ground and the Arduino pin connection is taken from the junction between LDR and Resister. For three LDRs there are three Arduino pin connections and connected to the analog input pins A0, A1 and A2 on the

https://doi.org/10.1051/e3sconf/202017001011

Arduino board. The sensor is mounted on the top of the solar panel to get the actual light from the sun.

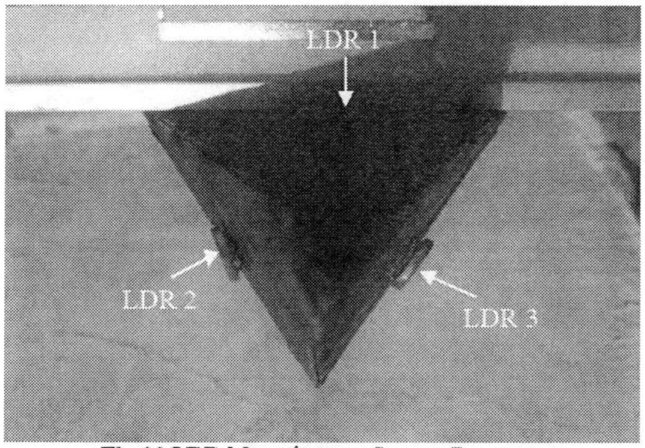

Fig.11 LDR Mountings on Sensor Geometry

9.2 Base Motor Mounting

The base motor is mounted vertically on the base plate to track the sun from East to West that is the daily motion of the sun. The base plate of the second motor is directly weld connected to the shaft of the base motor.

Fig.12 Base Motor Mounting

9.3 Second Motor Mounting

The second motor is mounted horizontally on the first base motor and it will track the sun from south to north which is the seasonal motion of the sun. On the shaft of the second motor, the solar panel is welded through a connecting rod.

Fig.13 Second Motor Mounting

9.4 Solar Panel Mounting

The solar panel is connected from the center with the connecting rod with the help of weld joint. The other end of connecting rod is connected to the second motor shaft with the help of weld joint.

Fig.14 Solar Panel Mounting

9.5 Arduino Uno board connections

The 12 Volt DC supply is given to the Arduino board through the wire from the battery unit. The three LDR output signal wires are connected at the A0, A1 and A2 analog input pins of the Arduino board.

The output signals of Arduino board which are given to the motors as input signals to them are taken from digital output pin numbers 9, 10, 11 and 12.

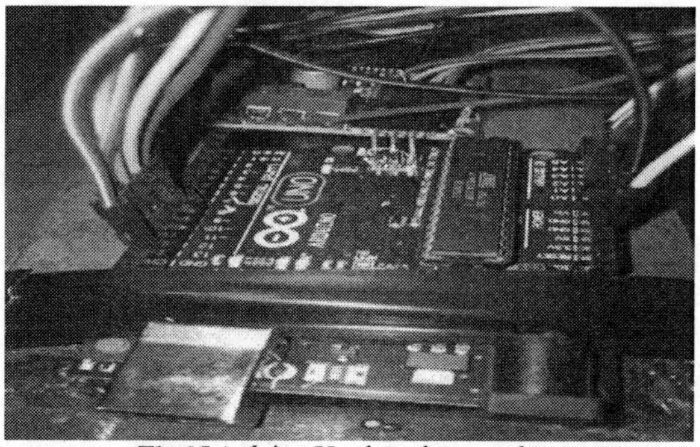
Fig.15 Arduino Uno board connections

9.6 Overall System

The overall system connections & assembly of the prototype of dual axis solar tracking system is as shown in figure 16.

Fig.16 Prototype of System

10. Summary

The manufactured dual axis solar tracking system is tested in solar light as well as in the normal tube light. In both the cases, LDR sensors perfectly detect the direction of maximum intensity light and accordingly rotate the solar panel to that direction. The input energy required to drive the system is as calculated above and which is 0.00875 Whr. The output energy can be calculated from the multi meter readings of voltage and current with respect to the time. Also the efficiency and the net tracking advantage of the system can be calculated from the following equations.
Therefore,

$$\text{Efficiency of the system} = \frac{\text{Output Energy}}{\text{Input Energy}} \quad (9)$$

$$\text{Net Tracking advantage} = \frac{E_t - E_{wt}}{E_{wt}} \quad (10)$$

Where,

E_t = Energy Produced with Tracking
E_{wt} = Energy Produced without Tracking

The manufactured dual axis solar tracking system tracks the current and exact position of sun and which can be cross checked by using Astronomical Equations. The Solar Panel moves from East to West in a day as well as from North to South in a year (according to the seasonal movement) in a direction of maximum light source i.e. Sun. The system uses only three LDRs. Thus, the complexity of tracking strategy is eliminated & system is made simpler.

11. Conclusion

The manufactured dual axis solar tracking system will increase the Energy efficiency considerably as it tracks the daily as well as the seasonal movements of the sun. Due to increase in efficiency, the size of solar panel & cost of power generation will reduce for the same power generation capacity in comparison with the fixed solar panel.

12. References

1. Y.Away and M.Ikhsan, ELSEVIER Automation in construction, (2017), **73**, pp. 175-183.
2. M.Rumbayan, M.D.Putro, IJESC, **6(11)**: (November 2017), pp.440-450.
3. S.Bazyari, R.Keypour, S.Farhangi, A.Ghaedi, K.Bazyari, JPEE, (2014), **2**, pp. 718-728.
4. J.K.Tharamuttam, A.K.Ng, ELSEVIER *Low Carbon Cities and Urban Energy Joint Conference WES-CUE*(2017), **143**, pp.629-634.
5. Y.Yao, Y.Hu, S.Gao, G.Yang, J.Du, ELSEVIER *RenewableEnergy*,(2014), **72**, pp. 88-98.
6. J.Juang and R.Radharamanan, *American Society for Engineering Education Conference of Zone 1*, (2014), pp.14-21.
7. M.K.Lokhande, IJCEM, (October 2014), **1**, pp. 122-133.
8. Government of India, Ministry of Power, annual report-2017-18 [https://powermin.nic.in/en/content/annual-reports-year-wise-ministry]
9. https://en.Wikipedia.org/wiki/Solar_tracker

Adaptive lighting comfort in the classrooms of educational building and student hostel rooms

Subbarao Yarramsetty[1], Nikhil Siva Deka[1], and Siva Kumar MVN.[1]*
[1]Department of Civil Engineering, NIT Warangal, Telangana State-506 004, India.

Abstract. The comfort level of occupants inside a building depends on various indoor environment factors like thermal comfort, indoor air quality, and indoor lighting quality. According to studies done on indoor environment quality of buildings, lighting plays a very critical role in determining the occupants comfort level. A proper indoor lighting is very important especially in the classrooms of an educational building as well as in student hostel rooms as it can be said as a vital element, controlling the psychology and comfort of the students and hence a deciding factor in the teaching-learning process. The lighting intensity and its distribution depends on the location of the interior partition as well as the design strategy of the exterior envelop. So, in order to obtain the most occupant friendly results, the impact of various strategies adopted should be evaluated in the early design period itself. This paper adopts a questionnaire-based field surveys which is done in the classrooms of National Institute of Technology Warangal and also in the hostel rooms of 1.8K Ultra Mega hostel of this institute. The main objective of this study includes. 1) To assess the indoor lighting comfort in the classrooms and hostel rooms of National Institute of Technology Warangal through a questionnaire-based lighting comfort study. 2) To identify preferences of subjects in the classroom environment with respect to lighting comfort and 3) To identify the comfort range of lighting intensity inside the classrooms and hostel rooms. SPSS software is used analyse the reliability of survey responses by finding the P value.

1 Introduction

The overall satisfaction of occupants inside a building can be improved favourably by enhancing the indoor air quality [1], thermal comfort [2] and indoor lighting quality [3]. Now a days with the increase in interest in health and also the environment, maintaining a proper comfortable environment inside a building is taken as an issue of importance by people in the twenty-first century to maintain health, while increasing productivity and efficiency of the occupants simultaneously. According to studies done on indoor environment quality of buildings, day lighting inside a room plays a very critical role in determining the occupants comfort level [4].

Lighting comfort can be defined as satisfaction of the residents in the given environment. Lighting quality is also dependent on the overall quantity of light. A good quality of lighting in general is contributed by a definite range of quantities of light. Various other parameters which affects the quality of lighting includes the illumination uniformity, good distribution of indoor light, colour characteristics of the lighting and room surfaces, flicker rate and glare [5]. There are many codes or guidelines for lighting design such as Chartered Institution of Building Services Engineers (CIBSE) Code for Interior Lighting [6], Illuminating Engineering Society (IES) Lighting Handbook [7] etc. that provide recommendations for some of the factors like uniformity and glare. A proper indoor lighting comfort is very important especially in the classrooms of an educational building as well as in student hostel rooms as it can be said as a vital element, controlling the psychology [8] and comfort of the students and hence a deciding factor in the teaching-learning process. A classroom requires the appropriate lighting environment that respond to the students as well as the staff sensitiveness. A continuous environmental stress can drain student's physical and mental resources which ultimately affects their performance [9]. Indoor environmental variables like distance from window, distance from board or projector screen, distribution of lighting inside the room, amount of interaction done in classroom etc. affects the lighting comfort sensation of the students. From studies it is found that the stress hormone known as cortisol are more regulated more efficiently by daylight than artificial lighting [10]. So, students should be always provided with access to daylight other than continuous exposure to electrical lighting systems in the building.

2 Area of Study

1.8K Ultra Mega hostel rooms and classrooms of Civil Engineering department of National Institute of Technology Warangal is chosen for this study. This study was done in month of September 2019. Warangal is one of the two metropolitan cities in the state of Telangana in

India. During September it mainly has clear and sunny sky with some occasional cloudy periods and rainy days in between. The study area in plan and the view of 1.8K Ultra mega hostel is shown in Fig. 1.

(a) (b)

Fig. 1. (a) NIT Warangal Campus, (b) 1.8K Hostel elevation

(a) (b)

Fig. 2. (a) Civil Engineering Department, (b) Classroom

The size of each classroom in which data is recorded are same. The vegetation covers near the classrooms are different creating somewhat different indoor lighting conditions for each classroom. The entrance view of Department of civil engineering division and the classroom arrangement is shown in Fig. 2. These classrooms are located in the 2nd floor of the Civil Engineering Department. 1.8K Ultra Mega hostel is a G+9 hostel building. The sizes of each room in the hostel are same but orientation of doors and windows and positions of the rooms vary according its location within the hostel regarding A wing (east facing) and B wing (west facing).Different rooms from different floors as well as different locations are chosen for the study so that varieties of data regarding indoor lighting comfort of students residing in these rooms can be observed.

3 Methodology

The lighting intensities (in lux) were measured continuously on hourly basis using the HTC LX-103 Digital Lux Meter instrument. The measurements were taken approximately at the working plane of the classrooms as well as hostel rooms. Data was collected from morning 8 AM to evening 5 PM for a week for the classrooms. The image of the digital lux instrument used for the study is shown in Fig. 3.

This study is carried out to assess their indoor lighting comfort conditions during the students' lesson hours. Data was taken at three positions within the classrooms i.e. near the window side, centre of the room and near the corridor side. The lighting intensity outside the room was also recorded. For the hostel rooms data was recorded for a month from 8 AM to 5 PM for different rooms located at different positions at both the east facing wing A and the west facing wing B which is shown in Fig. 4. In the hostel room data was taken near the window, farthest position from the window and also outside the room. Data recorded at different position in the room will help in analysing the light intensity distribution in the rooms.

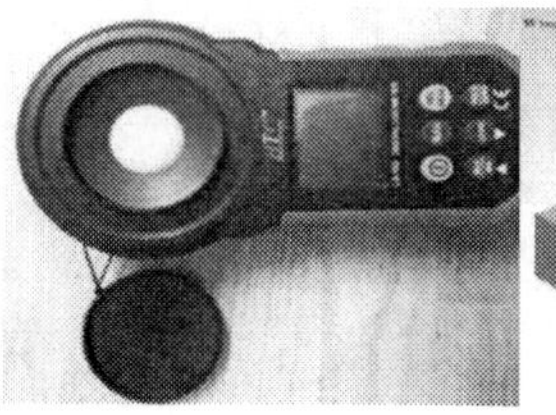

Fig. 3. HTC LX-103 Digital Lux meter

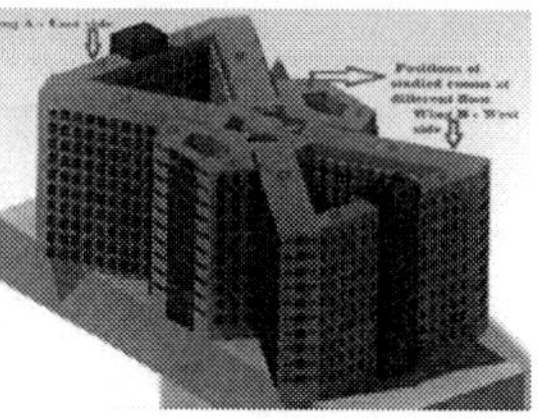

Fig. 4. Location of Study rooms – 3D view

Also, a questionnaire survey on various aspects of lighting in the selected rooms were conducted to study the subjective response of students to the indoor lighting comfort during daytime for classrooms as well as hostel rooms. The subjective response of the respondents and the daylight intensity recorded were compared to identify the comfort range of the students and also the different factors of discomfort regarding indoor lighting.

4 Results

4.1. Analysis of daylight intensity in the classrooms

All the studied classrooms are used from morning 8 AM to evening 5 PM for theoretical class purpose as well as seminar room for the students. The teaching learning process is delivered by using both white board and projector screen. So, a proper lighting environment is very necessary to accommodate all these facilities and commencing an efficient transfer of knowledge

These recorded lighting intensity variations with time are plotted in graph to analyse the pattern of variation which shows that there is a major difference in lighting intensity at different positions of the room. Near the window direct sunlight comes into the room and intensity is quite high but at the centre of the room and towards the corridor side intensity of daylight is poor. Except for a few hours, daylighting is well below the recommended comfortable lighting of 500 lux for classroom for a major part of the day. Again near the window daylight intensity is higher than 750 lux of upper limit value and caused discomfort among the students setting near the windows which was revealed after the analysis of the questionnaire survey. Fig. 5, Fig. 6 and Fig. 7 shows the indoor lighting intensity variations at different position of the classrooms in a more meaningful form to extract the pattern.

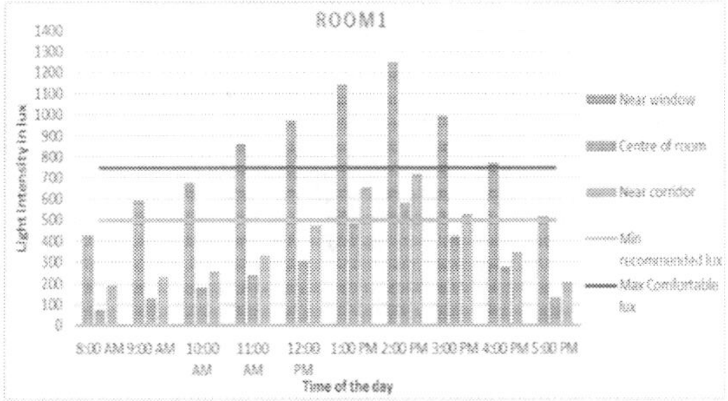

Fig. 5. Variation of indoor daylight intensity, Room # 1

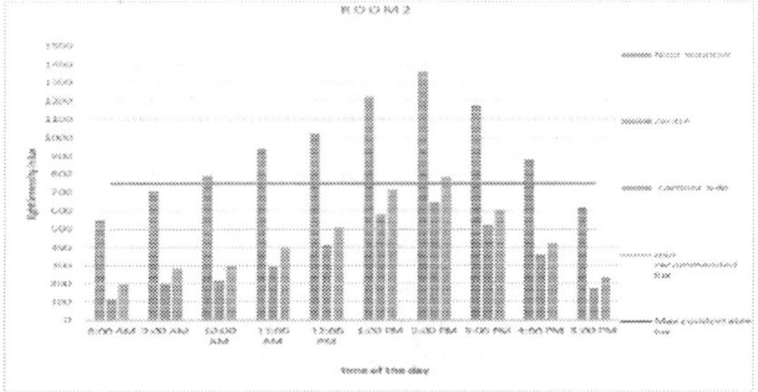

Fig. 6. Variation of indoor daylight intensity, Room # 2

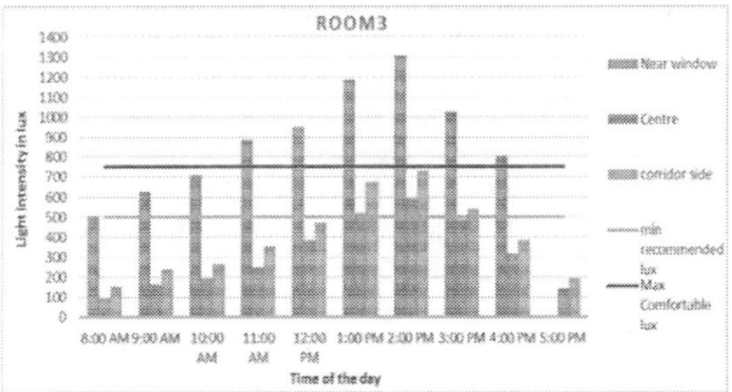

Fig. 7. Variation of indoor daylight intensity, Room # 3

It was observed that the overall indoor lighting of classroom # 2 is better than rooms 1 and 3. It is due to the fact that vegetation covers on the corridor side and window side of room 1 and 3 are much denser than room 2 therefore the light intensity outside the rooms are lower which in turn interferes with the indoor lighting of the respected rooms. Fig. 8 shows the variation of lighting intensity outside each classroom over the time.

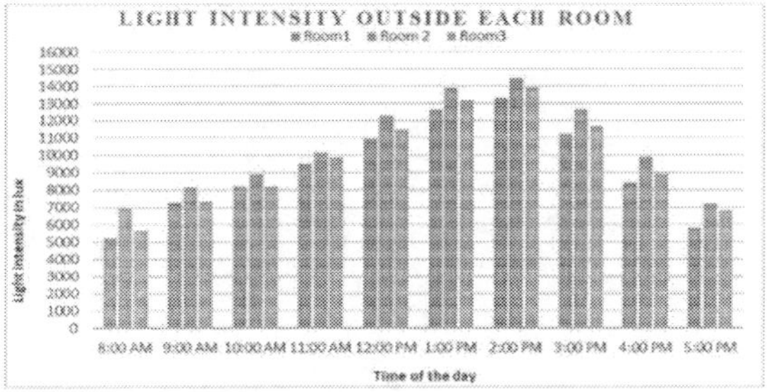

Fig. 8. Avg. Outside lighting intensity

Again, when the average lighting intensity of the day for different position within the room is calculated for 7 days of recorded value, it was found that daylight intensity varies widely for all the 7 days. Some days were bright enough to cause lighting discomfort like thermal irritation and glare whereas some days were dark and cloudy and artificial lighting was needed for smooth running of the classes. Fig.9, Fig10 and Fig.11 shows the variation of average lighting intensities of studied classrooms at different position over the 7 days period.

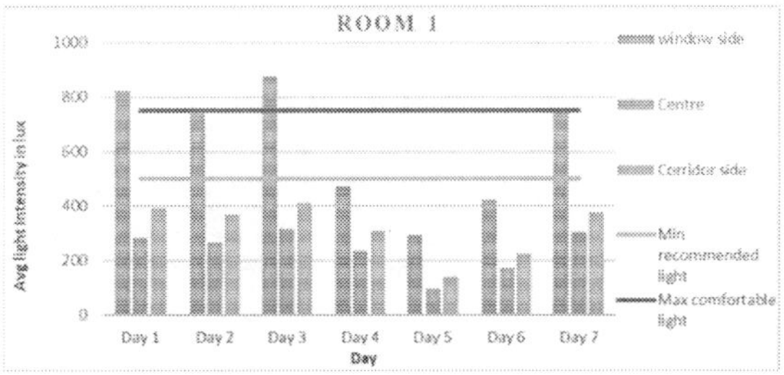

Fig. 9. Weekly Avg. daylight intensity variation, Room # 1

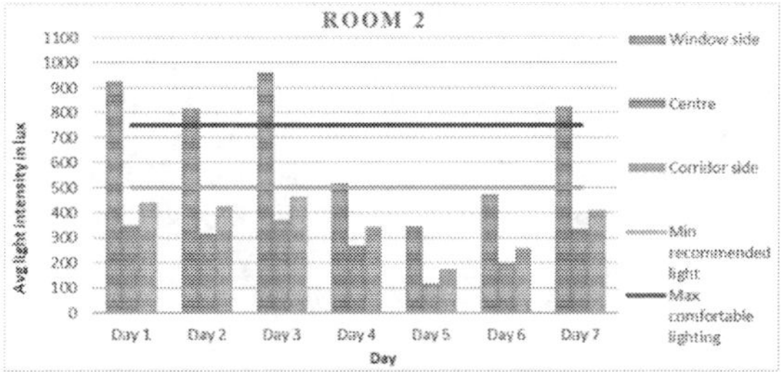

Fig. 10. Weekly Avg. daylight intensity variation, Room # 2

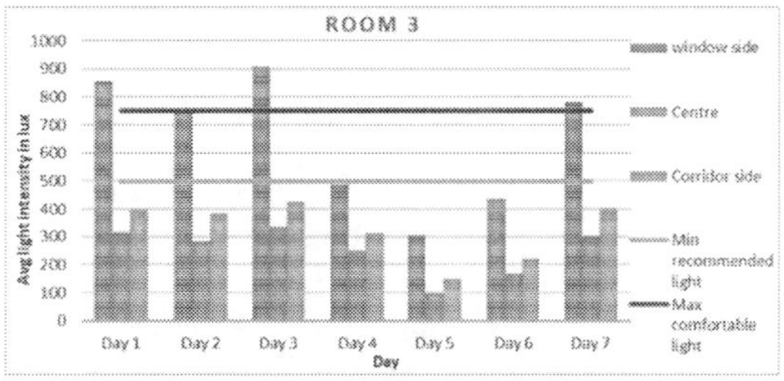

Fig. 11. Weekly Avg. daylight intensity variation, Room # 3

From the Fig. 9, Fig. 10 and Fig. 11 it is clear that for day 4, day 5 and day 6 which were cloudy days with occasional rain in between, the indoor daylight falls well below recommended intensity. Daylight intensity at all the positions especially at the centre of the room was very poor. Again, lighting intensities varies all over the day causing uncomfortable indoor environment. The indoor daylight intensity on a rainy day is compared with the comfort light intensity in Fig. 12.

https://doi.org/10.1051/e3sconf/202017001012

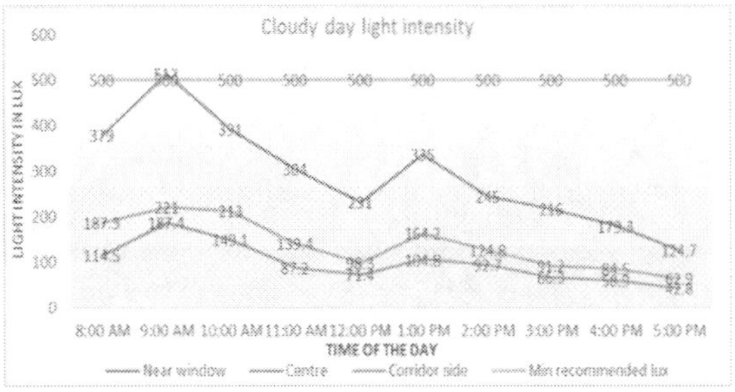

Fig. 12. Variation of daylight intensity on a rainy day

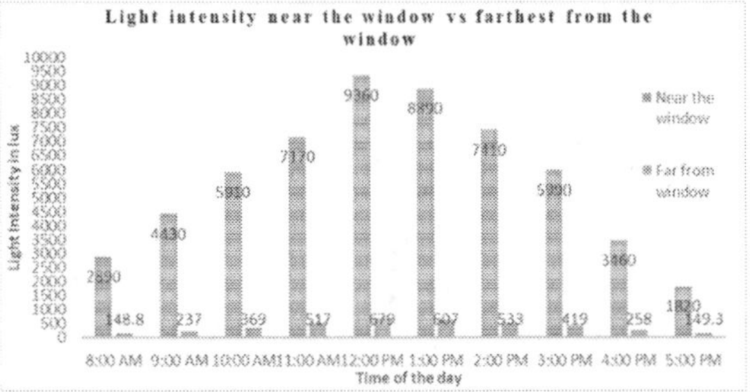

Fig. 14. Variation of daylight intensity at different position

4.2 Analysis of daylight intensity in the hostel rooms

The intensity of indoor daylight of the hostel rooms varies over different factors like time of the day, position of the room, floor on which the room is located etc. A number of 24 hostel rooms are studied over a period of 1 month in the month of September located at different positions and floors all over the hostel. It was found that position of the room plays a significant role in the indoor lighting environment of the hostel rooms.

When graphs are plotted from the recorded data it was seen that rooms located in the "A" wing (eastern facing) gets a good amount of morning light but decreases rapidly after noon whereas "B" wing (western facing) rooms have darker morning indoor environment as compared to A side rooms but gets max daylight between 1PM to 3PM in the afternoon hours. Fig.13 shows the graphical representation of variation of average daylight intensity of rooms located at Wing A and Wing B.

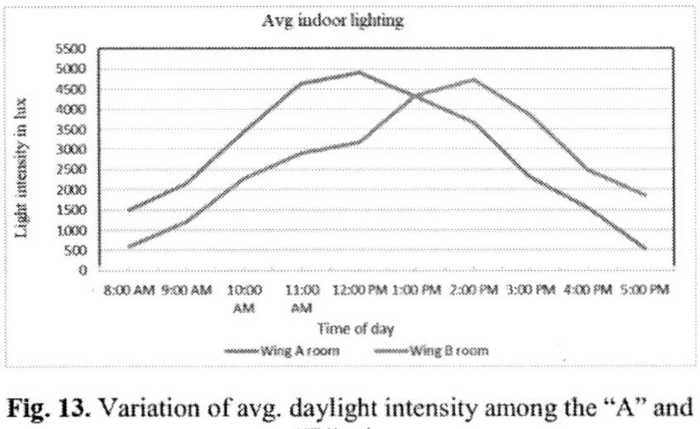

Fig. 13. Variation of avg. daylight intensity among the "A" and "B" wing.

Again, when the distribution of lighting intensity inside the room was analysed, it was found that lighting intensity at the farthest side from the window is lesser by a great margin from intensity near the windows. It means that despite a proper lighting near the windows, artificial lighting is still needed in the hostel rooms for a comfortable indoor environment. One of the reasons for this can be due the installation of windows, which are not directly on the wall but built outside the wall in a box like design. This design restricts daylight only to the areas near the window keeping the other parts comparatively darker. The variation of indoor light from window side to the farthest side over the time is shown in Fig. 14.

5 Analysis of the questionnaire survey

To check the reliability of the responses to confirm whether those responses can be used in this study to analyse the results, P value was found out by using the software SPSS. For a set of three variables, correlation of variable 1 was found out to be significant at the 0.01 level and the correlations of variables 2 and 3 are found to be significant at the 0.05 level. From this data it can be assumed that the responses of the survey is reliable enough to use in the analysis of the result for this study. For the questionnaire survey, a nine point scale was used for recording the responses of the occupants regarding the indoor comfort. For this scale, 1 refers to extremely unfavourable and 9 being the extremely favourable conditions of indoor environment. By analysing the responses through graphs it was found that majority of the occupants found the indoor lighting environment comfortable as most of the response values are in between 5 and 8. Daylight distribution was felt to be unfavourable by a few students specially who were seating far from the window sides. A large number of students reported to face uncomfortable indoor environments like glare, thermal irritation etc. with majority of the response values lying between 5 and 7 depicting a noticeable presence of uncomfortable environment but not on the extreme levels. Again regarding the effect of lighting on mood of the students, responses suggested that almost every students find the indoor daylight to affect their mood on some level. With majority of the responses lying between the ranges of 4 to 7, it can be analysed that indoor lighting environment plays a considerable role with the psychology of the occupants. Fig. 15 to Fig. 18 shows the graphical representation of the questionnaire survey data.

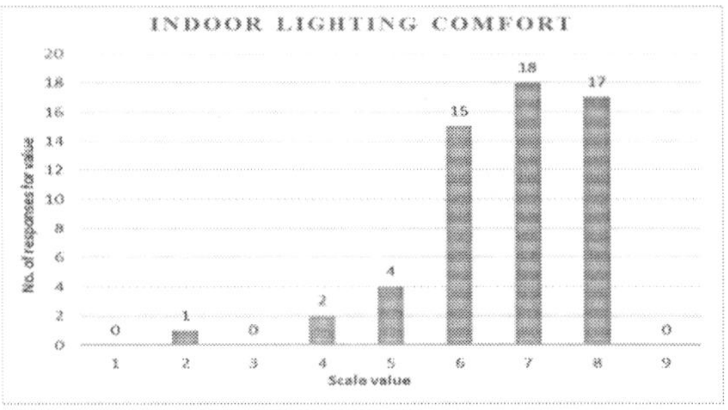

Fig. 15. Responses for indoor daylight comfort

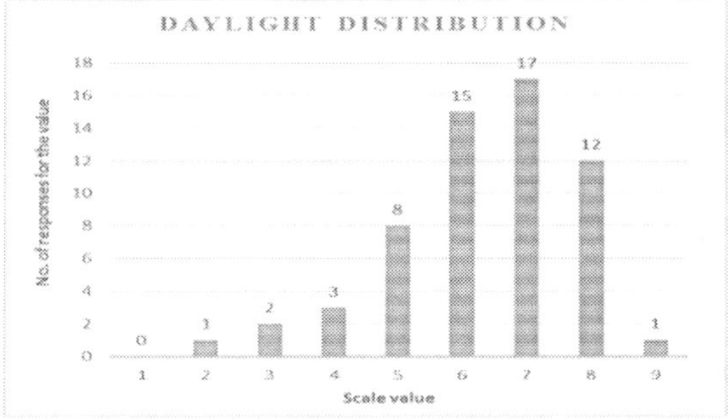

Fig. 16. Responses for indoor daylight distribution

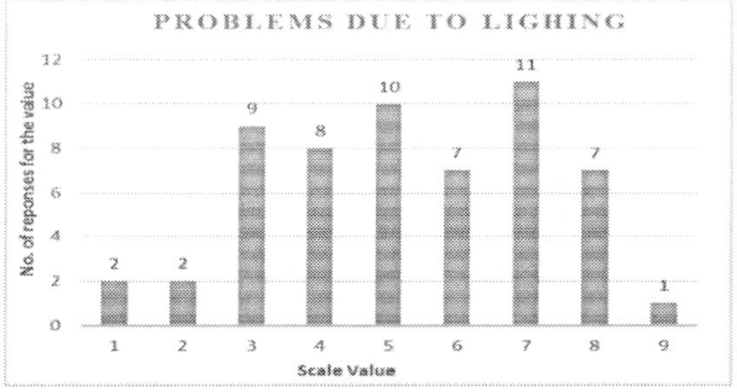

Fig. 17. Responses for problems caused by daylight

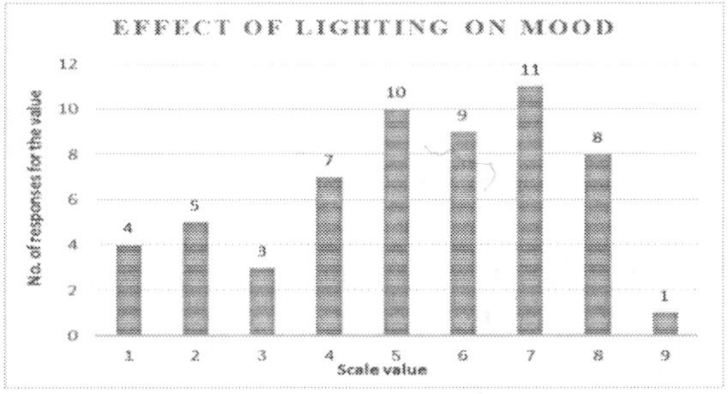

Fig. 18. Responses for effect of daylight on mood of the occupants

6 Conclusion

This study was done to analyse the adaptive lighting comfort in the classrooms of educational building and student hostel rooms. Daylight intensity was recorded in 3 classrooms for a week and also for 24 hostel rooms over a period of month. Also a questionnaire survey was distributed among students to collect their responses regarding the daylight sensation and indoor comfort environment. This responses were analysed with the help of graphs plotted from the recorded data. It was found that majority of the respondents found the daylight intensity between 250-400 lux comfortable which is below the recommended intensity of 500 lux for study purpose in classrooms. Need of artificial lighting was felt only below the intensity level of 200-250 lux. Students sitting at the window sides reported glare and thermal discomfort when the lighting intensity is recorded more than 800 lux. For the classrooms, the students sitting in the centre found the distribution of indoor lighting to be poor and felt the need of artificial lighting during early morning or late evening hours and also during the cloudy days. Majority of the students found the corridor side lighting environment to be more suitable. In the hostel rooms, students reported the distribution of daylight in the rooms to be unsatisfactory. Also some respondents reported thermal uncomforting at night due to high intensity of light in the room during the evening hour because of west facing orientation of the rooms. Students preferred daylight over artificial lighting as per their responses and also the moods of the students were reported to be dependent on indoor lighting environment. Students reported that dim indoor lighting than the comfortable intensity induces sleepiness during the study hours and uneven distribution of daylight, flickering of light causes lapse in concentration.

Indoor lighting is a major factor controlling the comfort level of occupants and plays even bigger role in educational buildings. Efforts should be made to make a room visually more comfortable to the occupants from the design phase itself. Curtains should be provided in all the classrooms to avoid undesirable bright light inside the room. Vegetation around the classrooms should not be allowed to grow too dense so that it may obstruct the natural daylight coming to the building. More efficient window designs should be applied to make the light distribution inside the room comfortable. Now a days with the help of BIM and Automation software, occupant friendly indoor lighting environment in classrooms and hostel room are easier to designs and implement. The scope for future research is high in this field as more and more people are giving emphasis on indoor comforting with each passing day.

References

1.J. Hummelgaard, P. Juhl, K. O. Sæbjörnsson, G. Clausen, J. Toftum, and G. Langkilde, Build. Environ. **42**, 4051 (2007).
2.A. Wagner, E. Gossauer, C. Moosmann, T. Gropp, and R. Leonhart, Energy Build. **39**, 758 (2007).
3.D. M. Franco-Duran and G. Mejia A, Proc. Constr. Res. Congr. 2016 2039 (2016).
4.J. A. Jakubiec and C. F. Reinhart, LEUKOS - J. Illum. Eng. Soc. North Am. **12**, 185 (2016).
5.T. M. Chung and J. Burnett, Indoor Built Environ. **9**, 335 (2000).
6.*CIBSE: Code for Interior Lighting. London,Chartered Institution of Building Services Engineers,1994* (1994).
7.*IESNA: Lighting Handbook, Ed 8. New York, Illuminating Engineering Society of North America, 1993* (1993).
8.A. T. P. So and L. M. Leung, Archit. Sci. Rev. **41**, 113 (1998).
9.P. Ricciardi and C. Buratti, Build. Environ. **127**, 23 (2018).
10.N. Gentile, T. Goven, T. Laike, and K. Sjoberg, Light. Res. Technol. **50**, 631 (2018).

Reduce energy consumption using efficient temperature Control inside building

Subbarao Yarramsetty[1*], Divyesh Sharma [1], and Siva Kumar MVN.[1]
[1]Department of Civil Engineering, NIT Warangal, Telangana State-506 004, India.

Abstract. One of the biggest problems we have in front of us today is climate change. However, this problem can be solved if we can take right steps individually. One of the major steps which can be taken is to reduce energy consumptions in the houses by automatic control of ACs according to users' comfort which will help in energy consumption. It is observed that a huge difference can be created by using automated temperature control system in respect to energy consumption. This paper contains study of indoor environment of the rooms in 1.8K Hostel and classrooms of Department of Civil Engineering in the Campus of NIT WARANGAL is done to analyse the comfort temperature inside the rooms for the efficient use of energy. Questionnaire survey-based study was adopted for this paper to analyse the range of thermal comfort of the students.

1 Introduction

In the current scenario, one of the biggest problems which is of concern on the Planet is the climate change[1]. The effect of Global Warming can be seen in different form either in the form of average increase in global atmospheric temperature or in the form of unpredictable weather. Buildings generally accounts a reasonable amount of energy consumption which means more CO2 emission [2]. As India a developing country, electricity consumption will keep on increasing as our population growth is high also, we are still having large energy production through non-renewable source which is resulting in energy consumption. In India building accounts for 35% of total energy consumption and is keep on increasing at 8% annually [3], which is affecting India's carbon foot print which is a matter of concern as the consumption will keep on increasing in future. If we know about how the temperature varies in that particular area or building, we can effectively utilize the consumption of energy.

Thermal comfort may be viewed as state of mind with respect to thermal satisfaction. Evaluation of thermal comfort is on the basis of codes like ISO, ASHRAE. In any accommodated space thermal adaptation can be characterize in three different processes behavioural, physiological accommodation and psychological training [4]. Indoor thermal comfort depends on proper ventilation, humidity level and performance of building envelop. Assessment of thermal comfort helps in finding quality of indoor environment and helps in optimum utilization of energy which is required for desired comfort level. A good thermal comfort has been defined as range of temperature as given by ASHRAE Standard 55, it also defines method to determine permissible temperature. Permissible Temperature ranges between 67 °F to 87 °F but the more accurate value depends on relative humidity. The standard specifies upper level of humidity of about 80% but has not specified any lower limit. It also specifies limit of relative humidity in occupied building to about less than 65% to reduce the condition for microbial growth[5].

For proper thermal comfort only, single parameter is not sufficient, as area having permissible temperature with high relative humidity and vice versa may provide discomfort to the occupant, it is therefore important to consider both parameter at a time and not in isolation. A proper temperature comfort especially in classroom is very important in order to minimize the artificial mode of providing comfort which will reduce the energy consumption which in turn reduce the CO_2 emission.

2 Area of Study

Location chosen for study 1.8K Ultra Mega Hostel and Postgraduate classroom of Department of Civil Engineering in the Campus of NIT Warangal. The geographical location of the campus is shown in Fig. 1. along with the front elevation of the 1.8k hostel. Warangal is located in the state of Telangana in Southern India. Temperature in Warangal varies between 12 to about 41 degree Celsius and average relative humidity ranges from 60% in June to about 77% in September. This study was done in the month of September 2019 which fall under ending of monsoon.

The elevation of the department of civil engineering is shown in Fig. 2(a), where the classrooms were located in the second floor. Fig. 2(b) shows the location of the study rooms in the 1.8K hostel. The size of rooms in 1.8K ultra mega Hostel are same but orientation of room is different depending on the wing they are located in. 1.8K is G+9 floor building which consist of two wings namely Wing A and Wing B at each floor. Wing A is East facing whereas Wing B is west facing. Different room with two room in each Wings were chosen for the study. Readings of temperature were taken from 8AM to 5PM at 1-hour interval covering one floor a day in the month of September. The red dot marked in Figure2(b) represent the location of reading taken in each wing.

(a) (b)

Fig. 1. (a) NIT Warangal Campus, (b) 1.8K Hostel elevation

(a) (b)

Fig. 2. (a) Civil Engineering Department, (b) Typical Floor plan of 1.8K Hostel

The size of each classroom in which data is recorded are same. The vegetation covers near the classrooms are different creating somewhat different indoor lighting conditions for each classroom. The entrance view of Department of civil engineering division and the classroom arrangement is shown in Fig. 2. These classrooms are located in the 2nd floor of the Civil Engineering Department. 1.8K Ultra Mega hostel is a G+9 hostel building. The sizes of each room in the hostel are same but orientation of doors and windows and positions of the rooms vary according its location within the hostel regarding A wing (east facing) and B wing (west facing).Different rooms from different floors as well as different locations are chosen for the study so that varieties of data regarding indoor lighting comfort of students residing in these rooms can be observed.

3 Methodology

The temperature of air and relative humidity were measured continuously at 1-hour interval using HTC-1 Digital Temperature Humidity meter at the selected location. Study is carried to determine the indoor comfort with respect to temperature and relative humidity with the changing environment condition classrooms.Fig.3 shows. HTC-1 Digital Temperature Humidity meter which is used for taking the reading. Readings were taken from 8AM TO 5PM at the occupied room in 1.8K Ultra Mega Hostel one floor at a time and at class room 8AM to 5PM for one week. Readings were taken at three different position namely near window, in the middle of room/class and outside the room. For the hostel room readings were taken at one floor per day with 4 rooms at each floor and two room at each wing. Average of readings at different position of room were taken to determine the average temperature of room at the particular time. Position of reading taken is marked by red mark in Figure-4 (a) and 4(b) for the class room and hostel room respectively.

Also, questionnaire survey in the selected room were done to examine the thermal comfort of the student in the room.

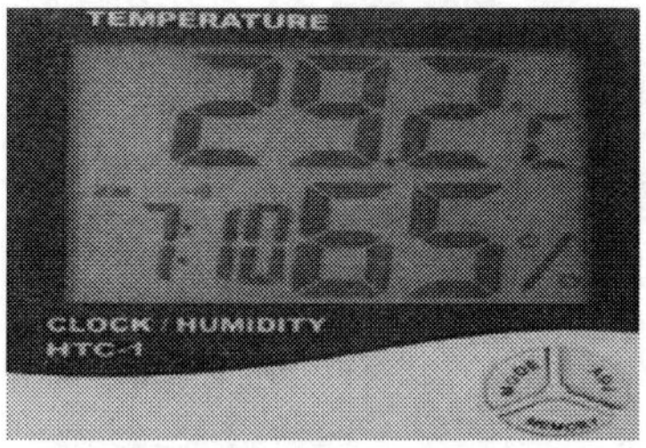

Fig. 3. HTC-1 Digital Temperature Humidity meter

(a) (b)

Fig.4. (a) Location of readings taken in hostel room, **(b).** Location of readings taken in classroom.

4 Results

4.1. Analysis of Temperature and Relative Humidity in Hostel room.

For rooms in hostel for the student to perform there self studies or other activity adequate thermal environment is a must, so that student has relax state of mind. Reading of temperature and humidity were taken from 8AM to 5PM. Figure 5 shows the variation of temperature of 4th floor at room 1 of wing A and wing B, it can be seen that at around 2pm temperature is maximum after which it start reducing. The red line marked in figure 5 shows the permissible temperature limit of 32°C. During the cource of the day maximum temperature rise is about 32°C which has just reach the permissible limit because the period of study is in mounsoon season hence temperature rise is not very high. Conscidering the humidity graph of figure 8, it can be seen that relative humidity is very high at begning of the day which rises to about 72%. Course of the day at which temperature and humidity at its peak is point of concern since high temperature combined with high humidity will not allow sweat do dry easily thus creat uncomfortable condition. As there is rise in temperature humidity in air gets reduce, but during monsoon season temperature is not very high that can provide relief from high humidity.

From the table1 we can see that as we go up the floor the average temperature is lower than the bottom floor this is because the heigher floor is open to atmosphere due to which there will be better circulation of air which provide lower temperature compared to floor below. As we can see from figure 6 Relative humidity in wing B of the floor is lower then that of wing B it is because of the fact that B wing is towards the open side of building while wing A face towards inside of building. It can also be seen from table1 that as time proceed towards maximum temperature temperature rise is seen between 1pm to 2pm and can be related to to the figure 7 that during the same time relative humidity is generally less compared to other time, it due to the fact that heigher the temperature generally lower the temperature which is a favourable case up to a permissible limit, since either heigher relative humidity or temperature beyond permissible limit creat uncomfortable condition.

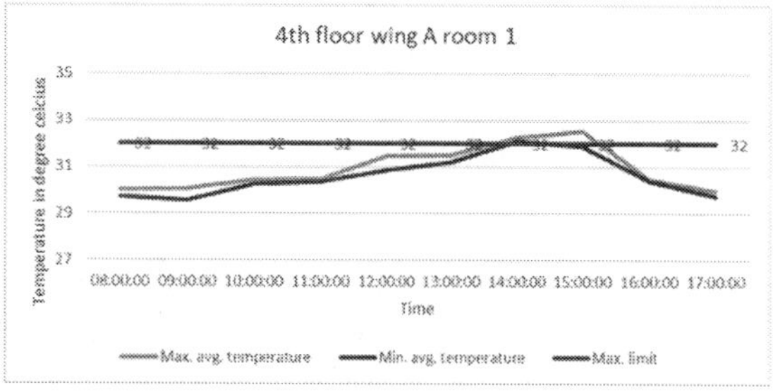

Fig 5: temperature variation of 4th floor wing A room 1

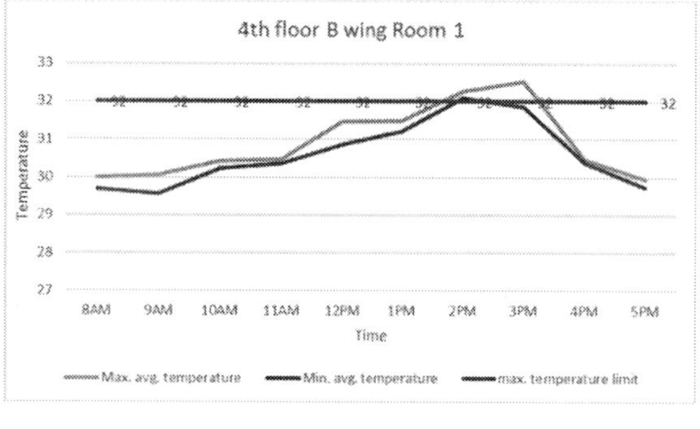

Fig. 6: temperature variation of 4th floor wing B room 1

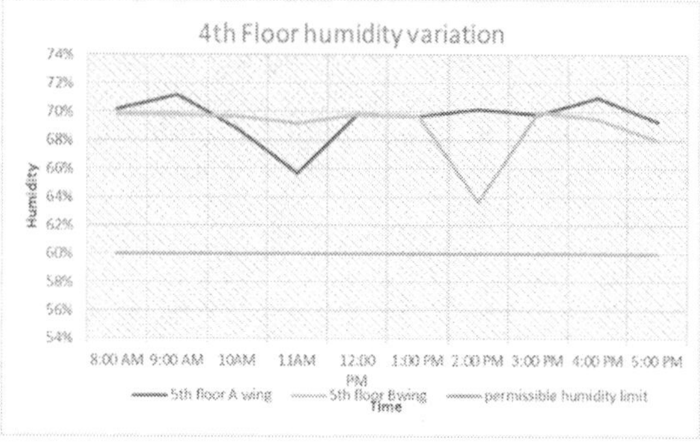

Fig 7: 4th floor relative humidity variation

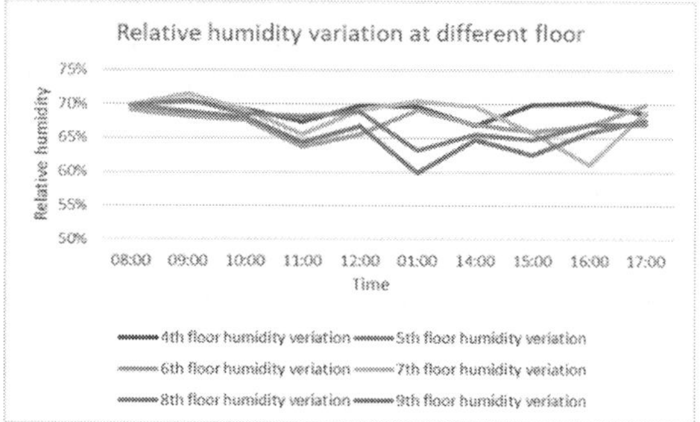

Fig.8: Graph showing humidity variation at different floor

Table1: average temperature variation of all floor at 1 hour interval

Floor / Time	4th floor temperature(°C)	5th floor temperature (°C)	6th floor temperature (°C)	7th floor temperature(°C)	8th floor temperature(°C)	9th floor temperature (°C)
08:00	29.98	29.22	29.03	30.01	29.22	28.67
09:00	30.13	29.87	30.67	29.89	29.87	29.1
10:00	30.38	30.93	30.68	30.71	30.93	29.68
11:00	30.95	29.76	31.71	31.19	29.76	30.57
12:00	31.38	29.98	31.31	31.35	29.98	30.68
01:00	31.59	29.49	30.28	30.66	29.49	30.13
14:00	32.34	29.85	30.49	30.18	29.85	30.12
15:00	32.27	29.32	30.63	32.27	29.32	29.98
16:00	31.13	29.41	30.52	32.78	29.41	29.84
17:00	30.52	29.48	30.09	30.84	29.48	29.72

4.2 Analysis of Temperature and Relative Humidity in the Class rooms

The class room functions with theoritical studies with black board, white board and projectorfor the study process, there fore it is necessary for the student to have relaxed state of mind for there performance, for this performance the be optimum adequate thermal indore environment is a must. As for the class room data is concerned, readings are taken at each class room from

8AM to 5PM in a day for 1 week. As we can see from figure 2(a) class room at second floor has some vegetation in front of the class which results in higher humidity and lower temperature than that corresponds to hostel. Because of vegetation, the energy which is required in the form of heat which reduce the humidity level is not available hence results in higher humidity level. As, we can see from figure 10 that temperature inside the room which is far from the window is low compared to the position near the window and outside the room, it is because of proper ventilation which is provided that results in proper air flow which results in lower temperature. As we can observe from figure 10 the temperature variation for class 1 and class 3 are very similar, this due to the condition around these classrooms are similar compared to the data in class 2 which is having higher vegetation in front of the class that result in different temperature and humidity condition.

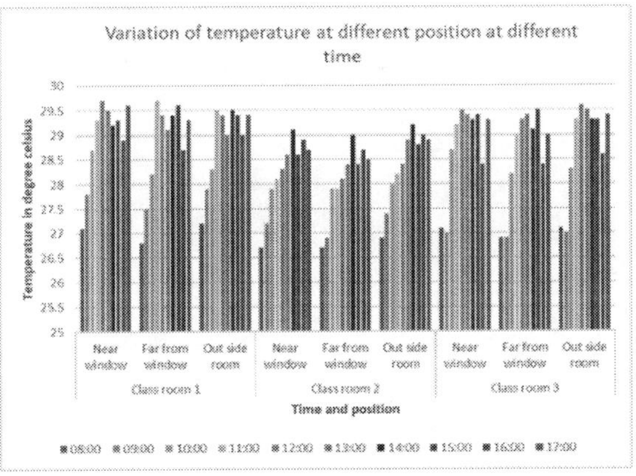

Fig.10: Temperature variation of class room at different position of class room at different time of a day.

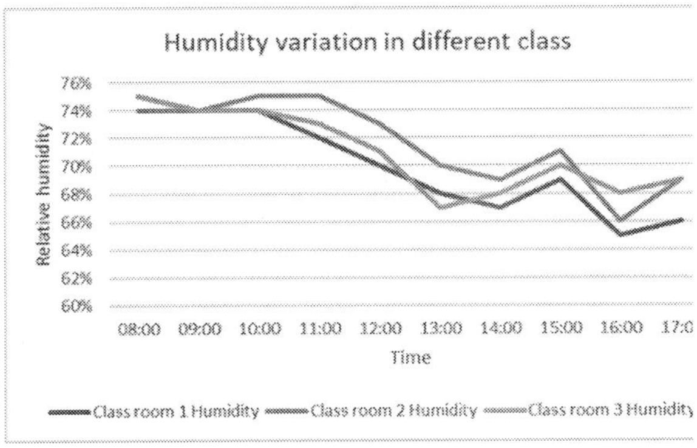

Fig.9: Graphical representation of humidity in different classroom.

5 Analysis of the questionnaire survey

A questionnaires survey was also done by distributing question by using google form to evaluate the thermal perception of the students. Question were asked about thermal perception, humidity perception and comfort criteria with each question marked on 9-point scale with extremely uncomfortable to extremely favourable as the extreme limit. After collecting the data from student, it was found that thermal perception of the students was found to fall under comfortable criteria. About 82% of the students marked the preference as under comfortable. In order to know the reliability of response to confirm the significance of the data P value is found using IBM- SPSS software. For the set of three variable, the correlation variable was found to be 0.01, which shows that the result obtained are reliable.

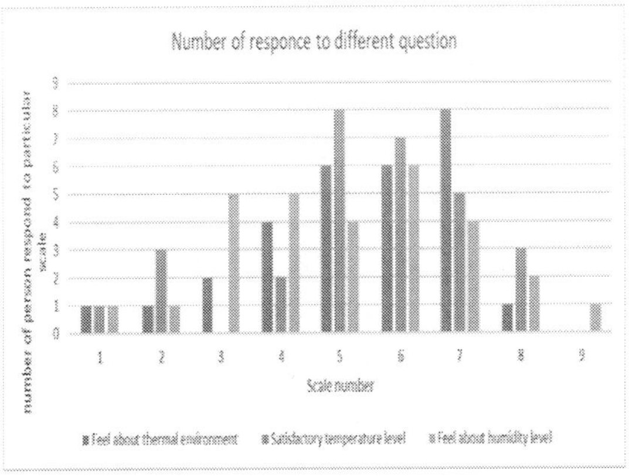

Fig.11 Number of responses for all question

6 Discussion and Conclusion

This study was conducted to understanding the thermal comfort of students and understanding the acceptable temperature in 1.8k Hostel and in classrooms in Campus of NIT WARANGAL. From the above study it is clear that solar radiation is main source for the room heating as the radiation can increase the temperature of rooms and can also reducing the humidity there by. It is observed that temperature of the floor above especially 9th floor in 1.8K hostel is lower compared to floor below this is because of better circulation of air since it is open to sky. From the survey it is found that even though temperature of the room is higher than the comfortable temperature response of the student was comfortable. It is mainly because of the fact that at the same time duration humidity was found relatively low which explains the response of the students in the survey. As we analyse from the study of the classroom, the vegetation cover is a must for controlling the indoor temperature of room which provide comfortable temperature required for students, which can be seen from the figure 10 that class 2 has relatively lower temperature compared to other class, which the support the role of vegetation cover in maintaining the thermal comfort of the students.(Added) With the provided good ventilation in the room almost uniform temperature and humidity can be maintained which can be depicted from the above study, good ventilation is therefore needed for providing comfortable temperature environment. By the use of proper ventilation and vegetation, energy consumption can be reduced as there will be minimum requirement of mechanical devices.

References

1. F. Giraldo-forero, F. Cardona-escobar, and E. Castro-ospina, **1**, 327 (2018).
2. H. Homayouni, C. S. Dossick, and G. Neff, in *Constr. Res. Congr. 2014 Constr. a Glob. Netw. - Proc. 2014 Constr. Res. Congr.* (2014).
3. R. Rawal, P. Vaidya, V. Ghatti, A. Ward, and T. Weidt, 313 (2002).
4. S. Rajkumar, L. R. Amirtham, and E. Horrison, **3**, 2 (2009).
5. ASHRAE, Renew. Sustain. Energy Rev. **80**, 1 (2011).

Comparative Investigations on Mahua Biodiesel-Diesel -Alcohol Low Percentage Blends with a VCR Diesel Engine

*Nikhil D Chavan[1], Amar P. Pandhare[1]**

[1]Sinhgad College of Engineering, Pune 411041, Maharashtra, India

Abstract. In this analysis mahua oil biodiesel isselected as one of the best alternative fuel to diesel.Transesterficiation processisusedforbiodieselproduction.Serious concern on the emissions, in particularthe nitrogen oxides (NOx), PM (particulate matters) and carbondioxide (CO_2)led to think about additives to improve the overall performance. In this investigation additives are used to improve the performance, combustion and emission characteristics of Mahua biodiesel (B) and Diesel (D) blends. B20 blend is used as basic blend to compare the performance.Ethanol (E) and Propanol (P) used as additives indiesel-biodiesel blend with 5%,10% and 15%and tested in VCR diesel engine. Ethanol shown better performance in particular as compare to Propanol. The reduction of NOx, CO, CO_2, HC emission in diesel engine and also increase performance in addition of 10% ethanol. From this conclude the result ethanol is alternative to improve efficiency of diesel engine by using the blended biodiesel.

1 Introduction

The diesel fuel is commercial and transportation fuel. The rapid increment in the demand of fossil fuel such as a diesel has a great impact on economy of countries. The transportation sector is identified as major polluting sector. The crude oil prices are fluctuating in the global market. Biodiesel has a capacity to cut down the lump sum of expenditure on fuel. Biodiesel is important alternative to fossil fuels. Different types of biodiesel are largely studied in the past.Different alcohols in different percentage in biodiesel can change the emissions of engine significantly. Mahua oil biodiesel is ecofriendly better performing biodiesel.Biggest problems with the use of biodiesel are decrement in performance, Increase in the emissions and incomplete combustion.Additives are added to increase the performance of biodiesel in diesel engine. These additives are environment friendly and useful for the growth of farmers. Second generation ethanol and propanol are used as additives.

2 Literature Review

Additives such as ethanol and 1- propanol added into diesel blends. By adding them decrease in NOx and soot emission is found. At the same time CO emissions are increased[1].Alcohols such as methanol, ethanol, propanol, butanol and all having 5% oxygen content are tested withwaste cooking oil diesel blends. It leads to reduction in emission of HC, NOx, CO, CO_2[2]. By using additives increase in oxidation stability and 4–5 °C change in cold flow properties achieved[3].Additives used in diesel–biodiesel blends are Propanol, n-butanol and 1-pentanol.Higher brake specific fuel consumption of D40B40Pro20 than ternary blends of D40B40nB20 and D40B40Pn20 at all engine loads[4]. D40B40Pro20, D40B40nB20 and D40B40Pn20 have the exhaust gas temperatures higher than that of the diesel–biodiesel blend[4]. Different oxygenates are considered for study. By adding 2-propanol in pseudo binary and ternary biodiesel density is measured at different temperatures. Excess molar value shows positive and negative effects on Performance. In this investigation Methanol (M)and Ethanol (E) used as additives. Reduction in emissions of NOx, CO, CO_2 HC takes place on addition of methanol in 10% and 15%[6].Biodiesel (20%) added with 5% and 10% additives shows better performance in emission reduction. Fuel properties are found to be similar to conventional fuel. [7,8].NOx and PM emission are reduced. Operating conditions and biofuel concentration are important parameters in reducing emissions.[9].B20D80 blend shows better performance than other blends[10,11].Mahua oil biodiesel blended with octanol reduces emissions by significant amount significant results are found[12,13].diesel–biodiesel–ethanol and bioethanol ternary blends are tested ternary blends shows better reduction in particulate matter other parameters are nearly same as like diesel[14,15].Mahua oil biodiesel shows better performance below 25% diesel blends. Less brake specific fuel consumption than the diesel. It has been proved by the results [16]. At the end Mahua oil biodiesel 20 % blend is significant and less

*Corresponding author: amarppandhare@gmail.com

studied. Hence mahua oil biodiesel is selected for study to improve the performance of engine by using additives. Propanol and ethanol in low percentage are selected as additives.

3 Material and method

The Mahua oil feedstock is collected from Indian biodiesel corporation Baramati. Mahua oil initially with free fatty acids having density 910 gm/cc which are need to be removed.This feed stock is then undergoes esterification and transesterification process.Under esreification 0.4% H2SO4 ACID catalyst by volume and10% methanol as a solvent were added. Mixture then heated at 55-60 0C and agitated at 550-600 rpm for 30 minutes. At the end of esterification fatty acid count reduced to the 5-6 mg from 17-18 mg of KOH. Density and viscosity of mahua oil then reduced by significant amount.Under transesterification esterified mahua oil mixed with 0.4% NaOH AND 10% methanol and agitated at 550rpm at 60 0C for 75 minutes.Setteling of mixture for 8 to 10 hrs resulted into 20% glycerol and 80% biodiesel. After purification and demoisturization at 110 0C for 20 minutes pure mahua oil biodiesel B100 obtained. Ethanol (E)and propanol(P)were added in 5%, 10%,15% in B20 biodiesel blend. Diesel with 20% Mahua oil biodiesel used as base fuel with best performance. Different blends to be tested of list is given in Table 1.

Table 1. List of blends to be tested

Blends	Diesel	Biodiesel	Propanol	Ethanol
Diesel	100%	-	-	-
D80B20	80%	20%	-	-
D75B20E5	75%	20%	-	5%
D70B20E10	70%	20%	-	10%
D65B20E15	65%	20%	-	15%
D75B20P5	75%	20%	5%	-
D70B20P10	70%	20%	10%	-
D65B20P15	65%	20%	15%	-

4 Experimental setup and methodology

The experimental setup consists of Kirloskar Computerized VCR engine of specification and details are given in Table 2 and Figure 1. Eddy current dynamometer is used to change the load from 0 to 16. Compression ratio kept constant at 18. Engine and dynamometer are water cooled there flow rate is measured by rotameter and calorimeter. Piezo sensor, load cell and temperature sensors are used for performance measurement. Exhaust gas analyzer is connected to measure emissions ofNOx, CO, CO_2, HC. Data acquisition is done by computer connected to the system.

The Diesel and blends D80B20, D75B20E5, D70B20E10 D65B20E15. D75B20P5, D70B20P10, D65B20P15 are tested respectively. During testing readings are taken when engine became totally steady. Loads are varied from 0 to 16.While compression ratio kept constant. Computerized system generated results for various blends.

Table 2. Diesel engine specification and details

Engine specification	Details
Type	1 cylinder 4 stroke
Bore dia x stroke length	87.5 x 110 mm
Cubic capacity	661 cc
Compression ratio	17.5
Piezo sensor Range	5000PSI,with low noise cable
Maximum speed	2000 rpm
Rotameter	40-400 LPH
Connecting rod length	234 mm
Over dimensions	W2000 x D2500 x H 1500 mm
Data acquisition device	Ni USB-5210, 16 bit ,250KS/s

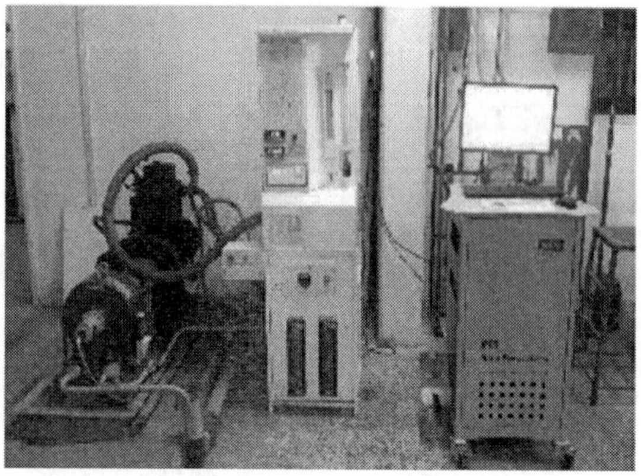

Fig. 1. Computerized VCR engine setup

5 Result and discussion

5.1 Performance

5.1.1 Brake power

Fig.2. Shows graph of Break power with Load. At low load condition nearly all blends shows same break power variation.As we know that the engine efficiency is depends upon the brake power and it is directly proportional to the brake power. From the graph we can concluded that the by using blends we can slight increase the efficiency of the engine.The power output is 2% increases for ethanol concentration blend and maximum for D70B20E10 and D65B20E15 Due to the complete combustion occur in Diesel engine.

5.1.2 Brake thermal efficiency

Fig.3. Shows graph of Break thermal efficiency with Load. At low load condition nearly all blends shows same break power variation. At higher load little change in result takes place.Brake thermal efficiency increases with increase in BMEP up to 2.92 bars beyond that small increases or decrease for various blends This result is beacase of additional lubricity and oxygen content in oxygenates additives , which results into the reduction in heat loss and active combustion process.Average 2% increment for D70 B20E10 as compare to diesel is seen as the load increases.D70B20E10 and D65B20E15 are coming out as best out of all blends as they are showing nearly 2% increment in the brake thermal efficiency.

5.1.3 Indicated thermal efficiency

Fig.4. Shows graph of Indicated thermal efficiencywith Load.The indicated thermal efficiency increases with addition of oxygenates both ethanol and propanol than pure diesel.The ethanol concentration blend is effective than the propanol concentration blend due to maximum oxygen content in ethanol concentration blend.Total increment in indicated thermal efficiency is about 3% for various loads.it is highest for the blend D70B20E10 and blend D65 B20E15.

5.1.4 Brake specific fuel consumption

Fig.5. shows graph of Specific fuel consumptionvs load condition.It is about 5% for various loads.Brake specific fuel consumption decreases with increase in load of the engine.The brake specific fuel consumption is lower for ethanol concentration blend due to lower viscosity than proanol, Hence the ethanol concentration blend consume less fuel than ethanol concentration blends. Brake specific fuel consumption is lower for the blend D70B20E10 AND D75B20E5. D70B20E10 shows less fuel consumption for all loads specially at higher loads.The Difference in specific fuel consumption is nearly about 1- 2%.

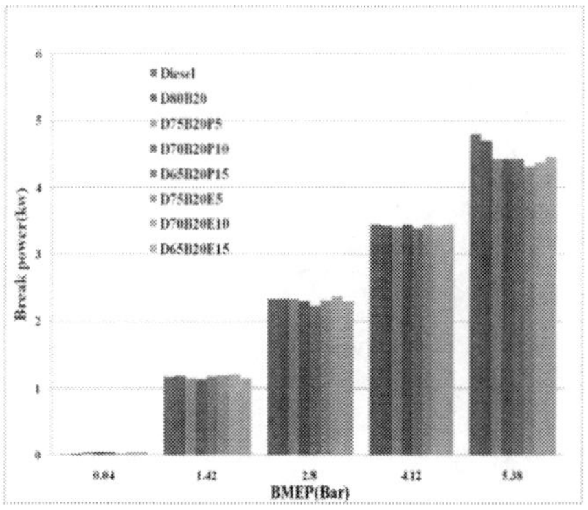

Fig. 2. Variation of Brake power with BMEP

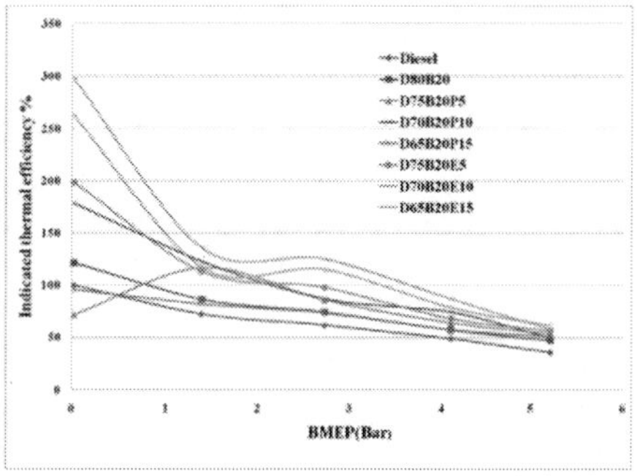

Fig. 3. Variation of brake thermal efficiency with BMEP

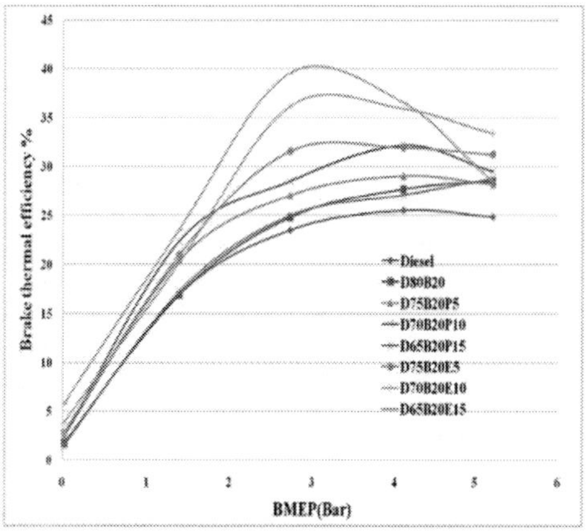

Fig. 4. Variation of indicated thermal efficiency with BMEP

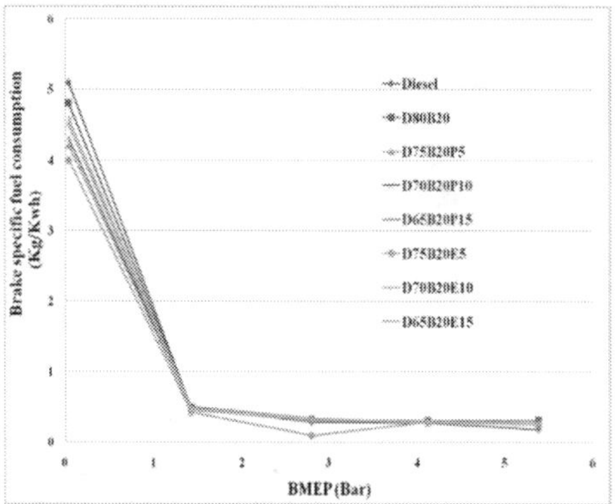

Fig. 5. Variation of Brake Specific fuel consumption with BMEP

5.2 Emmisions

Gas analyzer gives the results of emissions of CO, CO_2, HC and NOx are plotted on the graph.

5.2.1 Carbon monoxide (CO) emission

Fig.6. shows graph of carbon monoxide vs. changing load condition. CO is prodiced due to the incomplete combustion, which is formed due to scarcity of air or due to low gas temperature. CO emission decreases in addition of oxygenates in diesel.At higher load, the combustion of blends is almost complete due to the presence of oxygen which reduces the incomplete combustion of fuel in the region,results into to lesser CO emission. By more oxidation of CO, smallest possible emission of D70B20E10 is observed at higher loading condition. Average 3% decrement in CO emission is found for D70B20E10 which is highest as compare to other blends.

5.2.2 Carbon dioxide (CO_2) emission

Fig.7. shows graph of carbon dioxide vs. changing load condition.Fuel blend with ethanol and propanol has less CO_2 emission as compare to diesel fuel.It is obsereved that the emission of CO_2 increases with respective increase in load of the engine. Increasing percentage of ethanol in blend, decrease in CO_2 emission is observed because of higher oxygen content and higher carbon in mahua biodiesel.Decrement of CO_2 emission is 3% for all load conditions for the blend D70B20E10.

5.2.3 Hydrocarbon (HC) emission

Fig.8. shows graph of Hydrocarbon vs. changing load condition.Diesel and D75B20P5 have minimum HC emission at low load conditions. At higher loads D750B20P5 and ethanol blends have low HC emission.Overall less emission of HC is shown by D75B20P5 and D70B20E10. Reduction in emission of D70B20E10 is 3% as compare to B20D80.HC emission increases with addition of oxygenates than diesel at various loading condition. The result revealed that the HC emission for propanol concentration blends are less than ethanol concentration blends.HC emissions are found lowest for diesel due to less carbon and oxygen percentage as compared to various blends. The blends D70B20E10 and other shows contradictory results as compare to diesel.

5.2.4 Nitrous oxide (NOx) emission

Fig.9. shows graph of Nitrogen oxides vs. changing load condition.NOx emission decrease with increasing percentage of ethanol in the blend, this is due to the characteristic that lower heating value and latent heat of vaporization of ethanol as there is lesser peak temperature in the cylinder. In addition, this is due to the percentage of nitrogen in chemical composition of biodiesel and alcohol is less as compare to diesel. The NOx emission is decreases due to oxygenates added in biodiesel and diesel blend. The blend D70B20E10 and D65B20P15 shows minimum NOX emission as compare to other blends. It is nearly 5% decrement in NOX emission.

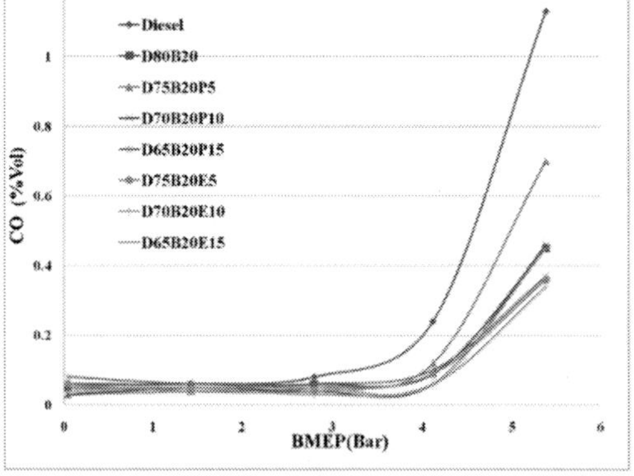

Fig. 6. Variation of CO with BMEP

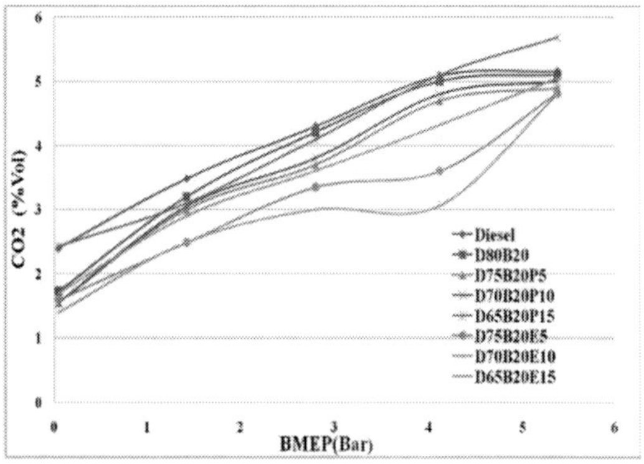

Fig. 7. Variation of CO2 with BMEP

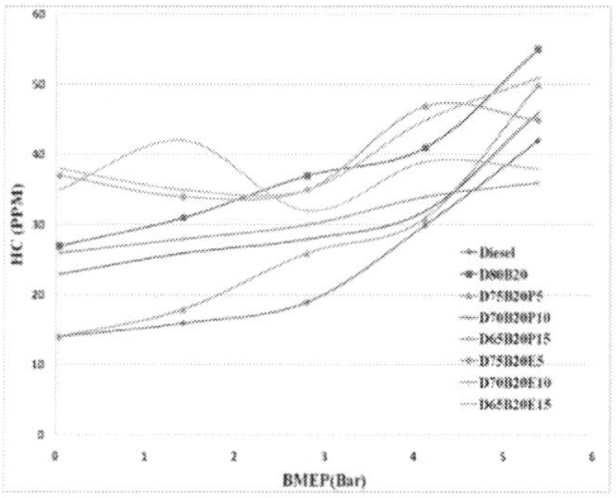

Fig. 8. Variation of HC with BMEP

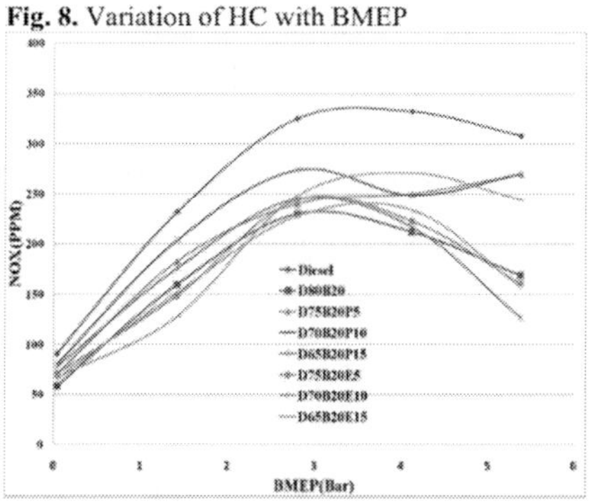

Fig. 9. Variation of NOX with BMEP

Conclusions

The experimental investigation concludes as given below

1 In performance the brake thermal efficiency for additives as a oxigenates is increased as compared to diesel. It is found maximum for the ethanol oxygenate blend (D70B20E10).

2 The brake specific fuel consumption is comparatively less as that of diesel fuel for D70B20E10 blend. It is 1-2% less for D70B20E10 blend.

3 The emission of CO is reduced by 2 % on addition of ethanol in diesel and biodiesel. It is lowest for D70B20E10 blend.

4 The emission of NOx is reduced by 5 % on addition of ethanol in diesel and biodiesel. It is lowest for D70B20E10 blend.

5 The emission of HC increases than diesel for both oxygenates and maximum for ethanol concentration blends.

6 From the experimental investigation, it can be concluded ethanol is found best oxygenate additives than that of propanol.

7 The effective blend found amongst the tested blend is D70B20E10.

References

S.Pinzi, JA.Soriano, 9th Int. Conf. on Applied Energy, ICAE2017, **142,** 849, (2017).

M. A.Ghadikolaei, C.S.Cheung, Ka-Fu.Yung, Energy. **164,** (2018).

I. Lawana,, W. Zhoua, Z. N. Garbaa, M. Zhanga,Z. Yuana, L.Chena, Ren. and Sustainable Energy Rev. **102,** 83, (2019).

A. Atmanli , Fuel **176,** 209, (2016).

A. Nayse, K. Sorate, Int. Engg. Res. Journal, **Special Edition,** (2017).

M. Yasin,R Mamat, A. Yusop, A. Aziz and G. Najafi,, Energy Procedia, **75**, 10, (2015).

1. H Li , S. Xia , H. Luo , P. Ma,. Fuel, **173,** 52, (2016).

M. Mofijur, M.G. Rasul, J. Hyde, Procedia Engineering, **105,** 658, (2015).

M. Parida, H. Joardar, A.. Rout, I. Routaray, B. Mishra, App. Thermal Engg. (2018),

S Rajkumar, J Thangarajab, Fuel, **240**, 101, (2019).

A Mahalingam , Y. Devarajan ,S. Radhakrishnan , S.Vellaiyan , B. Nagappan, Alexandria Engg. Jou. (2017)

2. M. Yasin , T Yusaf, R. Mamat, A. Fitri Yusop, Applied Energy **114,** 865, (2014).

R. Sakthivela, K. Rameshb, R. Purnachandrana,
P. Shameera, Ren. and Sustainable Ene. Reviews **82,** 2970, (2018)

S. Shahir,H. Masjuki, M. Kalam, A.Imran, A.Ashraful , Renewable and Sustainable Energy Reviews **48,** 62, (2015)

S.nandi, American Journal of Engineering Research **2,** 22.

AP.Pandhare, Journal of Renewable Energy, **11**, (2013)

Optimization and Techno-Economic Analysis of PV-Wind Power Systems for Rural Location in India.

Avinash Kaldate [1]*, *Amarsingh Kanase-Patil* [1], *and Shashikant Lokhande* [2]
[1]Department of Mechanical Engineering, Sinhgad College of engineering, Pune, India.
[2]Department of E&Tc Engineering, Sinhgad College of Engineering, Pune, India

Abstract. One downside to Green Energy is that it cannot be estimated. Therefore, determining the optimum planning and perfect working strategies for the resources to be included in the hybrid system is very important. HOMER software has been used in this research paper to solve the case study of the hybrid renewable energy system. Due to its extensive analytical capabilities and advanced prediction capabilities based on the sensitivity of variables, HOMER is one of the most used software for optimal planning purposes. A case study for the sizing of a renewable energy-based hybrid system is solved in this article, using the Hybrid Optimization of Multiple Energy Resources (HOMER) software. Photovoltaic panels (PV panels), wind turbines (WT), batteries, converters, electric charge and grid are used in case study. The results of the simulation are presented in graphical form and tabulated for better system visualization. The design of a system to supply 6.8 KWh/d whereas the peak is 1.04 KW electric loads has been performed using HOMER software. In order to allow the user to choose the most suitable option, a comparative analysis has made, showing the pros and cons of cases. Optimum construction conditions help to lower operating costs.

1 Introduction to HOMER software

Homer is free software created by the United States of America (USA) and the National Renewable Energy Laboratory (NREL). Using this software alternatives are planned and tested technologically and cost-effectively for off-grid and on-grid energy systems for centralized, stand-alone and distributed system generation. This enables the user to consider the accessibility of energy sources and other factors in a wide range of technology choices. HOMER was first developed under the Department of Energy Production (DOE) in 1993 to comprehend the trade-off between different energy production configurations [1]. NREL supplied free service to an increasing community of system designers who were interested in renewable energy after a few years of original design. HOMER has since remained free software, which has become a modelling tool for conventional and renewable energy technologies [2].

1.1 Principal Tasks of HOMER

1.1.1 Simulation

The simulation method is implemented in two respects. First, it tests the System's viability. This method is followed for two reasons. First, it is testing the viability of the system. If a system can properly compensate for electrical loads and take into account any other limitations imposed by the user, that would be considered feasible. Second, the cost estimate on the life of the system reflects the overall cost and overall Net Present Cost (NPC) of optimizing the hybrid renewable energy system installed and operated by HOMER PRO throughout its lifetime [3].

1.1.2 Optimization

The factors that make up the optimization system determine the combination of factors and the size or amount of the optimal value of such variables. In Homer, at the lowest total net present expense, the ideal or optimum device configuration meets the user-specified boundaries [9].

1.1.3 Sensitivity

Analysis of sensitivity enables estimation of the impact of uncertainty or conditions over which the planner has no influence over average wind speed or solar radiation [2]

2 HOMER Model

The planned hybrid energy systems include renewable sources such as PV, wind and biogas. Part of the battery is charged using a PV and wind system, but the generator is used in case of an emergency. The source bus was replaced by an AC, using a power converter. This integrated system is designed for off-grid electrification and on-grid electrification. A homer is assigned software for operational measurement. This software is used with various sensitive components for cost optimization. Various topics like load, global solar irradiation, wind speed etc. have been considered. HOMER's model organization is addressed in Fig.1.

*Corresponding author: **er.avinashkaldate@gmail.com**

2.1 Load Profile

Cited place for this analysis is Sarthak Niwas, A/p:Kashti, Tal: Shrigonda Dist: Ahmednagar . The latitude and longitude of location is $18^0 32.8' N$, and $74^0 34.9' E$ respectively. The load profile is acquired dependent on the basic requirements of appliances such as light,fan, refrigerator, TV and cooking. Depending on the climate, load varies with seasonal and monthly consumption. Total consumption is 6.8 kWh / d while peak consumption is 1.04 kW. Figure 2 depicts the load profile

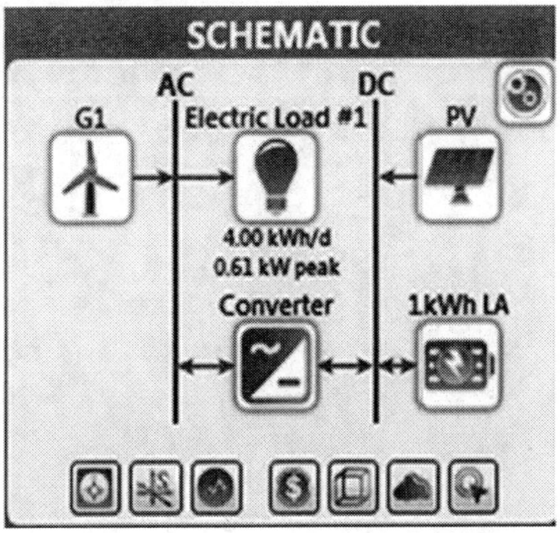

Fig. 1. Hybrid Energy System [16]

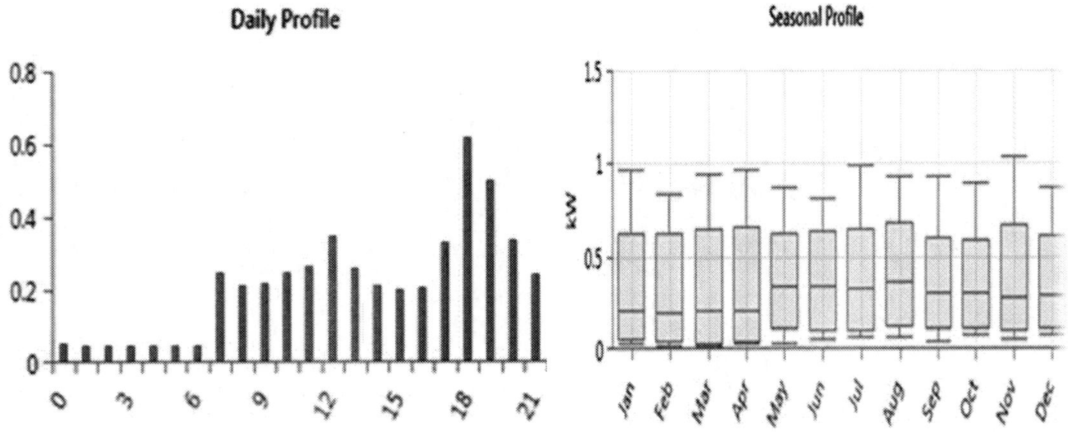

Fig. 2. Daily and seasonal profile [16]

2.2 Solar Resource Data

The solar resource has used for a rural site at a location of latitude and longitude is $18^{0}32.8^{'}N$, and $74^{0}34.9^{'}E$ respectively as shown in Figure 3. The solar radiation information was obtained from the website of NASA Surface Meteorology and Solar Energy. For this region, the average annual solar radiation is $5.33 Kwh/m^2/day$. Figure 4 shows a one-year solar resource profile. Maximum daily radiation occurs in a month of May which is 6.55 $Kwh/m^2/day$ and minimum daily radiation occurs in a month of July is 4.036 $Kwh/m^2/day$.

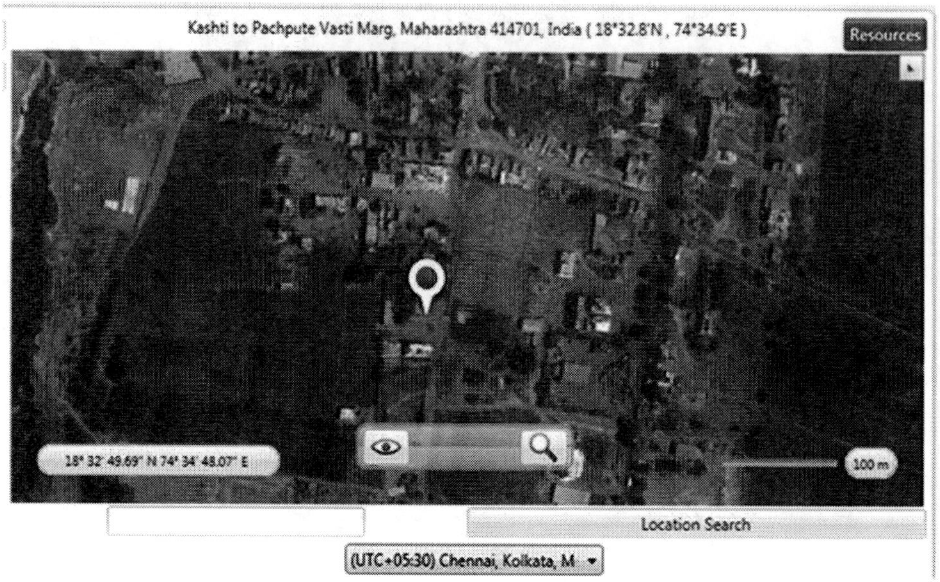

Fig. 3. location selected for case study [16]

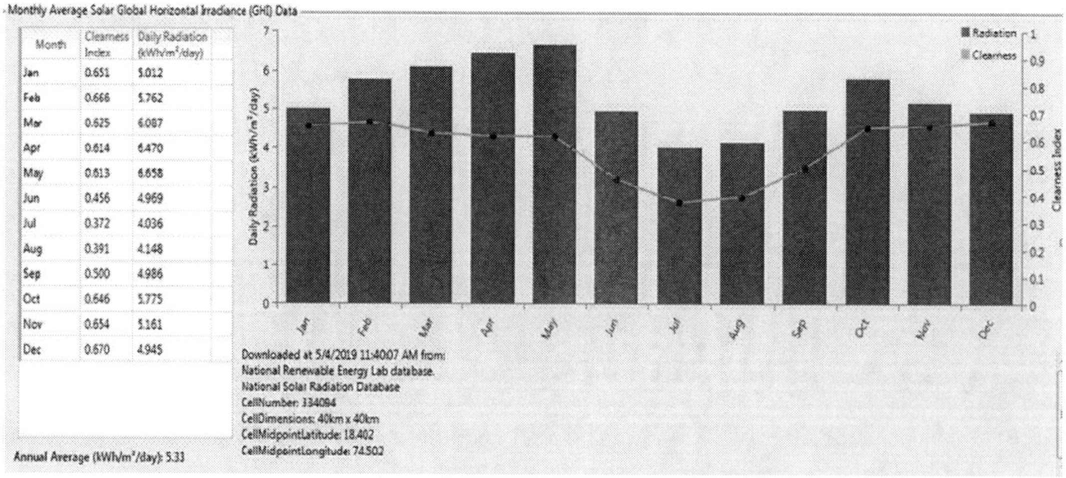

Fig. 4. Annual average solar radiation [16].

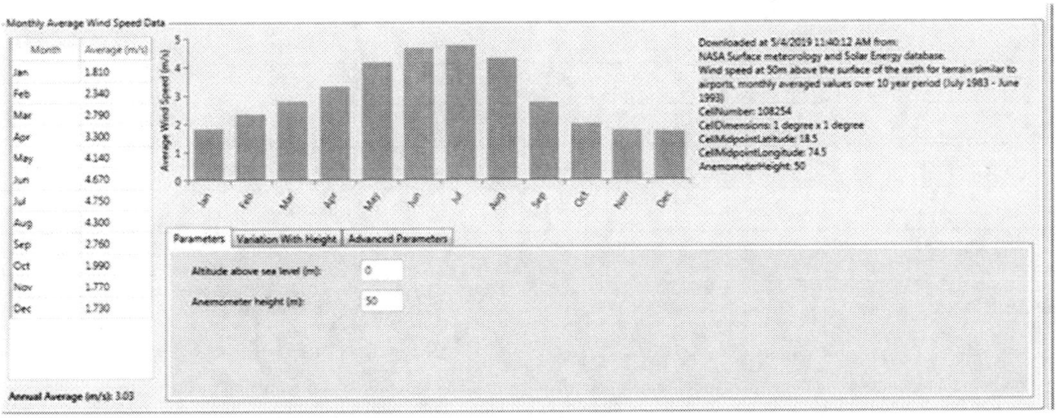

Fig. 5. Monthly average wind speed data [16].

2.3 Wind Resource data

Figure 5 shows the wind resource profile over a year period for a given region. This is acquired from the website of NASA surface metrology and solar energy. The anemometer is measured more than 50 meters above ground level and the wind represents an annual average speed of 3.03 m/s. Maximum wind speed occurs in a month of June is 4.67 m/s and minimum wind speed occurs in a month of December is 1.730 m/s.

3 Selection of Components

This project aims to reduce energy costs by using renewable energy systems such as photovoltaic panels / wind turbines. Optimization and COE are calculated using these two renewable power systems in HOMER software.

3.1. Sizes Considered for Optimization

Two renewable energy sources are used to generate electricity and to distribute charges. The sizes and prices of these renewable energy sources are set out in the table 1 and table 2. The information on wind turbines shows that five different size wind turbines were selected on the basis of load demand and size in the research area [7,11-14]. For each wind turbine rate, the factors that are useful in the mathematical modeling of the wind turbine are given in Table 1. Different parameters are considered such as rated output, rated wind speed, swept area, cut-in speed, Survival speed and cost per turbine etc.

Solar data is shown in Table 2. Four different size rating of PV panel are selected according to load demand. For the given study area, roof top area available for PV panel placement is 92 m^2. For each PV panel area is given in Table 2 and from this number of PV panel for each size to be calculated. Efficiency according to cost per PV panel is given in Table 2.

Table 1. Wind Turbine Data [10]

Rated output	5.1KW	4.2 KW	3.3 KW	1.8 KW	1.5 KW	650 W
Rated wind speed	11 m/s	11 m/s	10.5 m/s	10.5 m/s	10.5 m/s	10.5 m/s
Cut- in m/s, VC	2.7 m/s	2.7 m/s	2.7 m/s	2.7 m/s	2.7 m/s	2.7 m/s
Swept Area m2	21.4 m2	19 m2	16.4 m2	9.4 m2	9.2 m2	3.7 m2
Survival wind m/s	55m/s	55m/s	55m/s	55m/s	55m/s	55m/s
Cost per wind turbine	₹167500	₹134000	₹80400	₹67500	₹53600	₹40200

Table 2. PV cell Data [10]

Size	1 KW	2 KW	5 KW	10 KW
Area per panel	$9.2m^2$	$14.864\ m^2$	$27.87m^2$	$46.45m^2$
No. of PV panel	18	11	6	4
Efficiency	12%	16%	16 %	20%
Cost per panel	₹36000	₹72000	₹180000	₹360000

3.3 Converter

The converter is a system where the DC parts serve as AC loads or vice versa. It acts as a rectifier converting AC to DC, an inverter converting DC to AC, or both. The estimated capital cost of inverter is 3125 ₹/kW and replacement cost 3125 ₹/kW is calculated according to market survey. The normal life of the converter is considered 20 years.Battery and converter size costing shown in Table 3.

Table 3. Battery and converter size [10]

Component	Sizes	Cost
Converter	1kw	₹. 3125
Battery	1kwh	₹. 22600

4. HOMER Results

4.1 Flow chart of Homer optimization process.

The optimal system configuration can be determined by the components that need to be properly validated for each component of the system, and the strategies the system sends for use. The end goal is to identify a configuration that has low NPC [4]. Therefore, different system configurations and technical issues are simulated and create different combinations of system components. Finally, a list of those different configurations can be created and compared. The best viable pairs are separated, and the most viable pairs are considered, with the exception of the least viable pairs. The overall optimization process flow chart is shown in Figure 6.

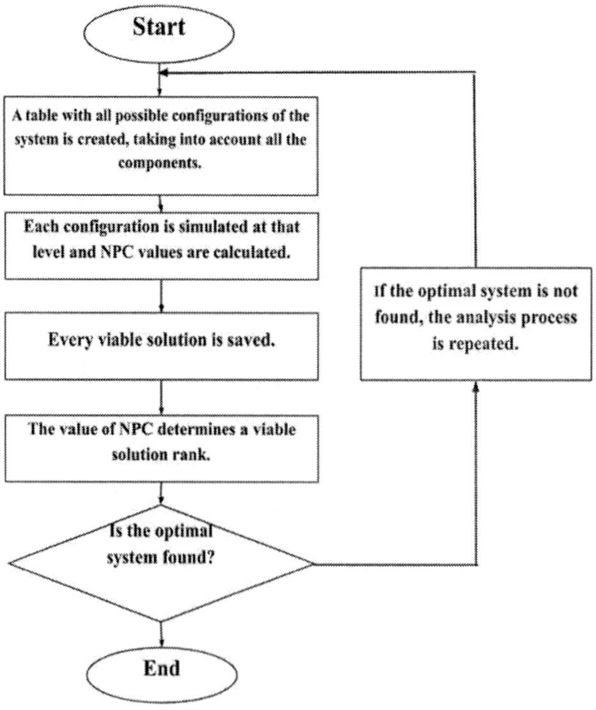

Fig. 6. Flow chart for optimization in Homer

Table 4. HOMER Results without Grid Connection [161].

PV	Wind	Battery	Converter	PV(kW)	NPC(₹)	COE(₹)	Operating cost(₹/yr)	Initial Capital cost(₹)	PV energy production(kWh/yr)
✔		✔	✔	5.86	₹647,671	₹34.34	₹17,682	₹419,083	9,688
✔	✔	✔	✔	6.44	₹663,100	₹35.15	₹18,844	₹419,493	10,650

4.1 Without Grid Connection

In this study, using NREL's Homer software, the selection and sizing of components of a stand-alone energy system has been carried out. Homer is a hybrid system design software that facilitates the design for stand-alone applications of electric power systems [4]. HOMER assists in designing a systematic power system to deliver the desired load. The ordering system takes into account the cost of the life-cycle. It will provide a efficient user interface and comprehensive device description with correct sizing [5,6]. The software conducts automated sensitivity analyzes for the hybrid system's response to core parameters such as cost of services and components [8]. The simulation has been performed considering the project life of 20 years. The hybrid renewable energy system is built into HOMER. It considers the broad rigor of chronological simulation and optimization in the model. All resources, including all capital, replacement possibilities, and annual maintenance costs, have been added as input to obtain the optimal size combination of different resources from Table 1, Table 2 and Table 3.

4.2 With Grid Connection

The hybrid renewable energy with grid system has been simulated in HOMER. For completion of system, battery was chosen per unit cost of ₹ 22,600. And the converter considered also has a power of 1 kW for this case. The Figure 7 and Table 5 shows HOMER developed a renewable energy grid hybrid model consisting of solar, wind as a renewable source, battery as a converter and grid power saver.

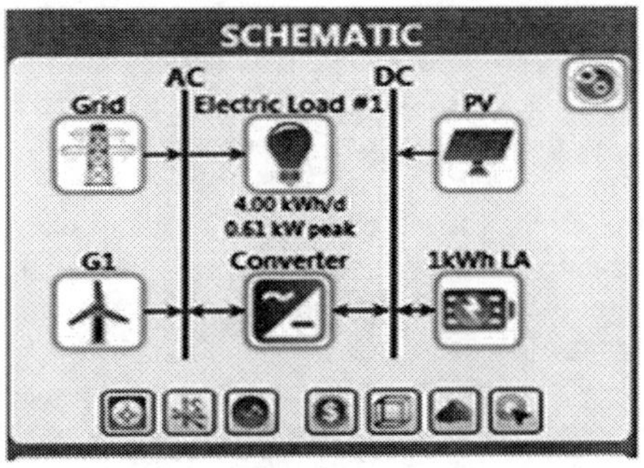

Fig. 7. Hybrid system with grid connection[16]

The result of the optimization in Table 5 shows that the combination takes into account both solar and wind, but among the values considered, the wind speed responsible here maximum power generation. This approaches the system's minimum NPC and lowest COE so it's acceptable. But if the wind speed is not above average, wind turbines should not be considered. The results of optimization without wind turbines are greater compared to wind turbines.

Table 5. HOMER Results with Grid Connection [16].

PV	Wind	Battery	Grid	Converter	NPC(₹)	COE(₹)	Operating cost(₹/yr)	Initial Capital cost(₹)
✔			✔	✔	-₹139,746	-₹1.35	-₹26,281	₹200,000
✔	✔		✔	✔	-₹112,028	-₹1.07	-₹26,071	₹225,000
✔		✔	✔	✔	-₹96,467	-₹0.930	-₹24,681	₹222,600
✔	✔	✔	✔	✔	-₹68,749	-₹0.659	-₹24,471	₹247,600
			✔		₹150,993	₹8.00	₹11,680	₹0.00
	✔		✔		₹177,257	₹9.39	₹11,778	₹25,000
		✔	✔	✔	₹194,478	₹10.30	₹13,285	₹22,739
	✔	✔	✔	✔	₹220,742	₹11.69	₹13,383	₹47,739

Cycle charging policy is a dispatch policy whereby the generator operates at full output power whenever it is required to operate to provide primary load. Surplus power goes towards lower-priority objectives such as reducing priority: servicing deferrable loads, charging battery bank. The plan behind sending a strategy is to send a generator running under load; it just produces enough energy to meet the primary load. Low-priority target, such as charging a battery bank for saving a deferrable load, are left to renewable energy sources.

5. Conclusion

Nature is rich in different sources, and abundant. But the capacity of the resources varies with the time and time of day. The lack of resources can be reduced by combining two or more resources by creating a hybrid energy system. For this purpose HOMER is capable of better optimisation. This paper represents the idea of creating a solar and wind hybrid energy storage system, taking into account the weather conditions latitude and longitudinal of that location ($18^0 32.8' N$, $74^0 34.9' E$). If the with-grid and without grid condition are compared with each other, it is observed that the capital cost in the with-grid condition is less than two lakhs. In terms of annual energy production with-grid condition produces 9688 KWh/year energy. But if sell excess energy through the grid, in with-grid cases you will make a profit. In the results of the simulation, it can be noted that the optimal solution can be achieved by incorporating PV panels, batteries and a converter into a hybrid grid connected system.

References

1. Y. Pavan Kumar and R. Bhimasingu, J. Mod. Power Syst. Clean Energy **3**, 1 (2015).
2. A. Kanase-Patil, R. Saini, and M. Sharma, Renew. Energy **35**, 1342 (2010).
3. M. Nehrir, C. Wang, K. Strunz, H. Aki, R. Ramakumar, J. Bing, Z. Miao, and Z. Salameh, IEEE Trans. Sustain. Energy **2**, 392 (2011).
4. A. Shiroudi, S. Taklimi, S. Mousavifar, and P. Taghipour, Environ. Dev. Sustain. **15**, 1389 (2013).
5. Y. Katsigiannis, F. Kanellos, and S. Papaefthimiou, Energy Syst. **7**,33 (2014).
6. O. Erdinc and M. Uzunoglu, Renew. Sustain. Energy Rev. **16**, 1412 (2012).
7. M. Bansal, D. Khatod, and R. Saini, ICROIT 2014 - Proc. 2014 Int. Conf. Reliab. Optim. Inf. Technol. **25** (2014).
8. M. Nehrir, C. Wang, K. Strunz, H. Aki, R. Ramakumar, J. Bing, Z. Miao, and Z. Salameh, IEEE Trans. Sustain. Energy **2**, 392 (2011).
9. A. Eltamaly and M. Mohamed, Math. Probl. Eng. **2014**, 16 (2014).
10. https://www.indiamart.com/ accessed on 4th May 2019
11. A. Kanase-Patil, R. Saini, and M. Sharma, *Optimum Utilization of Renewable Energy Sources in a Remote Area*, Presented in International Conference on Advances in Mechanical Engineering, Department of Mechanical Engineering, Sardar Vallabhbhai National Institute of Technology, Surat (2008).
12. A. Kanase-Patil, R. Saini, and M. Sharma, Renew. Energy **36**, 2809 (2011).
13. A. Kanase-Patil, R. Saini, and M. Sharma, Energy **36**, 5690 (2011).
14. A. Kanase-Patil, R. Saini, and M. Sharma, *Off Grid Electrification of Remote Rural Areas with Integrated Renewable Energy Systems*, Presented in 6th International Conference on Hydropower on the theme Hydropower supporting other renewable held at Tromso, Norway (2010).
15. A. Kaldate, A. Kanase-Patil, and S. Lokhande, *Review of SCADA based hybrid renewable energy source integration*, Presented in International Conference on Artificial Intelligence & Applications, College of Engineering. ,Roorkee (2019).
16. Site of HOMER Pro Available at:. https://www.homerenergy.com/products/pro/index.html : (accessed on 17 June 2019).

CFD based analysis of flow-driven vertical axis hydrokinetic turbines: blade hydrofoil and interactions effects

Zaid Hammoudi[1], *Ikram El Abbassi*[1,2], *Rafik Absi*[1,3,*], *A.-Moumen Darcherif*[1,3]
[1]ECAM-EPMI Graduate School of Engineering, LR2E-Lab, 13 Boulevard de l'Hautil, 95092 Cergy-Pontoise (France)
[2]L2MGC-Lab - EA 4114. [3]Quartz-Lab - EA 7393.

Abstract. In last years, interest in hydrokinetic energy conversion (HEC) technology has grown. However, HEC technology requires advancements to become successful for practical in-situ conditions. In this study we investigated the Darrieus type vertical axis hydrokinetic turbines (HKT). Simulations are carried out using commercial CFD package to study the hydrodynamic behavior of a three bladed turbine. Two HKT with same solidity equal to 0.17 were considered, a 1st with symmetric-NACA0018 blade hydrofoils and a 2nd with cambered-NACA4415 blades. Our simulations show a shift in instantaneous torque coefficients for two different tip-speed ratio (TSR) values. For TSR = 1, the 2nd HKT has a delay compared to the 1st HKT, while for TSR = 2 it has a lead. NACA0018 provides highest troque coefficients. For a hydrofarm use, it is important to consider the interactions between hydrokinetic turbines. We investigated the effect of streamwise distance on performance of a Darrieus turbine. A distance of about 11 times the diameter of HKT along the streamwise direction seems essential.

1. Introduction

Fuels and nuclear power represents more than 80% of the energy used in the world. These sources are easy to exploit, but the major problem is that they are not inexhaustible, and also cause environmental problems, such as greenhouse effect and nuclear waste. Most countries have already begun a transition to clean, renewable energy for their future. Non-conventional renewable energy (NCRE) sources can be classified into five families: solar, wind, biomass, geothermal and hydro or hydroelectric. In last years, interest in hydrokinetic energy conversion (HEC) technology has grown. However, hydrokinetic energy from tidal and river flows is currently little exploited around the world but generates great expectations due to its high potential for sustainable and predictable energy. The potential for total tidal energy in the world is estimated at about 3 TW, including 1 TW in accessible areas for the installation of energy capture devices [1]. To capture this kinetic energy, more than 50 devices can be used [2], in particular machines derived from wind turbine technologies, called tidal turbines [3] or hydrokinetic turbines (HKT), immersed along coasts or in rivers, or suspended on the free surface of the water [4].

There are mainly two types of turbines according to the position of the axis of rotation with respect to the current: horizontal axis (axial-flow) HKT and vertical axis (cross-flow) HKT. Darrieus type cross-flow hydrokinetic turbines (HKT) has some advantages over axial-flow turbines [4]. Different studies have been conducted to better understand cross flow hydrokinetic turbines HKT [5-7]. HEC technology requires advancements to become successful for practical applications. We need better understanding (1) individual turbines to allow optimization and (2) turbine wake and HKT interactions effects. Thanks to rapidly increasing computing power, Computational Fluid Dynamics (CFD) has become a powerful tool in engineering design and optimization. With adequate numerical methods, CFD provides accurate, cheap and quick results compared to a full-scale experimental study [8-9]. The aim of our study is to use CFD in order to provide an accurate description of the flow and therefore the best conditions for an optimization. Different commercial CFD package can be used, such as ANSYS CFX and Fluent [5-6]. In this study, we will use ANSYS Fluent for simulations of vertical turbines. We will introduce background related to HKT, define the different parameters and more adequate turbulence models. Finally, simulations with different parameters, geometries and configurations will be conducted.

2. Hydrokinetic turbines (HKT)

2.1 Principle

Hydrokinetic turbines are similar to wind turbines, they use the kinetic energy of flowing water to drive a generator [11]. However, the main difference is about the density of water which is about 850 times greater than air and flow velocities for a tidal or river flow tend to be an order of magnitude lower than flow velocities of

wind. Finaly, the Reynolds numbers are in the same range for both hydrokinetic and wind turbines.
Hydrokinetic turbines are on seabed or suspended under the free surface of the water. The rotor of the turbine turns under the effect of the current then drives an alternator that transforms mechanical energy into electrical energy.

2.2 Maximum extracted power: Betz's Law

The first efforts to assess the performance of rotating machines under the effect of a fluid flow (i.e. wind turbines) were done by [11] and [12]. The well know Betz's or Lanchester-Betz's law, which defines the maximum power that can be extracted from the stream flowing through an idealized « actuator disk » of section S. From the principles of conservation of mass and momentum, the power extracted from the water through the rotor is given by [11, 12]:

$$P = \frac{1}{2}(\rho\, S\, v)(v_1^2 - v_2^2) \tag{1}$$

where : v_1 et v_2 are respectively velocities in the front and downstream of the rotor, ρ the water density and $v = \frac{1}{2}(v_1 + v_2)$.
The totale power of a non-perturbed flow without rotor as for a cylinder of fluid with cross sectional area S is:

$$P_0 = \frac{1}{2}\rho\, S\, v_1^3 \tag{2}$$

The power coefficient is given by :

$$C_p = \frac{P}{P_0} = \frac{1}{2}\left(1 - \left(\frac{v_2}{v_1}\right)^2\right)\left(1 + \left(\frac{v_2}{v_1}\right)\right) \tag{3}$$

The curve of equation (3) riches a maximum for $v_2/v_1 = 1/3$, the maximale value is $C_p = P/P_0 = 16/27$. According to Eq. (3), no turbine can capture more than 16/27 (59.3%) of the kinetic energy of the flow (Ahmadi-Baloutaki *et al.* 2015). The factor 16/27 (0.593) is known as Betz's or Lanchester-Betz's coefficient. Practical utility-scale wind turbines achieve at peak 75% to 80% of the Lanchester-Betz's limit.

2.3 Tip Speed Ratio

The tip-speed ratio, or TSR for turbines is the ratio between the tangential speed of the tip of a blade and the actual flow velocity:

$$TSR = \lambda = \frac{\omega\, R}{v_\infty} \tag{4}$$

where ω is the rotor rotational speed in radians/second, R the rotor radius in meters et v_∞ is the flow velocity at an infinite point from the turbine in meters/second. The tip-speed ratio is related to turbines efficiency.

2.4 Power and torque coefficients

The torque coefficient is given by:

$$C_M = \frac{M_z}{\frac{1}{2}\rho\,(H\,\phi) R\, v_\infty^2} \tag{5}$$

where M_z the torque along the axis of rotation $\overrightarrow{oz}$, $S = H\,\phi$, H and ϕ are respectively the high and the diameter of the hydrokinetic turbine.
The power coefficient C_P is the ratio between the power drawn from the water by the hydrokinetic turbine and the power of the flow:

$$C_P = \frac{P}{\frac{1}{2}\rho\,(H\,\phi)\, v_\infty^3} \tag{6}$$

It is also possible to write the following relationship between power and torque:

$$P = M_z\, \omega \tag{7}$$

From the previous relationships, we can deduce that:

$$C_P = \lambda\, C_M \tag{8}$$

3. Results and discussions

We used statistical Reynolds Averaged Navier-Stokes (RANS) turbulence models. RANS models offer the most economic approach for computing turbulent industrial flows. Examples of RANS models are the different forms of k-ε or the k-ω two-equation models [13,14]. These models are based on the concept of eddy viscosity (turbulent viscosity) which is obtained by solving two additional transport equations. RANS models are suitable for many engineering applications and typically provide the level of accuracy required. Among the RANS models, the two-equation eddy-viscosity shear stress transport (SST) k-ω turbulence model combines a k-ω formulation in the inner parts of the boundary layer and a k-ε model a far from the wall [15]. In the present study, the flow is modeled using the k-ω SST model available in ANSYS Fluent.

To validate our method and results, we consider the cross-flow hydrokinetic turbine of Darrieus type of LEGI Laboratory (Grenoble). This HKT was used in the hydrodynamic tunnel of LEGI which consists in a rectangular section of 0.25x0.7m and a 1m length in the streamwise direction inserted in a closed hydraulic loop of 30m. Flow velocity are between 1 to 2.3 m/s. It corresponds to a blade Reynolds number between $1.7\text{x}10^5$ and $5\text{x}10^5$. Rotational speed is imposed by a synchronous generator connected to the turbine shaft. Turbine has 3 NACA0018 blade hydrofoils. The height H and the diameter D of the turbine are equal, H=Φ=175mm and blade chord C=32mm (figure 1.a) The solidity which is the fraction of circumference of rotor covered by blades is equal thereofore to 0.17. It represents a lateral blockage ratio (tunnel width/turbine width) of 4, and a vertical blockage ratio of 1.43.

It is important to select first the most appropriate mesh. Validation of our first results is done by comparisons with those of [6]. We will create the geometry with DesignModeler and mesh with Meshing. The methodology is explained in the next section.

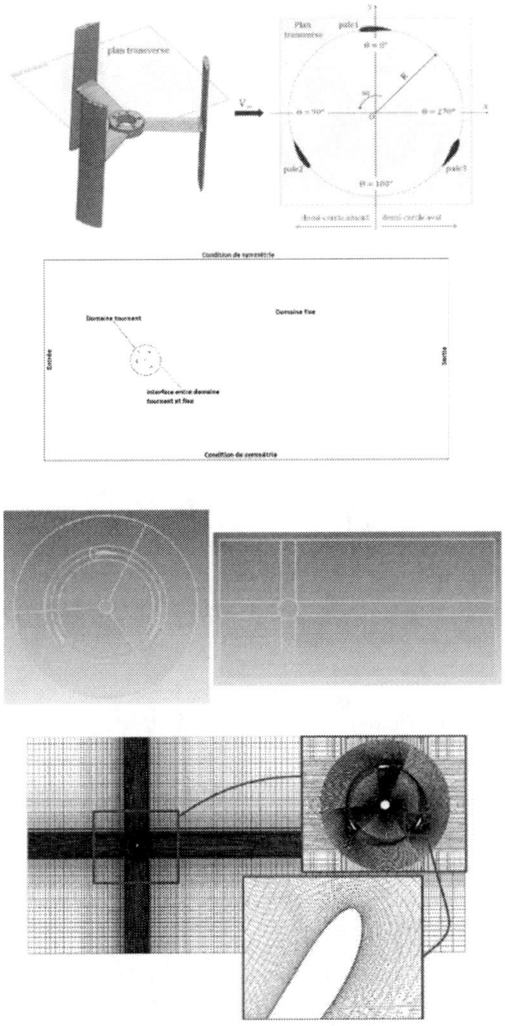

Fig. 1. Geometry and mesh, (Top) 3D view of the hydrokinetic turbine and its 2D section used for calculations, (Menchaca Roa 2011) (2nd) Geometry of a vertical hydrokinetic stream in an infinite medium, (3rd) Geometry under ANSYS Fluent, (Bottom) Structured mesh.

3.1. Geometry

The geometry is composed by two parts: 1) the first is rotating, of circular section with the horizontal sections of the three blades of the hydrokinetic turbine and its axis, 2) the second is fixed representing the medium surrounding the hydrokinetic turbine. Between the two, we apply a sliding mesh interface.

- The hydrokinetic turbine used has a diameter $\Phi = (2\ R) = 0.175$ m, an axis of diameter d = 0.022 m, a height h = 0.175 m, a blade chord C = 0.032 m and it is possible to change the shape of blade hydrofoil. The geometry is done with Design Modeler, the procedure is as follows:
- We import the coordinates of points constituting the blade hydrofoil NACAxxxx, then we transform them into a curve and apply functions "transformation of body" to move it to the desired place and duplicate it in a circular way to have the three blades.
- We draw all the other sketches of the geometry, sharing the two parts rotating and fixed in several surfaces to create several geometry bodies (Figure 1.c) to make a structured mesh, knowing that each part is created separately.
- At the end, we can gather the two pieces in one. By importing them, the piece fixes as a blocked body, and the rotating part as an unlocked body to apply fluent the sliding mesh technic and give the desired rotation.

3.2. Mesh

The mesh is made using Meshing software integrated in Ansys Workbench. Geometry already shared in geometry bodies. We create sizing on each edge, inflations around the walls of the blades and the axis of rotation and we apply the meshing method of faces to create a mesh in quadrilaterals (figure 1.d). Finally, we add the named selections for boundaries and the interface between rotating and fixed part [16].

Once the mesh created, we import it under Fluent and adjust the various parameters before starting the computation: First to create an interface between the rotating part and the fixed part: "Mesh Interface".

Then we can proceed as:

- Unsteady regime,
- We choose the turbulence model, $k - \omega$ (2eqn) in "Model" and SST in " $k - \omega$ model".
- In "Cell Zone Conditions", we put the desired rotation for the rotating zone by checking the "Mesh Motion" box and choose the axis of rotation and the angular velocity.
- We Define boundary conditions in "Boundary Conditions": Choose "symmetry" for the two horizontal boundaries of the domain. We choose "wall" for the three blades. For velocity inlet, velocity = 2.3 m / s, turbulence intensity I = 3%, characteristic length l = 0.002 as turbulence quantities. They were chosen by analogy with smooth pipes, since no experimental data are available [6]. Finally, we take for output "pressure outlet", the same parameters of turbulence as input.
- For computation of torque coefficient C_M given by equation (5), Fluent uses reference values, which must be inserted in "reference values", these values are the cross-sectional surface, the water density, the hydrokinetic turbine depth, a reference length, a reference velocity and the dynamic viscosity of water.
- We choose the resolution method,
- We set the residue values at 10^{-5}.
- And finally, before starting simulations, we define the time step, which is chosen as a degree of rotation [6],

to obtain the best compromise between precision and computational time, we increased the number of step times to reach a periodic solution with 100 max iterations per time step.

To obtain the better computation results, we need a mesh well adapted to the problem geometry. At initial step we used 3 different meshes: 1st mesh is unstructured, the 2nd is structured in part, the 3rd is structured and finer than the two others (about 180,000 points). We took this 3rd mesh for the simulations since the results at TSR = 2 show a good agreement with that of [6].

As shown in equation (8), to know the power factor C_P, we first need to compute the torque coefficient C_M, which is obtained by taking the average of instantaneous torque coefficient over a complete period. However, computations must be left until a periodic solution is obtained, as shown in FIG. 2. FIG. 3 presents instantaneous torque coefficients of the hydrokinetic turbine or of each of the individual bladesare.

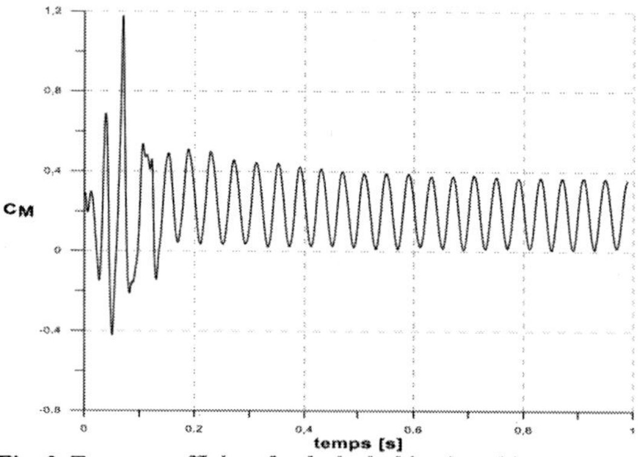

Fig. 2. Torque coefficient for the hydrokinetic turbine.

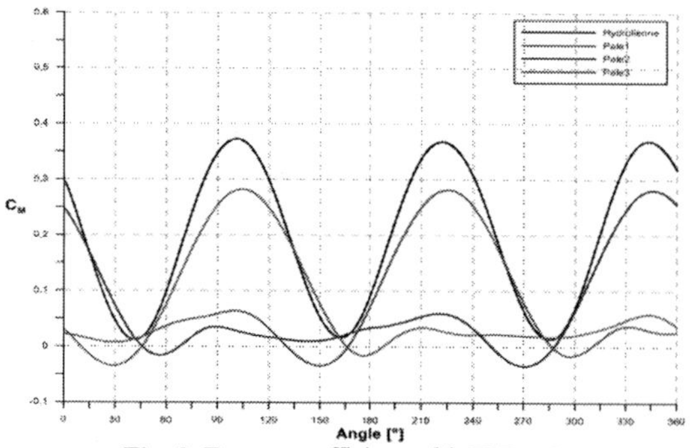

Fig. 3. Torque coefficient with TSR = 2.

3.3. Simulations with NACA4415 blade hydrofoil

After validating the simulations and the used mesh, the aim of this part is to investigate the effect of blade hydrofoil on the performance of the cross-flow hydrokinetic turbine of Darrieus type. We changed the symmetric-NACA0018 hydrofoils by cambered-NACA4415 blades to compare the performance of these two types of blade hydrofoils. When hydrokinetic turbines with fixed pitch blades show poor starting torque, it is possible to improve it by cambered blade profiles [4]. The simulation conditions are the same. Two simulations were carried out with TSR = 1 and TSR = 2.

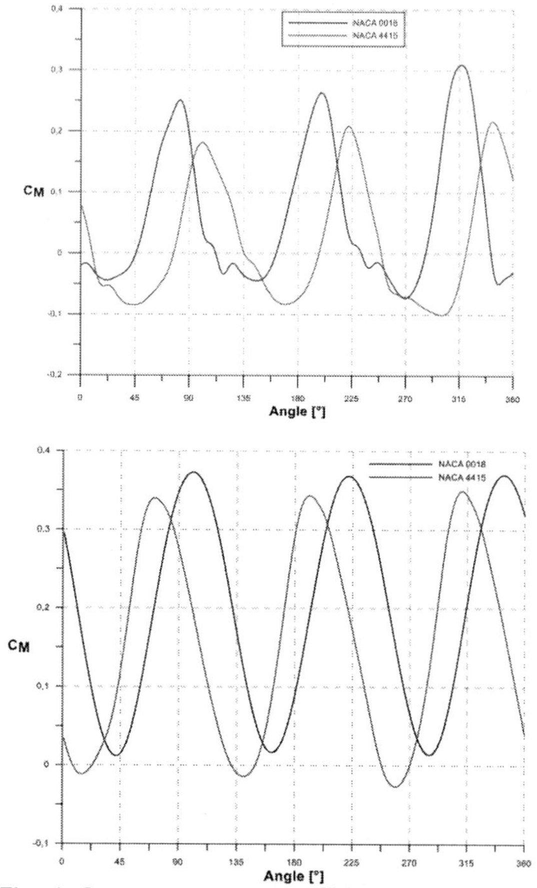

Fig. 4. Instantaneous torque coefficients: (Top) TSR=1, (Botton) TSR=2.

Figure 4 illustrates the difference in instantaneous torque coefficient for the two cases, between the hydrokinetic turbine (1) with NACA0018 blades and the hydrokinetic turbine (2) with NACA4415 blades. NACA0018 provides highest torque coefficients for the two TSR values and therefore a higher Power coefficient. However, we notice a shift of the hydrokinetic turbine (2) compared to the hydrokinetic turbine (1), it has a delay for TSR = 1, while for TSR = 2 it has a lead. The torque and therefore the power is more important for the hydrokinetic turbine (1), which gives a greater depression between the intrados and the extrados of the blades, thus a greater lift force, as well as a greater tangential force.

3.4. Effect of streamwise distance between Darrieus turbines toward hydro farm design

In marine environment, HKT are implanted in farms, such as wind turbines. The main shortcoming of HKT is the wake they create, which affects the power of other turbines downstream. The simple actuator disc model is unable to provide an adequate description. It can't reproduice blade tip vortices and wake rotation which provide a strong contribution to turbulence. Zhang et al. investigated near-wake wind turbine by PIV measurements [17]. They noticed a dislocation of coherent vortices after a few diameters and the important impact on turbulence production.

Figure 5 shows the change in the turbulence intensity caused by a single HKT. It is important to accurately predicte turblence quantities downstream a rotor since it is involved in the flow energy available for a 2nd HKT in hydrofarm configurations.

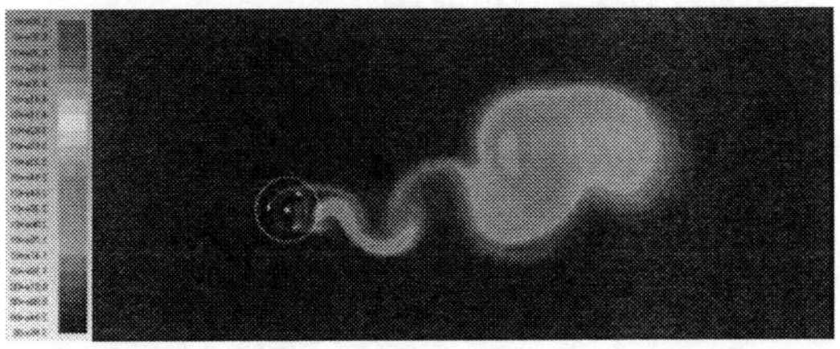

Fig. 5. Turbulence intensity (%).

In the case of two HKT in the streamwise direction, the second at downstream side has weak flow since the power density, which is the amount of power per unit volume, of the flow is less important compared to the 1st HKT at upstream side. This could be explained by the flow diversion after the 1st HKT due to loss in kinetic energy (Fig. 5). We aim to investigate the required distance for which the flow will recover his entire non-perturbed energy.

To understand the effect of the interaction between HKT, a simulation of two axially aligned turbines (NACA4415) separated by a distance A = 1m and then 2m, which represents respectively about 6 times and 11 times the diameter of the hydrokinetic turbine.

Figure 6 shows the effect of the first hydrokinetic turbine on the power of the second. The torque coefficient of the 2nd HKT is closer to that of the first one when X = 2m. For X = 1m, we notice a gap between torque coefficients of the two HKT.

Simulations show that the wake generated by upstream HKT has a high influence on the performance of the downstream HKT. The turbulence created by the 1st HKT dissipates away, so the further the 2nd HKT is distant, the closer their respective torque coefficients are. There is an optimal distance between two turbines axially aligned for which their torque coefficient becomes quite close. A stream wise distance of about X/D = 11 seems to be essential to have the least wake effect.

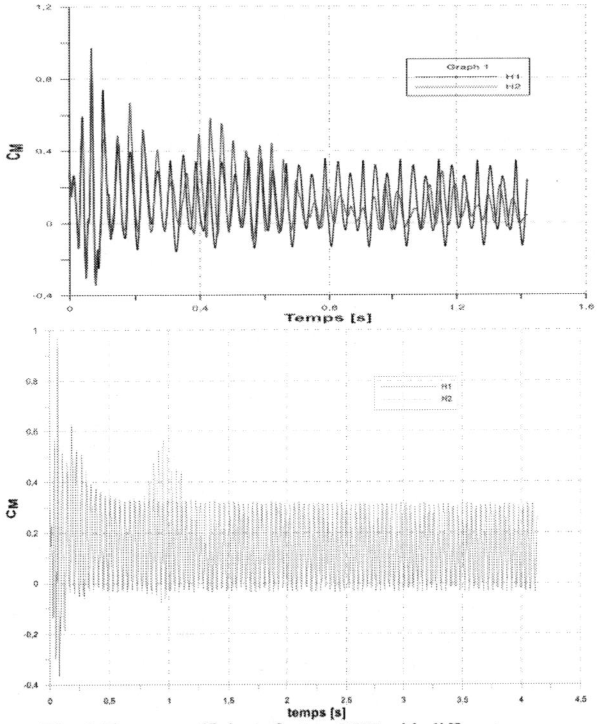

Fig. 6. Torque coefficients for two HKT with different streamwise distances: (top) X=1m and (Bottom) X=2m, respectively about 6 times and 11 times the diameter of the rotor..

4. Conclusions

The objective of the present study is to investigate: (1) the influence of hydrofoils on the performance of a Darrieus-type vertical-axis turbine and (2) the interaction between two axially aligned turbines. We performed two-dimensional simulations to study the behavior of the cross-flow Darrieus turbine. We outlined the steps for creating a geometry and mesh to apply the sliding mesh technique with ANSYS Design Modeler, Meshing, and Fluent.

Simulations were carried out for a HKT (1) with symmetric-NACA0018 and HKT (2) with cambered-NACA4415 blades in same conditions with TSR = 1 then TSR = 2. According to our results, HKT (1) with NACA0018 hydrofoils provides highest troque coefficients for the two TSR values and therefore more Power. However, we notice a shift in the instantaneous torque coefficient of the hydrokinetic turbine (2) compared to the hydrokinetic turbine (1), it has a delay for TSR = 1, while for TSR = 2 it has a lead. This result shows the importance of the TSR which can reverse the behavior for a HKT with a given blade hydrofoil. This behavior seems less important than the maximum torque coefficient; however it will require future investigations on possible mechanical or vibrating aspects.

In marine environment, HKT are implanted in farms, such as wind turbines. The main shortcoming of HKT is the wake they create, which impacts the power of other turbines downstream. We studied the case of two HKT aligned axially in the streamwise direction, separated by a distance respectively equal to about 6 and 11 times the diameter of the HKT. Simulations show that the wake generated by upstream HKT has a high influence on the performance of the downstream HKT. There is an optimal distance between two turbines axially aligned for which their torque coefficient becomes quite close. A stream wise distance of about X/D = 11 seems to be essential to have the least wake effect.

References

1. Twidell J., Weir T. Renewable Energy Resources, Chapter 13. Taylor & Francis, (2006).
2. Edenhofer O., Pichs-Madruga R, Sokona Y, Seybot K., Eickemeier P., Matschoss P., Hansen G., Kadner S., Schlömer S., Zwickel T., Von Stechow C. IPCC Special Report on renewable energy Sources and Climate change Mitigation. Prepared by working Group III of the Intergovernmental Panel on Climate change. (Cambridge University Press, 2011).
3. Behrens S., Griffin D., Hayward J., Hemer M.A., Knight C., MCGarry S., Osman P., Wright J. Ocean renewable energy, 2015-2050: an analysis of ocean energy in Australia. CSIRO. 212 p, (2012).
4. Kirke B.K. and Lazauskas L. Renewable Energy, **36**(3), 893-897, (2011).
5. Marsh P., Ranmuthugala D., Penesis I., Thomas G. Renewable Energy, **83**, pp 67-77, (2015).
6. Menchaca Roa A. PhD thesis, Université de Grenoble, France, (2011).
7. Patel V., Eldho T.I., Prabhu S.V. International Journal of Marine Energy, **17**, 110–135, (2017).
8. Bianchini A., Balduzzi F., Bachant P., Ferrara G., Ferrari L. Energy Conversion and Management, **136**, 318-328, (2017).
9. Shaheen M., Abdallah S., Sustainable Energy Technologies and Assessments, **19**, 125-135, (2017).
10. Bilgili M., Yasar A., International Journal of Green Energy, **14**(12), 1048-1056, (2017).
11. Lanchester F.W., J. Am. Soc. Nav. Eng., **27**(2), 509-510, (1915).
12. Betz A., Z., für Das. gesamte Turbinenwes., **26** (1920).
13. Absi R, Tanaka H, Kerlidou L, André A., Proc. 33rd International Conference on Coastal Engineering, 1-6 July 2012, Santander, Spain, 2772-2782, (2012).
14. Absi R., Proc. 27th International Conference on Coastal Engineering (ICCE), July 16-21, 2000, Sydney, Australia, American Society of Civil Engineers, 1268 – 1281, (2000).
15. El Gharbi, N., Absi, R., and Benzaoui, A., J. Appl. Fluid Mech., **5**, 63–70, (2012).
16. ANSYS INC (2013). ANSYS Fluent User's Guide, Release 15.0. 2620 p.
17. Zhang W., Markfort C.D., and Fernando P., Experiments in fluids, **52**(5), 1219-1235, (2012).

Effect of Scrubbing Liquid height on the performance of Self Priming Venturi Scrubber

Ashish M. Umbarkar[1,*], *Dinesh N. Kamble*[1], *Ashish Shelke*[2]
[1] Dept. of Mechanical Engineering, VIIT, Pune (India)
[2] University of Warwick, Coventry, (UK)

Abstract. The Containment Filtered Venting System (CFVS) is used to scrub the Particle matter & gaseous pollutants that are liberated during the severe accidents in nuclear power plants (NPP). This research work aims to an experimental study of the fluid dynamics of gas-liquid interaction of the circular cross-sectional self- priming VS at different experimental inputs. The performance of self-priming VS is indicated by the Pressure drop, Liquid to gas ratio & Collection efficiency. The effect of Scrubbing liquid height on these performance indicators is studied at different gas flow rates. The experimental result shows that the scrubbing liquid height greatly influences the performance of self-priming VS. It improves the Liquid to Gas ratio which is helpful to improve the collection efficiency.

1 Introduction

Post-Fukushima-Daiichi accident updating of passive safety devices was suggested as an urgent need to avoid the release of radioactive pollutants to the environment. By considering this updation the Filtered Containment Venting System (FCVS) used to remove the radioactive gaseous pollutants & PM from the gas flow which coming out from the NPP containment. The venturi scrubber is the main important part of FCVS system. It consists of three main segments as shown in fig. 1: convergent part due to its decreasing cross-section gas stream with particles passing through it gets accelerated & the velocity of the gas increases and pressure gets reduced. After that inside the throat, part gas achieves high velocity where the scrubbing liquid is get entrained inside the throat through orifices and due high gas velocity at this section the entrained scrubbing liquid gets atomized into small droplets.
The particle or gaseous pollutants get interact with these smaller droplets and due to the inertial impaction mechanism, the particles get stick to these droplets and get scrubbed[1]. And the divergent part where the increasing cross-section recovers the pressure of gas stream and reduces the high gas velocities. Due to a small cross-sectional area at the throat gas velocity increases and the pressure gets reduced heavily. Thus the pressure drop in the throat and the scrubbing liquid head outside of the throat cause the self-priming action of liquid at the throat through the orifices.
The study of the self-priming operation of venturi scrubber done previously by M. Lehner[2], Ali, et al.[3], Naoki HORIGUCHI et.al.[4] & Paridhi Goel et.al [5]. The conclusions showed that the rise in gas flow rate reduces the entrainment of scrubbing liquid inside the throat at constant scrubbing liquid height inside the tank. It was also reported that the increase in scrubbing liquid height increase the liquid entrainment at the throat at a constant gas flow rate.

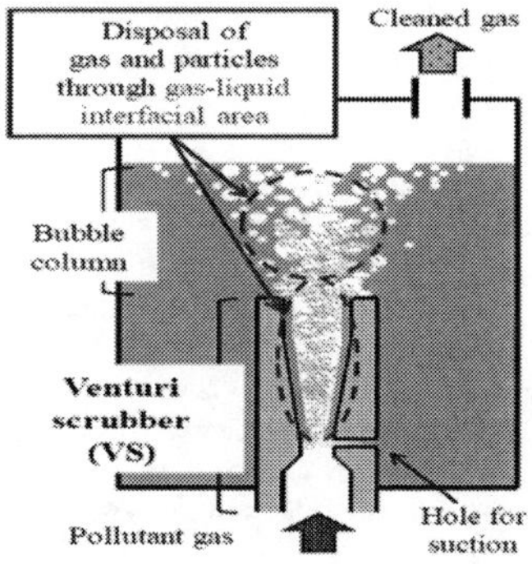

Fig. 1. Self Priming Venturi Scrubber[4]

The pressure drop across VS and collection efficiency are the important performance parameters of Self-priming venturi scrubber[6]. Also, the liquid to gas ratio (L/G ratio) is also reported as the factor which affects the collection efficiency[7]. Hence all these parameters get largely influenced by scrubbing liquid height. The present research work reports the analysis of the pressure drop across the self-priming venturi scrubber, L/G ratio & collection efficiency are the functions of the liquid height of scrubbing liquid inside the tank.

*Corresponding author: **ashishumbarkar9850@gmail.com**

2 Experimental Setup

All the experimental readings were taken on the circular venturi scrubber as shown in fig. 2 which was submerged inside the square cross-section scrubber tank of dimensions (300X300X1600mm). The scrubber tank was filled at three different liquid heights of 300mm, 550mm & 800mm. The compressed air at three different pressure i.e. 2 bar, 3bar, 4bar introduced at the inlet of the self-priming VS.

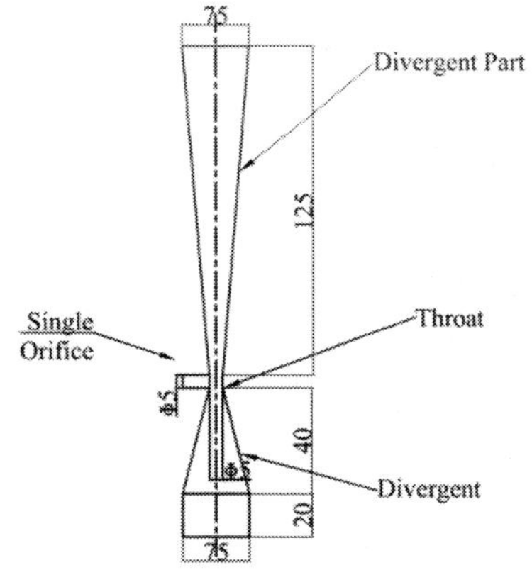

Fig. 2. Dimensions of Self-Priming VS

The fly ash of particle size (0.5 to 100 micron) as pollutants introduced in the compressed air stream before the inlet of venturi scrubber. For the measurement of pressure drop across venturi scrubber the use of pressure measuring devices has been done. To measure the airflow rate the Airflow rotameter was used to measure the airflow rate; also flow control valve provided for changing the flow rate of air. The measurement of the L/G ratio incorporated by using the secondary tank. Where the throat orifice hole was connected with the secondary tank and during measurement secondary and primary tank liquid heights kept at the same level. Other components of the setup are as shown in fig. 3

3 Results and Discussions

In this work, the effect of scrubbing liquid height above the throat on the pressure drop, L/G ratio & collection efficiency has been analyzed.

3.1 Effect of scrubbing liquid height on the pressure drop

The drop of pressure across the venturi scrubber has been calculated by differencing the pressure at the inlet of the convergent section and pressure at the outlet of the divergent section of venturi scrubber. The readings were taken at different liquid heights at 6 different pressure inlet conditions i.e. from 1 to 6 bar. Wherein pressure measurement at the outlet of VS calculated as follows: Pressure at the outlet of venturi = Atmospheric Pressure + Liquid Pressure head

Table 1. Pressure drop across venturi scrubber at different liquid heights

Pressure at Inlet	Pressure Drop in a bar at the different liquid height		
	300mm	550mm	800mm
2	0.75732	0.7328	0.70827
3	1.65732	1.6328	1.60827
4	2.65732	2.6328	2.60827
5	3.65732	3.6328	3.60827
6	4.55732	4.5328	4.50827

Table 1. Reveals that the increase in scrubbing liquid height decreases the Drop of pressure across the venturi scrubber.

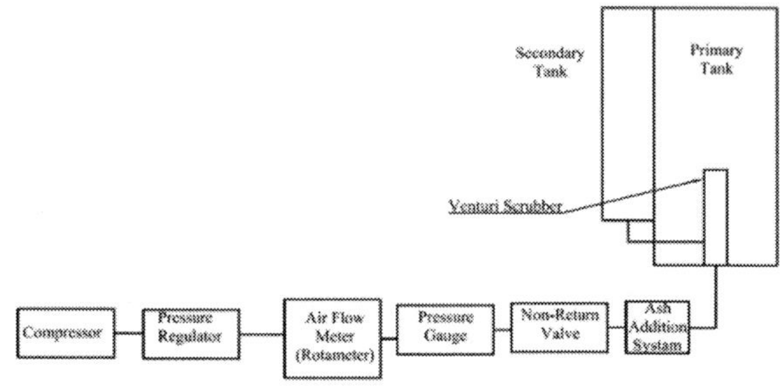

Fig. 3. Experimental Setup

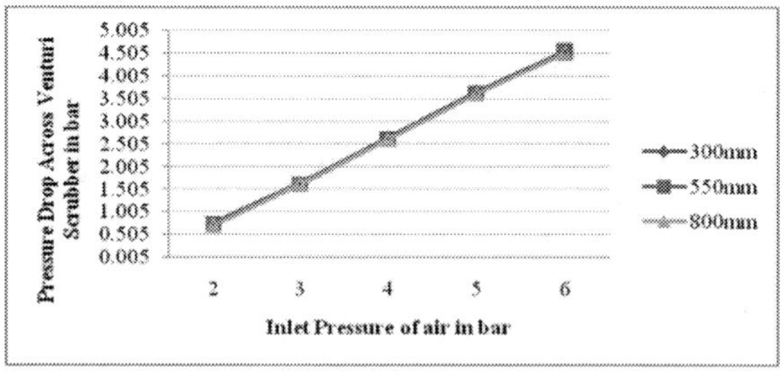

Fig. 4. Graph of Inlet Pressure vs Pressure drop at different scrubbing liquid heights

The fig. 4 shows that an increase in the inlet pressure to the venturi scrubber causes the increase in pressure drop across VS. This is caused due to the increase in throat velocity at the expense of pressure of air and the divergent section was unable to recover that loss of pressure at the end of venturi scrubber. Also, the pressure drop was slightly affected by the scrubbing liquid height it is found that pressure drop decreases with an increase in scrubbing liquid height.

3.2 Effect of scrubbing liquid height on Liquid to Gas ratio (L/G Ratio)

As described by M. Lehner[2] the venturi scrubber having two different types based on the scrubbing liquid supplied at the throat through orifice: forced feed injection, and self-priming VS. In the case of the force-feed injection method, the scrubbing liquid is injected through the orifice external source i.e. pumps, so that the liquid flow rate is not dependent on the gas flow rate. Whereas, in the case of self-priming venturi scrubber, the rate of scrubbing liquid flow rate is totally dependent on the gas flow rate because the liquid entrainment at the throat is totally dependent on the pressure drop at the throat section.

The liquid flow rate at the throat part is measured at different gas flow rates by using those readings the Liquid to gas ratio in kg/kg found out at different inlet pressure 2bar, 3bar, 4bar. Figure 5, 6,&7 depicts that with the increase in the gas flow rate at constant liquid height decreases liquid flow rate and due to this L/G ratio also decreased. But the increase in the liquid height of the tank at constant gas flow rate causes an increase in liquid flow rate which is, in turn, increase the L/G ratio. At high gas flow rates at the throat, the dynamic pressure of gas increases and this reduces the entrainment of liquid inside the throat through the orifice. But as the scrubbing liquid height increases, it increases the hydrostatic head required for suction of liquid at the throat and more amount of liquid gets sucked inside the throat.

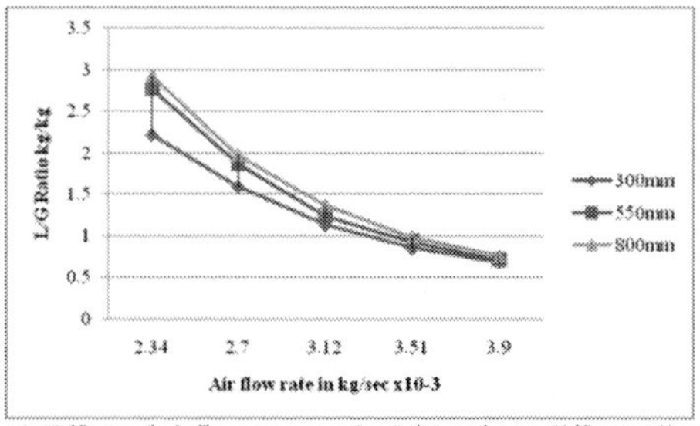

Fig. 5. Effect of airflow rate on the L/G ratio at different liquid heights at 2 bar

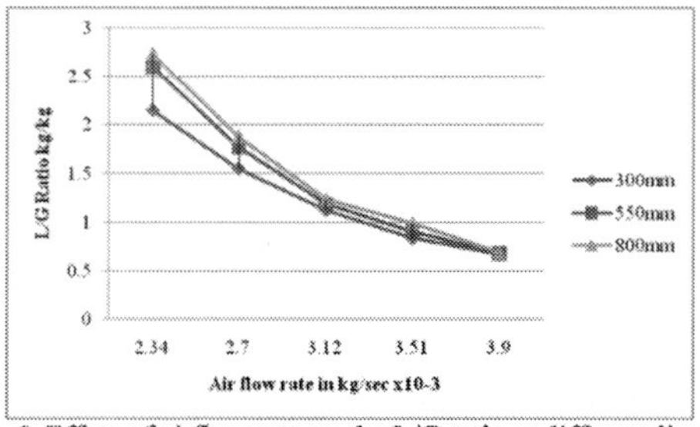

Fig. 6. Effect of airflow rate on the L/G ratio at different liquid heights at 3 bar

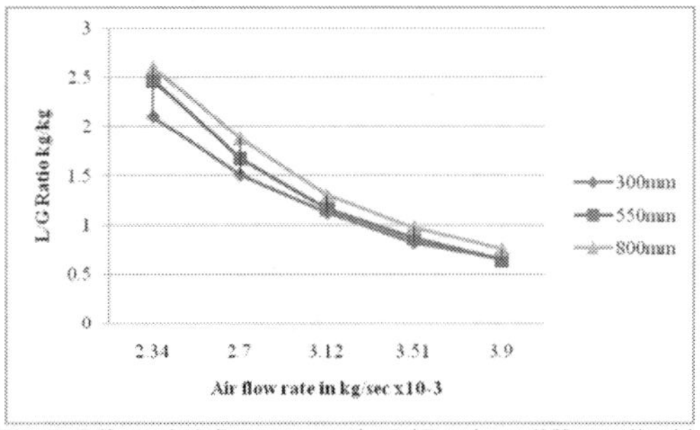

Fig. 7. Effect of airflow rate on the L/G ratio at different liquid heights at 4 bar

Table 2. Results of L/G ratio at different experimental conditions

Head of Water Level	Pressure in bar	Airflow rate in kg/sec X 10^{-3}				
		2.34	2.7	3.12	3.51	3.9
		L/G Ratio kg/kg				
30	2	2.218	1.59	1.133	0.855	0.684
	3	2.148	1.548	1.125	0.839	0.672
	4	2.103	1.507	1.125	0.824	0.654
55	2	2.772	1.87	1.242	0.933	0.712
	3	2.603	1.77	1.184	0.905	0.679
	4	2.475	1.675	1.161	0.867	0.641
80	2	2.925	1.973	1.364	0.983	0.758
	3	2.733	1.88	1.23	0.985	0.688
	4	2.602	1.883	1.299	0.974	0.755

Table No. 2 shows that the L/G ratio is gets affected by the inlet pressure of the gas. As the pressure at the inlet of venturi scrubber increases, it lowers the entrainment of liquid at the throat due to the increase in the throat pressure. Hence the decrease in entrainment reduces the L/G ratio.

3.3 Effect of Scrubbing liquid height on Collection efficiency

Paridhi et.al [5] studied the effect of scrubbing liquid height on the collection efficiency of the self-priming VS and found that as the scrubbing liquid height increases the collection efficiency also increases.

In the present work, the measured quantity of 100mg fly ash particles is introduced at the start of the venturi scrubber in the air stream. During the scrubbing process, this fly ash particle flows with airstream at the throat part of the venturi scrubber. Due to the low pressure at the throat part scrubbing liquid gets entrained inside the throat through an orifice. This entrained liquid gets atomized into tiny droplets due to the high velocity of the gas stream which breaks the liquid. These droplets provide the area for capturing the fly ash particles by the inertial impaction mechanism and scrub the pollutants from the air stream. After leaving the end of the venturi scrubber the air stream passes through the scrubbing liquid column in the form of bubbles and gets interacted with the scrubbing liquid in it. If the height of the liquid in the tank is maximum the interaction time for fly ash particle and scrubbing liquid increases and retention of fly ash in the scrubbing liquid increase. Then this retained fly ash gets settled at the bottom of the tank and the mass of it is measured for calculation of collection efficiency, hence higher collection efficiency is expected when higher scrubbing liquid height is used.

Table 3. Results of collection efficiency at different scrubbing liquid heights at the gas flow rate of 0.00234kg/sec

Water Head (mm) H	Fly ash Particles introduced (gm) M1	Fly ash Particles Collected (gm) M2	% Efficiency (M2/M1)*100
300	100	95.02	95.02
550	100	96.89	96.89
800	100	97.71	97.71

Table No. 3 shows that the increase in the height of scrubbing liquid inside the tank increases the collection efficiency at a constant gas flow rate of 0.00234kg/sec.

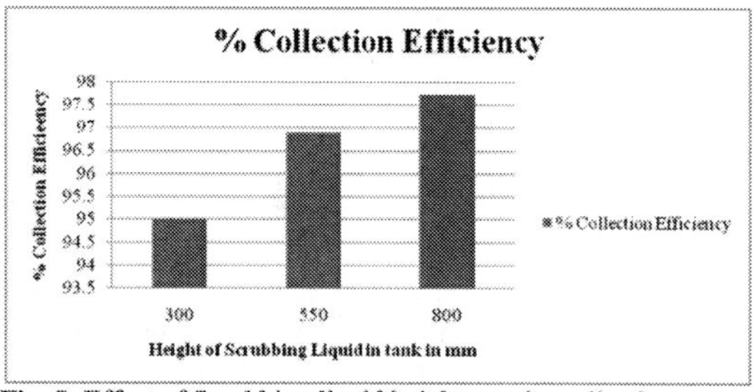

Fig. 8. Effect of Scrubbing liquid heights on the collection efficiency at 0.00243kg/sec gas flow rate

4 Conclusion

In this research experimental study of the effect of scrubbing liquid height in the tank on the pressure drop, L/G ratio and collection efficiency in self-priming venturi scrubber studied. The following conclusions are made from the present work;

1. The pressure drop across VS is slightly affected by the scrubbing liquid height, as it increases the pressure drop across venturi gets slightly reduced.
2. The liquid to gas ratio gets increased with increasing liquid height but gets reduced with increasing gas flow rate.
3. The collection efficiency of self-priming venturi scrubber is getting improved with increasing scrubbing liquid height in the tank.
4. It is also observed that the liquid entrainment at the throat is get reduced with increasing gas flow rate, thus it proves that the self-priming venturi scrubber works in a self-regulating way.

References

1. R. H. Boll, Ind. Eng. Chem. Fundam., **12**, 40, (1973)
2. M. Lehner, Aerosol Sci. Technol., **28**, 389, (1998)
3. M. Ali, Y. A. N. Changqi, S. U. N. Zhongning, G. U. Haifeng, and W. Junlong, Nucl. Eng. Technol., **45**, 203–210, (2013)
4. N. HORIGUCHI, H. YOSHIDA, A. KANEKO, and Y. ABE, Mech. Eng. J., **1**, TEP0026, (2014)
5. P. Goel, A. Moharana, and A. K. Nayak, Nucl. Eng. Des., **327**, 92, (2018)
6. Guerra, J. R. Coury, R. Be, and J. A. S. Gonc, Ind. Eng. Chem. Res., **51**, 8049, (2012)
7. V. Sekar, A. W. Gnyp, and C. C. S. Pierre, Ind. Eng. Chem. Fundam., **23**, 303, (1984)

Poly (N-Methyl Aniline)-Li Nanocomposite as an Electrolyte for Rechargeable Battery: In-situ Recipe

Nishigandh Pande[*1], *Adinath Jambhale*[1], *Dipika Jaspal*[2], *Jalinder Ambekar*[3], *Himanshu Patil*[1]
[1]School of Mechatronics Engineering, Symbiosis Skills and Professional University, Village – Kiwale, Adjoining Pune Mumbai Expressway, Pune - 412101, Maharashtra, India.
[2]Symbiosis Institute of Technology (SIT), Symbiosis International (Deemed University)(SIU), Gram: Lavale, Tal: Mulshi, Pune-412115, Maharashtra, India
[3]Centre for Materials for Electronics Technology (C-MET) Panchawati Road, Pashan, Pune- 411008, Maharashtra, India.

Abstract. The synthesis of cost-effective and safe polymeric nanocomposite materials has been a subject of interest and study for material science researchers. Poly (N-Methyl Aniline) –Li nanocomposite has been synthesized by a one-pot in-situ method and has been explored as an electrolyte in the battery. Poly (N-Methyl Aniline)-Li nanocomposite prepared, has been characterized by UV-visible, FTIR, FE-SEM, X-ray diffraction techniques. A mechanism of interaction of lithium-ion with nitrogen at the imine site has been proposed. The charge-discharge process of poly (N-Methyl Aniline) –Li nanocomposite, when used in the battery, has been discussed.

Keywords: *PNMA-Li nanocomposite, morphology, FESEM, battery*

1. Introduction

Long term performance and rechargeable ability play a vital role in lithium-ion battery technology. Nowadays, the Li^+ ion battery is a crucial component of the automotive industry [1]. Organic-inorganic composite layers have been developed to obtain a contemporary electrolytic material with enhanced properties [2, 3]. For the past few years, researchers are trying to revise the intercalation of conducting polymeric material by adding inorganic nanoscale materials as hosts. Substituted polyaniline has been chosen as a conducting polymeric material in the present research paper due to its optical and electrical properties [4-6].
The present article reports a facile synthesis method for developing poly(N-methyl aniline)-Li(PNMA-Li) nanocomposite as an electrolyte for battery, prepared by the in-situ green route technique, using the extract of *Plumeriarubra* flowers.[1]

***Corresponding author:** nishigandh.pande@gmail.com

2. Experimental Section

2.1 Materials and methods

Lithium ions were prepared in the nano form using *Plumeriarubra* flowers, after which the preparation of the poly(N-methyl aniline)-Li nanocomposite was carried out.

The extract of Plumeriarubra flowers were used to prepared lithium ion nanoparticles by thoroughly washing the petals with deionized water and extracting using soxhlet apparatus.

2.2 Synthesis of Lithium nanoparticles

20 mL of extract was used to synthesize nano lithium ions. The extract, along with 10 mL of 10% DL tartaric acid, was blended with 0.025 M $LiNO_3$in presence of ultrasonic waves at 28°C for a half-hour, to give fain pink colour of lithium nanoscalematerial [7].

2.3 Synthesis of Poly (N-methyl aniline)-Li nanocomposite

10 mL of Li+ ion nanomaterial was taken in 200 mL of a round bottom flask. N-methyl aniline (monomer) was dissolved in DL tartaric acid (acting as doping agent), and added to the container containing lithium nano ions with constant stirring. Ammonium

persulphate was added as an oxidizing reagent in the resultant mixture at 0 to 5 °C with continuous agitation. After 20 minutes of mixing the reaction mixture showed a dark emerald color indicative of the Poly (N-methyl aniline)-Li nanocomposite formed as the final yield.

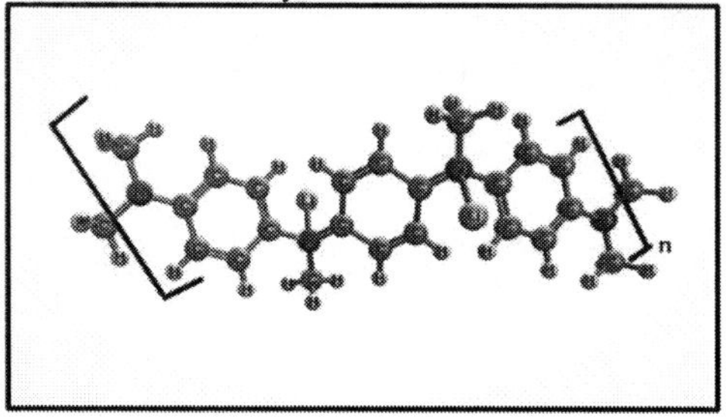

Fig.1. Structure of Poly (N- methyl aniline)-Li Nanocomposite

3. Characterization

The nanocomposite material synthesized was characterized by UV-visible spectroscopy. The absorption spectra were observed in the range of 200 to 800 nm wavelength. The nanocomposite was analyzed by X-ray diffraction with monochromatized Cu Ka (λ 1.5418 Å). The size distribution and morphology of nanocomposite were obtained by FE-SEM using JEOL-JSM-S4800. FTIR (Fourier transform infrared) spectral studies were carried out using Nicolet 67 FTIR spectrometer with KBr pellets used for analysis.

4. Result and Discussion

4.1 UV-Visible Spectroscopy

The nanocomposite showed absorption peaks at 265, 370, 450, and 800 nm. The absorption band of acidic range was observed at 265 nm. These absorption band at 370 nm emanated from π-π* electron transition for the benzenoid present in synthesized nanocomposite.The absorption peak at 450 showed formation of N=Q=N (salt formed due to doping) in nanocomposite. The absorption spectra moving towards higher level (800 nm) indicating the emeraldine state of nanocomposite [8].

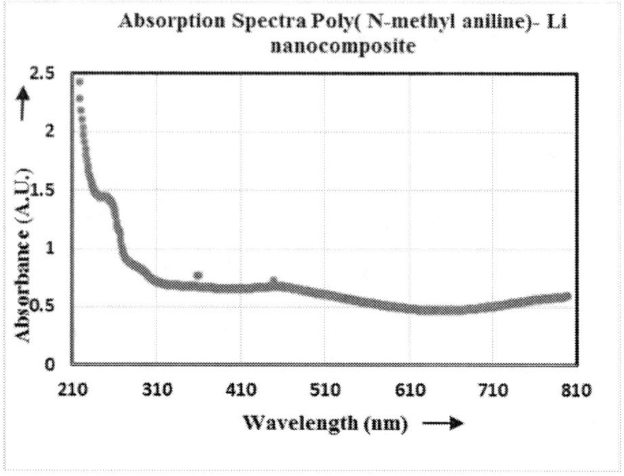

Fig.2. Poly (N- methyl aniline)-Li nanocomposite : UV-visible spectra

4.2 FTIR

The FTIR spectrum of Poly (N-methyl aniline)/Li nanocomposite has been portrayed in Figure 2. The spectrum shows a vibration band at ~3600-3000 cm^{-1} for –N-H stretching and 3000- 2800 cm^{-1} for C-H stretching, due to Poly (N-methyl aniline). Prominent peaks obtained at 1524 cm^{-1} and 1659 cm^{-1} were related to stretching vibrations due to C=N of the quinoid and benzenoid ring the C=C stretching, in the poly (N-methyl aniline)/Li nanocomposite. A stretching vibration at 1511 cm^{-1} was observed for the aromatic ring in Poly (N-methyl aniline)/Li nanocomposite. Signals at 1264 and 1131 cm^{-1} resulted from C-N stretching [4, 5]. The vibration at 551 $cm^{-1}$was attributed to the linear Lithium [9].

4.3 X-ray Diffraction

Figure 3 shows a broad peak at 2θ=22.03°, indicating a semi-crystalline polymeric material in the Poly (N-methyl aniline)/Li nanocomposite. Remaining peaks for Lithium were observed at 2θ = 18.96, 30.46, 35.44, 36.26, 40.72, 43.66, and 47.86, which were broad peaks indicative of the size of Lithium particles in nano sizes. The particle sizes of Lithium nanoparticles, as calculated using Debye Scherrer's equation (Eqn. 1), were ~ 50 nm[10].

$$D = \frac{0.9\,\lambda}{\beta.Cos\theta} \quad (1)$$

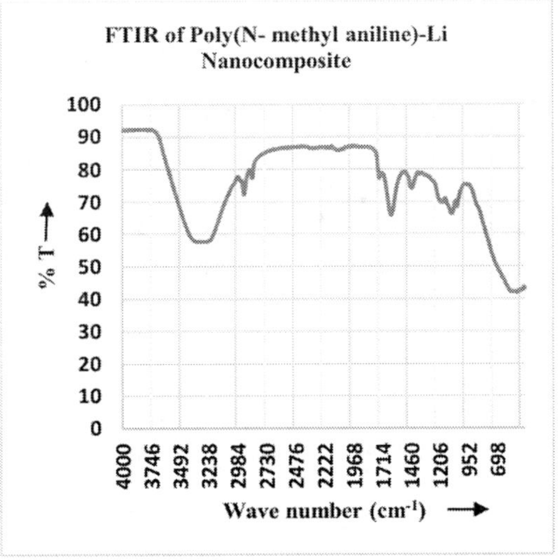

Fig.3. Poly (N- methyl aniline)-Li nanocomposite: FTIR spectra

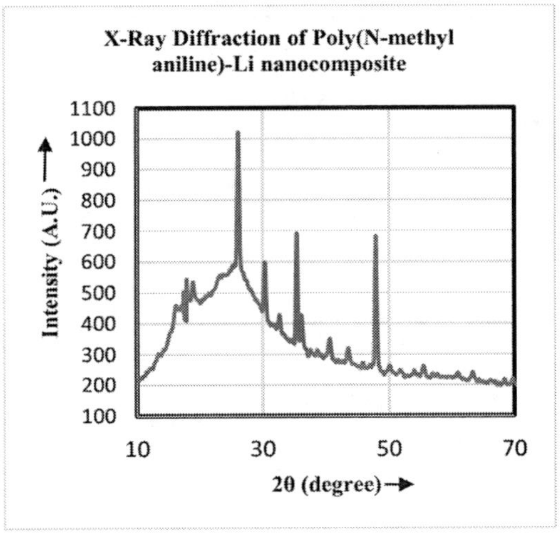

Fig.4. XRD of Poly (N- methyl aniline)-Li nanocomposite

4.4 FE-SEM (First Emission-Scanning Electron Microscopy)

The photograph of the Poly (N-methyl aniline)-Li nanocomposite (Figure 5) shows the morphology of the nanofibers and the polymer material as spherical shaped with agglomerated Lithium ions. The average diameter of the nanofibers (star-shaped) was in the range of 300±50 nm [11]. The agglomerated Lithium nanoscale materials show a spherical shape with a size of ~64 nm.

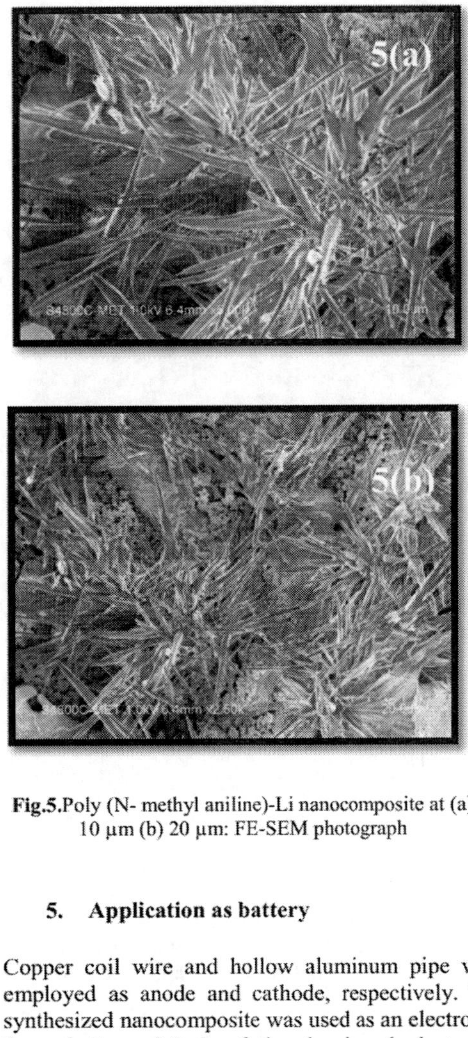

Fig.5. Poly (N- methyl aniline)-Li nanocomposite at (a) 10 μm (b) 20 μm: FE-SEM photograph

5. Application as battery

Copper coil wire and hollow aluminum pipe were employed as anode and cathode, respectively. The synthesized nanocomposite was used as an electrolyte for a battery. 2.5mL of the developed electrolyte (nanocomposite) was analyzed for a single cell battery. After the connection of two cells in series, current and time measurements were carried out using measuring equipment (multimeter). After contact in parallel, the battery showed a potential of 7.08 V, which remained the same even after 100 cycles. RC circuit diagram for charging- discharging is shown in figure 6.

The energy density of the developed battery was found according to Equation (2).

$$E = \int_{t1}^{t2} I.V.(t)dt (2)$$

Where,

E= Energy density

I= Discharge current

V= Cell Voltage

t_1 = Initial time of discharge

t_2= End time of discharge

The power density was calculated according to Equation (3).

$$P = \frac{E}{(t2-t1)} \quad (3)$$

The power density of the prepared battery was 119.233 W/m³, and was calculated using equation (2) and (3) [12].

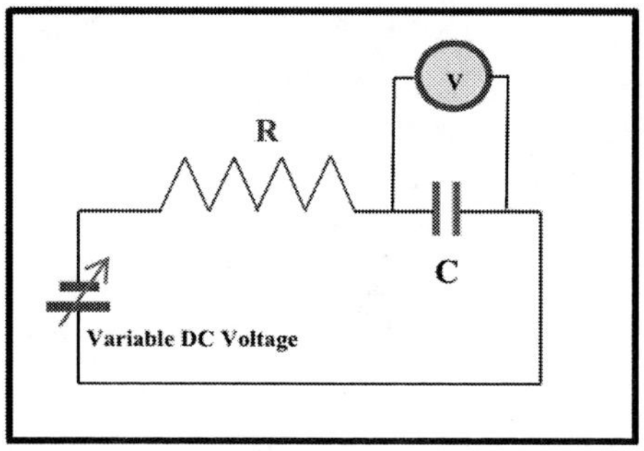

Fig.6. RC circuit for charging & discharging of battery

Figure 8 shows the discharge tests of battery for resistances 5, 10, and 27 KΩ, which indicates decreasing curves pattern for the higher resistance.

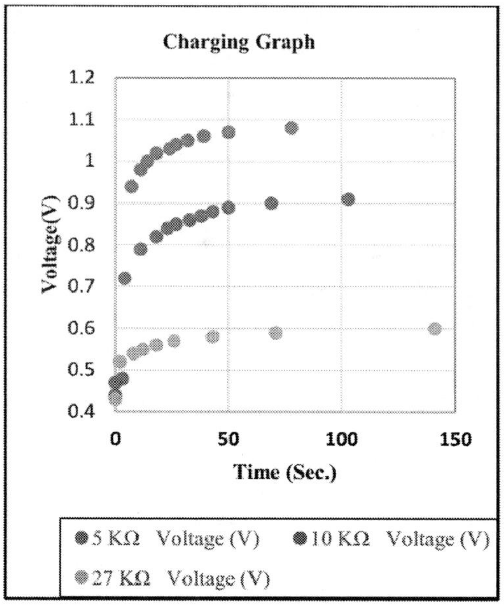

Fig.7. Charging graph of battery

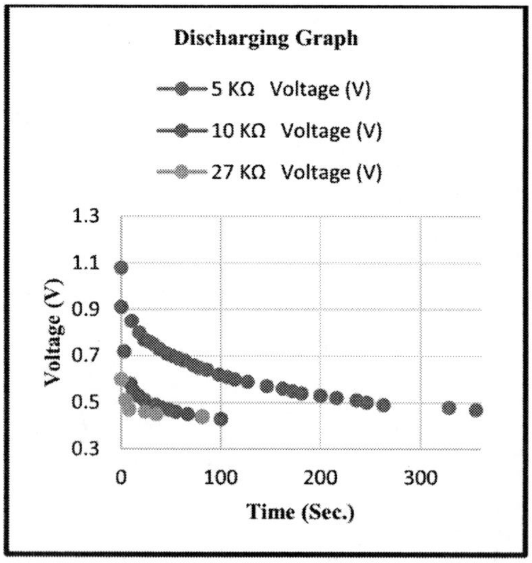

Fig.8. Discharge capacity of the battery

Fig.9. Photograph of (Green color)LED light glow by using battery

Conclusion

The developed nanocomposite material has given appreciable and outstanding results as an electrolyte for the battery. The networked structure of synthesized nanocomposite, with intermediate morphology of nanofibers (poly N-methyl aniline) and spheres (Lithium), was studied using FE-SEM analysis. The power density of battery was observed to be 119.233 W/m^{3}after 100 cycles of charging and discharging. The developed material, as an electrolyte, has shown immense potential to be a promising material for usage in batteries.

References

1. X. Han,M. Ouyang,L. Lu, J. Li,Y. ZhengZ. Li, J. power sources. **251**,1 (2014)
2. L.M. Bronstein, C. Joo, R. Karlinsey, A.Ryder, J.W. Zwanziger,Chem. Mater. , **13**, 10(2001)
3. E.M. Masoud, A.A. El-Bellihi, W.A. Bayoumy, M.A. Mousa, J. Alloy Compd.**575** (2013)
4. N.S.Pande, DipikaJaspal, J. Ambekar, Int. J. Nanosci.**16**,3 (2017)
5. N.Pande, D. Jaspal, A. Malviya, A. Warke, Inorg. Nano-Met. Chem.**47**,7(2017)
6. N.S. Pande, DipikaJaspal, A. Warke, V.V. Chabukswar, J. polym. Res. 10,4 (2016)
7. N.S. Pande, A. Jambhale, D. Jaspal, J. Ambekar, M. Kulkarni, Inorg. Nano-Met. Chem. 2470-1564 (2019)
8. E.I. Iwuoha, S.E. Mavundla, V.S. Somerset, L.F. Petrik, M.J. Klink, M. Sekota, P. Bakers, Microchim.Acta.**155**, (2006)
9. G. Aravind, M. Raghasudha, D. Ravinder, A. Gaffoor, V. Nathanial, J. Nanostruct. Chem. **5**, (2015)
10. N. Pande, D. Jaspal, J. Ambekar, V.P. Jayachandran, J. Chem. Pharm. **6** (2014)
11. B. Massoumi, N. Aalia, M. Jaymand, RSC Adv., **5** (2015)
12. A. Byeon ,A.M. Glushenkov ,B. Anasori , P. Urbankowski , J. Li , B.W. Byles, B. Blake, K.L. Van Aken , S. Kota, E. Pomerantseva, J.W. Lee , Y. Chen , Y. Gogotsi, J. Power Sources (2016)

Experimental Analysis of different shaped ribs on heat transfer and fluid flow characteristics

Sandeep S Kore[1]*, *Sunil V. Dingare*[2], *Satish Chinchanikar*[1], *Pravin Hujare*[1] *and Ashok Mache*[1]
[1] Vishwakarma Institute of Information Technology, Pune (India)
[2] MIT ADT University, Pune(India)

Abstract. There are number of applications such as gas turbines, solar air heating, electronics cooling and heat exchangers, where internal cooling passage is observed. For heat transfer augmentation inside these cooling passages different techniques are used like dimpled surface, wings and ribs. Ribs are used in most of the devices for internal cooling. The ribs disturb the boundary layer and increase the turbulent kinetic energy which enhances the heat transfer rate. Most of the researchers concentrate on square and rectangular shaped ribs. The cross section of the rib plays important role in the production of flow field. The shape of ribs affects on boundary layer separation, attachment and hot spots created. In the present paper heat transfer and fluid flow characteristics from rib roughened rectangular duct with different shapes of ribs were investigated. The experimental set up consists of rectangular channel of aspect ratio 4. The pitch to width ratio was varied as 5, 7.5 and 10 respectively. The Reynolds number was varied as 6000 to 30000. The ribs used for the investigations were square, house and boot shaped. From the investigations it is observed that boot shaped rib is having higher thermal performance than other two geometries

1 Introduction

The output in terms of efficiency & power of any thermal device increases by increasing turbine inlet temperature. In gas turbines efficiency can be improved by increasing turbine inlet temperature. The present turbine blades are exposed to temperature near about 1800K, which is higher than their melting temperature. To keep the temperature within working limit internal vanes and blades need to be cooled to lower the turbine blade temperature. For using any heat transfer enhancement technique it is important to know about boundary layer formation and higher temperature gradient spots. Otherwise higher temperature gradient cause thermal stresses and may cause decrease in life of turbine blade. The turbine blades and vanes are cooled by internal passage in which different turbulators are used to augment the heat transfer.

There are a wide range of cooling methods available and used in the gas turbine industry. Two categories exist when it comes to cooling methods, internal and external. An example of external cooling is the use of pushing air through blades and out, creating a cooler "film" over the blade and an example for internal cooling include passing relatively cooler air through the blade cavities. There are several specific methods of cooling, have been studied and implemented.

These methods take advantage of manufactured cavities, holes, etc. on the blade to improve heat transfer from the blade. With better heat transfer, higher inlet temperatures are able to be reached and thereby improving thermal efficiency and performance.

Film cooling uses coolant to create a boundary layer over the airfoil. On the airfoil itself, there are holes where the coolant, after entering the blade, is pushed out of these holes and creates this layer which acts as a barrier. This layer protects the blade and keeps it cooler than without the film.

Researchers have investigated internal cooling passage using various heat transfer augmentation techniques likes dimpled surface, twisted tape, flow blockage and ribs[1,2]. The various ribs investigated up till know are of square, isosceles triangular, reverse right-angle triangular, right angle triangular, right-angle trapezoidal, isosceles trapezoidal, and semicircular rib shapes [3]. But rib shapes like boot shape, house shape and pentagonal shape [3]were studied numerically. In present experimental work house shape and boot shaped ribs are investigated experimentally.

2 Experimental setup details

The experimental setup was designed keeping in view the validation of the smooth channels with considering results of Dittus–Boelter and modified Blasius correlations for smooth channels as shown in Fig.1

The experimental arrangement consists of 1 HP capacity blower to get the required flow rate. The test section is

* Corresponding author: sandeep.kore@viit.ac.in

manufactured from 10mm epoxy resin. The dimensions of the test duct are 1150mm x 100mm x 25 mm. The inlet and out sections are provided for hydro dynamically fully developed flow and to avoid eddies at outlet. To measure pressure drop across the test section two pressure taps are used. Fig.2 shows test section.

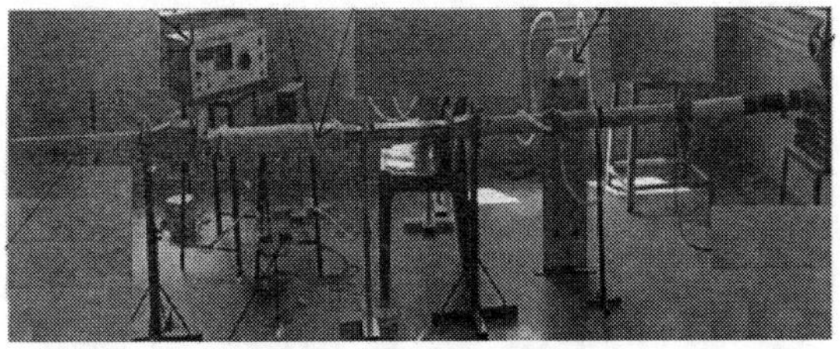

Fig.1 Experimental set up

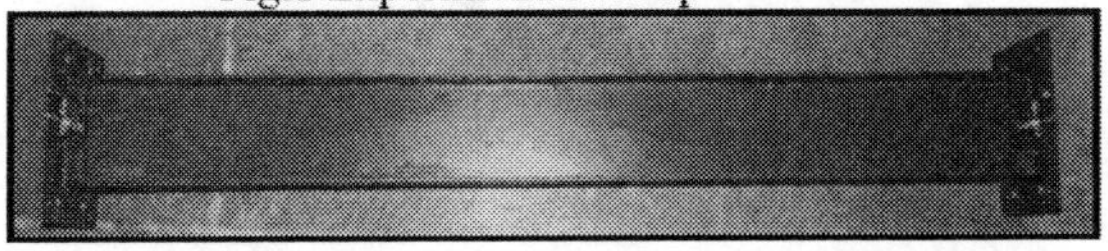

Fig.2 Test duct for experimental investigations

The test plate is of dimension 60 mm X 108 mm x 6mm. On the test plate ribs are pasted using metal glue. There are six 'T' types thermocouples are used to measure the temperature of test plate and two are used to measure the temperature of the air. The plate type of heater is used for constant heat flux boundary condition. Figure 3, shows the ribs used for experimental investigations. It includes square shaped, house shaped and boot shaped rib. The experimental analysis is carried out at 20mm, 40mm and 50mm pitch.

Fig.3. Schematics of ribs used in the present study

2.1 Data reduction

To measure the mass flow rate orifice meter is used connected to U tube manometer with water as manometric fluid. The following relation is used to find the mass flow rate.

$$\dot{m} = C_D a_0 \times \frac{\rho_{air} \times \sqrt{2gH_a}}{\sqrt{1-\beta^4}} \quad (kg/s) \quad (1)$$

Where,

C_D= Coefficient of discharge for orifice,
a_0= Cross sectional area of orifice (m^2),
β =d/D,diameter of pipe/ diameter of orifice =0.8,
g = Acceleration due to gravity (m/s^2),
H_a= Height of air column (m).

Heat transfer to air can be calculated using following relation,

$$Q = mC_p(T_{ai} - T_{ae})(Watt) \quad (2)$$

Where,

T_{ai}= Air temp at the inlet of the duct (°C),
T_{ae}= Air temp at the exit of the duct (°C),
m = Mass flow rate of air (kg/s^2),
Fig.1 Experimental set up
C_p= Specific heat of air (KJ/kg K),
Q = Convective heat transfer of air (W).

The heat transfer coefficient for the test section

$$h = \frac{Q}{A_s \cdot (T_s - T_{bm})}\left(\frac{W}{m^2K}\right) \quad (3)$$

Where

h = Convective heat transfer coefficient (W/m^2K),
As= Projected surface area of test surface,(m^2),
Ts=Average temperature of the test surface,
T_{bm}= Bulk mean temperature of air in the duct = (Tai + Tae / 2) (°C).

Nusselt number as,

$$Nu = \frac{hD_h}{K_{air}} \quad (4)$$

Where,

Nu = Average Nusselt number of the test surface,
D_h= Hydraulic diameter of the rectangular duct (m)=0.04,
K_{air}= Thermal conductivity of air (W/mK).

The pressure drop across the test duct is determined by,

$$\Delta P = (\rho_w * g * l_1 * \sin 1.25 * 13.5) \quad (5)$$

Where,

ΔP= Pressure Drop(Pa)
g = Acceleration due to gravity (m/s^2),
l_1 = Manometer reading (manometer scale)

$$f = \frac{(\Delta P)D_h}{2\rho_{air} L V_{air}^2} \quad (6)$$

Where,

f = Friction factor
D_h = Hydraulic diameter of the rectangular duct = 0.04 m.
ρ_{air} = Density of air (kg/m^3)
L = Length of test plate(m)

3 Results and Discussion

3.1 Validation of experimental set up

The experimental set up is validated with Dittus-Boelter correlation for heat transfer as shown in Fig.4.The percentage deviation between theoretical and experimental value is below 6%. For validation of friction factor modified Blasius correlation is used. The percentage deviation is below 9 % as shown in Fig.5

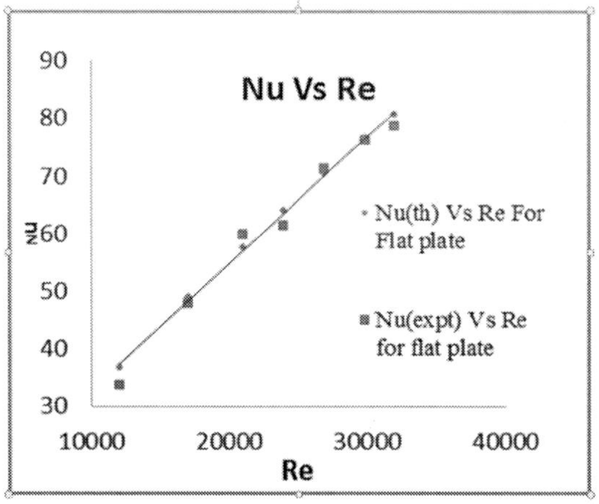

Fig.4. Validation of setup for heat transfer

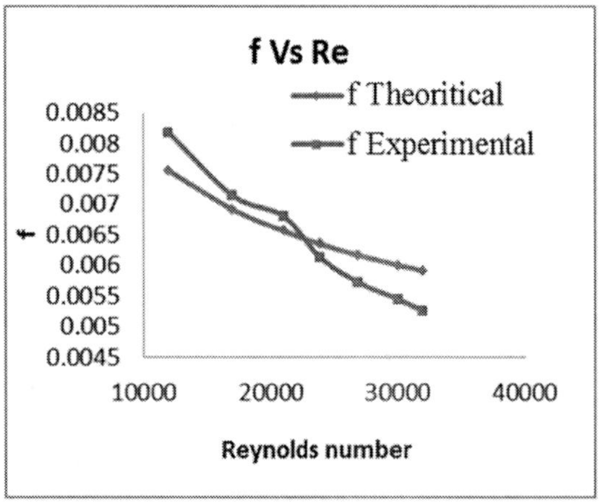

Fig.5. Validation of setup for friction factor

Figure 6, shows thermal performance for various shaped ribs. It is observed that at all Reynolds number thermal performance is higher. As Reynolds number increases thermal performance decreases from Reynolds number 18000 to 22000. This may be due to transitional flow, after this thermal performance increases.

All the ribs with 40mm pitch show highest performance. Compared to square, house shaped and boot shaped, performance of house shaped rib is higher. This is due to house shaped ribs reduces formation of recirculation zone behind and sharp edges causing the mixing of primary and secondary This results in turbulent mixing, which enhances the heat transfer rate. The performance of 20mm pitch is lower due to chocking of the flow.

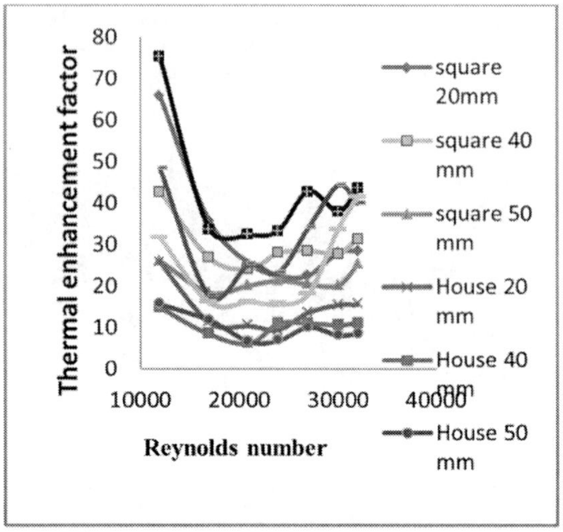

Fig.6. Thermal performance of different ribs

4 Conclusions

The experimental investigation with different shaped ribs is presented in this paper. The pitch between the ribs and different shapes of ribs are the performance parameters. Following conclusions can be drawn from experimental analysis. The results show that performance of boot strap rib is higher than square and house shaped rib.

1. The performance of house shaped rib is lowest among all the ribs. The difference is due to the formation of recirculation zone behind rib. Recirculation zone is lowest for boot strap rib compared to house shaped rib. The recirculation zone is necessary for the ejection and mixing of the fluid.
2. Thermal performance results states that with respect to both Heat transfer and friction loss analysis Ribs with 40 mm pitch shows consistent high performance whereas boot shaped ribs with 20mm pitch showed the highest performance
3. Comparing friction factor House shape ribs showed the lowest pressure drop whereas Boot shaped and Square ribs showed similar results.

References

1. Kore S.S., Yadav R.J. and Sane N.K., Investigations of Effect of dimple depth on heat transfer and fluid flow within rectangular

channel"Procedia Engineering 125, 1110-1117(2015)

2. Kore S.S., Patil Adik and Sane N.K., Experimental investigation of the effect of flow blockages on heat transfer and fluid friction in a round tube using wall-attached circular rings.,**50**,17-22(2019)
3. Mi-Ae Moon, Min-Jung Park, Kwang-Yong Kim Evaluation of heat transfer performances of various rib shapes" International Journal of Heat and Mass Transfer, **71,** 275–284 (2014)
4. Robert Robert Kiml, Sadanari Mochizuki Akira Murata, Effects of Rib Arrangements on Heat Transfer and Flow Behaviour in a Rectangular Rib-Roughened Passage: Application to Cooling of Gas Turbine Blade Trailing Edge, Journal of Heat Transfer,**123**, 675-681(2001)
5. P.R. Chandra and C.R. Alexander , J.C. Han,Heat transfer and friction behaviors in rectangular channels with varying number of ribbed walls, International Journal of Heat and Mass Transfer, **46,** 481–495(2003)
6. Hong-Min and Kwang-Yong Kim,Design optimization of rib-roughened channel to enhance turbulent heat transfer, International Journal of Heat and Mass Transfer, **47,** 5159–5168(2004)
7. R. Kamali and A.R. Binesh, The importance of rib shape effects on the local heat transfer and flow friction characteristics of square ducts with ribbed internal surfaces,International Communications in Heat and Mass Transfer **35,** 1032–1040(2008)
8. Pongjet Promvonge , Chinaruk Thianpong, Thermal performance assessment of turbulent channel flows over different shaped ribs, International Communications in Heat and Mass Transfer, **35** 1327–1334(2008).
9. E. Elsaadawy,H. Mortazavi M. S. Hamed, Turbulence Modeling of Forced Convection Heat Transfer in Two-Dimensional Ribbed Channels" Journal of Electronic Packaging,**130,**031011-1-031011-17(2008).
10. C. Thianpong , T. Chompookham, S. Skullong, P. Promvonge "Thermal characterization of turbulent flow in a channel with isosceles triangular ribs" International Communications in Heat and Mass Transfer **36,** 712–717(2009)
11. Akhilesh P. Rallabandi , Huitao Yang ,Je-Chin Han, Heat Transfer and Pressure Drop Correlations for Square Channels With 45 Deg Ribs at High Reynolds Numbers, Journal of Heat Transfer,**137,** 071703-1 - 071703-10(2009)
12. Teerapat Chompookham, Chinaruk Thianpong, Sutapat Kwankaomeng, Pongjet Promvonge, Heat transfer augmentation in a wedge-ribbed channel using winglet vortex generators, International Communications in Heat and Mass Transfer, **37,** 163–169(2010)
13. Huichun Liu, Jianhua Wang "Numerical investigation on synthetical performances of fluid flow and heat transfer of semiattached rib-channels" International Journal of Heat and Mass Transfer **54,** 575–583(2011)
14. Akhilesh P. Rallabandi ,Nawaf Alkhamis, Je-Chin Han,"Heat Transfer and Pressure Drop measurements for a Square Channel With 45 ° Round-Edged Ribs at High Reynolds Numbers, Journal of Turbo machinery, **133,** 031019-031029(2011).
15. M. I. Mohamed, R. Hoettiba, A. M. Saif, The Effect of the Corrugation Rib Angle of Attack on the Fluid Flow and Heat Transfer Characteristics Inside Corrugated Ribbed Passage, Journal of Heat Transfer, **133**, 081901-10(2011).
16. Linqi Shui, Jianmin Gao, Xiaojun Shi,Jiazeng Liu, Effect of duct aspect ratio on heat transfer and friction in steam-cooled ducts with 60 angled rib Turbulators, Experimental Thermal and Fluid Science, 123–134(2013)
17. Gongnan Xie, Jian Liu, Weihong Zhang, Numerical Prediction of Turbulent Flow and Heat Transfer Enhancement in a Square Passage With Various Truncated Ribs on One Wall, Journal of Heat Transfer, 2014, **136**,011902-1- 011902-11(2014)
18. Mi-Ae Moon, Min-Jung Park, Kwang-Yong Kim, Evaluation of heat transfer performances of various rib shapes" International Journal of Heat and Mass Transfer, **71**,275–284 (2014)

Experimental analysis of magnetic field effects on compressor energy saving cooling system

Rakesh Kumar Sidheshware[1,*], *S Ganesan*[1], *and Virendra K Bhojwani*[2]
[1]Sathyabama Institute of Science and Technology, Chennai 600119, Tamil Nadu, India
[2]MIT School of Engineering, MIT ADT University, Pune 412201, Maharashtra, India

Abstract. A research on different refrigerants with and without influence of the magnetic field is provided in this paper. The magnetic field degree viz is four gauss. The condenser exit row was 3000, 6000, 9000 and 12000 gauss. Magnetic fields at condenser exit lines are found to reduce the consumption of the compressor energy. Through applying magnet fields on R134a, compressor energy savings was observed up to 1.35%, 6.23%, 5.40% and 4.19%, respectively, in gauss levels of 3000, 6000, 9000 and 12000, while R152a savings in energy levels were up 13.10%, 15.59% and 25.86% and R407a saving in power up to 7.3%, 9.19%, 13.05% and 12.40%. The coolant R152a saves energy in contrast to other coolant compressors. As the magnetic field strength increases, a vapor refrigeration system decreases the energy consumption of the compressor.

1 Intoduction

Magnetic field effects on refrigerants are determined by its chemical composition and physical thermal properties such as viscosity, surface tension, thermal conductivity and heat. The highest measurable heat and thermal conductivity of coolants responds little and vice versa to the magnetic field. (Samuel M. Sami, R.J.Kita, 2005). The magnetic effect on the condensing outlet tube increases the cooling effect and reduces the compressor power. The compressor needs more power to remove liquid coolant from the compressor due to incomplete coolant evaporation. When the condenser has been magnetically applied, its cooling properties are modified and the evaporation frequency of liquid coolant is increased. The liquid coolant boiling in the compressor can therefore be reduced. (A Kotb, H.E.Saad2014).
The magnetic field effects are studied and its effect on vapor compression systems performance is analyzed in some modern refrigerant mixtures. (Samuel M. Sami, Shawn Aucoin, 2003). The impact of magnetic field on cooling mix R290 and R600a shows that magnetic field reduces energy cons-umption of the compressor and is dependent on strength of the magnetic field. The growing magnetic strength reduces compressor power and increases COP.
The effect of the magnetic field on the refrigerant mixes depends on the composition and boiling point of the mixture. (Kolandavel MANI, Velappan SELLADURAI, 2008). Vapor compression depends on the efficiency of the coolant. For high COP, liquid thermal conductivity should be high, liquid and molecular weight should be low, and the latentheat vapor density. The vapor compression device is very prone to the normal hot spot, fluid-viscosity, thermal conductivity, density of vapour and critical temperature. (Rotchana Prapainop,K O Suen, 2012).
As the magnet strength increases, the effect of the magnetic field on hydrocarbon, viscosity and surrounding pressure decreases with increasing mass flow rate and thus heat transfer in the condenser and evaporator. (HejunGuo, Zhizhong Liu, Yunchao Chen, Rujie Yao).
The air conditioning unit output under the different strength of the magnetic field is having a positive effect on COP. (Mr. Amod P. Shrotri. Mr. Nikhil S. Mane, Mr. Vishal P. Patil, Mr. Sawan A. Wani, 2015).

2 Apparatus and Procedure

Figure1 demonstrates the experimental set up of a magnet-applied vapor compression system for liquid line. A hermetically screened compressor is used during the experimental setup. The condensing system will change the cooling stage from vapor to liquid once the condenser is installed. Then the use of a capillary tube eliminates high-pressure liquid coolant. The agitator is used for a uniform water temperature in the water tank.

*Corresponding author: **rakesh1987thermal@gmail.com**

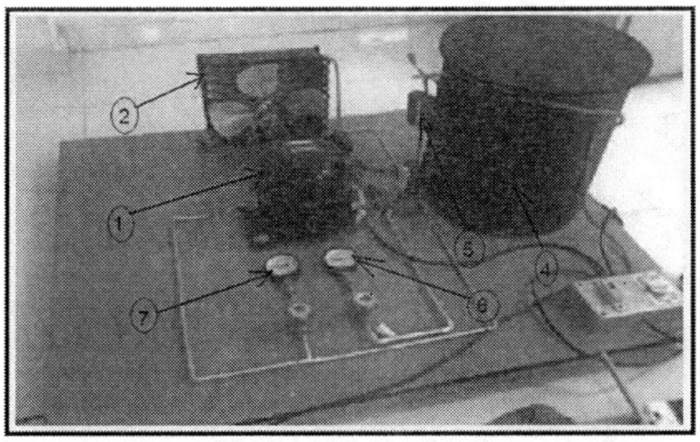

Fig.1. Setup of VCC system (1.compressor, 2.condenser 3.Capillary tube, 4.Evaporator, 5.Magnets, 6.Suction pressure gauge, 7.Discharge pressure gauge)

In this setup four magnetic elements are used with each of 3000 Gauss level. The magnet is located at condenser exit line in increasing order of 3000, 6000, 9000 and 12000 gauss level. For mounting the magnet at exit condenser row, magnet separator is used. Under the same operating environment, different refrigerant types will be checked.

Water is held at 24C for processing and cooled down to 8C in the water tank. At different points of the device, temperature readings are reported at a period of 10 minutes. For various gauzes, the separate reading is taken. For each coolant, the same process is repeated.

3 Results and discussion

The output of the vapor compression process is used as baseline information without adding a magnetic field. On the exit line of the condenser is applied the first magnetic intensity at a rate of 3000 gauss and reading is registered.

To research the effect on compressor energy when the magnetic force is applied on the condenser exit line and the increased magnetic power order.

3.1 Compressor Power for different refrigerants

Due to the incomplete evaporation of coolants, the vapour compression system compressor absorbs additional power. If any liquid refrigerant exits the compressor, the compressor absorbs more liquid coolant with more energy to boil. If magnetic field is applied to the condenser exit line, its vaporization increases the amount of liquid cold and reduces the energy consumption of the compressor. It rising compressor power by increasing the gauss rate.

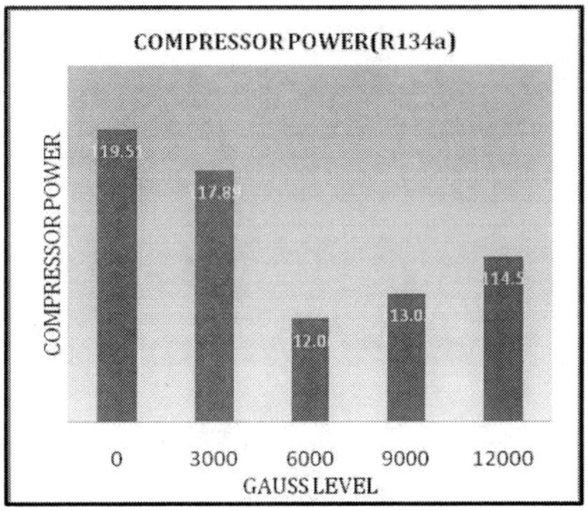

Fig. 2. Compressor Power (R134a)

Figure2 shows the power of the compressor versus the magnetic field strength for the R134a. The average compressor efficiency for gauss rates 3000 is reduced by 1.35% relative to the non-magnet field. Average compressor energy falls by 6.23 %, 5.40% respectively and 4.19 per cent for gauss rates 6000, 9000 and 12000.

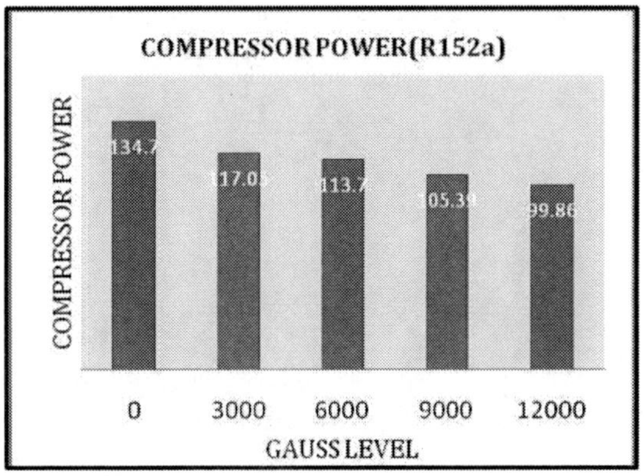

Fig. 3. Compressor Power (R152a)

The power vs. magnetic field strength for R152a is shown in Figure3. The rate of the gauss increases the compressor power. Average compressor power for gauss level 3000 decreased by 13.10 percent compared to the non-magnetic field. The average compressor power for gauss rates 6000, 9000 and 12000 is 15.59%, 21.75% and 25.86%, respectively.

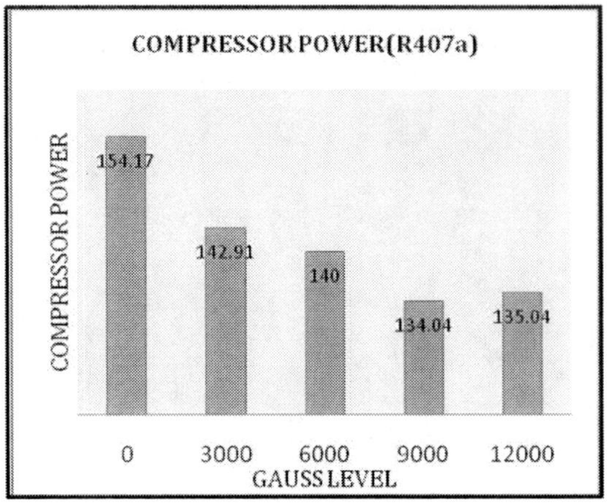

Fig. 4. Compressor Power (R407a)

The compressor power versus the magnetic field strength for the R407a is shown in figure 4. The average compressor strength of gauss level 3000 has been decreased by 7.3% compared to no magnet field. The average compressor capacity at Gauss rate 6000, 9000, 12000 and 13000 reduces respectively by 9.19%, 13.05% and 12.40%.

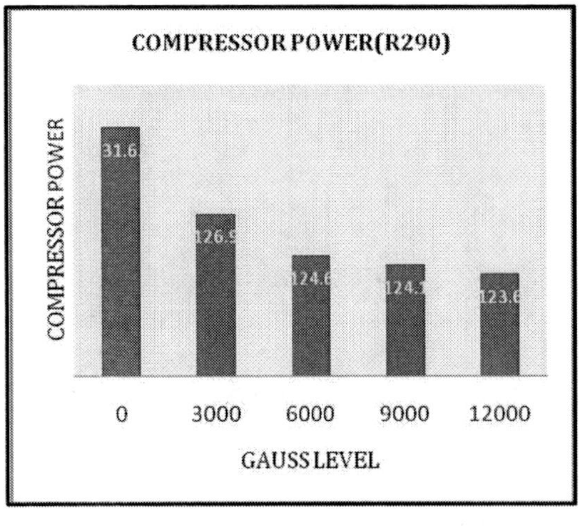

Fig. 5. Compressor Power (R290)

The compressor energy versus magnetic force of R290 is shown in Figure5. The rate of gauss increases the compressor power. The average compressor power of Gauss level 3000 is decreased by 3.60% compared to no magnetic field. The average compressor power for gauss rates 6000, 9000 and 12000 is decreased by 5.35%, 5.73% and 6.11% respectively.

4 Conclusions

During the test, the compressor energy savings of the compression cooling system was investigated under the influence of magnetic field using various refrigerants such as R134a, R151a, R407a, and R290.

The magnetic field has shown a positive effect on a decrease in vapor compression system power consumption.

The maximum power saving for R152a at 12000 gauss rate was 25.86 percent.

5 Reference

[1]. S.M. Sami, R.J. Kita, *Behavior of new refrigerant mixtures under magnetic field, IJES*, **29**, 1205-1213 (2005).

[2]. A Kotb, H.E. Saad, *Experimental Investigation for the Magnetic-Caloric Effect on the Refrigeration Cycle Performance, SJET*, **2(3A),** 339-347 (2014).

[3]. S.M. Sami, S. Aucoin, *Effect of magnetic field on the performance of new refrigerant mixtures, IJES*, **27**, 203-214 (2003).

[4]. K. Mani, V. Selladurai, *Energy saving with the effect of magnetic field using R290/600a mixture as substitute for $CFCl_2$ and' HFCl134a*, Thermal sci, **12**, 111-120 (2008).

[5]. R. Prapainop, K.O Suen (2012), *Effect of refrigerant properties on refrigerant performance comparison*, IJERA, **2**, 486-493 (2012).

[6]. H. Guo, Z. Liu, Y. Chen, R. Yao, *A study of magnetic effect on the physicochemical properties of individual hydrocarbons, Logistical Engineering College, Chongqing 400042, P.R. China(1994)*, 216-220.

[7]. A.P. Shrotri, N.S. Mane, V.P. Patil, S.A. Wani, *Magnetohydrodynamiccs emerging trend to enhance heat transfer in air conditioning system, IJSMR*, **1**, 41-48 (2015).

A Review of Recent Experimental and Theoretical Research of Dew Point Evaporative Coolers

Prashant Patunkar[1*], *Sunil Dingare*[2], *and Virendra Bhojwani*[1]
[1]Department of Mechanical Engineering, MIT School of Engineering, Pune (India)
[2]Department of Aerospace Engineering, MIT School of Engineering, Pune (India)

Abstract. Evaporative cooling is one of the techniques to lower the ambient air temperature according to the requirements. The technique includes the injection of moisture in the coming air, thus increasing its specific humidity. This technique can cool the air up to wet bulb temperature of the incoming air. In the recent past, the techniques are developed to lower the temperature of air without addition of moisture in the incoming air. These techniques demonstrate that the incoming air temperature can be approached near the dew point temperature of the incoming air. The present paper essentially focuses on the recent theoretical and experimental research of dew point evaporative coolers. Different flow configurations of heat exchange mediums, optimization of geometrical parameters, performance parameters (wet bulb and dew point effectiveness), energy efficiency ratio and energy saving potential are the key elements to study the prospects of dew point evaporative coolers over the conventional refrigeration systems operated by mechanical means. Dew point evaporative cooling technology is environmental friendly and has the potential in reducing the impact on earth due to global warming.

INTRODUCTION

The usage of conventional air conditioning system for residential, commercial and industrial applications contributes to the bulk consumption of electrical units. The consumption trends across world wide leads to climate change, which may further worsen the situations in the coming future. Therefore, it is the need of the hour to make available better options for the cooling pusposes. Evaporative cooling (direct and in direct) is considered an energy efficient and enivormental friendly technique. Direct evaporative cooling includes mixing of two streams, i.e. air and water, whereas in direct evaporative cooling both these streams are separated by heat exchange surface with wet and dry channels.

EXPERIMENTAL RESEARCH OF DEW POINT COOLERS

Liu et.al [1] investigated experimentally and theoretically the refrigeration behaviour of a highly efficient dew point evaporative cooler using most effective air and water flow arrangements. Different results obtained from the experiments reported the improvement of 29.3% in the cooler wet bulb efficiency. Increase in COP achieved by slightly greater than one third in contrast to the commercially available dew point air cooler of equal capacity. On account of usage of corrugated plates helped to achieve the cooling efficiency more that 10% as compared to flat plates. Further the refrigeration efficiency of dew point cooler in original configuration is only 62-67% that of a dew point evaporative cooler as compared to counter flow configuration. The paper reported that the refrigeration performance can be enhanced by enlarging the geometrical parameters such as length of the channel along with the air entrance length and minimizing the width and gap of the channel within a acceptable limit. The experimental set up designed with the corrugated plates is shown in Fig. 1.

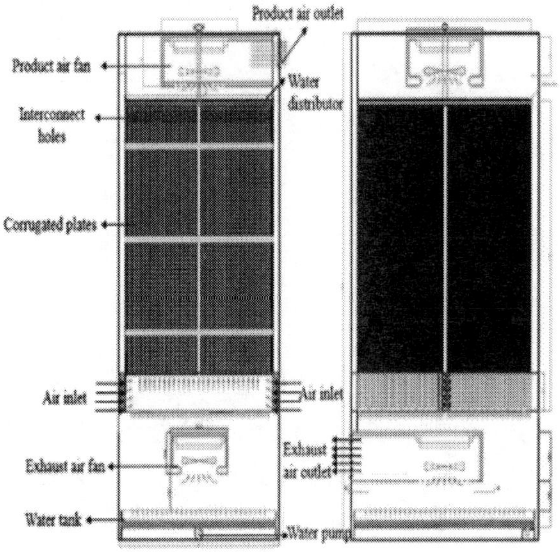

Fig. 1 Layout for Dew point evaporative cooler [1]

*Corresponding author: **prashant.patunkar@mituniversity.edu.in**

The detailed geometrical parameters of the prototype used are mentioned in Table 1.

Table 1. Different geometrical dimensions for prototype [1]

Geometrical Dimension	Value
Length of the channel	1.05 m
Width of the channel	0.716 m
Channel spacing	4.3 mm
Air entrance length	175 mm
Dry channels utilized	117
Wet channels utilized	117

The comparison of prototype and M30 cooler is mentioned in Table 2.

Table 2 Difference between prototype and M30 [1]

Type of arrangement	M30	Prototype
Preliminary air flow rate (m^3/h)	4522 3349	2719 2972
Actual air flow rate (m^3/h)	2363 1615	1188 1223
Flow ratio	0.523 0.482	0.437 0.445
Inlet air dry bulb temperature (^0C)	37.78 37.78	38.6 38.1
Inlet air wet bulb temperature (^0C)	21.11 18.33	20.28 22.03
Product air temperature (^0C)	22 20.94	17.8 20.9
Cooling capacity (kW)	11.77 10.08	10.83 9.65
Wet bulb efficiency	94.5% 88.2%	114% 107%
COP	28.9 31.8	42.8 33.3

The tests were carried in constant temperature and humidity environment. The experimental results obtained the wet bulb efficiency of 114% and dew point efficiency of 68.4%, refrigeration effect of 10.8 Watts and highest coefficient of performance of 42.8 for standard operating conditions.

Frank Bruno [2] presented the findings achieved after testing a prototype cooler operated in commercial as well as residential purposes over varied sets of atmospheric conditions. Two different coolers operated at other sites were tested for the purpose to evaluate the output characteristics of the devices working in commercial as well as residential applications in the presence of varied atmospheric conditions. For the commercial application the device could cool the air from more than 40^{0}C to nearly 15^{0}C. The total data was obtained for the period of 44 days where in the daily average ambient air temperature ranged between 22.5^{0}C and 40. 3^{0}C. The daily average humidity ranged between 10% and 55%. Fig. 2 shows the variation of Energy efficiency ratio with the difference of dry bulb and dew point temperature.

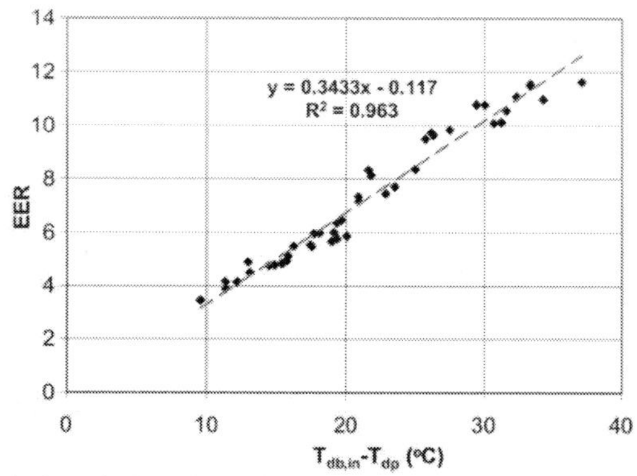

Fig.2 Variation of energy efficiency ratio with difference between dry bulb and dew point temperature of atmospheric air (commercial building) [2]

The linear relationship was obtained between the energy efficiency ratio and difference between dry bulb inlet and dew point outlet temperature. The daily average wet bulb and dew point effectiveness of the prototype with respect to dew point temperature is shown in Fig. 3.

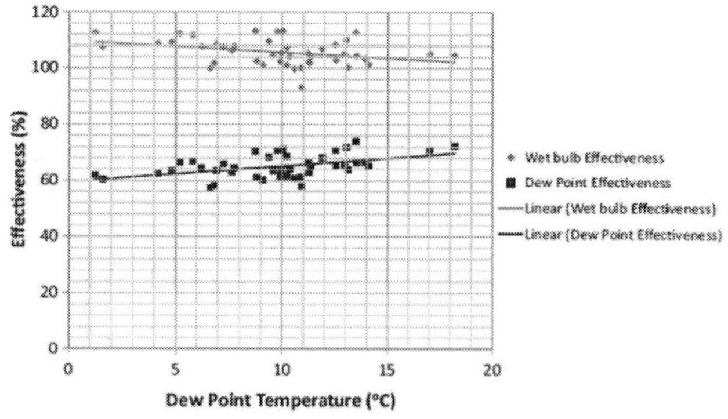

Fig.3 Variation of wet bulb and dew point effectivenesses with dew point temperature (commercial application) [2]

Average values for wet bulb and dew point effectiveness obtained were 106% and 65% respectively for the commercial application. Fig. 4 shows the variation of energy efficiency ratio with respect to dew point temperature and Fig. 5 shows the daily average wet bulb and dew point effectivenesses with respect to atmospheric dew point temperature.

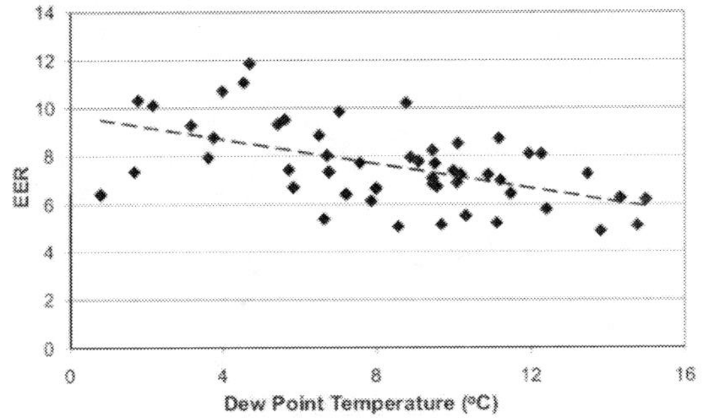

Fig.4 Variation of energy efficiency ratio with dew point temperature of atmospheric air (household application) [2]

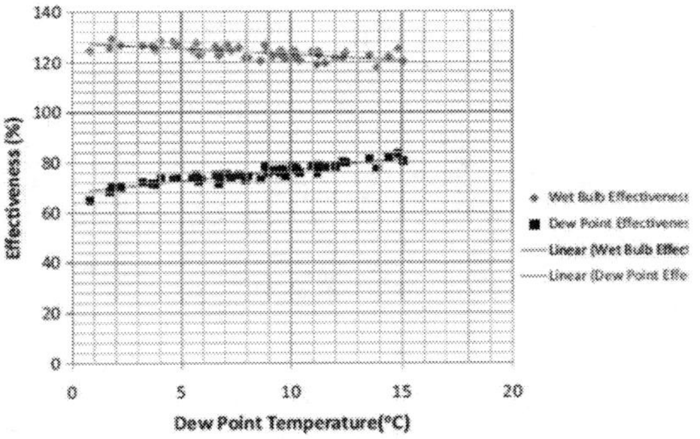

Fig.5 Variation of Wet bulb dew point effectivenesses with dew point temperature [2]

The EER observed to increase as the dew point temperature reduces. The average wet bulb and dew point effectivenesses obtained were 124% and 75% respectively for residential installation. The analysis was also performed to study the potential energy saving from the cooler for commercial building application. The empirical correlation was developed between the energy efficiency ratio, condenser temperature and air temperature on the evaporator. The results were obtained for the configuration of counter flow indirect evaporative cooler for commercial as well as household installations. The yearly energy saving was proposed in the range of 52 and 56%.

The thermodynamic performance of a innovative device as cooler was investigated by Jia et.al. [3] A mini size counter cross flow device coated with nylon fibers over wet channel portion was investigated. The significance of the operational parameters on thermodynamic performance was investigated. The findings obtained indicated that, device applied with coating of nylon fibers delivered enhanced cooling output. The combination of polystyrene nylon fibre delivered the dew point efffectiveness of 46.7% to 78.6%. The effectiveness of this combination was in good agreement with counter flow evaporative cooler and higher than that of cross flow evaporative devices. The schematics of the arrangement is shown in Fig.6.

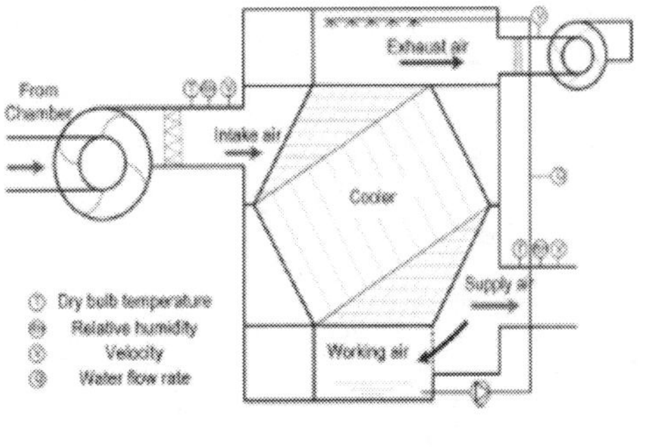

Fig. 6 Arrangement for experimental apparatus [3]

Experimental apparatus consists of an air distribution arrangement, a water circulation arrangement along with arrangement for heat transfer. The test set up operated with supply air flow rate and working -to-intake air ratio (WIR) at 85m^3/hour and 0.4 respectively. The inlet air temperature was adjusted to 29.5^{0}C, 32.5^{0}C and 35.5^{0}C and the specific humidity of 16 g/kg. The variation of dew point effectiveness with respect to intake air temperature for different wet surface materials is shown in Fig.7.

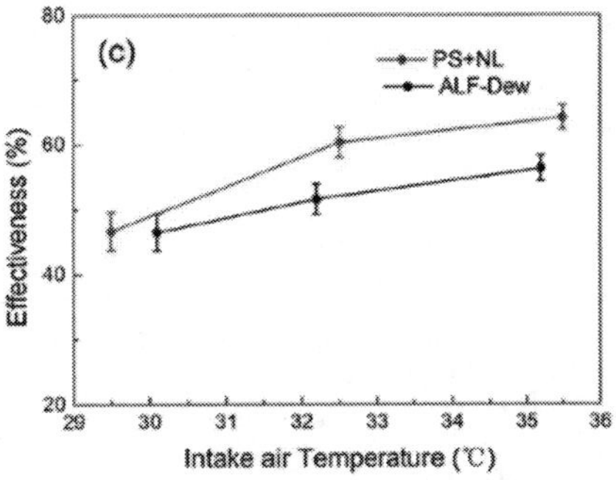

Fig.7 Dew point effectiveness versus intake air temperature for various substances [3]

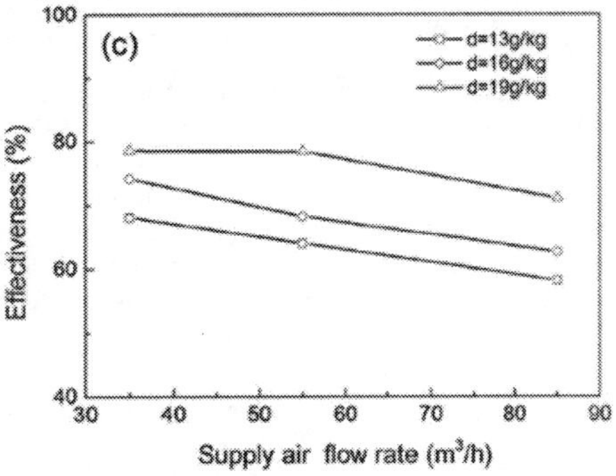

Fig. 8. Variation of effectiveness with supply air flow rate [3]

The effect of intake air flow rate on the dew point effectiveness is shown in Fig. 8. It is observed that the dew point effectiveness of the cooler decreases slightly with enhanced intake air flow rate which may be due lower contact period of air and wet surface due to increased air flow rate. The parameters such as air temperature, relative humidity and air velocity were examined. Dew point effectiveness was determined mathematically. The impact of wet surface material, intake air flow rate, work to intake air ratio (WIR), air intake temperature, intake air humidity on the thermal performance of the device was analysed. The suggested device showed the capability to make it compact and weight of the device with simultaneously constant volume of air.

Lin et.al. [4] designed and engineered a counter flow dew point evaporative cooler and the assembled experimental set up is shown in Fig.9. The size of device has dimensions as 0.8 m x 0.18 m x 4.0 m (L x W x H).

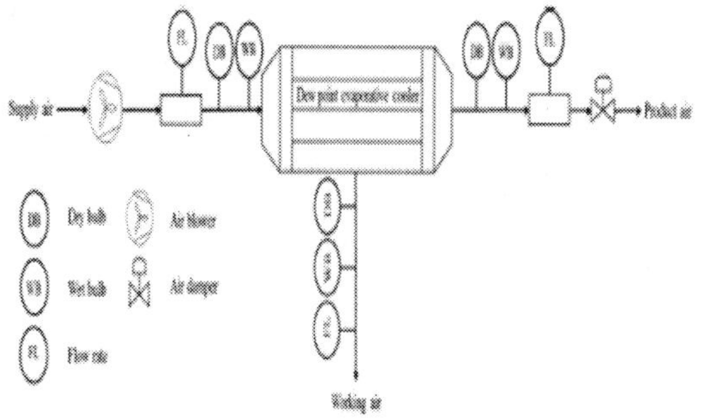

Fig. 9 Experimental set up for the device [4]

The testing parameters for the set up are mentioned in Table 3.

Table 3 Testing parameters for the set up [4]

Parameter	value	Range
Length of the Channel	0.8 m	-
Width of the Channel	0.18 m	-
Height of the Channel	4.0 mm	-
Intake air temperature	32.5^{0}C	27.2-40.3
Intake air moisture content	10.0 g/kg	9.8-19.6
Intake air velocity	2.0 m/s	1.9-2.1
Operating air ratio	0.33	0.31-0.33

Pakari and S. Ghani [5] developed and tested a prototype of cooling system with the prime objective for validation of the numerical model of the device. Fig. 10 shows a schematic of the device. The trials were taken on the device by regulating the intake air temperature, intake air moisture content, and intake air velocity. The regulation of extraction ratio was carried with the help of standby blower fixed at the outlet of the wet channel.

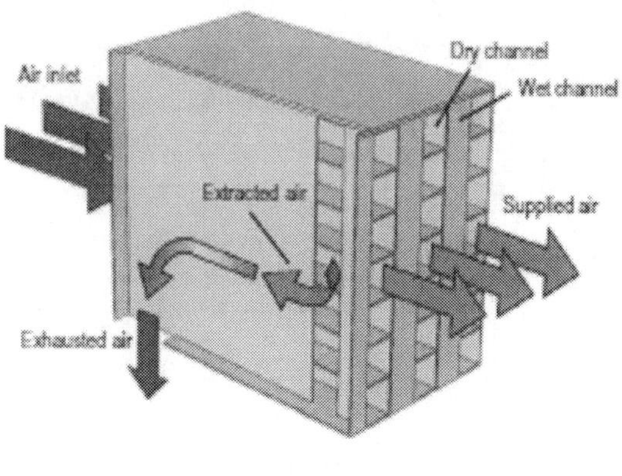

Fig. 10 Flow arrangement for the test device [5]

The technical specifications of the experimental set are mentioned in Table 4.

Table 4. Dimensional parameters for test device [5]

Parameter	Value
Dry channel thickness	2 mm
Wet channel thickness	2 mm
Length of the channel	0.5 m
Height of the channel	0.3 m
Heat exchanger thickness	0.22 m
Thickness of plate	0.2 mm
Paper thickness	0.15 mm

Fig. 11 shows layout of the test device.

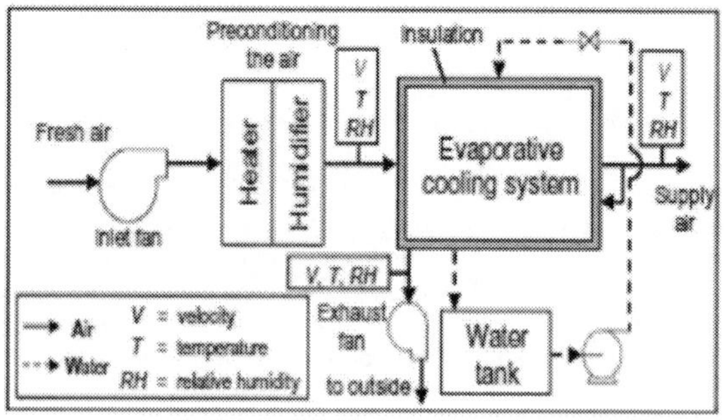

Fig. 11 Layout for the test device [5]

The exit temperature and moisture content for the device were tested in 3 cities for a year and changes of monthwise exit air temperature and moisture content obtained between 30^{0}C and 20 to 70 % respectively.

THEORETICAL RESEARCH OF DEW POINT COOLERS

The fundamental analysis of heat and mass transfer for the counter-flow dew point evaporative cooler was investigated by Lin et.al. [4]. The simulation and experimental analysis has been done for the objective of validating at the preliminery stage, the convection and mass transfer phenomenon of the device. The value of the convection heat and mass transfer coefficients were obtained by log mean temperature/humidity difference technique. The simulation of heat and mass transfer response of the device has been analysed with the help of 2 dimensional model. Various parameters such as product air temperature, convection heat and mass transfer coefficients and dimensionless numbers were simulated with various parameters as shown in Table 5.

Table 5 Simulation parameters for the device [4]

Parameter	Value	Range
Length of the channel (m)	0.8	0.5-1.2
Thickness of plate and film (mm)	0.25	0.15-0.55
Channel height (mm)	4.0	2.0-6.0
Supply air temperature (^0C)	30	27.5-40.0
Inlet air moisture (g/kg)	13.3	8.0-18.0
Inlet air velocity (m/s)	2.0	1.0-3.0
Working air ratio	0.33	0.1-0.9

The study also revealed that factors which made significant impact on the heat and mass transfer performance were Re_D, H/L, δ/H and π. A 2 dimensional CFD model of the device is obtained with fundamental balance mathematical relations. The theoretical results obtained may be utilized for different real life examples.

Sohani et.al. [6] proposed an optimization methodology to boost improving potential of devices. The system output was modeled throughout the year and then, hour basis readings for inlet parameters was used and diversified optimization was carried out after alternate hour. The objective functions focussed were cost of running, coefficient of performance and rate of water flow. The hour, month and yearwise findings were analyzed and put forth. Based upon four different cliamatic conditions investgated, the results showed substantially enhanced scope for the hour basis optimization technique to escalate the objective functions 18.7, 19.6 and 36.2% on an average. The hourly optimization strategy (HOS) is proposed, which was so far untapped for the improvement of potential of the device.

Akhlaghi et.al.[7] develpoed a regression model to predict the thermodynamic performance of the device. The proposed model helps to establish the relationship of various parameters with operating parameters such as, flow and dimensional parameters as mentioned in Table 6.

Table 6 Geometrical and water parameters for the proposed model [7]

Parameter	Value
Length of the channel	1.2 m
Width of the channel	0.348 m
Pitch	0.005 m
Stacks	200
Water temperature	16^0C
Water circulation rate	18 kg/s

The operational parameters taken into consideration are, inlet air conditions, which includes moisture content, flow rate in addition to working air fraction over the inlet air, where as refrigeration capacity, COP, pressure drop, dew point, wet bulb effectiveness were considered as operating conditions. The gemoetric characteristics selected are height of the channel, its interval and number of stacks in heat and mass exchange. The selected 8th degree model produced highest comparative difference for the refrigeration capacity, COP, pressure drop, dew point along with wet bulb effectivenesses were 6.1%, 7.54%, 0.07%, 3.54% and 2.53% respectively. The introduction of the MPR into dew point cooler technique imparts crucial significance for characterization of the specific feature to device. The proposed polynomial has the potential of further extension to determine optimised dimensional as well as flow characteristics correlated with the device not only for varied climatic but also for the different operating situations.

Pakari and S. Ghani [5] developed the regression models so as to predict the out put characteristics, help for the design and optimization of the device. The detailed analysis includes regression models, analytical equations, which in turn correlates various inlet conditions, such as working and dimensional terms and properly chosen exit performances of the device by applying numerical tests in addition to response surface technique. The parameters taken into consideration are temperature of the intake air, moisture content, air intake velocity along with extraction ratio. The dimensional terms include length and width of refrigeration stsyem's passages.

Various results obtained from the numerical model and experimental measurements for the outlet air temperature are shown in Fig. 12. The results match within the range of 10%. The average discrepency observed has been about 5.2%. The output behaviour parameters taken into consideration are temperature of the air, moisture content in addition to wet bulb efficiency.

The proposed regression model helps to analyse the exit parameters for the refrigeration systems in the absence of any priority for the solution differential equations.

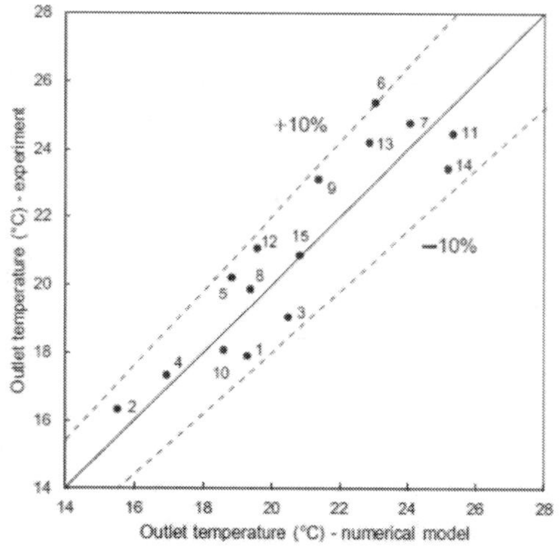

Fig.12 Variation of the trend in exit temperatures obtained by numerical and experimental technique [5]

The regression models possesses resonable accuracy which turns to be beneficial for the application in energy oriented codes for the thermodynamic response projection of the cooling devices.

Liu et.al. [8] proposed the numerical model which was subsquently cross checked experimentally with the similar results obtained from various investigations. The significance of the model extended to investigate the different working parameters such as intake discharge of air, air ratio, intake temperature of water, discharge of water along with temperature and moisture content for the intake air related to refrigeration capacity on the modified construction of the device. The proposed numerical model achieves similar results with experimental results obtained in contrast to the various researchers with variation of less than 5% obtained by experimental and numerical techniques.

Ali Pakari and Saud Ghani [9] studied the thermodynamic behaviour of refrigeration device by numerical as well as experimental technique. The proposed one dimensional model of heat and mass exchange for the cooler is analysed and compared with developed three dimensional model. Findings obtained from experimental set up achieved the wet bulb efficiency for refrigeration cycle attained figures close to 125%. Further investigation revealed that, one dimensional model proposes relatively less exit temperatures nearly 1.86% as compared with three dimensional model. The proposed exit temperatures for refrigeration cycle obtained by both models validates the experimental findings close to 10% and 8.5 % respectively. One dimensional model appropriately indicates thermodynamic response for refrigeration cycle on account of faster computing speed and available facilities in contrast to three dimensional model. From the various factors considered, the extraction ratio makes the significantly appreciable impact on the wet bulb efficiency. Wet bulb efficiency is directly proportional to the extraction ratio.

CONCLUSION

The present paper reviews on the recent experimental and theoretical research of dew-point evaporative coolers. Such coolers have the significant energy saving potential for residential and commercial applications. The performance of the dew point evaporative cooler is governed significantly by operational and geometrical parameters. Those parameters can be further used for the experimental investigations and numerical validation for a small capacity domestic air conditioning system based on the usage of non-conventional energy for the year round climate.

REFERENCES

1. Y. Liu, Y.G. Akhlaghi, X. Zhao, J. Li, Experimental and numerical investigation of a high-efficiency dew-point evaporative cooler, Energy & Bulidings 197 (2019) 120-130.
2. F. Brunó, On-site experimental testing of a novel dew point evaporative cooler, Energy & Buildings 43 (2011) 3475-3483.
3. L. Jia, J. Liu, C. Wong, X. Cao, Z. Zhang, Study of the thermal performance of a novel dew point evaporative cooler, Applied Thermal Engineering 160 (2019) 114069.
4. J. Lin, D.T. Bui, R. Wang, K.J. Chua, On the fundamental heat and mass transfer analysis of the counter-flow dew point evaporative cooler, Applied Energy 217 (2018) 126-142.
5. A. Pakari, S. Ghani, Regression models for prediction of counter flow dew point evaporative cooling systems, Energy Conversion and Management 185 (2019) 562-573.
6. A. Sohani, H. Sayyaadi, M. Zeraatpisheh, Optimization strategy by a general approach to enhance improving potential of dew point evaporative coolers, Energy Conservation and Management 188 (2019) 177-213.
7. Y.G. Akhlaghi, X. Ma, X. Zhao, S. Shittu, A statistical model for dew point air cooler based on the multiple polynomial regression approach, Energy 181 (2019) 868-881.
8. Y. Liu, J.M. Li, X. Yang, X. Zhao, Two-dimensional numerical study of heat and mass exchanger for a dew-point evaporative cooler, Energy 168 (2019) 975-988.
9. A. Pakari, S. Ghani, Comparison of 1D and 3D heat and mass transfer models of a counter flow dew point evaporative cooling system : Numerical and experimental study, International Journal of Refrigeration 99 (2019) 114-125.

Thermal Analysis of Vertical Plate Fin Heat Sink

Anilkumar Sathe[1*], *Sudarshan Sanap*[2], *Sunil Dingare*[3], *Narayan Sane*[4]
[1,2,3]MIT School of Engineering, Rajbaug, Pune, Maharashtra, India
[4]Walchand College of Engineering, Maharashtra, India.

Abstract: Heat Sinks are widely used to remove the heat from the components which are generating heat during their functioning. Overheating causes malfunctioning of the components as well as it is responsible for reducing their life. Free convection is very common way of heat transfer from the heat sink considering power requirement, pressure drop and cost of the forced convection. This paper presents the thermal analysis of vertical plate fin heat sink by theoretical and experimental method at variable heat input. The results are obtained by taking experimental observations and are validated with already existing correlations suggested by various researchers in the literature.
Keywords: Heat Sink, Natural Convection, Fin Array

1 Introduction

Fins are used when present surface area is not sufficient for the removal of generated heat in the components with available temperature difference between surface and surrounding. In various electronic systems such as relays, mother boards of personal computers, audio systems, LED lamps etc. as shown in Fig. 1, heat is generated during their working. Thermal management of these components is very essential in order to avoid their failure as well as to improve their life span. One of the effective techniques of cooling the electronics components to equip heat sink on their surface to increase their effective surface area and thereby to increase the heat dissipation rate. Heat sinks are the array of number of fins arranged together in order to increase the surface area available for heat transfer. In majority of applications passive cooling is preferred considering its low cost and trouble free service. Though air has low specific heat, low conductivity but because of its free availability, low cost and ease of maintenance still it is preferable choice in most of the cases. While using the heat sink the shape of the fin, its orientation, material, size and fin spacing plays very important role in transferring the heat. Various researchers studied the air flow in natural convection over vertical surface of the fin plate and proposed various correlations.

Fig. 1 Heat Sinks

2 Literature Survey

Hong and Chung [1] carried out experimental and numerical analysis of vertical plate fin array in natural convection heat transfer. They found out optimum spacing for maximum heat transfer rate and further studied the effect of varying Prandtl number on thermal performance. Chang et al. [2] examined numerically the thermal performance of vertical fin array with and without dimples. They concluded that average value of Nusselt number (Nu) is the strong function of Rayleigh number (Ra). They tested four configurations of dimpled fin array and found out the best dimple arrangement for higher heat transfer rate. Awasarmol and Pise [3] experimentally studied rectangular fin array with and without perforation effect at different angles of orientation. They concluded that perforations on the fin surface increases heat transfer rate with advantage of reduction in weight. In their study they found out optimum perforation diameter. Naserian et al. [4] studied the effect of v type fin array on a vertical base experimentally and numerically. Results for plate fins are validated with Churchill and Chu and McAdam correlations. Haghighi et al. [5] carried out experimental investigation of vertical fin array and compared the results with combination of plate fins and pin fins arranged between two consecutive plate fins of heat sink. The results show that combination of plate and pin fin have lower thermal resistance and higher heat transfer compared to only plate fins heat sinks. Heat transfer enhancement of new-designed heat sinks is about 10-41.6% higher, compared to normal pin-fins. Effect of fin spacing on the thermal resistance was also studied.

3 Objective

In the present work the vertical symmetrical heat sink is manufactured and tested in natural convection mode with various heat inputs in order to study the effect of heat input on the thermal performance of the heat sink. The results obtained experimentally are validated with already existing correlations. Many of the researchers validated their

https://doi.org/10.1051/e3sconf/202017001022

experimental results by using correlations applicable for vertical plate where characteristic dimension is the vertical length of the plate. But in the present work validation is done with correlations suggested for vertical rectangular fin array where characteristic dimension is the fin spacing. Finally, thermal performance is measured in terms of convective heat transfer coefficient.

4 Experimental set up

The dimensions for the fin for experimental work are width 150 mm, height 200 mm and thickness 2 mm. Spacing between the fins is created by spacer block of width 50 mm, height 200 mm and thickness 16.5 mm. the thickness of the block serves as fin spacing. All fins and spacer blocks are made up of aluminium material. Spacer blocks and fin plates are having three central holes of 18 mm diameter along middle line to insert three stainless steel cartridge type heater rods. Two small holes are also provided on top and bottom sides of fin to insert tie rods to held complete assembly of fins and spacer blocks together as shown in figure. The overall size of the vertical fin array is 150 mm width, 200 mm height and 150 mm breadth. After assembling the fin array system (FAS), the effective length of the fin i.e. base to tip distance is measured as 50 mm.

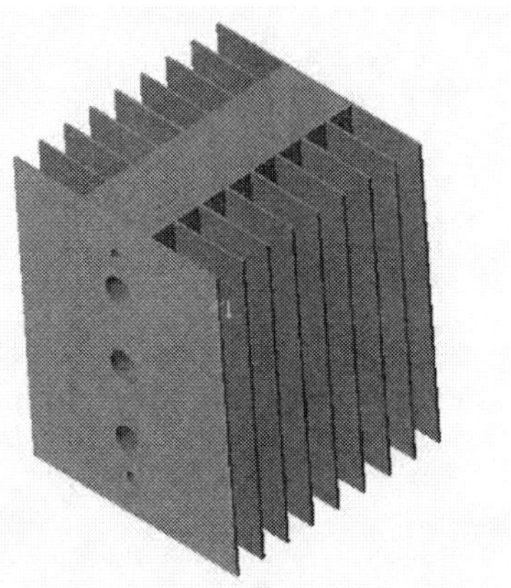

Fig. 2 Vertical Plate Fin Heat sink

The assembly is covered by bakelite plates from both the sides in order to reduce side heat loss. Bakelite plate size is equal to size of fin plate. Number of fins is taken as 9 and spacer blocks are 8. K type thermocouples with accuracy of 0.1^0C are used to measure the temperatures at different locations of vertical heat sink. Total 6 no. of thermocouples are arranged on the spacer blocks and 6 no. of thermocouples are arranged on the fin plate surface. Two thermocouples are arranged on side bakelite covers and one thermocouple is in the enclosure to record ambient temperature. The arrangement of all thermocouples is as shown in Fig. 3.

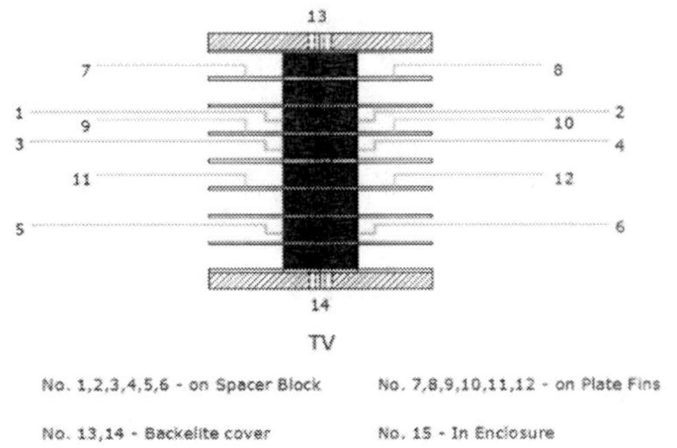

Fig. 3 Top view of the heat sink with Thermocouples positioning

In order to provide undisturbed surrounding, a rectangular box type enclosure from plywood is manufactured which is surrounding the arrangement. The total assembly of finned system is held vertical in hanging position with the help of metal wire supports so that natural convection current of hot air will setup around the fin surface.

The control panel is designed with arrangement of dimmerstat, display of heater wattage, temperature indicator with selector switch and mains switch. The power input connections to the heater is given through the control panel. All thermocouples are connected to control panel for displaying temperature on indicator.

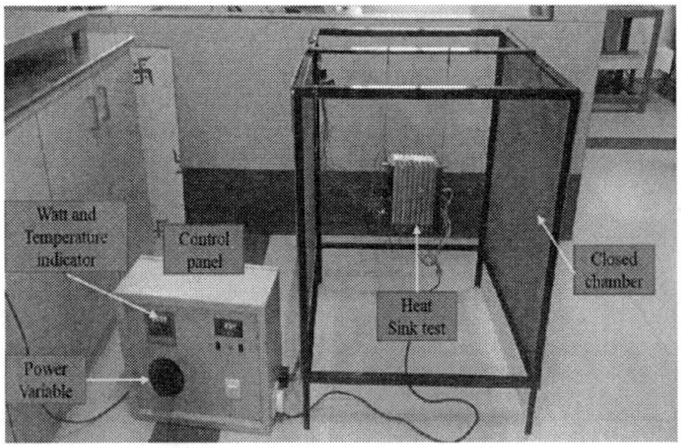

Fig. 4 Experimental Setup

5 Data reduction

a) Temperature of finned surface is found out by taking average of all thermocouple reading mounted on base plate and fin surface.
b) Temperature of surrounding is measured by thermocouple reading mounted in surrounding inside the enclosure.
c) Area is calculated for the finned surface.
d) The variable heat input is generator with the help of dimmer and heat supply is determined by wattage reading given by wattmeter.
e) The various heat losses are calculated including side convective and radiation loss.
f) By using Newton's law of cooling, the convective heat transfer coefficient is determined.

6 Validation of Experimental Setup and Procedure

6.1 Experimental Method:

1. Area calculation of system of fin array (FAS) for experimental analysis:
   A_b = Area of base
   A_f = Area of Fins
   A = Total area considered for analysis = $A_b + A_f$
   s = Fin spacing i.e. thickness of spacer block
   H = Fin height i.e. vertical length of FAS
   W = Width of FAS
   B = Breadth of FAS
   t = Thickness of fin
   l = Fin lenth i.e. base to tip distance
   $A_b = (8 \times s \times H) \times 2$
   $= (8 \times 0.0165 \times 0.2) \times 2 = 0.0528\ m^2$
   $A_f = (8 \times l \times H) \times 2 \times 2 + (9 \times t \times H) \times 2$
   $= (8 \times 0.05 \times 0.2) \times 2 \times 2 + (9 \times 0.002 \times 0.2) \times 2$
   $= 0.3272\ m^2$
   $A = 0.0528 + 0.3272 = 0.38\ m^2$
2. Average surface temperature of FAS (T_s)
   $$T_s = \frac{T_1+T_2+T_3+T_4+T_5+T_6+T_7+T_8+T_9+T_{10}+T_{11}+T_{12}}{12}$$
   $T_s = 41.4\ ^0C = 314.4\ K$
3. Surrounding temperature (T_∞)
   $T_\infty = T_{15} = 28.3\ ^0C = 301.3\ K$
4. Radiation heat loss (Q_r)
   $$Q_r = \sigma.A.\epsilon.(T_s^4 - T_\infty^4)$$
   Assuming emissivity of comercial aluminim surface (ϵ) = 0.09
   $Q_r = 5.67 \times 10^{-8} \times 0.38 \times 0.09 \times (314.4^4 - 301.3^4)$
   = 2.96 W
5. Heater Input (Q_H)
   Q_H = Wattmeter Reading = 25 W
6. Effective heat transfer rate for convection (Q)
   $$Q = Q_H - Q_r$$
   $$Q = 25 - 2.96 = 22.04\ W$$
7. Convective heat transfer coefficient (h)
   Newton's law of cooling is given by mathematical expression:
   $$Q = h \times A \times (T_s - T_\infty)$$
   $$h = \frac{Q}{A \times (T_s - T_\infty)}$$
   $$h = \frac{22.04}{0.38 \times (41.4 - 28.3)} = 4.42\ W/m^2\ ^0C$$

6.2 Theoretical Method

After calculating the experimental value of heat transfer coefficient, the value is compared with already existing correlations for vertical plate available in literature.

6.2.1 Correlations for vertical plate fin array

The correlations used for comparing the results are

a) McAdams correlation
   $Nu = 0.59\ (Ra)^{1/4}$ for $10^4 < Ra < 10^9$ ----------- [1]

b) Churchill and Chu's first relation (for laminar and turbulent flows)
   $$Nu = \left[0.825 + \frac{0.387(Ra)^{1/6}}{\left[1+\left(\frac{0.492}{Pr}\right)^{9/16}\right]^{8/27}}\right]^2 \text{ for } 10^{-1} < Ra < 10^{12} \text{ --------[2]}$$

c) Bar-Cohen Correlation
   $$Nu = \frac{h.s}{k} = \left[\frac{576}{\left(\frac{Ra_s.s}{H}\right)^2} + \frac{2.873}{\left(\frac{Ra_s.s}{H}\right)^{0.5}}\right]^{-0.5} \text{ -------------------- [3]}$$

d) Elenbaas Correlation
   $$Nu = \frac{h.s}{k} = \frac{1}{24}\frac{Ra_s.s}{H}\left[1-\exp\left(\frac{-35}{\frac{Ra_s.s}{H}}\right)\right]^{\frac{3}{4}} \text{ --------------- [4]}$$

6.2.2 Calculations

1. $\Delta T = T_s - T_\infty = 41.4 - 28.3 = 13.1\ ^0C$
2. Mean film temperature (T_{mf})
   $$T_{mf} = \frac{T_s+T_\infty}{2} = \frac{41.4+28.3}{2} = 34.85\ ^0C = 307.85\ K$$
3. Coefficient of thermal expansion (β)
   $$\beta = \frac{1}{T_{mf}} = \frac{1}{305.67} = 0.00324\ K^{-1}$$
4. Properties of fluid (air) at mean film temperature (Calculated from property table of air)
   a. Kinematic viscosity (ϑ) = $1.65 \times 10^{-5}\ m^2/s$
   b. Prandtl Number (Pr) = 0.70
   c. Thermal conductivity (k) = 0.0267 W/m^{0}C
5. Grashoff Number (Gr)
   $$Gr = \frac{\beta.g.\Delta T.L_c^3}{\vartheta^2}$$
   $$Gr = \frac{0.00324 \times 9.81 \times 13.1 \times 0.2^3}{(1.65 \times 10^{-5})^2} = 1.23 \times 10^7$$
6. Rayleigh Number (Ra)
   $$Ra = Gr.\ Pr$$
   $$Ra = 1.23 \times 10^7 \times 0.70 = 8.62 \times 10^6$$
7. Nusselt Number (Nu)
   By using correlations given by equation 1,2,3 and 4 the Nusselt number values are obtained.
8. Convective heat transfer coefficient (h)
   $$h = \frac{Nu.k}{L_c}$$

Table 1 Values of h by various correlations for 25 W

Correlation	Mc Adam	Churchill and Chu's - first	Bar-Cohen	Elenbaas	Expt.
h (W/m^2 ^{0}C)	4.27	3.99	4.21	4.23	4.42

Similarly, calculations are extended for 50, 75, 100 and 125 W and experimental and theoretical results are plotted on the graph in order to justify the agreement between them.

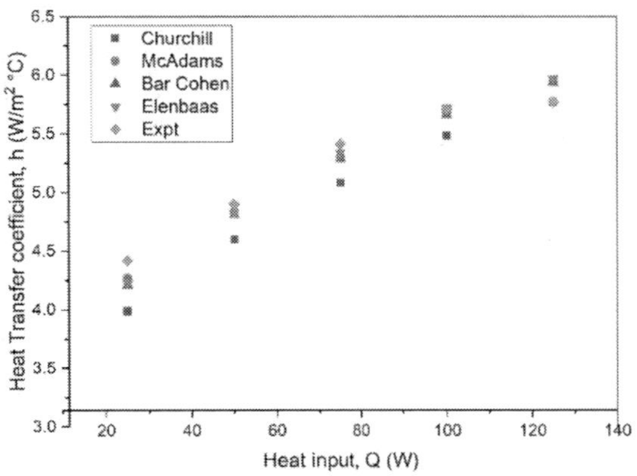

Fig. 5 Validation of the Experimental setup

The plot examinations indicate that the values of convective heat transfer coefficient (h) obtained from experimental observations are in good agreement with already existing correlations with relative deviation between 3-10 %. These results prove the validity of the experimental setup and experimental procedure.

7 Result of Plane Fins

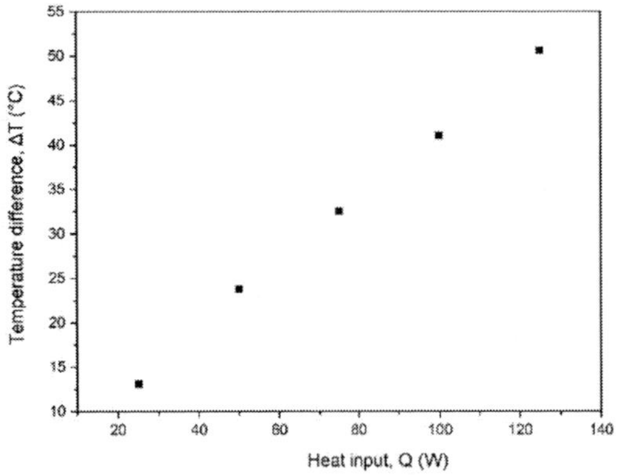

Fig. 6 Plot of temperature difference (TD) vs heat input (Q)

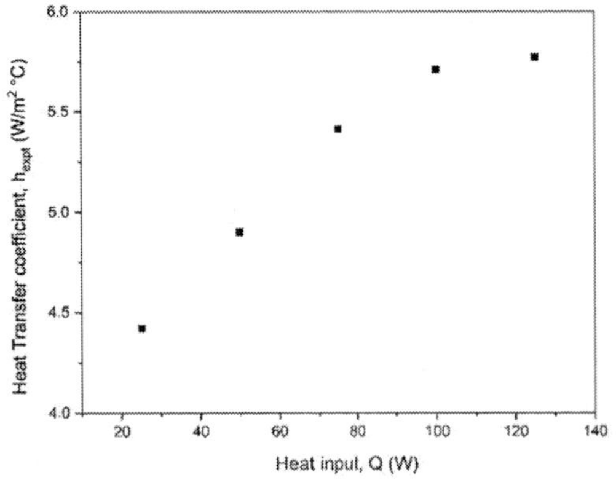

Fig. 7 Plot of heat transfer coefficient (h) vs heat input (Q)

It is clear from the Fig. 6 and Fig. 7 that heat input supplied at the base of FAS provides driving potential to create natural convection currents of air through the space available between two consecutive fins. With increase in heat input the temperature difference between fin surface and surrounding goes on increasing. The convection current becomes dominant with increase in heat input results into more degree of buoyancy force and higher convective heat transfer coefficient.

An uncertainty analysis was done to investigate the error in experimental setup which is a sum of individual uncertainty of experimental parameter used in the analysis. The total uncertainty found in the experimental setup and analysis is below 2%.

Conclusion

1. The experimental results of convective heat transfer coefficient are in good agreement (Maximum deviation less than 10%) with already existing correlations in the literature.
2. Thermal performance measured in terms of convective heat transfer coefficient for vertical heat sink is the function of heat input. As the heat input increases, the natural convection current augments as a result of density difference.

References:

1. S. Hong, B. Chung, Int. J. Therm. Sci., **101**, 1-8 (2016)
2. S. Chang, H. Wu, D. Guo, J. Shi, T. Chen, Int. J. Heat Mass Tran. (2016)
3. U. Awasarmol, A. Pise, Exp. Therm. Fluid Sci., **68**, 145-154 (2015)
4. M. Naserian, M. Fahiminia, H. Goshayeshi, J. Mech. Sci. Technol., **27**, 2191-2197 (2013)
5. S. Haghighi, H. Goshayeshi, M. Safaei, Int. J. Heat Mass Tran. **125**, 640-647 (2016)
6. W. Elenbaas, Physica, **9**, 1-28 (1942)
7. A. Bar-Cohen, W. Rohsenow, J. Heat Trans, **106**, 116-123 (1984)

Solar photovoltaic module performance characterisation using single diode modeling

Javed K. Sayyad[1,*] *and Paresh S. Nasikkar*[1]
[1]Symbiosis Institute of Technology, Symbiosis International (Deemed University), Lavale, Pune, 412115 Maharashtra (India).

Abstract. Single or double diode electrical modeling of SPV module gives valuable results which will help to identify the exact behavior of SPV module under the normal operating condition. Accurate modelling of SPV module will also help to calculate internal resistances (R_s, R_{sh}) and parasitic of the SPV module. The main contribution of this work is the stepwise simplification of the current equation of single and double diode electrical model of SPV module. Then the single diode model of SPV module having 36 SPV cells in series is simulated in LTspice simulator. Simulated results are compared with labelled electrical parameters which shows close proximity to the labelled parameter at particular values of series and shunt resistance. This paper also presents the effect of variation in series resistance (R_s) and shunt resistance (R_{sh}) on the performance of the SPV module under normal operating condition. The dependency of SPV electrical parameters (I_{max}, V_{max}, P_{max}, FF, η) with the variation of series resistance (R_s) and shunt resistance (R_{sh}) is simulated, and the effects are discussed in details.

1 INTRODUCTION

In 1839, A French Physicist, E. Becquerel has investigated and observed Photovoltaic (PV) effect which converts incident light energy into electrical energy. Now a days Solar Photovoltaic (SPV) energy is most evolving and fast growing technique to fulfills the need of electricity of mankind [1-3]. The SPV cell is the basic unit of SPV module. SPV Cells are connected in series to get an output of the desired value of voltage. Similarly, the dimension of SPV cells will decide the current rating pf SPV module.

The parameters like module temperature (T_m), incident solar irradiation (G), and load impedance states the conversion efficiency (η) of photovoltaic module [4-7]. The major disadvantage of commercially available SPV module is its low conversion efficiency (η < 20%) and the non–linearity of the output I-V curve [8, 9]. In normal operating condition, meteorological data such as solar irradiance (W/m^2), ambient temperature (^0C), humidity (*%Rh*), and wind velocity (*mph*), etc. varies unpredictably. This uncertain variation affects current, voltage and hence power output generated by the SPV module [10, 11].

Manufacturers provide labelled electrical parameters (specification given on backside of SPV panel) and performance characteristic curves of SPV module for optimized power generation. This parameters are captured indoor under Standard Test Condition (STC) where $G = 1000\ W/m^2$, $T_m = 25^0C$ and *AM* (Air Mass) = 1.5 [12]. Commonly, SPV cell is modeled using a single diode model circuit equivalence [13]. To improve precision and accuracy, it can be modeled using a double diode model circuit equivalence [14]. Proper electrical modelling of SPV module will help to predict accurate performance actual operation condition [15].

In section 2, a need for SPV module performance characterisation is described with some definitions of significant parameters. Single and double diode modeling is compared with mathematical derivations of currents in section 3. In section 4, the single diode model of ASP-12-75W_p SPV module is simulated in SPICE simulator. In section 5, the results of the simulation are presented and compared for variation in series and parallel resistance. The conclusion of the work is presented in section 6.

2 CHARACTERISATION

Performance characterisation of SPV module/cell by tracing I-V and P-V curves is helpful while designing new SPV power plant. It also useful for scheduling Operation and Maintenance (O&M) of the SPV power plant [16]. Fig. 1 shows that during curve SPV cell/module is swept under different load conditions using variable resistor [17]. SPV cell is an active device which generates energy from incident solar light. Fig. 2 shows that, the I-V and P-V curve of an illuminated SPV cell.

I-V and P-V characteristic curve is used for extracting some important performance parameters like short circuit current (I_{sc}), open circuit voltage (V_{oc}), Maximum power (P_{max}), voltage at maximum power (V_{max}), current at maximum power (I_{max}), conversion efficiency (η) and field factor (*FF*) [18].

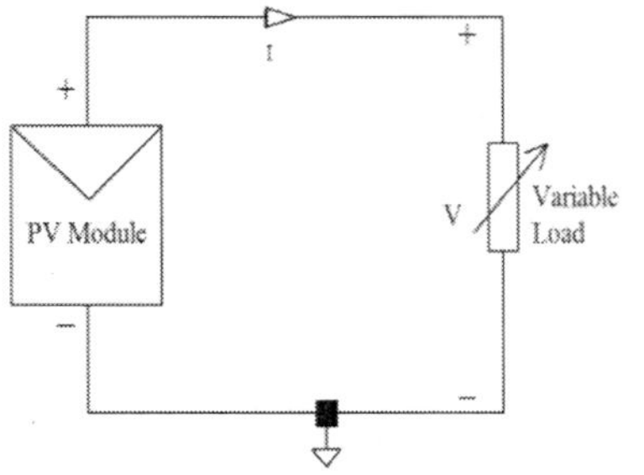

Fig. 1. Characterisation of SPV module

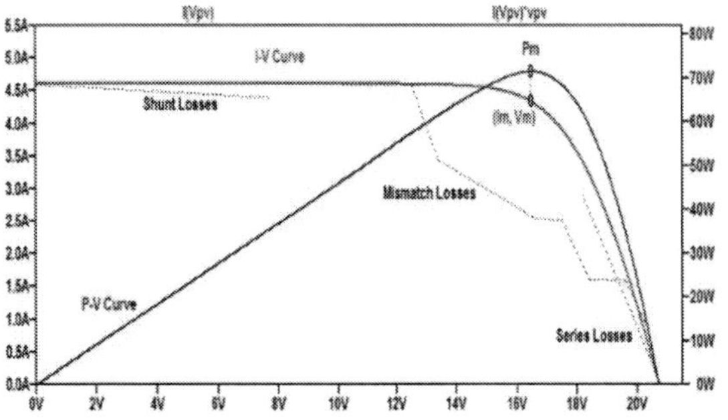

Fig. 2. I-V and P-V Curve

2.1 Significance of Series Resistance (R_s)

Slope near V_{oc} is determined by series resistance (R_s) as shown in Fig. 2 and the value of R_s is calculated by using Equation 1;

$$R_s = -\left(\frac{dV}{dI}\right)_{V=V_{oc}} \tag{1}$$

Resistance due to SPV cell metallization, SPV cell solder bonds, emitter–base region, cell interconnected bus-bars and junction box terminations implies finite value of series resistance in electrical equivalence circuit modeling. The ideal value of *Rs* is 0 [19]. The value of I_{sc} is reduced due to high value of R_s, which shows inverse relation between them.

2.2 Significance of Shunt Resistance (R_{sh})

The ideal value of R_{sh} is infinite but finite value of R_{sh} due to manufacturing defects , device degradation and parallel high conductivity paths through or on the solar cell edges results slope near I_{sc} [19, 20]. This effect is shown in Fig. 2 and the value of R_{sh} is calculated by using Equation 2;

$$R_{sh} = -\left(\frac{dV}{dI}\right)_{I=I_{sc}} \tag{2}$$

The low value of R_{sh} leading to higher power loss at output [21].

3 MODELING OF SPV CELL

SPV cell is modeled as a current source (I_{pv}) in parallel with a P-N junction diode in a single diode electrical equivalent circuit model [22]. SPV cell behaves like an only diode if there is no incident light present (dark condition $G = 0$) to generate any current. With an increase in the intensity of incident light, the current generated by the SPV cell increases [23]. Fig. 3 shows the only diode, and the current flowing through it is given by Equation 3;

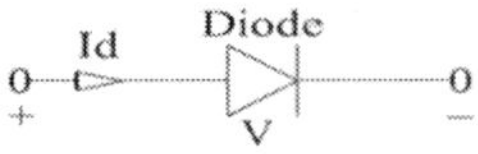

Fig. 3. Single Diode

$$I_d = I_o\left(e^{\frac{qV}{akT_m}} - 1\right) \tag{3}$$

put;

$$v_t = \frac{akT_m}{q} \tag{4}$$

Equation 5 becomes;

$$I_d = I_o\left(e^{\frac{V}{v_t}} - 1\right) \tag{5}$$

a is ideality factor of diode which is ranging between 1 and 2 (given in datasheet of diode). Considering a = 1 at 300 K temperature, value of V_t becomes 25.86 mV.

3.1 Single Diode Modeling (Ideal)

Fig. 4 shows ideal SPV cell single diode model (Considering $R_s = 0$ and $R_{sh} = \infty$);

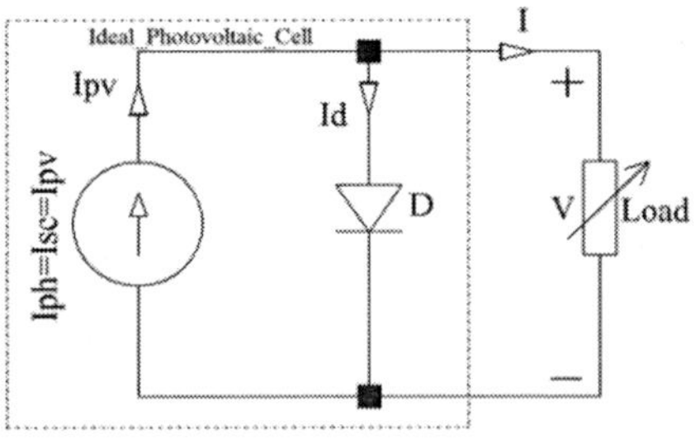

Fig. 4. Simplified Single Diode Modeling of SPV cell (Ideal)

The equation of current flowing through the load is given by Equation 6;

$$I = I_{pv} - I_d \tag{6}$$

$$I = I_{pv} - I_o\left(e^{\frac{V}{v_t}} - 1\right) \tag{7}$$

Fig. 5 shows practical SPV cell single diode model (Considering $R_{sh} = \infty$);

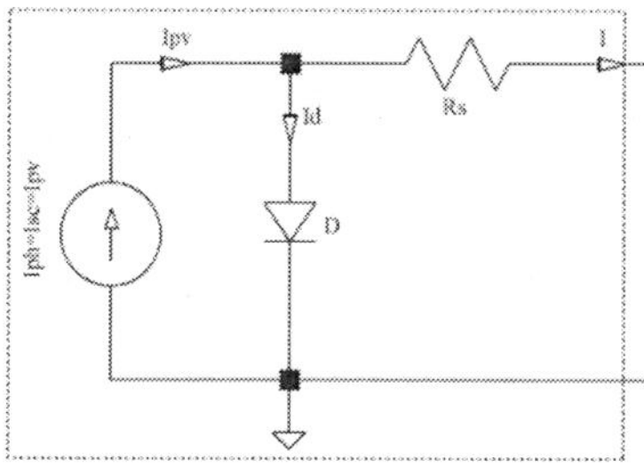

Fig. 5. Single Diode Modeling of SPV cell with R_s

The equation for practical SPV cell considering R_s only is given by Equation 8;

$$I = I_{pv} - I_o\left(e^{\frac{(V+I R_S)}{v_t}} - 1\right) \quad (8)$$

Fig. 6 shows practical SPV cell single diode model considering finite values of R_s *and* R_{sh}.

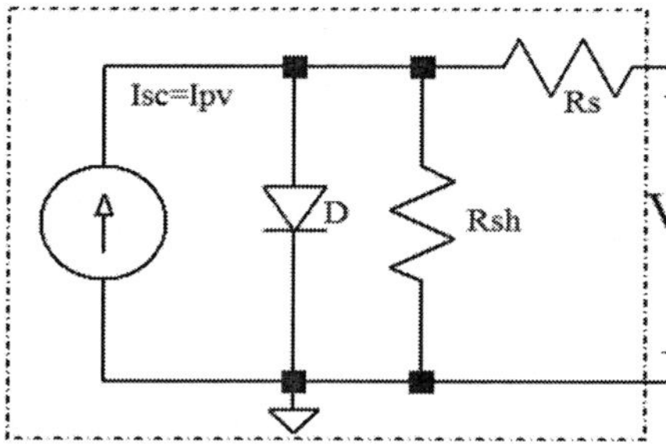

Fig. 6. Single Diode Modeling of SPV cell with R_s and R_{sh}

Considering R_s and R_{sh}, the single diode model equation becomes;

$$I = I_{pv} - I_d - I_{Rsh} \quad (9)$$

$$I = I_{pv} - I_o\left(e^{\frac{(V+I R_S)}{v_t}} - 1\right) - \frac{V+IR_S}{R_{sh}} \quad (10)$$

Commonly, 36 or 72 cells are connected in series to get desired voltage value at output of SPV module. Different current rating of SPV module depends on physical size of each cell. Cell is bigger in dimension has more current rating. Considering N_s as a number of cells are connected in series, equation of practical SPV module becomes;

$$I = I_{pv} - I_o\left(e^{\frac{(V+I R_S)}{N_S v_t}} - 1\right) - \frac{V+IR_S}{R_{sh}} \quad (11)$$

This single diode modeling with R_s and R_{sh} is also known as five parameter model (I_o, η, R_s, R_{sh}, I_{pv}) [13]

3.2 Double Diode Modeling

To increase the precision and accuracy of the equivalence circuit model, many researchers have added several diodes and resistances in series or in parallel with the current source [24]. Fig. 7 shows that, double diode model of the SPV cell considering effect of R_s and R_{sh}. Double diode model considers the effect of recombination of parallel diode [23]. At low value G and low value of T_m, this model will give significant and more accurate results. Double diode model has better ability to give mode precise and sharp I-V and P-V characteristics curve under normal as well as partial shading operating condition [25].

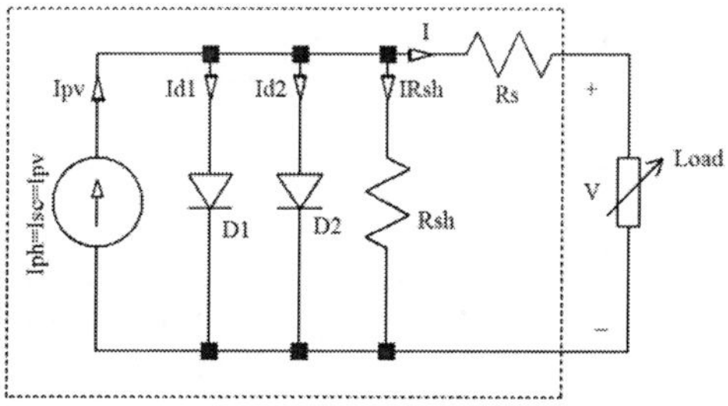

Fig. 7. Double Diode Modeling of SPV cell with R_s and R_{sh}

The current flowing through load using double diode model is given by Equation 12;

$$I = I_{pv} - I_{d1} - I_{d2} - I_{Rsh} \quad (12)$$

$$I = I_{pv} - I_o\left(e^{\frac{(V+I R_S)}{N_{s1} v_{t1}}} - 1\right) - I_o\left(e^{\frac{(V+I R_S)}{N_{s2} v_{t2}}} - 1\right) - \frac{V+IR_s}{R_{sh}} \quad (13)$$

3.3 Summary of comparison between single and double diode modeling

Table 1. Comparison of single and double diode model

Quantities	Single Diode	Double Diode
Computation time for parameter extraction	Shorter	Longer
Unknown parameters	Less	More
Accuracy	Good	Better
I–V and P–V characteristics	Smooth	Sharp
Consider recombination effect	No	Yes

4 SPICE SIMULATION OF SINGLE DIODE MODELING

Labelled parameter of ASP-12-75W_p SPV module is used as a reference which are calculated under standard test condition, 1000W/m^2 and T=25^0C given in Table 2.

 https://doi.org/10.1051/e3sconf/202017001023

Table 2. Labelled Specification of SPV Module

ASP-12-75W_p SPV Mono–crystalline (1000 W/m^2, 25°C)	
Electrical Parameter	**Specification at STC**
Maximum Power (P_{max})	75 W_p
Voltage at Maximum Power (V_{max})	17.60 V
Current at Maximum Power (I_{max})	4.26 A
Open Circuit Voltage (V_{oc})	21.5 V
Short Circuit Current (I_{sc})	4.60 A
Number of series connected SPV cell (N_s)	36

Single diode model of complete SPV module having 36 SPV cells (N_s = 36) in series is simulated in SPICE simulation software assuming all cells are identical and having equal solar radiation. Photon current (I_{pv} = I_{sc}) flowing through SPV cell is directly proportional to input incident light energy [12]. Leakage current of diode is considered as 1 nA. Current controlled current source (CCCS) having current gain 1 is used which gives same current at output of voltage controlled voltage source (VCVS). VCVS is used to multiply single cell voltage with number of cells in series in the SPV module (N_s*E_{pv0}). Considering uniform operating condition, multiplication of SPV cell output with 36 will shows that there is no possibility under consideration of partial shading condition in simulation. Fig. 8 shows the single diode model of SPV module.

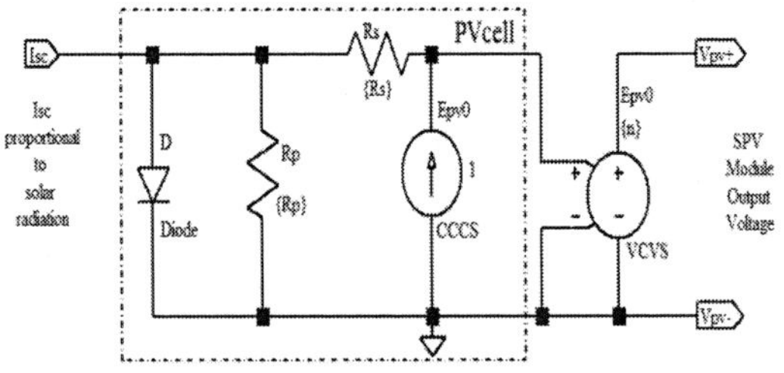

Fig. 8. Simulated single diode model (X1)

Current source of 4.60 A represents I_{sc} of SPV module (ASP-12-75W_p) which is proportional to solar radiation incident on SPV module. Fig. 9 shows the final equivalence model of SPV panel with X1 is the solar panel symbol assuming internal resistance of voltmeter ($R_{internal}$) is 0.01 Ω.

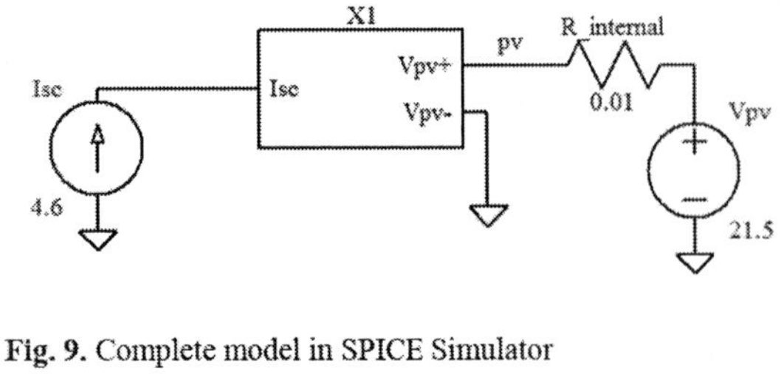

Fig. 9. Complete model in SPICE Simulator

5 RESULTS AND DISCUSSIONS

In order to validate single diode modeling and SPICE simulation for above SPV module, the specified values of the ASP-12-75W_p SPV module are compared with simulation results for different combination of R_s and R_{sh}. In this section, effect of variation of R_s and R_{sh} on characteristic curve is observed carefully. The I–V and P–V characteristic curves from SPICE simulation of single diode model of SPV module are presented in this section. The effect of R_s and R_{sh} variation on SPV module characteristics curve is validated.

5.1 Effect on R_s

In this case, the series resistance R_s of single diode model was varied in multiple of 10 from 0.0001Ω to 1Ω with all other parameters kept constant (R_{sh} = 1KΩ). The results (I– V and P–V curves) are presented in Fig. 10 to Fig. 14. From this figures, it may be noted that with increase in series resistance (R_s), the slope near V_{oc} decreases which effectively decreases output power of SPV module and it is tabulated in Table 3.

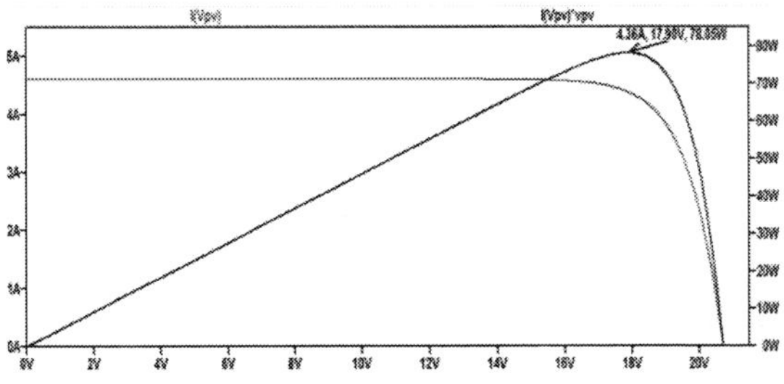

Fig. 10. I-V and P-V Curve with $R_s = 10^{-4}\,\Omega$ and $R_{sh} = 1\ K\Omega$

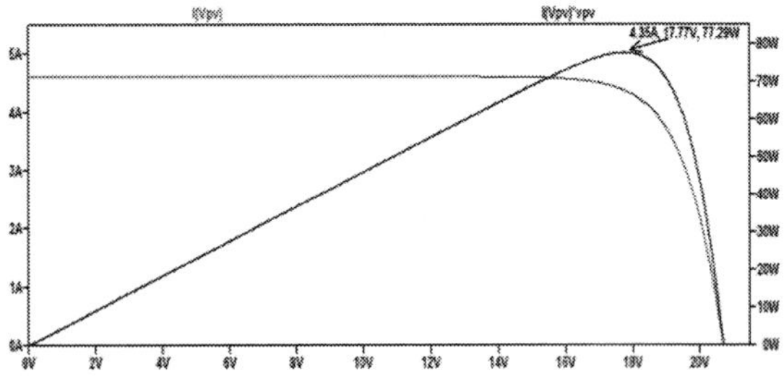

Fig. 11. I-V and P-V Curve with $R_s = 10^{-3}\,\Omega$ and $R_{sh} = 1\ K\Omega$

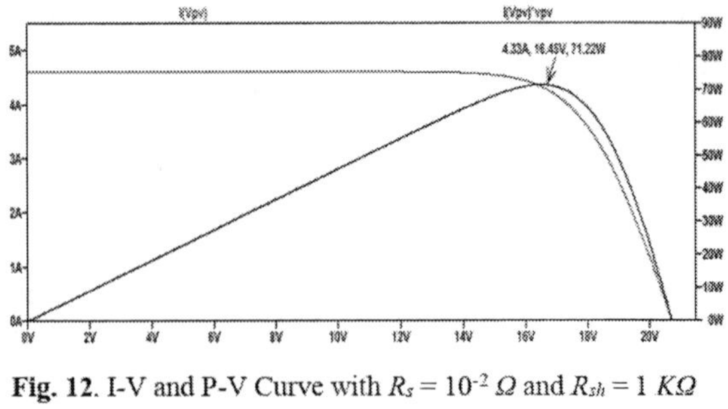

Fig. 12. I-V and P-V Curve with $R_s = 10^{-2}\,\Omega$ and $R_{sh} = 1\ K\Omega$

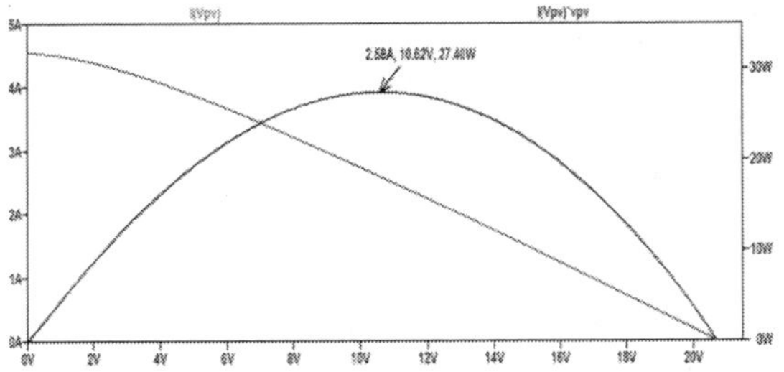

Fig. 13. I-V and P-V Curve with $R_s = 10^{-1}\ \Omega$ and $R_{sh} = 1\ K\Omega$

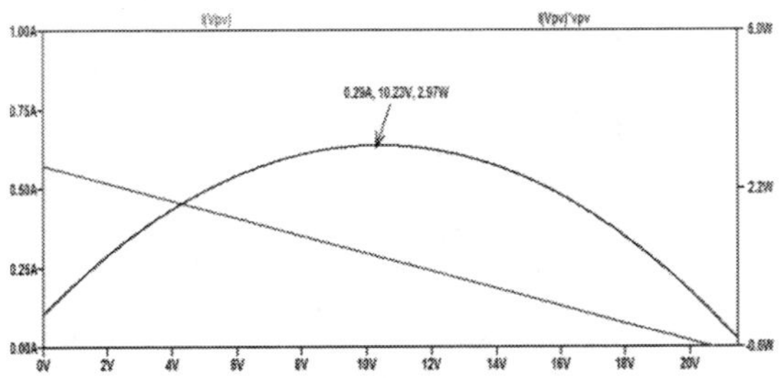

Fig. 14. I-V and P-V Curve with $R_s = 1\Omega$ and $R_{sh} = 1\ K\Omega$

Table 3. Effect of R_s Variation

	Value				
R $R_s\ (\Omega)$	10^{-4}	10^{-3}	10^{-2}	10^{-1}	1
$R_{sh}\ (\Omega)$	1 KΩ				
$I_{max}\ (A)$	4.36	4.35	4.33	2.68	0.29
$V_{max}\ (V)$	17.90	17.77	16.45	10.62	10.23
$P_{max}\ (W_p)$	78.05	77.29	71.22	27.40	2.97

5.2 Effect on R_{sh}

In this case, the shunt resistance (R_{sh}) of single diode model was varied in multiple of 100 from 0.1Ω to 10$M\Omega$ with all other parameters kept constant ($R_s = 0.01\Omega$). The results (I–V and P–V curves) are presented in Fig. 15 to Fig. 19. From this figures, it may be noted that with increase in shunt resistance (R_{sh}) the slope near I_{sc} changes and output power increases which is tabulated in Table 4.

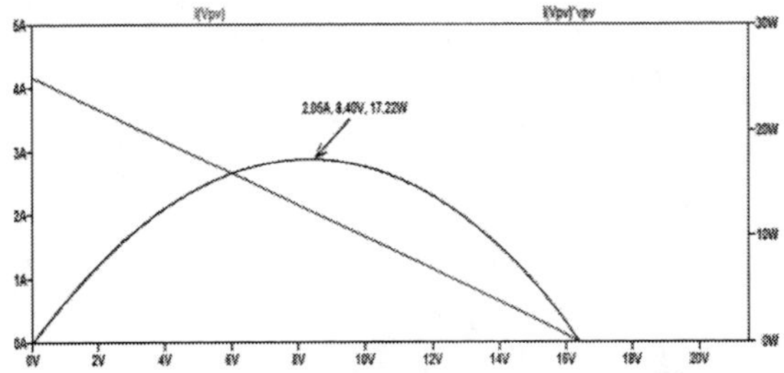

Fig. 15. I-V and P-V Curve with $R_{sh} = 0.1\Omega$ and $R_s = 0.01\Omega$

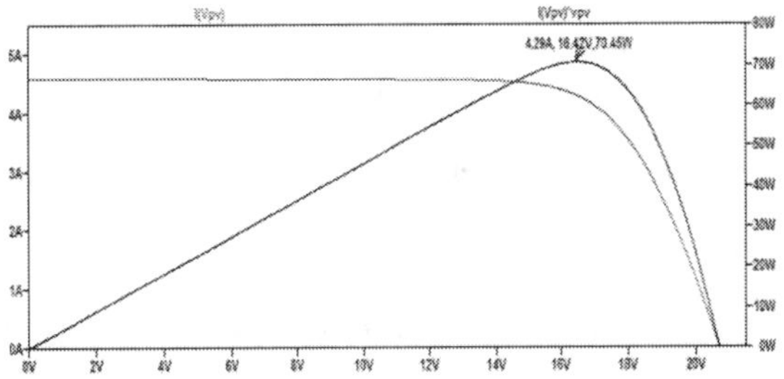

Fig. 16. I-V and P-V Curve with $R_{sh} = 10\Omega$ and $R_s = 0.01\Omega$

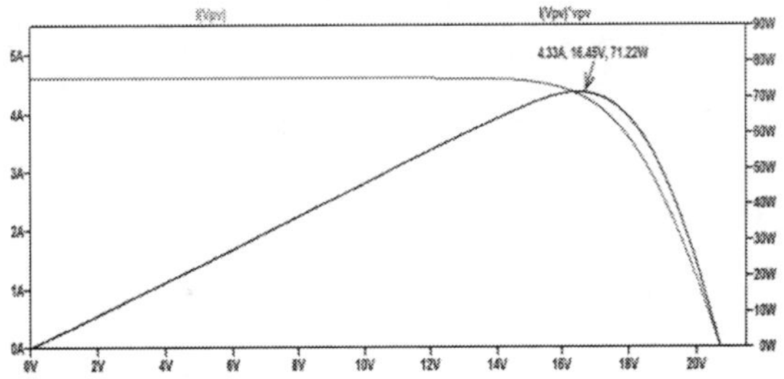

Fig. 17. I-V and P-V Curve with $R_{sh} = 1K\Omega$ and $R_s = 0.01\Omega$

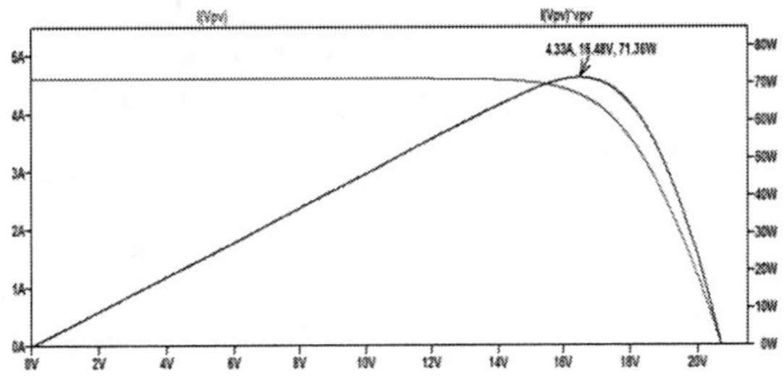

Fig. 18. I-V and P-V Curve with $R_{sh} = 100K\Omega$ and $R_s = 0.01\Omega$

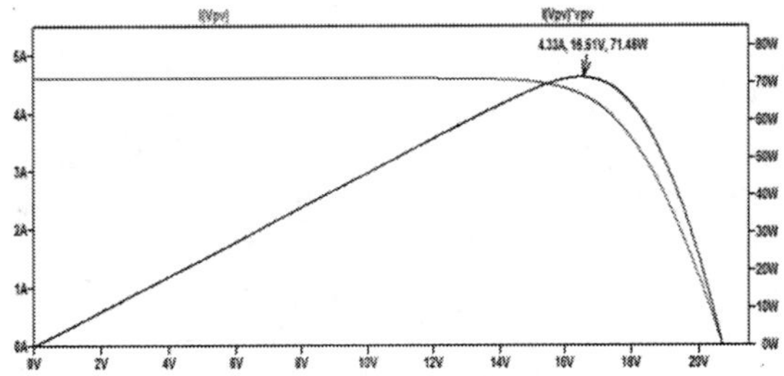

Fig. 19. I-V and P-V Curve with $R_{sh} = 10M\Omega$ and $R_s = 0.01\Omega$

Table 4. Effect of R_{sh} Variation

	Value				
$R_s\ (\Omega)$	0.01				
$R_{sh}\ (\Omega)$	0.1	10	10^3	10^5	10^7
$I_{max}\ (A)$	2.05	4.29	4.33	4.33	4.33
$V_{max}\ (V)$	8.40	16.42	16.45	16.48	16.51
$P_{max}\ (W_p)$	17.22	70.45	71.22	71.36	71.46

6 CONCLUSION

A detailed comparative study of single and double diode electrical modeling was carried with their advantages and disadvantages. The ideal value of series resistance(R_s) is 0, but finite value shows the change in the slope of a curve near V_{oc}. Similarly, the ideal value of shunt resistance (R_{sh}) is ∞, but finite value shows the change in slope of curve near I_{sc}. The results presented in the above section clearly shows that as R_s increases the output power decreases. Similarly increases in R_{sh} value, the output power increases. The implemented single diode model of SPV module gives results having close proximity with a labelled specification of ASP-12-75W_p SPV module with a low value of R_s and high value of R_{sh} under normal operating conditions at a constant temperature and constant solar irradiance

References

1. Kurtz, S., *Opportunities and challenges for development of a mature concentrating photovoltaic power industry.* 2012: Citeseer.
2. Friedman, D.J., et al., *Toward 100 Gigawatts of concentrator photovoltaics by 2030.* IEEE Journal of Photovoltaics, 2013. **3**(4): p. 1460-1463.
3. Branker, K., M. Pathak, and J.M. Pearce, *A review of solar photovoltaic levelized cost of electricity.* Renewable and sustainable energy reviews, 2011. **15**(9): p. 4470-4482.
4. Ibrahim, H. and N. Anani, *Variations of PV module parameters with irradiance and temperature.* Energy Procedia, 2017. **134**: p. 276-285.
5. Siddiqui, R. and U. Bajpai, *Deviation in the performance of solar module under climatic parameter as ambient temperature and wind velocity in composite climate.* International Journal of Renewable Energy Research, 2012. **2**(3): p. 486-490.
6. Xiao, C., et al., *Impact of solar irradiance intensity and temperature on the performance of compensated crystalline silicon solar cells.* Solar Energy Materials and Solar Cells, 2014. **128**: p. 427-434.
7. Hishikawa, Y., et al., *Precise Outdoor PV Module Performance Characterization Under Unstable Irradiance.* IEEE Journal of Photovoltaics, 2016. (5): p. 1221-1227.
8. Kandemir, E., N.S. Cetin, and S. Borekci, *A comprehensive overview of maximum power extraction methods for PV systems.* Renewable and Sustainable Energy Reviews, 2017. **78**: p. 93-112.
9. Vorster, F. and E. Van Dyk, *Current-voltage characteristics of high-concentration, photovoltaic arrays.* Progress in Photovoltaics: Research and Applications, 2005. **13**(1): p. 55-66.
10. Engeland, K., et al., *Space-time variability of climate variables and intermittent renewable electricity production – A review.* Renewable and Sustainable Energy Reviews, 2017. **79**: p. 600-617.
11. Kahoul, N., et al., *Evaluating the reliability of crystalline silicon photovoltaic modules in harsh environment.* Renewable Energy, 2017. **109**: p. 66-72.
12. Kawagoe, K., Y. Hishikawa, and N. Yamada, *Outdoor Direct STC Performance Measurement of PV Modules Based on a Sun-Shading Technique.* IEEE Journal of Photovoltaics, 2017. **7**(6): p. 1725-1730.
13. Lineykin, S., M. Averbukh, and A. Kuperman. *Five-parameter model of photovoltaic cell based on STC data and dimensionless.* in *2012 IEEE 27th Convention of Electrical and Electronics Engineers in Israel.* 2012.
14. Bellia, H., R. Youcef, and M. Fatima, *A detailed modeling of photovoltaic module using MATLAB.* NRIAG Journal of Astronomy and Geophysics, 2014. **3**(1): p. 53-61.
15. Bana, S. and R.P. Saini, *A mathematical modeling framework to evaluate the performance of single diode and double diode based SPV systems.* Energy Reports, 2016. **2**: p. 171-187.
16. Amiry, H., et al., *Design and implementation of a photovoltaic I-V curve tracer: Solar modules characterization under real operating conditions.* Energy Conversion and Management, 2018. **169**: p. 206-216.
17. Warner, T.H. and C.H. Cox, *I-V CURVE TRACER EMPLOYNG PARAMETRIC SAMPLNG*, USA, Editor. 1984: USA.
18. Macabebe, E.Q.B., C.J. Sheppard, and E.E. van Dyk, *Parameter extraction from I–V*

characteristics of PV devices. Solar Energy, 2011. **85**(1): p. 12-18.

19. Meyer, E.L. and E.E.v. Dyk, *Assessing the reliability and degradation of photovoltaic module performance parameters.* IEEE Transactions on Reliability, 2004. **53**(1): p. 83-92.
20. Rummel, S. and T. J. McMahon, *Effect of cell shunt resistance on PV module performance at reduced light levels.* Vol. 353. 1996: American Institute of Physics. 581-586.
21. Tamrakar, V., S.C. Gupta, and Y. Sawle. *Study of characteristics of single and double diode electrical equivalent circuit models of solar PV module.* in *International Conference on Energy Systems and Applications 2015.* 2015. Dr. D. Y. Patil Institute of Engineering and Technology, Pune, India.
22. Durgadevi, A., S. Arulselvi, and S.P. Natarajan. *Photovoltaic modeling and its characteristics.* in *2011 International Conference on Emerging Trends in Electrical and Computer Technology.* 2011.
23. Masmoudi, F., F.B. Salem, and N. Derbel. *Single and double diode models for conventional mono-crystalline solar cell with extraction of internal parameters.* in *2016 13th International Multi-Conference on Systems, Signals & Devices (SSD).* 2016.
24. Ishaque, K., Z. Salam, and H. Taheri, *Simple, fast and accurate two-diode model for photovoltaic modules.* Solar Energy Materials and Solar Cells, 2011. **95**(2): p. 586-594.
25. Yin, O.W. and B.C. Babu, *Simple and easy approach for mathematical analysis of photovoltaic (PV) module under normal and partial shading conditions.* Optik, 2018. **169**: p. 48-61.

Design of Monopile Foundation for Offshore Wind Turbine

Tushar Raktate[1,*], *Rohan* choudhary[1]
[1]Department of Civil Engineering, Sandip University, Nashik, Maharashtra, India

Abstract: Offshore wind turbines are currently considered as a reliable source of renewable energy. Pre-feasibility study includs calculation of preliminary dimensions of the offshore wind turbine structure used to be perform for preliminary costing to achieve at the commercial capacity of project. The main objective of study is to perform preliminary configuration for commercial viability and approximate size of the foundation pile structure. Design nomograms and equations are derived for preliminary design of monopile founded wind turbines situated at offshore of Gujarat. Parametric studies are carried out on various configurations of hollow monopile by changing water depths and properties of soil. A nonlinear static analysis of substructure is carried out considering aerodynamic and hydrodynamic forces for various structural and soil parameters. The design of sub structure wind turbine is based on API (American petroleum institute) standards. An example problem involving the design of foundations for The proposed area is located 23-40 km seaward side from the Pipavav port at Gulf of Khambhat off Gujarat coast. The site is easily accessible from the Pipavav and Jaffrabad Port, is taken to demonstrate the proposed calculation procedure. The data used for the calculations are obtained from publicly available sources.
Keywords: OWT, support structures, monopile, regression analysis, nomogram.

1. Introduction

Environmental preservation and an integrated perspective of energy use can be achieved by the utilization of renewable energy resources. Total Renewable energy resources account for 12.2% of India's installed capacity, with 70% contribution from wind sector. Presently, the focus is moving towards Coastal wind energy due to some favorable factors like higher wind speed, low visual intrusion and noise, easy transportation of higher capacity wind turbine, Minimum transmission costs for seashore cities, conservation of useful land and over exploitation of potential onshore sites. India currently ranks as the 5th largest onshore wind market in the world and is yet to develop its offshore wind potential.

The MNRE (ministry of new renewable energy) has recognize the Coastal regions of Rameswaram and Kanyakumari in Tamilnadu, the gulf of Kutch and the gulf of Kambhat in the state of Gujarat, as potential zones in India for installation of OWTs. This study investigates the sub-structure design of OWT using ESLs and nonlinear analysis. The novelty in the approach lies in relating important site parameters local to the individual mill to the preliminary dimensioning of the wind turbine foundation.

Almost 75% of the OWT in Europian countries are supported due to their geological character, offshore wind foundations represent a significant part of a project's essentiel expenditure, so the optimiation of model is may leads to considerable savings. By Considering this dffrent parametrs are related to foundation technology selection. i.e.water depth, turbine class based om MW, site condition, cost, installation vessels availability and local fabrication facilities between others. Typical foundations are descibed in following fig. (1)

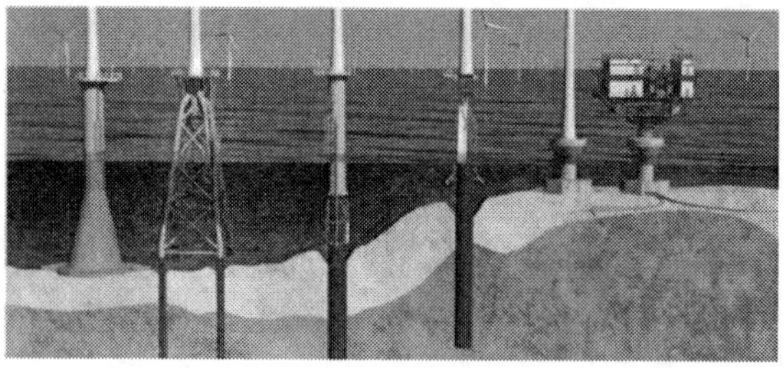

Fig. 1. Types of OWT Foundations design by COWI (From Left :(a) - Concrete gravity based, (b) - Jacket, (c/d) - monopile, (e) -WTG, (f) -Gravity based Substation)

Monopile : Monopiles are used in several wind farm worldwide. This are used upto 40 m deep into water, and for 6-8 MW tubine capacity. Also suitable for different soil condition, and single steel pile is to be embedded into soil, which is connected to WTG through Transition piece.A monopile foundation have 50-80m long steel pile of 5-8 m diameter of pile which is generally driven into soil.Above that "Transition Piece (TP)" is fitted, TP contains appurtenances including internal working

* Corresponding author: **tusharraktate13@gmail.com**

platform, J-tube for cable protection, boat landing and much more.

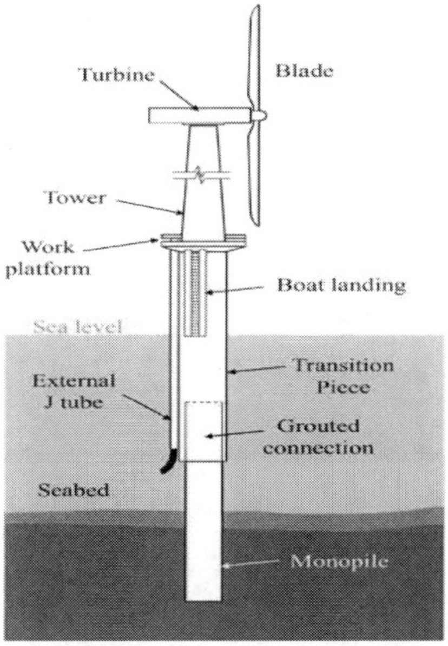

Fig. 2. Component Part of OWT

2. Methodology

The study involves nonlinear static analysis using SACS (structural analysis computer system) and MLRA (multivariable linear regression analysis) using statistical tools like MS Excel and Minitab.

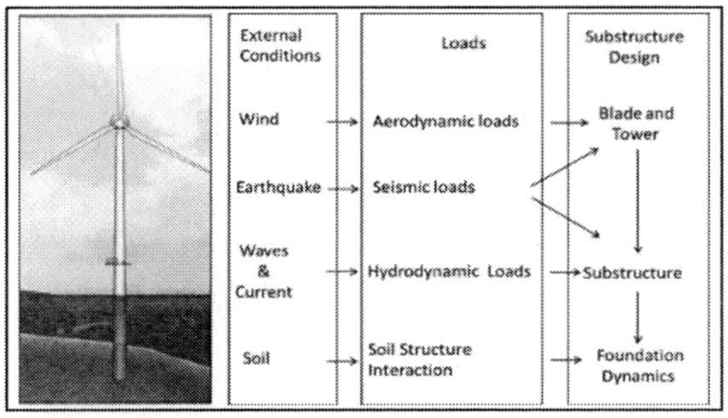

Fig. 3. Substructure design based on external condition

The substructure is design based upon (Aerodynamic, hydrodynamic and Pile-Soil Interaction). Aerodynamic load depend on site condition and turbine. In this paper Aerodynamic load are carried out by using [IEC 61400-2, (2006)] for 5MW wind turbine. The wind load on substructre and tower is estimated based on indian condition [IS 875 part 3, (2015)]. Stokes 5th order wave theory used to find wave kinematics based on water depth and wave parameters. Calculation of forces for structure is based on morison's equation for combination of wave and current. Earthquake load calculation is to be carried out by using [IS 1893 part 2, (2002)] by using response spectrum method.

2.1 The Monopile Supported Wind Turbine Model

The reference wind turbine used in the study is 5 MW NREL wind turbine configuration, as the data are available from the published literature NIWE and the turbine capacity is similar to those adapted in the site Gulf of khambhat gujrat. The modeling, analysis, and design are carried-out using SACS. Fig. 9 shows the SACS model of 5 MW NREL wind turbine. Homogeneous layer of silty sand is considered up-to a depth of 60 m from the mudline. A density 7,850 kg/m3 for steel pile and transition piece is used in the study. Tables 1 and 2 show the properties of 5 MW NREL wind turbine structure

Table 1: Properties of 5MW (NREL) wind turbine

Rating	5 MW
Rotor orientation, Configuration	Upwind, 3 Blades
Control	Variable speed, Collective pitch
Drivetrain	High speed, Multiple stage gearbox
Rotor, Hub Diameter	126 m, 3 m
Height of Hub	90 (m)
Cut-in, Cut-Out, Rated wind speed	3 (m/s), 25(m/s),11.4(m/s)
Cut-in, Rotor speed	6.9 (rpm), 12.1 (rpm)
Rated trip speed	80 (m/s)
Overhang, Tilt of shaft and procone	5m, 5 (deg), 5.5 (deg)
Nacelle Mass	240,000 kg
Rotor mass	110,000 kg
Tower Mass	347,460 kg

Table 2: Tower Properties

Base Diameter of Tower	6 (m)
Base thickness of Tower	0.028 (m)
Top diameter of Tower	3.86 (m)
Top thickness of Tower	0.018 (m)
Modulus of Elasticity	210 GPa
Modulus of Rigidity	80.7 GPa
Steel density	8,500 kg/m^3
Height above ground	87.5 (m)
Total mass	347,460 kg
Structural - Damping Ratio	1%

2.2 Nonlinear Static Analysis

Wind is predominant for offshore structure. Calculation of Aerodynamic forces require information related site condition, along with wind shear and mean wind speed. The aerodynamic load on turbine are carried out by using FAST developed by NREL based on BEM (blade element momentum) theory, for standard 5 MW NREL wind turbine. The predominant frequency for the aerodynamic thrust force is obtained through FFT (Fast Fourier Transform) for various wind conditions (Fig.

10). Related information is taken by IEC (61400 part 3-2015) The aerodynamic load in time-domain is converted to ESL using DAF (Dynamic amplification factor). DAF is a dimensionless quantity which describe the number of time deflection or stress that is multiplied to deflection and stresses caused by static load when dynamic load is to be applied to the structure.

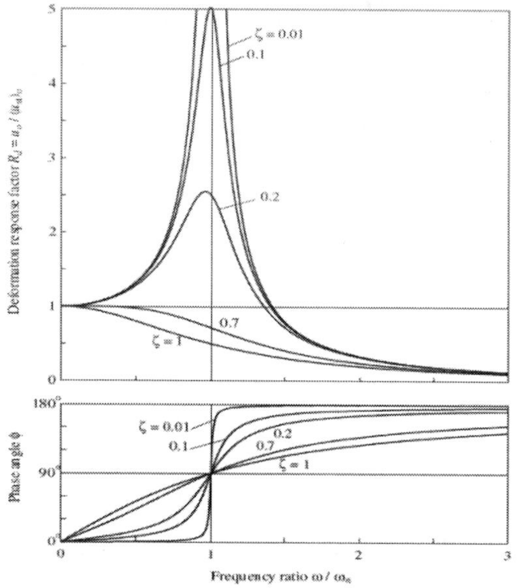

Fig. 4. DAF for damped SDOF system subjected to harmonic force.

Foe single degree of freedom subjected to harmonic motion with viscous damping, DAF is to be calculated based on following eq (1). The graphical representation of DAF vs. frequency ratio for damped Single degree of freedom is excited by harmonic force is shown in fig. (4)

$$DAF=\frac{Xdyn}{Xst}=\frac{1}{\sqrt{[(1-(w/w_n)2)2+(2\times\epsilon\times w/w_n)2]}} \;\ldots\ldots\ldots\; (1)$$

The maximum force of thrust is 1.2 MN at the rotor level of the turbine of self weight is 3434 kN is obtained for ECD+R wind condition. Extreme wind speed value of 50 m/s for gulf of Kambhat and gulf of Kutch regions is obtained using IS 875-part 3-1987 code provision. The total wind force on the tower is calculated using SACS (Connected edition) tool based on API standards.

For wind load calculation structure is divided into different geomatrical shapes and wind load on this geomatrical shape is calculated by using eq. (2)

$$F = 1/2\times\rho\times C_e\times U^2\times A \;\ldots\ldots\ldots\ldots\; (2)$$

The variation of wind intensity is generlly in time and space. Statistical wind properties like (mean wind speed and standard deviation of speed) taken over duration with 1 hour variation. The 1 hour means wind speed U(z) at height z (ft) above the sea level which is calculated based on Eqs. (3) and (4)

$$U_{(z)} = U_O \times [1 + C \times \ln(\frac{z}{32.8})] \;\ldots\ldots\ldots\ldots\; (3)$$

$$C = 5.73 \times 10^{-2} \times (1 + 0.0457 \times U_O)^{1/2} \;\ldots\ldots\ldots\ldots\; (4)$$

The hydrodynamic load due to extreme waves of period 12 S and breaking wave height of 0.78 times the Water depth at site is calculated basd on Morison's equation (Eq. (5)). The equation express wave forces as a inertia forces which is proportinal to partical acceleration and a Non linear force is proportinal to square of the particle velocity.

$$F = C_M\frac{\rho\pi D^2 A}{4} + C_D\frac{\rho D v|v|}{2} \;\ldots\ldots\ldots\ldots\; (5)$$

Pile Soil interaction, analyzes behavior of pile supported strcture structure which is subjected to one or more static load condition. Pile finite deflecton ("P-delta" effect) and nonlinear behavior along with directions bot transverse and longitidunal axis are considered. The program uses distributed spring type foundation model and uses finite difference technique to solve pile model which is represented beam column on nonlinear elastic foundation. Beam-column elements model structural members to resist both axial and bending actions. The structure resting above pile is represented as a linear elastic model. Soil exibits a nonlinear behavior for axial and transverse loads, therefore interactive procedure is used to calculate pile effect on deflection on structure. The analysis is generally based on the theory of subgrade reaction. The pile is discretized into segments with nodes at each end of the segments, and the soil around the pile is replaced by a series of discrete springs. Fig. 5 shows a typical pile-soil model.

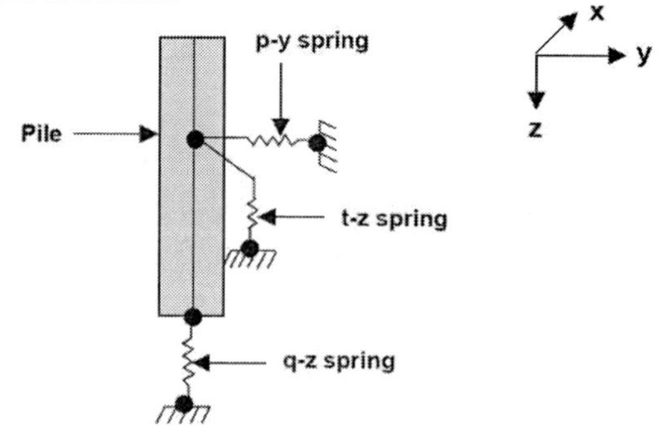

Fig. 5. Typical pile-soil model.

The equation governing the pile-soil interaction for an elastic pile (based on Bernoulli-Euler beam theory) is given in Eqs. (6) and (7).

$$EI\frac{d^4y}{dx^4} + P_x\frac{d^2y}{dx^2} - p - w = 0 \;\ldots\ldots\ldots\ldots\; (6)$$

and substituting,

$p = -k_y$ .... (7)

2.3 Structural Design

The analysis and design is to be carried out on 64 different models of different water depth, current velocity and angle of internal friction. The input variable used for analysis include water depth varying from 10 m to 30 m, current velocity varying from 0.5 m/s to 2 m/s and angle of internal friction of soil varying from 25° to 40° (Table 3). A homogeneous layer of soil is considered throughout the length of the pile for simplicity. The output variables include diameter (D), thickness (t), and minimum length of the pile (L). Thickness of the pile is obtained as a function of diameter, based on API standard.

Criteria of servicability is based on tolerance requirments for design of wind turbine and is described by "Turbine Manufacturer Requirements". Basically this should be turbine specific, inclusing size and hub height. Typically this tolerance are specified in some code of practice or a design specification given by client that should be given by the manufacturer. Maximum pile head rotation after installation (as per DNV-GL code) is specified with 0.25° limit on "Tilt" at necelle level. Maximum rotation at pile head is limited to 0.22° and the UC (unity check) ratio is limited to 0.8.

Table 3: Design parameters

Angle of internal friction of sand	25°, 30°, 35°, 40°
Water depths	25, 30, 35, 40 (m)
Water height	78% of water depth
Wave period	12 (S)
Current speed	0.5, 1, 1.5, 2.0 (m/s)
Wind speed	50 (m/s)

2.4 Statistical Analysis of Monopile foundation

MLRA (multivariable linear regression analysis) is a statistical analysis used for calculating design parameters. It explains the linear relationship between one dependent variable (Y) and two or more independent variables (X). multivariable linear regression analysis is generally uses least squares method to estimate the parameters. The least square method finds optimum when sum (S) of squared residuals is minimum (Eq. (8)). A residual is calculated by using difference between the actual value of the dependent variable (Y) and the value predicted by the model (Y').

$S = \sum_{i=1}^{N}(y_i - y_i')^2$ .... (8)

The prediction of unknown dependent variable is accomplished by the following equation (Eq. (9)):

$y_i' = b_0 + b_1X_{1i} + b_2X_{2i} + \ldots + b_kX_{ki}$ .... (9)

For this study, multivariable linear regression analysis is performed on a 64 different sample size to obtain the equation or nomograms for pile diameter and for pile length. A sample size is a part of the population chosen for the study. Sample size is increases ($n \geq 30$), shape of sampling distribution is approximately normal irrespective distribution of population. It is stated that the minimum number of samples for multiple regression analysis be at least ten times the total number of variables involved in the study.

The calculation of affected parameters for the dependent variables is based on the p-value. The predictor has low P-value is to be addition to the model because changes in the predictor value are related to change in the response value. Conversely, a larger (insignificant) p-value gives that changes in the predictor are not corelated with changes in the response. In other words, p-value indicates the probability of insignificance of a variable in the prediction model. A maximum p-value of 5% is allowed in the model. MLRA mainly involves the following steps:

- Based on the data set, establish a mathematic model to estimate the unknown parameters. Least square method is a common method of estimation.
- Credibility of the obtained relations is tested by p-value of predictors in F-test of the overall significance.
- Introduce predictors with significant influence into the model, and eliminate those with no significant influence.
- Predict the pile outer diameter and pile length using the derived relations.

3. Results

3.1 Multivariable Linear Regression Analysis

The multivariable regression analysis is performed on the sample data using MS Excel's data analysis tool and the design equations are obtained. These equations are tested with new samples. The equation for obtaining the optimum outer diameter (D) of monopile (Eq. (10)) with coefficient of determination, R^2 of 94.50% is:

$D=4.252+0.0732\times WD+0.110\times C$............ (10)

where,
WD: Water depth (m)
CV: Current velocity (m/s)

A equation for calculating the pile outer diameter for a given water depth and current velocity is based on the API formula (Eq. (11))

$t = 6.35 + (D/100)$ ....(11)

The equation for obtaining the minimum (min.) pile length (L) of monopile (Eq. (12)) with R^2 of 86.50% is:

$L = 56.532 - 1.394 \times SA + 1.135 \times MI$........(12)

where,
SA: Angle of internal friction of soil (deg.)
MI: Moment of inertia (m^4)

3.2 Validation

The predicted model is validated by using eight different cases which are taken randomly between 25-40 m. Table 4 and 5 shows the corresponding results from SACS and predicted design equations for pile length and pile diameter.

Table 4: Validation for pile diameter

No.	WD	CV	SA	PD1	PD2	Er1
1	25	0.5	25	6.14	5.30	-0.84
2	27	0.7	27	6.31	5.80	-0.51
3	29	0.9	29	6.47	6.70	0.23
4	31	1.1	31	6.64	6.15	-0.49
5	33	1.3	33	6.81	6.25	-0.56
6	35	1.5	35	6.98	6.40	-0.58
7	37	1.7	37	7.15	6.65	-0.50
8	40	1.9	40	7.39	7.15	-0.24

WD: Water depth (m)
CV: Current velocity (m/s)
SA: Angle of internal friction of soil (deg)
PD1: Pile diameter using equation (m)
PD2: Pile diameter from SACS (m)
Er1: Error in pile diameter (%)

Table 5: Validation for pile length

No.	WD	CV	SA	MI	PL1	PL2	Er2
1	25	0.5	25	5.95	28.44	26.10	-2.34
2	27	0.7	27	6.63	26.42	24.45	-1.97
3	29	0.9	29	7.31	24.40	22.30	-2.10
4	31	1.1	31	8.09	22.50	21.00	-1.50
5	33	1.3	33	8.93	20.67	15.00	-5.67
6	35	1.5	35	9.84	18.91	16.50	-2.41
7	37	1.7	37	10.81	17.22	12.40	-4.82
8	40	1.9	40	12.30	14.73	10.50	-4.23

WD: Water depth (m)
CV: Current velocity (m/s)
SA: Angle of internal friction of soil (deg)
MI: Moment of inertia (m4)
PL1: Minimum pile length using equation
PL2: Minimum pile length from SACS
Er2: Error in pile length (%)

The variation of predicted values for pile outer diameter and pile length from the mathematical model with respect to the original values from SACS analysis is represented graphically in Figs. 6 and 7.

It is seen that the predicted pile diameter values are very much close to the original values with a maximum error of 0.84%. The predicted values for minimum pile length are found to be slightly deviating with that of the original results from SACS analysis with a maximum error of 5.67%. This may be due to the following reasons: The assumption that the soil layer throughout the depth of the pile has the same uniform properties introduces a certain amount of error in the model. However, consideration of various soil type layers in the analysis increases the complexity of the problem by involving several parameters.

3.3 Worked-Out Example

Consider a location where water depth is 25 m, current velocity of 0.5 m/s and medium sand with angle of internal friction of 25°. By substituting the values in the proposed design equations, outer diameter of pile (D) and minimum pile length (L) are obtained.

$D = 4.252 + 0.0732 \times WD + 0.110 \times CV$
$D = 4.252 + 0.0732 \times 25 + 0.110 \times 0.5$
$D = 6.140$ m
$L = 56.532 - 1.394 \times SA + 1.135 \times MI$
$MI = (\pi/64) \times (D^4 - d^4)$
$MI = (\pi/64) \times (D^4 - (D - 2t)^4)$

Thickness of pile (t) = 6.35 + (D/100) with all units in mm, based on API standards.

$t = 6.35 + (6140/100) = 67.75$ mm
$= 6.775$ cm
$MI = (\pi/64) \times ((6.14)^4 - (6.14 - 2 \times 0.06775)^4)$
MI = 5.95 m4
$L = 56.532 - 1.394 \times 25 + 1.135 \times 5.95$
$L = 28.44$ m

Hence, a pile of outer diameter 6.14 m, thickness 6.77 cm and minimum length of 28.44 m is required for the given site conditions. The length mentioned here is the minimum desired length to satisfy the rotation conditions.

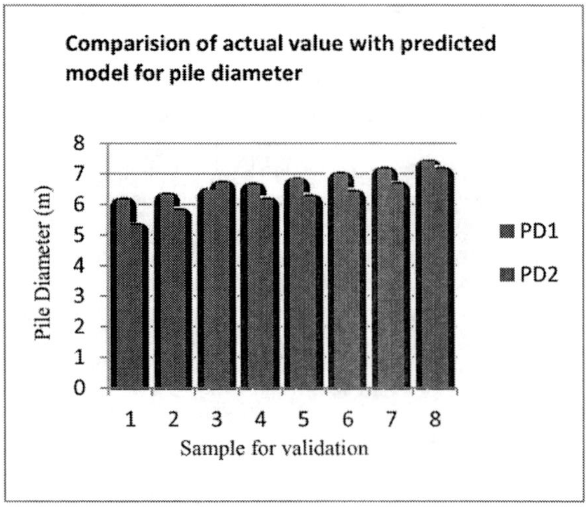

Fig. 6. Graphical representation of predicted values with actual values of pile diameter

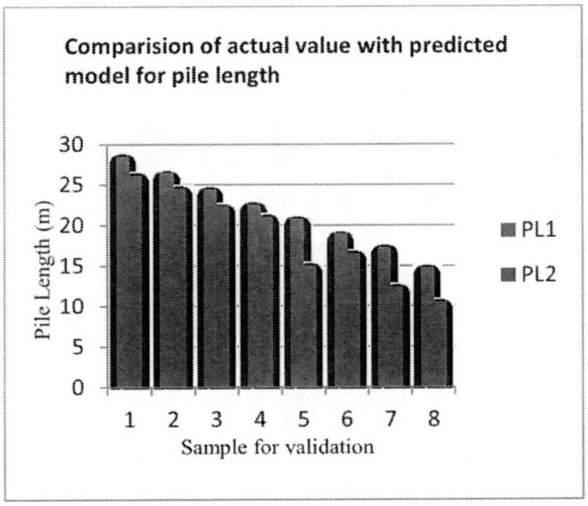

Fig. 7. Graphical representation of predicted values with actual values of pile length

4. Conclusion

This research study gives equations and nomograms for quick preliminary design of monopile foundation for offshore wind turbine using simple design methodology. This study is based on extreme environmental conditions of Gujarat, India. Sometimes, these equations and nomograms are useful for other site having similar environmental conditions and turbine characteristics. The obtained design equations give a clear estimate for the preliminary study on the offshore monopile design for a given site. Hence, the effort as well as time required for pre-feasibility study are reduced by this simplified design methodology. The response of monopiles under dynamic load is not properly understood and guidance in design codes is not given properly. If dynamic design is incorrect, monopile may tilt in the long term. If the tilt is greter than the allowable limit, the wind turbine may be need to shutdown. Monopile design is generally carried out by using API design procedure calibrated for flexible pile design where the pile is pretend to fail by plastic hinges. Future scope include development of prediction model by considering different layers of soil along the pile length and finite element model to generate p-y curves for large diameter piles which are to be further involve in the monopile design for developing the prediction model. The large diameter calibrated p-y curves formulations for sand and clay under monotonic and dynamic conditions will be contribute substantially to reducing conservatism in OWT substructure design.

Acknowledgement

The authors wish to thank and acknowledge the support extended by the Sandip School of engineering and tehnology, Nashik.

References

1. "India Wind Energy Outlook." Global Wind Energy Council, (2012)
2. "Denmark Breaks its Own World Record in WindEnergy."(2016) /.http://www.euractiv.com/section/climate-environment/news/denmark-breaks-its-own
3. Alluri, S. K. R., Gujjula, D., Phani Kumar, S. V. S., and Ramana Murthy, M. V. "Pre-feasibility Study on Fixed Platforms for Offshore Wind Turbine in India." In Proceedings of the 18th Asia-Pacific Conference on Wind Engineering, Chennai, India. (2013)
4. A. M. David.,"Offshore Wind Market and Economic Analysis-Annual Market Assessment." Document Number: DE-**EE005360**, U.S. Department of Energy. (2013)
5. Bhattacharya, Subhamoy. *"Challenges in Design of Foundations for Offshore Wind Turbines."* IET Engineering and Technology Reference, **ISSN 2056-4007**, (2014)
6. API recommended practice 2A-WSD (RP 2A-WSD), "Recommended Practice for Planning, Designing and Constructing Fixed Offshore Platforms—Working Stress Design"
7. NIWE, "FOWPI - wind turbine, Layout and AEP" (June 2019)
8. NIWE, "FOWPI – Advisory Foundation Concept Design" (June 2018)
9. Watsamon Sahasakkul, Hieu Nguyen, Ali Sari, "An Improved Methodology for Design and Analysis of Offshore Wind Turbines Supported by Monopiles" (2016)
10. Sanjeev Malhotra, "Selection, Design and Construction of Offshore Wind Turbine Foundations structure" (2017)
11. Harvey J. Burd, Byron W. Byrne, Ross A. McAdam, Guy T. Houlsby, Chris M. Martin, William J.A.P. Beuckelaers, "Design aspects for monopile foundations " (September 2017)
12. B.C. O'Kelly, "The Offshore wind turbine foundations analysis and design" (May 2013)
13. Ishwariya Shrikanth, Satya Kiran Raju Alluri1, Krishnaveni Balakrishnan, M. V. Ramna Murthy and M. Arockisamy, "simplified design methodology for monopile foundation" (2013)
14. WILLIAM A. SCHAFFER "Monopile Foundation for Offshore Wind Turbine Simulation and Retrofitting" (2017)
15. R Damiani, K Dykes, G Scott, "A comparitive study of offshore wind support structures with monopiles and jackets for structure" (2015)

Tribological investigation on oil blended with Additive using Response surface methodology

Tushar Gadekar[1,*], *Dinesh Kamble* [1]
[1]Vishwakarma Institute of Information Technology Pune 411048, India

Abstract.

Friction and wear in dynamic parts is the primary reason for energy loss in gearbox lubrication system and this can be optimized by utilizing modified lubricant. The tribological nature of gearbox system is critically affected by factors such as type of lubricant, loading & speed etc. In latest years, multiple advanced oil and modern tribological techniques & instruments have been utilized to investigate behaviour of oil like pin on disc, Fourball tester etc. This paper presents comparative investigation of oil blended with additive for two different conditions using prediction model & RSM. The design of experimentations has been conducted by using response surface methodology. The value of inputs parameters such as concentration, load & sliding velocity ranges from 0.5 to 5 %, 60 to 100 N and 0.65 to 1.5 m/s, respectively are utilized to evaluate the outcomes of coefficient of friction and specific wear rate. At the end results from Prediction equations are compared with experimental literature based outcomes to signify the effect of parameters like blend %, load & Sliding speed. The Coefficient of friction model showed 47.57 % more closer outcomes as compared to the Specific wear rate model for specific variation of unknown parameters for pin on disc setup in oil.

1 Introduction

Lubricants are broadly used in the manufacture and engineering zones, along with consumed in automotive and several products to advance the softness of motion between two faces of material and therefore shrink the friction & wear factors in between these studied faces. The prime function of oil is to drop friction, avoid wear of material, soak up heat, avoid rust as well as reduce pollution. Also, it's necessary that a lubricant would be capable to circulate easily adequate to take in and leave generated heat. The additives present in oil advance and reinforce the life of lubricants in various lubrication systems like engine, gearbox, bearing etc.

1.1 Importance of Problem

The low rotating speed of the main shaft with in a wind turbine needs the use of a turbine gearbox in order to increase the rpm of the secondary shaft until the true angular velocity values used for the generator are reached. The gearbox in wind turbine system has dominant maintenance as compared to other component .One of the key issue associated with wind turbine system is the optimal viscosity and anti-scuffing parameter reached at temp. beyond 80 $^{0\,C}$ [1] .Friction and wear reduction is critical problem in such application due to its impact on the energy consumption and maintenance costs. High quality lubrication is required in such applications to overcome the problem. Current assessments proved that 33 % of the fuel energy in car is lost due to friction in an engine, trans- mission, tires and brakes, and only 21.5% of the energy is used to move the car [2]. One of a chief losses happening in an IC engine is due to friction between its dynamic engine parts. This loss is considerable and approximately 15% of the total loss of energy and has a direct impact on the efficiency and durability of the machine.

1.2 Current Status

Recent researchers are concentrated on refining performance of lubricating oil by referring different additives. The influence of add-ons on lubricant performance has been investigated by using pin on disc type tribometer. On the other side there is less investigation on influence of Composite additives on tribological performance of oil in heavy load system like Wind turbine gearbox. Also some of researcher investigated impact of ionic liquid as add-ons on gearbox oil using tribometer but the drop in friction parameter not observed significantly in study.

2 Literature review

The literature study on experimental examination of lubrication nature of various lubricants blended with add-ons & influence of it on COF, wear, viscosity are reviewed and reported under.

* Corresponding author: **tdgadekar13.scoe@gmail.com**

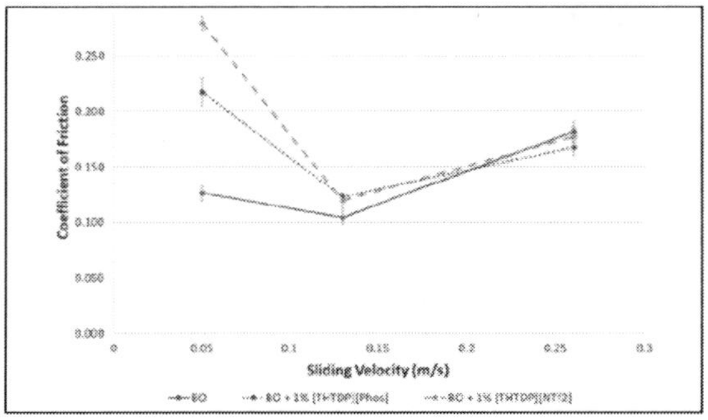

Fig. 1. Coefficient of friction model behavior for oil blended with Ionic Liquid [3]

R. Monge et al. [1] have investigated friction & wear performance of Ionic liquids as add-ons in gearbox oils. In an investigation two distinct ionic liquids 5 wt. % concentration are utilized for examination in wind turbine lubrication system. The Investigation result has showed that used additive have a less friction altering feature but a significant wear reducing performance.

The Lubrication skill of ionic liquids as add-ons in a bio-oil for gearbox lubrication system has been investigated for various sliding velocity [3]. An occurence of the ionic liquid add-on effectively decreases a volume loss in track, particularly under larger speed & reduces less Coefficient of friction as shown in Figure 1. Marques et al. [4] have studied efficiency of a gearbox lubricated with wind turbine gear oils. The examination showed that amongst the investigated gear lubricants the Glycol grounded oil has showed healthier than any further of a selected gear lubricants. Greco et al. [5] have reported wear & Friction performance of boron based surface treatment and Nano-particle lubricant additives for gearbox of wind turbine. The authors have suggested a combined way to include electrochemical boriding in synchronisation with the use of Nano additives.

A. Greco et al. [6] have reported material friction &wear parameter in turbine Structures. This article reports contact failures of bearings from wind turbine gearboxes. The investigation showed that microstructural alteration has been seen closer to the surface. The experimental study on oil blended with single additive by using various types of tribometer is presented for cases like Gearbox, Engine, Bearing etc [2, 3, 8-16].

M. Husnawan et al. [17] have developed friction force Prediction Equation for oil base stock inclosing palm olein & anti-wear additive. Tribological experiment was carried on four ball tester & model force developed by using wear load, speed & operating time. The Response surface methodology (RSM) has been utilized statistical model development.

M. H. Sakinah et al. [18] have presented the use of DOE method in the investigation of the Tribological outcome of Palm Cooking lubricant mixed in Engine oil. The DOE has been assembled by using Response surface methodology to reduce experimental conditions and to develop a mathematical model between parameters such as rotational speeds, volume concentration, and applied loads. Analysis of variance (ANOVA) test has been carried out to check the adequacy of the empirical models developed.

Yashvir Singh et al. [23] have optimized tribological characteristics of seed oil using response surface methodology. The tribological investigation has been conducted by using pin on disc instrument .also statistical modelling of particular wear rate & COF parameters has been done by using experimental results.

3 Experimental Setup

The experimental test Setup & Parameters for tribological investigation of oil blended with additive are reported under.

3.1 Parameter of investigation

After a literature survey implementation it is revealed that several fundamental factors have been utilized in a Lubrication performance measurement. The performance measurement mainly includes investigation of friction coefficient, viscosity, wear etc. A detail summary of tribological performance measurement is shown in fig 2.

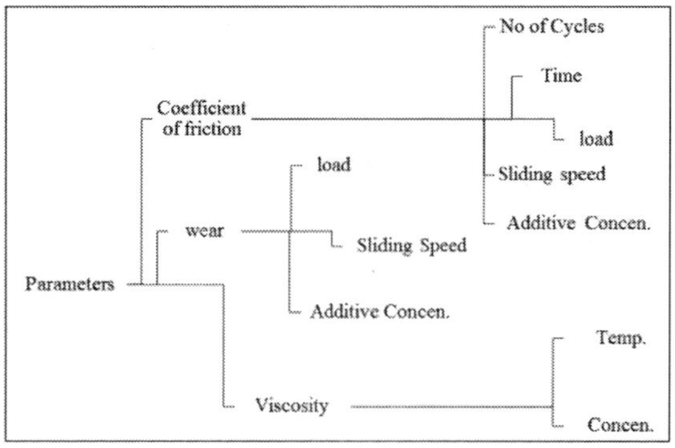

Fig. 2. Parameters in tribological study of oil

3.2 Pin on disc instrument

The Pin-on-disc type configuration is used to investigate coefficient of friction & wear in sliding contact as shown Figure 3.

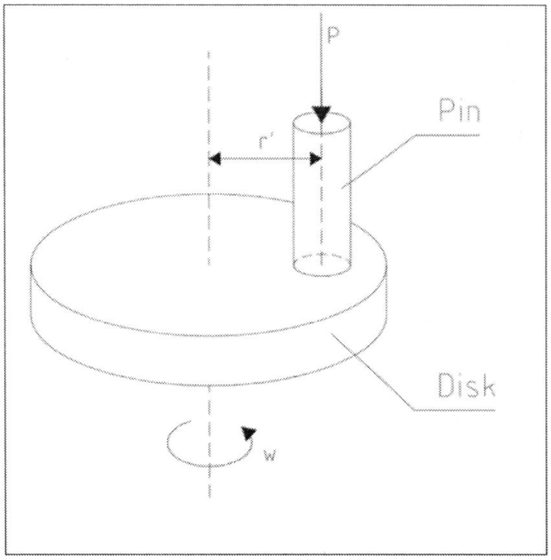

Fig. 3. Pin on disc setup

The configuration includes cylinder-shaped pin counter to a rotating circular disc by a load 'L'. For a pin-on-disc Arrangement two specimens are required namely disc & pin of standard dimensions .The standards & disc dimension for experimentation on pin on disc instrument is mentioned in Table 1.

Table 1. Pin on disc test standard

Pin diameter range in mm	Disc specimen diameters in mm	Disc specimen Thickness in mm	Standard
2 to 10	30 to 100	2 to 10	G99

3.3 Coefficient of friction & Wear rate model for Pin on Disc test

The quadratic models for specific wear rate and Coefficient of friction for Pin on disc setup is given below,

$$COF = 0.039497 - (4.76850\times 10^{-4}) \times B) + (4.475 \times 10^{-5} \times L) - (3.93760\times 10^{-3} \times S) - (5.0\times10^{-8}\ B\times L) + (1.2\times 10^{-5})\times B\times S) + (2.78750 \times10^{-5} \times B^2) - (5.00 \times10^{-9})\times L^2) - (2.400\times10^{-5})\times S^2) \quad (1)$$

$$Wear = 1.37377\times10^{-5}) - (4.6299 \times 10^{-7})\times B) + (6.8257 \times 10^{-8} \times L) - (3.3852\times 10^{-6}\times S) - (2.45 \times10^{-11} \times B \times L) + (1.18 \times 10^{-8}\times B\times S) - (8.20560 \times10^{-10} \times L \times S) + (2.9357\times 10^{-8}\times B^2) - (1.05993\times10^{-10} \times L^2) - (1.34229\times10^{-7})\times S^2) \quad (2)[23]$$

Where, B is the blend percentage, L is the load in N, S is the sliding velocity in m/s

4 Design of Experiment

The Design of experiment for gearbox oil blended with additives is mentioned under. The experimental run suggested by Central composite design of response surface approach has been presented.

4.1 Response surface methodology

Response surface methodology is the method to choose numerous relations between progression parameters and tribological parameters and discover the influence of these progression parameters on the combined responses.Fig.4.

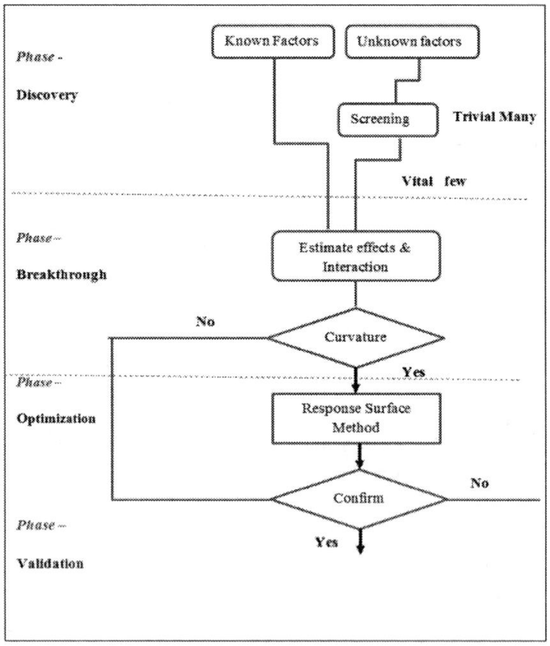

Fig. 4. Steps in RSM

RSM is deal with the usage of factorial design .In RSM the crucial effect of a factor is well-defined as the deviation in response initiated by a variation in a level of the factor considered like Concentration or Speed or Load etc., while the other parameters are retained constant.

4.2 DOE by Response surface methodology

The experimental trial suggested by CCD Approach of response surface approach has been developed on Design expert ® Software. Table 2 shows the appropriate levels of variables referred to design the factors for a tribological experiment on pin on disc Tribometer.

Table 2. values of independent parameters

Parameters	-1	1
Load in (N)	60	100
Concentration in wt. %	0.5	5
Sliding velocity in m/s	0.65	1.5

In DOE three parameters considered as a known parameter namely Load, Concentration ,sliding velocity, which has 2 levels each hence for CCD with 3 center points 17 runs obtained from Design expert ®.Table 3 shows the Design of experiment using by Response surface methodology .

Table 3. Design of experiment using CCD

Run	Load in N	Concentration	Sliding Velocity
1	60	5.00	1.50
2	114	2.75	1.08
3	80	2.75	0.36
4	80	6.53	1.08
5	60	5.00	0.65
6	100	5.00	1.50
7	100	5.00	0.65
8	80	2.75	1.08
9	46	2.75	1.08
10	60	0.50	1.50
11	100	0.50	0.65
12	80	2.75	1.79
13	80	2.75	1.08
14	80	2.75	1.08
15	60	0.50	0.65
16	100	0.50	1.50
17	80	-1.03	1.08

5 DOE vs. Experimental literature based results

The design of experiment & experiment literature based results of tribological parameter for Coefficient of friction & specific wear rate model is presented under.

5.1 Coefficient of friction

The Coefficient of friction model results for various 17 trials of independent parameters are presented in Figure 5. In literature based inputs the maximum value of Coefficient of friction model is 0.0406 & minimum value is 0.0248.

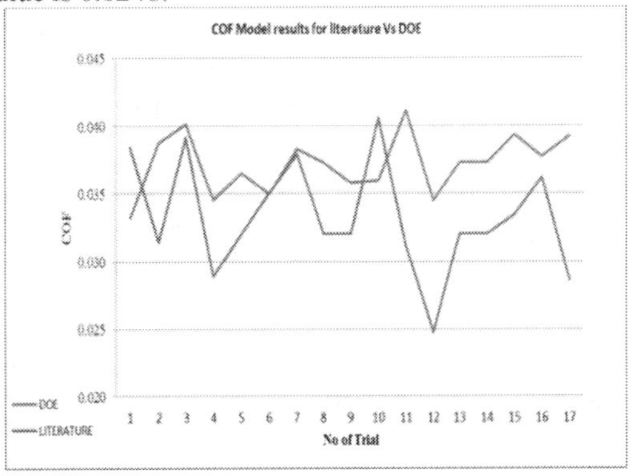

Fig. 5. Coefficient of friction values for DOE & Literature

In CCD based inputs the maximum value of COF model is 0.0411 & minimum value is 0.033 shown in Fig.5.In COF based results analysis most of the values of DOE varies in parallel pattern as that of Literature values.

5.2 Wear

The wear outcomes for both inputs shown in Figure 6. In literature based inputs the maximum value of specific wear rate in (mm^3/Nm) model is 1.69×10^{-5} & minimum value is 9.71×10^{-8}.

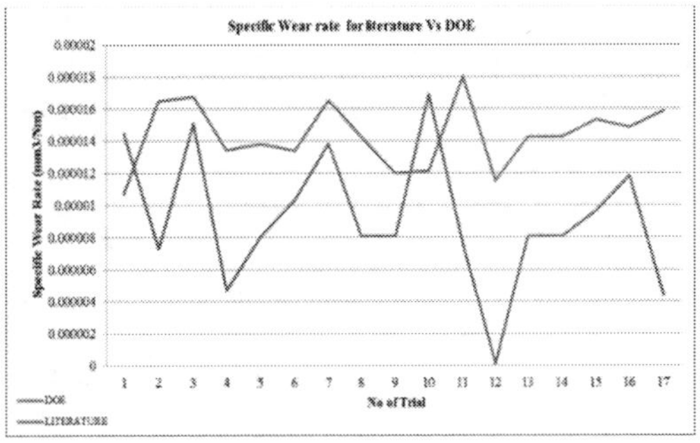

Fig. 6. Wear values for DOE & Literature inputs

In DOE based inputs the maximum value of specific wear rate in (mm^3/Nm) model is 1.80×10^{-5} & minimum value is 1.072×10^{-5}. In Specific wear rate based results analysis most of the values of DOE varies in in extreme pattern as that of Literature outcomes.

6 Conclusion

Wear and friction drop is critical issue in gearbox systems and it can be resolved by using high performance modified Lubricant. The Prediction model results for literature & CCD based inputs are compared & investigated by referring 17 trials for various conditions for Pin on disc setup. On the origin of present comparative investigation particular important findings are given as:

1. Response surface methodology (CCD) & literature based values are compared for pin on disc setup and the comparative examination revealed that 23.5 % closer values to experimental reference values in COF model, the specific wear rate model showed 11.7 % closer values to experimental values. .

2. In COF model the maximum value of Coefficient of friction is 0.0411 & 0.0406 for CCD & Experimental literature based inputs respectively. Also the minimum value of COF is 0.0248 found in experiment literature outcomes.

3. In specific wear rate model the extreme variation in curves are observed for 17 trials of tribological examination.

4. The extreme specific wear rate value for CCD based inputs is 1.80 x 10^{-5} mm^3/Nm & the minimum value is observed for trial 12.

5. The Coefficient of friction model showed 47.57 % more closer outcomes than Specific wear rate model for specific variation of unknown parameter like load, blend % & sliding velocity.

References

1. R. Monge, R.González, A.Hernández Battez, A.Fernández-González, J.L.Viesca, A.García, Wear, 50-63(2015).
2. Z. Y. Xu, K. H. Hu, Y. K. Cai, F. Huang and C. L. Han, 1-7 (2014);.
3. Edward Cigno, Christina Magagnoli, Michael S. Pierce, P. Iglesias, Wear, 756–765 (2017).
4. Pedro M.T.Marques, Carlos M.C.G.Fernandes, Ramiro C.Martins, Jorge H.O.Seabra, Tribology International, **71**,7-16 (2014).
5. A. Greco, K. Mistry, V. Sista, wear, **271**,1754-1760 (2011).
6. A. Greco, S.Sheng, J.Keller, A.Erdemir, wear, **302**; 1583-1591(2013).
7. Naveenkumar Rajendhran, Siva Palanisamy, Arunkumar Prabhakaran Shyma, Rajendran Venkatachalam, Tribology International, 156–168 (2018).
8. Carlos M.C.G.Fernandes, Pedro M.T.Marques, RamiroC.Martins, Jorge H.O.Seabra,Tribology International, (2014).
9. Carlos M.C.G.Fernandes, Pedro M.T.Marques, Ramiro C.Martins, Jorge H.O.Seabra, **88**, 309-316(2015).
10. Meena Laad,Vijay Kumar S. Jatti, Journal of King Saud University – Engineering Sciences, (2016).
11. W. Gregory Sawyer, John C. Ziegert, Tony L. Schmitz & Timothy Barton, Iranian Polymer Journal, 284-290(2006).
12. Lin Zhang, Dapeng Feng, Bin Xu,Tribol Lett,95-101 (2009) .
13. M.S. Charoo , M.F. Wani , Tribology Industry,156-162 (2016) .
14. Fatima Leonor Guzman Borda, Sylvio Jos_e Ribeiro de Oliveira, Leticia Maria Seabra Monteiro Lazaro, Albino Jos_e Kalab Leir, Tribology International, 52-58(2018).
15. Kai Gao, Qiuying Chang, Bin Wang , Ningning Zhou , Tao Qing, (2018).
16. A. Vadiraj, G. Manivasagam, K. Kamani, V.S. Sreenivasan, Tribology in Industry, (2012):1; 3-10
17. M. Husnawan, M.G. Saifullah, H.H. Masjuki, Tribology International, 74-81 (2007).
18. M. H. Sakinah, A. K. Amirruddin, K. Kadirgama, D. Ramasamy, M. M. Rahman, & M.M. Noor, Advances in Tribology, 1-11(2016).
19. Sheida Shahnazar, Samira Bagheri & Sharifah Bee Abd Hamid, International journal of hydrogen energy, 1-18(2015).
20. Mohammad Hemmat Esfea, Ali Akbar Abbasian Aranib & Saeed Esfandehb, Applied Thermal Engineering, 493-506(2018).
21. Ting Luo,Xiaowei Wein, Haiyan Zhao,Guangyong Cai, Xiaoyu Zheng, ceramics international, 10103-10109(2014).
22. Zhenyu J. Zhang, Dorin Simionesie and Carl Schaschke, lubricants, 44 -65(2014).
23. Yashvir Singh, Paramvir Singh, Abhishek Sharma, Praveen Choudhary, Amneesh Singla, Nishant Kumar Singh, Egyptian Journal of Petroleum, 1145–1155(2018).

Application of additives with gasoline fuel: A review

Mayur Jadhav[1,*], *Swati* Jadhav[1], *and Supriya* Chavan[2]
[1]Government College of Engineering, Karad, Maharashtra, India
[2]Indian Biodiesel Corporation, Baramati, Maharashtra, India.

Abstract: Alternative fuels have become very important nowadays and proving their importance in both positive ignition engines and compression ignition engines, so as to reduce the consumption of gasoline and diesel respectively. Also, blending of these alternative fuels (additives) into gasoline has been proved important in improving various performance parameters and reducing emission parameters. Various studies have been carried out in the field of using additives in gasoline to see the effects on performance parameters like Brake power, Brake thermal efficiency, Volumetric efficiency and brake specific fuel consumption. Also, this study emphasizes reducing the emissions to the ambient. The aim of this review is to compare and study various additives which can be blended with gasoline and study their effect on performance and emission characteristics of Spark ignition engines.

1 Introduction

Fossil fuels serve as one of the primary energy sources for the world today. World energy demand is increasing day by day due to rapidly increasing population. Prices of fossil fuels are increasing due to increasing demand which is impacting the economy of nations where fossil fuel production is not significant. Fossil fuels are hazardous for the environment and are causing the effects like acid rain, ozone layer depletion and the greenhouse effect. Therefore, worldwide research in the field of alternative fuel sources for engines is carried out.

Gasoline consists of C-4 to C-12 hydrocarbons and it is a mixture of alkanes, alkenes and cycloalkanes. Generally used chemical formula for gasoline is $C_{6.97}H_{14}$.

Over the years, rules and regulations regarding the emission standards are changing and demand reduction in emissions from automobiles. From April 2020 BS-VI (Bharat stage VI) emission standards will be implemented in India which require reduction in NO_x and CO emissions as compared to BS-IV (Bharat Stage IV) standards. Hydrocarbon (HC) emission limit has been kept the same in BS-IV as well as BS-VI standards. NO_x emission limits are reduced by 25% for light duty vehicles (M1, M2, N1 Class I, N1 Class II, N1 Class III, N2) as per BS VI standards as compared to BS IV standards whereas CO emission limits are kept unchanged. For three wheelers NO_x emission limits are kept unchanged and CO emission limit is reduced by 53% in BS VI as compared to BS IV. For two wheelers, NO_x emissions are reduced by 85% for class 1 and 2-1, 82% for class 2-2 and 70% for class 3-1, class 3-2. CO emission limits are reduced by 29% for class 1 and 2-1, 49% for class 2-2 and by 49% for class 3-1, 3-2. With the help of additives it is observed that there is improvement in performance and reduction in emissions. Hence we can comply to BS VI emission standards with the use of various additives.

There are various additives that are added to gasoline for boosting the performance and reducing emissions. Additives are mainly characterised into two categories Eco-friendly additives (bio-additives) and Chemical based additives. Bio-additives mainly consist of Ethanol, Turmeric leaf oil and algae oil. Whereas Chemical based additives are mainly categorised as follows:

a) Oxygenates: Propyl alcohol, Butyl alcohol, Isopropyl alcohol, gasoline grade t-butyl alcohol.

b) Ethers: Methyl tertiary-butyl ether, Tertiary-amyl methyl ether, Tertiary-hexyl methyl ether, Ethyl tertiary butyl ether, Tertiary-amyl ethyl ether.

c) Antiknock agents: Dimethyl methyl phosphate, Toluene, Isooctane, Triptane, etc.

d) Nanoparticles: Fe_2O_3, TiO_2 etc.

e) Fuel additives in general: Ethers are used for running the engine in difficult to start engines. Nitro-methane is used for high performance in racing.

2 Discussions

Schematic of the experimental setup used by various researchers for performance, combustion and emission characteristics evaluation of fuel blends for an SI engine is shown in figure. 1. Experimental setup consists of a variable compression ratio SI engine fuelled with blends of gasoline and other additives. Five gas analyzer is used for measuring exhaust gases like CO_2, CO, NO_x and HC.

*Corresponding author: **jadhavmayur455@gmail.com**

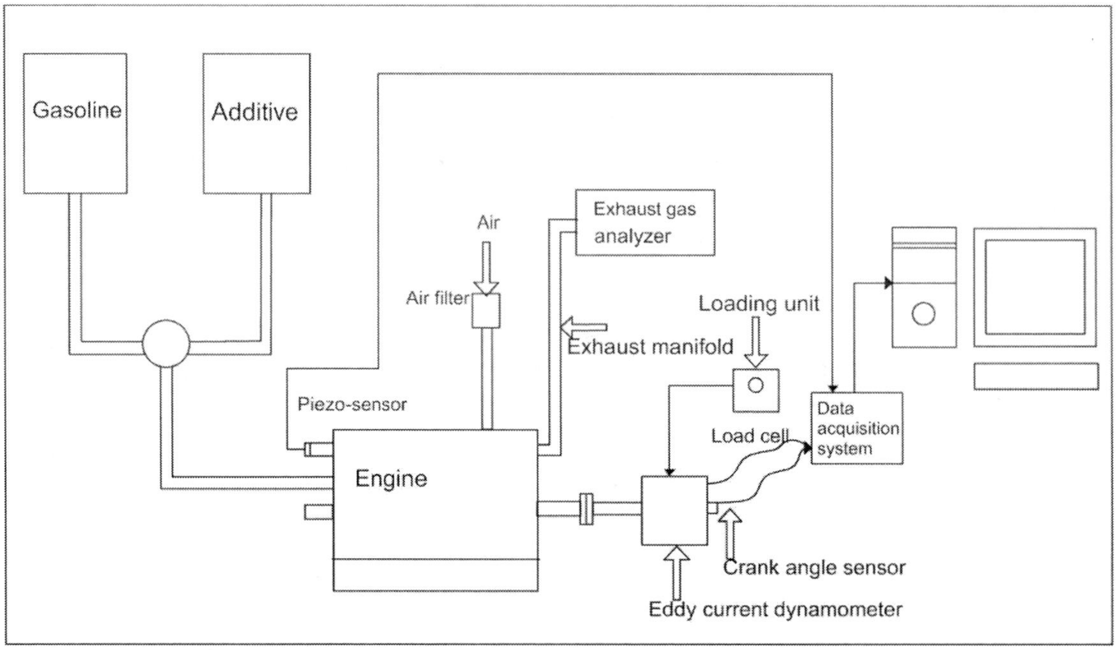

Figure. 1. Experimental schematic for performance, combustion and emission investigations on SI engine setup

2.1 Literature review

There are various researches carried out related to additives in gasoline. Brief overview of previous research along with different additives is given below:

Manish Saraswat and Nathi Ram Chauhan [1] investigated the use of butanol and algae oil blends. Compression ratio, blending percentage and load were varied and performance, emission parameters are determined. The studies indicated enhancement in power, torque and HC, CO, NO emissions were reduced with the use of butanol as additive. On the contrary, upsurge in nitrous oxide, carbon monoxide could be seen. For algae oil Brake specific fuel consumption, nitrous oxides and carbon monoxide emissions are scaled up whereas brake thermal efficiency is reduced.

M. Mourad and K. Mahmoud [2] investigated spark ignition engine performance and emission with ethanol/ butanol-gasoline mixtures. Gasoline is blended with butanol and ethanol in the blending percentage of 2%, 5%, 10%, 15% and 20%. Performance and pollutant emissions were measured at varying engine speeds at low loads applied to the engine. The results show that mitigation in exhaust gases were measured at variable engine speeds at low loading. The results show that there was a mitigation of exhaust gases from engine for Carbon monoxide and Hydrocarbons. Also there was a slight scale down in engine power which was 11% lower than neat gasoline. Also scale down in specific fuel consumption was recorded.

Ashraf Elfasakhany [3] investigated emission and performance of SI engines using ternary bio-ethanol, iso-butanol and gasoline blends. It was observed that ternary blends were responsible for reduction in emissions of UHC and CO while also mitigating performance compared to pure gasoline fuel. However by increasing the percentage of additives (>10%) engine performance shows considerable improvement in performance as compared to neat gasoline.

Mustafa Koç et al. [4] studied the effects of ethanol blends (E0, E50, E85) in unleaded gasoline at different compression ratios. The results indicated that ethanol blended with gasoline increases performance parameters such as engine torque, power and increases fuel consumption and reduces Carbon monoxide, nitrous oxides and hydrocarbons emissions. It was also found that if ethanol blended with gasoline then higher compression ratio without knock can be achieved.

Ashraf Elfasakhany [5] investigated performance and exhaust emissions from SI engine operated with ethanol-gasoline, methanol-gasoline and methanol-ethanol-gasoline blend (3%, 7% and 10% each additive). Results show that when SI engine is operated with ethanol-methanol-gasoline blends then CO and UHC emissions were considerably reduced in comparison to base fuel. Methanol-gasoline blends recorded lowest CO, UHC emissions whereas ethanol-gasoline blends exhibit moderate CO and UHC emissions. Highest volumetric efficiency and torque were recorded in case of ethanol-methanol blends whereas highest power output was recorded with ethanol-gasoline blends.

Mehmet Ilhan Ilhak et al. [6] studied engine performance, emission and combustion analyzes in SI engine fuelled by gasoline-acetylene blends. The testings were performed at 1500 rpm engine speed and variable loads. To prevent knocking phenomena, the addition of acetylene to gasoline reduced brake thermal efficiency at rich stoichiometric condition and increased CO emissions whereas brake thermal efficiency was increased by adding acetylene to lean mixtures and decreased CO emissions. Also addition of acetylene increased combustion limits, reduced UHC emission and increased NO_x emissions.

Yu Li et al. [7] studied combustion, emission and performance characteristics of MPFI SI engine fuelled with methyl alcohol, ethyl alcohol and butyl alcohol-gasoline blends under various blending percentage and engine loading conditions. The combustion phasing was improved using alcohol due to higher burning velocity, which subsequently reduced brake thermal efficiency. ethyl alcohol and methyl alcohol blends produced lowest UHC and NO_x emissions.

Hazim Sharudin et al. [8] investigated the effects of iso-butanol mixing on SI engine operated with methanol-gasoline blends. methanol blend (5%) was blended with iso-butanol additive for 5%, 10%, 15%. Experiments were conducted at 100% load conditions with speed ranging between 1000-2500 rpm. The results show that 15% of iso-butanol in methanol-gasoline (M5) blend indicated scaling up in brake power, brake thermal efficiency and exhaust gas temperature as compared to other blends. In terms of engine emissions, above mentioned blends gave significant mitigation in carbon monoxide and hydrocarbons emissions as compared to neat gasoline.

I Gravlos et al. [9] studied emission parameters of SI engine operating on different alcohols blended with gasoline fuel. The alcohol component in the blends consisted of different alcohols like methyl alcohol, ethyl alcohol, propyl alcohol, butyl alcohol and pentyl alcohol. The parameters under test were engine speed, CO, CO_2, NO and HC emissions. During variable load tests, it was observed that CO and HC levels in engine emissions were scaled down with the use of alcohol-gasoline blends. NO emissions were higher as compared to base fuel.

Eran Alptekin and Mustafa Canakci [10] studied that performance and emission parameters of solketal-gasoline fuel blend in SI engines. Octane number, brake specific fuel consumption were increased as well as CO_2 and NO_x emissions were increased with the use of solketal blends as compared to neat gasoline. Hence, it was concluded that solketal can be used to improve octane number and as an oxygen providing fuel.

Mohsen Ghazikhani et al. [11] experimentally studied exhaust emitted and performance characteristics of a 2-stroke spark ignition engine fuelled with gasoline blended with ethanol. In this study performance characteristics such as delivery ratio, efficiency of trapping and scavenging efficiency were studied. Also, emissions of carbon monoxide, carbon dioxide and nitrous oxides and hydrocarbons were under consideration. In this experiment, gasoline was blended with ethanol in three different percentages of 5% to 15%. Results were recorded for speed range of 2500 rpm to 4500 rpm. Due to vaporization of ethanol scavenging efficiency enhanced along with delivery ratio. Also, for 5% ethanol in gasoline brake specific fuel consumption decreased even though ethanol has lower calorific value. Carbon monoxide emissions reduced by 71% for 15% ethanol blended gasoline sample. Hydrocarbon emissions decreased by 6% for 5% ethanol, 12% for 10% ethanol and around 18% for 15% ethanol. For 15% ethanol blended gasoline sample highest NO_x reduction of 80% is observed and around 38% reduction in NO_x for all other samples is observed.

Huseyin Serdar Yucesu et al [12] studied the effect of ethanol in gasoline fuel on operating performance and exhaust with different operating compression ratio. Blending percentages of 10%, 20%, 40%, 60% were used and the results were compared with neat gasoline. Experiments were performed with six different compression ratios of 8:1, 9:1, 10:1, 11:1, 12:1 and 13:1. It was observed that as the compression ratio increased, torque increased and with increase of blending percentage, torque is enhanced slightly for particular compression ratio. In the case of exhaust emissions , fuels containing 40% and 60% ethanol had a large effect on the emissions. The maximum reduction was obtained with 40% and 60% ethanol blend at 2000 rpm speed. Average reduction of CO emissions were close to 11% for both 40% and 60% blends. Also hydrocarbons decreased was also observed which was comparatively higher than carbon monoxide.

Rinu Thomas et al. [13] experimentally studied the effect of varying compression ratio on emission and performance parameters of spark ignition engine operated with gasoline and n-butanol blends at different loading conditions. Parameters under consideration were brake thermal efficiency, exhaust gas temperature, emitted pollutants. Experiments were conducted at compression ratios of 7, 8.5 and 10. The fuels used in study are pure gasoline and 20% n-butanol blended in gasoline. It was observed that thermal efficiency was improved at higher compression ratios. Combustion efficiency improves at higher temperature and pressure obtained at higher compression ratios. It was noticed that comparatively better results for brake thermal efficiency were obtained for 20% blend of n-butanol than neat gasoline. It was concluded that fuels having comparatively good anti-knocking properties will enhance brake thermal efficiency at higher compression ratio. As compression ratio is increased , reduced exhaust gas temperatures were recorded. This can be related to better conversion of heat energy to useful work. Higher in-cylinder temperature, pressure and stoichiometrically rich mixtures at maximum load resulted in enhanced exhaust gas temperature. It was observed that supply of stoichiometrically rich mixture at higher load results in incomplete combustion of fuel. This increases hydrocarbon and carbon monoxide emissions. The use of 20% blend reduces carbon monoxide emission at higher compression ratio and higher load. NO_x emission is observed to be increased but that can be tackled by effective use of exhaust gas recirculation.

M. Bahattin Celik et al. [14] studied the effect of methanol fuel on single cylinder SI engine for variable compression ratios. In this experiment, methanol was used as as fuel at high compression ratio to enhance the performance and decrease the emissions. Knocking was observed at a compression ratio of 8 for gasoline while it was not observed for methanol fuel. The knock was determined from cylinder pressure-time curves. It was found that carbon monoxide, carbon dioxide and NO_x emissions were reduced significantly with very less changes in brake power at CR=6:1. With increase in CR from 6:1 to 10:1 with methanol as a fuel, engine brake power and thermal efficiency increased. Also exhaust emissions were reduced with an increase in compression ratio.

Zhi Tian et al. [15] studied the various parameters in turbocharged direct injection spark ignition engine. Blending percentage of 10% and 20% each was used for methanol, ethanol and butanol. Neat gasoline was also tested simultaneously. It was observed that as the rpm was increased, there was an increase in torque for all the samples. It was observed that alcohol as an additive in gasoline increases brake torque as well as heat release rate. Highest increase in brake torque and heat release rate was observed for 20% methanol blend followed by 20% ethanol blend respectively. It was also found that adding alcohol increases thermal efficiency of the engine but it also results in increase in fuel consumption. 20% methanol has the highest thermal efficiency and fuel consumption. Also, decrease in carbon monoxide and carbon dioxide emissions is observed with the addition of alcohol.

Changwei Ji et al. [16] investigated performance and combustion parameters of spark ignition engine fuelled with blends of dimethyl ether and gasoline. It was found that by increasing dimethyl ether content in fuel, fuel consumption could be reduced upto maximum 28%. Also by using a blend of dimethyl ether and gasoline, HC and NO_x emissions were reduced. With the blending of dimethyl ether in gasoline, combustion period, flame development and propagation durations were reduced. Also results were obtained that there is an increase in CO emissions as dimethyl ether content is increased.

M. Mourad and Khaled R. M. Mahmoud et al. [17] investigated performance parameters and emissions of SI engine fueled with propanol blended with gasoline blends. Neat gasoline and blends of 5%, 10%, 15% and 20% of propanol were used. Engine tests were carried out in the speed range of 1000 rpm to 5000 rpm.at constant load. The results show that engine torque is increased marginally and brake specific fuel consumption is decreased if the propanol blended gasoline is used. 15% propanol blend gives best results among all the samples.

Gopinath Dhamodaran et al. [18] studied operating performance, exhaust and combustion characteristics of multi point fuel injection SI engine system fuelled with gasoline blended with di-isopropyl ether. Blending percentage of 10%, 20% and 30% di-isopropyl ether fuel blend in gasoline was used. The experiments were performed in the speed range of 1400 rpm-2800 rpm under two different loads. Results of neat gasoline were compared and analyzed with blends of di-isopropyl ether and gasoline. Di-isopropyl ether acts as an oxygenate and provides oxygen for the complete combustion of fuel inside the combustion chamber. Hence, when SI engine was operated with DIPE-gasoline blends, thermal efficiency was enhanced. Also, heat release rate and pressure inside the cylinder was increased. CO and HC emissions were reduced considerably by 25% and 27% less than neat gasoline whereas CO_2 and NO_x emissions gasoline. It was observed that by adjusting ignition timing, it is possible to reduce NO_x emissions.

A. Keskin and M. Gürü [19] studied the effect of blending of ethanol and propanol with neat gasoline and using it for SI engine. Exhaust and noise of the engine were parameters under consideration. Blending percentage of 4% to 20% with intervals of 4% on a volume basis were used for ethanol and propanol blending. Carbon monoxide and hydrocarbons emissions were decreased whereas NO emissions increased as the blend percentage increased for ethanol and propanol. Carbon dioxide emissions were increased as the blend percentage increased. Noise levels were comparatively lower with ethanol-gasoline blend. However for propanol-gasoline blend, noise level increased with blend percentage.

Alper Calam et al. [20] studied engine operating performance and exhaust of engine fuelled with gasoline blended with fusel oil. In this study, neat gasoline and fusel oil-gasoline blended samples are investigated for variable ignition timing. Blend percentages of 5%, 10%,15%, 20%, 30% and 50% were used with pure gasoline. It was observed that torque increased whereas fuel consumption reduced with ignition timing. Exhaust gas temperature reduced for all the samples with decrease in ignition timing. Effective efficiency increased with ignition timing for all the samples.

Ahmed N. abdalla et al. [21] studied performance parameters and exhaust emissions of SI engine fuelled with gasoline-fusel oil blend. Fusel oil is a by-product produced from fermented agricultural products. Pure gasoline and gasoline-fusel oil blends of 5%, 10%, 20%, 30% and 50% were used. It was observed that brake power increases for all the blends with speed and it is higher than gasoline for higher speed and 60% open throttle. Brake specific fuel consumption of 10% blend was higher than 20% blend and neat gasoline. 10% blend has shorter combustion period compared to 20% blend and neat gasoline, hence engine power was observed to be increased. NO_x emissions were lower for blends as compared to gasoline while CO_2 and HC emissions are increased for blends.

2.2 Concluding remarks

From above literature review following remarks can be made:

(1) Alcohols like methanol, ethanol, propanol and butanol reduce emissions of HC and CO whereas they increase emissions of CO_2 and NO_x in most of the cases. Also performance parameters like brake thermal efficiency is slightly reduced. Methanol and Ethanol blended with gasoline allows to achieve higher compression ratio without knocking phenomena. blending of ethanol has two important effects on fuel blend properties: An increase in octane rating and decrease in Calorific value. The first effect dominates performance upto 10% blending and after that second effect starts to dominate. Low reid evaporation pressure of alcohols enables blended gasoline to be stored and transport safely.

(2) Algae oil is not desirable additive as it increases emissions of NO_x and CO as compared to neat gasoline. Also, it increases fuel consumption and reduces thermal efficiency.

(3) If acetylene is added to lean mixtures then it increases brake thermal efficiency and reduces CO emissions. Also, acetylene addition reduces UHC and increases NO_x.

(4) With the blending solketal-gasoline fuel, octane number and brake specific fuel consumption increases. Also, CO_2 and NO_x emissions were hiked with the use of solketal blends as compared to neat gasoline. Hence, solketal can be used to improve octane number.

(5) By addition of dimethyl ether fuel consumption can be reduced marginally. HC and NO_x emissions were reduced and CO emissions were increased. For di-isopropyl ethers-gasoline blends thermal efficiency was improved. CO and HC emissions were lowered as compared to neat gasoline whereas CO_2 and NO_x emissions were increased for blends as compared to neat gasoline. Ethers are carcinogenic and are responsible for water contamination hence they are banned in several countries to use as an additive.

(6) Fusel oil-gasoline blend increased engine power, CO_2 and HC emissions and NO_x emissions were reduced as compared to neat gasoline.

3 Recommendations

Many things should be taken into consideration before selecting the best alternative fuel or additive for gasoline on a large scale. Researchers have discussed various factors to be considered for additive or alternative fuel selection:

(1) Optimisation in performance and emission parameters for SI engine when operated with additive should be taken into consideration. Performance should not degrade with the use of additive as compared to gasoline. Most likely performance should be enhanced. Emissions should reduce with the use of additives.

(2) Effect of additives on the engine should be determined. Additives should have similar properties as that of gasoline so that modification in the current engines is not required.

(3) Environmental effects should be considered. Additives should not contaminate the environment and should not be harmful for human health. Methyl tert-butyl ether, aromatic alcohols are carcinogenic.

(4) Cost of additive should not exceed that of base fuel. Price of the resultant blend should be affordable,

(5) Availability and accessibility of additive on a large scale should be checked.

(6) Alcohols production on large scale is possible and are widely used today for blending in gasoline. Bio-ethanol is the best option among all the alcohol as it is a renewable fuel.

(7) Other renewable bio-additives are being ignored and not given much importance. Bio-additive like Turmeric leaf oil (Curcuma longa L. oil) should be deeply studied as it is produced from agricultural waste of turmeric leaves.

4 Economic study for bio-additives

Usage of ethanol upto 10% is permitted in India as per national policy but still many petroleum companies are not blending ethanol in gasoline. As petroleum production is not significant in all the countries, many countries are dependent on other nations for their petroleum needs. As the demand is increasing day-by-day due to exponentially increasing population, many developing nations are facing adverse effects on the economy due to heavy imports. If these countries begin to blend bio-additive in gasoline, they can save upto atleast 10% gasoline and expenditure. Also using bio-additives like ethanol, turmeric leaf oil help to boost the local economy. Use of ethanol in gasoline has increased demand for ethanol hence local sugar factories have got extra income source. As ethanol is produced from waste product in sugar factories, hence production of ethanol has proved beneficial for factories. Use of , turmeric leaf oil in gasoline will ensure extraction of oil from turmeric leaves which is a waste product in the field. By selling turmeric leaves the income of farmers will increase and it will improve the economic condition of farmers.

5 Conclusions

Additives can be used in the engine without modification. The emissions using additives can be satisfactorily reduced and performance parameters are improved in most of the additives. Bio-additives like ethanol and turmeric leaf oil should be used instead of chemical based additives as they are safer to the environment and efficient in improving performance and reducing emissions. Also bio-additives can be beneficial in improving the local economical conditions.

References

1. M. Saraswat, N. R. Chauhan, Engg Sci and Tech, an I Journal, **23**, 92-100 (2020)
2. M. Mourad, K. Mahmoud, Renewable Energy, **143**, 762-771 (2019)
3. A. Elfasakhany, J of Cleaner Production, **139**, 1057-1067 (2016)
4. M. Koç, Y. Sekmen, T. Topgul, H. S. Yucesu, Renewable energy, **34,** 2101-2106 (2009)
5. A. Elfasakhany, Engg Sci and Tech, an I Journal, **18**, 713-719 (2015)
6. M. İlhak, S. Tangöz, S. Akansu, N. Kahraman, Energy, **175**, 434-444 (2019)
7. Y. Li, J. Gong, Y. Deng, W. Yuan, J. Fu, B. Zhang, App Thermal Engg, **115**, 53-63 (2017)
8. H. Sharudin, N. Abdullah, G. Najafi, R. Mamat, H. H. Masjuki, App Thermal Engg, **114**, 593-600 (2017)
9. I. Gravlos, D. Moshou, Th. Gialamas, P. Xyradakis, D. Kateris, Z. Tsiropoulos, Renewable Energy, **50**, 27-32 (2013)
10. E. Alptekin, M. Canakci, App Thermal Engg, **124**, 504-509 (2017)
11. Mohsen Ghazikhani, Mohammad Hatami, Behrouz Safari, Davood Domiri Ganji, Prop and Power Research, **2(4)**, 276-283 (2013)
12. Huseyin Serdar Yucesu, Tolga Topgul, Can Cinar, Melih Okur, App Thermal Engg, **26**, 2272-2278
13. Rinu Thomas, M. Sreesankaran, Jeevan Jaidi, Dileep, M. Paul, P. Manjunath, Perspectives in Science, To be published
14. M. Bahattin Celik, Bulent Ozdalyan, Faruk Alkan, Fuel, **90**, 1591-1598 (2011)
15. Zhi Tian, Xudong Zhen, Yang Wang, Daming Liu, Xiaoyan Li, Fuel, **259**, 1-12 (2020)
16. Changwei Ji, Lei Shi, Shuofeng Wang, Xiaoyu Cong, Teng Su, Menghui Yu, Energy, **126**, 335-342 (2017)
17. M. Mourad, Khaled R.M. Mahmoud, Energy, **149**, 741-749 (2018)
18. Gopinath Dhamodaran, Ganpathy Sundaram Esakkimuthu, Yashwanth Kutti Pochareddy, Fuel, Article in press
19. A. Keskin and M. Gürü, Energy Sources Part A, **33**, 2194-2205 (2011)
20. Alper Calam, Yakup İçingür, Hamit Solmaz, Hasan Yamık, Int J of Green Energy, **12**, 767-772 (2015)
21. Ahmed N. Abdalla, Omar I. Awad, Hai Tao, Thamir K. Ibrahim, Rizalman Mamat, Ali Thaeer Hammid, Energy Sources, Part A: Recovery, Utilisation, and environmental effects

Optimization of electric vehicle based on triboelectric nanogenerator

Sumit Bhamre[1,], Sainath Mali[1], and Chittaranjan Mane[1]*
[1]JSPM's Jayawantrao Sawant College of Engineering, S. No. 58, Handewadi Road, Hadapsar, Pune 411028, Maharashtra, India.

Abstract. Triboelectric Nanogenerator is a newly invented energy harvesting technology which converts external mechanical energy into electricity based on universally accepted triboelectric principle. In this research paper we have developed an easy way to produce Triboelectric Nanogenerator by using easily available material such as Kapton, Polyethylene Terephthalate (PET Plastic Sheet) and Aluminium Foil unlike more complex material such as Polydimethylsiloxane (PDMS) used by the core researchers. We have also designed and developed a sheet of Triboelectric Nanogenerator and implemented it on electric vehicle and observed the scavenging of friction energy from rolling tyres and also calculated that how much energy Triboelectric Nanogenerator can produce. This successful demonstration of Triboelectric Nanogenerator on wheel and its theoretical calculations provides a promising solution to scavenging the wasted energy of the vehicle and use this wasted energy to increase the battery capacity and range of electric vehicle.

1 Introduction

Intensive research efforts have been dedicated to the creation of alternative energy technologies due to the rapidly increasing use of fossil fuel and global energy demand in the last decades. The harvesting of environmental mechanical energy is regarded as an effective and promising way to meet energy requirements due to its large abundance and omnipresence. In 2006 a nanogenerator was developed to collect mechanical energy from different mechanical energy sources, including fluid or air flows, ultrasonic waves, deformations, rotations and human activities using nano-composites or piezoelectric nanomaterials. In recent years triboelectric nanogenerator (TENG) has been developed as a new scavenging theory based on the conjunction of triboelectric friction and electric induction

As a power source for a variety of applications, Triboelectric Nanogenerator has quickly shown great promise, including powering of portable electronics, self-allocated sensors, charging lithium ion batteries, electron deposition, electron degradation and electrolyzing water.

The potential of Nano generator to scavenge waste mechanical energy in many different ways is one of the most appealing merits. Friction is a big loss of energy in a rotary mechanism, the most common example being rolling tires on the ground. Studies performed by the US Energy Department showed that the rolling resistance in cars (without including break loss) absorbs 5–7 percent of the energy. In expanding the range of electric vehicles, extracting this type of energy would be promising solution.

2 Introduction to Triboelectric Nanogenerator

In 2012, the Zhong Lin Wang group developed the Triboelectric Nanogenerator (TENG) to convert small-scale mechanical energy by a combination of triboelectrification and electrostatic induction into electricity. TENG has unprecedented success and a paradigm shift with a power rating of up to 500 W/m2 and an instantaneous conversion efficiency of 70 percent [1]. TENG is lightweight, low density, low cost and can be made with organic materials and neither magnets, nor coils are used for TENG. Most notably, TENG works best with low frequencies (< 5–10 Hz) as opposed to the traditional electromagnetic generator; therefore it is a unique alternative for low frequency body and sea wave (blue energy) energy harvesting.

A Triboelectric Nanogenerator (TENG) can also be utilized to actively detect dynamic and static arising from mechanical agitation using voltage and current output signals respectively.

An energy harvesting device, triboelectric nanogenerator, transforms mechanical external energy into electrical energy through a conjunction of triboelectric effect and electrostatic induction.

In terms of this power generation unit, in the internal circuit, the triboelectric effect generates potential due to a transfer of charging among two fine organic / inorganic films with an opposite tribo-polarity; electrons move

https://doi.org/10.1051/e3sconf/202017001027

between two electrodes connected to the back of the films to balance the potentials in the external circuit.

The triboelectric effect is an electrification caused by touch, when a material is electrically charged by the use of friction with a separate material. Triboelectric effect is a common cause of electrostatics every day. In relation to the contact material, the signed charges to be borne by a substance depend on their relative polarity.

This new energy technology offers a number of other benefits, in addition to unparalleled output efficiency, includes low manufacturing and production costs, great strength and reliability and environmental friendliness. The Triboelectric Nanogenerator is ideal for the processing of all kinds of mechanical energy that is available but wasted in our daily lives, including human movement, vibration, mechanical triggers.

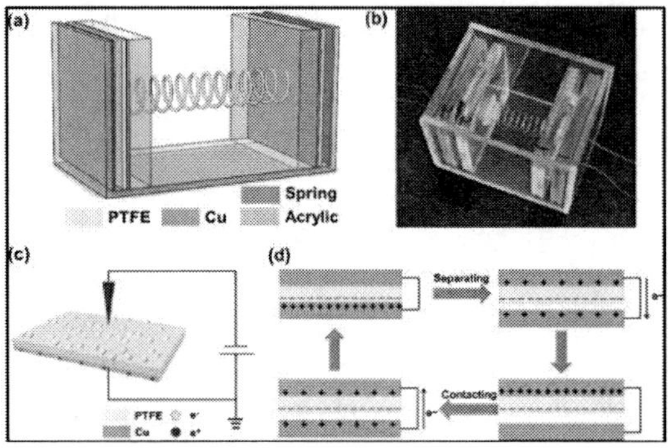

Fig. 1. (a) Illustration of spring type triboelectric nanogenerator (b) A spring-assisted TENG produced computer photograph (c) Schematic method of electron injection to produce negative charges on the PTFE sheet (d) The spring-assisted TENG device's operating principle.

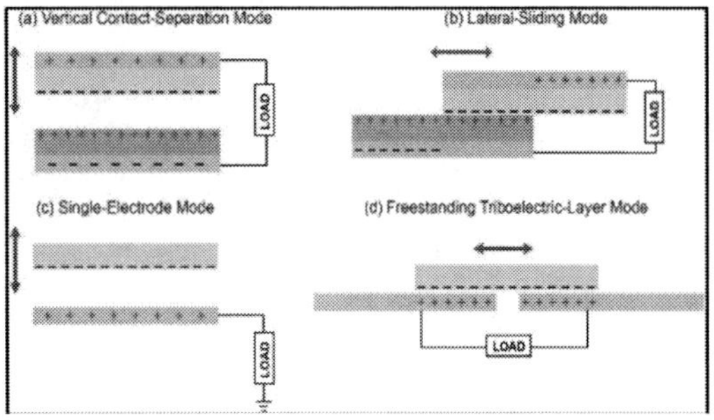

Fig. 2. Types of mode of operating Triboelectric Nanogenerator.

3 Fabrication of Triboelectric Nanogenerator

Triboelectric Nanogenerator is a very hard manufacturing process because it requires a piece of sandpaper, then tapped to the bottom of a plastic polystyrene board, then a 10 to 1 ratio of liquid PDMS and cross-connecting system is mixed and then inserted uniformly into the sandpaper. After that a film is taken off the sandpaper and a portion of the Cu foil is applied to the flat side of the thin film PDMS to conduct pores with a rough surface of thin film and PDMS thin film, serving as a sandpaper electrode [2].

So, to overcome this intricate process we developed a new type of Tribo-generator. We have created it by using materials such as Kapton, Polyethylene Terephthalate (PET Plastic Sheet) and Aluminium Foil. In which Kapton acts as an electron donor, Polyethylene Terephthalate (PET Plastic Sheet) acts as charge acceptor and aluminium acts as charge collector.

Following diagram shows the demonstration of Triboelectric Nanogenerator which we have created;

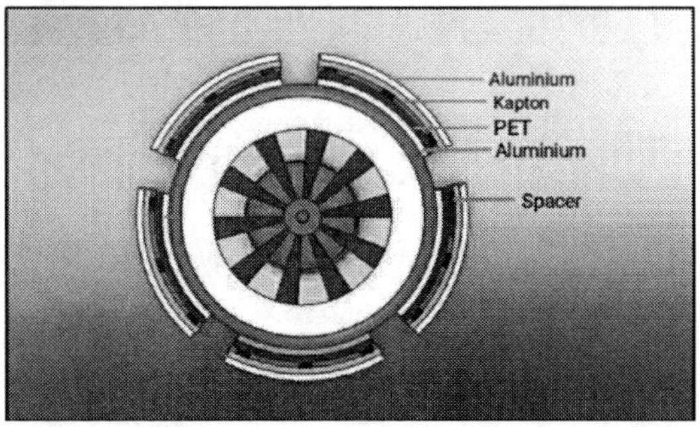

Fig. 3. Implementation of Triboelectric Nanogenerator on wheel.

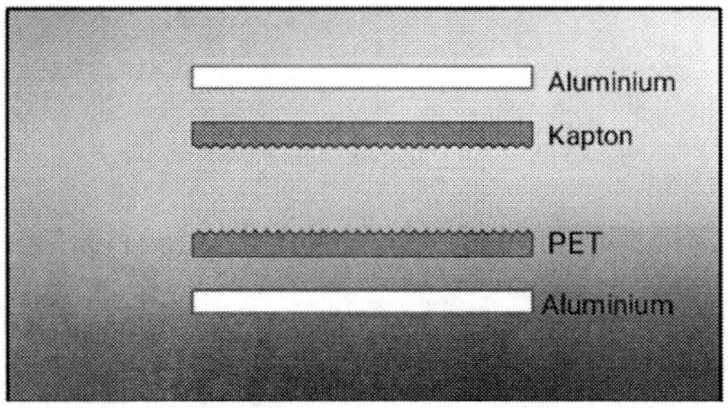

Fig. 4. Two dissimilar material, 1st one created by Aluminium and Kapton and 2nd one created by Polyethylene Terephthalate (PET) and Aluminium.

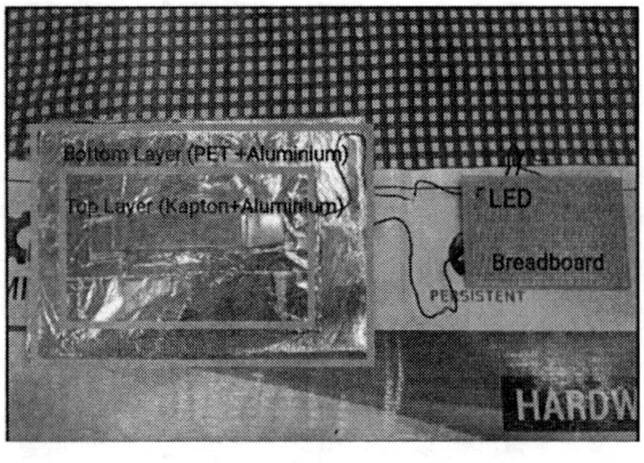

Fig. 5. Actual scale model of the Triboelectric Nanogenerator.

When the 2 dissimilar surface comes in contact and again gets discontent there is simultaneous charge transfer between the two dissimilar materials i.e. in between Kapton and PET, and aluminium acts as an electron charge collector.

 https://doi.org/10.1051/e3sconf/202017001027

As shown in Fig. 5. The charge collected on the aluminium is transferred to the breadboard and then to the LED which shows the charge transfer circuit completion.

4 Results and Discussions

We have taken the results on the digital oscilloscope and every time due to variable load the reading were observed to be different, readings were observed between the range of 5V to 16V (V_{pp}), following are the real time graphical demonstration of digital oscilloscope;

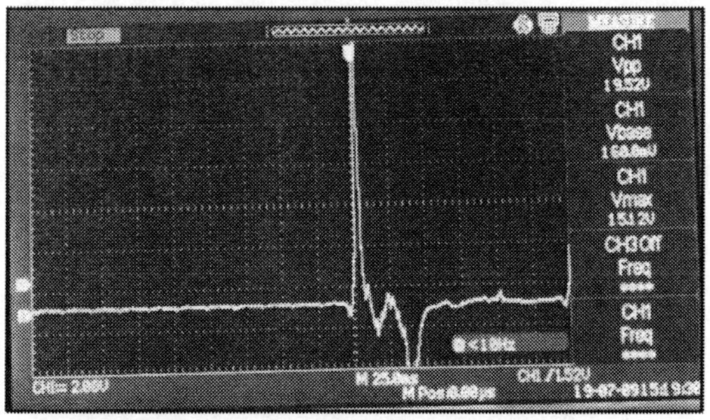

Fig. 6(a). Output result with V_{pp} value 19.52v

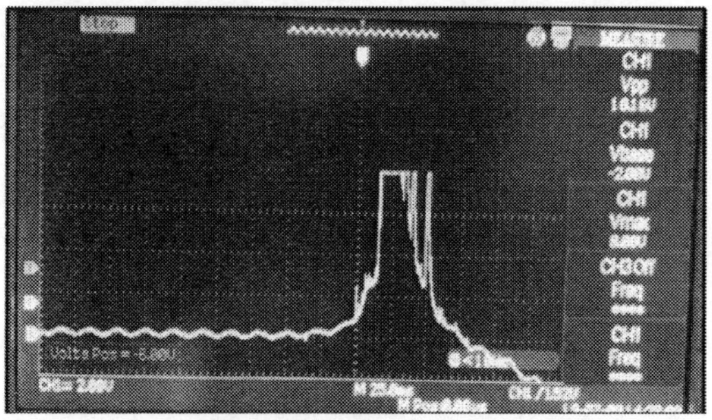

Fig. 6(b). Output result with V_{pp} value 10.16v

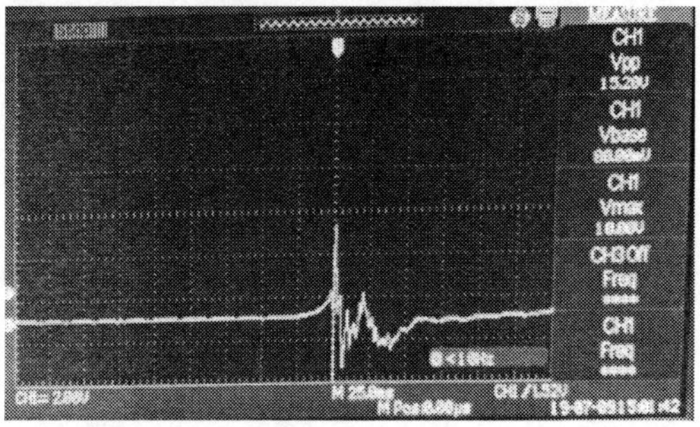

Fig. 6(c). Output result with V_{pp} value 15.20v

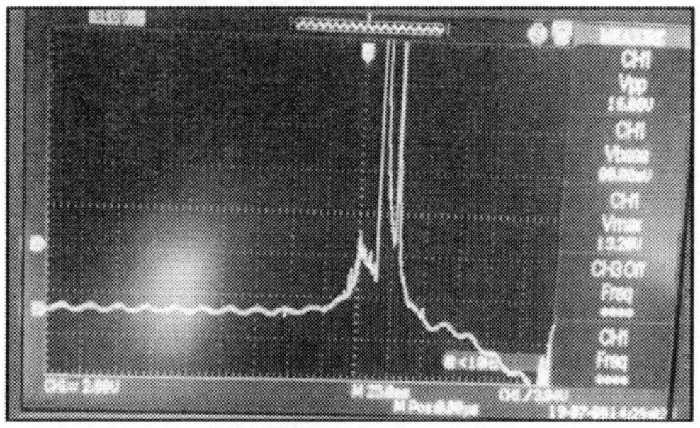

Fig. 6(d). Output result with V_{pp} value 16.88v

5 Calculations

As of now the maximum instantaneous power output is achieved to be 14.92 mW [3] by the Triboelectric Nanogenerator and the conversion efficiency is achieved up to 10.4% [3] of the total energy harvested.

So, assuming that we have implemented these Triboelectric Nano generators on the vehicles tyre as shown in fig. 3, following are the total energy calculation harvested by the 4 wheels in one complete rotation.

Tyre Size: 175/65 R14

Radius of the tyre: 291.55 mm

Overall width: 174 mm

Now, circumference of the tyre

$= \pi \times D$

$= \pi \times 583.1$

=1831.8626 mm

Outer surface area of the tyre =circumference x width

=1831.8626 x 174

=318744.1057 mm^2

Now let us consider thickness of outer surface be 1 mm;

Therefore, volume of outer surface

=318744.1057 mm^2 x 1 mm

=318744.1057 mm^3

Now, it was found that power density of 14.92 mW/cm^3 can be produced with the convergence efficiency of 10.4%

So, the total number of strips can be placed on single tyre
=318744.1057 mm^3 / 1 cm^3

=318744.1057 mm^3 / $(10)^3$ mm^3

=318.7441057 (Number of strips)

https://doi.org/10.1051/e3sconf/202017001027

Therefore, total power produced by 318.744 strips

= 14.92 mW x 318.744

=4755.6620 mW

=4.7556 W

Therefore, power produced by all four tyres in one single revolution is =4 x 4.7556 W

=19.022 W

By considering the conversion efficiency of 10.4% of the triboelectric Nano generators;

Obtained power

=19.022 x (10.4/100)

=1.9783 W

Hence, the total power output in one revolution of tyre is theoretically obtained to be approximately 2 Watts.

6 Conclusion

In brief, we have demonstrated an innovative design of TENG using materials such as Kapton, Polyethylene (PET plastic sheets) and Aluminium based on the triboelectric principle This design has a unique ability to scavenge the friction energy which is generally wasted between rolling wheels and the ground.

The successful modification of Triboelectric Nanogenerator has demonstrated a promising process capability to scavenge the wasted friction energy between the tires and the ground from a running vehicle and thus improves efficiency or the total mileage per charge of electric vehicles.

References

1. Z.Wang, L.Lin, J. Chen, S. Niu, Y.Zi, *Triboelectric Nanogenerators*, xxxiii (2016).
2. Y.Mao, D. Geng, E. Liang, X. Wang, Nano Energy, 228 (2015).
3. Y.Mao, D. Geng, E. Liang, X. Wang, Nano Energy, 233 (2015).

Comprehensive study on smart cooling techniques used for batteries

Amey Ekbote[1,*], Mathew Karvinkoppa[1], Virendra Bhojwani[1,] and Naveen Patil[2]
[1] MIT School of Engineering, Rajbaug Loni, Pune-412201, India
[2] Vellore Institute of Technology, Vellore, Tamil Nadu, India

Abstract. Electric vehicles are important in today's world to reduce pollution. The demand for electric vehicles is increasing day by day. The major component is the battery for an electric vehicle which gives the power to drive motor and drives the vehicle. Continuous operation of the vehicle causes the battery to heat and while heating there are some flammable gases released which may cause a fire. The heating of batteries reduces the performance of the vehicle and reduces the efficiency, therefore there is a need for cooling techniques to keep the temperature of batteries below the critical temperature for safe operating conditions. The present study emphasizes various cooling techniques used for the battery thermal management system. Cooling improves the performance of the battery and reduces the temperature of the battery. It helps in maintaining the temperature of the battery at the desired level. Before cooling the battery, it is necessary to study the thermal behavior of the battery. Various aspects of the thermal behavior are also reported in the paper and the problems associated with the time required for charging the batteries are also discussed.

1. Introduction

Conventional vehicles cause pollution leading to global warming which is affecting our daily life [1]. The only solution to reduce pollution due to transportation is by using eco-friendly electric vehicles. Electric vehicles do not harm the environment which does not operate on conventional fuels.

Electric vehicles have a battery that stores the energy and drives the motor shaft of the vehicle. In most cases, lead-acid or lithium-ion batteries are used. Lithium-ion batteries take less time than lead-acid batteries for charging. There is an increase in internal heat generation occurs due to continuous usage of battery causing the rise in the temperature of the system and the performance of the battery reduces. The continuous usage of battery may lead to failure and a small change in the wiring diagram may lead to a short circuit too. The batteries generating heat consume more power and it reduces the economy of the vehicle. The parameters that affect the performance of battery due to a rise in temperature are given below:

- Electrochemical reaction in the battery
- Vehicle mileage
- Charging time
- Reliability
- Life of battery[a]

To cool down the temperature of batteries and keep the temperature below the desirable for the effective working of them, various cooling systems are studied in this topic. The cooling system plays a vital role in removing heat from batteries. Cooling systems reduce the temperature of the battery. The various cooling techniques are developed to reduce the temperature of the battery. One of the advantages of using a cooling system is the performance of the vehicle gets improved and increases the life of the battery. Air cooling, liquid cooling, and the use of phase change material (PCM) are some of the cooling techniques used in batteries. According to the concept of heat transfer heat travels from the high-temperature areas to the low-temperature area. In the current study, the battery thermal management system and preheating of battery is also discussed. Battery thermal management plays a chief role in heat analysis. Battery thermal management aims to distribute the temperature uniformly. The goals of battery thermal management system are (a) to maintain temperature of the battery (b) uniform temperature distribution (c) to maximize the heat dissipation

2. Battery Thermal Management System

Battery thermal management system (BTMS) [24] is used for maintaining temperature conditions in electric vehicles. A thermal imbalance affects the performance of

*Corresponding author: ameyekbote9@gmail.com

the battery. Heating causes a reduction in the life span of cells. It is observed that hot cells degrade quickly compared to cold cells. Arrhenius law states that the chemical reaction in battery increases exponentially with the battery cell temperature. The lithium-ion battery system should have the maximum cell temperature below 40 °C and the cell temperature difference below 5 °C. Due to this life span of the battery is maintained. The aim of thermal management is maintaining uniform temperature [24].

A systematic approach has been shown while studying thermal management. Following are steps for designing and evaluating a BTMS are given below:

1. Specifications for modules and packs should be identified under different climates like geometric entities, dimensions, number of modules, and positions, based on the vehicle integrators requirements should be identified.
2. Calculate overall heat generation and overall heat capacity from the module at different discharge cycles and temperature.
3. Determine the steady-state temperature of modules using energy balance principles under operating conditions. By using these fans or pumps power requirements can be quoted with a selection of coolant fluids.
4. Carry out Finite Element Analysis to understand the thermal behavior of modules. Using Computational Fluid dynamics heat transfer rate can be obtained and systems power requirements can be evaluated. The impact can be studied on modules.
5. Design a battery module with fans and cooling materials and carry out the detailed experiment. The analysis must include cost estimation, FEA validation, maintenance, ease of operation, and comparison with an alternative system.
6. Build a system of the module and test it under different climatic conditions. Observe performance at various flow rates.
7. Carry out the optimization of the system. Several factors are taken into consideration such as battery performance and life, impact on vehicle performance, cost, and ease of maintenance while doing optimization.

2. Cooling methods

Cooling takes place under temperature difference. Cooling can be done using air or liquid as coolant media. Heat transfer plays a key role in the cooling of the battery. Heat transfer includes conduction, convection, and radiation. Conduction takes place between two solid particles one having more energy and the other one having less energy. The heat transfer rate in conduction is more compared to convection and radiation. heat conduction occurs due to two effects: the vibrational motions of the molecules positioned at relatively fixed positions periodically induce the lattice vibrational waves called a lattice, and the energy gets traveled via the free flow of electrons in the solid. Convection is a mode of heat transfer in which heat exchange takes place between solids and liquids or gases. Convection is classified as natural convection and forced convection. In natural convection, the heating body is exposed to natural phenomena while in forced convection external device is used to remove heat. Radiation takes place by electromagnetic waves i.e change in molecular configuration or atomic structure. In radiation, there is no intervening medium required. These modes of heat transfer play a vital role in the removal of heat while cooling. Cooling can be classified as active cooling or passive cooling. The active cooling method uses a fan to remove heat while in passive cooling natural cooling takes place. Cooling methods are described below

2.1 Air cooling

The heat transfer medium plays a prime role in removing heat. For the removal of heat, air can be used as a medium. Air cooling is a method for dissipating heat [2]. It is done by increasing the surface area or increasing the airflow over the object to be cooled, or both. Mathew and Hotta [25-26] carried out numerical studies to cool integrated circuits which acts as a hot spot. They used natural, forced, and mixed convection methods for cooling these hotspots generated in IC chips. They observed the effect of cooling using a fluent solver. These cooling methods were found to be effective for cooling circuits. The methodology adopted by them is also feasible for battery modules.

2.2 Liquid Cooling

In some cases, air cooling is not preferable as it cannot access some parts present inside vehicles [28]. Liquid cooling is preferred when the temperature of the ambient condition is more than 35° C, in an enclosed system where less volume of air enters inside the parts, and when there is a higher heat flux removal rate. Liquid cooling consists of fluid which removes heat from surrounding. In this type, the flow rate of liquid can be increased and recirculate. Designing of a liquid system is simple and liquids have high thermal conductivity (HTC) and high heat transfer coefficient (HHTC) [28]. Sometimes liquid cooling systems have problems like leakage, corrosion, extra weight, and condensation, and even cavitation that may take place in tubes. There are some factors which are taken into the selection of fluid mentioned below

- High specific heat
- Low viscosity
- High latent heat of vaporization
- High boiling point
- Low freezing point
- Non-corrosive and non-inflammable

Liquid cooling systems are of two types viz. direct cooling and indirect cooling systems. Indirect cooling systems, the cooling system has contact with components. While in the case of the indirect cooling system they don't have contact with components. Liquid cooling systems have closed-loop and open-loop systems, depending on whether the liquid is discharged or sent again into the system after it is heated. In open-loop systems, fluid is passed through the cooling system and discharged into a drain when it is heated. The hot fluid in closed-loop systems is cooled using a heat exchanger and recirculated to the system again. Patil and Hotta [27] used different fluids for studying the cooling effect on integrated circuits. They used fluids like FC-72, water and they carried out analysis using software Ansys Icepak. Patil and Hotta [28] used water, FC-72, dielectric fluids, nanofluids for studying the cooling effect on electronic modules. They studied the characteristics of these working fluids and these can remove heat at high temperatures.

2.3 Phase Change Materials (PCMs)

Phase changing materials are the ideal material for thermal applications. These materials store and release energy while melting and freezing processes [9]. When freezing occurs a large amount of energy is released by these materials. These materials absorb energy during the melting process. Generally, these materials are used in passive thermal management and developed as an alternative to active cooling. The advantage is that they absorb a large amount of heat. Phase change materials are more efficient in the forced convection method. PCMs play a key role in the battery thermal management system and its shape is simple, light, and requires less space. Mathew and Hotta [31-32] used phase changing material for studying temperature characteristics. They used different PCMs to cool down the temperature of IC chips mounted on the SMPS board. PCMs showed effective cooling and it can be used for cooling battery modules. The studies conducted by authors [26-32] mainly focuses on identifying the hot spot at the board level. These hots spots are further cooled down using various cooling techniques as discussed. Thes cooling techniques are also feasible at the battery module level and the temperature of the modules can be reduced.

3 Background of cooling studies

Various researchers have carried studies on battery cooling techniques. The main objective is to increase the performance of the battery. Cost optimization and weight reduction is also a major factor. Sometimes system weight is also taken into consideration while designing a system. A background study carried out by researchers is given in Table 1.

Table 1. Summary of the literature pertains to cooling of Batteries

Author	Coolant	Parameters
Ahmad A. Pesaran and Matthew Keyser [1]	-	• Various tests were performed to carry out the thermal analysis of the battery. • Battery modules used for testing were valve-regulated lead-acid, lithium-ion, and nickel-zinc and heat generation capacity and thermal imaging carried out.
Ahmad A. Pesaran [2]	Air	• Developed a model for thermal analysis • Lump capacitance model for temperature analysis • Finite elements to observe effects on the model
Ahmad A. Pesaran [3]	Air, liquid	• 60 Hz AC power supply was given to a 12 volt,13 Ah battery for heating • Experiments were carried at -40°C where warming of battery took place • By conducting experiments transient behavior and preheating of the battery was studied
Anthony Jarrett, Yong Kim∗ [4]	Air	• 1 mm × 160 mm × 200 mm dimensions of cold plate used for analysis • CFD analysis was carried out • Optimization was studied using parameters coolant pressure drop, temperature uniformity, and average temperature
Rajib Mahamud, Chanwoo Park [5]	Air	• Carried out CFD and developed the lumped capacitance model • Carried out a study on reciprocating air flow method
Rajib Mahamud and Chanwoo Park [6]	Air	• Lump capacitance method is studied • An analysis is done by considering biot number <0.1 • They used lithium-ion cell which is constructed using multiple layers and Material of cell is non-homogenous and non-isotropic • The lumped-capacitance thermal model was used to calculate the core, average and surface temperatures of the cell under battery duty cycles • They kept convective heat transfer coefficients 151.9, 1519, 7600, and 15200 W/m^2-K that were corresponding to the Biot numbers of 0.1, 1, 5, and 10, respectively. • Accurate results were obtained using CFD modeling

Thanh-Ha Tran, et al [7]	Air, liquid	• They used 14 cylinder cells battery modules having 6.5 Ah of capacity, and dimensions 38 mm in diameter and 142 mm in height • Heat pipe cooling system was used to cool the battery • They used 4 modules, each module consists of 7 tubes with 50 helical grooves • Using heat pipe power consumption and noise level can be reduced
M.C. Browne et al [8]	Phase changing material (PCM)	• They studied the effects of temperature on photovoltaic cell • They reviewed about phase change materials and studied their parameters • They also studied the effects of phase change materials
Ziye Ling, et al [9]	Air	• The battery was tested under two conditions viz .completely passive system and hybrid passive system • In a completely passive system battery was surrounded by RT44HC composite while in hybrid system battery with phase changing material was exposed to forced air convection. • They carried out a comparison between the performance of passive and active thermal management systems • Using PCM with an active cooling system increases the cooling efficiency
Shahabeddin K. Mohammadian et al [10]	Liquid electrolyte and water	• The prismatic structured battery was used to observe thermal management • Heat generation causes failure and heat dissipation in battery • Internal and external cooling methods were used for cooling battery • Internal cooling is differentiated with the external one • They observed temperature distribution using numerical and graphical analysis • The internal cooling method reduces the temperature inside the battery and also improves temperature distribution.
Bahman Shabani and Manu Biju [11]	Air	• Transient conduction approach was used for uniform temperature distribution • Various governing equations were used to carry heat analysis • The finite element method was used to study behavior. • Lumped Capacitance Thermal Approach, Numerical and Analytical Thermal Models and Equivalent Circuit Thermal Models methods are used and deep analysis is carried out.
Dafen Chena et al [12]	Liquid	• Air cooling, liquid cooling(direct and indirect), and fin cooling techniques were used to see cooling effects. • A prismatic cell was made using software ANSYS/Fluent which is based on electrochemical reaction theory and its performance was studied • Dielectric mineral oil is used in direct liquid cooling. • 169mm × 197mm × 1mm was the dimensions of the fin • The average temperature increase was observed at the end of discharge reach a minimum of 4°C so that all cooling methods were observed • Fin cooling adds weight while air cooling consumes more power
Lan et al [13]	Air	• Mini channels were tested at different flow rates and directions and results were calculated • 173 x 168 x 39 mm were dimensions of battery module while the dimension of channel 3 x3 mm • It was tested at 15 °C and 35 °C and analyzed • This system was found superior for battery thermal management system
S.Panchal*, et al [14]	Water	• An aluminum-laminated battery was used which contains $LiFePO_4$ cathode material • The rating of the battery was 20 Ah.

		• Thermocouples and heat flux sensors were used to detect temperature variation. • The dual cold plate approach was used and studied cooling at different temperatures from 5° to 35°C. • As current discharge increases heat generation increases
L. H. Saw, et al [15]	Air	• The battery pack was used consisting of 24 pieces of 38,120 cells • The heat generated was measured using the calorimeter • Various mass flow rates were applied to analyze air cooling • They found a correlation between Nusselt number and Reynolds number
Chen et al [16]	Air	• The battery pack was used consisting of 24 pieces of 38,120 cells • The heat generated was measured using the calorimeter • Various mass flow rates were applied to analyze air cooling • They found a correlation between Nusselt number and Reynolds number
Wang et al [17]	Air	• Thermal stability of components of a lithium-ion battery is being studied • Heat generation and thermal runaway has been studied • Working and arrangement of lithium-ion battery has been explained • They investigated the thermal effects of battery on various passenger vehicles • They investigated the performance of battery at sub-zero temperature
S. Panchal et al [18-19]	Water + Air	• Testing was done at temperatures between 5°C and 25°C using air cooling and water cooling method. • Using lab view software parameters were recorded like charge and discharge current, charge and discharge voltage, and sampling. • By increasing the temperature discharge rate increases. • The cooling patterns obtained from the numerical results were stable
	Water	• Mini -channel cold plates are used for cooling of battery modules • Water is used as fluid to act as coolant and water was set at a temperature 5°C,15°C and 25°C • Computational Fluid Dynamics is studied using Reynolds Averaged Navier Strokes equation. • Temperature distribution and velocities were compared in minichannels
Jian Xu et al [20]	Ethylene glycol and water	• Mini-channel cooling system was designed for battery cooling • Nail penetration technique was used to avoid circuit failure • The dimensions of each cell were 180 mm X 130 mm X 50 mm and the rating was 55 Ah. • They conducted an experiment and model was validated and carried out numerical analysis • Fluids ethylene glycol was compared with water
Abdul Haq Mohammed [21]	Water	• Two types of cold plates were used one with s-type and other with pin-type design • Surface temperature and pressure drop were taken into consideration • The flow rate was kept 0.2 L/min and 0.3 L/min in s-type and pin type 4.1°C and 4.2°C temperature drop was observed • High-pressure drop values occurred in S-type design compared to the pin type design.
Ben Ye, et al [22]	Air	• 131mm X 65mm X16mm was a dimension of the battery. • The comparison was done between experimental analysis and CFD temperatures • They carried out cost optimization
Ahmed pearsan [23]	Air	• Carried out finite element analysis • Then used Infrared photography to obtain thermal images • liquid crystal thermography to obtain thermal images • selected air at 25°C for cooling the module •

A.A. Pesaran et al [24]	Air	• Steps for studying battery thermal management system are given • For the ideal design of the battery , the BTMS approach is useful • Various methods are discussed to observe uneven temperature distribution • Air cooling is preferred over liquid cooling as it is difficult to design

4. Emerging Smart technologies in cooling

New cooling technologies impress the market today. The major factor is cost optimization and improving the performance of the system. These techniques play a vital role in the cooling of the battery. After carrying out extensive study researchers have developed smart cooling technologies. By using these concepts temperature of the battery can be controlled. Some techniques are shown below:

4.1 Cabin Air Cooling

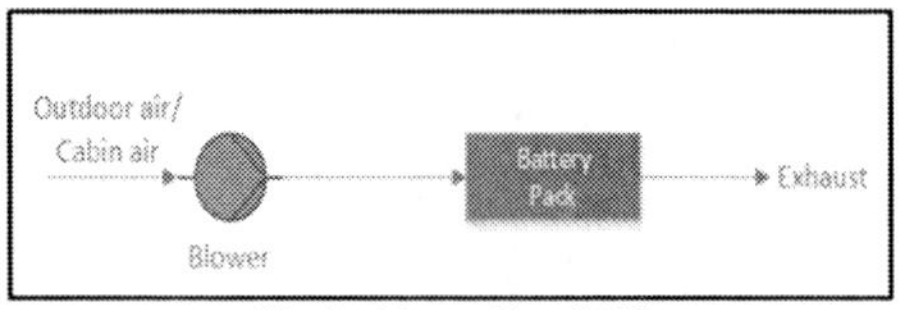

Fig 1. Air cooling [33]

This system is used in the Toyota Prius [33]. Preconditioned air from air conditioning is used for battery cooling. Airflow takes place over a surface and channels through the battery.

4.2 Independent Air Cooling

This system is similar to cabin air cooling. This system consists of an evaporator that removes heat from the battery. Refrigerant gets evaporated at low temperature and absorbs heat. Thus battery gets cooled below cabin temperature.

4.3 Direct refrigerant cooling

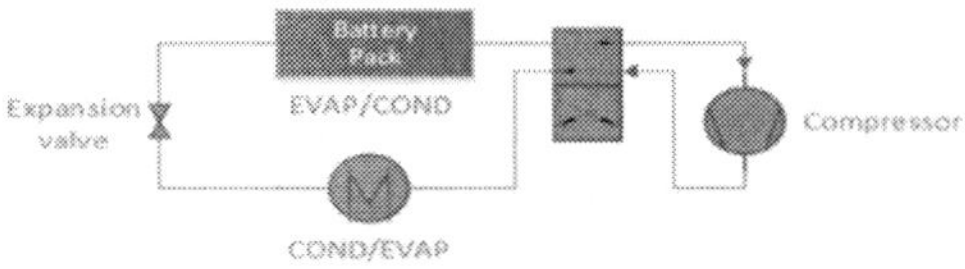

Fig 2. Direct Refrigerant cooling [33]

Direct refrigerant connects an evaporator plate in parallel with the current evaporator of vehicle air condition. Evaporator plates have direct contact with the battery plate.

4.4 Thermoelectric Cooling

In the thermoelectric cooling conversion of electric voltage takes into temperature difference using thermoelectric modules. Heat transfer takes place directly by consuming electricity. It can be used for heating applications just by reversing poles.

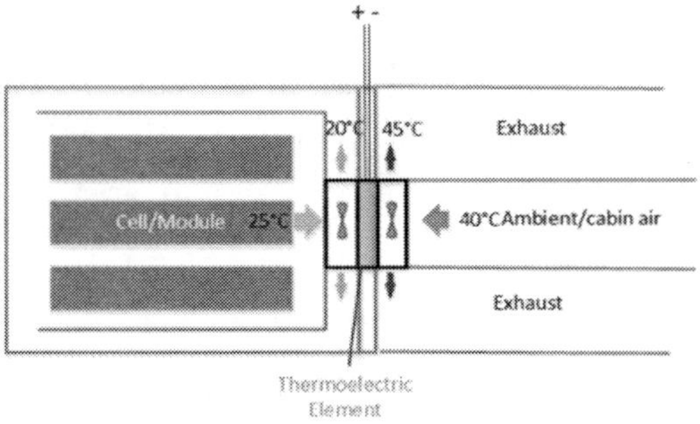

Fig 3. thermoelectric cooling [33]

4.5 Heat Pipe Cooling

The heat pipe is another solution for cooling batteries. The heat pipe is an envelope of pipes having a capillary powder structure with sintered copper powder. It is used as an evaporator which absorbs heat while operation and dissipates heat in the condenser and becomes liquid again.

4.6 Combinational liquid cooling system

This system has four modes: bypass with heater modes, passive cooling, and active cooling This system can be used as passive cooling as well as active cooling. Passive cooling is most preferred as it has simple architecture and has low power consumption

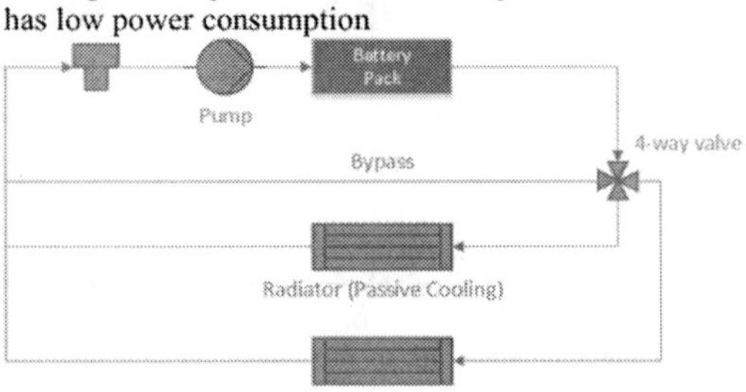

Fig 5. combinational System [33]

5 Preheating Techniques

All batteries create problems during cold climatic conditions[3]. Electrochemical processes become slow and overall internal resistance increases in the winter season. Due to slow initialization, starting of the vehicle becomes hard. Some steps are given below The battery core could be heated internally by applying electricity (electrochemical components)[24].

- External heating can be done for a module using an electric heater
- Each cell in a module can be heated externally using an electric heater.
- Internal heating can be done for each cell using a hot fluid.
- Internal Hot fluid heating method can be used for preheating of each module.

6 Conclusion

Electric vehicles are major key players in the market. Batteries are key components for power generation. Cooling systems are necessary to improve performance. Cooling systems reduce the heat generation of the system. They always try to cool the system without affecting the internal system thus increasing the mileage of vehicles. Various types of cooling systems are discussed. Every cooling system has its uniqueness. The major problem is the cost of the system gets increased. Hence researchers have also studied cost optimization. Due to this cost of the system gets reduced. These new emerging technologies play a vital role in the cooling system. The battery reactions get easily initiated. Battery Thermal Management is useful for studying temperature effects. The systematic approach for temperature distribution is also discussed. The currents study will help the researcher to choose the appropriate battery cooling technique considering all the crucial factors.

References

[1] A. A. Pesaran and M. Keyser, "Thermal characteristics of selected EV and HEV batteries," in Sixteenth Annual Battery Conference on Applications and Advances. Proceedings of the Conference (Cat. No.01TH8533), Long Beach, CA, USA, 2001, pp. 219–225.

[2] A. A. Pesaran, "Battery thermal models for hybrid vehicle simulations," Journal of Power Sources, vol. 110, no. 2, pp. 377–382, Aug. 2002.

[3] A. Pesaran, A. Vlahinos, and T. Stuart, "Cooling and Preheating of Batteries in Hybrid Electric Vehicles," p. 7, 2003.

[4] A. Jarrett and I. Y. Kim, "Design optimization of electric vehicle battery cooling plates for thermal performance," Journal of Power Sources, vol. 196, no. 23, pp. 10359–10368, Dec. 2011.

[5] R. Mahamud and C. Park, "Reciprocating airflow for Li-ion battery thermal management to improve temperature uniformity," Journal of Power Sources, vol. 196, no. 13, pp. 5685–5696, Jul. 2011.

[6] R. Mahamud and C. Park, "Spatial-Resolution, Lumped-Capacitance Thermal Model for Battery Power Cycle Analysis," presented at the SAE 2011 World Congress & Exhibition, 2011, pp. 2011-01-1362.

[7] T.-H. Tran, S. Harmand, and B. Sahut, "Experimental investigation on heat pipe cooling for Hybrid Electric Vehicle and Electric Vehicle lithium-ion battery," Journal of Power Sources, vol. 265, pp. 262–272, Nov. 2014.

[8] M. C. Browne, B. Norton, and S. J. McCormack, "Phase change materials for photovoltaic thermal management," Renewable and Sustainable Energy Reviews, vol. 47, pp. 762–782, Jul. 2015.

[9] Z. Ling, F. Wang, X. Fang, X. Gao, and Z. Zhang, "A hybrid thermal management system for lithium ion batteries combining phase change materials with forced-air cooling," Applied Energy, vol. 148, pp. 403–409, Jun. 2015.

[10]S. K. Mohammadian, Y.-L. He, and Y. Zhang, "Internal cooling of a lithium-ion battery using an electrolyte as coolant through microchannels embedded inside the electrodes," Journal of Power Sources, vol. 293, pp. 458–466, Oct. 2015.

[11]B. Shabani and M. Biju, "Theoretical Modelling Methods for Thermal Management of Batteries," Energies, vol. 8, no. 9, pp. 10153–10177, Sep. 2015.

[12] D. Chen, J. Jiang, G.-H. Kim, C. Yang, and A. Pesaran, "Comparison of different cooling methods for lithium-ion battery cells," Applied Thermal Engineering, vol. 94, pp. 846–854, Feb. 2016.

[13] C. Lan, J. Xu, Y. Qiao, and Y. Ma, "Thermal management for high power lithium-ion battery by mini-channel aluminum tubes," Applied Thermal Engineering, vol. 101, pp. 284–292, May 2016.

[14] S. Panchal, I. Dincer, M. Agelin-Chaab, R. Fraser, and M. Fowler, "Experimental and theoretical investigations of heat generation rates for a water cooled LiFePO 4 battery," International Journal of Heat and Mass Transfer, vol. 101, pp. 1093–1102, Oct. 2016.

[15] L. H. Saw, Y. Ye, A. A. O. Tay, W. T. Chong, S. H. Kuan, and M. C. Yew, "Computational fluid dynamic and thermal analysis of Lithium-ion battery pack with air cooling," Applied Energy, vol. 177, pp. 783–792, Sep. 2016.

[16] K.-H. Chen, T. Han, B. Khalighi, and P. Klaus, "Air Cooling Concepts for Li-Ion Battery Pack in Cell Level," in Volume 1: Aerospace Heat Transfer; Computational

Heat Transfer; Education; Environmental Heat Transfer; Fire and Combustion Systems; Gas Turbine Heat Transfer; Heat Transfer in Electronic Equipment; Heat Transfer in Energy Systems, Bellevue, Washington, USA, Jul. 2017, p. V001T09A001, doi: 10.1115/HT2017-4701.

[17] Q. Wang, B. Jiang, B. Li, and Y. Yan, "A critical review of thermal management models and solutions of lithium-ion batteries for the development of pure electric vehicles," Renewable and Sustainable Energy Reviews, vol. 64, pp. 106–128, Oct. 2016.

[18] S. Panchal, R. Khasow, I. Dincer, M. Agelin-Chaab, R. Fraser, and M. Fowler, "Numerical modeling and experimental investigation of a prismatic battery subjected to water cooling," Numerical Heat Transfer, Part A: Applications, vol. 71, no. 6, pp. 626–637, Mar. 2017.

[19] S. Panchal, R. Khasow, I. Dincer, M. Agelin-Chaab, R. Fraser, and M. Fowler, "Thermal design and simulation of mini-channel cold plate for water cooled large sized prismatic lithium-ion battery," Applied Thermal Engineering, vol. 122, pp. 80–90, Jul. 2017.

[20] J. Xu, C. Lan, Y. Qiao, and Y. Ma, "Prevent thermal runaway of lithium-ion batteries with minichannel cooling," Applied Thermal Engineering, vol. 110, pp. 883–890, Jan. 2017.

[21] A. H. Mohammed et al., "Design and Comparison of Cooling Plates for a Prismatic Lithium-ion Battery for Electrified Vehicles," presented at the WCX World Congress Experience, 2018, pp. 2018-01-1188.

[22] B. Ye, M. Rubel, and H. Li, "Design and Optimization of Cooling Plate for Battery Module of an Electric Vehicle," Applied Sciences, vol. 9, no. 4, p. 754, Feb. 2019.

[23] A. Pesaran, A. Vlahinos, and S. Burch, "Thermal Performance of EV and HEV Battery Modules and Packs," p. 13.

[24] A. A. Pesaran, S. Burch, and M. Keyser, "An Approach for Designing Thermal Management Systems for Electric and Hybrid Vehicle Battery Packs," p. 18.

[25] M. V. Karvinkoppa and T. K. Hotta, "Numerical investigation of natural and mixed convection heat transfer on optimal distribution of discrete heat sources mounted on a substrate," IOP Conference Series: Materials Science and Engineering, vol. 263, p. 062066, Nov. 2017.

[26] V. K. Mathew & Tapano Kumar Hotta (2020) Experiment and numerical investigation on optimal distribution of discrete ICs for different orientation of substrate board, International Journal of Ambient Energy, DOI: 10.1080/01430750.2020.1712255

[27] V. K. Mathew and T. K. Hotta, "Numerical investigation on optimal arrangement of IC chips mounted on a SMPS board cooled under mixed convection," Thermal Science and Engineering Progress, vol. 7, pp. 221–229, Sep. 2018.

[28] K. Mathew and N. Patil, "Convective Heat Transfer on the Optimum Spacing of High Heat Dissipating Heat Sources—A Numerical Approach," in Emerging Trends in Mechanical Engineering, L. Vijayaraghavan, K. H. Reddy, and S. M. Jameel Basha, Eds. Singapore: Springer Singapore, 2020, pp. 73–84.

[29] N. G. Patil and T. K. Hotta, "Role of working fluids on the cooling of discrete heated modules: a numerical approach," Sādhanā, vol. 43, no. 11, Nov. 2018.

[30] N. G. Patil and T. K. Hotta, "A Review On Cooling Of Discrete Heated Modules Using Liquid Jet Impingement," Frontiers in Heat and Mass Transfer, vol. 11, Jul. 2018.

[31] M. V K and T. K. Hotta, "Role of PCM based mini-channels for the cooling of multiple protruding IC chips on the SMPS board - A numerical study," Journal of Energy Storage, vol. 26, p. 100917, Dec. 2019, doi: 10.1016/j.est.2019.100917.

[32] Mathew V Karvinkoppa and T.K.Hotta, "Transient analysis of Phase Change Material for cooling of discrete modules under mixed convection,'' Journal of Energy and Environmental Sustainability, vol.8 oct 2019

[33] J. Li and Z. Zhu, "Battery Thermal Management Systems of Electric Vehicles," p. 79.

[34] School of Chemistry and Chemical Engineering, Henan Normal University, Xinxiang, Henan Province, 453007, China and S. Yang, "A Review of Lithium-Ion Battery Thermal Management System Strategies and the Evaluate Criteria," Int. J. Electrochem. Sci., pp. 6077–6107, Jul. 2019.

Speed scaling technique integrated in scheduling of production and maintenance under energy constraints using genetic algorithms

Sadiqi Assia[1, 2*], El Abbassi Ikram[2], El Barkany Abdellah[1], Darcherif Moumen[2] and El Biyaali Ahmed[1]

[1]Laboratoire Génie Mécanique, Faculté des sciences et Techniques de Fès, Université Sidi Mohammed Ben Abdellah, B.P. 2202 - Route d'Imouzzer - Fès, Maroc
[2]ECAM-EPMI, 13 Boulevard de l'Hautil, 95092 Cergy-Pontoise (France)

Abstract. Recently, energy scarcity and environmental pollution have become increasingly serious problems, while the scheduling of energy savings is attracting increasing attention from industries to minimize energy consumption. This article deals with the problem of scheduling production and maintenance under energy constraints in the flow shop. Two mixed binary integer programming models are provided to derive optimal scheduling to minimize total energy consumption (TEC). To solve this problem, we use an approximate method as a genetic algorithm (GA).

1 Introduction

It is certainly indisputable that the climate changes induced by this increase in the concentration of greenhouse gases will have multiple consequences that are still difficult to identify. However, they are expected to cause regional and global changes in temperature, precipitation and other climate variables, which could result in global changes in soil moisture, rising mean sea level and the prospect of more severe episodes of extreme heat, flooding and drought.

Excessive emissions of greenhouse gases and in particular carbon dioxide produced by the combustion of fossil fuels. Although fossil fuels have been widely used for electricity production, rational use of energy will contribute to a significant reduction in carbon emissions.

Manufacturing industry has always been a major consumer of energy. As a result, modern government regulations have forced manufacturers to adopt energy-saving measures [1]. In particular, the researchers understood that scheduling could play an important role in reducing the energy consumption of manufacturing processes. Therefore, operational research tools and metaheuristic research methods, such as genetic algorithm [2], particle swarm optimization [3] and simulated annealing [4], are also used for the design of a cleaner production system.

The purpose of this article is to establish two models for integrated scheduling of production and maintenance tasks, in order to minimize total energy consumption. We use an approximate method as a genetic algorithm (GA).

2 PROBLEM DESCRIPTION

This section addresses the problems of integrated scheduling of fixed and flexible production and maintenance jobs with the integration of the machine's dynamic speed variation technique into the non-switching flow shop, where the objective is to minimize total energy consumption (TEC).

Table 1. The notation used to formulate the problem

Indices:	
i	Index of jobs
j	Index of machines
k, k'	Indices of the position in each machine
l	Index of the batch in each machine
s	Index of speed level
Sets:	
J	set of jobs; $J = \{J_1, J_2, \dots, J_n\}$
M	set of machines; $M = \{M_1, M_2, \dots, M_m\}$
O_i	set of the operations of J_i; $O_i = \{O_{i,1}, O_{i,2}, \dots, O_{i,m}\}$ $O_{i,j}$ is the operation of J_i in M_j
$BH_{l,j}$	set of the batches in the lth batch in machine M_j
Parameters :	
n	Number of jobs
m	Numbre of machines
S	Numbre of speed level
$w_{i,j}$	The required workload of the job i $(1 \leq i \leq n)$ in the machine j $(1 \leq j \leq m)$
V_s	Processing speed of a machine operating at a speed level s.
$P_{i,j}$	The processing power of each operation $O_{i,j}$ in machine j.
TBF_j	The period between two consecutive unavailability intervals in M_j
PM_j	The maintenance time of the machine M_j
Z	An infinite integer
C_{max}	The makespan, which corresponds to the maximum execution time of all jobs
P_0	The common power, consumed by

* Corresponding author: **Sadiqiassia@gmail.com**

	auxiliary equipment and installations
P E C	The total energy consumption of the treatment $O_{i,j}$
C E C	The common energy consumption of the auxiliary installations in the flowshop
T E C	The total energy consumption in the "flow shop".

- All jobs are available at the beginning of the scheduling horizon,

- The optimal job sequences are not necessarily identical when the unavailability constraint is taken into account, even for flow shop workshops with two machines. Thus, the non-displacement scheduling is considered in this study.

- The machines can handle jobs at different speed levels, which are selected from a set of levels S.

- Either v_S $(1 \leq s \leq S)$ represents the processing speed of a machine operating at the speed level s.

3 Modelling of the problem $Fm|nr - FI|_{TEC}$

The first case, where unavailability intervals are set periodically, is referred to as follows:

- Unavailability intervals or maintenance periods are fixed periodically, is noted $Fm|nr - FI|_{TEC}$.

- The time between two consecutive unavailability intervals in M_j is TBF_j and the maintenance time is PM_j.

- The lth unavailability interval can be calculated as follows: $lth = [l \ TBF_j + (l-1) PM_j \ , \ l(TBF_j + PM_j)]$.

Minimize

$$\sum_{i=1}^{n}\sum_{j=1}^{m}\sum_{k=1}^{n}\sum_{s=1}^{S}\frac{w_{i,j} \cdot P_{i,j,s}}{v_s} \cdot x_{i,j,k,s} + P_0 \cdot C_{max}$$

S.t

$$\sum_{i=1}^{n} x_{i,j,k,s} = 1 \qquad ;\forall j,k,s \qquad (1)$$

$$\sum_{k=1}^{n} x_{i,j,k,s} = 1 \qquad ;\forall j,i,s \qquad (2)$$

Constraints (1) and (2) ensure that each job can only be placed at one position in the sequence of each machine, and that only one job can be placed at each position, respectively.

$$\sum_{s=1}^{S}\frac{w_{k,j}}{v_s} = \sum_{i=1}^{n}\sum_{s=1}^{S} x_{i,j,k,s} \cdot \frac{w_{i,j}}{v_s} ;\forall k \ , \ \forall j \ ,\forall s \qquad (3)$$

The constraint (3) specifies the processing time of the operation $O_{k,j}$.

$$C_{k,j} = S_{k,j} + \sum_{s=1}^{S}\frac{w_{k,j}}{v_s} \qquad ; \forall j > 1 \ , \ \forall k \quad (4)$$

The constraint (4) establishes the link between the start time and the end time of $O_{k,j}$.

$$S_{1,1} = 0 \qquad (5)$$

$$S_{k,j} \geq C_{k-1,j} \qquad ; \forall j, \ \forall k \geq 2 \quad (6)$$

Constraints (5) and (6) ensure that the kth task cannot start before the end time of the previous task in M_j.

$$(2 - x_{i,j,k,s} - x_{i,j+1,k',s})Z \geq S_{k,j} + \sum_{s=1}^{S}\frac{w_{i,j}}{v_s} - S_{k',j+1} \quad ; \forall j < m \ , i,k,k',s \qquad (7)$$

The constraint (7) ensures that the start time of the task i in M_{j+1} is not earlier than its end time in M_j.

$$C_{k,j} = e_{k,j} + (PM_j + TBF_j)\sum_{k'=1}^{k-1} g_{k',j} + (PM_j + TBF_j)\sum_{k'=1}^{k-1} q_{k',j} ; \ \forall j, \ \forall k \geq 1 \qquad (8)$$

$$C_{1,j} = e_{1,j} + (PM_j + TBF_j)\ q_{1,j} ; \ \forall j \qquad (9)$$

Constraints (8) and (9) represent that the arrival time of $O_{k,j}$ is equal to the sum of the elapsed time of $O_{k,j}$ and the times of the previous batches $O_{k,j}$, including actual and empty batches.

$$q_{k,j} \leq g_{k-1,j} \qquad ; \forall j, \ \forall k > 1 \quad (10)$$

The constraint (10) ensures that no empty batches exist between $O_{k-1,j}$ and $O_{k,j}$ when $O_{k-1,j}$ is not the last task of a real batch.

$$e_{k,j} \geq \sum_{s=1}^{S}\frac{w_{k,j}}{v_s} \qquad ;\forall k \ , \ \forall j \qquad (11)$$

The constraint (11) ensures that the elapsed time of the task is greater than its processing time.

$$e_{k,j} \leq TBF_j \qquad ;\forall k \ , \ \forall j \qquad (12)$$

The constraint (12) ensures that no jobs can be processed during unavailability intervals.

$$e_{k+1,j} \geq e_{k,j} + \sum_{s=1}^{S}\frac{w_{k+1,j}}{v_s} - g_{k,j} Z; \forall k < n, \ \forall j (13)$$

The constraint (13) ensures that the elapsed time of $O_{k+1,j}$ is greater than the sum of the processing time of $O_{k+1,j}$ and the elapsed time of $O_{k,j}$ if they're in the same batch.

$$C_{max} = C_{n,m} \qquad (14)$$

The constraint (14) confirms that the makepan is equal to the completion time of the machine's nth task.

$x_{i,j,k,s}$, $g_{k,j}$ are of the variables binary (15)

$q_{k,j}$ is a complete (16)

$\frac{w_{k,j}}{v_s}$, $C_{k,j}$, $S_{k,j}$, $e_{k,j}$ ≥ 0 (17)

4 Modelling of the problem $Fm|nr-CI|_{TEC}$

Minimize *Fct Obj*

S.t

Constraint sets (1)-(5) and constraint set (7) are the same with the model 1.

$$S_{k,j} \geq C_{k-1,j} + PM_j \cdot y_{k,j} \quad ; \forall j, \forall k \geq 2 \quad (18)$$

The constraint (18) ensures that the operation $O_{k,j}$ must not start before the sum of the end time of the immediate previous job in M_j and the possible maintenance time.

$$b_{1,j} = 0 \quad ; \forall j \quad (19)$$

$$y_{1,j} = 0 \quad ; \forall j \quad (20)$$

Constraints (19) and (20) specify the initial state of the system.

$$a_{k,j} = b_{k,j} + \sum_{s=1}^{S} \frac{w_{k,j}}{V_s} \quad ; \forall k, \forall j \quad (21)$$

$$b_{k+1,j} = a_{k,j} (1 - Y_{k+1,j}) \quad ; \forall j, k \leq n \quad (22)$$

Constraints (21) and (22) specify the age of the machines before and after each operation $O_{k,j}$, respectively.

$$a_{k,j} \leq TBF_j \quad ; \forall k \quad (23)$$

$$\frac{w_{k,j}}{V_s}, C_{k,j}, S_{k,j} \geq 0 \quad (24)$$

5 Genetic algorithm for solving the models

The genetic algorithm begins with a step called genesis in which a generation of initial Pop_Size size population of individuals is generated. For each individual generated, a cost function is calculated in order to define the adaptation score of the individuals during the selection process. These individuals evolve through the application of crossing according to a probability Pc. Subsequently, the children obtained undergo an inversion at the level of the genes with a probability of mutation Pm. These three phases of evolution allow with a great chance to produce a new population better than that of the previous generation.

With each new generation, the new populations are increasing and a loop is made as long as the evaluation considers that the solution is not yet optimal. Figure 1 shows the general mechanism of genetic research.

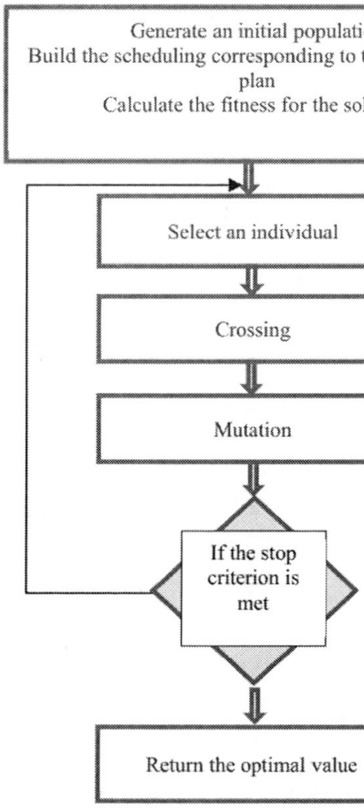

Figure 1. General mechanism of genetic algorithms

The pseudo-code of genetic algorithm :

In order to solve our problem of integrating the planning and ordering of production systems, we present in the following the algorithm of the figure 2 which contains the main stages of the genetic algorithm.

The research begins by initializing a population of individuals who will be called Popsize (population size), Pc is the percentage of growth, Pm is the percentage of mutation, max_iteration is the maximum number of iterations. The genetic algorithm stops when the maximum number of iterations is reached. Among the major difficulties of the genetic algorithm is the good choice of the values of the Popsize, Pc, Pm and max_iteration parameters.

```
Algorithm: the genetic algorithms pseudo-code
A : Initialize Pop size, Pc, Pm and max_iteration ;
B : Generate an initial population P (n) which represents a sequencing of jobs on the resources;
C : Calculate the value of fitness (TEC);
D : Select the best fitness for the population using the tournament selection method
E : Until the number of iteration is reached:
        For each sequence from 1 to Pop size :
        Choose randomly 2 sequences of P (n).
                If random <= Pc :
                        Apply the 2 point cross and produce 2 children.
                        If random <= Pm :
                         Apply the shift change  mutation ;
                Copy the children into the new  population ;
                Evaluate the new  population.
F : If the stop criterion is reached, complete and return the optimal value found otherwise return to step D.
```

Figure 2. The genetic algorithms pseudo-code

3 Conclusion

This work concerns the integrated scheduling of production and maintenance tasks under energy constraints. To our knowledge and according to the literature review, the documents dealing with production and maintenance scheduling do not take into account energy constraints. In the literature, two types of unavailability constraints are examined. In the first case, the unavailability intervals are fixed periodically and known in advance. In the second case, the intervals are flexible and the continuous working time of the machines cannot exceed the maximum authorised time.

We solved these problems with two models for programming mixed binary integers. MILPs are provided with fixed and flexible maintenance types to derive optimal schedules in order to minimize total energy consumption (TEC). The performance of the proposed mixed binary integer programming models is evaluated based on the genetic algorithm (GA). A comparative study of the results provided with those available in the literature proved the performance of the model developed.

References

1. EIA, 2013. International Energy Outlook 2013 (Online). www. eia . gov / forecasts / ieo / pdf / 0484 (2013) . pdf . Accessed on 5 May 2018.
2. CSY, 2016. China Statistical Yearbook 2015 (Online). www. stats. gov .cn / tjsj / ndsj / 2016/ indexeh. htm . Accessed on 5 May 2018.
3. Wang S, Wang X, Yu J, Ma S, Liu M, Bi-objective identical parallel machine scheduling to minimize total energy consumption and makespan, Journal of Cleaner Production (2018), doi: 10.1016/j.jclepro.2018.05.056.
4. Ding J., Song S., Zhang R., Chiong R., Wu C. (2016a) Parallel machine scheduling under time-of-use electricity prices: new models and optimization approaches. IEEE Transactions on Automation Science and Engineering, 13(2): 1138-1154.

A Study on Performance Enhancement of Parabolic Trough Collector

Aditya Bawane[1*], *Sanjay Lakade* [1], *and Virendra Bhojwani*[2]
[1]Pimpri Chinchwad College of Engineering, Pune, India
[2]MIT-ADT University, Pune, India.

Abstract. Solar energy is available in abundant quantity which can be utilized for thermal and power generation applications. The maximum solar energy extraction for application is challenging. This review focuses on the performance enhancement of parabolic trough collector. Heat transfer through absorber tube, various nanofluids with concentration is stated. Thermal efficiency increases due to the use of parabolic collector with booster reflector and glass cover over the system performance. The complex 3D analysis (ANSYS) gives accurate distribution of heat flux over the absorber tube.
Keywords: Parabolic trough collector, secondary booster reflector.

1 Introduction

Many technologies have been developed to make use of solar radiation to reduce the dependency on the exhaustable energy sources. Sun releases energy in the form of radiation which is 1.74×10^7 W at the upper layer of the earth. 51% of these energy reaches to earth surface which is huge [1]. Solar energy is free, but capturing, processing and storaging costs are high. There are many methods to utilize the solar energy in low to high temperature range. Among these different solar collector are flat plate collector, evacuated tube collector, parabolic trough collector (PTC), compound collector, linear Fresnel reflector, parabolic dish collector, heliostat field collector.

The flat plate collectors are used for applications such as domestic water heating for low temperature (120-140^{0}C).PTC and CPC are used to steam generation plant. PTC has parabolic shaped mirror, absorber and receiver tube, and tracking mechanism. It can reach the temperature range of 350-400 ^{0}C. CPC can be used for low and medium temperatures of 60-240 ^{0}C [2] with high efficiency and low heat loss. Linear Fresnel reflector have long rows of mirror which direct the solar radiation to the downward facing fixed receiver.

2 Parabolic Trough Collector

In PTC, Solar energy incident on parabolic shaped mirror then it reflected to the receiver tube through glass covered absorber tube. The fluid that circulates through the receiver tube is heated and used for additional applications such as steam production. PTC uses single axis or two axis tracking system. Mostly two axes system are arranged in north-south direction for tracking the sun and it is moved along east-west direction for high optical efficiency. Single axis tracking system with receiver tube is arranged in east-west direction, which results in lesser efficiency as compared to dual axes tracking. PTC with have 60-70% thermal efficiency with evacuated receiver tube which is 10% higher than non-evacuated tube.

2.1 Thermal Analysis of PTC Receiver

The heat transfer through receiver tube occurs through three different heat transmission modes – conduction, convection and radiation.

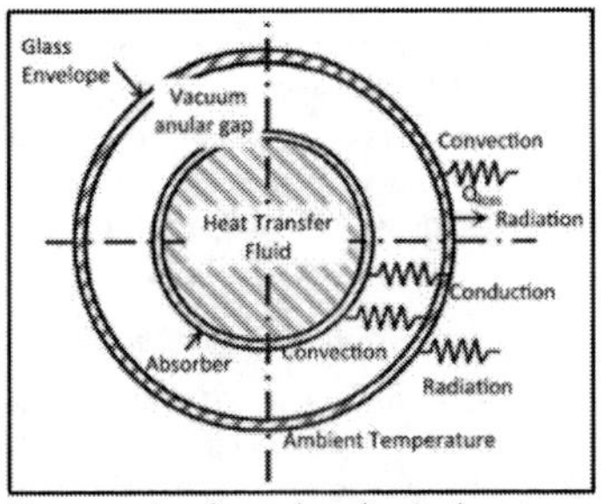

Fig. 1. Cross-sectional receiver for the heat transmission

The transfer of convection heat take place whenever medium changes and it greatly depends upon geometry, surface roughness and fluid properties [3]. Heat transmission by conduction through glass and receiver tube is related by Fourier law: [4]

$$q_{pipe\,(in-out)} = \frac{2\pi K_c}{\ln\left(\frac{D_{pipe,in}}{D_{pipe,out}}\right)}(T_{in} - T_{out}) \quad (1)$$

Where K_c is the thermal conductivity of receiver (W/m ^{0}C).

Heat transfer by convection between the fluid transfering heat and inside surface of receiver tube can be expressed as ;

*Corresponding author: aditya23892@gmail.com

$$q_{f,conv} = \pi D_{pi} h_f (T_{pi} - T_{fluid}) \quad (2)$$

Where h_f is the coefficient of convection heat transfer at the internal diameter of tube (W/m °C); D_{pi} is the diameter (internal) of the receiver (m); T_{pi} is internal surface temperature of tube (°C) and T_{fluid} is the average temperature of the circulating fluid (°C).
Radiation does not require any heat transfer medium, it can be transmitted, absorbed or reflected. The heat transfer due to radiation between the envelope of glass and the receiver pipe considering diffused reflection and radiation is given by equation (3)

$$q_{pi-po,rad} = \frac{\sigma \pi D_{pi,o}(T_{po}^4 - T_{gi}^4)}{\frac{1}{\varepsilon_{po}} + \left(\frac{(1-\varepsilon_{gi})D_{po}}{\varepsilon_{gi} D_{gi}}\right)} \quad (3)$$

3 Heat Transfer Enhanancement

3.1 Nanofluids

The increased heat transfer for the base substance with insert in the laminar region was increased 20-300% as compared to regular receiver [5]. This is due to the turbulence enhances the diffusion of energy and hence the heat transfer. TiO_2/De- ionized water with nano fluid increases the output of PTC. TiO_2 will achieve an improvement in maximum efficiency of 8.66 % (57%) higher than the solution of water base at a volume concentration of 0.2% having mass flow of 0.0667 kg/s [6]. These nanofluids have an energy factor of 9.51 % higher than water. Distilled and deionized water and alumina nanoparticles (Al_2O_3) were used in volume fraction of 1% and 3%. The addition of nanoparticles increases the thermal properties of the fluid, thereby allowing energy exchange between the working fluid and the tube to be increased. For 3% volume fraction of nano-fluid, the maximal efficiency was 52.4 % and 40.8 % for water for various ambient conditions and incident angles. Nano fluid had a maximum efficiency of 57.7 % for nanofluid and 46.5 % for water for 1 % volume fraction of nanofluid. The maximum efficiency was found for the small angle of incident [7]. For a small solar radiation value, the outlet nanofluid temperature was higher than water.

3.1.1 Different Geometry for Receiver Tube

The absorber tube of PTC was connected with twisted tubes which resulted in enhancement of energy exchange. Five reviver tubes were used for receiver tube each one with overlapped flows of two types. The first one is the flow-through twisted inner tubes of different hydraulic diameters (single, double and triple). Second with reverse flow via an external tube. The measurements showed a significant improvement in the energy exchange of the single twisted tube (STT) and a greater increase in overlapped the triple twisted tube (TTT) for the single receiver tube. This is because the pressure difference for TTT is greater than STT [8].

Fig. 2. Twisted tubes [8]

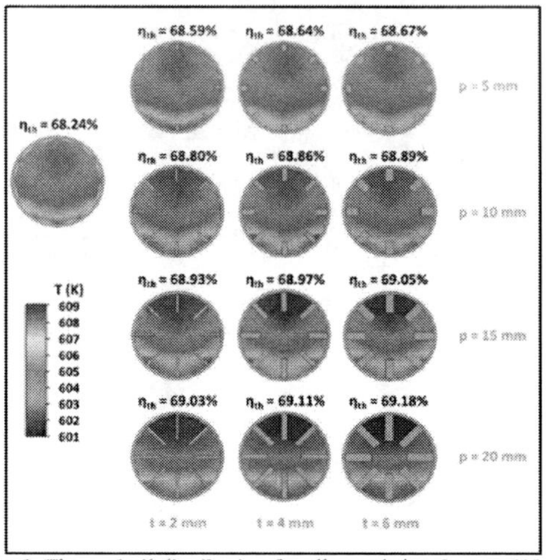

Fig. 3. Thermal oil distribution for all tested absorbers at 600 K and 150 L/min [9].

Twelve absorber tubes are compared with smooth tube with internally finned length of 5 mm to 20 mm and a thickness of 2 to 6 mm. As the number of fin increases the efficiency will increase but losses occurs at high pressure. The fin's length is significantly higher than its thickness [9]. However thickness have significant effect on pressure losses.

3.2 Receiver and Optical Efficiency

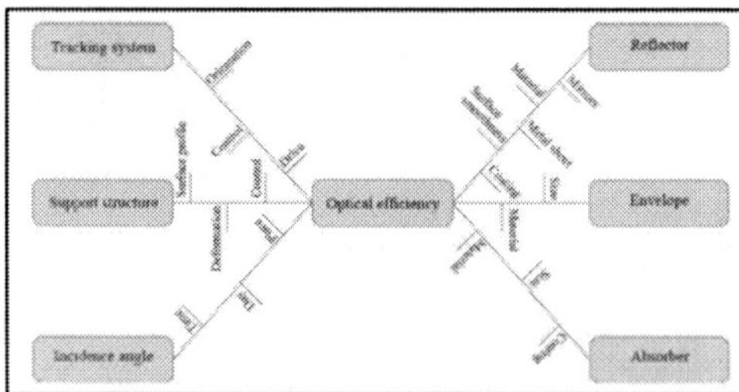

Fig. 4. Parameters affecting the optical efficiency [16].

Silver and Aluminium mirrors are most commonly used materials for mirrors having ideal reflectance of 0.99 [16]. Fig. 5. Shows the parameter affecting the optical efficiency. The unit length cost for receiver decreases when size of solar field expands from 2.8 hectares to 10 hectares (almost halved about $1400 to $760). Efficiency depends on reflective area, aperture area and solar field area. For large capacity plants it is possible to cut down the cost in each sector. The labour cost for mechanical work can be drops by 48% for 10 hectares and 75% for 160 hectares. Where as civil cost remain same. Copper can be used as receiver tube material for PTC considering thermal stresses and bending phenomenon [10]. Optical performance reduces with respect to time and glass breaks. To prevent this risk, the thickness of metal ring should be minimal enough and coefficient of thermal expansion between the glass cover and the metal ring should be lower than 7% [11]. The best method to eliminate the conduction losses from receiver tube is creating vacuum between glass cover and receiver tube. Borosilicate glass and stainless steel are good material for vacuum. Getter is necessary to maintain vacuum [10]. Data from existing PTC plant shown 55% of failure due to broken glass and 29% due to vacuum loss. The receiver cost is 30% of plant cost having payback period of 1-5 years. The bent receiver tube may touch glass cover and damage it [12]. In order to concentrate thermal photovoltaic collectors, increase in inlet temperature increases exergy efficiency but reduces thermal and electrical efficiencies [13].

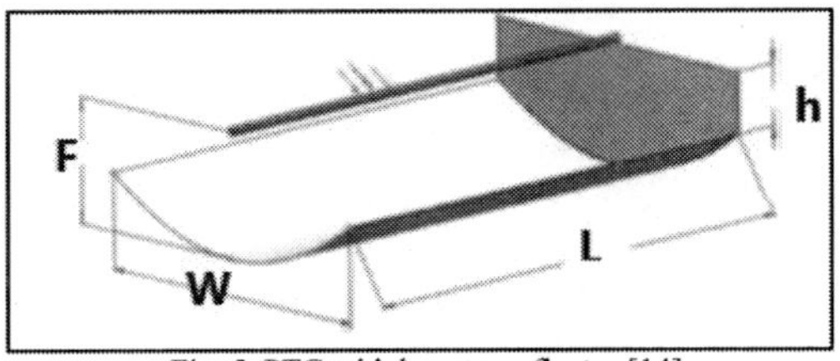

Fig. 5. PTC with booster reflector [14]

A secondary vertical reflector is placed at the edge of collector to enhance the optical performance. This reduces the losses at the end of receiver of PTC by reflecting back to receiver. With higher solar incident angle, the optical enhancement with booster reflector is higher. With a ratio of 0.236 (focal), the annual enhancement is 21.7 %. In winter it is greater and smaller in summer as a percentage. When the system operates at high temperatures, the thermal improvement is higher [14]. The flat glass coverup over the PTC increases the thermal output by 9% [15]. This is due to the air trapped inside the envolope heat up the inside air and reduces the convection losses.

4 CFD Analysis

To predict thermal performance, it is useful to analyze PTC by 1D and less complex 2D. The complex 3D analysis is most suitable and accurate and it provides all information about actual distribution of heat flux on the receiver, which cannot be achieved by 1D or 2D model. Fig.7. a) and b) shows the distribution of temperature and heat flux on receiver pipe. For selection of optimum receiver tube properties 3D analysis is useful for prediction of minimum and maximum temperature. For this purpose CFD analysis is useful for creating and meshing the geometry, defining the boundary conditions and heat transfer of fluid flow. MTRC method used to provide an non homogenous solar flux distribution over the heat collecting element for optical analysis [17].

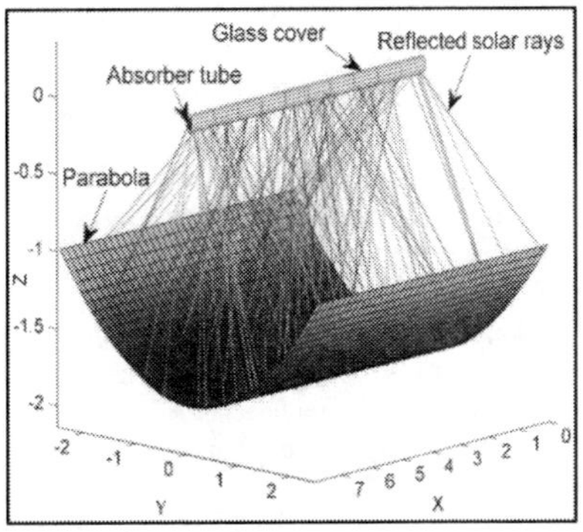

Fig. 6. Solar concentration mechanism obtained using 3D analysis in parabolic trough collector [17]

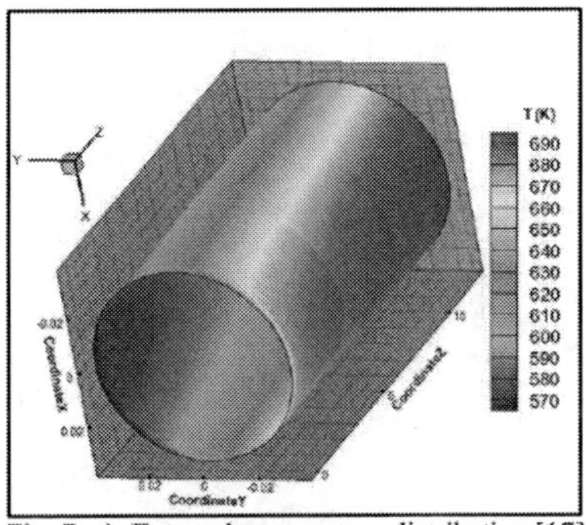

Fig. 7. a) External temperature distribution [17]

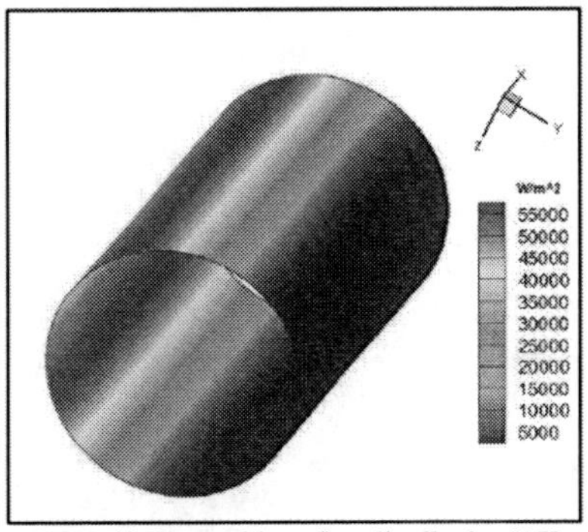

Fig. 7. b) External heat flux distribution [17]

5 Conclusion

This study focuses on mathematical modeling and techniques used to improve the energy exchange in PTC. The following conclusion are drawn from review:

a) In laminar flow regimes with inserts, the heat transfer improvement is significant. Nanofluids can increase the energy factor to 9% as compared to water.

b) Increase in fin numbers in receiver tube increases the efficiency but it results into higher pressure losses. Fin thickness have significant effect on pressure losses.

c) PTC with a secondary booster reflector raises the annual enhancement by 21.7 % with a focal ratio of 0.236.

d) Glass cover over the PTC increases the thermal efficiency by 9%.

e) The complex 3D model is most accurate for actual heat flux distribution over a receiver tube.

References

1. W. Fuqiang, C. Ziming, T. Jianyua, Y. Yuan, S. Yong, L. Linhua, *Progress in concentrated solar power technology with parabolic trough collector system: A comprehensive review*, Renewable and Sustainable Energy Reviews **79** (2017)
2. V. Pranesh, R. Velraj, S. Christopher, V. Kumaresan, *A 50 year review of basic and applied research in compound parabolic concentrating solar thermal collector for domestic and industrial applications*, Solar Energy **187** (2019).
3. A. Abdulhameda, N. Adam, M. Ab-Kadir, A. Hairuddin, *Review of solar parabolic-trough collector geometrical and thermal analyses, performance, and applications*, Renewable and Sustainable Energy Reviews **91** (2018)
4. L. S. Conrado, A. R. Pulido, G. Calderon, *Thermal performance of parabolic trough solar collectors*, Renewable and Sustainable Energy Reviews **67** (2017)
5. H. M. Sanse, Renewable and Sustainable Energy Reviews (to be published)
6. J. Subramani, P.K. Nagarajan, Omid Mahian, Ravishankar Sathyamurthy, *Efficiency and heat transfer improvements in a parabolic trough solar collector using TiO_2 nanofluids under turbulent flow regime*, Renewable Energy (2017)
7. B. Rios MS, Rivera-Solorio, G. Cuéllar AJ, *Thermal performance of a parabolic trough linear collector using Al_2O_3/H_2O nanofluids*, Renewable Energy (2018)
8. Ra'ad K. Mohammed Aldulaimi, *An Innovative Receiver Design for a Parabolic Trough Solar Collector Using Overlapped and Reverse Flow: An Experimental Study*, Arabian Journal for Science and Engineering, (2019).
9. E. Bellos, C. Tzivanidis, D. Tsimpoukis, *Thermal enhancement of parabolic trough collector with internally finned absorbers*, Solar Energy · (2017)
10. S. Akbarzadeh, M. Sadegh Valipour, *Heat transfer enhancement in parabolic trough collectors: A comprehensive review*, Renewable and Sustainable Energy Reviews **92** (2018)
11. Ze-Dong Cheng, Xue-Ru Zhao, Ya-Ling He, *Novel optical efficiency formulas for parabolic trough solar collectors: Computing method and applications*, Applied Energy **224** (2018)
12. Mohammed Soudani, Kamal E. Aiadi, D. Bechki and S. Chihi, *Experimental and theoretical study of Parabolic trough collector (PTC) with a flat glass cover in the region of algerian sahara (Ouargla)*, Journal of Mechanical Science and Technology (2017)
13. M. Valizadeh, F. Sarhaddi, M. Adeli, *Exergy performance assessment of a linear parabolic trough photovoltaic thermal collector*, Renewable Energy **138** (2019)
14. Evangelos Bellos, C. Tzivanidis, *Investigation of a booster secondary reflector for a parabolic trough solar Collector*, Solar Energy **179** (2019)
15. M. E. Soudani, K. E. Aiadi, D. Bechki and S. Chihi, *Experimental and theoretical study of Parabolic trough collector (PTC) with a flat glass cover in the region of algerian sahara (Ouargla)*, Journal of Mechanical Science and Technology (2017)
16. Mokheimer EM, Dabwan YN, Habib MA, Said SA, Al-Sulaiman FA. *Techno-economic performance analysis of parabolic trough collector in Dhahran, Saudi Arabia*. Energy Convers Manage (2014)
17. B. Agagna, A. Smaili, Q. Falcoz, O. Behar, *Experimental and numerical study of parabolic trough solar collector of MicroSol-R tests platform*, Experimental Thermal and Fluid Science (2018).

A thermostatted model for a network of energy sources: Analysis on the initial condition

Marco Dalla Via[1,2,*], *Carlo Bianca* [1,2], *Ikram El Abbassi*[2], *and Abdelmoumen Darcherif* [1,2]
[1]ECAM-EPMI, Laboratoire Quartz EA 7393, 13 Boulevard de l'Hautil, 95092 Cergy-Pontoise (France)
[2]ECAM-EPMI, Laboratoire LR2E, 13 Boulevard de l'Hautil, 95092 Cergy-Pontoise (France)

Abstract. The energy multisource network is a complex system characterized by the interactions between the energy sources. Recently the thermostatted kinetic theory has been proposed for the modelling of a hybrid energy multisource network with storage. The present paper is devoted to the presentation of a thermostatted kinetic theory model for a network composed of a non-renewable and a renewable energy source. The storage system is modelled by introducing an outer force field. In particular the modelling interest is addressed to the analysis on the initial condition of the distribution functions which describe the two energy sources.

Keywords. Thermostatted kinetic theory, energy source, storage system, initial condition

1 Introduction

Currently a special attention is paid to the modelling of the energy source network, in order to face the increasing energy demand [1,2]. This issue has attracted researchers coming from different fields. Indeed, the inclusion of the renewable energy sources in the power grid is a challenging task which is necessary to realize the so called "energy transition"[3, 4]. The renewable energy sources are characterised by a certain intermittency [5, 6] which needs to be considered in order to optimize the management of the hybrid multisource network. The introduction of storage systems enables facing the low reliability of the renewable intermittent energy sources [7].

During the last few decades, the concept of "smart grid" [8-10] has emerged to depict the features of the modern energy network. In particular, the interactions and the exchange of information between the sources is the fundamental tool with which the network manages to supply the required energy to the users.

Among the different mathematical approaches, that have been considered for the modelling of the energy distribution network, one has: ODE-based models [11, 12], complex networks [13-15] and optimization approaches [16, 17]. In particular, a number of studies have identified the energy network as a complex system [18, 19], because of its interconnected character.

Lately, the paper [20] has proposed the thermostatted kinetic theory [21] as modelling tool for the energy multisource network. The thermostatted kinetic theory has already been employed as general paradigm for the modelling of complex systems in biology [22, 23] and within the pedestrian dynamics framework [24].

According to the thermostatted kinetic theory approach, one divides the complex system, i.e. the multisource network, into different subsystems, i.e. the energy sources. The sources of energy are represented by distribution functions over the activity variable, i.e. the produced energy value, which expresses the behaviour and the "strategy" of the subsystems. The binary interactions between the sources drive the time evolution of the network which is depicted by nonlinear integro-differential partial equations. The modelling theory also considers the introduction of an outer force field. The action of the force on the energy sources affects the energy supply and enables to model the charging of the storage system. A specific energy quantity is not yielded to the users and it is used to fill the storage system. The outer force field is balanced by introducing a thermostat term which controls the global activation energy and ensures the existence of a nonequilibrium steady state [25, 26]. The computation of the thermostat term follows the papers [27, 28] where the authors use the Gaussian thermostat in the molecular dynamics simulations.

The authors of this work have proposed in [29], and furtherly developed in [30], a thermostatted model for a hybrid power grid. Specifically, the attention has been focused on the sensitivity analysis on the parameters and in the modelling of the time dependent performance of each source of energy (renewable and non-renewable).

The present paper deals with a deeper analysis on the initial condition for the aforementioned model. Specifically, different initial conditions are considered in order to enlighten their effect on the evolution over the energy variable of the distribution functions and on the charging of the storage system. The computational

* Corresponding author: **m.dallavia@ecam-epmi.com**

analysis shows that the distribution functions mostly evolve in the neighborhood of the initial condition. Besides, the charging of the storage system is slower if the initial condition is characterized by low values of energy. The contents of this work are exposed by means of three more sections. The second section presents the thermostatted model for the grid composed of one source of non-renewable energy and one source of renewable energy. The third section deals with the analysis on the initial datum of the distribution functions which represent the two sources of energy. Finally, the fourth section proposes some conclusions.

2 The thermostatted kinetic theory model

This section presents the thermostatted model for the network of energy sources with storage proposed in [29]. The model is derived within the paradigm described in [21]. The network $\mathcal{N}$ is composed of two sources of energy: a source of non-renewable energy NR and a source of renewable energy R. According to the proposed modelling framework, every single energy source is associated to a quality parameter $u_i \in [0,1]$ for $i \in \{1,2\}$: $u_1 = 1/3$ is related to the low quality energy source NR and $u_2 = 2/3$ is related to the high quality energy source R. The energy sources are assumed to be two functional subsystems whose description is realized by a vector of distribution functions $\boldsymbol{f} = (f_1(t,w), f_2(t,w))$, where:

$$f_i(t,w): [0,+\infty[\times D_w \to \mathbb{R}^+, i \in \{1,2\}, \quad (1)$$

and the continuous variable $w \in D_w \subset \mathbb{R}^+$ denotes the value of the furnished energy. Specifically, $f_i(t,w)dw$, for $i \in \{1,2\}$, indicates the number of users who are receiving a value of energy $w \in [w, w+dw]$, produced by the energy source u_i, at time t. Bearing all above in mind, $f_1(t,w)$ describes the energy production of the source of non-renewable energy NR and $f_2(t,w)$ describes the energy production of the source of renewable energy R.
It is assumed that an outer positive force field $\boldsymbol{F} = (F,F) \in (\mathbb{R}^+)^2$ acts on the energy sources: This allows the modeling of the charging of the storage. The whole system is moved out of the equilibrium due to the action of the outer force field. Thus, $\boldsymbol{F}$ is balanced by a thermostat term $\alpha[\boldsymbol{f}](t)$. The thermostat aims at controlling the global activation energy of the network and therefore grants the reaching of a nonequilibrium steady state.
The description of the multisource grid at the macroscopic scale is obtained by considering the distribution functions moments. In detail, the global moment of order *(p,q)* writes as follows:

$$\mathbb{E}^{[F]}_{(p,q)}[f](t) = \sum_{i=1}^{2} u_i^p \int_{D_w} w^q f_i(t,w)dw, \quad \forall p,q \in \mathbb{N}. \quad (2)$$

In particular, the moment of order *(0,0)* $\mathbb{E}^{[F]}_{(0,0)}[f](t)$ denotes the number of users who receive energy. The moment of order *(0,1)* $\mathbb{E}^{[F]}_{(0,1)}[f](t)$ denotes the average quantity of energy which is yielded to the users. The moment of order *(2,2)* $\mathbb{E}^{[F]}_{(2,2)}[\boldsymbol{f}](t)$ denotes the global activation energy of the system. In the case $\boldsymbol{F} = \boldsymbol{0}$, the notation $\mathbb{E}^{[F=0]}_{(p,q)}[\boldsymbol{f}](t)$ simplifies to $\mathbb{E}_{(p,q)}[\boldsymbol{f}](t)$.
The following master equation balances the inlet and outlet flows into the microscopic volume *[w,w+dw]* and gives the time derivative of the distribution functions:

$$\partial_t f_i(t,w) + F\partial_w\left(\left(1 - \alpha_{(2,2)}[\boldsymbol{f}](t)w\right)f_i(t,w)\right) = J_i[\boldsymbol{f}](t,w), \quad i \in \{1,2\}. \quad (3)$$

The functional parameter $\alpha_{(2,2)}[\boldsymbol{f}](t)$ refers to the thermostat term employed to control $\mathbb{E}^{[F]}_{(2,2)}[\boldsymbol{f}](t)$, and writes as follows:

$$\alpha_{(2,2)}[\boldsymbol{f}](t) = \frac{\mathbb{E}^{[F]}_{(2,1)}[\boldsymbol{f}](t)}{\mathbb{E}^{[F]}_{(2,2)}[\boldsymbol{f}](t)}. \quad (4)$$

The reader interested in the formal derivation of the thermostat (4) is addressed to [29].
The functional operator $J_i[\boldsymbol{f}](t,w) = G_i[\boldsymbol{f}](t,w) - L_i[\boldsymbol{f}](t,w)$, for $i \in \{1,2\}$, models the interactions between the energy sources. The operator $J_i[\boldsymbol{f}](t,w)$ is divided into the gain term $G_i[\boldsymbol{f}](t,w)$ and the loss term $L_i[\boldsymbol{f}](t,w)$, which write as follows:

$$G_i[\boldsymbol{f}](t,w) = \sum_{l=1}^{2}\sum_{j=1}^{2}\int_{D_w} \eta_{lj}(w_*,w^*)\,\mathcal{B}^i_{lj}\,\mathcal{A}_{lj}(w_*,w^*;w) \times f_l(t,w_*)f_j(t,w^*)dw_*dw^*, \ \mathrm{i} \in \{1,2\}, \quad (5)$$

$$L_i[\boldsymbol{f}](t,w) = f_i(t,w)\sum_{j=1}^{2}\int_{D_w} \eta_{ij}(w,w^*)\,f_j(t,w^*)\,dw^*, \quad \mathrm{i} \in \{1,2\}. \quad (6)$$

The functional parameters model the phenomena described below:

- $\eta_{lj}(w_*,w^*): D_w \times D_w \to \mathbb{R}^+$, for $l,j \in \{1,2\}$, models the rate of the exchanges of information between the energy source u_l, yielding the value of energy w_*, and the energy source u_j, yielding the value of energy w^*. The interaction rate is constant. It is assumed that the two distinct energy sources have an interaction rate which is smaller than the autointeraction rate: $\eta_{lj}(w_*,w^*) = \alpha + \beta\delta_{lj}$, with $\alpha,\beta \in \mathbb{R}^+$, where δ_{lj} refers to the delta of Kronecker.
- $\mathcal{A}_{lj}(w_*,w^*;w): D_w \times D_w \times D_w \to \mathbb{R}^+$, for $l,j \in \{1,2\}$, represents the probability that the binary interaction involving the energy sources u_l and u_j, which yield the values of energy w_* and w_*, respectively, makes the energy source u_l to yield the value of energy w. The modeling of $\mathcal{A}_{lj}(w_*,w^*;w)$ is based on the following statements: The low values of energy should be furnished by the non-renewable energy source NR and the high values of energy should

be furnished by the renewable energy source R. Therefore, the probability density $\mathcal{A}_{lj}(w_*, w^*; w)$ reads:

$$\mathcal{A}_{lj}(w_*, w^*; w) = \delta(w - d_{lj}(w_*, w^*)), \quad (7)$$

where δ refers to the delta of Dirac and

$$d_{lj}(w_*, w^*) = \begin{cases} w_* - \lambda_1 \text{ if } l = 1,\ j = 2 \\ w_* + \lambda_2 \text{ if } l = 2,\ j = 1 \\ w_* \quad\quad \text{if } l = j, \end{cases} \quad (8)$$

and $\lambda_1, \lambda_2 \in [0,1]$. The parameters λ_1, λ_2 grant the production of different values of energy.

• $\mathcal{B}^i_{lj}$, for $i, l, j \in \{1,2\}$, represents the probability that the production of the energy source u_i is boosted due to the binary interaction involving the energy sources u_l and u_j. The probability density $\mathcal{B}^i_{lj}$ reads:

$$\mathcal{B}^i_{lj} = \begin{cases} \epsilon_{lj} & \text{if } i = l \\ 1 - \epsilon_{lj} & \text{if } i \neq l \end{cases} \quad (9)$$

where $\epsilon_{lj} \in [0,1]$ are the activation parameters. Specifically, the energy source u_l has probability ϵ_{lj} of being activated, for $l, j \in \{1,2\}$.

Given a specified initial datum $\boldsymbol{f}^0 = (f_1^0(w), f_2^0(w))$, the mathematical model (3) is a system of nonlinear integro-differential equations containing 9 phenomenological parameters. The analysis performed in [21] has proved that the corresponding initial boundary value problem has a solution and that this solution is unique. The sensitivity analysis on the parameters has been performed in [29] with special attention to the activation parameters ϵ_{12}, ϵ_{21} and the heterogeneity parameters λ_1, λ_2. Moreover [29] has shown the relation between the magnitude F of the outer force field $\boldsymbol{F}$ and the charging of the storage system. The quantity $\boldsymbol{E}[\mathbf{F}, \mathbf{f}](t)$ denotes the energy placed into the storage system and reads:

$$\boldsymbol{E}[\mathbf{F}, \mathbf{f}](t) = \mathbb{E}_{(0,1)}[\boldsymbol{f}](t) - \mathbb{E}^{[F]}_{(0,1)}[\boldsymbol{f}](t) \quad (10)$$

and the equation (2) gives the expression of $\mathbb{E}_{(0,1)}[\boldsymbol{f}](t)$ and $\mathbb{E}^{[F]}_{(0,1)}[\boldsymbol{f}](t)$.

3 The computational analysis on the initial condition

This section presents the computational analysis on the initial condition $\boldsymbol{f}^0 = (f_1^0(w), f_2^0(w))$ for the model (3). Specifically, the attention is addressed to the effects of different initial conditions on the evolution over the energy variable of the distribution functions, in the case $F = 0$, and on the charging of the storage system, in the case $F \neq 0$.

The parameters presented in the previous section are set as done in [29]: $\alpha = 0.6$, $\beta = 0.1$, $\epsilon_{11}=0.5=\epsilon_{22}$, $\epsilon_{12}=0.3$, $\epsilon_{21}=0.8$, $\lambda_1=0.5$ and $\lambda_2 = 0.6$. In particular, $\epsilon_{12} < \epsilon_{21}$ boosts the production of the source of renewable energy R, increasing the quality of the energy furnished to the users. The magnitude of the parameters λ_1 and λ_2 has been chosen to promote a diversified production of energy. The sensitivity analysis on the functional parameters can be found in [29].

The numerical integration of the model (3) is performed by means of an order 4 Runge-Kutta method (MATLAB).

The following initial conditions are considered:

$$a)\ f^0_{i,a}(w) = \begin{cases} 2/\sqrt{\pi}\, e^{-(4(w-1.8))^2} \text{ if } i=1 \\ 2/\sqrt{\pi} e^{-(4(w-1.2))^2} \text{ if } i=2 \end{cases} \quad (11)$$

$$b)\ f^0_{i,b}(w) = \begin{cases} 2/\sqrt{\pi} e^{-(4(w-2.3))^2} \text{ if } i=1 \\ 2/\sqrt{\pi} e^{-(4(w-1.7))^2} \text{ if } i=2 \end{cases} \quad (12)$$

$$c)\ f^0_{i,c}(w) = \begin{cases} 2/\sqrt{\pi} e^{-(4(w-1.3))^2} \text{ if } i=1 \\ 2/\sqrt{\pi} e^{-(4(w-0.7))^2} \text{ if } i=2 \end{cases} \quad (13)$$

The three initial conditions are shown in the Figures 1, 2 and 3. Each initial condition is defined by means of two Gaussian functions with $\sigma^2 = 1/32$. The initial distribution $f^0_{1,j}(w)$ attains higher values of w (values of energy) than the initial distribution $f^0_{2,j}(w)$, for $j \in \{a, b, c\}$. Accordingly, the three initial conditions are characterized by the fact that the source of non-renewable energy NR supplies the most part of the energy $\mathbb{E}^{[F]}_{(0,1)}[\boldsymbol{f}](t)$ which is used by the users. For what concerns the number of users, the three initial conditions are such that the two energy sources equally share the total number of users $\mathbb{E}^{[F]}_{(0,0)}[\boldsymbol{f}](t)$. The initial condition a) is characterized by the production of medium values of energy (in the middle of the energy domain D_w), the initial condition b) is characterized by the production of high values of energy (in the right side of the energy domain D_w), the initial condition c) is characterized by the production of low values of energy (in the left side of the energy domain D_w).

3.1 The $F=0$ case

This section is devoted to analyzing the evolution over the energy variable of $f_1(t,w)$ and $f_2(t,w)$, for the initial conditions $\boldsymbol{f}^0_a = (f^0_{1,a}(w), f^0_{2,a}(w))$, $\boldsymbol{f}^0_b = (f^0_{1,b}(w), f^0_{2,b}(w))$ and $\boldsymbol{f}^0_c = (f^0_{1,c}(w), f^0_{2,c}(w))$ in the case $F = 0$. As it is shown by the Figures 4, 5 and 6, the evolution over the energy variable is characterized by the fact that the distribution functions become wider and overlap. This phenomenon is related to the parameters λ_1 and λ_2 which prompt the heterogeneity of the values of the produced energy. The overlapped distribution functions indicate that some users receive a mix of energy produced by both sources. The choice of the magnitude of the parameters ϵ_{12} and ϵ_{21}, namely $\epsilon_{12} < \epsilon_{21}$, promotes the activation of the source of renewable energy R whose distribution function $f_2(t,w)$ grows more than the distribution function $f_1(t,w)$ of the source of non-renewable energy NR. It must be stressed that the distribution functions hardly leave the region of the energy domain where the initial condition is located. This means that the energy production is strongly dependent on the initial produced energy values. If at t^0 the two energy sources are producing low, medium or high values of energy, they will continue to produce low, medium or high values of energy as time passes.

The dependence on the allocation of the initial condition is related to the nature of the kinetic theory framework where the interactions between the subsystems drive the dynamics. Bearing the mathematical model (3) in mind, the intensity of the interactions is proportional to the overlapping between the distribution functions. Therefore, the regions of the energy domain, which are not covered by the distribution functions, are involved by the dynamics only if the distribution functions translate. The translation towards energy values, which are not covered by the initial condition, depends on the magnitude of the parameters λ_1 and $\lambda_2$and promotes a diversified production of energy.

3.2 The $F \neq 0$ case

The analysis presented in this section focuses on the energy $\boldsymbol{E}[\boldsymbol{F},\boldsymbol{f}](t)$, see the equation (10), put into the storage system. The magnitude of the outer force field is set as $F = 0.1$. The analysis performed in [29] has shown that, as F increases, more energy $\boldsymbol{E}[\boldsymbol{F},\boldsymbol{f}](t)$ is put into the storage system. Moreover, it has shown that the thermostatted force field makes the distribution functions shift towards low values of energy. Consequently, the users receive a reduced quantity of energy $\mathbb{E}_{(0,1)}^{[F]}[f](t)$.

As the Figure 7 shows, the charging of the storage system is practically the same with the initial condition $\boldsymbol{f}_a^0$ and $\boldsymbol{f}_b^0$. At the beginning, the allocation of energy with the initial condition $\boldsymbol{f}_b^0$ is slower than the allocation of energy with the initial condition a) and then the former becomes faster than the latter. The initial condition $\boldsymbol{f}_c^0$ is characterized by a different augmentation of the energy $\boldsymbol{E}[\boldsymbol{F},\boldsymbol{f}](t)$: The allocation of energy into the storage system is much slower than with the other two initial conditions. This is related to the fact that the charging of the storage system begins by taking the high values of energy, see [29]. Therefore, since the initial condition $\boldsymbol{f}_c^0$ is characterized by low values of the energy variable and the previous section has shown that the distribution functions evolve in the neighborhood of the initial condition, the action of the outer force field is weakened. It can be now explained why the charging of the storage is not efficient with $\boldsymbol{f}_c^0$: The energy sources provide a little amount of energy to the users and therefore there is a small amount of energy available for the storage system.

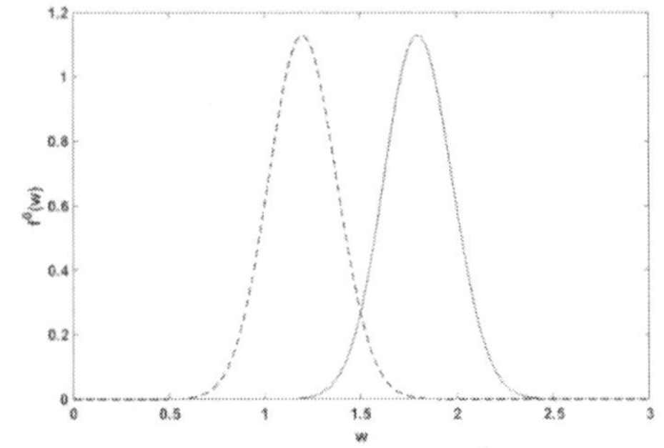

Fig. 1. The initial distribution functions $\boldsymbol{f}_a^0$. The solid line refers to the energy source NR and the dashed line refers to the energy source R.

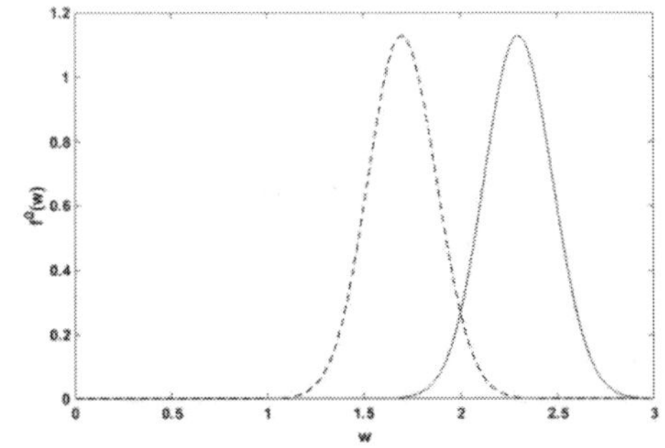

Fig. 2. The initial distribution functions $\boldsymbol{f}_b^0$. The solid line refers to the energy source NR and the dashed line refers to the energy source R.

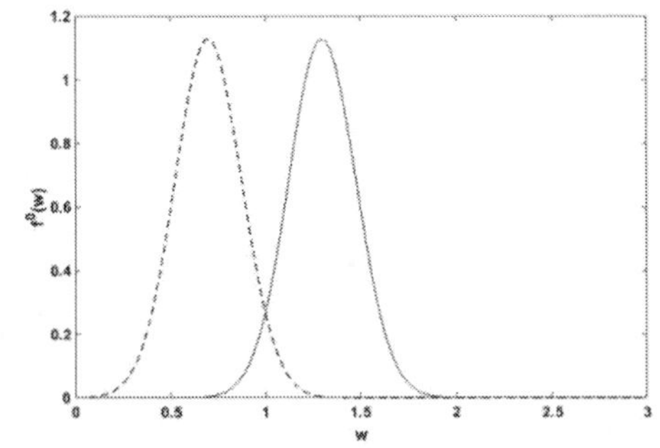

Fig. 3. The initial distribution functions for the case $\boldsymbol{f}_c^0$. The solid line refers to the energy source NR and the dashed line refers to the energy source R.

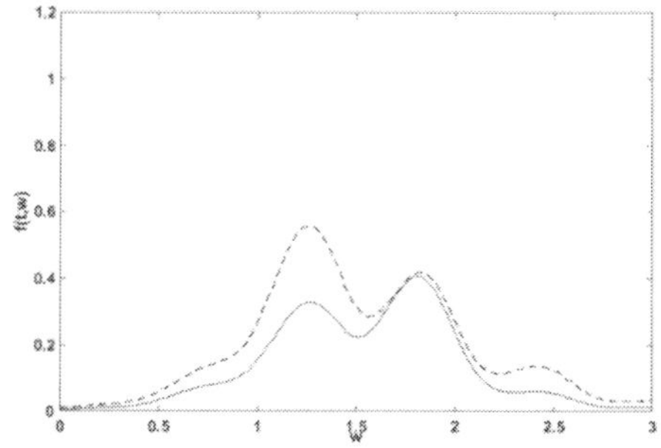

Fig. 4. The evolution over the energy variable of the distribution functions at time $t = 30$ for the initial datum $\boldsymbol{f_a^0}$. The solid line refers to the energy source NR and the dashed line refers to the energy source R.

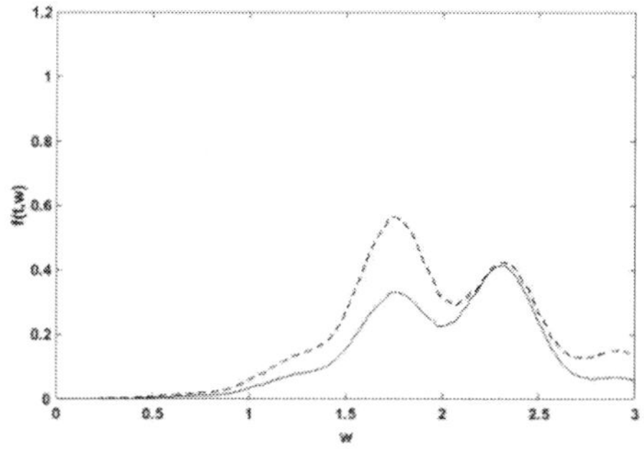

Fig. 5. The evolution over the energy variable of the distribution functions at time $t = 30$ for the initial datum $\boldsymbol{f_b^0}$. The solid line refers to the energy source NR and the dashed line refers to the energy source R.

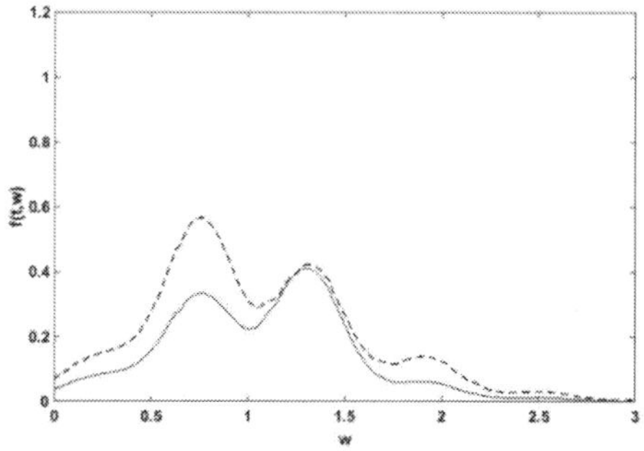

Fig. 6. The evolution over the energy variable of the distribution functions at time $t = 30$ for the initial datum $\boldsymbol{f_c^0}$. The solid line refers to the energy source NR and the dashed line refers to the energy source R.

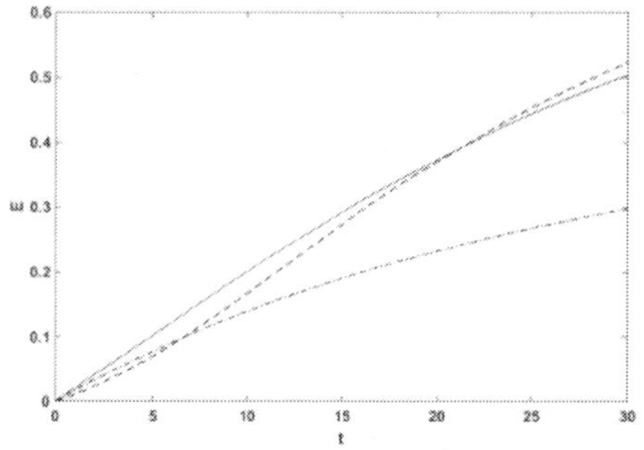

Fig. 7. The charging of the storage system (energy $\boldsymbol{E}[\boldsymbol{F},\boldsymbol{f}](t)$) for $F = 0.1$. The solid line refers to the initial datum $\boldsymbol{f_a^0}$, the dashed line refers to the initial datum $\boldsymbol{f_b^0}$ and the dashed-dotted line refers to the initial datum $\boldsymbol{f_c^0}$.

4 Conclusions

This work has performed the analysis on the initial condition for the mathematical model treated in [29]. The model is derived within the thermostatted kinetic theory framework exposed in [21] for the modelling of a grid of hybrid energy sources with storage. According to the mathematical framework, a distribution function *f(t,w)* describes each source of energy, while an outer force field $\boldsymbol{F}$ models the charging of the storage system. The mathematical model needs a thermostat term α for the purpose of balancing the outer force field and ensure the existence of a nonequilibrium steady state. A system of integro-differential equations depicts the time derivative of the distribution functions.

In particular, the present work has considered a source of non-renewable energy and a source of renewable energy. The authors have investigated the effects of the initial condition on the way the distribution functions evolve in the energy variable and on the charging of the storage system. Three different initial conditions have been considered, characterized by medium, high and low values of the energy variable w. In the case $\boldsymbol{F} = \boldsymbol{0}$, the computational analysis has shown that the distribution functions evolve in the neighbourhood of the initial condition because of the interaction-driven dynamics of the multisource grid. In the case $\boldsymbol{F} \neq \boldsymbol{0}$, the computational analysis has shown that, if the initial condition is characterized by low values of energy, the charging of the storage system is slower.

The analysis performed in this paper can be generalized: Different initial conditions can be considered, characterized by a different shape and by a different sharing of the total number of users, i.e. $\mathbb{E}_{(0,0)}^{[F]}[\boldsymbol{f}](t)$, and of the average quantity of energy supplied to them, i.e. $\mathbb{E}_{(0,1)}^{[F]}[\boldsymbol{f}](t)$, between the energy sources. Besides the analysis on the initial datum can be performed for the model presented in [30] where the authors have introduced time dependent activation parameters.

References

1. S. Jebaraj, S. Iniyan, Renew. Sust. Energ. Rev. **10**, 218-311 (2006)

2. P.H.J. Nardelli, N. Rubido, C. Wang, M.S. Baptista, C. Pomalaza-Raez, P. Cardieri, M. Latva-aho, Eur. Phys. J. Special Topics **223**, 2423-2437 (2014)

3. B.D. Solomon, K. Krishna, Energy Policy **39**, 7422-7431 (2011)

4. A.S. Dagoumas, N.E. Koltsaklis, Appl. Energy **242**, 1573-1587 (2019)

5. G. Notton, M.-L. Nivet, C. Voyant, C. Paoli, C. Darras, F. Motte, A. Fouilloy, Renew. Sust. Energ. Rev. **87**, 96-105 (2018)

6. F. Romanelli, Eur. Phys. J. Plus **131**, 53 (2016)

7.Y. Yang, S. Bremner, C. Menicats, M. Kay, Renew. Sust. Energ. Rev. **91**, 109-125 (2018)

8. S. Blumsack, A. Fernandez, Energy **37**, 61-68 (2012)

9. X. Fang, S. Misra, G. Xue, D. Yang, IEEE Commun. Surv. Tutor. **14**, 944-980 (2012)

10. M. Ourahou, W. Ayrir, B. EL Hassouni, A. Haddi, Math. Comput. Simulation https://doi.org/10.1016/j.matcom.2018.11.009

11. G. Filatrella, A.H. Nielsen, N.F. Pedersen, Eur. Phys. J. B **61**, 485-491 (2008)

12. R. Carareto, M.S. Baptista, C. Grebogi, Comm. Nonlinear Sci. Numer. Simulat. **18**, 1035-1046 (2013)

13. A. Sciré, I. Tuval, V.M. Eguíluz, Europhys. Lett. **71**, 318-324 (2005)

14. G.A. Pagani, M. Aiello, Physica A **392**, 2688-2700 (2013)

15. B.C. Wang, M. Sechilariu, F. Locment, Math. Comput. Simulation **91**, 119-133 (2013)

16. R. Baños, F. Manzano-Agugliaro, F.G. Montoya, C. Gil, A. Alcayde, J. Gómez, Renew. Sust. Energ. Rev. **15**, 1753-1766 (2011)

17. K.-H. Chang, G. Lin, Simul. Model. Pract. Theory **52**, 40-51 (2015)

18. S. Mei, X. Zhang, M. Cao, *Power Grid Complexity*, (Springer, 2011)

19. C. Peterman, S.B. Amor, A. Bui, *Advances in Knowledge-Based and Intelligent Information and Engineering Systems*, 149-158, (Frontiers in Artificial Intelligence and Applications, IOS Press, 2012)

20. M. Dalla Via, C. Bianca, I. El Abbassi, A.-M. Darcherif, *A hybrid thermostatted kinetic framework for the modeling of a hybrid multisource system with storage*, submitted (2018)

21. C. Bianca, Appl. Math. Sci. **6**, 651-660

22. C. Bianca, J. Riposo, Eur. Phys. J. Plus **130**, 159 (2015)

23. C. Bianca, L. Brézin, Int. J. Biomath. **10**, 1750072 (2017)

24. C. Bianca, C. Mogno, Eur. Phys. J. Plus **133**, 213 (2018)

25. G.P. Morriss, C.P. Dettmann, Chaos **8**, 321-336 (1998)

26. O.G. Jepps, L. Rondoni, J. Phys. A **43**, 133001 (2010)

27. W.G. Hoover, A.J.C. Ladd, B. Moran, Phys. Rev. Lett. **48**, 1818-1820 (1982)

28. D.J. Evans, W.G. Hoover, B.H. Failor, B. Moran, Phys. Rev. A **28**, 1016-1021 (1983)

29. M. Dalla Via, C. Bianca, I. El Abbassi, A.-M. Darcherif, Appl. Math. Model. **78,** 232-248 (2020)

30. M. Dalla Via, C. Bianca, I. El Abbassi, A.-M. Darcherif, Eur. Phys. J. Plus **135**, 198 (2020)

E3S Web of Conferenced **170**, 02001 (2020) https://doi.org/10.1051/e3sconf/202017002001
EVF'2019

Petri Net-Based Approach for "Cyber" Risks Modelling and Analysis for Industrial Systems

K. Labadi[1,2,3,*] , *A.-M. Darcherif*[1,2,3], *I. El Abbassi*[1,2,4] , *and S. Hamaci*[1,2,3]
[1]*ECAM-EPMI, 13 Bld de l'Hautil, 95092 Cergy-Pontoise, France*
[2]*LR2E Laboratoire de Recherche en Energétique et Eco-Innovation Industrielle*
[3]*Quartz-Lab (EA 7393) –* [4]*L2MGC (EA 4114) - France*

Abstract. Today, industrial systems are large, complex, and increasingly vulnerable. Specifically, due to the current digital transformation, the industry 4.0 creates crucial cyber-risks and cyber-security challenges. In this context, risk modelling and impact analysis has become a crucial research topic. Based on the formal modelling and performance analysis power of Petri Nets (PN), this paper represents a summary of our methodological approach for "risk" modelling and "impact" analysis of cyber vulnerabilities and / or other critical events. The applicability of the developed approach is demonstrated on a real-life industrial system.

1 General introduction

The current fourth industrial revolution affecting manufacturing systems (Industry 4.0), supply chains and logistic systems (Logistics 4.0) creates crucial cyber risks and cyber-security challenges [8]. In era of Industry and Logistic 4.0, the organizations are hyper connected with their smart devices and smart networks. Unfortunately, this creates cyber-attacks vulnerabilities and opportunities for cyber criminals to infiltrate networks via the weakest links.

As other critical events (technology failures, supplier bankruptcy, disturbances, disruptions, disasters ...), a cyber-attack can affect many functions across the network such as production process, transportation organisation, data systems and, more generally, different performances of the system such as customer service, production level, product quality, costs and profits. Figure 1 represents the typical impact of a critical event on the performance of a system.

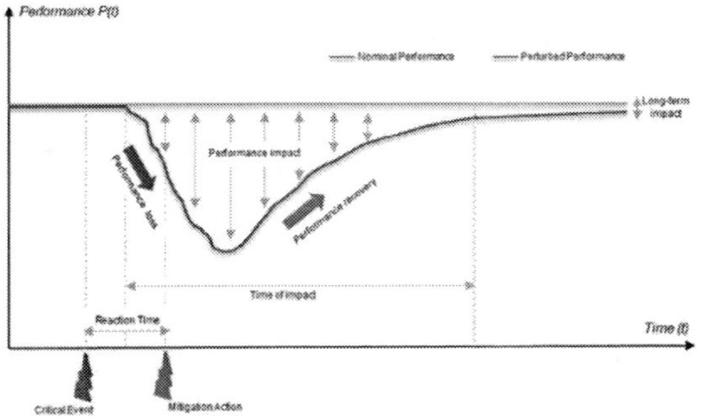

Fig. 1. Typical performance evolution under critical event(s)

Attacks can cause physical damage at facilities, disruption of the information, financial and material flows of the global system. They generally have a low probability and the potential for a large loss. Attacks come in many forms by inserting malicious hardware or software such as malware insertion within any system and malicious integration of counterfeit components during system design, system development or system management.

In practice, cyber security consists of all the technologies that keep computer and data systems [2]. Cybersecurity should no longer be viewed as a function of information technology or information security alone. It is a complex discipline that needs to combine between traditional cyber security and industrial management fields. It is concerned with mitigating risks both to products, services, information, process, and technology related to systems. Except for the technological aspects of the problem, industrial managers need methods and tools for control, analysis, and risk mitigation of their cyber vulnerabilities.

Motivated by this crucial context and problem, this paper suggests a promising modelling and analysis methodology for cyber risk and security purposes. Based on the formal modelling and performance analysis power of Petri Nets [7] for complex discrete event systems, including manufacturing systems, supply chains and logistic systems, this work demonstrates their potential for modelling, impact analysis, and risk mitigation of their critical events including cyberattack vulnerabilities. The developed technique based on the "Nominal" and "Perturbed" Petri net models is presented in a summary way (section 3) and an industrial manufacturing and packaging of pharmaceutical products is used to demonstrate the methodology (section 4).

* Corresponding author: k.labadi@ecam-epmi.com

2 DES and Petri Nets

In this work, the systems are viewed as "Discrete Event Systems" (DES). Many real world systems, including supply chains, logistic and manufacturing systems, can be considered as DES not because of their intrinsic characteristics, but because of the aspects of their behaviour that we want to emphasize. In this field, Petri Nets (PN) [7] have arisen as a practical formalism for modelling DES, widely used thanks to their graphical and mathematical foundation, ready to be exploited for modelling, simulation, theoretical analysis, as well as performance evaluation. They were applied not only to address modeling and performance analysis issues but also to study other specific topics such as logistics optimization, production scheduling and planning, system control and supervisory, diagnosis and fault detection in many discrete event systems [1, 4, 6, 9]. However, although the literature of Petri nets is very plentiful and have been widely used in various domains, their potential and applicability for cyber risk modelling and analysis still to be developed. Today, study, design and management of any industrial system or organization without considering risk evaluation and analysis issues lead to inconsistent networks. Besides some adequate technological solutions, industrial managers need methods and tools for modelling, analysis, and testing of their cyber risk and security issues. This work constitutes part of this emerging issue.

For readers not familiar with Petri nets, some of their basics concepts are given in the following. As illustrated in Figure 2, as a basic definition, a Petri net is bipartite directed graph composed of places (represented as circles), transitions (represented as rectangles or bars), and directed arcs used to connect between places and transitions. A place may contain tokens and the current marking (state) of the modeled system is specified by the number of tokens in each place. Each transition usually used to model an activity whose occurrence is represented by its firing. A transition can be fired only if it is enabled, which means that all preconditions for the activity are fulfilled (there are enough tokens available in the input places of the transition). When the transition is fired, tokens will be removed from its input places and added to its output places. The number of tokens removed or added is determined by the weight of the arc connecting the transition with the place.

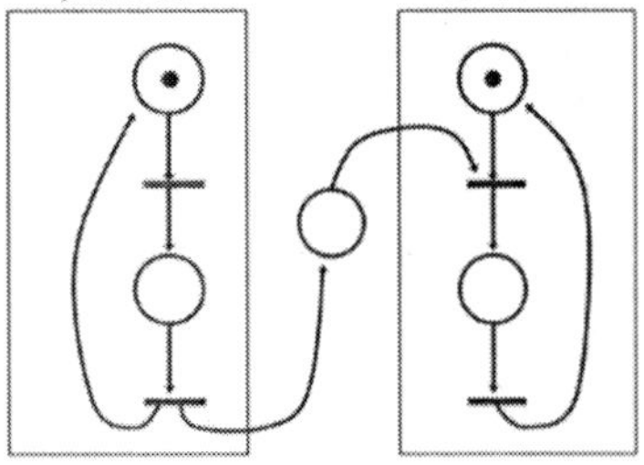

Fig. 2. Simple example (Producer / Consumer)

The signification of the places, transitions and arcs in a Petri net depends on the system modeled. As a simple example, Figure 2 shows a Petri net which models a producer-consumer system. The model consists of three different parts: a production part on the left, a consumption part on the right, and a buffer in the center which allows up to four items to be produced in the system at any time.

The concept of Petri nets has been originally proposed without any notion of time. For performance evaluation and analysis, "Time" is introduced in Petri nets. Generally, two types of transitions can be used to model discrete events: (i) immediate transitions, with zero firing delay and (ii) timed transitions with deterministic or stochastic firing delay [3]. Furthermore, the performance evaluation is based on the temporal evolution of the marking (state) process of the modeled system. If a system is modeled by a stochastic timed PN [3], its analysis can be performed based on the underlying stochastic process of the net. The analytic approach is feasible particularly when the stochastic process has a finite number of states and is analytically tractable under some assumptions. For large and complex systems, simulation techniques and associated software tools may be required.

3 "Nominal" and "Perturbed" Petri Net Based Approach

Based on the formal modeling and performance analysis power of (stochastic) Petri nets for complex discrete event systems, including industrial systems, supply chain and logistic systems, in this section, we develop a "Nominal" and "perturbed" Petri Net approach for "risk" modeling, "impact" analysis and risk mitigation of cyber vulnerabilities and / or other critical disruption events in such dynamical systems.

The proposed approach, synthesized in Figure 3, contains three different steps:

(1) Modelling step, (2) Analysis step and (3) Mitigation step, which can be used in a closed-loop configuration for risk mitigation purposes of the system network. More details of the methodology are given in the following:

- **(Step 1)** – Based on the modelling power of (Stochastic) Petri Nets, the system is modelled firstly as a "Nominal Petri Net model" (without any critical events), and secondly, we introduce one or several perturbation(s) event(s) into the nominal model to obtain its associated "Perturbed Petri Net model".

- **(Step 2)** – Based on the quantitative and/or qualitative methods associated to the Petri Net formalism, the modelled system can be studied through its "Perturbed PN model" in order to analyse the impact of the critical perturbation(s) inserted into the initial model. Both qualitative and quantitative analysis can be performed with the two models for impact study.

— **(Step 3)** – Based on the analysis of the perturbed model, actions and mitigation solutions can be proposed (new system design, new parameters, human error reduction …). Then, the "newly" system can be analysed as often as necessary by the steps above.

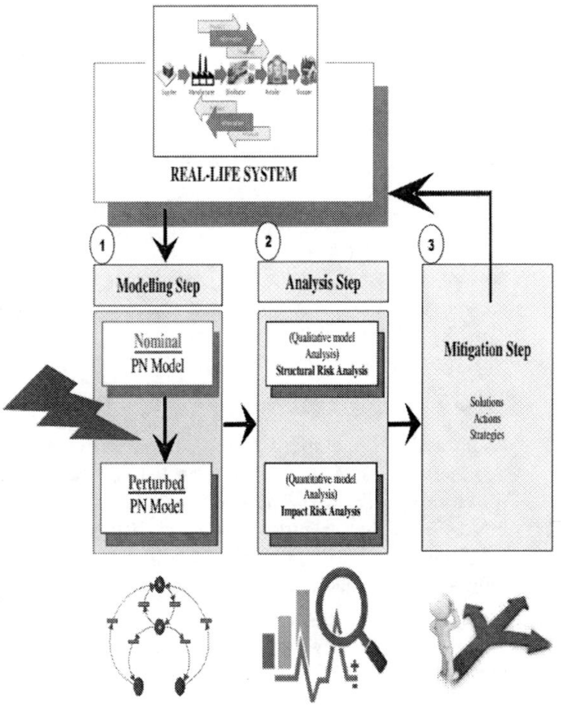

Fig. 3. The Nominal and Perturbed Petri Net Approach.

The main question is "How to introduce (model) a perturbation such as an attack event into the Nominal Petri net model that generates the Perturbed Petri Net model?"

— According to the practice, any attack (Hardware, Software, Process …) and/or critical event can cause operational and / or structural disruption across the system (machine breakdown, disruption of the flow, stock out, lateness, blocking, cancellation of operations, transport interruption …).
— "Perturbation(s)" can be inserted at any location in the Nominal Petri net model according to the critical event(s) to be represented (modelled). Indeed, parametric and/or structural modifications can then be introduced into the nominal model via its three different components: transitions, places and arcs.

4 Industrial Application

To demonstrate the applicability of the proposed approach, we consider a real-life industrial system in Figure 4. The system represents a pharmaceutical manufacturing and packaging system.

Fig. 4. A pharmaceutical manufacturing and packaging system

4.1. Presentation of the industrial system

The industrial system is composed of 7 automated machines (P_i) connected to each other by pallet conveyors (C_{ij}). All the system is controlled by PLC (Programmable Logic Controllers) and equipped with a computer system for monitoring and supervision (S) of different ranges and orders of the production. As schematized in Figure 5, the machines are structured around several conveyors, and three closed-circuits connected by pneumatic jacks. They are used to move pallets from conveyors and vice versa. At the end of the packaging process, two other pneumatic jacks are used for unloading of bottles from pallets and packaging them as finished products.

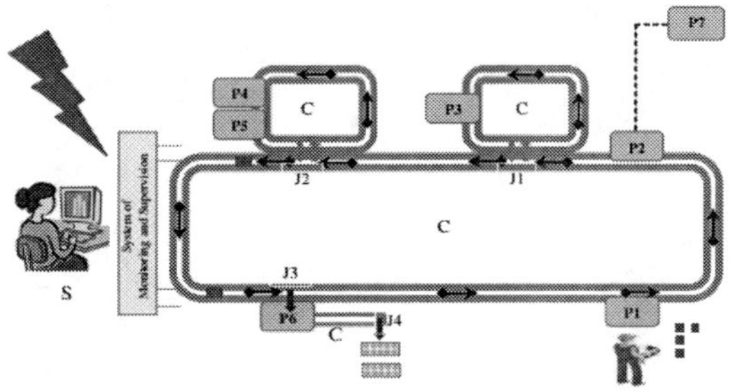

Fig. 5. Structure of the manufacturing and packaging system

Table 1. Components of the industrial system

S	System of Monitoring and Supervision
P1	Operator station (loading of pallets and empty bottles)
P2	Tablet counting/filling machine
P3	Granule dosing/filling machine
P4	Bottle capping machine
P5	Labeling machine for bottles
P6	Unloading and packaging machine
P7	Pharmaceutical tablet manufacturing machine
C1-C4	Convoyors between the differents machines
J1-J4	Input / output jacks for paletts

To more understand the general process of the system and the scheduling of its activities, Table 1 gives its different components and automated machines. Several ranges of manufacturing and packaging can be performed by the system according to their programs to choose by using the monitoring and supervision part of the system.

4.2. Nominal Petri Net model of the system

The nominal Petri net model of the industrial system is shown in Figure 6. The model reproduces its real structure represented in Figure 5.

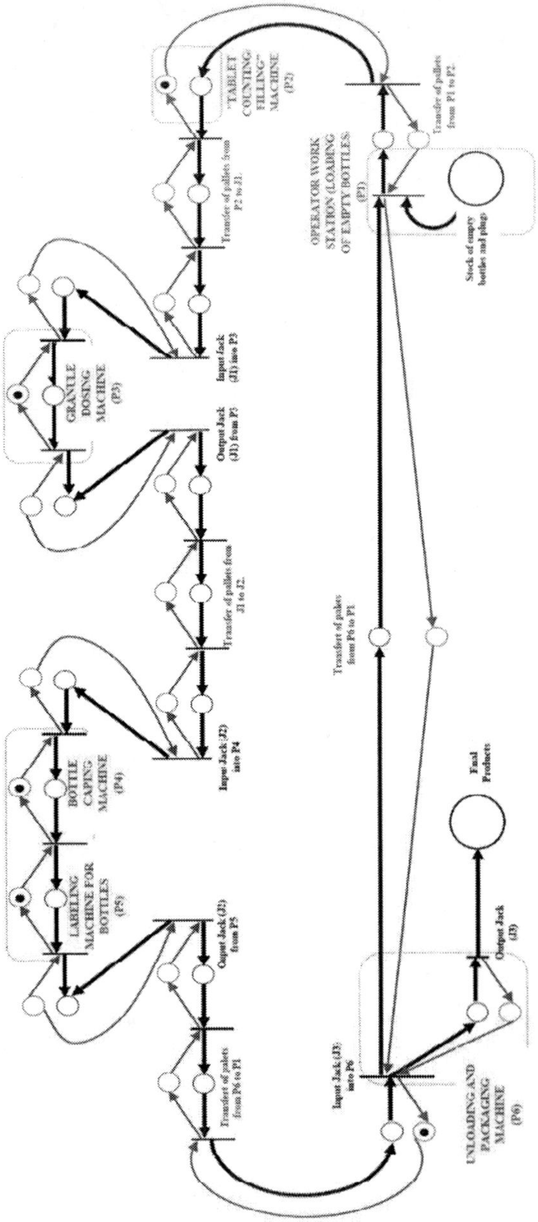

Fig. 6. The "Nominal" Petri Net model of the system

A set of parameters relating to capacities of the different conveyors between machines (Table 2) and the average operating delays (OT) of the different machines are given for three different production ranges: G1, G2, G3 (see Table 3).

Table 2. Capacities of conveyors

Conveyor (C_{ij})	Capacity [pallets]
$C_{(P1-P2)}$	20
$C_{(P2-J1)}$	13
$C_{(J1-P3)}$	26
$C_{(P3-J1)}$	9
$C_{(J1-J2)}$	21
$C_{(J2-P4)}$	26
$C_{(P5-J2)}$	9
$C_{(J2-J3)}$	22
$C_{(P6-P1)}$	32

Table 3. Operating delays of three different ranges

Machines	TO(G1)	TO(G2)	TO(G3)
P1	6	6	6
P2	14	30	-
P3	8	-	18
P4	8	8	8
P5	7	7	7
P6	15	15	15
P7	15	15	15

In addition, the speed of the conveyors is adjustable. So, the different ranges will be tested according to two different speeds: V1 = 15.0 cm/second and V2 = 28.8 cm/second. Thus, all travel times of pallets between the machines can be determined according to the different lengths (or capacities) of the conveyors. Then, by combining the three production ranges and the two conveyor speeds, the simulation results of this study can be given for six configurations denoted by GiVj (i = 1 to 3, and j = 1 to 2).

4.3. Performance evaluation of the Nominal PN model

The performance evaluation and analysis of the system is performed by discrete event simulation based on its associated nominal Petri net model represented in Figure 6. The production rate of the system is considered in this study.

For each configuration (G_iVj, Np), this performance indicator can be expressed in two different ways:

- As the number of products that can be produced during a given unit of time.

$$PR_1(G_iV_j, N_P) = \frac{1}{N_{Rep}}\left(\left(\sum_{k=1}^{N_{Rep}} M(P_{PROD})_k\right)\Big/T_{Sim}\right)$$

- As the amount of time it takes to produce one unit of a product.

$$PR_2(G_iV_j, N_P) = \frac{1}{N_{Rep}}\left(T_{Sim}\Big/\left(\sum_{k=1}^{N_{Rep}} M(P_{PROD})_k\right)\right)$$

Where:

$PR_1\,(G_iV_j,\,Np)$	Average number of products per unit of time.
$PR_2\,(G_iV_j,\,Np)$	Average production time of a unit of product
T_{Sim}	Total time of simulation.
N_{Rep}	Number of replications.
k	1 … N_{Rep} (k-th replication).
$M(P_{PROD})_k$	Number of tokens, at the end the k-th simulation, of the place P_{PROD} that represent the stock of finished products.

The discrete simulation results are represented graphically in Figures 7 and 8. Clearly, from a given number of pallets (Np*), the production rate of the system is optimal (see Table 4).

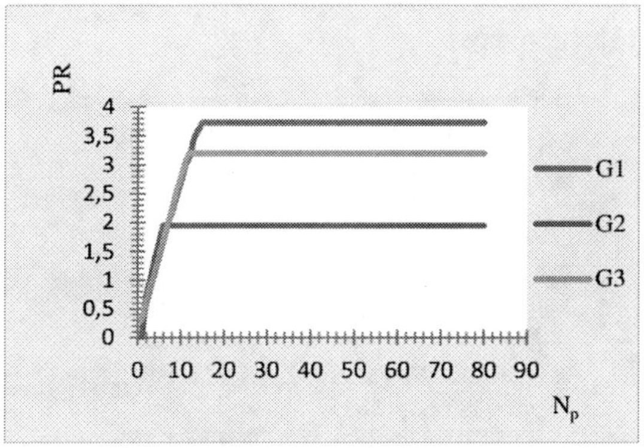

Fig. 7. Production rate evolution (case with V1)

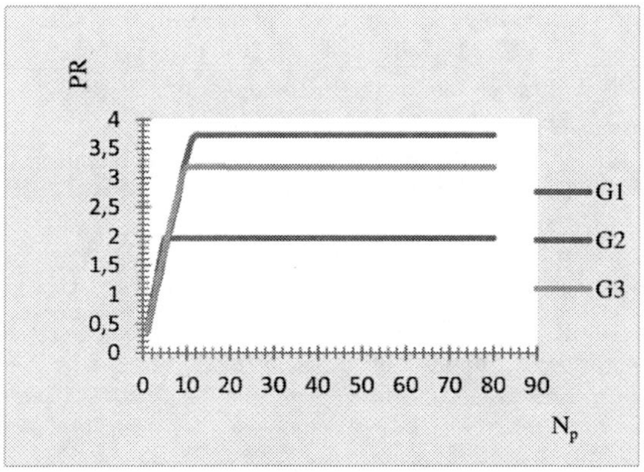

Fig. 8. Production rate evolution (case with V2)

Table 4. Optimal production rates of the industrial system

Ranges	PR*	N_P*
G1V1	3,725	15
G2V1	1,946	6
G3V1	3,195	12
G1V2	3,730	12
G2V2	1,965	5
G3V2	3,185	9

4.4. Modelling and Performance evaluation of the Perturbed PN model

Here, we develop the case of the perturbed model in order to evaluate the risk of a computer intrusion that could affect the transmission network between the monitoring system and all of its programmable logic controllers. The risk is due to the absence of Firwalls for industrial security. The perturbed PN model is represented in Figure 9, where we can distinguish between two PN modules described as follows:

- The first part (subnet 1) represents the nominal part of the industrial system studied previously. The global manufacturing system is represented by the stochastic transition T_{PROD}. Its firing leads to the finished products deposited in P_{PROD}. The stochastic firing delay to be associated to the transition T_{PROD} corresponds to the average production rate determined by using the nominal model (see Table 4).

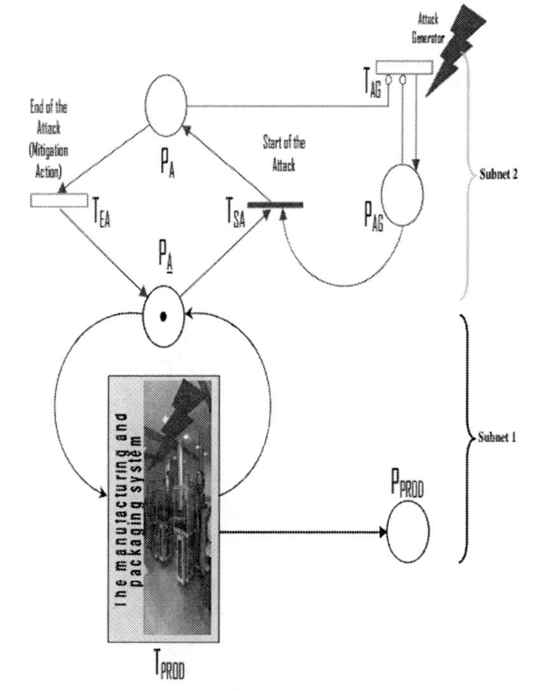

Fig. 9. The "Perturbed" Petri Net model

- The second part (subnet 2) represents the attack generator of the system. The stochastic transition T_{AG} means to generate randomly an attack. It can be set to generate on average "x" attacks/ unit of time. In this study, the perturbed system will be analyzed for several scenarios: 1 attack / year, 2 attacks/ year ...). By firing the transition T_{AG}, the generation of an attack will be indicated by marking the place P_{AG} with one token (i.e. $M(P_{AG}) = 1$). The use of the two inhibitor arcs connecting the transition T_{AG} with the places P_A and P_{AG} serves to avoid another attack, while the system is already subject to an ongoing attack. When $M(P_{AG}) = 1$, the attack starts immediately by firing the transition T_{SA}. Indeed, this immediate transition is enabled if only if: $M(P_{AG}) = 1$ and $M(P_A) = 1$. The immediate firing of the transition T_{SA} (start of the attack) leads to: $M(P_A) = 1$ to indicate the current attack of the system; and $M(P_{\underline{A}}) = 0$ that makes the transition T_{PROD} not enabled (interruption of the production as long as the attack is untreated). After a stochastic delay that depends on the mitigation action, the transition T_{EA} can be fired to reverse the marking of the place P_A and P_A as $M(P_A) = 0$ and $M(P_{\underline{A}}) = 1$ indicating the end of the attack and the recovery of production with nominal conditions.

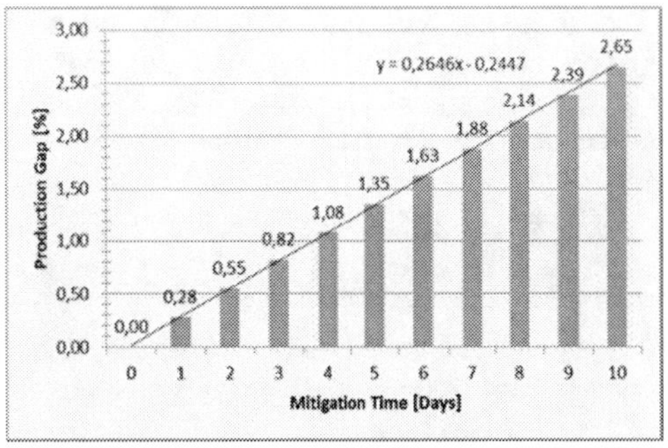

Fig. 10. The impact of the attack on Production Gap [%] according the mitigation delay.

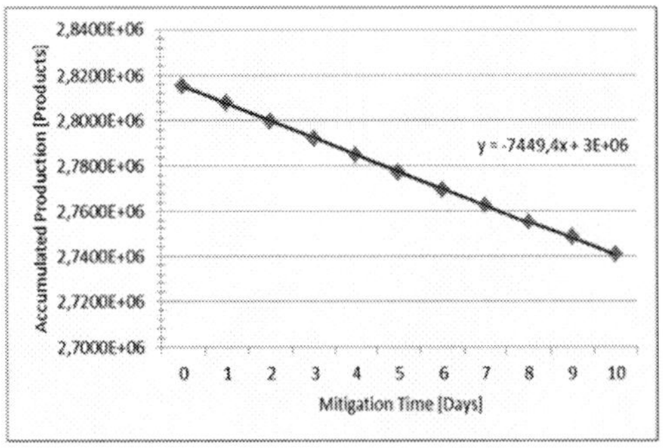

Fig. 12. The impact of the attack on Accumulated Production [Products] according the mitigation delay.

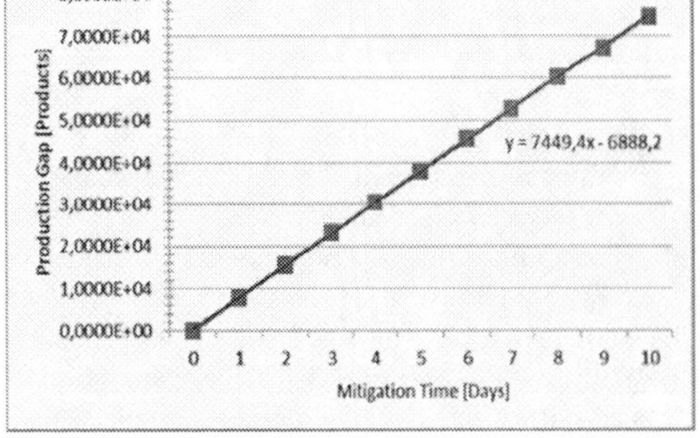

Fig. 11. The impact of the attack on Production Gap [Products] according the mitigation delay.

Finally, some discrete event simulation results obtained for the range G1V1 are given graphically in Figures 10-11-12. They represent a quantitative analysis of the impact of a computer intrusion (T_{AG}) on the productivity of the industrial system (T_{PROD}). The results are presented for ~1 Attack / Year and for several scenarios in terms of the mitigation time.

5 Conclusion

Based on the formal modelling and performance analysis power of Petri Nets (PN), this paper represents a summary of our methodological approach for "risk" modelling and "impact" analysis of cyber vulnerabilities and / or other critical events. The applicability of the developed approach is demonstrated on a real-life industrial system.

References

1. H. Chen, L. Amodeo, F. Chu, and K. Labadi, , IEEE TASE, Vol. 2, No. 2, pp.132–144 (2005).
2. Derui Ding, Qing-Long Han, Yang Xiang, Xiaohua Ge, Xian-Ming Zhang, Vol. 275, pp. 1674-1683 (2018).
3. P.J. Haas, "Stochastic Petri Nets: Modelling, Stability, Simulation". Springer-Verlag, New York, (2002).
4. A.A. Kadri, K. Labadi, I. Kacem, EJIE, Vol. 9, No. 5, pp. 638-663, (2015).
5. K. Labadi, T. Benarbia, J.B. Barbot, and A. Omari, IEEE TASE, Vol. 12, No. 4, pp. 1380-1395, Oct. (2015).
6. K. Labadi, H. Chen, L. Amodeo, IEEE SMC, Part C, Vol. 37, pp.4515–4520 (2007).
7. Murata, Tadao, In: Proceedings of the IEEE, Vol. 77, No. 4, pages 541-580, (1989).
8. William Knowles, D. Prince, D. Hutchison, J. Ferdinand Pagna Disso, K. Jones, IJCIP, Vol. 9, Pages 52-80 (2015).
9. R. Zurawski, and Zhou Mengchu, IEEE TIE, Vol. 41, No. 6, pp.567–583 (1994).
10. K. Labadi, H. Chen, L. Amodeo, Journal Européen des Systèmes Automatisés, JESA, Vol. 39, n° 7, pp. 863–886. (2005).

Precipitation estimation by a multi-threshold method using cloud optical and microphysical properties from MSG / SEVIRI data

Mourad Lazri[1,*], *KarimLabadi*[2], *Jean Michel Brucker*[2] *and Soltane Ameur*[1]
[1]Laboratoire LAMPA, Université Mouloud MAMMERI, Tizi-Ouzou (Algérie)
[2]ECAM-EPMI/QUARTZ-Lab (France)

Abstract. The aim of this paper is the estimation of precipitation in northern Algeria using satellite data. To do this, a multi-threshold method based on the exploitation of the clouds optical and microphysical properties is developed. Depending on the availability of the MSG / SEVIRI (Meteosat Second Generation/Spinning Enhanced Visible and Infrared Imaging) satellite channels, the database is divided into daytime data and nighttime data. In the learning phase, daytime and nighttime two-dimensional thresholds are determined from comparisons between the MSG / SEVIRI satellite data of rainy season 2010/2011 and the corresponding data from the Sétif meteorological radar. Using linear regression, an empirical relationship between the SEVIRI spectral data and the precipitation intensities from the radar is determined.To estimate rainfall, the determined empirical relationship is applied to the validation dataset collected during rainy season 2011/2012. To evaluate the method, a comparison of the estimation results with reference radar data is performed. The results show that these estimations are in good correlation with those measured by radar.

1 Introduction

Due to its importance to human civilization, drinking water has become a major challenge for the world's population because of the inequitable distribution of this water between different regions of the globe. To this end, the study of water resources is one of the major scientific priorities.

These last years, drinking water resources are starting to become scarce and the use of water is constantly growing. Faced with this growing use, the implementation of a reasonable and rational management plan for these resources is urgent. A considerable part of these resources comes from precipitation whose spatio-temporal distribution is highly variable. Knowledge of rainfall fields in time and space therefore contributes to better management of these water resources.
Algeria is one of the countries most affected by this water deficit. Indeed, in this region, the drinking water that is exploited comes mainly from rainfall, which is mainly concentrated in the north of the country. The north of Algeria is characterized by a humid and cold climate, and at the same time hot and dry [1].

The measurement of rainfall amounts can be carried out by rain gauges or by remote sensing means such as radar and meteorological satellites. The rain gauges measure the precipitations fallen locally with a great precision. However, to cover large areas, the establishment of a network of rain gauges is difficult to achieve, if not impossible especially for areas difficult to access. The use of radars for the direct measurement of rainfall intensities has become widespread in order to remedy the point measurements of rain gauges. Despite this success, radar coverage is still limited and cannot cover large areas. As a result, the use of optical instruments on board meteorological satellites has become essential. The satellite has a wider field of observation. However, these satellites do not provide direct measurements of precipitation intensities.

Developed studies have exploited measurements in infrared channels from satellite, which connects the height of the clouds top height and cloud top temperature to the rain rate [2]. These methods are adapted to convective systems [3]. Stratiform precipitation is not identifiable in the infrared because is relatively warm at the top and generate low to moderate rainfall intensities. To address this issue, in the case of mixed situations, precipitation estimation techniques from the clouds optical and microphysical properties are used [4] to [6]. The new generation of satellites gives rich information about these properties. This has made it possible to better relate the cloud water path to rain rates, significantly improving the classification and precipitation estimation [4].

As part of this work, we will present a multi-threshold technique for precipitation estimation using the correspondences betweenMSG multi-spectral data and the Radar data. Because of the complexity of the Mediterranean climate, information about theclouds optical and microphysical properties is used here to account for both convective and stratiform systems. This article is structured as follows: in Section 2, we describe the study area and data used. Section 3 contains the methodology of developed scheme. In section 4, the

*Corresponding author: m_lazri@yahoo.fr

results of estimation are presented. Finally, in section 5, we closed this article with summary and conclusions.

2 Presentation of the study area and data

2.1 Study region

The study area of 800 * 500 km² is situated at North of Algeria. We have chosen the part which is covered by the Sétif Radar observations (figure 1). The geographic coordinates are 34 ° N to 38 ° N for latitude and 2 ° W at 9 ° E for longitude.

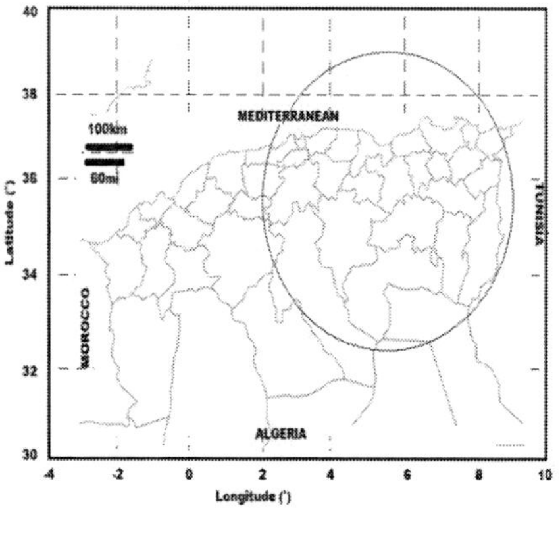

Fig. 1. Northern Algeria:Study area indicated by the circle.

2.2 Data used

This study exploits the data pairs of the MSG/SEVIRI multispectral radiances and the Sétif radar observations. Radar data is used for calibration and validation of the method. The radar provides direct and instantaneous measurements of precipitation intensities, hence the choice of this type of data.

Two rainy seasons are considered in this study; data collected during November 2010 to February 2011 is used for learning the developed technique. It includes 1463 scenes of daytime and nighttime precipitation. The dataset from November 2011 to February 2012 is used for the evaluation and validation of the developed technique. It includes 1120 scenes of daytime and nighttime rainfall collected during the indicated period. To properly exploit the spectral channels of SEVIRI, the nighttime data and daytime data are used separately.

2.2.1 Radar data

The radar managed by National Office of Meteorology is placed at 36 ° 11'N, 5 ° 25'E and at an altitude of 1700 m.the temporal resolution and spatial resolution are respectively 15 min and 1 km. A radar imageis constituted of 512 × 512 pixels collected in PPI (Plane Position Indicator). The number of classes for this Radar is 16 classes corresponding to different intensities of reflectivity [04, 12, 18, 22, 26, 30, 34, 38, 42, 46, 50, 54, 58, 62, 66, 70 (dBZ)]. The Radar reflectivity (Z) expressed in $((mm^6 m^{-3})$ can be given in dBZ.

2.2.2Satellite data

This data obtained from the Spinning Enhanced Visible and Infrared Imager (SEVIRI) instrument, it is a radiometer-imager aboard the MSG (Meteosat Second Generation) satellite. The SEVIRI radiometer provides 12 images at wavelengths between 3.8 and 14 µm. The temporal resolution is 15 minutes and the spatial resolution is 3 * 3 km² at the point below the satellite in all channels with the exception of the HRV channel which has a resolution of 1 * 1 km². In the case of northern Algeria, the spatial resolution is 4 * 5 km² due to its geographical position and the spherical shape of the Earth.

For our study, the selected satellite channelsare: VIS0.6, NIR1.6, IR3.9, WV7.3 and IR10. The combination of its spectral characteristics makes it possible to provide information about clouds microphysical and optical properties that are decisive for the quantification of precipitation intensities.

In this context, the Cloud Water Path (CWP) which depends both on the size of the water and/or ice droplets $r_e$and the optical thickness$\tau$of the clouds is used. Indeed, a cloud of important optical thickness with large droplets gives significant precipitation. The relationship is given by equation 1 [4]-[7]:

$$CWP = \frac{2}{3}\rho\, \tau\, r_e \qquad (1)$$

ρ : represents the density of the water/ice in the clouds in(g/m^3).

Several approaches based on "Look-Up-Table (LUT)" for retrieving optical thickness and effective particle radius of clouds from a radiative transfer model [8]. The results are simultaneously comparison between satellite observed reflectances with LUT of simulated reflectances for given particle sizes and optical thicknesses. For this reason, we used CWP which is implicitly derived from the MSG/SEVIRI to estimate the precipitation.

A set of combinations obtained from the different spectral channels are selected:

The parameter $\Delta T_{IR3.9-IR10.8}$gives information on the effective radius of particles and the optical thickness of clouds [9]-[4]. In the IR3.9 channel, scattering is significant for particles with smaller radius, reducing cloud emissivity. Whereas the larger particles have a higher emission. The$\Delta T_{IR3.9-IR10.8}$ reaches the highest values for optically thin clouds of small/large particles. Medium to high $\Delta T_{IR3.9-IR10.8}$values are obtained for high optical thickness with a large particles. These values of $\Delta T_{IR3.9-IR10.8}$are lower for clouds with a thin optical thickness. Small $\Delta T_{IR3.9-IR10.8}$are obtained for small particles with thick clouds.

The parameter $\Delta T_{IR3.9-WV7.3}$ as for the $\Delta T_{IR3.9-IR10.8}$provides information on the effective radius of particles and the optical thickness of clouds [10].

The combination of $\Delta T_{IR3.9-IR10.8}$ and $\Delta T_{IR3.9-WV7.3}$characterizes the water capacity of clouds (CWP) (Fig. 2). As a result, cloudy areas with a high CWP value correspond to higher precipitation intensities. Cloudy areas with a low CWP value result in low rainfall intensities. These two parameters, namely $\Delta T_{IR3.9-IR10.8}$ and $\Delta T_{IR3.9-WV7.3}$ are used during the nighttime. Based on parameter thresholds, we determined several levels of CWP (fig. 2).

The channels $VIS0.6$ and $NIR1.6$ are used to obtain information about the effective radius and the optical thickness of the clouds respectively [5][4]. High reflectance values of VIS0.6 show a high optical cloud thickness and low reflectance values of NIR1.6 correspond to large particles in the cloud. It should be noted that these two channels are available during the day. The combination of the two channels ($R_{VIS0.6}$ and $R_{NIR1.6}$) characterizes the CWP (Fig. 3). To determine the levels of CWP, thresholds are used.

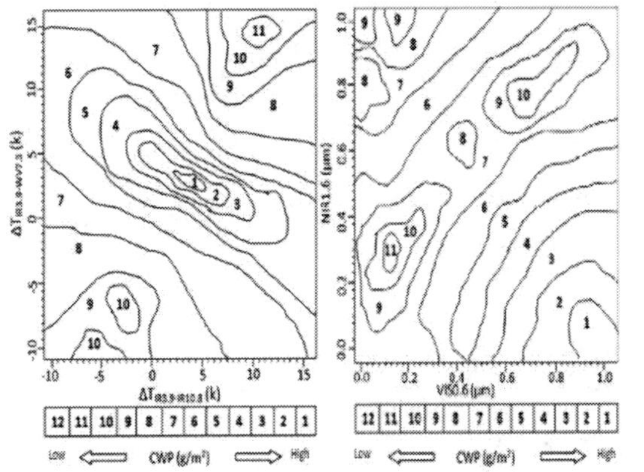

Fig.2. Cloud water path(CWP) versus $\Delta T_{IR3.9-IR10.8}$ and $\Delta T_{IR3.9-WV7.3}$ (left), Cloud water path (CWP) versus VIS0.6 and NIR1.6 (right) [11].

The following Table illustrates the multi-spectral parameters used in this work.

Table 1. Multi-spectral parameters used.

	Spectral parameters	Definition
Night time	$\Delta T_{IR3.9-IR10.8}$	brightness temperature difference between $IR3.9$ and$IR10.8$
	$\Delta T_{IR3.9-WV7.3}$	brightness temperature difference between $IR3.9$ and $WV7.3$
Day time	$R_{NIR1.6}$	Reflectance of the canal$NIR1.6$
	$R_{VIS0.6}$	Reflectance of the canal $VIS0.6$

2.2.2 Spatio-temporal correspondence between radar and satellite

The original spatial resolution of Sétifradar data (1 x 1 km^2) was re-projected and resampled to spatial resolution 4x5 km2, which corresponds to the resolution of the data of SEVIRI in the study area (northern Algeria). Due to the small area of overlap observed by the satellite and Radar, the spatial resolution is fixed at 4x5 km^2.

The comparison of the two images may be hampereddue to parallax offsets and co-location errors. This parallax offsets occurs because SEVIRI observes the study area at an oblique angle. Thus, to find the maximum correlation between the SEVIRI images and radar images, a radar pixel has been replaced by the average of the 5x5 neighborhood pixels for the high-level clouds, and the average of the 3x3 pixels for the low level clouds.

2 Methodology

In order to estimate precipitation using satellite data, we used a linear regression established between precipitation intensities and the CWP parameter (Fig.5). Due to the inadequacy of the IR3.9 channel during the day and to the unavailability of VIS0.6 and NIR1.6 at nighttime and, two models were developed, a daytime model and a nighttime model.

The method is based on the use a multi-threshold of spectral parameters to identify the CWP levels. The description of the multi-threshold method (MULTh) is presented in Fig. 3.

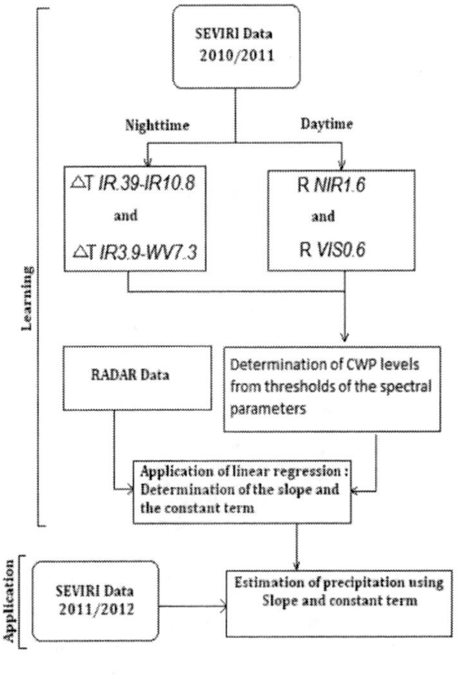

Fig.3. Flowchart of the developed method.

During the nighttime, we traced the variations of the parameters $\Delta T_{IR3.9-IR10.8}$ and $\Delta T_{IR3.9-WV7.3}$according to the radar reflectivities (fig. 4). Similarly, the RVIS0.6 and RNIR1.6 parameter changes as a function of radar reflectivity are plotted in Fig. 4.

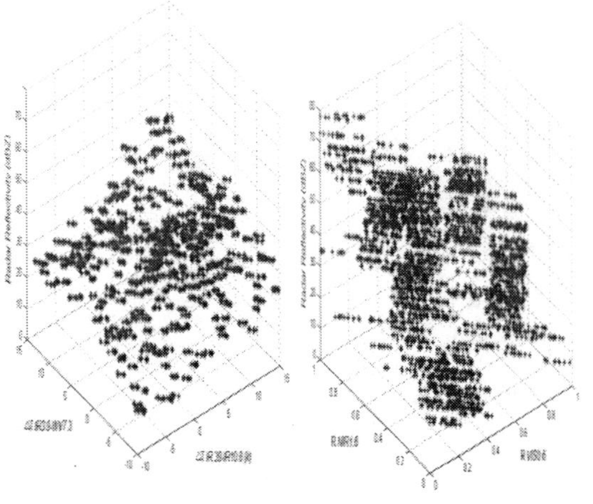

Fig. 4. Radar reflectivities versus $\Delta T_{IR3.9-IR10.8}$ and $\Delta T_{IR3.9-WV7.3}$ (left), Radar reflectivity versus $RVIS0.6$ et $RNIR1.6$ (right).

The fig. 4 shows that there is a relationship between the spectral parameters and the radar reflictivities. To relate the CWP to precipitation intensities, we have represented CWP levels as a function of radar reflectivities (Fig. 5).

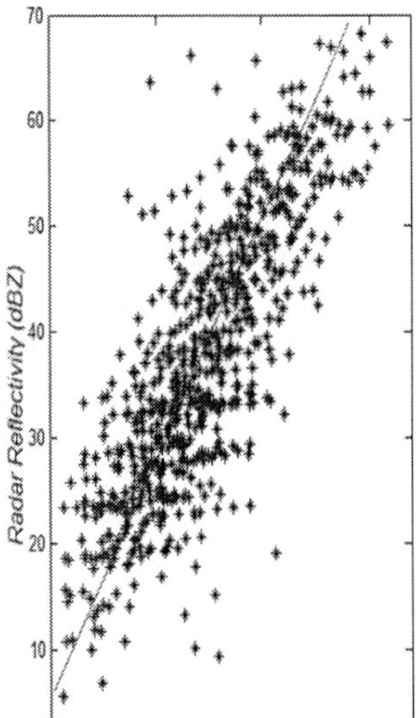

Fig.5. Radar reflectivities versus CWP.

Using linear regression, we were able to establish a mathematical relationship. The relationship between CWP and radar reflectivity is given by equation 2:

$$R(dBZ) = S \cdot CWP + c \qquad (2)$$

Where R is the radar reflectivity, S is the slope and is the constant term. In our case, we found S and c equal to 5.2 and 6.4 respectively.

4 Results of rainfall estimation

In this section, we calculated cumulative precipitation on monthly scales during the 2011/2012 rain period. To do this, we used the CWP levels obtained from the SEVIRI data and the two parameters S and c determined in the linear regression. Using equation 2, these estimates at the study area and at monthly time intervals (November, December, January and February) were calculated. The results of these estimates and the radar measurements are shown in Fig. 6.

To show the performance of the method, we calculated a set of evaluation parameters equations (3) to (6).

— The correlation coefficient

$$r = \frac{\sum_{i=1}^{N}(E_i-\bar{E_i})(V_i-\bar{V_i})}{\sqrt{\sum_{i=1}^{N}(E_i-\bar{E_i})^2}\sqrt{\sum_{i=1}^{N}(V_i-\bar{V_i})^2}} \qquad (3)$$

— The Root Mean Square Difference

$$RMSD = \sqrt{\frac{1}{N}\sum_{i=1}^{N}(E_i - V_i)^2} \qquad (4)$$

— The bias

$$Bias = \frac{1}{N}\sum_{i=1}^{N}(E_i - V_i) \qquad (5)$$

— The Mean Absolute Difference

$$MAD = \frac{1}{N}\sum_{i=1}^{N}|E_i - V_i| \qquad (6)$$

Where:

E_i : The estimated value.

V_i : The value of the radar validation data.

$\bar{E_i}$: The average value of the estimates.

$\bar{V_i}$: The average value of the radar validation data.

The evaluation parameters obtained are shown in the Table 2. The MULThmethod was also compared to the WkNN technique (weighted k *nearest neighbor*) implemented by Bensafi et al, (2019) [12] using the same database. The results of comparison are indicated in table 2.

The Fig. 6 shows that the fluctuations are larger in December and January than in November and February. This is corroborated by the correlation coefficients given in Table 2. The correlation coefficients vary between 0.86 and 0.89. The month of January has the highest correlation coefficient (r = 0.89), this is due to the presence of significant precipitation, especially those of stratiform type.

The values of the evaluation parameters (RMSD, Bias and MAD) show that the estimates obtained are close to the actual measurements. The best estimates are obtained for the months of December and January compared to the months of November and February. The precipitation scenes observed during these two months are dominated by stratiform precipitation. While convective rainfall is most observed during the months of November and February. The elaborate technique shows interesting performances for stratiform precipitation.

Table 2 shows the evaluation parameters for precipitation estimation obtained by the application of the WkNN technique and the application of the MULTh technique. According to the comparison results, the MULTh techniqueshows superior performances compared to the WkNN techniques.

The application of the MULTh technique to the SEVIRI data has improved the results of the precipitation estimation by taking into account the different types of cloud. The use of the CWP has contributed to better estimate rainfall. The MULTh technique estimates better the stratiform precipitations. This remarkable performance for stratiform precipitation is due to the dominance of water content in this type of cloud contrary to the ice content

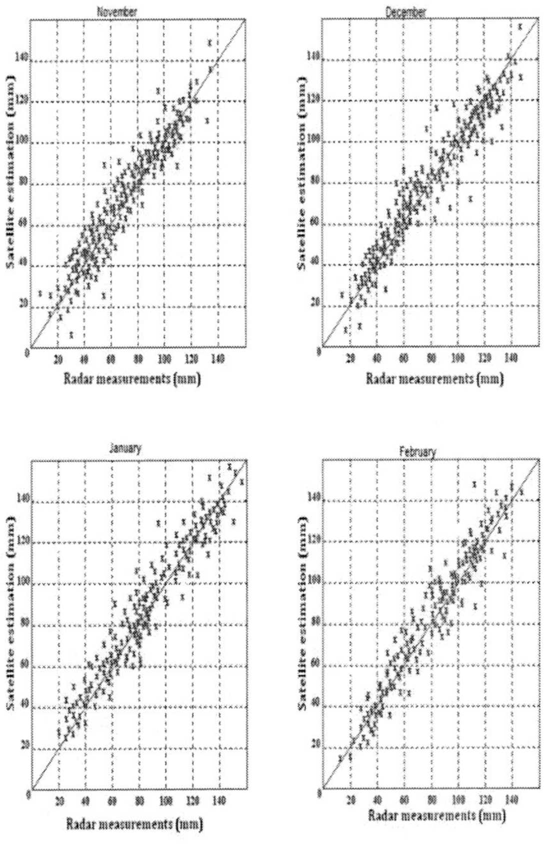

Fig.6. Satellite estimation versus radar measurements

Table 2. Parameters of evaluation of the cumulative precipitation estimated by the techniques WkNN and MULTh.

	WkNN Technique				MULThTechnique			
	r	RMSE (mm)	Bias (mm)	MAE (mm)	r	RMSE (mm)	Bias (mm)	MAE (mm)
November	0.85	15.45	-4.52	12.58	0.86	14.7	-3.6	10.2
December	0.87	12.58	2.85	8.25	0.88	10.5	2.5	7.6
January	0.88	11.62	2.94	7.29	0.89	09.1	2.4	7.3
February	0.87	17.23	-3.74	10.58	0.87	16.6	-5.7	11.3

5 Conclusion

The aim of this work is to develop a precipitation estimation technique applied to the northern Algeria using MSG satellite data. In this context, the technique aims to improve rainfall estimates, on the one hand, and extend these estimates to areas not covered by meteorological radar and / or rain gauges. The developed scheme is based on the exploitation of the optical and microphysical properties of clouds. The advantage of such a technique is that it can be applied both for convective systems and stratiform systems. The MULTh method compares CWP levels to precipitation intensities. A linear relationship has been established using linear regression.

Using this method, estimates of totals precipitation are determined. These estimates are compared to radar measurements. The results indicate a good correlation between the satellite estimates and the Radar measurements. The results also show that best estimates are obtained for stratiform precipitation. In this type of precipitation, precipitation intensities are proportional to CWP levels.

The results of the MULTh technique are also compared to the results obtained by applying the WkNN technique. The results showed a clear improvement for precipitation estimation using the MULTh method. However, in the case of convective precipitation estimates, the developed method showed a slight underestimation due to the presence of ice clouds. The MULTh method seems to be more interesting for the extratropical regions where advective-stratiform precipitation is dominant.

The perspective of this work is that it would be interesting to integrate the Cloud Ice Path to consider clouds with ice in its top, such as cumulonimbus.

References

1. Trigo, R., Xoplaki, E., Zorita, E., Luterbacher, J., Krichak, S.O., Alpert, P., Jacobeit, J., Sàenz, J., Fernàndez, J., Gonzàlez-Rouco, F., Garcia- Herrera, R., Rodo, X., Brunetti, M., Nanni, T., Maugeri, M., Turkex, M., Gimeno, 18 L., Ribera, P., Brunet, M.,

Trigo, I.F., Crepon, M., Mariotti, A., Chapter 3 Relations between variability in the Mediterranean region and mid-latitude variability. In: Developments in Earth and Environmental Sciences, vol. 4. Elsevier, pp. 179–226 (2006).

2. Hong, Y., Hsu, K.-L., Sorooshian, S., Gao, X., Precipitation estimation from remotely sensed imagery using an artificial neural network cloud classification system. J. Appl. Meteorol. 43, 1834–1853 (2004).
3. Ebert, E.E., Janowiak, J.E., Kidd, C., Comparison of near-real-time precipitation estimates from satellite observations and numerical models. Bull. Am. Meteorol. Soc. 88, 47–64 (2007)
4. Thies, B., Nauss, T., & Bendix, J., Precipitation process and rainfall intensity differentiation using Meteosat Second Generation SEVIRI data. Journal of Geophysical Research, 113, 19 (2008).
5. Lazri, M., Ameur, S.,.A satellite rainfall retrieval technique over northern Algeria based on the probability of rainfall intensities classification from MSG-SEVIRI. J. Atmos. Sol. Terr. Phys. 147, 106–120 (2016).
6. Levizzani, V., Satellite rainfall estimations: new perspectives for meteorology and climate from the EURAINSAT project, Ann. Geophys., 46, 363–372, http://www.ann-geophys.net/46/363/2003/ (2003).
7. Nauss, T., Kokhanovsky, A.A., Assignment of rainfall confidence values using multispectral satellite data at mid-latitudes: first results. Adv. Geosci. 10, 99–102 (2007).
8. Roebeling, R.A., Holleman, I., SEVIRI rainfall retrieval and validation using weather radar observations. J. Geophys. Res. 114 (2009).
9. Lazri, M., Ameur, S., Mohia, Y., Instantaneous rainfall estimation using neural network from multispectral observations of SEVIRI radiometer and its application in estimation of daily and monthly rainfall. Adv. Space Res. 53, 138–155 (2014).
10. Schmetz, J., Pili, P., Tjemkes, S., Just, D., Kerkmann, J., Rota, S., Ratier, A., 2002. An introduction to Meteosat Second Generation (MSG). Bull. Am. Meteorol. Soc. 83, 977–992 (2002).
11. Lazri, M., Ameur, S., Brucker, J.M., Testud, J., Hamadache, B., Hameg, S., Ouallouche, F., Mohia, Y., Identification of raining clouds using a method based on optical and microphysical cloud properties from Meteosat second generation daytime and nighttime data. Appl. Water Sci. 3, 1–11 (2013).
12. Bensafi N., Lazri M., Ameur S., Novel WkNN-based technique to improve instantaneous rainfall estimation over the north of Algeria using the multispectral MSG SEVIRI imagery, Journal of Atmospheric and Solar-Terrestrial Physics, doi.org/10.1016/j.jastp.2018.12.004 (2019).

A review on some investigations into sustainable machining techniques

Shrikant Gunjal[1, *], *and Sudarshan Sanap*[1]
[1]MIT ADT University, Loni Kalbhor, Pune, Maharashtra- 412 201, India.

Abstract. Machining plays a crucial role in process of economy through the way it contributes around 5% of the country's total economy. The method of machining has been changed through the last periods due to the competition within the market to urge more profit and High Speed Machining (HSM) plays the vital role to realize the equivalent. Hence, many researchers are working on reduction of the machining cost and consumption of energy. Hard to cut materials have long list of practical applications, which is the main reason to consider their research and development aspects. Thus, it's important to review the consequences of process parameters mainly like cutting speed, feed rate and depth of cut so as to determine the correlation with reference to key measuring parameters like tool life, surface roughness and tool wear. The authentication of this approach must be proved in future, using different experimental data as high speed machining leads to extreme machining temperature, sudden tool failure and subsequent adverse effects on overall machining performance in all the aspects. Sustainable machining techniques have the effective answer to avoid the adverse consequences of HSM. This paper emphasizes upon the review of application of various sustainable machining techniques like Minimum Quantity Lubrication (MQL), Cryogenic Technique, and Compressed Gas into the high-speed machining of hard to cut materials. Further, the appliance of vegetable based lubricants is highlighted for better comparison.

1 Introduction

It is well known fact that the manufacturing industry contributes significantly in terms of overall development worldwide. Most of the countries have observed the increase in industrial energy usage and production house contributes majorly in terms of CO2-emissions. With no surprise, modern society is striving for tailor made solutions across the product range. [1]. The thought of sustainability [2] is to overcome the issues that face from various segments, as it is important that many manufacturers adopt sustainable practices. Sustainable manufacturing consists of protection of environment, profit and social benefits of all these manufacturing areas. Manufacturing in alignment with sustainability and optimum machining aspects will serve the benefits of production house [3]. The idea of cleaner and sustainable production is related to creating services and products, utilizing procedures and frameworks that preserve the natural resources as well as the energy. Such model ought to be monetarily feasible, safe furthermore, energizing for workers and alluring for the consumers [4]. This research emphasize on the investigations into various parameters of machining and its effects on the sustainability.

1.1 Sustainable machining

Sustainable manufacturing requires the preservations of the energy as well as materials. It is related to ecological effect, energy utilization, operational security, labors wellbeing, management of waste, and cost of production. Sustainability is significant in the process of machining [5]. The sustainability of a production process is often achieved by integrating all the tiny factors so as to achieve sustainable machining. The important market demand has got to be considered in dependence to cost; quality that has got to consider while the delivery of the ultimate product. Along with major manufacturing/machining attributes; waste reduction is important area of consideration by the virtue of efficient utilization of resources [6, 7].

2 Literature review

The literature review focus on the authors contribution that have made the research and experiment to promote sustainable practices in order to consume less energy and enhance surface integrity. High speed machining is an important manufacturing process and widely used for extrusion dies, automobile, medical and aeronautics industry applications. Considerable developments are happening in the area of machining of various hard and conductive materials under different machining conditions [8, 9].

Sustainable machining techniques viz. MQL, compressed gas, cryogenic etc. has shown environment friendly direction for better future in all the aspects. With wide research work; it has been observed that, sustainable machining techniques help to reduce cutting forces, temperature and overall machining cost along with improved surface finish, tool life and overall machining performance even in hard to cut materials (titanium alloy, nickel alloy, AISI series/ hardened steel materials) [10-12]. An exhaustive literature review in this regard showing the contributions of various researchers is a source of vision to look towards latest developments in the area. This is an attempt to categorize the various researches made by different researchers for the sake of deciding the direction or path of research for new researchers.

2.1 High- speed machining and optimization

Pawade et al. [13] have discussed the application of a novel approach called Jaya algorithm to optimize method parameters in turning of AISI S7 material at higher speed. The algorithm does not need any algorithm-specific control parameters. Tool nose radius, cutting speed, feed rate, and depth of cut are the important method parameters considered. Surface roughness and cutting force observed to be influenced by these parameters. L18 mixed Taguchi orthogonal array was used while the mathematical prediction models were developed for output using regression method.

Khedekar et al. [14] have observed that time affects majorly over thickness variation during the experimentation on nickel chrome in view of its wide range of application in plating. Shinde et al. [15] performed the experimental work to optimize the method parameters in the process of powder-mixed electrical discharge machining. They used fine abrasive particles of aluminium, silicon and silicon carbide powder during experimentation which was performed on laboratory developed set-up. Response surface methodology (RSM) was used to assess the process performance.

Kale et al. [16] found that performance of copper electrode is better over brass during machining of Inconel 718. Response surface analysis and D-optimal method is used for optimization of process parameters to find the minimum tool wear and maximum material removal.

2.2 Cutting tool techniques based on cryogenic treatment

Cryogenic is a potential machining technique of the cutting tool cooling and part, while performing an operation under consideration. Cryogenic machining basically refers to delivering the cryogenic fluid to the cutting surface during the machining which undergoes the extreme temperature during the operation and where there's change within the material characteristics. Coolant used is typically liquefied nitrogen at –196°C. Nitrogen is safe, non-combustible, and noncorrosive gas which can be stored [17, 18]. Rubio et al. [19] noted that, cold compressed air which is subsequent to cryogenic technique; improves the surface finish by the virtue of temperature reduction in the cutting zone. Pereira et al. [20] performed the experimental investigations into AISI 304 turning and found the improvement in tool life by 50% under cryogenic application. Also, they noted the possibility 30% increase in cutting speed compare to dry machining. Schoop et al. [21] implies the liquid nitrogen (LN2) at -195.8 temperatures during high speed machining of Ti-6Al-4V material. They noted the use of cryogenic cooling allow lowest tool wear < 10 μm at higher cutting speed (240 m/min.) and reduced surface roughness compared to flood machining. Pande and Patil [22] performed the experimentation to check the effect of method parameter followed by analysis of cutting speed and feed correlation with reference to surface finish. ANFIS model and statistical validation are found to be in agreement of experimentation.

Pusavec et al. [23] assessed the cryogenic machining for a nickel alloy; Inconel 718. They assessed the attributes of surface integrity for distinct combinations of lubrication conditions. The outcomes demonstrate that cryogenic machining procedures can be actualized to enhance the surface integrity quality in view of improved quality of final product. Pusavec et al. [24] also assessed the machining of Inconel 718 using RSM by developing a predictive model for MQL, cryogenic, dry and cryo-lubrication machining processes. ANOVA was used to validate the models. From the results it could be observed that the lubrication/cooling significantly affected the performance of machining. Pusavec et al. [25] further presented a machining observation of Ni-alloy (high-temperature) which indicated that the cost of tooling contributed majorly to the cost of overall production.

2.3 Cutting tool techniques based on surface structuring

Baisane et al. [26] found the higher surface roughness and lower material removal rate under the consideration of Al2O3p. Taguchi method is used during the machining of composite materials. Bhople et al. [27] noted that cutting speed and depth of cut are major parameters affecting surface roughness and main cutting force respectively during machining of austempered ductile iron. Surface plot used to analyze the effects of feed rate, depth of cut and speed on surface finish, microhardness and cutting force. The validation of these models was done by ANOVA. Jadhav et al. [28] have performed drilling experimentation on Al6061/ Al2O3 aluminium metal matrix composite to study the effect on surface finish. Design of experiment was done by Taguchi method to know the effects of cutting speed, point angles of drill bits, feed rate and reinforcement percentage of Al2O3p on cylindricity, mean force, torque and surface roughness.

Nizamuddin et al. [29] have compared the karanja oil with conventional cutting fluid during the orthogonal cutting of AISI 1045 steel and found better results with eco-friendly oil over the conventional fluid. Nilesh and Brahmankar [30] have experimented to incorporate the particles dimensions along with development of a replacement model for surface roughness. Experimental results are found to be in close agreement with predictions of this model. The nonlinear estimation results while machining of these composites projects the importance or properties like coefficient of thermal expansion, thermal diffusivity, melting temperature and heat of fusion. Pawade et al. [31] have noted the significant reduction in cutting force using honed plus chamfered cutting edge during machining of Inconel 718 super alloy at higher speed with PCBN cutting tool.

2.4 Cutting tool techniques based on cooling and lubrication

Wang et al. [32] performed the experimental evaluation during grinding under the application of MQL using different vegetable oil types. They performed the experiment on Ni- alloy (GH4169) and noted that use of vegetable oils namely castor, peanut, sunflower and soybean oil; help to achieve superior results in comparison of flood lubrication. Key ingredients of vegetable oils like ricinoleic and oleic acids which have low friction coefficient and high binding energy; helps in producing excellent lubrication properties they observed. On the equivalent line, Sunday et al. [33] concluded that use of MQL method using Bio- lubricant in various machining processes is an effective way to tackle environmental concerns, health issues and price related to the lubricant application during machining operation.

Sachin and Nilesh [34] have developed new cutting fluid during machining of M2 steel turning operation under MQL and compared it with traditional oil. Results show that surface roughness obtained under alove vera is better as compared to mineral oil. Further, reduced tool wear noted under alove vera application. Shrikant and Nilesh [35] have investigated that bio-lubricant canola oil performance is much closer to the synthetic oil results during the turning of hardened AISI 4340 material. Gunjal et. al. [36] found the similar kind of closer results with bio-lubricant coconut oil in comparison to synthetic oil during experimental investigation and noted less than micron variation in results of chip thickness study.

Pande and Patil [22] found that the application of vegetable oil under MQL had better lubrication in comparison to flood and mineral lubrication. They utilized liquid paraffin oil for assessing and comparing the lubrication properties of sunflower oil, castor, palm, rapeseed, maize, peanut and soybean as the MQL base oils. On the same line, Karkade and Nilesh [37] have performed turning experimentation on titanium alloy under cryogenic (CO_2) gas cooling. The comparative investigation was undertaken on surface integrity and process forces. Surface finish was noted to enhance up to 28 % compared to dry cutting.

Thakur et al. [38] have analyzed machinability characteristics during Inconel 718 machining under dry and MQL conditions. Collected chips were analyzed using X-ray diffraction analysis. Results show that MQL application is far better than that of dry machining. Elshwain et al. [39] reviewed the coolant-lubricants based on gas. They concluded that in comparison to conventional cooling and dry cutting for nickel-alloys and machining titanium, most of the attributes of the gas based cutting fluids enabled a considerable improvement. Jozic et al. [40] have performed the Taguchi's design of experiment for optimizing the cutting environments and process parameters during the process of machining. It is seen that feed per tooth, cutting time and cutting condition applied a critical impact on different reactions. Therefore, the adequacy of this system was confirmed by an experimental test. The concerned study affirms that air cold system is often effectively applied when it's important to satisfy recently established attributes.

2.5 Comparison table

Table 1. Researchers contribution

Ref. No.	Method	Type of Metals & Machining Procedure	Evaluation metrics
4	Taguchi	AIMg3-Milling	Surface roughness, power consumption
5	Taguchi, ANOVA	Magnesium Alloy-Turning	Lubrication, surface roughness and finish
24	ANOVA, RSM	Nickel alloy, Dry machining, high speed machining, cryogenic	Cooling lubrication and tool life
25	Genetic algorithm	Nickel alloy, Dry machining, cryogenic	Energy consumption, safety measures, and cost
38	Taguchi	Steel 42CrMo4-Milling	Surface finish and cutting force

Conclusions

The concerned review paper is an attempt to summarize the research effort to drive sustainable machining future for all the relevant machining operations. Though sustainable machining techniques has shown promising results to all the materials even at high speed; the subsequent challenge like complexity reduction and structuring for continuous improvement to orientate manufacturing towards excellence has to be evaluated effectively. This shows the sustainable machining as a potential approach for the modern manufacturing world which prompt for tool efficiency and low energy consumption for quality and overall machining performance.

References

1. Enomoto, T., & Sugihara, T. Improvement of anti-adhesive properties of cutting tool by nano/micro textures and its mechanism. *Procedia Engineering*, 19, 100-105 (2011).
2. Rahim, E. A., Ibrahim, M. R., Rahim, A. A., Aziz, S., & Mohid, Z. Experimental investigation of minimum quantity lubrication (MQL) as a sustainable cooling technique. *Procedia CIRP*, 26, 351-354 (2015).
3. Gajrani, K. K., Suvin, P. S., Kailas, S. V., &Sankar, M. R.. Hard machining performance of indigenously developed green cutting fluid using flood cooling and minimum quantity cutting fluid. *Journal of Cleaner Production*, 206, 108-123 (2019).
4. Fratila, D., and Caizar, C., "Application of Taguchi Method to Selection of Optimal Lubrication and Cutting Conditions in Face Milling of AlMg3," J. Cleaner Prod., 19(6–7), pp. 640–645 (2011).
5. Eker, B., Ekici, B., Kurt, M., and Bakır, B. "Sustainable Machining of the Magnesium Alloy Materials in the CNC Lathe Machine and Optimization of the Cutting Conditions," Mechanics, 20 (3), pp. 310–316 (2014).
6. Gajrani, K. K., & Sankar, M. R. Sustainable Machining with Self-Lubricating Coated Mechanical Micro-Textured Cutting Tools. In Reference Module in Materials Science and Materials Engineering. Elsevier (2018).
7. Revuru, R. S., Posinasetti, N. R., VSN, V. R., & Amrita, M. Application of cutting fluids in machining of titanium alloys—a review. The International Journal of Advanced Manufacturing Technology, 91 (5-8), 2477-2498 (2017).
8. Borlepwar, P. T. and Patil, D. N., Recent advances in electrical discharge machining process: a review. In *IVth international conference on production and industrial engineering, CPIE* (2016).
9. Bandapalli, C., Sutaria, B. M., & Bhatt, V. D., High speed machining of Ti-Alloys–A critical review. In *Proceedings of the 1st International and 16th National Conference on Machines and Mechanisms (iNaCoMM2013)* (pp. 18-20) (2013).
10. Sharma, A. K., Tiwari, A. K., & Dixit, A. R. Effects of Minimum Quantity Lubrication (MQL) in machining processes using conventional and nanofluid based cutting fluids: A comprehensive review. Journal of Cleaner Production, 127, 1-18 (2016).
11. Park, K. H., Suhaimi, M. A., Yang, G. D., Lee, D. Y., Lee, S. W., & Kwon, P. Milling of titanium alloy with cryogenic cooling and minimum quantity lubrication (MQL). International Journal of Precision Engineering and Manufacturing, 18 (1), 5-14 (2017).
12. Gill, S. S., Singh, J., Singh, H., & Singh, R. Investigation on wear behaviour of cryogenically treated TiAlN coated tungsten carbide inserts in turning. International Journal of Machine Tools and Manufacture, 51 (1), 25-33 (2011).
13. R. S. Pawade, A. S. Awale, & Brahmankar, P. K. Application of Jaya Algorithm in Optimization of High Speed Turning of AISI S7 Tool Steel (2016).
14. Khedekar, D., Gosavi, V., Gogte, C. & Brahmankar, P., Optimization of Process Parameters of Nickel–Chromium Electroplating for Thickness Variation using Genetic Algorithm. In *International Conference on Communication and Signal Processing (ICCASP 2016)*. Atlantis Press (2016,).
15. Shinde, R., Patil, N., Raut, D., Pawade, R. & Brahmaknakr, P., Experimental Investigations into Powder-Mixed Electrical Discharge Machining (PMEDM) of HCHCr D2 Die Steel. In *International Conference on Communication and Signal Processing (ICCASP 2016)*. Atlantis Press (2016).
16. Kale, S., Khedekar, D., Brahmankar, P. & Sadaiah, M., Some Investigations into the Electrical Discharge Machining of Inconel 718 Alloy using Copper and Brass Electrodes. *In International Conference on Communication and Signal Processing (ICCASP 2016)*. Atlantis Press (2016).
17. Kalia, S. Cryogenic processing: a study of materials at low temperatures. Journal of Low Temperature Physics, 158 (5-6), 934-945 (2010).
18. Gill, S. S., & Singh, H. Cryogenic treatment of materials: cutting tools and polymers. In Polymers at Cryogenic Temperatures (pp. 245-273). Springer, Berlin, Heidelberg (2013).
19. Rubio, E. M., Agustina, B., Marín, M., & Bericua, A. Cooling systems based on cold compressed air: A review of the applications in machining processes. *Procedia engineering*, 132, 413-418 (2015).
20. Pereira, O., Rodríguez, A., Fernández-Abia, A. I., Barreiro, J., & de Lacalle, L. L. Cryogenic and minimum quantity lubrication for an eco-efficiency turning of AISI 304. Journal of Cleaner Production, 139, 440-449 (2016).
21. Schoop, J., Sales, W. F., & Jawahir, I. S. High speed cryogenic finish machining of Ti-6Al4V with

polycrystalline diamond tools. Journal of Materials Processing Technology, 250, 1-8 (2017).

22. Pande, P. P. and Patil, N. G., Investigations into Machining of Inconel 718 by Using Adaptive Fuzzy Based Inference System. International Journal of Engineering Research, 3(5) (2014).
23. Pusavec, F., Hamdi, H., Kopac, J., & Jawahir, I. S. Surface integrity in cryogenic machining of nickel-based alloy—Inconel 718. Journal of Materials Processing Technology, 211 (4), 773-783 (2011).
24. Pusavec, F., Deshpande, A., Yang, S., M'Saoubi, R., Kopac, J., Dillon, O. W., & Jawahir, I. S., "Sustainable Machining of High Temperature Nickel Alloy—Inconel 718: Part 1—Predictive Performance Models," J. Cleaner Prod., 81, pp. 255–269 (2014).
25. Pusavec, F., Kramar, D., Krajnik, P., & Kopac, J., "Transitioning to Sustainable Production—Part II: Evaluation of Sustainable Machining Technologies," J. Cleaner Prod., 18 (12), pp. 1211–1221 (2010).
26. Baisane, V., Patil, N., Lahane, S., Pawade, R. & Brahmankar, P., December. Investigations into Wire Electro-discharge Machining of A6061/Al2O3p Composites. In *International Conference on Communication and Signal Processing (ICCASP 2016)*. Atlantis Press (2016).
27. Bhople, N., Patil, N. and Mastud, S., The Experimental Investigations into Dry Turning of Austempered Ductile Iron. *Procedia Manufacturing*, 20, pp.227-232 (2018).
28. Jadhav, S. S., Kakde, A. S., Patil, N. G. & Sankpal, J. B., Effect of cutting parameters, point angle and reinforcement percentage on surface finish, in drilling of AL6061/Al2O3p MMC. *Procedia Manufacturing*, 20, pp.2-11 (2018).
29. Nizamuddin, M., Agrawal, S. M. and Patil, N., The Effect of Karanja based Soluble Cutting Fluid on Chips Formation in Orthogonal Cutting Process of AISI 1045 Steel. *Procedia Manufacturing*, 20, pp.12-17 (2018).
30. Patil, N. G. & Brahmankar, P. K., Semi-empirical modeling of surface roughness in wire electro-discharge machining of ceramic particulate reinforced Al matrix composites. *Procedia CIRP*, 42, pp.280-285 (2016).
31. Pawade, R. S., Joshi, S. S., Brahmankar, P. K. & Rahman, M., An investigation of cutting forces and surface damage in high-speed turning of Inconel 718. Journal of Materials Processing Technology, 192, pp.139-146 (2007).
32. Wang, Y., Li, C., Zhang, Y., Yang, M., Li, B., Jia, D., & Mao, C. Experimental evaluation of the lubrication properties of the wheel/workpiece interface in minimum quantity lubrication (MQL) grinding using different types of vegetable oils. Journal of cleaner production, 127, 487-499 (2016).
33. Lawal, S. A., Choudhury, I. A., & Nukman, Y. A critical assessment of lubrication techniques in machining processes: a case for minimum quantity lubrication using vegetable oil-based lubricant. Journal of Cleaner Production, 41, 210-221 (2013).
34. Agrawal, S. M. and Patil, N. G., Experimental study of non-edible vegetable oil as a cutting fluid in machining of M2 Steel using MQL. *Procedia Manufacturing*, 20, pp.207-212 (2018).
35. Gunjal S. U. and Patil N. G., Experimental investigations into turning of hardened AISI 4340 steel using vegetable based cutting fluids under minimum quantity lubrication. *Procedia Manufacturing*, 20, pp.18-23 (2018).
36. S. U. Gunjal, S. B. Sanap & N. G. Patil, Role of cutting fluids under minimum quantity lubrication: An experimental investigation of chip thickness, *Materials Today: Proceedings*, (2020).
37. Karkade, H. B. & Patil, N. G., Comparative investigations into high speed machining of AB titanium alloy (Ti–6al-4v) under dry and compressed Co2 gas cooling environment. In AIP *Conference Proceedings* (Vol. 2018, No. 1, p. 020009). AIP Publishing (2018).
38. Thakur, D. G., Ramamoorthy, B. & Vijayaraghavan, L., Influence of minimum quantity lubrication on the high speed turning of aerospace material superalloy Inconel 718. International Journal of Machining and Machinability of Materials, 13 (2-3), pp.203-214 (2013).
39. Elshwain, A. E. I., Redzuan, N., &Yusof, N. M. Machinability of nickel and titanium alloys under of gas-based coolant-lubricants (CLS)–A review. International Journal of Research in Engineering and Technology, *2*(11), 690-702 (2013).
40. Jozic', S., Bajic', D., & Celent, L., "Application of Compressed Cold Air Cooling: Achieving Multiple Performance Characteristics in End Milling Process," J. Cleaner Prod., 100, pp. 325–332 (2015).

Friction welding of Aluminium Alloy 6063 with copper

Yashwant Chapke[1,*], *Dinesh Kamble*[1] *,and Saoud Md. Salim Shaikh*[2]
[1]Vishwakarma Institute of Information Technology, Pune 411048, India
[2]Sinhgad Academy of Engineering, Pune 411048, India.

Abstract Friction welding process is a forging welding process in which work piece are joined due to heat produced by friction between two joining surfaces and upset pressure is applied by non-rotating work piece. Joining of aluminum alloy with dissimilar material is important research area to focus on as maximum aircraft structures havexx Aluminum alloy frame and aerospace designers familiar with Aluminum alloy and its design considerations. After comparison of mechanical properties and application of light weight alloys aluminum alloys, tungsten, stainless steel and copper, copper selected as dissimilar material to join with Aluminum alloy AA6063. AA 6063 also known as architectural alloy selected based upon its properties. This dissimilar joint of AA6063 and Copper has application in electrical conductors as copper is good electrical conductivity and used in maximum electrical conductors. In this research work AA6063 joined with Copper successfully using Rotary Friction Welding process. Through process study effective process parameters like Friction Pressure, Upset Pressure, Spindle Speed, and Friction Time identified and their effect on weld joint strength were studied.Testing for measuring UTS of friction welded joint conducted. Using DOE tool optimized set process parameters for friction welding identified and their effect on weld joint strength studied experimentally. Maximum UTS of 222.787 MPa for Friction welded joint achieved, bend test also performed on friction welded samples.

1. Introduction

Rotary friction welding is used for joining of round parts like engine valves, steering rod, shafts and other parts in aerospace application. Bimetallic valves are produced using RFW process.

RFW process is solid state joining approach hence limitation of fusion welding of joining different material of different melting point overcome using Friction Welding. In this friction welding process despite of different melting point both AA 6063 and Copper joined which was not feasible earlier by fusion welding process. Aerospace Industry use Aluminum alloy as primary material for structure and frame due to its light weight and mechnaical properties hence Aluminium alloy selected as one material for this research work [1].
Also in aerospace, automobile, marine, defense equipment appliances for joining various copper electrical parts to Aluminum Alloy alloy frame. Mainly in aerospace structures frames made up of Aluminum Alloy and various other metals parts to be joined to Aluminum Alloy frame using friction welding.

2. Literature Review

Various Researchers aimed at achieving maximum UTS for friction welded joint through parametric optimization of friction welding process.[1-17] While doing parametric optimization following parameters are considered by researchers,

Friction Time (FT), Friction Pressure (FP), Spindle Speed (SS), Burn of length (BOL)[1-9]. Further sub branching of friction welding parameters can be done in form of upset time, upset pressure, breaking time [1-7]. Researchers focused on study of microstructure of friction weld joint under electron microscope and micro structural characterization done for friction welded dissimilar metal joints. Researchers also stated interested observations by microstructure study of joints in regards to new layer formation and grain structure at welded joint. Mumin Sachin et al found Friction Time, Friction Pressure have direct effect on tensile strength of joint[5]. C. Meengam et al joined semi solidAA7077 using RFWprocess[2]. L.Zhou et al in their research work considered Rotational Speed as dominant parameter and studied its effect on mechanical properties, on microstructure of Ti-6Al-4V. They have observed along

* Corresponding author: **yashchapke@gmail.com**

with increase in rotational speed hardness in weld zone decreases. In conclusion remark made by researcher through this study in regards to defect free welds are produced by employing rotational speed ranging from 400 to 600 rpm. Researchers through this research work found that 6061 T6 Aluminum can be successfully joined with steel AISI 1018 Steel getting tensile strength in range of 170 Mpa to 250Mpa for welded joint. J. Romero et al in their research work have done investigation on effect of forging pressure on microstructure and micro hardness of welding joint Ti-6Al-4V [11].

2.1 Difficulty in joining AA6063 with Copper

Aluminum Alloy 6063and copper has different melting point temperature , Aluminum Alloy 6063 has 610°C Melting Point where as Copper has 1024°C Melting temperature.
Melting Temperature difference and other factors like formation of intermetallic layer and oxides makes conventional fusion welding of Aluminum Alloy 6063 copper joining obsolete and impractical.

Table 1.0
Material Comparison

Material	*Mechanical Properties*	*Chemical composition*	*Industrial application*
Stainless steel SS304	Toughness Durability Corrosion resistance Cryogenic resistance Ductility	SS304 minimum of 18% chromium 8% nickel, combined with a maximum of 0.08% carbon	Aerospace Automotive oil & gas industries kitchenware consumer products
Titanium alloy	Very high tensile strength and toughness at high temperature also	Titanium alloy has 6% aluminum and 4% vanadium, 0.25 % maximum iron,0.2% oxygen and remaining titanium	Military Aircraft Spacecraft Bicycle Connecting rod Consumer electronics Medical
Copper	Soft Malleable Ductile High thermal electrical conductivity	Copper available in nature in directly usable metal form	Widely used nonferrous metal. Electronics Electric power Machineries

Stainless Steel, titanium, Copper and Magnesium materials are identified for joining with aluminum alloy. Material comparison done based upon its material properties, chemical composition and industrial application.

Parameters have dominant effect on tensile strength of friction welded joint identified through study.
Following process parameters identified related to RFW process,

Friction Pressure (FP), Upset Pressure (UP), Friction Time (FT) and Spindle Rotation Speed (SS). Following inferences were drawn based upon literature study and pilot experimentation.

1. High Spindle Speed results in good tensile strength of joint
2. High upset pressure will increase hardness of joint
3. Increasing Friction Pressure value results in increase in tensile strength of joint.

3. Experimentation

3.1 Design of Experiments (RSM)

Three levels low medium and High identified for selected process parameters and Using Response Surface Methods (RSM) optimum set of process parameters obtained and used for conduction of experiments.
For performing DOE Design Expert Tool used and optimum values of Friction Time (FT) 1 second, Spindle Rotation speed (SS) 1800 RPM Friction Pressure (FP) 39 and 48 Kg/mm^2 and Upset pressure 97 kg/mm^2 obtained.

3.2 Experimentation

Total Nine Trial runs conducted on Rotary Friction Welding Machine. Optimized process parameters sets used for conduction of experiments as per parameters sets obtained from DOE tool. In first two trails Aluminum Alloy 6063couldn't joined with copper alloy due to less Friction Pressure (39Mpa and 48Mpa) and upset pressure (80Mpa and 97MPa). Then for next four trail run as per process parameter obtained from design expert tool and expert advice Friction Pressure and Upset Pressure. Again four more trails are taken using and maximum tensile strength achieved through variation of process parameters. Experimentation performed on Rotary Friction welding Machine having maximum forge force 120 KN, maximum spindle speed of 2000 RPM.

3.2.1 First trail run Experiments

In this trail run Friction Pressure is used 39 MPa and 48 MPa respectively, Upset Pressure used 80 MPa and 98 MPa and Spindle Speed maintained at 1800 RPM.

Fig.1.0 First two trailsrunAA6063- Cu Failed

Table 2.0

Process Parameters Set 1 (Failed to join)

Parameters	***Run 1***	***Run 2***
Spindle length L1	141.6	141.6
Slider side L2	141.2	141.2
Total= L1+L2	282.8	282.8
Final length (after welding)	264.2	266.2
Loss(actual measured)	18.6	16.2
Shrinkage range	18	18
Friction time t1b	0.5	1
Break delay t2	0.1	0.6
Upset delay t3	0.3	0.1
Upset time t4	2	3
Soft friction	0	25
Upset pressure@13kg/mm2	80	97
Feed	75	75
Rpm	1800	1800

Trail run taken with two sets of process parameters. In first run Aluminium Alloy 6063 not welded with copper and this runs become void run. Following process parameters used for first trail run and it is observed that selected values of Friction Pressure needs to be increased for successful weld joint between Aluminium Alloy 6063 and Copper.Aluminium Alloy 6063 not welded with copper and this runs become void run. Following process parameters used for first trail run and it is observed that selected values of Friction Pressure needs to be increased for successful weld joint between Aluminium Alloy·6063 and Copper.

3.2.2 Second and third trail run Experiments

With second set of parameters four runs were conducted under second and third trail run. From Parameter set 2 sound friction welded joints produced. But after welding final length received was less hence more losses observed. To reduce losses with same set further trail taken.

In third trial four more trail run with same set of parameters used in second trail run performed. Here actual measured loss was less than second trail run. Final length received after weld also was more than previous trail run which reduces the losses and saving material.

Second and third trial run total eight experiments performed with following set of parameters given in Table No. 3.0

Table 3.0

Process Parameters set 2(successful joining)

Parameters	**1**	**2**	**3**	**4**
Spindle length	140	141.6	142.3	140.8
Slider side L2	140.7	141	140.8	140.1
Total= L1+L2	280.7	282.6	283.1	280.9
Final length welding)	268.4	267.2	276.4	277.5
Loss(actual measured)	12.3	15.4	6.7	3.4
Friction time t1	1	1	1	1
Break delay t2	0.6	0.6	0.6	0.6
Upset delay t3	0.1	0.1	0.1	0.1
Upset time t4	3	3	3	3
Soft Friction P	25	25	48	48
Upset Pressure	97	97	97	97
Feed	75	75	75	75
Rpm	1800	1800	1800	1800

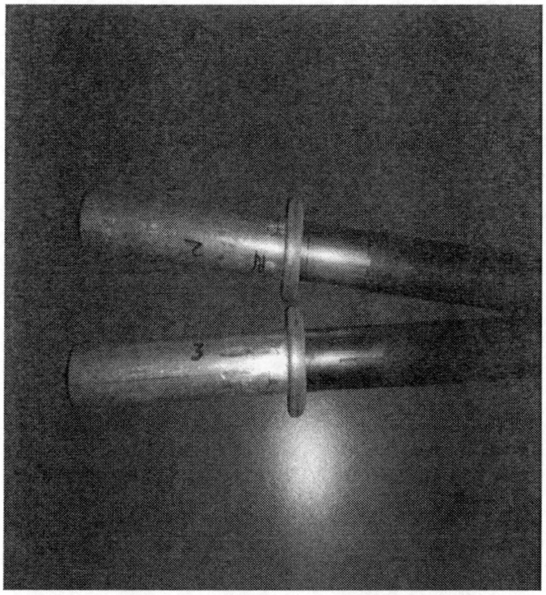

Fig.2.0 Experimental run 2 (More Losses)

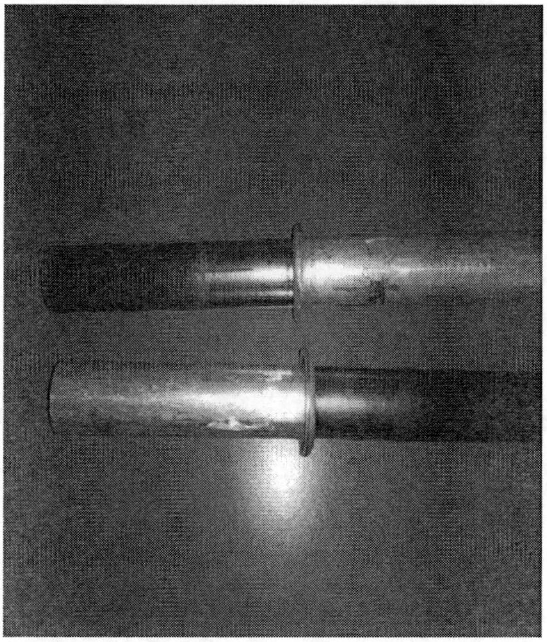

Fig. 3.0 Experimental run 3 (less losses)

AA6063 successfully welded with copper using friction pressure of 48 Kg/mm^2 and upset pressure of 97 kg/mm^2at Friction time of 1 sec and upset time 3 sec. and spindle rotation speed SS 1800 RPM.

3.3 Ultimate Tensile Strength (UTS) Testing

Friction welded joints obtained from process parameters set 2 and set 3 are tested for measuring its UTS. For measuring UTS testing specimen prepared as per ASM standards specimen diameter set to 12.5mm and testing performed on friction welded joint.

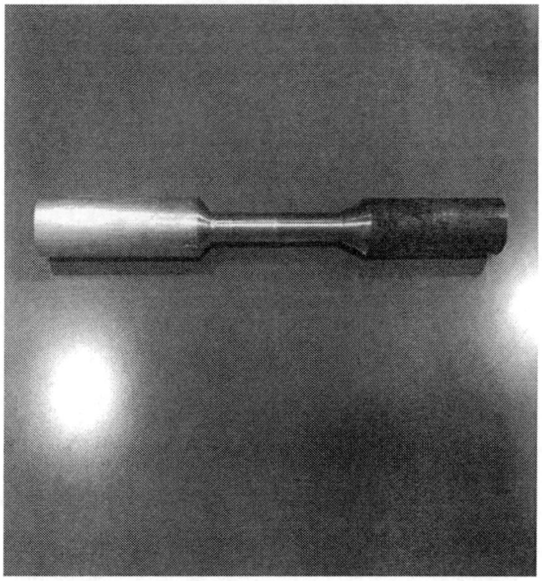

Fig. 5.0 Tensile Test Specimen

3.4 Bending test

Fig. 6.0 Bending Test Specimen

Bending test is performed for investigating the strength of IMC layers formed at welded joint interface. Bending test performed on samples welded using Friction Time of 1 Sec, Upset Pressure 97 kg/mm^2 and Friction Pressure of 48 Kg/mm^2. Bending test performed at 90° all the specimens welded at given set of process parameters tested for above bending test. All Samples failed the bending test.

4 Results and Discussion

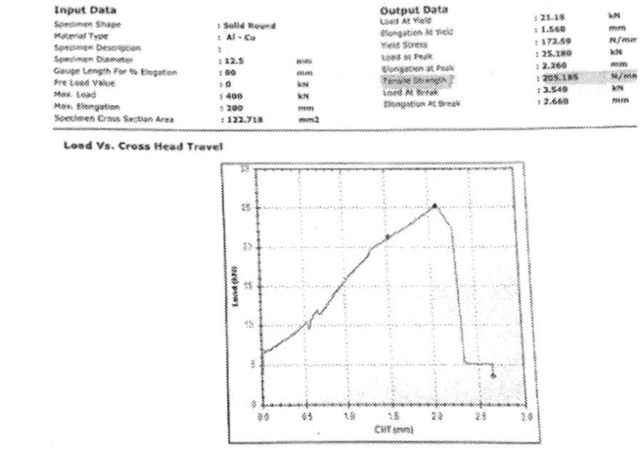

Fig.8.0 UTS test report of second trail run (205.185Mpa)

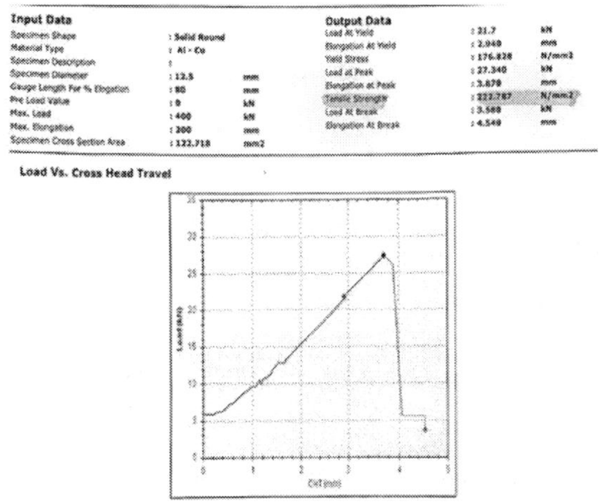

Fig.9.0 UTS test report of second trail run(222.7Mpa)

4.1 Effect of Process Parameters

Weld joint strength Increases with increase in Friction Time and Upset Pressure up to certain limit. Insufficient weld time results in weak friction welded joint. For first trail run Friction Time (FT) 0.5 Seconds failed to join Al6063 with Copper, increase in Friction Time will form adequate Friction Welded joint. This is proved from later two set of experimentations.

Upset pressure directly proportional to UTS of weld joint up to certain limit as shown in Fig. 8 and Fig.9. From this study it has been observed that Variation in Upset Pressure and Friction Time are contributing dominantly as compared to other process parameters of RFW process.

Friction Time (FT), Friction Pressure (FP) and Upset Pressure (UP) varied keeping other parameters constant and through this research work inferences can be drawn that this parameters are dominant process parameters and have more influence on weld strength. For joining dissimilar material using RFW and obtaining adequate tensile strength process parameters should be selected from above range to achieve good tensile strength of joint.

5 Conclusions

1. Experimentation shows that less friction time (0.5 Sec) forms inadequate IMC layer results in failed friction weld joints.

2. Ultimate Tensile Strength of 205.185 MPa obtained for friction welded specimen welded with optimised process parameters given in Table No.2.0.

3. For Friction welded joint welded with process parameters given in Table 2.0 More flash produced at joint. Due to excess flash formation considerable losses are observed (actual measured losses 12.3mm, 15.4mm, 15.3mm, 16.6mm) resulted in final length of work piece after welding (268.4 mm, .

4. Ultimate Tensile Strength of 222.787 MPa obtained with third set of process parameters given in Table 3.0. This is maximum strength achieved as in respect to base metal strength. Obtained UTS value clearly states that tensile strength of friction welding joint can be increased by increasing friction and Upset pressure up to threshold value.

5. Less flash formation takes place when Al6063 joined with Copper using Process Parameter set given in Table 3.0. In this welding less losses observed (actual measured losses 4.3mm, 5.2mm, 6.7mm, 3.4mm) resulted in Final length of work piece after welding (278.4mm, 276.3, 276.4, 277.5 mm).

6. In this research work UTS of 205.185 MPa and 222.787 MPa observed this strength is greater than previous research work studied related to joining of dissimilar metals to Aluminium alloy using Rotary Friction Welding Process.

Authors would like to express their sincere gratitude towards FWT Friction Welding Technology and Mr.YatinTambe for providing support for conduction of Experiments.

References

1. Radosław Winiczenkoa, Olgierd Gorochb, Anna Krzynska , Mieczysław Kaczorowski 2017, Friction welding of tungsten heavy alloy with Aluminum Alloy 6063alloy , Journal of Materials Processing Technology **246**, 42–55, (2017).
2. C. Meengam, S. Chainarong, P. Muangjunburee, Friction Welding of Semi-Solid Metal 7075 Aluminum Alloy 6063Alloy, *Materials Today Proceedings,* **4** ,1303–1311, (2017).
3. M. Meisnar, S. Baker, J.M. Bennett, A. Bernad, A. Mostafa, S. Rescha, N. Fernandes, A. Norman Microstructural characterization of rotary friction welded AA6082 and Ti-6Al-4V dissimilar joints, *Materials and Design,* **132**, 188–197, (2017).
4. Tolga Dursun, Costas Soutis, Recent developments in advanced aircraft Aluminum Alloy 6063alloys, *Materials and Design* **56** (2014) 862–871, (2014).
5. MuminSahin, Joining of Aluminum Alloy 6063and copper materials with friction welding, *Springer-Verlag London Limited,* (2010).
6. R.Paventhan,P.R.Lakshminarayanan, V Balasubramanian, Optimization of Friction Welding Process Parameters for Joining Carbon Steel and Stainless Steel, *Science Direct International Journal of Iron and steel*, **19**: 66-71,(2012).
7. Wenya Li, Achilles Vairis, Michael Preuss&Tiejun Ma, Linear and rotary friction welding review , *Taylor &francis International Materials Reviews*,(2016).
8. Nirmal S. Kalsi, Vishal S. Sharma, A statistical analysis of rotary friction welding of steel with varying carbon in workpieces, *Springer-Verlag London Limited*,(2011).
9. S.D. Meshram,T. Mohandas, G. Madhusudhan, Reddy Friction welding of dissimilar pure metals, *Journal of Materials Processing Technology*,184 330–337.
10. J.Romero, M.M. Attallah,M. Preuss, M. Karadge S.E. Bray Effect of the forging pressure on the microstructure and residual stress development in Ti–6Al–4V linear friction welds, *Science direct Acta Materialia*, **57**, 5582–5592, (2009).
11. EmelTaban, Jerry E. Gould, John C. Lippold, Dissimilar friction welding of 6061-T6 Aluminum Alloy 6063and AISI 1018 steel: Properties and microstructural characterization, *Materials and Design* **31** , 2305–2311, (2010).
12. Mumin Sahin, Evaluation of the joint-interface properties of austenitic-stainlesssteels (AISI 304) joined by friction welding, *Mater. Des.* **28**, 2244–2250, (2007).
13. Fukumoto S., Tsubakino H., Okita K., Aritoshi M., Tomita T., Friction welding process of 5052 aluminium alloy to 304 stainless steel. *Materials Science and Technology,* **15(9)**, 1080-1086,(1999).
14. Li P,Li X, Xiong J, Zhang F, Liang L, A study of the mechanisms involved in initial friction process of continuous drive friction welding, *Journal of Adhesion Science and Technology* **29(12),** 1246-12,(2015).
15. Lee WB, Bang KS, Jung SB, Effects of intermetallic compound on the electrical and mechanical properties of friction welded Cu/Al bimetallic joints during annealing, *J Alloy Compd,* **390(1–2)**, 212–219,(2005).
16. M. Sahin, Evaluation of the joint-interface properties of austenitic-stainless steels (AISI 304) joined by friction welding, *Mat. Design,* **28(7)**, 2244–2250, (2007).
17. C. Shanjeevi, S. Satish Kumar, P. Sathiya, Multi-objective optimization of friction welding parameters in AISI 304L austenitic stainless steel and copper Joints, *Proc. Inst. Mech. Eng.* Part B **230**, 449–457, (2014).

Statistical Method to Extract Evoked Potentials from Noise

Besma Benchabane[1], *Moncef Benkherrat*[2]* *and Salah Djelel*[3]
[1]LASA, Badji Mokhtar-Annaba University, Algeria
[2]ECAM-EPMI, 13 Boulevard de l'Hautil, Cergy-Pontoise, France
[3]Université 20 août 1955, Skikda, Algeria

Abstract. Evoked Potentials are induced by visual or auditory stimulation. The Evoked Potentials represent transient electrical activities of some limited brain regions. The signal-noise ratio (SNR) of the EPs is typically around -10 dB. In order to study brain activities related to information processing in the brain, one has to "extract" the single EPs from the noise. We propose a method does not require a priori information concerning the characteristics (time, frequency) of the signal and does not use a template. The method proposed in this work use the wavelet transform associated with a statistical test.

1 Introduction

The evoked potentials (EPs) are discrete signals embedded in the spontaneous Electroencephalography activity (EEG). Their extraction from noise requires a repetition of the recording. The Visual or auditory stimulation trigger the acquisition system and then collect the "evoked potential". The evoked potential differs from spontaneous nerve activity (EEG) because it is synchronous with a triggering "event". In practical terms, the signal triggering the event is used to the acquisition of the evoked signal. The evoked potential (PE) is defined as a transient variation of the electrical potential of a limited region of the brain relative compared to another electrically neutral region. The EP is captured by an electrode placed in the electric field emitted by the active structure and compared to the potential detected by a so-called "reference" electrode. When the reference electrode captures encephalic nerve activity, the sensor system is called bipolar. On the other hand, when the reference electrode is located on an area without cerebral activities for example on the ear lobe, the sensor system is called monopolar.

In the best case, the evoked potential (PE) of which we have just seen the interest is captured very far from its source with an amplitude so small that it does not exceed ten micro-volts. Moreover, it is embedded in a continuous brain activity (the EEG is higher than 100 microvolts) also captured by the electrodes. The PE is sometimes lower than the background of the amplifiers.

It is therefore necessary to extract PE from background noise before examining its characteristics. The classical method used for 40 years is average method. This method consists of the average of the synchronous successive responses.

The Evoked Potential is a brain activity that evolves according to the attention of the subject, therefore the average is insufficient to study it satisfactorily. The average is not representative of the reaction of the subject during the accomplishment of a task and the difficulty to realize it. In order to extract the single variations of the recorded signals, it is then necessary to look at alternative methods more relevant than the averaging.

For this purpose, some techniques have been proposed. They consider the event-related potential as a stationary signal. These methods use parametric modelling [1-3]. Some researchers [4, 5] apply a digital filter to the EP signal in order to improve his SNR. The frequency band is optimized from a spectral analysis of a series of averaged response subsets. Benkherrat [6] applies a series of successive digital filters called a bank of filters to improve the SNR of the single ERPs signals. This method combines the properties of the signal in the frequency and time domain to build the reference signal. It first calculates the average of the records, and then divides the average signal into potentially useful segments. Subsequently, it calculates the spectrum for each segment, which uses it to build the band-pass of the filter associated with the studied segment.

The method proposed in this work use the wavelet transform associated with the statistical method. Our method does not require a priori information concerning the characteristics (time, frequency) of the signal and does not use a template like the methods mentioned above.

The performances of the method will be presented after its application on the synthetic and real signals.

First, we defining the multiresolution wavelet transform. Second, we will describe our method which combines a statistical test and multiresolution wavelet transform.

2 Multiresolution Wavelet Transform

The aim of the multiresolution wavelet transform [7, 8] is to describe, by providing information on the local

*Corresponding author: m.benkherrat@ecam-epmi.com

regularity, the evolution of the signal over time at different time scales. The multiresolution wavelet transform is based on a simple principle. We consider a function whose main characteristic is to be of limited duration in time. This basic function is called the analyzing wavelet. The wavelet transform consists, at a given time position, in compressing or expanding the analyzing wavelet by a scale factor, and then calculating the product between the wavelet and the signal at each scale factor. The wavelet coefficient is higher when the variation occurring in the signal is at the same scale as the wavelet. Thus, the variation of very short duration will be detected at a very small scale and conversely. The variation of long duration will be detected at a high scale.

The Multiresolution Wavelet Transform is written as follows:

$$T_s(a,b) = \frac{1}{\sqrt{a}}\int_{-\infty}^{+\infty} \psi\left(\frac{t-b}{a}\right) s(t)\, dt \tag{1}$$

Where ψ is the analyzing wavelet, a is the scale and b is time. The applications of the wavelet transform are numerous: multi-scale analysis, the detection of singularities in a signal, compression of the signal and denoising. The last application is the one that interests us in this work.

3 Algorithm of decomposition

We use the Mallat decomposition algorithm [9]. It allows a rapid decomposition of the signal at different scales, as well as a reconstruction of the signal. The decomposition is carried out by a cascading algorithm proceeding by successive filtering.
The signal to be analyzed is projected on two orthogonal sub-spaces.

$$f = T_{-1}(f) + D_{-1}(f) \tag{2}$$

Where $T_{-j}(f)$ is called the approximation of the level -1, and $D_{-j}(f)$ is called the detail of level -1.
By noting (respectively) the sequence of the coefficients of the approximation (respectively of the detail) of level -j, one passes to the next level by:

$$T_{-j-1,k} = \sum_{n \epsilon Z} T_{-j,n}\, h(n-2k) \quad \forall\, k \in Z \tag{3}$$

$$D_{-j-1,k} = \sum_{n \epsilon Z} D_{-j,n}\, g(n-2k) \quad \forall\, k \in Z \tag{4}$$

Where h and g are defined by:

$$h(x) = \sqrt{2}\int_{-\infty}^{+\infty} \Phi(2s-x)\, \Phi(s)\, ds, \quad \forall x \in R \tag{5}$$

$$h(x) = \sqrt{2}\int_{-\infty}^{+\infty} \Phi(2s-x)\, \psi(s)\, ds, \quad \forall x \in R \tag{6}$$

We passing from the sequence T_{-j} to the sequence T_{-j-1} (respectively D_{-j} to D_{-j-1}) by filtering then under-sampling (decimation) by keeping only the even terms of the result obtained.

4 Algorithm of denoising

The problem is to distinguish the noisy coefficients from the coefficients related to the useful waves. The proposed method is based on two principles. The first is the standard deviation of the noise coefficients is larger than the standard deviation of the signal coefficients. The second, the orthogonal wavelet transform, compresses the signal energy into a relatively small number of large coefficients; it arranges the signal in some compartments. The energy of the noise is dispersed over the whole transform, and gives small coefficients; in wavelet space, noise and signal dissociate [10]. By averaging the coefficients of the wavelet decomposition, the coefficient of variation (CV) can be calculated. The coefficient of variation [11] is the ratio between the standard deviation of the coefficients of the wavelet decomposition and the average of these coefficients. Thresholding can be applied as a function of the value of the coefficient of variation to separate the signal from the noise. A noisy coefficient gives a large CV and inversely for a less noisy coefficient. We note the recorded signals $y_{1,N}, y_{2,N}, y_{3,N} \dots y_{M,N}$ with M the total number of signals and N the number of samples per signal. Each signal is decomposed into five levels of detail D_j $(j = 1:5)$ and the approximation T_5. We denote by $C_i(n)$ the coefficients of the wavelet decomposition at the level i and n represents the number of the coefficient. Let CM_i^n the average of the coefficients at the discrete instant n for the level on the M signals.

$$CM_i^n = \frac{1}{N}\sum_{n=1}^{N_i} C_i(n) \tag{7}$$

With N_i the number of coefficients of level i. We can estimate the standard deviation σ_i for each level i and for all coefficients at the discrete instant n of the M trials. When the standard deviation is high the coefficients of the number at the discrete instant n are noisy and conversely for less noisy coefficients containing a useful signal. We assumed that the distribution of the coefficients for each number n follows a normal distribution, hence the thresholding condition [12]:

$$C_{i(n)} = \begin{cases} C_i(n) & if\ |CM_i^n| \geq \dfrac{1.96\,\sigma_i}{\sqrt{M}} \\ & 0 \end{cases} \tag{8}$$

 https://doi.org/10.1051/e3sconf/202017002005

4 Simulation data

In order to assess the performance of our method, we constructed 60 simulation signals. Each signal contains 512 samples and a sequence of waves. The latency and amplitude of these waves are varied to simulate the variation of amplitude and latency of the waves of the event-related potentials. To generate the noise, we use the random generator with uniform distribution. We build 60 different sequences of the noise. This noise is added to the synthetic signal with different SNR. The SNR varies between 0 dB and -10 dB. To evaluate the performance of the method described above; we compared it to digital low pass filtering. We used a Butterworth low pass digital filter. To compare the two methods: wavelet method and low pass digital filtering. We used two parameters. The first parameter is the improvement of the signal-to-noise ratio (SNR). The second parameter is the mean squared error (MSE). The MSE is calculated between noisy signals and denoised signals by using our method and digital low pass filter. In figure 1, the improvement of the SNR is higher for wavelet method (red curve) than for low pass digital filtering (blue curve). The difference between the two curves is 4 dB. The maximum of the SNR enhancement is 6 dB for the wavelet method while it is less than 0 dB for the low pass digital filtering method. These results show the superiority of the wavelet method compared to digital filtering. Figure 2 compares the MSE of the low pass digital filter (blue curve) and the wavelet method (red curve). The MSE is lower for wavelet method than that of low pass digital filtering. This result confirms the previous results and shows the superiority of our method.

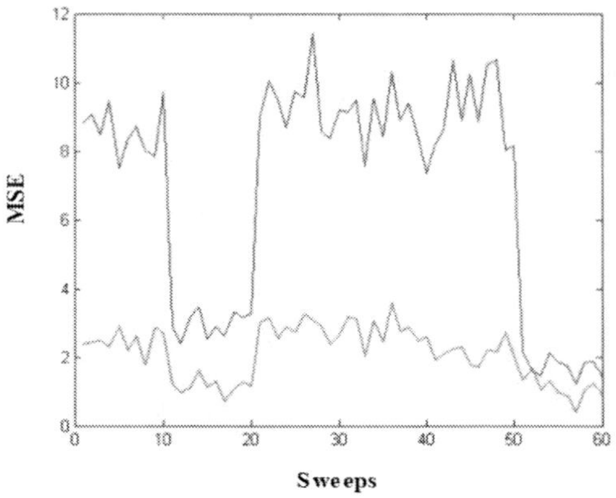

Fig. 2. Comparison of the MSE between wavelet method (red curve) and digital filtering (blue curve).

The signals are digitized with a sampling frequency of 256 Hz. Figure 3 represents the average of the signals denoised by using our method (blue curve), low pass digital filtering (red curve) and the average of the noisy signals (black curve). The number of signals averaged is 103. The signal filtered by the digital low pass filter is less smooth. We can also see that the pre-event interval (-0.5 s to -0.2 s) that contains the noise has been removed. The post-event interval (0.4s up to 0.5s) was also removed. This proves that the wavelet method (standard deviation thresholding) retains only the useful signal between 0s and 0.5s.

5 Real data

We have tested the wavelet method (standard deviation thresholding) and low pass digital filtering on real signals. These signals were recorded using a visual stimulation protocol.

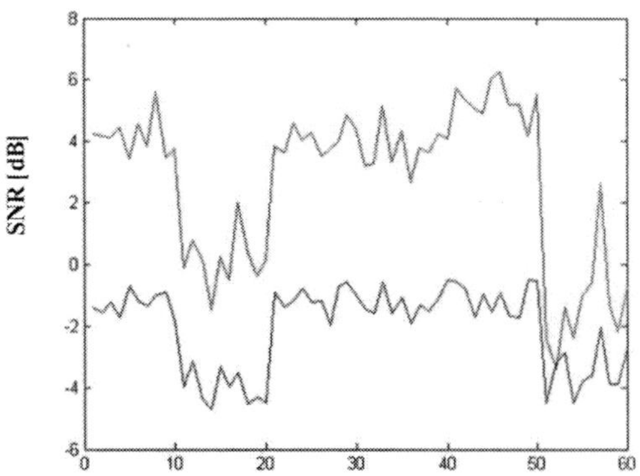

Fig. 1. Comparison of the improvement of the SNR between the wavelet method (red curve) and the digital filtering (blue curve).

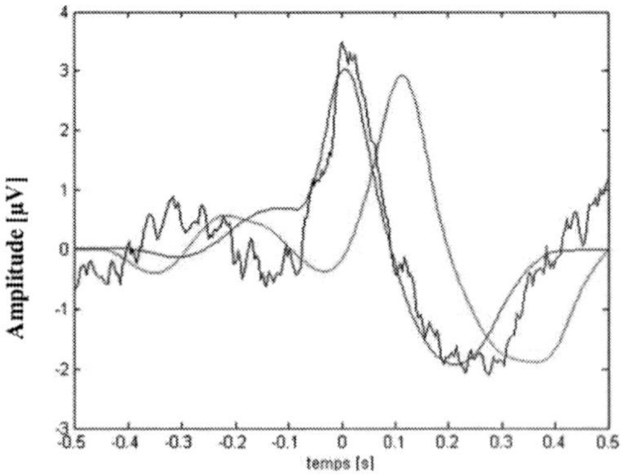

Fig. 3. The average of the noisy signals (black curve), the average of the denoised signals by using wavelet method (blue curve) and by using digital filtering (red curve).

6 Conclusion

Estimation of individual signals is a difficult problem because of the lack of information about signal

characteristics and low signal-to-noise ratio. We have proposed a method that can be applied without a priori. This method is based on the coefficient of variation. In this study, we show that the wavelet method (standard deviation thresholding) is more robust than the low pass filtering method in estimating of the single ERPs. Using two parameters of comparison, the mean squared error and the improvement of the signal-to-noise ratio. These two parameters demonstrated the superiority of the wavelet method compared to the classical method of the digital filtering. The low pass digital filtering method is not suitable to extract ERPs signals because of the overlap of the ERPs signal spectrum and noise (EEG). The enhancement of the SNR of the individual signals in the case of the standard deviation threshold method will allow us to estimate the average with a smaller number of signals. The temporal and frequency localization thresholding method can be applied in the case of signals whose temporal characteristics are stable along a recording session. The standard deviation thresholding method is simple to implement and can be extended to other signal processing applications.
Digital signal processing is an important stage for application, like the automatic diagnosis of neurological disease. The biosignal processing is very useful for telemedicine of the future. Filtering and denoising the neurophysiological signals is an integrate part of system like brain-computer interface (BCI). The BCI is useful for severely disabled people. This technology participates in the social integration of people with disabilities and therefore to build the city of the future with more integration of these people.

References

1. S. Cerruti, G. Chiarenza, D. Liberati, P. Mascellani, G. Pavesi, ' A Parametric method of identification of single-trail event-related potentials in the barin', IEEE Trans. Biomed. Eng., **35**, 701-710, (1988)

2. S. Cerruti, G. Baselli, D. Liberati, G. Pavesi, ' Single sweep analysis of visual evoked potentials trough a model of parametric identification', Biol. Cybern., **56**, 111-120, (1987)

3. M. Hansson, T, Gänsler, G. Salomonsson, 'Estimation of single event-related potentials utilizing the Prony method', IEEE Trans. Biomed. Eng., **43**, 973-981, (1996)

4. J.R. Boston, P.J. Ainslie, 'Effects of analog and digital filtering on brain stem auditory evoked potentials', Electroencephal. Clin. Neurophysiol., **48**, 361-364, (1980)

5. J. Fridman, E.R. John, 'Application of digital filtering and automatic peak detection to brain stem auditory evoked potential', Electroencephal. Clin. Neurophysiol., **53**, 405-416, (1982)

6. M. Benkherrat, 'Analyse et traitement des potentiels évoqués cognitifs en temps réel', thèse université Paris XII, (1998)

7. Y. Meyer, S. Jaffard, O. Rioul, ' L'analyse par ondelettes', Pour la science, **119**, septembre (1987)

8. O. Rioul, P. Duhamel, ' Transformée en ondelettes discrètes et continues : comparaison et algorithmes rapides', Actes du treizième colloque Gretsi, Juan Les Pins, 193-196, septembre (1991)

9. S. Mallat, 'A theory for multiresolution signal decomposition : the wavelet representation', IEEE Trans. Pattern Anal. Machine Intell, **2**, 674-693, (1989)

10. B.B. Hubbard, ' Ondes et ondelettes la saga d'un outil mathématique', Pour la science diffusion Belin, (1995)

11. D. Schwartz, 'Méthodes statistique à l'usage des médecins et des biologistes', Flammarion (1972)

12. M. R. Spiegel, 'Statistique', série schaum, (1993)

Acquaintance with Natural Language Processing for Building Smart Society

Ravindra Parshuram Bachate[1*], *Ashok Sharma*[1]
[1]School of Computer Science & Engineering, Lovely Professional University, Phagwara, Punjab, India.

Abstract. Natural Language Processing (NLP) deals with the spoken languages by using computer and Artificial Intelligence. As people from different regional areas using different digital platforms and expressing their views in their spoken language, it is now must to focus on working spoken languages in India to make our society smart and digital. NLP research grown tremendously in last decade which results in Siri, Google Assistant, Alexa, Cortona and many more automatic speech recognitions and understanding systems (ASR). Natural Language Processing can be understood by classifying it into Natural Language Generation and Natural Language Understanding. NLP is widely used in various domain such as Health Care, Chatbot, ASR building, HR, Sentiment analysis etc.

1. Introduction

Languages which are spoken by human being called as a Natural Languages. There are many resources which are generating data in natural languages on daily basis. For example, Facebook, twitter, Instagram, blogs etc. are generating a big data which is very much difficult to process with traditional approaches. Today's world is data driven and we must work on techniques which will improve the living standard of human being. To achieve this, Natural Language Processing (NLP) plays a vital role because we will deal with user's expressions and feelings expressed on various platforms. A lot of research has been done on Natural Language Processing (NLP) but it is limited to some widely used languages only e.g. English. If we think about the India, around 70% of Indian population lives in rural area where they are not having good understanding of English language. If we want to make their life better using NLP, we need to work on regional languages like Marathi, Hindi, Punjabi and so on. In India there are 22 official languages are spoken and more than 1000 dialects are spoken [1]. These regional and dialect languages belongs to different language families like Indo-Aryan, Dravidian, Austric, Tibeto-Burman and other [1].

To work with these family of languages, we need to understand the representation of it. Natural Language can be described in two parts – Natural Language Processing and Natural Language Understanding. Natural Language Processing deals with many tasks such as Named Entity Recognition (NER), part of speech tagging (POS), text categorization, syntactic parsing, conference resolution, machine translation and question answering. Whereas Natural Language Understanding deals with relation extraction, semantic parsing, question and answering, sentiment analysis, summarization, paraphrase and natural language interface and dialogue agents. While building any ASR system or text to speech system or speech to text system, NLP and NLU is used.

2. Historical Review of Natural Language Processing

Historical review of natural language processing will help us to understand journey of progress of natural language processing in last 10 decades. In 1906, Professor Saussure thought about natural language as a science and since then people started doing research on it. He did a research about natural language and delivered lectures on it during 1906 to 1911. After his death, his colleagues Albert Sechehaye and Cherles bally written a paper entitled "Cours de Linguistique" in 1916. Alan Turing in 1950 written a revolutionary paper on Thinking Machine. In 1957, Noam Chomsky written a book on Syntactic structure[2]. In 1964, US established "Automatic Language Processing Advisory Committee (ALPAC)" and initiated research project in 1966 in collaboration with National Research Council on NLP and AI. Until 1980s, people were using handwritten rules for processing natural language, but it is difficult to write handwritten rules every time. To deal with this problem, statistical models were introduced in 1990s. N-gram became very powerful due to its use in recognizing and tracking clumps of linguistic data.

LSTM Recurrent Neural Network model introduced in 1997 and wide use of this algorithm for voice and text processing started around 2007. In 2001, first feedforward neural network was proposed by Yoshio Bengio. Apple

*Corresponding author: **bachateravi@gmail.com**

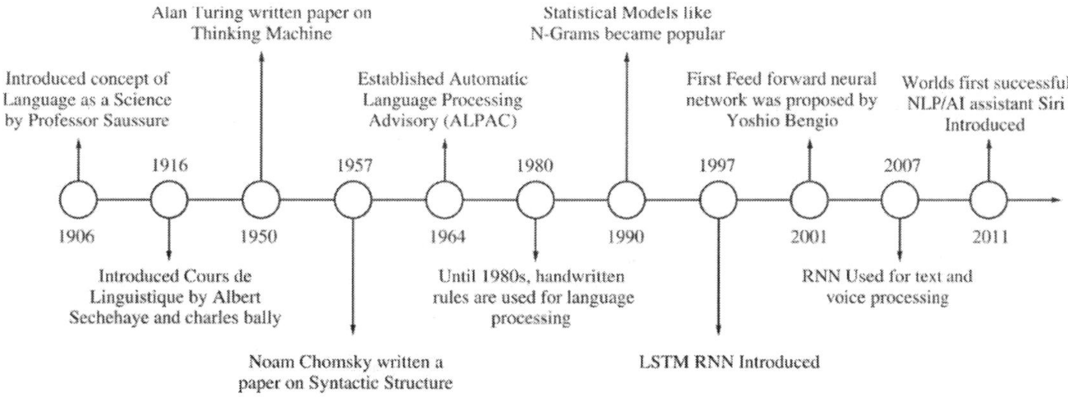

Fig.1: NLP History

introduced Siri which became world's first NLP/AI based assistant. Later other companies also introduced assistant application such as Googles Assistant, Amazon's Alexa, Microsoft's Cortona etc. Artificial Intelligence and Machine learning branches adds significant contribution in expanding application area, effectiveness and efficiency of natural language processing for the society. Automatic Speech Recognition and Understanding is a branch of Natural Language Processing which is widely used in applications such as Siri, Alexa etc.

3. NLP Architecture

Understanding Natural Language is not easy due to various kind of ambiguity exists while processing natural language as many factors influence on it such as gender, surrounding, voice clarity, mic used etc. Natural Language Processing is carried out in various phases or tasks which are described in Figure 2. Each phase having its own problems occurred while processing natural languages due to ambiguity present in input.

3.1 Lexical Analysis / Phonetical Analysis

Input for this phase can be text or voice. Based on type of input, respective analysis will be done in this phase. Text is taken as an input for lexical analysis and it generates tokens for the next phase. For phonetic analysis, sound is taken as an input and it classifies sound based on its physical properties which considers transmission, articulation and reception of speech sounds. Phonetic analysis is required to convert speech into text. Phonetic variation is a big challenge as it varies as user changes[3].

3.2 Morphological Analysis

In morphological analysis, basically it analyses both words and non-words. After analysing, it is segregated into two parts i.e. words and non-words such as semi colon. Morphology is used to describe the relationship between word forms and lexical forms[4].

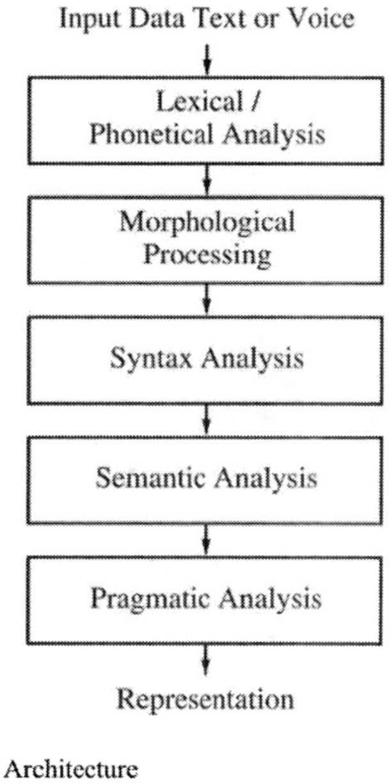

Fig. 2: NLP Architecture

3.3 Syntax Analysis

Syntax analysis involves two tasks-grammar checking and parsing input[4].The first task i.e. grammar checking involves checking correctness of input and second task deals with parsing input statement and arranging their relationship among the word. Parser generates three kind of structure.

3.4 Semantic Analysis

Semantic analysis deals with finding the meaning of statements. It helps to understand the meaning of natural language so that it can be used for other domains such as Artificial Intelligence, Machine Learning, Data Analysis, Speech Understanding etc. Same statement can be interpreted more than one meaning while doing semantic analysis.

3.5 Pragmatic Analysis

The same statement may have more than one meaning found in semantic analysis while interpreting its meaning. Now it is difficult to choose which meaning is applicable into respective context of that statement. The last phase of Natural Language Processing (NLP) chose the correct meaning of the statement with respect to context of use.

4. NLP for Smart Society

Due to NLP's strong ability of making machines to work independently, it is widely used in all possible areas or domains in the world. The applications of NLP are shown in Figure 3.

4.1 Question & Answering

Question and Answering is one of best application of Natural Language Processing. It will be greatly useful for smart societies to have such system to whom we can ask any question and get a answer in fraction of second. Applications such as Siri, Google Assistant, Cortana and Alexa are the best examples of developing QA system. Developing QA framework needs intense NLP work to be done to achieve the greater accuracy[5]. Q & A Systems belongs to AI and NLP domains. Such systems are designed to answer humans' questions by machine. It is useful for building Human Computer Interaction Systems.

4.2 Automatic Speech Recognition

Building Automatic Speech Recognition (ASR) Systems requires natural language processing. To analyses and understand the speech is a challenging task. To deal with this, natural language processing is used in ASR systems. Applications like Siri uses ASR for recognizing the speech of user while developing Human Machine Interaction Systems[6]. ASR systems are used in various products where commands to that product is supposed to given by spoken language such as operating music systems, smart home systems, handling different car operations etc.

4.3 Machine Translation

Machine translation is an application of NLP which translates one language text into another language e.g. Marathi to English and vice versa by keeping meaning of statements same[7]. Basically, there are two types of Machine Translations – Bilingual and Multilingual. Bilingual Machine Translation systems translates between two languages only whereas Multilingual systems translates text into more than one language. Such translations help people of different regions who speak different languages but still wants to communicate with each other.

4.4 Chatbot

A chatbot can be defined as a software program implemented using Artificial Intelligence and Natural Language Processing intended to interact with human like a human. Chatbot gives humans answer on chat and user doesn't feels like that it is a computer program. Many domains including banking, e-learning, telephone uses chatbot to give 24 * 7 support to their customer. Home automation can be done using IoT and Chatbot such as Amazon Alexa, Mi which internally uses Natural Language Processing[8].

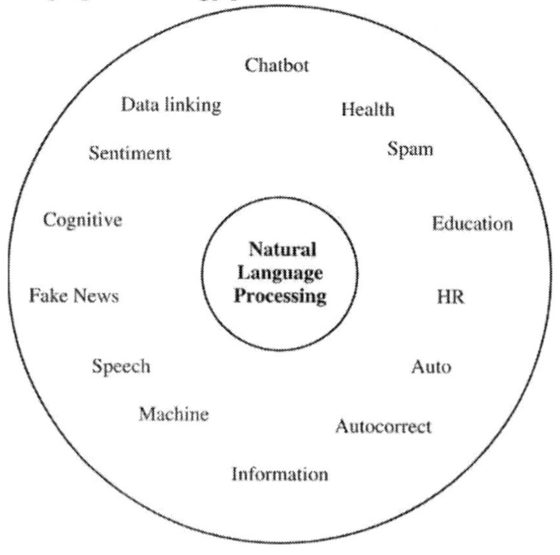

Fig. 3: NLP Applications

4.5 Education

Natural Language Processing can bring a very positive change in education system[9]. As knowledge in the world is exist in various languages, NLP can make that knowledge available and understandable to students with the help of NLP's different applications. To improve e-learning systems in education, NLP will help a lot as knowledge will be available to seekers in their own spoken language. And it is proved that seekers can understand and remember the things, if they learn it in their own language.

4.6 Fake News Detection

Due to tremendous growth of using social media, reaching to people gets more easier than ever. It has both pros and cons too. People having negative mindset posts fake news using social media platforms such as Facebook, twitter, Instagram etc which spreads hatred[10]. Finding fake news on social media is a priority task for social media companies as they are partially responsible for it. So fake news detection is one of the applications of NLP which is used in all social media companies to detect and remove fake news from their platforms.

4.7 Sentiment Analysis

As name suggest, this is used to identify the sentiment of people on various platforms. To improve any business or wants to understand users view about the product or system, Sentimental analysis plays a vital role. Natural language processing is used to understands the reviews of user and then sentiments of user are identified. Sentimental analysis is used to judge whether people liked movie or not. Also, it can be used for identifying satisfaction level of users.

4.8 Health Care

NLP can be widely used in health care sector. To learn adverse effect of drugs or to understand the purposes of drug Natural language processing can be used[11].NLP also can be used for various purposes in health care domain such as clinical trial matching, clinical documentation, clinical decision support, to develop computer assistance system for patients etc.

4.9 Spam Detection

Now days, people are getting many spam mails in their mail box which distracts the user form important mails and sometimes users get trapped in fraudulent activities. Spam mails can be defined as a bulk mail sent to large number of audiences in an intention to do the marketing of their products or to do fraud by collecting users' details[12]. So, it is a big challenge for all mail service providers to protect their users from such spam mails. By implementing natural language processing in mail program, mail service provider can find and segregate spam mails from important mails.

4.10 Text Auto Correction

Natural language processing has another application which corrects the spelling if typed wrong. It helps people to write the content without any fear and tension as it reduces the risks of doing spelling mistakes. It improves the correctness of document. NLP also can be used for auto completion of statement based on the analysis of previous typing history. It reduces cognitive load and typing work of user.

5. NLP Tools

There are many NLP Tools available for its implementation. These tools are listed technology wise in figure 4. Main three languages considered here to make a list of NLP tools i.e. Python, Java and Node which are most popular and widely accepted programming languages in the world. Python includes NLP tools such as NLTK, Spacy, TextBlob, PyTorch, Texacy. Java has CoreNLP, StanfordNLP, OpenNLP and CogCompNLP whereas Node has Compromise, Nlp.js, Retext and Natural NLP tools.

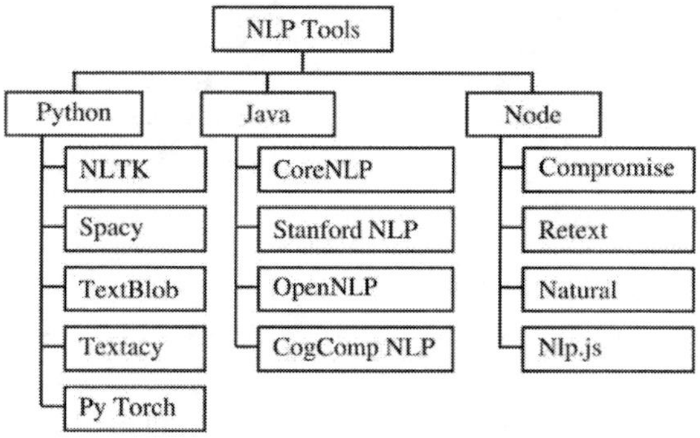

Fig. 4: NLP Tools

6. Conclusion

This paper discusses the Natural Language Processing in all possible aspects of it for understanding its use for making our society smart. Natural Language Processing research work and its implementation for regional languages which are spoken in respective part of India will help to improve the different domains associated with respected area. The various applications of NLP will help to build smart society or city with its effective and efficient use while developing products for it.

References

1. MHRD, Indian Linguistic. (1995).
2. N. Chomsky, *Syntactic Structures.* PARIS: MOUTON PUBLISHERS, (1957).

3. D. Williams and P. Escudero, "Detecting phonetic variation versus phonemic differences," pp. 265–269.
4. A. Reshamwala, D. Mishra, and P. Pawar, "Review on natural language processing," no. February, (2013).
5. G. Bouma, I. Fahmi, J. Mur, and G. Van Noord, "Linguistic knowledge and question answering," vol. **46**, pp. 15–40, (2005).
6. R. Weiscbedel, J. Carbonell, W. Lehnert, M. Marcus, and R. Perrault, "White Paper on Natural Language Processing," pp. 481–493.
7. M. T. Makwana and D. C. Vegda, "Survey: Natural Language Parsing For Indian Languages," (2015).
8. C. J. Baby, F. A. Khan, and J. N. Swathi, "Home automation using IoT and a chatbot using natural language processing," 2017 Innov. Power Adv. Comput. Technol. i-PACT 2017, vol. 2017-January, pp. 1–6, (2018).
9. K. M. Alhawiti, "Natural Language Processing and its Use in Education," vol. **5**, no. 12, pp. 72–76, (2014).
10. F. Cardoso, A. Cristina, and B. Garcia, "Can Machines Learn to Detect Fake News ? A Survey Focused on Social Media," vol. **6**, pp. 2763–2770, (2019).
11. T. Ly et al., "Evaluation of Natural Language Processing (NLP) systems to annotate drug product labeling with MedDRA terminology," J. Biomed. Inform., vol. **83**, no. April, pp. 73–86, (2018).
12. D. Khurana, A. Koli, K. Khatter, and S. Singh, "Natural Language Processing: State of The Art, Current Trends and Challenges," no. August (2017).

Anomaly Detection for Predictive Maintenance in Industry 4.0- A survey

Pooja Kamat[1,*] *and Rekha Sugandhi*[2]
[1] Symbiosis Institute of Technology, Symbiosis International (Deemed University),Lavale, Pune, Maharashtra, India
[2] MIT School of Engineering, MIT-ADT University, Loni-Kalbhor, Pune, Maharashtra, India

Abstract. Maintenance and reliability professionals in the manufacturing industry have the primary goal of improving asset availability. Poor and fewer maintenance strategies can result in lower productivity of machinery. At the same time unplanned downtimes due to frequent maintenance activities can lead to financial loss. This has put organizations' thought process into a trade-off situation to choose between extending the remaining functional life of the equipment at the risk of taking machine down (run-to-failure) or attempting to improve uptime by carrying out early and periodic replacement of potentially good parts which could have run successfully for a few more cycles. Predictive maintenance (PdM) aims to break these tradeoffs by empowering manufacturers to improve the remaining useful life of their machines and at the same time avoiding unplanned downtime and decreasing planned downtime. Anomaly detection lies at the core of PdM with the primary focus on finding anomalies in the working equipment at early stages and alerting the manufacturing supervisor to carry out maintenance activity. This paper describes the challenges in traditional anomaly detection strategies and propose a novel deep learning technique to predict abnormalities ahead of actual failure of the machinery.

1 Introduction

1.1 IIoT- Industrial Internet of Things:

IoT or Internet of Things has boomed in recent years. IoT allows machines to communicate with each other over the Internet in real-time. One of the application areas of IoT in the industrial domain is called as Industry 4.0. Currently, Industry 4.0 is undergoing a revolution with innovation at its core for those who wish to revamp the organizational processes in the industry [1]. This has led to the upsurge of the Industrial Internet of Things (IIoT) technology wherein industry equipment is largely driven by sensors and sensor data. For any industry, whether it is an automobile or thermal or any manufacturing industry, it runs on the oil of data that the manufacturer can use to gain meaningful insights into his businesses (e.g. to identify a fault via predictive analytics). Also, data can be considered as a critical asset for structural health checks of the equipment used in the industry.

However, in most of the industrial checks, it is difficult to identify alert rules, since the processes are too complex, and dynamic. Today, inadequate maintenance techniques can reduce an equipment's overall productive capacity by five to twenty percent [2]. Studies carried out recently also show that machine downtime which is unplanned or has occurs at last minute due to some faults costs industrial manufacturing organizations an estimated $50 billion each year [3]. Predictive maintenance is one of the major goals of Industry 4.0 and it depends majorly on Anomaly detection. Anomaly detection or outlier detection is a set of major techniques with an aim to identify anomaly patterns that deviate from normal behavior. At the same time, manufacturing processes generate a large volume of data that needs to be monitored in real-time for generating timely anomaly alerts. This paper outlines the need for anomaly detection, summarizes the current research work in this domain and propose a deep-learning based technique for identifying anomalies in an efficient manner.

The rest of the paper is structured in the following way: The next section describes the evolution and need for Predictive Maintenance followed by Section 2 describing the current state-of-art methodology for Anomaly detection in Predictive Maintenance. Section 3 describes the challenges faced in detecting anomalies for predictive maintenance and Section 4 describes some of the current research work done in the area of anomaly detection. In Section 5 the authors propose a novel anomaly detection framework using the power of Artificial Intelligence, its benefits and also present some of the available datasets as case studies. At the end the authors present a conclusive summary about the need of using AI for predictive maintenance.

1.2 Predictive Maintenance how, what and why?

* Corresponding author: **pooja.kamat@sitpune.edu.in**

 https://doi.org/10.1051/e3sconf/202017002007

It's a common practice for manufacturers who have been carrying out maintenance techniques for years according to their companies' maturity in dealing with machine failures. Figure 1 describes how maintenance activity has evolved with changing technology:

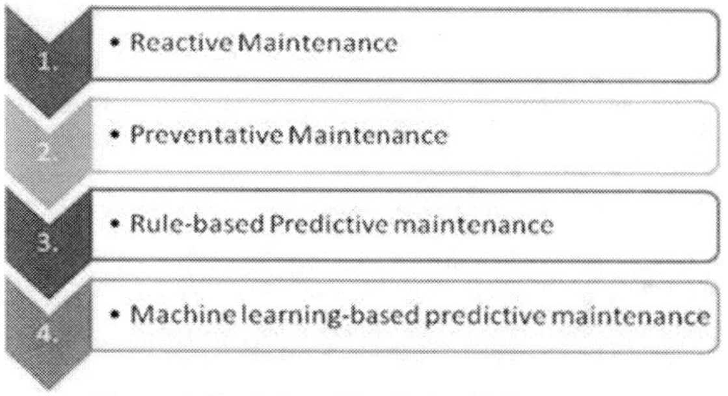

Figure 1: Evolution of Predictive Maintenance

Predictive maintenance has evolved in a big way that can be characterized as follows:

- **Reactive Maintenance** – This basically allows you to repair after the failure has occured.
- **Preventative maintenance** - This involves carrying out regular maintenance activities to avoid failures. However, many of the times it is done without the actual need of doing it. For example replacing an equipment part after 6 months when it could have run successfully for at least 10 more cycles.
- **Rule-based predictive maintenance** – Maintenance is carried out on basis of hard-coded threshold rules and alert is sent in case a measurement crosses the thresholds.
- **Machine learning-based predictive maintenance** – Makes the use of advanced analytics and machine learning techniques to predict when the next failure would take place and accordingly carry out maintenance ahead of time.

Some of the benefits of Predictive Maintenance (PdM) are as follows:

- Avoid unplanned downtime
- Reduce planned downtime
- Increase productivity
- Optimize the use of maintenance resources
- Improve asset health and performance
- Increase customer satisfaction

2.Anomaly Detection for Predictive Maintenance:

2.1 Need for Anomaly Detection

One of the biggest issues faced by the manufacturing industries is their processes can identify only 20 % anomalies beforehand [4]. This means that most of the time the anomalies go undetected and the industry are unable to handle them due to ineffective anomaly detection techniques.

Anomaly detection is the process to identify variables or items that do not belong to an expected pattern in the same dataset and is usually unobservable to human eye. Such anomalies which we can term as early signs of failure usually result into equipment breakdown or lead to faults in the working of the equipment. Anomaly detection is tough because:

- The current knowledge is not enough to define a class.
- Often, no examples are available in the data to describe the anomaly accurately.

Example: A worn-out roll bearing that leads to the breakdown of the machine.

2.2 Anomaly Detection for Condition Monitoring in Machines:

A rotating machine such as a pump or a steam turbine or a non-rotating equipment such as a distillation column will eventually start deteriorating due to regular use and starts generating abnormalities which we can term as 'anomalies' here. That state should not be considered as a case of total shutdown but should be considered as a warning that the machine is not in its optimal state and requires to undergo a maintenance activity.

Condition Monitoring or Statistical Process Control which was introduced in 1924, is an effective statistical technique for quality control and assurance in manufacturing domain. Data in the form of machine sensor measurements are recorded in real-time during the manufacturing process and a graph is then plotted with predetermined threshold limits that reflect the capability of the process. The data points that falls within the thresholds indicates that everything is normal and working operating as expected with some natural variation that is expected as part of the process. If data points fall outside of the threshold limits, this signifies an assignable issue that should be addressed at the earliest before faults occur. In this way, SPC is an efficient method to detect anomalies at early stages. Figure 2 describes the typical lifecycle of condition monitoring in the manufacturing industry.

https://doi.org/10.1051/e3sconf/202017002007

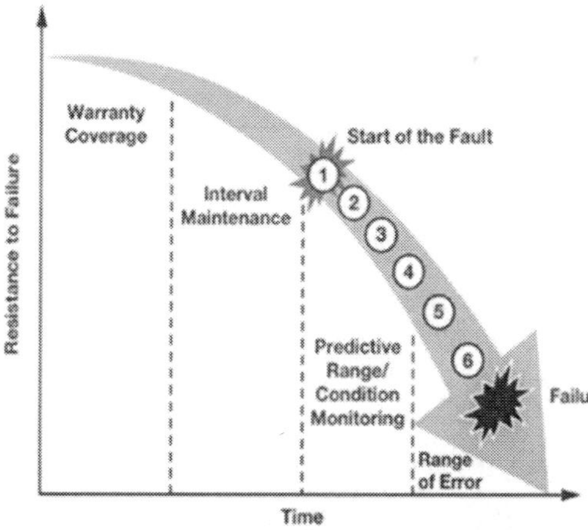

Figure 2: Condition Monitoring in machines [5]

The author Thomas Brand (2017) in his article points out the use of some of the current analytical techniques such as time-based and frequency-based techniques for condition monitoring using vibration sensors [5]. The time domain based statistical analysis gives more idea about the vibration degree of the system over a period and checks whether it is within the permissible limits. The statistical analysis based on frequency allows monitoring of vibration in the form of the fundamental frequency along with harmonic components for fault diagnosis. The accurate tracking of the basic frequency is especially important because the effective values and other statistical parameters can deviate with speed over time. If the parameters deviate significantly from the last measurement, the basic frequency must be verified to avoid false alarms.

3. Challenges faced by Anomaly Detection:

Researcher Pramit Choudhary (2019) in his article *Introduction to Anomaly Detection* points out some of the issues in Anomaly detection using traditional statistical methods [6]:

- The statistical data contains noise which could be alerted as an anomaly, because the difference between normal and abnormal behavior is mostly not specific.
- The characteristics of anomalous or normal may frequently change hence the average thresholds will change frequently.
- The anomaly pattern is mostly based on seasonality. Hence this requires more dynamic techniques to understand the change in anomaly trends.

Predictive Maintenance (PdM) due to its varied benefits such as significant savior in maintenance costs, reduction in downtimes, etc. has become a sought-after goal for many manufacturers.

Traditionally manufacturers have been carrying out PdM using SCADA systems manually hard-coding thresholds limits, setting static alert rules and other configurations. This kind of semi-manual technique fails to consider the dynamic and complex behavioral patterns of the machinery or is unable to relate to the contextual data relating to the manufacturing process at large. For example, a sensor on an automobile machinery may suddenly record a rise in temperature. A rigid rule-based system would not take into account that this is due to the fact that the machine is undergoing sterilization and would further trigger a false-positive alarm.

Flovik, V. (2018) in his article "How to use machine learning for anomaly detection and condition monitoring" mention challenges faced by the traditional methods of anomaly detection [7]:

- Generation of false positives:

The process of having hardcoded anomaly alarm limits tends to send a huge number of false alerts for conditions which in actual are normal situations for a specific sensor. This unnecessarily wastes the time and efforts of the supervisor overlooking the process and also impacts the availability of the sensor.

- Missing anomaly alarms:

There might also be cases of missing alarms for situations that are actually problematic but are not alarmed. This is a more serious situation as the damage is already done and would lead to repair costs and lost production.

The above two challenges have the same root cause as follows as Flovik V.(2018) states:

"The health of a complex piece of sensor equipment cannot be reliably judged based on the analysis of each measurement on its own" [7]

In Figure 3, author R.Silipo (2019) clearly states that anomaly detection problems do not offer a supervised dataset with labeled classes such as a signal from a normally functioning system and a signal from a system with an analogy. In this case, we can only train a machine learning model on a training set with "normal" examples and use a distance measure between the original signal and the predicted signal to trigger an anomaly alarm [8].

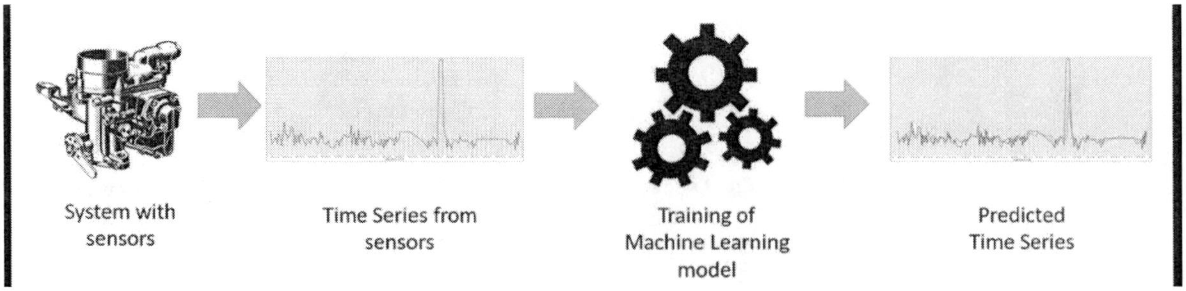

Figure 3: Time Series prediction for anomaly detection [8]

4.Current Research Work in Anomaly Detection

Tang, Z. et al. proposed a unique anomaly detection method for a long- width cable- stayed bridge [9]. Here the descriptions of anomaly patterns were studied using the convolutional neural network (CNN). The authors have implemented a semi-supervised deep learning approach for anomaly classification and further research is needed for unsupervised classification.

Li, X.,et al. in their paper had chief emphasis on data augmentation techniques while carrying out anomaly detection. This was done to get high accuracy even for a small original training dataset as it is difficult to obtain accurately labeled data in real industries [10]. It was observed that individual augmentation techniques achieved better accuracy than when they were used in combination.

Graß, A. et al. proposed a novel Generic Anomaly Detection for Production Lines (GADPL), for detecting anomalies for a real reflow oven using old sensor data which was recorded over a period of more than seven years [11]. However the work did not study the association of anomalies within multivariate data. The work needed further improvement on how one can correlate varied dimension information within a particular dissimilar measure to find anomalies.

Żabiński, T. et al.in their paper propose a platform that utilizes a toolset for intelligent condition monitoring systems for novelty detection and classification depicting an example of CNC milling tool [12]. However the diagnosis of CNC tool head imbalance was not explored in real time.

Liang, Y. C. et al. proposed to use fruit fly optimization (FFO) algorithm to optimize the threshold values based on historical data to achieve more accurate anomaly detection [13]. However different data sources such as vibration, force data, etc. were not fused for better prediction. Deep learning algorithms were not explored.

5. Deep Learning Framework for Anomaly Detection:

There is a strong research needed in case of detecting a true anomaly by considering a blend of several sensor measurements to get an accurate and real warning of the situation as depicted in Figure 4.

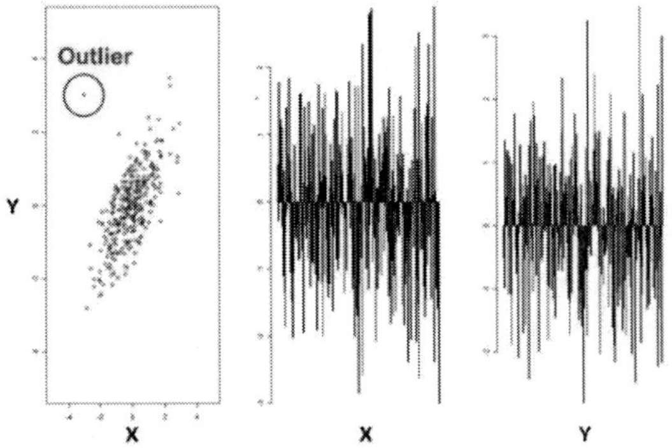

Figure 4: Outlier detection using multivariate anomaly detection techniques [7]

5.1 Proposed Deep Learning technique:

The chief goal of predictive maintenance is to detect anomalies and failure patterns and provide early alerts to the manufacturing personnel. The authors of this paper propose a novel deep learning technique which is proactive in terms that will "learn" and study patterns leading to anomalies and faults in machinery beforehand and proactively warn the personnel about possible issues well ahead of time. Figure 5 describes the steps of the proposed technique. This will benefit the personnel in two ways:

- The warnings and alerts can enable efficient maintenance of the faulty components

https://doi.org/10.1051/e3sconf/202017002007

- The resources required for maintenance can be optimally utilized.

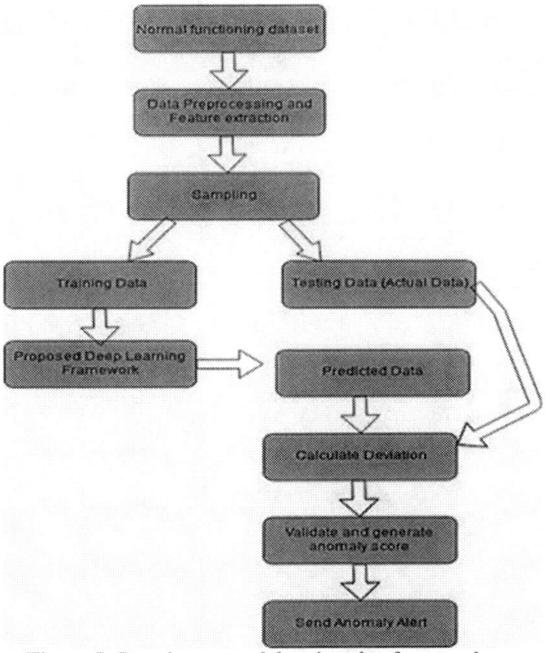

Figure 5: Steps in proposed deep learning framework

5.2 Benefits of proposed Deep Learning technique for Anomaly Detection

The authors propose to use the combination of AutoEncoders and Long Short Term Memory (LSTM) deep learning model variants for effective anomaly detection in manufacturing domain. An autoencoder is an unsupervised artificial neural network and is used to "learn" from input data. Autoencoders are mostly useful for noise removal in data and for dimensionality reduction which is the case in most of the manufacturing sensor data. Autoencoders are helpful for data representation and visualization in situations where the input data has high dimensionality and cannot be plotted easily. Using the dimensionality reduction feature of Autoencoders we can explore the data in 2D or 3D space. Figure 6 depicts a typical Autoencoder neural network.

However most of the manufacturing sensor data are sequential in nature and are collected over a longer period. Prediction over a longer sequence is difficult due to varying lengths of the input sequence. The time-based ordering of the observations can also make it challenging to consider only required features suitable for use as input to supervised deep learning models.

Chalapathy, R., & Chawla, S. (2019) and Mohammadi, M. et. al propose the use of Autoencoders and LSTM deep learning variants for anomaly detection [15][16].

The LSTM (Long short-term memory) architecture consists of an Encoder-Decoder system and can support input sequences of variable length and generate output sequences accordingly. LSTM has the capability to understand the dynamics and complex interactions within the time-series based ordering of input sequences and at the same time store this information in memory.

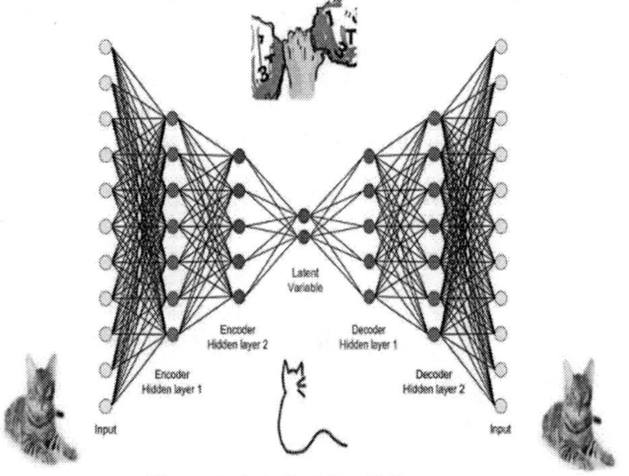

Figure 6: AutoEncoders [14]

5.3 Case Studies/Available Datasets:

Here the authors describe some of the publicly available sample datasets in the domain of manufacturing where the proposed algorithm can be utilized effectively.

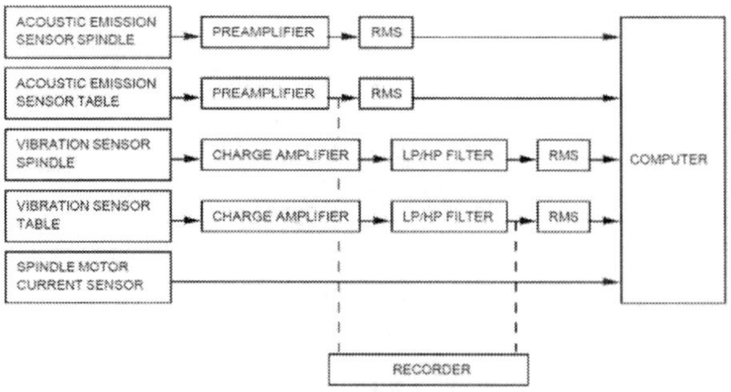

Figure 7: Experimental Setup of Milling machine [17]

5.3.1 Mill Data Set

The data in this dataset describes experiments conducted from various runs on a milling machine under different operating conditions. The primary motive of this experimentation was to investigate tool wear. Data was captured using three types of sensors mainly the current sensor, acoustic emission sensor and vibration sensor. Readings were recorded at different instances of time[17]. Figure 7 depicts the experimental setup of the milling machine.

Need for anomaly detection in investigating tool wear:

In case of milling machine, a sharp tool is always desired to maintain a high-quality product with accurate dimensions and smooth surface finish. A dull one may deform the surface to a great extent and the cost of repair would be high. A dilapidated tool would in turn cause more friction resulting in higher cutting temperatures and other unwanted effects. Therefore, in this case, tool wear can be considered as a true anomaly condition which has to be controlled at the right time before the damage is done. Some of the implications of tool wear are:

- flank wear which can be caused due to friction between the tool and the workpiece.
- The speed of cutting leading faster damage
- intuitive rounding of the cutting edge

Field Name	Description
case	Case Number (1-16)
run	Counter for experimental run in each case
VB	Flank wear measured in VB
time	Duration of experiment
DOC	Depth of cut (does not vary for each case)
feed	Feed (does not vary for each case)
material	Material (does not vary for each case)
smcAC	AC spindle motor current
smsDC	DC spindle motor current
vib_table	Table vibration
vib_spindle	Spindle vibration
AE_table	Acoustic emission at table
AE_spindle	Acoustic emission at spindle

Figure 8: Dataset description of sensors used in milling experiment[17]

The above Figure 8 describes the sensor parameters used in the milling experiment. By analyzing the threshold values of these multiple sensors using deep learning models, an alert regarding possible tool wear can be predicted.

5.3.2 Turbofan Engine Degradation Simulation Data Set:

The turbofan or fanjet is a category of airbreathing jet engine that is widely used in aircraft propulsion as shown in Figure 9. In this case study, engine degradation simulation was carried out using C- MAPSS (Commercial Modular Aero-Propulsion System Simulation). Four different conditions were simulated for different groupings of functioning conditions and fault modes. The simulator captured recordings of various sensor channels to diagnose fault evolution [18].

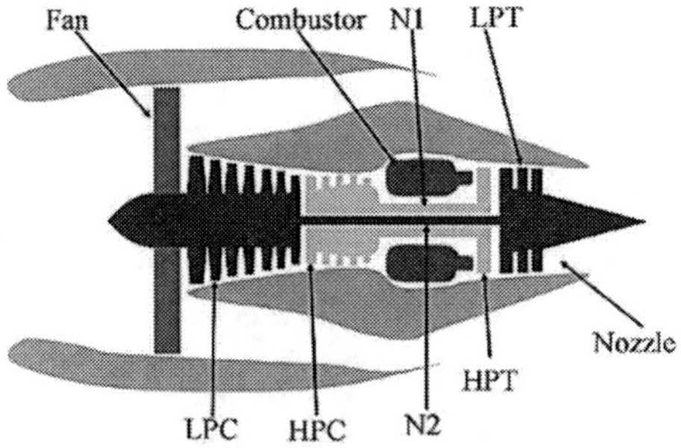

Figure 9: Simplified diagram of engine simulated in C-MAPSS [18]

C-MAPSS is a MATLAB based tool for simulation of large aircraft engines. It provides the user a numerous editable input parameter for controlling the flight engine statistics.

Figure 10 list down some of the input parameters of this simulation engine: The inputs consist of fuel flow and a dataset of 13 health-checking parameter inputs which the user can simulate to study the effects of faults and degradation in any of the engine's five rotating components (Fan, Low Pressure Compressor, High Pressure Compressor, Low Pressure Turbine, High Pressure Turbine) [18].

https://doi.org/10.1051/e3sconf/202017002007

Name
Fuel Flow
Fan efficiency modifier
Fan flow modifier
Fan pressure-ratio modifier
LPC efficiency modifier
LPC flow modifier
LPC pressure-ratio modifier
HPC efficiency modifier
HPC flow modifier
HPC pressure-ratio modifier
HPT efficiency modifier
HPT flow modifier
LPT efficiency modifier
HPT flow modifier

Figure 10: Turbofan Engine Degradation Simulation Input Dataset [18]

Symbol	*Description*	*Units*
Parameters available to participants as sensor data		
T2	Total temperature at fan inlet	°R
T24	Total temperature at LPC outlet	°R
T30	Total temperature at HPC outlet	°R
T50	Total temperature at LPT outlet	°R
P2	Pressure at fan inlet	psia
P15	Total pressure in bypass-duct	psia
P30	Total pressure at HPC outlet	psia
Nf	Physical fan speed	rpm
Nc	Physical core speed	rpm
epr	Engine pressure ratio (P50/P2)	--
Ps30	Static pressure at HPC outlet	psia
phi	Ratio of fuel flow to Ps30	pps/psi
NRf	Corrected fan speed	rpm
NRc	Corrected core speed	rpm
BPR	Bypass Ratio	--
farB	Burner fuel-air ratio	--
htBleed	Bleed Enthalpy	--

Figure 11: Turbofan Engine Degradation Simulation Output Dataset [18]

Need for Anomaly Detection for investigating Equipment wear:

The above scenario describes the simulation data for aircraft engines throughout their usage. An engine unit may be deployed under different usage conditions from one flight to another. The volume and speed of damage accretion will be different for each engine. It is assumed that the amount of damage accumulated will vary depending on various factors during a particular flight and hence, one must depend on sensor information recorded during each flight. This scenario describes deterioration due to equipment wear based on the usage trend of the engines and may not necessarily be due to any particular fault cause. This justifies the need to use deep learning to study anomalies in the various sensor measurements to monitor the flight engine health during or after the flight journey.

6.Conclusion and Summary:

Maintenance and reliability professionals in the manufacturing industry are posed with huge challenges, but the primary goal of any manufacturing organization remains constant throughout i.e to improve asset availability.

Anomaly detection in particular, and predictive analytics as a whole, cannot be an afterthought - it must be implemented in your business goals and planning from the beginning. Extensive research is needed in the development of effective anomaly detection techniques which not only alert manufacturers well ahead of time but also abstain them from carrying out unnecessary maintenance activities resulting in wastage of resources. The current statistical approaches fail to consider combination of various parameters that would lead to anomalies in the working of the equipment. The proposed deep learning framework overcomes this by detecting abnormalities and checking correlations while searching for anomaly patterns across thousands of data feeds which is not possible by manual approach. Also, most of the current research work consider semi-supervised approach for classifying anomalies. The proposed framework makes use of autoencoders variant of deep learning which is an unsupervised learning technique. This will help uncover new classes of anomalies. Artificial Intelligence empowers the Industry 4.0 due to its capability to analyse huge volume of data in real time and generate timely alerts for issues that may arise. The framework will assist in the evaluation of the asset health and behaviour and the system as a whole and help identify component degradation prior to its malfunctioning.

7. Future Scope:

The authors propose to apply this technique in real world manufacturing scenarios in order to check its accuracy in predicting anomalies on real streaming time series data. The algorithm can also be compared against other deep learning model variants for better accuracy.

References:

1. S. T. N. R. Y. Zhong, X. Xu, E. Klotz, Engineering **3(5)**, 616-630. (2017)
2. Deloitte Insights. (2019). A whitepaper on Making maintenance smarter [last accessed on September. 2019].
3. How Manufacturers Achieve Top Quartile Performance. (2017, March 6) [last accessed on September 2019].
4. Loon, R. (2019). How Industrial IOT is Influenced by Cognitive Anomaly Detection - The Digital Transformation People [last accessed 15th July. 2019].
5. Demands on Sensors for Future Servicing: Smart Sensors for Condition Monitoring by Thomas Brand, 2017 [last accessed on September 2019].
6. Choudhary, P. (2019). Introduction to Anomaly Detection, Towards Data Science [last accessed on 1st September, 2019].
7. Flovik, V. (2018). How to use machine learning for anomaly detection and condition monitoring, [last accessed 5th September 2019].

8. R. Silipo,2019 "IoT Anomaly Detection 101: Data Science to Predict the Unexpected", Dark Reading, [last accessed: 23- Sep- 2019].
9. Z. Tang, Z. Chen, Y. Bao, and H. Li, Struct. Control Heal. Monit **26**,(2019)
10. X. Li, W. Zhang, Q. Ding, and J. Q. Sun, J. Intell. Manuf. **31**, 433–452 (2020).
11. A. Graß, C. Beecks, and J. A. C. Soto, Unsupervised Anomaly Detection in Production Lines,TIA Vol. **9** (pp 18-25) Springer Berlin Heidelberg, (2019)
12. T. Zabiński, T. Maoczka, J. Kluska, M. Madera, and J. Sęp, in Procedia CIRP **79**, 63-67, (2019)
13. Y. C. Liang, S. Wang, W. D. Li, and X. Lu, Engineering, **5(4)**, 646– 652, (2019)
14. Extreme Rare Event Classification Using Autoencoders In Keras: Preparing For GRE [last accessed on October 1, 2019]
15. Chalapathy, R., & Chawla, S. (2019). Deep learning for anomaly detection: A survey. arXiv preprint arXiv:1901.03407.
16. M. Mohammadi, A. Al-Fuqaha, S. Sorour, and M. Guizani, IEEE Commun. Surv. Tutorials **20**, 2923--2960. (2018)
17. A. Agogino and K. Goebel (2007). BEST lab, UC Berkeley. "Milling Data Set ", NASA Ames Prognostics Data Repository
18. A. Saxena and K. Goebel (2008). "Turbofan Engine Degradation Simulation Data Set", NASA Ames Prognostics Data Repository.

Modern information and telecommunication systems for medical purposes

Oleksii Serheiev-Horchynskyi[1,*]

[1]NTUU "Igor Sikorsky KPI", Inst. for Applied System Analysis, Dept. of Sys. Design, 37 Prosp. Peremohy, 03056, Kyiv, Ukraine

Abstract. The analysis and classification of modern information and telecommunication systems (ITS) for medical purposes were carried out. Historically established areas of application of medical telecommunication systems were distinguished. Classes of communication and data related to modern medical telecommunication systems were analysed. The role of standards and protocols for data transmission in the process of interaction between the hardware and software components of medical ITSs was defined. Standards for transmitting various types of information were listed; the exploration of the standards for telemedicine information transmission was carried out. Various structural-logical schemes of medical ITSs implementation were illustrated; components of the schemes were reviewed. The importance of development and implementation of a unified interaction model for various medical ITS components, that will allow adapting to extreme characteristics of data transmission physical environment, was figured out.

1 Introduction

The classes of medical ITSs and structural-logical schemes of their implementation vary due to a wide range of used medical data types [1; 2], specifications of the implementation of data transmission protocols, parameters of the hardware implementation, and so on. The underlying issue of effective exploitation of ITSs from the beginning of their development up to the present time is the noise immunity of data transmission [1].

It is clear that in emergencies, specialized medical systems become strategically crucial in the struggle for human life. The general objective of remote medicine is to provide medical services based on ITSs of various classes, demanding that the quality of medical data exchange at all levels including registration, transmission, and storage of medical data is guaranteed [1; 2].

For remote areas, the issue of transmission of telemedicine information about a person's condition over a distance is essential and involves the provision of storage and processing of the medical data [1; 3]. The use of remote access technologies in medical ITSs results in improved quality and speed of providing medical services, especially in areas of social disasters with a damaged medical infrastructure.

2 Classification of Medical Data and Telecommunication Systems

2.1. Areas of Medical ITSs Application

The following historically established main areas of application of information and telecommunication technologies in medicine can be distinguished [3-6]:

- telemedicine consultations;
- patient video surveillance;
- telemedicine of urgent conditions, and emergencies;
- tele-education;
- telesurgery and operative prevention;
- space telemedicine.

The most common application of medical ITSs is remote consultations, which can use both static (video, audio, text) and dynamic (conferencing) information.

2.2. Medical Data

Modern medical conferencing can be segmented into two classes: consultations (point-to-point mode) and councils (multipoint mode), providing for direct communication between the patient, doctor, and consultant in near real-time mode. In both cases, the transmission of medical data requires a stable process at high levels of interference.

Medical data cover a wide range of critical information needed to make informed decisions about a patient's health state. Assessments of the patient's condition are usually performed based on requests to the electronic database of the history of diseases, the results of recent clinical tests, visual data about the patient's condition (radiograms, computer tomograms, ultrasound images, and so on).

The static resources, which are analysed in off-line mode, don't require high-speed communication channel and specialized hardware [1]. Unlike static, providing a conference communication (based on dynamic resources) requires the installation of special equipment (digital video and audio equipment for data acquisition and

*Corresponding author: alexey.sergeev@ymail.com

 https://doi.org/10.1051/e3sconf/202017002008

transmission) with access to broadband communication technologies, the use of specialized software and specialized data exchange protocols [3].

Medical ITSs (systems for dynamic monitoring) allow registering the physiological and biometric parameters of the patient's body and transmitting them to the base station for further analysis. Such systems are already used for patients who require frequent medical screening and continuous physiological monitoring [1; 3].

For data transmission, current medical ITSs use 3G, 4G, Wi-Fi, WiMAX, LTE standards. Satellite systems, including the International Maritime Satellite-Communication System, are used to deliver critical data from remote areas [3].

3 Standards

3.1. Principles of Information Transmission

Essential telemedicine information can be classified as follows [6-8]: documents in electronic form; visual images; recordings of electrical signals; multimedia information; information in the video conferencing mode. The interaction and data exchange between components of ITS are governed by the standards and protocols.

The names of the standards include the abbreviations of the groups, organizations or institutions involved in their development [8]. For instance, ASC X12N deals with external standards for the exchange of electronic documents, ASTM E31.11 – exchange of laboratory test data, IEEE P1157 – exchange of medical data (MEDIX), ACR-NEMA DICOM – exchange of video/audio data.

Modern medical ITSs use various medical and communication hardware, specialized communication interfaces and protocols [9]. There is an urgent need to develop and implement a unified model of interaction of medical equipment, allowing adapting to any characteristics of the physical environment for data transmission.

3.2. Standards for Information Transmission

Below are essential and most commonly used standards for the transmission of information, taking account of their adaptation to medical ITS [10].

Standards for telemedicine data transmission developed by the International Organization for Standardization (ISO):

- ISO 5218 (1977) – exchange of data, representation of the gender of a person;
- ISO 1000 (1981) – units of measurement, SI system;
- ISO 2955 (1983) – processing of data for the representation of SI units and other units in systems with a limited set of characters;
- ISO 8072 (1986) – networking standards;
- ISO 8601 (1988) – data elements and exchange formats, date and time representation;
- ISO 8859 (1988) – data processing, a table of 8-bit single-byte character codes.

Standards developed by the American National Standards Institute (ANSI):

- ANSI X3.30 (1985) – representation of a calendar and ordinal date;
- ANSI X3.4 (1986) – a table of character codes, the American Standard Code for Information Interchange (7-bit ASCII table);
- ANSI X3.43 (1986) – representation of the local time of day for data exchange in information systems;
- ANSI X3.50 (1986) – representation of SI units, traditional U.S. units and other units in systems with a limited set of characters;
- ANSI X3.51 (1986) – representation of universal time, local time shifts and U.S. time zones for data exchange.

Standards developed by the American Society for Testing and Materials (ASTM):

- ASTM E31.12 (1990) – representation of clinical tests;
- ASTM E1467 (1994) – transmitting digital neuro-physiological data in computer systems;
- ASTM E1394 (1997) – exchanging data between diagnostic equipment and computer systems;
- ASTM E1381 (2002) – transmitting low-level data between clinical laboratory and computer systems.

The specific standards for the exchange of medical data are described below.

3.3. HL7 Standard

HL7 (Health Level 7) – approved in 1996 the United States' national standard for the exchange of medical data in electronic form [11; 12], see Fig. 1.

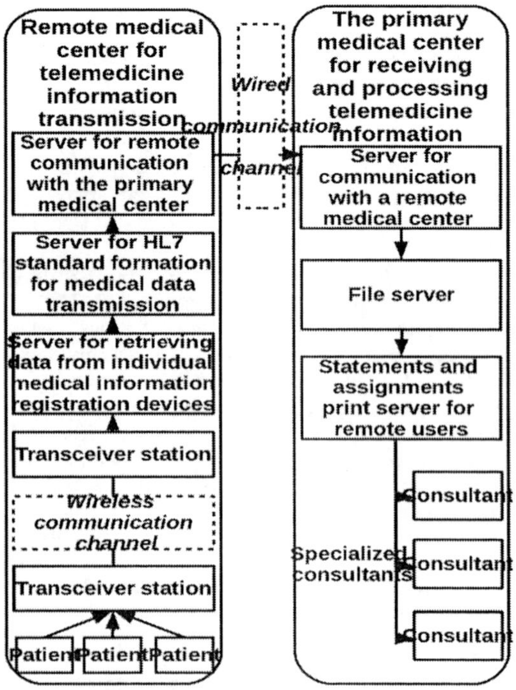

Fig. 1. Structural-logical scheme of organization of medical ITS using the HL7 standard.

The objectives of the standard HL7 were: to provide the interoperability of automated systems in health care facilities; to unify the exchange of external data; to standardize data exchange between medical ITS (standardization of software interfaces); to support digital data exchange across a wide range of communication environments [12].

The term "Level 7" refers to the top level of the Open Systems Interconnection (OSI) model. The standard includes the definition and structure of data transmission, synchronization of medical data exchange, and standard principles of communication systems' provision based on the noise-resistant data encoding/decoding.

The HL7 standard defines the mechanism of addressing to the interfaces of various medical systems for exchanging patient's medical data, namely: patient reception or registration, referencing to specialized consultants, processing of various types of requests, planning of medical supplies, recording the results of the treatment process, conducting clinical observations, patient monitoring, and so on.

3.4. DICOM Standard

Another transmission standard that is widely used by medical equipment and software producers is DICOM (Digital Imaging and Communications in Medicine), was developed by the National Electrical Manufacturers Association and continues to improve [11; 12], see Fig. 2.

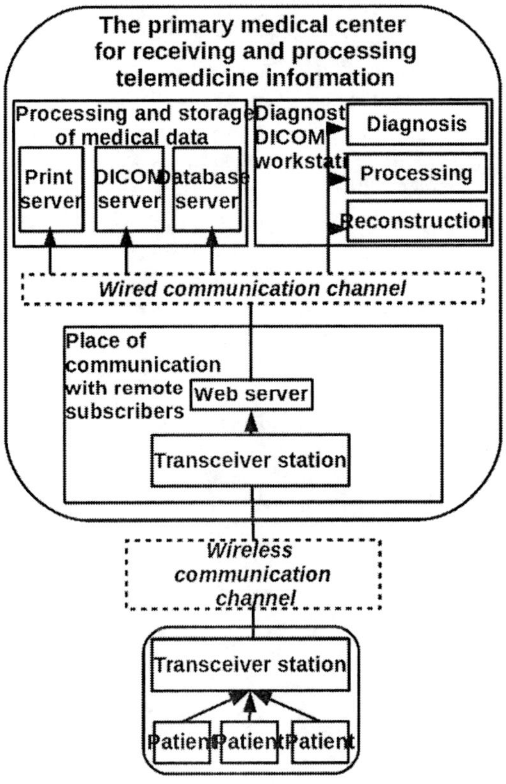

Fig. 2. Structural-Logical Scheme of Implementation of Medical ITS Based on DICOM Protocol.

The DICOM standard allows one to create, store, transmit and print selected images, sequences of images, information about the patient, medical institution, the medical staff that conducts treatment, existing equipment, and so on (see Table 1).

ITSs based on DICOM protocol are client-server structures consisting of interlinked components and software [12]:

- medical DICOM equipment, playing the role of DICOM clients;
- web server to support the client-server structure;
- workstation for medical survey and data processing (DICOM automated diagnostic station, single or a branched system);
- DICOM peripheral equipment.

The DICOM standard provides a digital communication system between DICOM server and diagnostic equipment, including [12]: x-ray equipment (RF, CF, DF, RG, MG, CR); angiographic complexes (XA, DSA); computed tomography (CT, PET); magnetic resonance tomography (MR); isotope scanners (NM); ultrasound scanners (US, EC); microscopes (DM); endoscopes (ES).

Table 1. DICOM Services.

№	Name of the DICOM service	Function of the DICOM service
1	Store	Processing and storage of visual and other telemedicine information
2	File	Data storage on data carriers for digital data exchange
3	Storage Service Class Provider	Software (client-server architecture) for organizing DICOM network
4	Storage Service Class User	Customer software for interacting with DICOM network
5	Query/Retrieve	Exchange of patients' history and lists
6	Modality Worklist	Creation of lists of conducted and necessary patients' health tests
7	Print	Provision of interaction between software and specialized peripheral DICOM equipment for printing telemedicine information

The DICOM standard enables the integration of medical equipment of different classes, including DICOM scanners, DICOM servers, workstations, and DICOM printers into a single medical information and communication network.

4 Interaction Models

Several different classifications of medical ITSs can be distinguished: by the principle of operation or scope of application and by the level of placement in the health care system. There are the following levels [3; 4]: 1 – the level of the country's institutional system (hospitals, clinics, mobile laboratories, and so on), 2 – the essential (clinical) level (doctors of various fields and profiles), 3 – the local level of the medical system (profiled and specialized medical services and regional management entities), 4 – the regional level of the country's activities (regional institutions).

At the essential (clinical) level, medical ITSs can be segmented into two main groups [3; 12]:

– Remote counseling, diagnosis, and staff training – the patient's automated workstation is connected to *N* consultants' workplaces.

– Remote monitoring and registration of patients' vital functions – the consultant's automated workstation is connected to personal medical devices of *N* patients.

Despite the diversity of medical ITSs, they are characterized by typical operations related to registration, transmission, processing, and analysis of data. In general terms, according to [13-15], the structure that processes and transmits information in standard medical ITS can be described as follows.

Analog video or audio signals from any information source (diagnostic equipment, scanners) are received on the input of medical ITS. The signal is sampled based on the frequency characteristics of the equipment and the network, then quantized and becomes a subject to noise-resistant encoding and compression, regarding the network data rate requirements. The signal is then fed to the data-transmitting unit that provides modulation of a certain type, which further becomes the subject of optimal filtering on the receiver side.

The modulated information signal, spreading over the data channel, is distorted by an interference of artificial or natural origin. These interferences contain, as fluctuating, focused or pulsating components and distort the message. Taking account of that the information message is distorted during transmission, an important decision for the ITS functioning is to combine optimal filtering and noise-resistant decoding units in the receiver.

5 Conclusion

The provision of high quality of ITS functioning is possible using the noise-resistant methods and structures of the network organization. The need to ensure the efficient transmission of large volumes of medical data in noisy conditions with low signal-to-noise ratios requires the improvement of existing and the development of new methods and models for transmitting and receiving data.

Mentioned circumstances cause the efficiency of the functioning of medical ITS and its ability to provide desired medical services in emergencies. The development and implementation of a unified interaction model for various medical ITS components, which will allow adapting to extreme characteristics of data transmission physical environment, is a vital scientific and practical issue.

References

1. E. Coiera, *Guide to Medical Informatics, the Internet and Telemedicine* (Chapman and Hall Medical, London, 1997)
2. V. M. Bezruk, I. P. Nikolaev, D. V. Chebotareva, Information and Telecommunication Signal Recognition Technologies in Telemedicine, Bulletin of Lviv Polytech. National Univer.: Inform. Sys. and Net., **805**, 17-22 (2014)
3. V. N. Konyukhov, *Fundamentals of Telemedicine Systems* (Samara State Aerospace University, Samara, 2012)
4. A. V. Vladzimirskiy, *Telemedicine in the System of Health Care Management and Organization* (Digital PH, Donetsk, 2012)
5. N. A. Lysov, Y. L. Minayev, V. S. Vopilin, A. A. Supilnikov, A. V. Chebykin, *Medical Informatics* (Reaviz, Samara, 2013)
6. Y. E. Lyah, A. V. Vladzimirskiy, *Introduction to Telemedicine. Series: Essays of Biological and Medical Informatics* (Lebed, Donetsk, 1999)
7. S. S. Furuie, et al., Managing medical images and clinical information: InCor's experience, IEEE Transactions on Inform. Tech. in Biomed., **11**, 1, 17-24 (2007)
8. V. A. Gerasevich, *Computer for Doctor* (BHV-Petersburg, St. Petersburg, 2004)
9. A. G. Faibushevich, V. D. Protsenko, *Application of Clinical Informatics Methods in Complex Studies and Treatment of Patients* (PFUR, Moscow, 2008)
10. J. Quinn, *HL7*, http://www.hl7.eu/HL7v2x/v27/std27/ch01.html, accessed 2019/10/16 (2011)
11. J. Rodrigues, *Health Information Systems: Concepts, Methodologies, Tools, and Applications* (IGI Global, Hershey, 2010)
12. V. N. Kazakov, V. G. Klimovitskiy, A. V. Vladzimirskiy, *Telemedicine* (Nord, Donetsk, 2002)
13. G. F. Konakhovich, I. O. Machalin, O. Yu. Puzyrenko, *The Theory of Electrical Communication*, (Interservice, Kyiv, 2013)
14. O. N. Novoselov, A. F. Fomin, *Fundamentals of the Theory and Calculation of Information-Measuring Systems* (Mashinostroenie, Moscow, 1991)
15. R. W. Heath Jr., *Introduction to Wireless Digital Communication: A Signal Processing Perspective* (Prentice Hall, Upper Saddle River, 2017)

Internet of Things (IoT) based outdoor performance characterisation of solar photovoltaic module

Paresh S. Nasikkar[1,*], *and Javed K. Sayyad*[1]
[1]Symbiosis Institute of Technology, Symbiosis International (Deemed University), Lavale, Pune, 412115 MH (India).

Abstract. Outdoor performance characterisation of the Solar Photovoltaic (SPV) module is essential to while designing and commissioning a new SPV power plant. The health of SPV plant is monitored using a conventional method which underutilises the workforce and resources used for Operation and Maintenance (O&M) of the SPV power plant. Outdoor performance characterisation of SPV module using reliable, compact, portable, and economical Current-Voltage (I-V) curve tracer having IoT capability and auto sweep capability is presented in this paper. The capacitive load method is used for I-V curve sweep, and the result is compared with the resistive load method. In this paper, the advantages of using a capacitive load method over resistive load method are observed and experimentally validated. The ease of using IoT feature makes this proposed I-V Curve Tracer (IVCT) device more reliable to trace Current-Voltage (I-V) curve and Power-Voltage (P-V) curve for outdoor performance characterisation of SPV module.

1 INTRODUCTION

Limited availability of fossil fuel, cost effectiveness due technology upgradation in SPV panel manufacturing and consumer awareness towards use of green technology has exponentially increases the use of SPV. Globally growing need of energy can be fulfilled by grid tied SPV power plant [1]. It can be easily generated and used in household, commercial and industrial applications [2]. The major disadvantage of commercially available SPV module is its low conversion efficiency ($\eta < 20\%$) and the non-linearity of the output I-V curve [3, 4].

Commissioning the SPV power plant on the remote location where unshaded space is largely available is the prime concern of the power plant design engineer [5]. Performance analysis and continuous monitoring of this power plant during operation is the necessity now a day. It adds financial burden over power plant owner. More accurate power plant monitoring system helps to schedule regular maintenance which will increase life span of the power plant [6]. It also helps to find faults during operation because of uncertain change in metrological parameters [7-9]. Internet of Things (IoT) based outdoor performance characterisation is the necessity to monitor the plant continuously from remote location [10].

In this paper analysis of SPV system is evaluated using IoT based I-V Curve Tracer (IVCT). This developed IVCT device with capacitive load method gives high degree of flexibility and scalability in comparison with resistive load is shown and validated experimentally.

1.1 Performance Characterisation of SPV Module

Tracing I-V and P-V curve for performance characterisation is predominant method than other existing method. These curve help to extract important electrical parameters like short circuit current (I_{sc}), open circuit voltage (V_{oc}), maximum current (I_m), maximum voltage (V_m), maximum power (P_m), series resistance (R_s), shunt resistance (R_{sh}), field factor (FF), efficiency (η) etc. which shown in Fig. 1 [11].

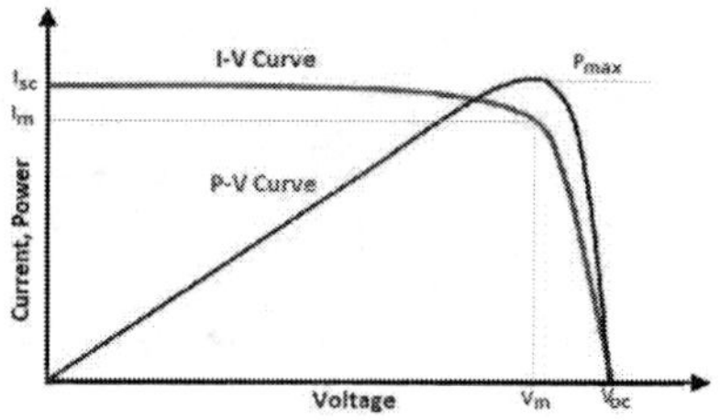

Fig. 1. I-V and P-V curve

Conventionally for curve tracing, SPV module is loaded from zero resistance path to infinite resistance path using variable resistance. SPV module is active device it generates energy and IVCT device acts as load which sweeps curve using internal load circuitry.

*Corresponding author: **paresh.nasikkar@sitpune.edu.in**

Conductivity of silicon material increases as increases in solar irradiance makes SPV module to generate more current at output. For change in temperature, the series resistance increases which makes more voltage drop across series resistance and less voltage at output of SPV module.

2 PROPOSED TECHNIQUE FOR IVCT

Major building blocks of the proposed IVCT device is SPV modules, Sensor, load Unit and Raspberry Pi 3B model as a central processing unit. Detailed block diagram of proposed IVCT is shown in Fig. 2.

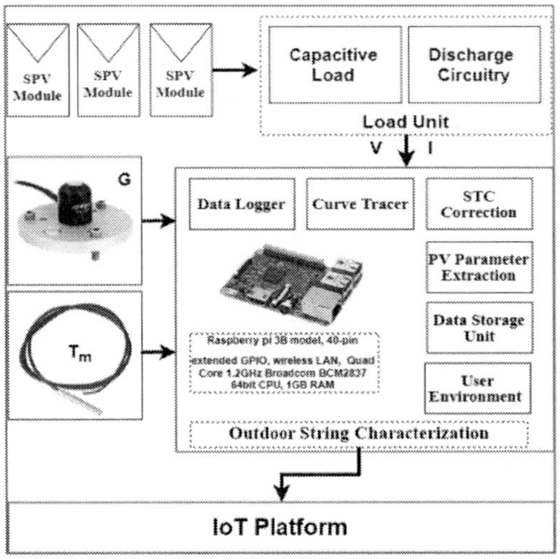

Fig. 2. Detailed block diagram of Proposed IVCT

2.1 SPV Module

Eight SPV modules were installed on the terrace of the solar photovoltaic lab for testing designed IVCT. These modules are manufactured by Akshaya solar power Pvt. Ltd. and there STC technical specifications electrical characteristics are as tabulated in Table 1. For this research ASP-12-60W_p SPV modules is the device under test and installed on terrace as shown in Fig. 3.

Fig. 3. ASP-12-60W_p installed on terrace of SPV Lab

Table 1: Technical specification of ASP-12-60W_p

Symbol	Parameter	Value
P_{max}	Maximum Power	60 W_p
V_{max}	Maximum Voltage	17.6 V
I_{max}	Maximum Current	3.41 A
V_{oc}	Open Circuit Voltage	21.5 V
I_{sc}	Short Circuit Current	3.68 A
α	Temperature coefficient of I_{sc}	+0.0009/^{0}C
β	Temperature coefficient of V_{oc}	-0.004/°C

2.2 Sensors

In order to obtain correct measurement of voltages during operation, the SPV panel voltage needs to be divided which is accomplished by a resistive voltage divider circuit. This provides voltage isolation and protects other circuitry. This voltage is then read by the external 12 bit analog to digital converter ADS1015 and send back to raspberry pi.

DC Current can be measured using small shunt resistor or different current sensors like ADS712, LEM DC current transducer DK 20 C5 U etc. [12]. We have used ACS712 IC based hall effect current sensor which is cost effective and easily available. ACS712 has 3 variants, 5A, 20A, 30A having different current measuring capability. This sensor is bidirectional and can sense current in both directions. With no load connected sensor shows 2.5V at the output pin. Sensitivity of ACS712 5A is 185mV/A. This signal is directly feed to external ADC1015 which converts analog signal to digital and send it to the Raspberry pi.

For light irradiance measuremtes, Apogee SP110SS pyranometer sensor is used. It is calibrated by manufacturer and it has sensitivity 0.2mV/Wm^{-2}. This senor is aligned with SPV module to measure light irradiance accurately as shown in Fig. 4. DS18B20 temperature sensor is placed backside of the SPV module to measure of module temperature continuously.

Fig. 4. Apogee SP110SS Pyranometer on terrace

2.3 Load Unit

2.3.1 Capacitive Load

To obtain characterisation curves, it is important to control the current generated by the SPV module. To perform this, one has to vary resistance (load) over entire range in a short span of time T_{scan} from zero to infinity. It will measure the points of the I-V curve from the short-circuit current to the open-circuit voltage. This research uses one of the methods to emulate as a variable resistance is by using capacitor as a load. Results for the resistive load is also taken to validate the results. Capacitor load provides natural sweep and able to perform well under Partial shading condition (PSC) [13-15]. Capacitive method is useful to develop I-V curve tracer for high power rated photovoltaic modules.

2.3.2 Discharge Circuitry

To start the new sweep, charged capacitor during curve tracing need to be discharged first for accurate measurement. It requires resistive (energy sink) elements to discharge capacitor safely within a short span of time. We have used heat sinkable resistor which has Aluminium Housed direct heat sink mooting. Five resistors of 300ohm, 100W specification is used in series or parallel as per requirement to safely discharge the capacitor before and after I-V curve tracing is shown in Fig. 5.

Fig. 5. Heat Sinkable High Wattage Resistor

Total time duration required to discharge the capacitor (τ) safely is given by Equation 1 [16]

$$\tau = 5 * R * C \tag{1}$$

2.4 Raspberry Pi 3B as Central Processing Device

The Raspberry Pi 3B was programmed with python programming to monitor both the current and the voltage of the SPV module. Real time measurement of metrological parameters like light radiance and module temperature is measured using sensor and used for STC correction. Flow chart of the proposed device working is as given in Fig. 6. Clear sky condition is checked using light irradiance data to start the device.

2.4.1 Modifying measured parameters to STC condition at varying temperature and light irradiance (STC correction)

Impact of varying meteorological parameters will give the erroneous reading of instantaneous current and voltage. To reduce the effect, measured electrical parameters need to be corrected with STC conversion [8, 17]. STC conversion for of measured parameters are calculated using Equation 2 and 3 which are given below,

$$V_{STC} = V_{meas}\left\{1 + \left[\beta * ln\left(\frac{G_{STC}}{G_{meas}}\right) + [\beta * (T_{STC} - T_{meas})]\right]\right\} \tag{2}$$

$$I_{STC} = I_{meas}\left\{1 + \left[\alpha * |(T_{STC} - T_{meas})| * \left(\frac{G_{STC}}{G_{meas}}\right)\right]\right\} \tag{3}$$

Where, V_{STC} and I_{STC} is voltage and current at STC (labelled specification) respectively, V_{meas} and I_{meas} is the voltage and current measured during I-V tracing respectively, G_{STC} is the light irradiance at STC (1000W/m^2), T_{STC} is module temperature at STC (25^{0}C), G_{meas} and T_{meas} is the light irradiance and module temperature measured using pyranometer and temperature senor during I-V sweep respectively, α and β is the thermal coefficient of the short-circuit current (V/^{o}C) and thermal coefficient of the open-circuit voltage (A/^{o}C) respectively.

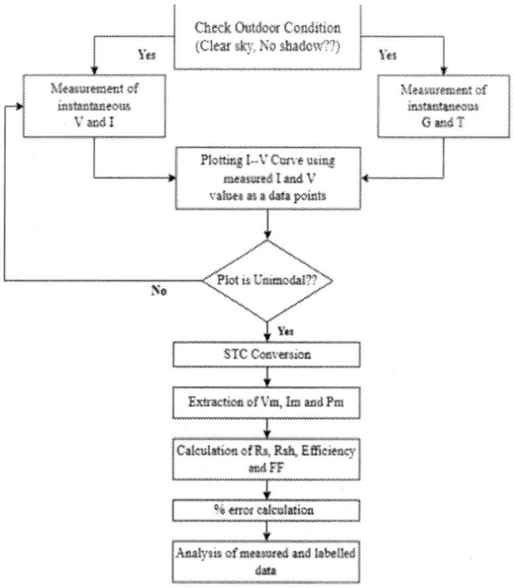

Fig. 6. Flow chart of python program

2.4.2 *Data Logger*

The instantaneous values of I, V, G, and T_m can be feed directly to external ADS1015 for data conversion from analog to digital. This Digital data is given to Raspberry pi which acts as central processing unit in the proposed IVCT device.

Measurement of high current and high voltage requires proper isolation during measurement. Galvanic isolation is used to protect main circuitry and Raspberry pi microprocessor from unwanted surge and high current due to short circuit. It will also help to reduce noise and provide safety for technician/engineer during performance characterisation.

3 RESULT AND DISCUSSION

Fig. 7 shows the actual hardware test setup in the SPV research lab. The reference values specified by the manufacturer on the back side of the SPV module are utilized in the analysis for error and accuracy calculation.

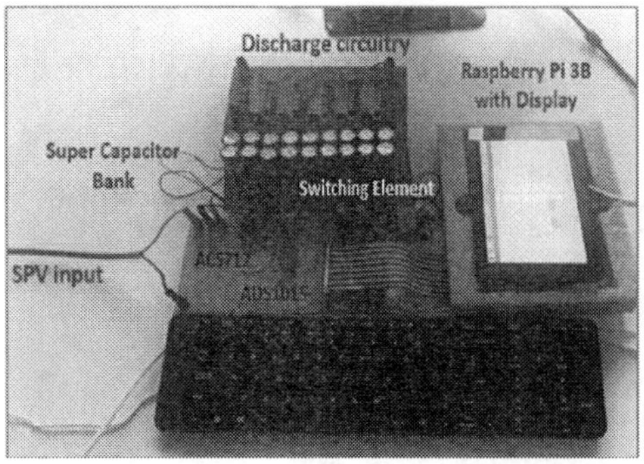

Fig. 7. Actual Hardware of Proposed IVCT

3.1 Result of conventional method (R Load)

Conventional R load method is tested and measurement has been taken out for validation. Rheostat of 300ohm is used as resistive load to vary load resistance from minimum to maximum manually. This method is simple, cost-effective and shows significant results for low operating solar power modules. But this method has certain limitations. It unable to track fast changing (non-uniform) operating and climatic condition. For large operating power, resistances become heavy, bulky and costly [18-20]. For power solar modules resistive load shows inherent losses and dissipates heats during operation [20].

Throughout all the analysis of R load, the sample frequency is depending on the operator expertise. SPV module electrical parameters extracted from captured I-V and P-V curve of ASP 12-60Wp on different time instants on the same day is tabulated in Table 2,3,4.

Table 2: Result analysis for 12:00PM Time stamp

Time Stamp	12:00			
Parameters	Measured	STC Corrected	% Error	% Accuracy
V_{oc} (V)	20.7	21.998	2.3	97.7
V_{max} (V)	14	14.878	18.3	81.7
I_{sc} (A)	3.67	3.5966	2.3	97.7
I_{max} (A)	3.49	3.4202	0.3	99.7
P_{max} (W_p)	48.86	50.886	17.9	82.1
G_{meas} (W/m^2)	720			
T_{meas} (^{0}C)	41			

Table 3: Result analysis for 13:00PM Time stamp

Time Stamp	13:00			
Parameters	Measured	STC Corrected	% Error	% Accuracy
V_{oc} (V)	20.4	20.661	4.1	95.9
V_{max} (V)	16.1	13.974	25.9	74.1
I_{sc} (A)	3.18	3.373	9.1	90.9
I_{max} (A)	2.76	3.375	1	99
P_{max} (W_p)	44.436	47.162	27.3	72.7
G_{meas} (W/m^2)	626.25			
T_{meas} (^{0}C)	48			

Table 4: Result analysis for 14:00PM Time stamp

Time Stamp	14:00			
Parameters	Measured	STC Corrected	% Error	% Accuracy
V_{oc} (V)	20.5	23.314	7.8	92.2
V_{max} (V)	15.6	15.768	11.6	88.4
I_{sc} (A)	2.3	3.507	4.9	95.1
I_{max} (A)	2.14	3.335	2.2	97.8
P_{max} (W_p)	33.384	52.586	14.1	85.9
G_{meas} (W/m^2)	650			
T_{meas} (^{0}C)	57			

3.2 Result of Proposed method (C Load)

For DC input, capacitor during charging offers low resistance path and after full charging it offers infinity resistance path to flow of current. It sweeps curve naturally from I_{sc} to V_{oc}. IoT based IVCT is tested on the same time stamp and results are tabulated in Table 5,6,7. Below results shows that, measured electrical parameters of SPV module using capacitive load based IVCT closely reassembles the reference values. Measured parameters are taken from I-V curve and corrected using STC correction formula. External ADC is used with gain 8 and 3300 samples per second for better resolution of I-V curve trace.

Table 5: Result analysis for 12:00PM Time stamp

Time Stamp	12:00			
Parameters	Measured	STC Corrected	% Error	% Accuracy
V_{oc} (V)	20.52	21.806	1.4	98.6
V_{max} (V)	15.292	16.251	7.7	92.3
I_{sc} (A)	3.467	3.536	3.9	96.1
I_{max} (A)	3.12	3.182	6.7	93.3
P_{max} (W_p)	47.711	51.711	13.8	86.2
G_{meas} (W/m^2)	720			
T_{meas} (^{0}C)	41			

Table 6: Result analysis for 13:00PM Time stamp

Time Stamp	13:00			
Parameters	Measured	STC Corrected	% Error	% Accuracy
V_{oc} (V)	18.806	20.501	4.6	95.4
V_{max} (V)	15.287	16.665	5.3	94.7
I_{sc} (A)	3.4611	3.576	2.8	97.2
I_{max} (A)	3.184	3.289	3.5	96.5
P_{max} (W_p)	48.674	54.811	8.7	91.3
G_{meas} (W/m^2)	626.25			
T_{meas} (^{0}C)	48			

Table 7: Result analysis for 14:00PM Time stamp

Time Stamp	14:00			
Parameters	Measured	STC Corrected	% Error	% Accuracy
V_{oc} (V)	18.982	21.379	0.6	99.4
V_{max} (V)	15.252	17.178	2.4	97.6
I_{sc} (A)	3.598	3.757	2.1	97.9
I_{max} (A)	3.381	3.531	3.5	96.5
P_{max} (W_p)	51.567	60.656	1.1	98.9
G_{meas} (W/m^2)	650			
T_{meas} (^{0}C)	57			

All the entries of measured current voltage, light irradiance and module temperature is uploaded on ThingSpeak IoT platform. This results can be analysed anytime from anywhere. We can program threshold values of individual parameter to initiate alarm or emergency signalling during operation.

7-inch touch screen display of Raspberry Pi 3 B model is used with graphical user interface programmed using python library to ease user handling experience. Raspberry pi 3B model has inbuilt Wi-Fi module which help to connect IVCT to network and share data over cloud or IoT platform wirelessly. Screenshot of uploaded data on ThingSpeak IoT platform is shown in Fig. 8.

Fig. 8. G and T_m data on ThingSpeak IoT Platform

4 CONCLUSION

The main circuit is based on a capacitive loading which allow auto sweep of characterisation curves of the solar photovoltaic modules having maximum power up to 60W_p. The designed IV curve tracer is easily scalable and can trace the I-V curve upto 50V open circuit voltage and 10A short circuit current. Error in P_{max} is upto 15% due to the impact of error in current measurement. This shows the internal resistive and parasitic effect of components in the circuit while capturing current continuously.

The design and implementation of a low-cost, portable and scalable I-V curve tracer for SPV module is presented in this paper. This device able to examine and monitor SPV modules performances continuously, and it becomes very useful, particularly when the SPV modules are installed at the rooftop or in a rural area. IoT capability of designed IV curve tracer helps to speed-up data capturing, monitoring and subsequent analyses of SPV power plant from a remote location. The bidirectional communication capability of IoT will help to tackle emergency situation easily and will also help to take necessary measures to reduce power/energy losses. Analysis of data on the cloud help to schedule maintenance and reduces the O&M cost drastically.

ACKNOWLEDGEMENT

This research did not receive any specific grant from funding agencies in the state, commercial, or non-commercial sectors. Nonetheless, the authors would like to thank the Department of Electronics and Telecommunication Engineering, Symbiosis Institute of Technology, Symbiosis International (Deemed University), Pune, Maharashtra, India for providing the space to set up the Solar Photovoltaics laboratory.

References

1. B. Shivakumar, and K. Sudhakar, Performance evaluation of 10 MW grid connected solar photovoltaic power plant in India, Energy Reports, **1**, 184-192 (2015).
2. A. Sulaiman, F. Inambao, and G. Bright, Solar Energy as an Alternative Energy Source to Power Mobile Robots, *Advances in Intelligent Systems and Computing,* **274**, 957-969 (2014).
3. E. Kandemir, N. S. Cetin, and S. Borekci, A comprehensive overview of maximum power extraction methods for PV systems, Renewable and Sustainable Energy Reviews, **78**, 93-112 (2017).
4. F. J. Vorster, and E. E. Van Dyk, Current-voltage characteristics of high-concentration, photovoltaic arrays, Progress in Photovoltaics: Research and Applications, **13**(1), 55-66 (2005).
5. S. Fadhel, C. Delpha, D. Diallo, I. Bahri, A. Migan, M. Trabelsi, and M. F. Mimouni, PV shading fault detection and classification based on I-V curve using principal component analysis, Application to isolated PV system. Solar Energy, **179**, 1-10 (2019).

6. S. Sarikh, M. Raoufi, A. Bennouna, A. Benlarabi, and B. Ikken, Photovoltaic system fault identification methodology based on I-V characteristics analysis, AIP Conference Proceedings, **2123**(1), 0200371-02003714 (2019).
7. M. R. Maghami, H. Hizam, C. Gomes, M. A. Radzi, M. I. Rezadad, and S. Hajighorbani, Power loss due to soiling on solar panel : A review, Renewable and Sustainable Energy Reviews, **59**, 1307-1316 (2016).
8. E. L. Meyer, and E. E. van Dyk, Assessing the reliability and degradation of photovoltaic module performance parameters, IEEE Transactions on Reliability, **53**(1), 83-92 (2004).
9. H. Ibrahim, and N. Anani, Variations of PV module parameters with irradiance and temperature, Energy Procedia, **134**, 276-285 (2017).
10. H. B. Chi, M. F. N. Tajuddin, N. H. Ghazali, A. Azmi, and M.U. Maaz, Internet of things (IoT) based iv curve tracer for photovoltaic monitoring systems, Indonesian Journal of EE&CS, **13**(3), 1022-1030 (2019).
11. E. Q. B. Macabebe, C. J. Sheppard, and E. E. Van Dyk, Parameter extraction from I–V characteristics of PV devices, Solar Energy, **85**(1), 12-18 (2011).
12. Y. Erkaya, H. S. Illa, C. Conway, S. Dhali, and S. Marsillac, Development of a string level fault detection system for solar tracking applications, *IEEE 40th Photovoltaic Specialists Conference* (2014).
13. M. M. Mahmoud, Transient analysis of a PV power generator charging a capacitor for measurement of the I–V characteristics, Renewable Energy, **31**(13), 2198-2206 (2006).
14. S. Filippo, J. Ahmad, A. Ciocia, P. Di Leo, A. F. Murtaza, and M. Chiaberge, Capacitor charging method for I–V curve tracer and MPPT in photovoltaic systems, Solar Energy, **119**, 461-473 (2015).
15. R. Ahmad, F. Murtaza, U. T. Shami, Zulqarnain, and F. Spertino, An MPPT technique for unshaded/shaded photovoltaic array based on transient evolution of series capacitor, Solar Energy, **157**, 377-389 (2017).
16. T. H. Warner, and C. H. Cox, A high power current-voltage curve tracer employing a capacitive load, Solar Cells, **7**(1), 175-181 (1982).
17. Y. Erkaya, P. Moses, and S. Marsillac, On-site characterization of PV modules using a portable, MOSFET-based capacitive load, *IEEE 43rd Photovoltaic Specialists Conference* (2016).
18. E. E. van Dyk, A. R. Gxasheka, and E. L. Meyer, Monitoring current–voltage characteristics and energy output of silicon photovoltaic modules, Renewable Energy, **30**(3), 399-411 (2005).
19. E. Duran, M. Piliougine, M. Sidrach-de-Cardona, J. Galan and J. M. Andujar, Different methods to obtain the I-V curve of PV modules : A review, *33rd IEEE Photovoltaic Specialists Conference* (2008).
20. A. Q. Malik, and S. J. B. H. Damit, Outdoor testing of single crystal silicon solar cells, Renewable Energy, **28**(9), 1433-1445 (2003).

Intelligent Vehicle Speed Controlling and Pothole Detection System

Aditya Anand[1,*], *Rushiraj* Gawande[1], *Prathamesh* Jadhav[1], *Rima* Shahapurkar[1], *Anjali* Devi[1], *and NikhilKumar*[1]
[1]*JSMP'S Jayawantrao Sawant College of Engineering, Pune, Maharashtra, India.*

Abstract- Speed is the single largest killer on India's roads. The higher the speed, the greater the impact and the more the chances of grievous injury and death. This is why speed management is something which needs to be seriously considered. Here is an approach to build a system which reduces the number of accidents due to driver's negligence of over speeding. We are proposing a Dynamic Speed Limiter with the help of machine learning algorithm which will help in reducing the accidents caused due to over speeding and rash driving of vehicles on road. The proposed system set the maximum speed of vehicle with the help of sign board available on the roads which define the safe driving speed of vehicle for that area or road. Also to solve the potholes we are using ML algorithm and mobile phone accelerometer to detect the potholes, accelerometer vibrations are set. After detecting the vibrations, the location (i.e. longitude and latitude) is marked on the GPS, commuters will get information about potholes.

Keywords- Speed Limiter, Camera, Image Processing, Sign-Board, CNN, SVM, OpenCV, MCU (Raspberry Pi).

1. Introduction

According to the World health organization, around the world approximately 1.35 million die every year due to road accidents. Injuries caused on the road are one of the sixth leading causes of death in India with a higher percentage of hospitalization, loss of lives, disabilities and socio- economic losses to the mankind. We have come up with the solutions of two of the major causes of accidents.

Firstly, speeding, speeding endangers everyone on the road. According to Central Ministry of Road Transport and Highway India[1], in 2017 speeding on road was reason for death of 9717 peoples accounting for more than a quarter (26%) of all the traffic caused fatalities. Speeding endangers not only the life of the speeder but also all the commuters around them. To solve this cause, we are designing and implementing a dynamic speed limiter through image processing technique. As, the speed-limit sign board recognition system has two separate tasks: first is the detection stage and second is the recognition stage. The detection stage is responsible for detecting the sign board from the image, and in the recognition stage the values specified in the sign board will be identified. In the recognition stage, template matching or learning- based classification techniques are usually used[2,3]. For learning- based classification, Artificial Neural Networks and Support Vector Machines (SVMs) have been widely investigated . Recently, Convolutional Neural Network (CNN) has been used to recognize traffic signs and has high accuracy[4] , however, huge computation resource is usually required either in training or test phases. In this paper, we introduce an automatic detection and recognition system of speed- limit sign board that can handle different conditions of lighting and blurriness in images[5,6]. The system uses a detection method based on MSER detection and recognizes speed-limit signs with Histogram of Oriented Gradient (HOG) features by a SVM classifier. Further this limit is provided input to

*Corresponding author: anandaditya59@gmail.com

the speed limiter through MCU. The speed limiter then sets the maximum speed of the vehicle which restricts the vehicle to over speed.

Second one is the pothole; potholes are also one of the major reasons of accidents[1]. According to the data shared by states to Centre Uttar Pradesh, India logged most such deaths at 987. In the year 2017, 4250 people died nearly on the roads which were under construction. To solve this cause, we are using ML algorithm and mobile phone accelerometer to detect the potholes, accelerometer vibrations are set. When a user passes on the potholes for the first time,

Fig. 1. A typical speed limit sign board

Accelerometer in mobile phones detects the pressure of potholes based on the vibration. After detecting the vibrations, the location (i.e. longitude and latitude) is marked on the GPS, commuters will get information about potholes via updated Maps and also the location will be sent to the corporation to fix the potholes.

2. Literature Survey

According to the traditional Systems and prior research, the methodologies and technologies used to control the vehicle speed using the sign board where the speed limit is mentioned but we can still exceed that limit and move with a higher speed. This is the major reason for fatal accidents and loss of lives.

Following we have mentioned some important researches which have some advantages and limitations which are been surveyed.

Table. 1. Literature Survey of some papers

Sr. no	Title of Paper	Author Name	Advantages	Techniques Used	Drawbacks
1.	Automatic Recognition of Speed Limits on Speed-Limit Signs by Using Machine Learning	Shigeharu Miyata	This study describes a method for using a camera to automatically recognize the speed limits on speed-limit signs	Machine Learning with local binary pattern recognition, Image Processing with HSV, Neural Network	number extraction was not successful due to signs with faded colors, signs that were backlit, or signs that were partially hidden by trees or other objects
2.	Wireless Speed Controlling Module For Two Wheelers	1. P.R. Sonawane 2. Vinayak Kaviskar 3. Abhinav Kumar 4. Debabrata Karmakar	Intimate the driver about the zones and to automatically maintain the speed is accomplished by means RF technology.	micro controller along with a RF transmitter, speed sensor, RF receiver, Servo motor	Implemented system works only for two wheelers
3.	Real Time Pothole Detection using Android Smartphones with Accelerometers	1. Artis Mednis 2. Girts Strazdins 3. Reinholds Zviedris 4. Georgijs Kanonirs	The paper is describing a mobile sensing system for road irregularity detection using Android OS based smart-phones.	Android OS based smart-phone, accelerometer sensors	System only detects the potholes without alerting the users.
4.	RoADS: A Road Pavement Monitoring System for Anomaly Detection Using Smart Phones	1.Fatjon Seraj 2.Berend Jan van der Zwaag 3. Arta Dilo 4. Tamara Luarasi	In this paper we proposed a system that detects road surface anomalies using mobile phones equipped with inertial sensors:	Smartphone orientation inside the vehicle, accelerometer, anomaly detection supervised machine learning techniques	time consuming process of manual labeling

 https://doi.org/10.1051/e3sconf/202017002010

			accelerometers and gyroscopes.		
5.	Vehicle Collision Avoidance with Dynamic Speed Governor	1. Kashyap A R 2. Mohankrishna K	The proposed system is based on microcontroller for collecting data related to speed and transmitting it through a transceiver to a base station that analyzes the transmitted data and takes decisions related to speed limit .	Microcontroller, Software: MPLAB IDE, CAD Software: EAGLE, Ultrasonic sensors	Traces are placed manually as done in the traditional method where you change the path of the trace every time you click the mouse.

3. Research Elaboration

3.1 System description:

We have come up with the solutions of two of the major causes of accidents. Firstly, speeding, speeding endangers everyone on the road. According to Central Ministry of Road Transport and Highway, in 2017 speeding on road was reason for death of 9717 peoples accounting for more than a quarter (26%) of all the traffic caused fatalities. Speeding endangers not only the life of the speeder but also all the commuters around them.

To solve this cause, we are designing and implementing a dynamic speed limiter through image processing technique. The camera installed in front of the vehicle captures real time images of the speed limit sign board and then MCU receives the images as input for image processing. With the help of image processing detection and recognition algorithm from the captured image we detect the speed limit of that particular area. Further this limit is provided input to the speed limiter through MCU. The speed limiter[7] then sets the maximum speed of the vehicle which restricts the vehicle to over speed.

The flowchart in Figure 2 shows the complete process of speed controlling using image processing the speed limit signboard.

Second one is the pothole; potholes are also one of the major reasons of accidents. According to the data shared by states to Centre Uttar Pradesh logged most such deaths at 987. In the year 2017, 4250 people died nearly on the roads which were under construction.

To solve this cause, we are using mobile phone accelerometer to detect the potholes, accelerometer vibrations are set.

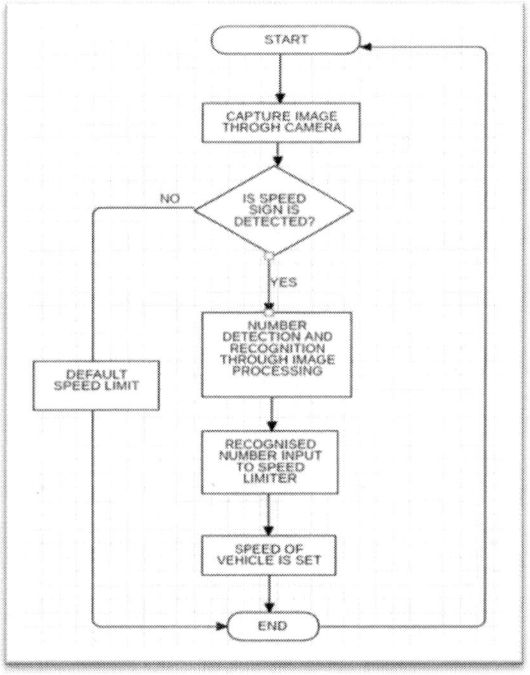

Fig. 2. Flowchart of Speed Control System

When a user passes on the potholes for the first time, Accelerometer in mobile phones detects the pressure of potholes based on the vibration. After detecting the vibrations, the location (i.e. longitude and latitude) is marked on the GPS, commuters will get information about potholes via updated Maps and also the location will be sent to the corporation to fix the potholes. The flowchart in the figure 3 shows the complete process of pothole detection using mobile accelerometer[8,9] and GPS to mark location of pothole on the Map.

https://doi.org/10.1051/e3sconf/202017002010

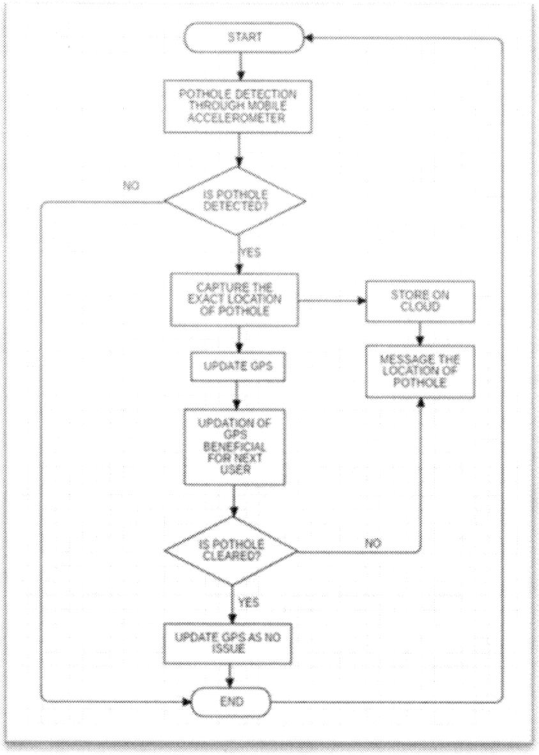

Fig. 3. Flowchart of Pothole Detection System

3.2 System Architecture:

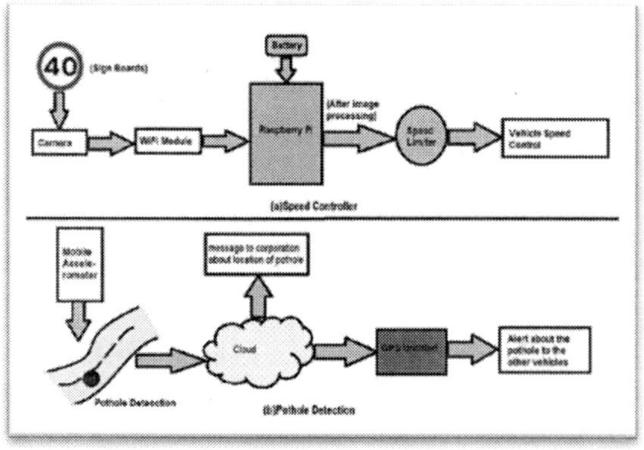

Fig. 4. Proposed Architecture

In the proposed speed controlling system the speed-limit sign board detection and recognition system works in three different stages: first preprocessing of the image, second is detection of the sign board, and third is the recognition of speed limit in the sign board. The first step, in the preprocessing stage is responsible for enhancing the input image taken from the live video so that the speed-limit sign board in real time can be more obvious. The second step, the detection stage, uses MSER detection to find speed-limit sign candidate regions. Finally, the last step, in the recognition stage is responsible for extracting the HOG features from the candidate regions and feeds it to a SVM and the class of a speed-limit sign will be classified.

Before detection, the input image needs to be processed for the detection stage to work properly and efficiently. Since the input to the MSER computation requires grayscale image, we need to convert the input RGB image to a grayscale one. Also, the speed-limit signs are designed to have bright red colors, so we can take advantage of this feature and emphasize the pixels that have dominant red channel values. Therefore, we adapt the red color enhancement approach. For every pixel x = [x_R, x_G, x_B] in the sign board image, where x_R, x_G, and x_B are the red channel, green channel, and blue channel of the image, in the image the red color enhancement is provided by

$$f_R(x) = \max\left(0, \frac{\min(x_R - x_G, x_R - x_B)}{x_R + x_G + x_B}\right) \tag{1}$$

And the final grayscale image can be obtained by

$$Red(x) = \begin{cases} f_R(x), \; x_R \geq x_G \; and \; x_R \geq x_B \\ \quad and \; x_G/(x_R - x_G) \leq T_R \\ \quad 0, \; otherwise \end{cases} \tag{2}$$

Where T_R is a parameter that obtained from experiment results. The grayscale image Red(x) obtained from this way has the advantage that the outer circle of sign board in red color signs becomes more apparent.

In the detection part, the system will detect MSERs and find potential speed-limit sign regions among them. The main advantage of this approach is that the detection of MSERs is robust under different environmental conditions, such as lighting and weather of the scene or the complexity of the background. MSERs are regions in an image that maintain their shapes when the image is thresholded at several levels. The detection of MSERs is performed on a grayscale image that comes from the result of the preprocessing stage.

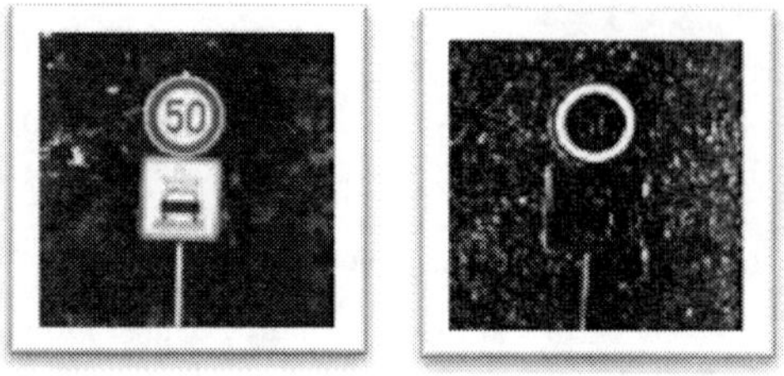

(i) (ii)

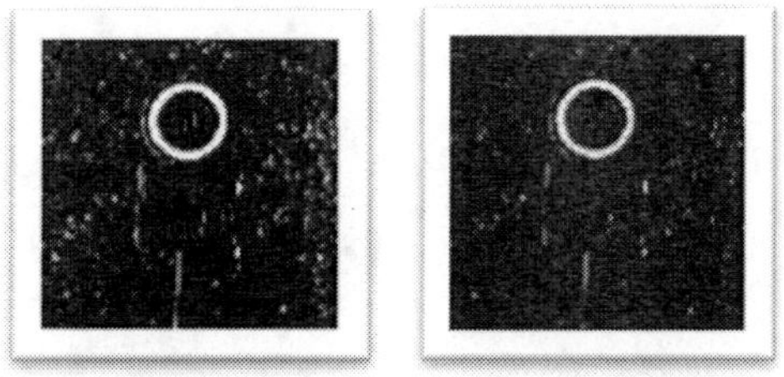

(iii) (iv)

Fig. 5. The speed-limit sign maintains its shape under different thresholds.
(i.) Original image. (ii.) The preprocessed image. (iii.) Threshold = 150. (iv.) Threshold = 250.

The recognition stage is responsible for verifying a candidate region as a speed-limit sign board. If the candidate region is actually a speed-limit sign board then, the speed-limit values present on the sign board will be determined. Otherwise, the candidate region verified by recognition stage will be discarded. In the recognition stage, the HOG features of the sign board will be extracted from the candidate region. Since speed-limit signs have strong appearance—circle shapes and high-contrast edges, HOG features are very suitable to capture these characteristics. HOG feature is computed for each candidate region and a vector will be generated. For speed and simplicity, SVM is used as a tool for the recognition stage.
After detection of the speed we set the that speed limit to the vehicle by changing the limit of speed limiter by providing it the limit by using raspberry pi. Thus, the vehicle maximum speed is set automatically by using the speed limit sign board and speed limiter.
In the proposed pothole detection system using mobile phone accelerometer to detect the potholes, accelerometer vibrations are set. When a user passes on the potholes for the first time, Accelerometer in mobile phones detects the pressure of potholes based on the vibration sensed by the mobile phone and recorded by the change in coordinates of the accelerometer. After detecting the vibrations, the location (i.e. longitude and latitude) is marked on the map by fetching the location by using mobile phone GPS, then after commuters will get information about potholes via updated Maps and also the location of the pothole will be sent to the corporation by GSM service of mobile phone to fix the potholes.

4. Conclusion

This paper proposes a model for the Intelligent Speed Controlling and Real time Pothole Detection and monitoring system, after detection of signboard the vehicle maximum speed will be set by using speed limiter so as to limit the vehicle maximum speed to the speed limit decided for that road and a mobile phone will be used to detect the pothole and alert the users and concerned authorities about the location of pothole . Thus, proposed system ensures the safety of user in terms of controlling the over speeding, pothole notification on map and suggesting alternative path, that he drives in an area and safety to all of the people on road around them.

5. References

[1]. K. K. Sandhu, IndiaToday, *Over 9300 deaths, 25000 injured in 3 years due to potholes*, (2018).

[2]. A. D. Lattner, A. Miene, and O. Herzog, Springer, *A Combination of Machine Learning and Image Processing Technologies for the Classification of Image Regions*, **Adaptive Multimedia Retrieval pp 185-199,** (2003)

[3]. E. Forson, Towards Data Science, *Recognising Traffic Signs With 98% Accuracy Using Deep Learning*, (2017).

[4].S. Doshi, Towards Data Science, *Traffic Sign Detection using Convolutional Neural Network*, (2019)

[5]. R. Biswas, H. Fleyeh, M. Mostakim, IEEE, *Detection And Classification Of Speed Limit Traffic Signs*, **10.1109/WCCAIS.2014.6916605**, (2014)

[6]. A. A. Dalve, S. S. Shiravale, International Journal of Computer Applications , *Real Time Traffic Signboard Detection and Recognition from Street Level Imagery for Smart Vehicle*, **Volume 135 – No.1**, (2016)

[7]. R. J. Hanowski, G. Berrgoffen, J. S. Hickman, F. Guo, ResearchGate, *Research on the Safety Impacts of Speed Limiter Device Installations on Commercial Motor Vehicles: Phase II*, **FMCSA-RRR-12-006**, (2012).

[8]. H. Song, K. Baek, Y. Byun, ResearchGate, *Pothole Detection using Machine Learning*, **10.14257/astl.2018.150.35**, (2018).

[9]. A. Mendis, G. Strazdins, R. Zviedris, G. Kanonirs, IEEE, Real time pothole detection using Android smartphones with accelerometers, **10.1109/DCOSS.2011.5982206**, (2011)

Design and Fabrication of Solar Powered Vapour Absorption Refrigeration System

Varun Yadav[1]*, *Supradeepa* Panual G[1]*, *Neeraj* Yadav[2]*, *Ratnam* Bordia[1], *Rohini* Soni[1] and *Rinkesh* Khandey[1]
[1]Dept. of Mechanical Engineering, Medicaps Institute of Tech. and Mgmt., Indore, M.P., India
[2]Dept. of Mechanical Engineering, Medicaps University, Indore, M.P., India

Abstract. Engineering is all about the application of knowledge and ideas for continuous development in society. In today's world, there is a strong need for an environment-friendly refrigerating system, therefore, our focus is on a solar powered vapour absorption refrigeration system. This project focuses on a cooling system that minimizes the dependency over electricity and to show our ability to save our resources for future generations. The objective of this work was to design and fabricate a vapour absorption refrigeration system, using LiBr-H_20, as the refrigerants and powered by solar energy. Performance Evaluation of the system has been done on the basis of different operating conditions and parameters like, solar irradiance, collector, generator, condenser and evaporator temperature. The COP of the system was obtained as 0.1 and the capacity was 0.01 TR. Since it's an ab-initio development it will be a unique one in terms of understanding and underlying engineering. The system is an eccentric one that can be operated by multiple heat sources like solar energy, biomass etc. without much change in the design. This system can be used to develop an Air Conditioner, Refrigerator or a Chiller.

1. Introduction

Vapour Absorption Refrigeration Systems (VARS) works on the refrigerant absorption principle. It belongs to the category of vapour cycles similar to Vapour Compression Refrigeration Systems (VCRS). However, unlike VCRS, heat is used as an input for the absorption systems. Hence these systems are also called heat operated or thermal energy-driven systems.

The most commonly used pair of refrigerants are Ammonia-Water and LiBr-Water. For Instance, V.K Bajpai's work included a VARS system using NH_3 as refrigerant and H_2O as absorbent. The COP of the system was found be around 0.58, when used with flat plate solar collector [1]. In previous research it has been found that the system with LiBr solution performs better than the system using ammonia solution [2]. This system has been designed as a single effect absorption system as opposed to a double effect system although double effect system has a significantly higher coefficient of performance because of the simplicity and cost-efficiency of a single effect system [3] [4].

There are various ways to provide heat energy as input in the system, the most common methods in case of Solar Energy are Flat Plate Solar Collector, Parabolic Solar Collector and Evacuated tube type Solar Collector. Ajay et al studied the performance of a Vapour Absorption based Air Conditioning System, powered by a parabolic solar collector, with LiBr- H_20 being the working fluids. They obtained around 3-4°C drop in cabin (evaporator) temperature over the whole working time of the system [5].

Due to high thermal efficiency, higher attainable temperature and less footprint Evacuated tube type collectors are widely used. S Arora et al, studied the thermal analysis of evacuated tube solar collectors, in relation to VARS system [6]. Their work showed that the incident flux of solar radiation and heat transfer coefficient of the air gap in tubes, affects the amount of heat transferred to working fluid. Also, V Mittal et al analysed the effects of collector temperatures on the coefficient of performance (COP) of the system [7].

2. Working Principle

VARS has similar working components, except generator and absorber, to that of VCRS. Generator, absorber and pump in VARS work similarly to the compressor present in VCRS. An illustrated single effect VARS system is shown in Fig. 2.1. It should be noted that in LiBr solution, water act as refrigerant and LiBr act as the absorbent. The role of absorbent is to enable the flow of refrigerant from the absorber to the generator by absorbing the vapours of refrigerant and forming a solution that can be easily pumped to the generator while consuming significantly less electricity used by the electric VCRS system to compress the refrigerant.

The input energy in the form of thermal energy (1) is added through hot steam and water from a solar water heater/electric heater/gas burner. The thermal energy vaporises the water in LiBr solution present in the

*Corresponding authors: varunyadav1910@gmail.com, gspanual@gmail.com and neeraj.yadav@medicaps.ac.in

generator to steam. The high-pressure steam (2) flows to the condenser leaving strong LiBr solution in the generator. This pressurised steam then rejects heat to the surrounding and condense to pressurised water (3), which is throttled through an expansion valve to low pressure and temperature liquid water. This low-temperature liquid water flows to the evaporator (4), absorbs the heat from the load and evaporates to vapour state. These water vapours are then sucked in the absorber by the strong LiBr solution (5), which is throttled down from the generator (6), to form weak LiBr solution and reject heat in the process. This weak LiBr solution is then pumped back to the generator (7) to complete the cycle.

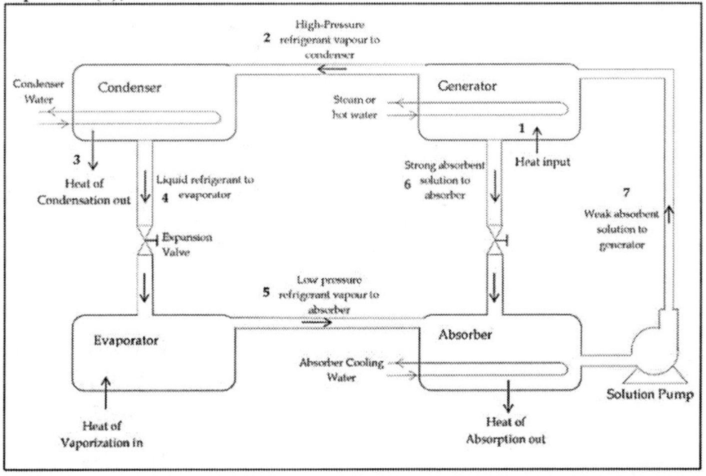

Fig. 2.1. Single Effect Vapour Absorption System

3. Design of Solar Collector

The collector designed here is Evacuated tube type Solar Collector. The geographical coordinates of the place for which it is designed are (Latitude: 22.6216°; Longitude: 75.8034°). The month of operation is assumed to be May. Thus, calculating declination angle and hour angle

$$\delta = 23.45 \sin[360/365\ (284 + N)] \quad (1)$$

N = 135 (15th May); Solar daytime = 8 hours (9AM-5PM)

Substituting values in the above equation

$\delta = 18.8°$

Hour Angle = ω

$\omega = 15°$ (Solar Time-12)

$\omega = 37.5°$

γ = Latitude = 22.62°

Assuming, Solar Constant = I_{sc} = 1367 W/m^2

Average Global Horizontal Irradiance = 7.48 kWh/m^2/day

Thus, zenith angle can be calculated as,

$$Cos\ z = sin\,\gamma\ sin\,\delta + cos\,\gamma\ cos\,\delta\ cos\,\omega \quad (2)$$

$$\text{Cos z} = \sin(22.62)\sin(18.80) + \cos(22.62)\cos(18.80)\cos(37.50)$$

$$\text{Cos z} = 0.38 \times 0.32 + 0.92 \times 0.95 \times 0.79$$

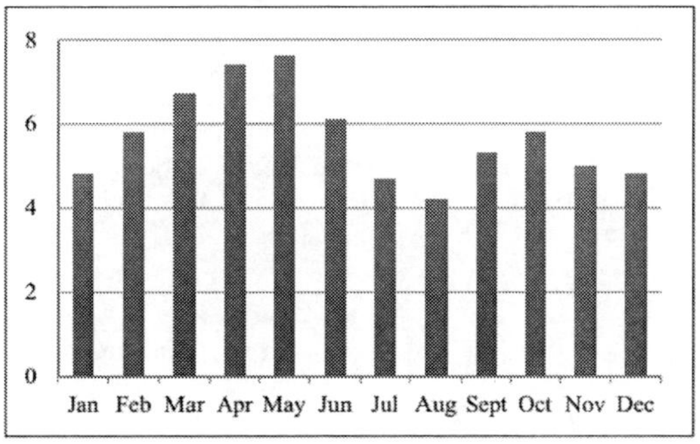

Fig. 3.1. Average Global Horizontal Irradiance (kWh/m2) vs. Month

$Cos\ z = 0.82$

$z = 35.2°$

$$I_z = I_{sc}\ e\ (-c\ (sec\ z)\ s) \quad (3)$$

The value of solar constant is, I_{sc} = 1367 W/m^2

Here, cs = 0.2424

Let,

60% η

Thus, we get, I_z = 609.66 W/m^2

But practically

$I_z = 700\ W/m^2$

Since, $T_a = 37^\circ C$, $T_g = 92^\circ C$, $m_w = 5$ kg & c = 4.186 kJ/kgK

Hence heat required for heating the water from T_a to T_g in an hr

$$Q_i = m_w \times c \times \left(\frac{T_g - T_a}{3600}\right) \quad (4)$$

$$Q_i = 5 \times 4.18 \times \left(\frac{92 - 37}{3600}\right)$$

$$Q_i = 0.32\ kW$$

And,

$$Q_i = I_z \times A \quad (5)$$

Effective area, $A = \frac{319.3}{700} m^2$

$$A = 0.456\ m^2$$

Area of 1 tube = $(\pi \times d \times l) \times 0.52$

$= \pi \times 58 \times 10^{-3} \times 1.8 \times 0.52$

$= 0.171$

Number of tubes $= \frac{Effective\ area}{Area\ of\ 1\ tube} = \frac{0.456}{0.171} \approx 3$

4. Design of VARS system

4.1. Heat absorbed in Generator (Q_G):

A Generator is a heat exchanger to give heat energy as input to the VARS system. In this system, we have used a shell and tube type heat exchanger due to its simple and cost-effective structure. A considerable amount of research has been done on the boiling of liquids but data on LiBr solution is not extensive. It has been found that the average heat transfer coefficient lies between 1600 W/m^2 K and 7500 W/m^2 K [8].

$$Q_G = hA(T\infty - Ts) \quad (6)$$

$$Q_G = 3200 \times (3.14 \times 0.02 \times 0.32) \times (95 - 90)$$

$$Q_G = 323.55\ W$$

4.2. Heat absorbed in Evaporator (Q_E):

Dimensions of evaporator cabin: 48x36x25 cm^3; Thickness of evaporator cabin: 2 cm; Thermal conductivity: 0.04 W/mK.

The cabin is made up of polystyrene which is a thermal insulator with low thermal conductivity. Assuming laminar flow around all the surface, we can use simplified equation for air:

Vertical Plate

$$h_v = 1.42 \times (\Delta T/L)^{1/4} \quad (7)$$

Cabin temperature = $T_e = 10^\circ C$

Inside surface temperature = $T_i = 17^\circ C$

Outside Surface temperature = $T_o = 35^\circ C$

Ambient temperature = $T_a = 37^\circ C$

$$h_{va} = 1.42 \times \left(\frac{37 - 35}{0.48}\right)^{1/4} = 2.03\ W/m^2K$$

$$h_{vb} = 1.42 \times \left(\frac{17 - 10}{0.48}\right)^{1/4} = 2.77\ W/m^2K$$

Similarly, for horizontal plate

$$h_h = 1.32 \times (\Delta T/L)^{1/4} \quad (8)$$

$$h_{ha} = 1.32 \times \left(\frac{37 - 35}{0.36}\right)^{1/4} = 2.03\ W/m^2K$$

$$h_{hb} = 1.32 \times \left(\frac{17 - 10}{0.36}\right)^{1/4} = 2.77\ W/m^2K$$

Heat absorbed from Front and Back walls of evaporator:

$$Q_{e1} = \frac{2 \times (T_a - T_e)}{[(1/h_{va}) + (l/kA_1) + (1/h_{vb})]} \quad (9)$$

$$Q_{e1} = \frac{2 \times (37 - 10)}{[(1/2.03) + (0.02/(0.04 \times 0.12)) + (1/2.77)]}$$

$$Q_{e1} = \frac{2 \times 27}{5.02} = 10.76\ W$$

Similarly, for Top & Bottom walls of evaporator:

$$Q_{e2} = \frac{2 \times (T_a - T_e)}{[(1/h_{ha}) + (l/kA_2) + (1/h_{hb})]} \quad (10)$$

$$Q_{e2} = \frac{2 \times (37 - 10)}{[(1/2.03) + (0.02/(0.04 \times 0.17) + (1/2.77)]}$$

$$Q_{e2} = \frac{2 \times 27}{3.75} = 14.41\ W$$

Side walls:

$$Q_{e3} = \frac{2 \times (T_a - T_e)}{[(1/h_{va}) + (l/kA_3) + (1/h_{vb})]} \quad (11)$$

$$Q_{e3} = \frac{2 \times (37 - 10)}{[(1/2.03) + (0.02/(0.04 \times 0.09) + (1/2.77)]}$$

$$Q_{e3} = \frac{2 \times 27}{6.41} = 8.43\ W$$

$$Q_E = Q_{e1} + Q_{e2} + Q_{e3} \quad (12)$$

$$Q_E = 33.59\ W$$

Coefficient of Performance:

$$COP = \frac{Q_E}{Q_G} = \frac{33.59}{323.55} = 0.10$$

Tonnage of Refrigerator:

Heat due to infiltration,

$$Q_i = \frac{\left[q \times \left(\frac{1}{v_1}\right) \times (h_1 - h_2)\right]}{60} kW \quad (13)$$

$$q = \frac{L \times W \times H \times A_c}{60} m^3/min \quad (14)$$

$A_c = A_{c1} + A_{c2} = 0.5 + 2 = 2.5$ Refer [9,10]

$$q = \frac{0.48 \times 0.36 \times 0.25 \times 2.5}{60} = 1.8 \times 10^{-3}\ m^3/min$$

Outside Conditions:

At $T_a = 37°C$ & RH = 10%

$h_1 = 47.22$ kJ/kg; $v_1 = 0.89$ m³/kg

Inside cabin conditions:

At 10°C & RH = 60%

$h_2 = 21.58$ kJ/kg; $v_2 = 0.81$ m³/kg

$$Q_i = 1.8 \times 10^{-3} \times \left(\frac{1}{0.89}\right) \times (47.22 - 21.58)\ kJ/min$$

$$Q_i = 0.87\ W$$

Since,

$$Total\ Cooling\ Load\ (TCL) = Q_E + Q_i \quad (15)$$

$$TCL = 33.59 + 0.87 = 34.46\ W$$

$$Tonnage\ of\ Refrigeration = \frac{TCL}{3.52 \times 10^3} \quad (16)$$

$$= \frac{34.46}{3.52 \times 10^3} = 0.01\ TR$$

5. Construction

5.1. Condenser:

Air-cooled fin type condenser (Fig. 5.1) with 0.375mm outer diameter (OD) and 225mm length of copper tube was used to condense the refrigerant and cool it down further. The refrigerant flows inside the copper tube and fins are soldered on the surface of the copper tube so that it increases the overall contact surface area for heat transfer. Finned type condenser has better efficiency than bare tube type condenser and is economical than shell and tube type condenser which makes it convenient to use and maintain.

Fig. 5.1. Air-Cooled fin type condenser

5.2. Evaporator:

The chilled refrigerant after throttling enters the insulated polystyrene evaporator casing (Fig. 5.3), of length 48mm, breadth 36mm and height 25mm, in the copper tube (OD = 0.625mm). This copper tube (Fig. 5.2) is bent so that the length of the copper tube inside the evaporator casing could be increased which in turn increases the effective surface area for heat to be transferred from the load eventually to the refrigerant.

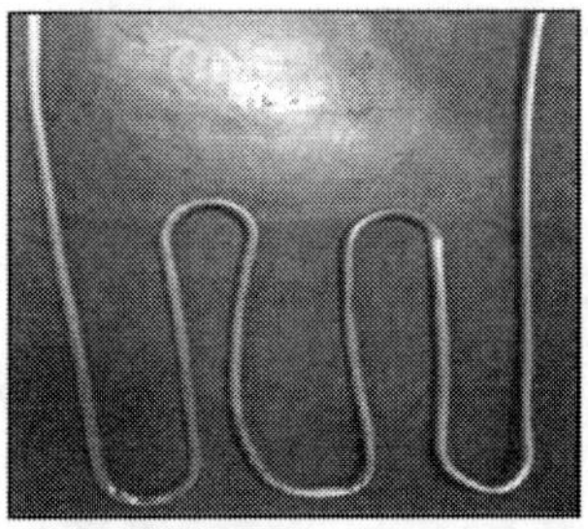

Fig. 5.2. Evaporator coil

Fig. 5.3. Evaporator Casing

5.3. Absorber:

When water from the evaporator enters the absorber, it is absorbed by the absorbent, this process is exothermic and hence heat is released. This heat increases the temperature in the absorber. Several pieces of research have observed that with the increase in temperature of absorber the COP of the system reduces significantly [11], hence the most widely used method of absorption of vapor refrigerant into a falling film of solution over cooled horizontal tubes is used to reduce the temperature inside the absorber.

A copper tube of (OD = 0.625mm) with precisely drilled holes is placed on top of an array of horizontal copper tubes (OD = 0.375mm) bent from a single pipe such that all of them are connected. The strong LiBr solution flows through the tube at the top and drips down the holes on to the array of horizontal absorber tubes with cooling water flowing through them. The refrigerant flows to the evaporator through a tube to the absorber, where it is absorbed by the absorbent in the solution. This whole setup is enclosed in non-reactive and anti-corrosion high-density polythene cylindrical container.

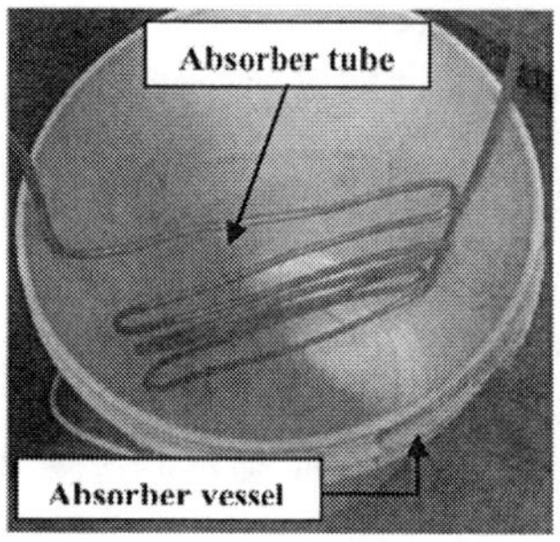

Fig. 5.4. Absorber

5.4. Generator:

The generator is made with galvanized iron plates welded together to form a cuboidal box (Fig. 5.5) of dimension 15x15x30mm. LiBr solution is heated in this box with the help of hot water, from the solar collector, flowing inside the helix-shaped copper tube (Fig. 5.6) of OD = 0.375mm. The box is insulated with a thick asbestos sheet to maintain the high temperature inside the generator.

Fig. 5.5. Generator vessel

Fig. 5.6. Generator heating coil

5.5. Evacuated Tube Solar Collector:

Three evacuated tubes of length 1.8m and OD 58mm are placed parallel to each other on a stand. A small container is fitted at the top of the collector with the top end of each of the three tubes connected to the container (Fig. 5.7). The tubes are placed at an angle of 10^{0} (around 20^{0} less than the latitude) to maximise the annual energy collection [12].

Table 1. General Parameter of Evacuated Tube Collector

Parameters	Value
Tube length	1800mm
Tube Diameter, (Outer)	58mm
Absorptive Coating Material	Cu/SS-ALN
Glass Material	Borosilicate Glass 3.3
Vacuum Range	$<= 5 * 10^{-3}$ Pa
Total No. Of tubes used in the collector	3

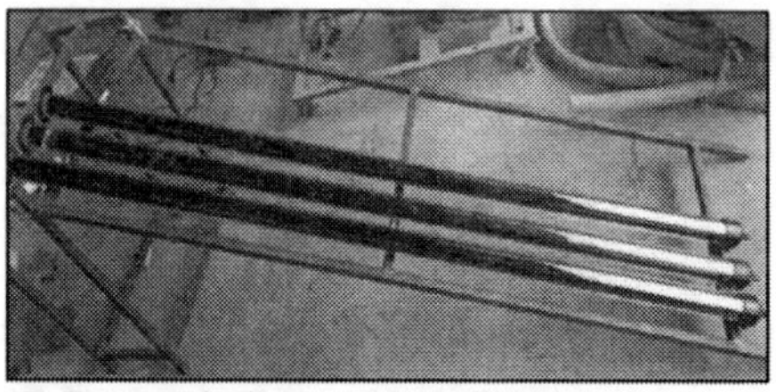

Fig. 5.7. Evacuated tube type solar collector

Fig. 5.8. Experimental setup of VARS using solar thermal energy

6. Observation & Result

The evacuated tube type solar collector was tested for around 2.5hrs, on an average sunny day (04/04/2018) at 11:30 am and the temperature corresponding to the time taken was measured and tabulated below.

Table 2. Collector fluid temperature vs Time

S. No.	Temperature (⁰C)	Time (in min)
1	32.7	0
2	35.6	4
3	40.9	8
4	42.1	12
5	43.6	16
6	49.2	24

7	55.6	32
8	61.1	40
9	65.8	48
10	72.3	60
11	78.2	72
12	81.8	84
13	84.5	96
14	86.8	108
15	89.2	120
16	91.6	132
17	93.8	144
18	94.8	156

Evaporator temperature corresponding to the generator temperature was measured at 3 millibar pressure (maintained with the help of a vacuum pump). In the course of testing for 4 hrs, a maximum drop of 20°C was achieved (Fig. 6.2).

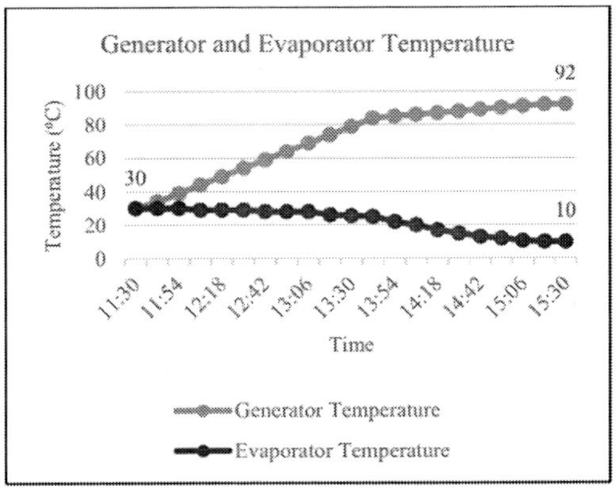

Fig. 6.2. Generator and Evaporator Temperature

7. Conclusion

A vapour absorption refrigeration system was designed and fabricated, with LiBr-H_20 as the working fluid and to be able to be powered by renewable and sustainable heat energy sources. An evacuated tube solar collector was designed for delivering the required power to the system. It was found that the collector was able to reach a temperature of around 92°C in a couple of hours, which in turn facilitated the system such that, a temperature difference of the order 18-20°C was found in the evaporator (cabin) as compared to the ambient temperature. The COP of the system was calculated to be 0.1 and the capacity of the system was computed to be 0.01 TR.

Based on the observation it's conclusive that this system can be used as a viable sustainable alternative to conventional refrigeration systems. The low running cost is an additional cost-benefit for the consumer and since it requires heat energy as input it can be used in remote areas with limited access to electricity. This unit offers an approach to provide billions of people with access to low-cost refrigeration medium while contributing significantly to curb climate change.

Authors thank the Director and HoD of Mechanical Engineering of Medicaps Inst. of Tech. and Mgmt for their support. We thank Dr. V. Ganesan (Director, UGC-DAE CSR Indore) for his encouragement., M.P. Saravanan, Dr. M.M. Patidar, Mr. N. Vijayakumar and the members of Cryogenics Lab of CSR Indore deserve special thanks for their guidance and support.

References

1. V. Bajpai, in *Proceedings of the World Congress on Engineering* (2012), pp. 4-6
2. I. Horuz, International Communications in Heat and Mass Transfer 25, 711 (1998)
3. L. Dominguez-Inzunza, J. Hernández-Magallanes, M. Sandoval-Reyes, and W. Rivera, Applied Thermal Engineering 66, 612 (2014)
4. C. B. Dorgan, C. E. Dorgan, and S. P. Leight, ASHRAE Journal 37, (1995)
5. N. Ajay Sankar and D. S. Sankar, International Journal of Innovative Research in Science, Engineering and Technology (An ISO 3297: 2007 Certified Organization) 4, (2015)
6. S. Arora, S. Chitkara, R. Udayakumar, and M. Ali, Journal of Petroleum and Gas Engineering 4, 74 (2011)
7. V. Mittal, K. Kasana, and N. Thakur, Journal of Energy in Southern Africa 17, 65 (2006)
8. H. Varma, R. Mehrotra, K. Agrawal, and S. Singh, International Communications in Heat and Mass Transfer 21, 539 (1994)
9. A. Handbook and others, ASHRAE Handbook, Fundamentals (2007)
10. K. Karthik, International Journal of Emerging Technology and Advanced

Engineering 4, (2014)

11. B. Babu and G. M. P. Yadav, International Journal of Engineering Research and Management (IJMERM) ISSN 2349 (2015)

12. M. Jamil Ahmad and G. N Tiwari, The Open Renewable Energy Journal 2, (2009)

Heat balance analysis in electric arc furnace for process improvement

Sandeep Ohol[1,*], *Mathew VK*[1], *Savita Shinde*[1] *and G. Balachandran*[2]
[1] MIT School of Engineering, MIT ADT University, Pune (India)
[2] JSW Bellary, (India)

Abstract. The current study deals with optimizing the melting process used in electric arc furnace by heat balance equations. Heat balance is a very important aspect in an arc furnace in which the energy input consists of electrical energy [65%], chemical energy [25%] and exothermic reaction heat [10%]. This energy is optimized with the charge mix, charge quantity, fluxes, fuel used, and O2 used in the burners. The present model considers all these aspects and gives heat distribution in the process. The model spreadsheet gives a reasonable prediction in terms of metal yield, composition, and energy consumption. The model also predicts the amount of iron oxidized in the process. The mass and heat balance model is a useful tool for process analysis and improves the process efficiency of electric arc furnace steelmaking.
Keywords Energy consumption analysis, Electric Arc Furnace, Steel scrap, Heat balance, Optimization

1 Introduction

The use of electric arc furnaces (EAF) for steelmaking has grown dramatically in the last few decades. Presently, around 36 % of the steel produced is made by the electric arc furnace route and this is expected to further increase to 50% in the next few years. Electric Arc Furnace (EAF) is the most common way to recycle steel from scrap. By melting the scrap in a furnace with the help of electrodes and an electric current, new, functional steel can be produced from old products. Once the scrap has been melted and refined to the desired composition and temperature, the molten steel is tapped into a ladle for secondary treatment and casting. The international iron and steel institute classifies EAFs based on power supply per ton of furnace capacity. Power classification ranges are shown in the following Table 1[1]

Vladimir Viktorovich Pavlov and Oxana Sergeevna Logunova conducted the study of charging material at large integrated steel factory. They found that by using alternative material charge, consumption of specific energy is changed[1] Jens Wendelstorf and Karl-Heinz Spitzer developed a process reactor model which will be helpful to predict tapping temperature and meltdown ststus[2]. Dr. Bernd Kleimt and et al. conducted the energy balance of electric arc furnace and developed one dynamic model which will be helpful for online observation and dynamic process control[3]. Dragoljub Gajic and et al. developed one model which predicts the value of electrical energy consumption as a function of the chemical composition the charging mix[5]. Jens Wendelstorf developed online model which demonstrate meltdown of solid material and temperature development in the furnace[6]. Marcus Kirschen et al. conducted mass and energy balance of EAF melting process which gives information about chemical energy input and total energy demand of particular EAF process[7]. Bake lee and Il Sohn reviewed the energy innovations for EAF steel making route and found that preheating of scrap will lower the energy consumption by 90 kWh/t. Bucket type and twin shell preheaters will also be effective in lowering overall power consumption by 60 kWh/t[8]. Biau lo et al. analysed flow characteristics of ferrite-flows and established an energy intensity production optimization model for production system in iron and steel industry. They suggested Unified resource deployment is the basic guarantee of the implementation of the optimization scheme. Therefore, resource sharing should be fully considered during the iron and steel company process design stage. The advanced energy saving technologies and measures, which favors production route re-optimizes, should be adopted in actual production process[9].

Using ultra-high power furnaces it is possible to increase productivity, reduce tap to tap time, and reduce electric energy consumption [1]

*Corresponding author: sandeepashu31@gmail.com

Table 1 Electric power classification in the electric arc furnace

Range of power (KVA/t)	Furnace type
100-200	Low power
200-400	Medium power
400-700	High power
700-1000	Ultra-high power

2 Experimental work

2.1 Experimental set up

For this study, the electric arc furnace at Kalyani Carpenter Special Steels Limited, (KCSSL) Pune is used. The EAF at KCSSL has a capacity of 35 MT and it is equipped with a 22 MVA transformer. The details of the furnace used for the study is shown in following Electric Arc Furnace at KCSSL.

- •Make: ABB
- •Melting capacity: 35 MT
- •Transformer Power: 22 MVA
- •Tapping: Slide gate assembly
- •Two virtual lance burners
- •Bucket Charging: Minimum Two charges per heat
- •Continuous DRI feeding Conveyor
- •Charge Mix: In house scrap; forging flashes, HMS, Pig iron, DRI

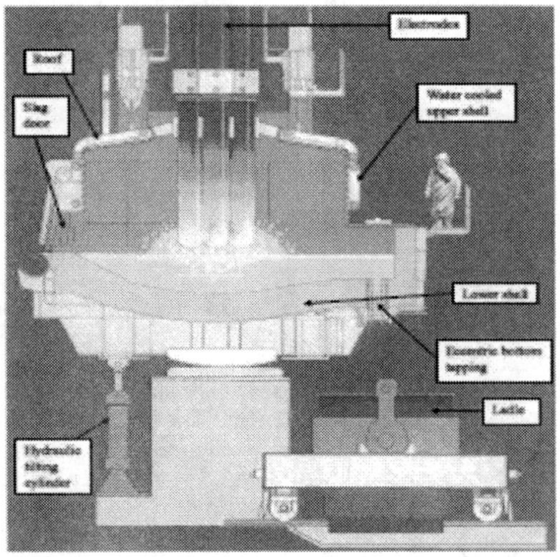

Fig. 1 Typical Electric Arc Furnace

There were one batch of experimentation that was carried out in the furnace

2.2 Melting Technique

Initially, the EAF roof is opened and about half of the total charge is charged into the furnace. After the initial charge, the roof is replaced on the furnace top and the furnace is powered on with a graphite electrode to generate electric arc. During the melting period, Virtual Lance Burners (VLB) run as per the computer preset program. The current and voltage selection is initially set and a profile for the melting set as per designed experimental requirements. After the initial lot is melted, the roof is opened and the second charge of another 30% of charge is added. After the second charge, the furnace is powered on and the VLB switched on. When all the 80% of charge is melted, a temperature reading is taken. When the temperature was greater than 1570°C The first metal sample for chemical analysis is taken. This is followed by the final 20% charge addition consisting mainly of DRI, using a conveyor belt feeder. At the various stages of the above operation, metal and slag samples were taken for chemical analysis. The process events were continuously recorded from start to end. On attainment of desired chemical composition, the heat is tapped into a ladle for secondary refining. The typical processing cycle carried out in the experimental batch is shown in Table 2 The heat input, in terms of electric arc, was gradual. The various processing times at various stages such as charging, melting, refining, sampling, foaming, and tapping are given.

Table No. 2 EAF process details for the experimental batches

Batch 1		Batch 2		Batch 3	
Time (Min.)	Process Details	Time (Min.)	Process Details	Time (Min.)	Process Details
0 – 3	The first charge added	0-3	First charge added	0-3	First charge added
3 – 23	First charge melting	3-52	Furnace melting	3-23	First charge melting
23 – 26	The second charge added	52-55	Second charge added	23-26	Second charge adeed
26 – 44	Second charge melting	55-74	Second charge melting	26-44	Second charge melting
44 – 45	First slag sample is taken	74-75	DRI feeding started	44-45	First slag sample taken
45 – 46	DRI feeding started	75-77	First slag sample taken	45-46	DRI feeding start
46 – 50	Temperature and the first metal sample is taken for analysis	77-79	Temperature and first metal sample taken for analysis	46-50	Temperature and first metal sample taken for analysis

50 – 55	Second slag sample is taken/Foaming and temperature raising	79-81	Second slag sample taken	50-54	Second slag sample taken. Foaming and temp. raising
55 – 56	Third slag sample is taken	81-86	Second and third slag sample taken	54-55	Third slag sample taken
56 – 58	Final coke injection	86-89	Foaming and temp. Raising	55-58	Final coke injection
58 – 59	DRI feeding ends	89-91	Forth slag sample taken	58-59	DRI feeding ends
59 – 62	Tapping temperature recorded	91-94	Final coke injection, DRI feeding ends	59-60	Tapping
62 – 65	Fourth slag sample is taken/ Tapping starts	94-96	Tapping	61-62	Fourth slag sample taken, Tapping start

The process schedule for Batch-1 and Batch-3 were almost same. In Batch 2 a prolonged processing was carried out, where the melting was carried out in a progressive manner. The arcing was very gradual for initial charge melting till 80% of the charge. Parameters such as power consumed during the process due to electric energy input, fuel combustion through lance burners and carbon injection is shown in Table 3

Table No. 3 EAF input data for three batches

Parameter	Batch 1	Batch 2	Batch 3
Charge mix, (MT)	39.50	36.00	37.50
Electric energy consumption, (kWh/MT)	384	424	428
VLB oxygen consumption (NM^3/MT)	33.19	45.04	44.08
VLB Fuel consumption, (Lit/MT)	0.0045	0.0057	0.0052
Carbon injection, (kg/MT)	0.017	0.009	0.011

The metallic and slag output from the furnace is shown in Table 4

Table No. 4 EAF output data for three batches

Parameter	Batch 1	Batch 2	Batch 3
Theoretical metallic output, MT	36.612	34.894	34.421
Actual metallic output, MT	37.420	32.900	33.820
Metallic loss, MT	-0.807	1.994	0.6
Theoretical slag output, MT	6.273	4.920	5.396
Actual slag output, MT	6.320	5.500	5.670
Slag loss, MT	-0.046	-0.579	-0.273
Yield %	94.73	91.38	90.18

The theoretical energy required to convert solid steel to molten condition is in the range of 350 kWh/ton. The EAF steelmaking is only 55 to 65 % efficient and as a result the total equivalent. energy input is usually in the range of 400 to 550 kWh/ton in modern operations. The energy input consists of more than half the energy input for the EAF is from electrical energy. Chemical energy from fuel contributes about 40% and 10% can come from the exothermic reaction heat. The chemical energy can be from the calorific value of the fuel the rate of heating is controlled by the fuel and oxygen burning rate. The output energy distribution shows that 60% of heat energy is contained in the superheated steel, slag carries about 10%; other heat losses are about 30%. Minimizing this can save process energy.

Table No. 5 Distribution of heat (in %) from input to output in a typical EAF

Heat Input		**Heat Output**	
Parameter	**%**	**Parameter**	**%**
Electric energy	56-60	Liquid Steel	55-60
Chemical energy	30-40	Cooling water	8-10
Burners	5-10	Miscellaneous	1-3
		Slag	8-10
Total	**100**	**Total**	**100**

Factors such as raw material composition, power input rates, and operating practices can alter the above balance. In operations utilizing a large amount of charge carbon or high carbon feed materials, up to 60 % of the energy contained in the off-gas may have high calorific due to large quantities of uncombusted carbon monoxide. The recovery of this energy in the EAF could increase energy input by 8 to 10 %. Thus, it is important to consider such factors when evaluating the energy balance for given furnace operation.

3 Governing equations

3.6 Governing equations for mass balance

1) Governing equation for this mass balance model is as

$\Sigma M_{input} = \Sigma M_{output} + \Sigma M_{ls}$ ....(1)

where $\Sigma M_{input} = M_{Scr} + M_{flx} + M_{ck} + M_{dm} + M_{bd} + MO_2^{blown}$ ....(2)

$\Sigma M_{output} = M_{st} + M_{sl} + M_{dust} + M_{gases}$ ....(3)

2) Iron balance

$W_{Fe}^{DRI}.M_{DRI} + W_{Fe}^{Scr}.M_{Scr} + W_{Fe}^{Flx}.M_{Flx} = W_{Fe}^{St}.M_{St} + W_{Fe}^{Sl}.M_{Sl} + W_{Fe}^{dust}.M_{dust}$ ......(4)

3) Silicon balance

$W_{Si}^{DRI}.M_{DRI} + W_{Si}^{Scr}.M_{Scr} + W_{Si}^{Flx}.M_{Flx} = W_{Si}^{St}.M_{St} + W_{Si}^{sl}.M_{sl}$.........................(5)

Total

$M_{Scr} + M_{DRI} + M_{Flx} + M_{O2}^{blown} + M_{coke} = M_{st} + M_{sl} + M_{dust} + M_{gases}$.............................(6)

Where

W_i^j = Weight percent of specie i in phase j M_{input} = Mass of all input scrap material.

M_{output} = Mass of all output material

M_{ls} = Mass of losses

M_{Scr} = Mass of scrap

M_{Flx} = Mass of fluxes

M_{ck} = Mass of coke

M_{dm} = Mass of deoxidizing material

M_{bd} = Mass of fuel

M_{O2}^{blown} = Mass of O_2 blown into the furnace

M_{st} = Mass of steel

M_{sl} = Mass of slag

M_{dust} = Mass of dust M_{DRI}

Mass balance analysis of steel production process has been carried out according to the input component, and using mass conservation law as mentioned in above equations.

4 Results and discussion

Heat balance is another important aspect of an electric arc furnace which enables optimization of process conditions to minimize the energy consumption. This enables energy management from the charge point of view. The broad energy parameters for three experimental batchs is shown in Table 6

Table No. 6 Heat balance for the experimental batches

Fuel used Kg	Fuel heat in kWh/MT	Electric heat in kWh/MT	Chemical heat kWh/MT	Total energy consumed kWh/MT
170	47.02	384.00	238.08	622.08
187	58.83	424.00	227.24	649.42
176	53.86	428.00	287.80	706.88

The Fig.2 shows that fuel used is moderately higher for Batch-2. The electrical heat input increases from Batch-1 to Batch-3. The heat input due to stoichiometric exothermic heat generated by various steel making reactions is moderately higher for Batch-3 compared to the other two batches. The overall heat used in the process increases from Batch 1 to Batch 3 progressively. It is seen that average 63% of heat comes from electrical energy and average 37 % of heat from fuel burnt in the furnace and the exothermic reaction heat varies from 7 to 9% due to exothermic reaction. This is important as minimising electrical energy saves energy cost, and this energy itself is produced with fuel combustion elsewhere. Hence, usage of fuel in furnace to generate heat is good. The present experiments shows 34 to 40% and possibility of enhancing this should be explored. However, this also increases emissions in the factory and the cost of this and impact on surrounding have to be assessed. The exothermic heat is <9%. To maximise this may be increasing carbon containing charge such as pig iron may be of helpful. The impact of this on the energy saving and productivity has to be assessed. Further pig iron is more expensive than scrap. The model developed can now be run to predict the theoretical power generation for a given charge mix.

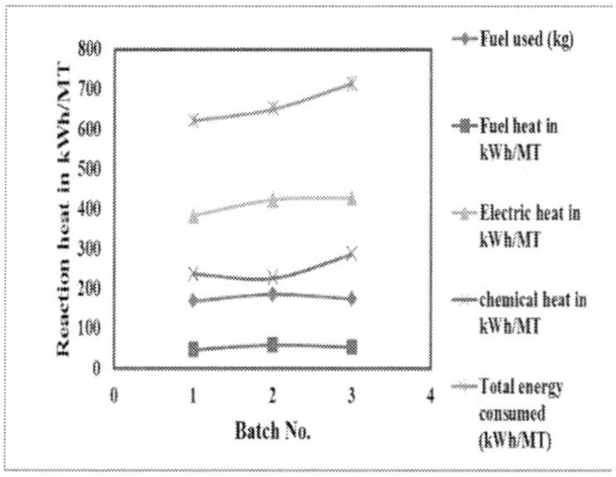

Fig. No. 2 Heat balance in the experimental batches

5. Summary and Conclusion

Three experimental trials were carried out in electric arc furnace at KCSSL to understand the process behaviour at various stage of processing. The elemental and slag constituents were tracked by studying the chemistry of the raw materials input and the output metal and slag obtained in the furnace. The model does the energy balance in the process and it is possible to optimise the electric energy consumption and fuel based energy and exothermic heat inputs. Using the model, it is now possible to reduce the process time, fuel, electric energy and oxygen consumption.

Acknowledgement

We thank the organisation KCSSL for providing the experimental facility.

References

1 Jeremy Jones, The EPRI centre for material production, C M R I, Pittsburgh, PP **1** (1997)

2 F. P. Edneral, vol. 1 Electric steelmaking, Mir publishers, Moscow, PP **17** (1979)

3 Richard J. Fruehan, AISI steel foundation, 11th edition, steelmaking and refining volume, chapter **10,** (1998)

4 Yuri N. Toulouevski and Iiyaz Y. Zinurov, Scientific basis for selection, Springer, pp **1**, (2010)

5 Ali Akbar Mottahedi et al, Int. J. of chemTech re, Vol.1,No1, pp **62** (2009)

6 F. Grobler and R.C.A. Minnitt, The Jr. of South-African inst. of mining and metallurgy, (1999)

7 R. J. fruehan, Y. Li, L. Brabie, Carnegie Mellon University, Pittsburgh.

8 Richard J. Fruehan, the AISI steel foundation, 11th edition, steelmaking and refining volume, chapter **2** (1998)

9 Jernkontoret, Sweden. [online]. Available: http://www.jernkontoret.se(2006, May 17).

10 Jacob paulraj, Steel world, pp **26** (2007)

11 Marcus Kirschen, Victor Risonarta, Herbert Pfeifer, Energy 34, pp **1065**, (2009)

12 Ismail Ekmekci, YasarYetisken, UnalCamdali, Int. Jr. of iron and steel re., 14(05) pp **01,** (2007)

13 Karl R. Haapala and John W. Sutherland(20), Annals of the CIRP vol. pp **1**, (2007)

Parametric Design and Comparative Analysis of a Special Purpose Flexure Spring

Suraj Bhoyar[1*], *Virendra Bhojwani*[1] *and Sudarshan Sanap*[1]
[1]Dept. of Mechanical Engineering, School of Engineering, MIT-ADT University, Pune, India.

Abstract. Flexure springs play pivotal role to ensure reliability and durability of the cryocooler compressors. The critical application of flexure spring intends to eliminate the wear, frictional losses and significantly vibrations. A parameter study was performed considering the design variables of a flexure spring to meet the requirements of the cryocooler compressors viz. complexity of the geometric configuration, radial and axial stiffness, fatigue life, material, etc. The parametric design approach was deployed using finite element analysis (FEA), to validate the same by experimental data to analyse the performance of the flexure spring. This paper also presents a comparative analysis of the design variables in terms of axial and radial stiffness, induced stress as well as the manufacturing methods' considerations to optimize the design of a special purpose flexure spring for the cryocooler compressors. Furthermore, experimental study of the flexure spring was carried out in order to maximize the stroke.

1 Introduction

Cryocooler originates from the phrase cryogenic cooler which provides active cooling down to cryogenic temperatures. Cryocoolers finds versatile applicability across pulse tube, Stirling, GM, Joule Thompson industrial applications, etc. The wide applications are discussed in the following table 1:

Table 1. The wide applications

Domain of applications	Applications of Cryocoolers
Space missions	Observing in the infrared, gamma-ray and x-ray spectrums, Satellite imaging
Police and Security	Infrared sensors for night vision goggles; surveillance
Defense	Infrared sensors for missile guiding control used in tactical applications; satellite based Infrared sensors assisting surveillance and monitoring
Commercial	Used in Cryopumps for semiconductor manufacturing and fabrication; High temperature superconductors for telecom applications; Semiconductors for high speed computing devices; Programmable Focusing Mechanism
Medical	Cooling magnets for MRI scanning systems; SQUID magnetometers for heart and neural studies; Liquefaction of oxygen for warehouse storage at hospitals; Cryogenic catheters and cryosurgery applications
Transportation & Logistics	LNG integration for fleet vehicles; Super conducting magnets in maglev trains/metros
Energy Supplies	Liquified Nitrogen for storing surplus energy; Infrared sensors for thermal loss detection; Super Conducting magnetic energy storage for peak shaving and power reconditioning; power applications (motors, transformers, etc.)
Agriculture and Biology	Storage of biological stems, cells and specimens for experimentation and research

Justifying the versatility of cryocoolers in general, the salient features are high resolution, high precision, high sensitivity, high power density, low power and low noise. The major concern for the applications of cryocoolers relies on the reliability for the development of tactical cryocoolers.

Cryocoolers requires high reliability in performance and high operating life. The linear motors are preferred over the rotary compressors because rotary compressors apply large radial forces and cause large amount of wear and also require lubrication. Lubrication is the common pain point in cryocoolers. For maintenance free operation, longer life for cryocoolers, a special purpose flexure spring is designed. The introduction of the Flexure spring technology enables more reliability. Flexure springs play pivotal role to ensure reliability and durability of the cryocooler compressors. The critical application of flexure spring intends to eliminate the wear, frictional losses and significantly vibrations. The flexure spring operates by bending of spiral arms.

The flexure spring boasts the following advantages:

- Simple and inexpensive.
- Compact, lightweight, low friction.
- Manufactured by Chemical itching, wire cut EDM, WJM.
- Long life and highly reliable.
- Noiseless during operation.

*Corresponding author: suraj.bhoyar@mituniversity.edu.in

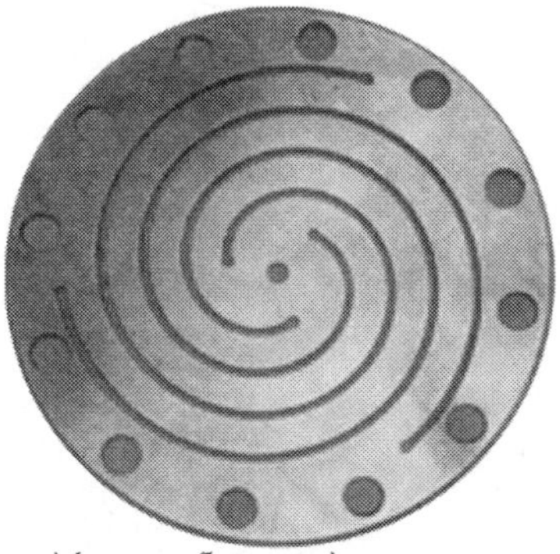

Fig. 1 A special purpose flexure spring

Fig.1 shows the special purpose flexure spring.
A parameter study was performed considering the design variables of a flexure spring to meet the requirements of the cryocooler compressors viz. complexity of the geometric configuration, radial and axial stiffness, fatigue life, material, etc. The parametric design approach was deployed using finite element analysis (FEA), to validate the same by experimental data to analyse the performance of the flexure spring.

2 Research Motivation

A flexure spring mostly relies on the material which can be repeatedly flexed without disintegrating; so, the selection of the material becomes critical for the same. Most of the metals will tend to experience fatigue with repeated flexing, and will ultimately snap. Another point for selection of Flexure springs revolves around very low but uniform friction.

The essence of special type flexure spring lies in the fact that it acts a spring in axial direction (exhibiting low stiffness) and it acts as a bearing in radial direction (exhibiting high stiffness).
The parametric study was conducted to meet life requirements of cryocooler compressors especially for space applications. The parameter study approach was deployed to identify the optimum geometrical configuration; and it was achieved through the CAE package and validated through the experimental results.
As mentioned, the preferable way to achieve high operating life and reliability is by refraining physical contact between the piston and cylinder encompassed in the compressor. The supporting piston mass at the front and back side with high radial stiffness using flexure springs would be the most competent way to accomplish the same.

3 Literature Survey

The very concept of a flexure spring was put forward by A. Wolf et al. in 1938 and later granted with the patent for it [1]. The application was intended for capturing the Earth's Vibrations using mounting in a vibration detector. University of Oxford, late in 1981, had employed flexure springs in a Stirling cryocooler first time for the research purpose [2]. In 1992, Using Finite Element Method, the analysis was carried out by Wong et al. and proposed optimization of a three spiral flexure bearing. The same was validated with the experimentation. The researchers found that the radial stiffness of flexure bearing gets decreased with axial displacement and the maximum stresses get induced at the spiral cut ending [3].
Marquardt et al. presented that the ratio of radial to axial stiffness of the flexure spring could be considered to choose the appropriate geometrical configuration in correlation to the design parameters for flexure bearings [4].
The correlation was further insisted and presented by Wong et al.to provide the conclusion on static and dynamic testing using Finite Element Method for a three spiral cut flexure geometry. The results were validated through experimentation and concluded that the dynamic stresses should be considered for the fatigue analysis [5].
Analysis of a three-spiral flexure spring by Finite Element Analysis approach was done by Gaunekar et al. and proposed that axial displacement leads to increase in the stresses. Also, they pointed out that the axial stiffness and radial stiffness of the geometry illustrated a linear behaviour when plotted against the axial displacement of the spring. The researchers had developed normalised graphs for three parameters

1. Maximum stress,
2. Radial stiffness and
3. Axial stiffness with normalized axial displacement.

The findings were validated using experimentation with a fibre–optic interferometer technique and a simple dead weight methodology. The major findings were that the axial and radial stiffness of the flexure geometry is directly proportional to spring material thickness, which resulted in normalised graphs which would be stepping stone for the flexure bearing design considerations and optimization [6].
Z. S. Al-Otaibi and A. G. Jack considered a flexure bearing design for a linear motor supported with experimental investigation and validation of the results. The research carried out projected that an increase in spiral angle decreases the stresses and thereby axial stiffness of the disc, while the axial stiffness increases with the thickness of the flexure geometry [8]. These research outputs were also later reinforced by C.J. Simcock et.al who performed iterations using strain gauges and studied a flexure spring to identify bearing stresses [9].
Recently, Amit Jomde et al. had analysed the various parameters to find equivalent stress and stiffness for beryllium copper UNSC 17200 material. the researchers considered the parameters focusing on performance and service life at lower operating stresses and higher stiffness for the given application [10].
Yingxia et al. had proposed finite element simulation for the spring grooved with spiral slits under deformation. The research findings were carried out for the deformation

pattern, the stress distribution, and the natural frequency of the flexure spring with three spiral slots [11].

The performance of Oxford and triangle flexure bearings were investigated by Wenjie Zhou et al. The researchers had put forward that the triangle arm flexure bearings were responsible for larger radial/axial stiffness ratio, and the size of linear compressors could be miniaturized considering the adequate parameters [12].

4 Methodology

This research work presents a comparative analysis of the design variables in terms of axial and radial stiffness, induced stresses considerations to optimize the design of a special purpose flexure spring for the cryocooler compressors. Furthermore, experimental study of the flexure spring was carried out in order to maximize the stroke of displacement.

This research investigation has considered the following design approach:

1. To achieve higher mechanical spring stiffness – compact size
2. To accomplish lower operating stresses and infinite fatigue life
3. To identify Optional material with better mechanical properties
4. To find best geometrical configuration

The most important design considerations for the special purpose flexure spring were:

- Fatigue Strength
- Radial Stiffness
- Axial Stiffness

Practically, a single unit of flexure spring disc would not perform and deliver the total desired stiffness. So, a stack of flexure springs was used in a parallel arrangement for the application. it means that multiple flexure springs were kept in parallel to achieve the desired stiffness. The total stiffness was the sum of individual discs arranged through the arrangement.

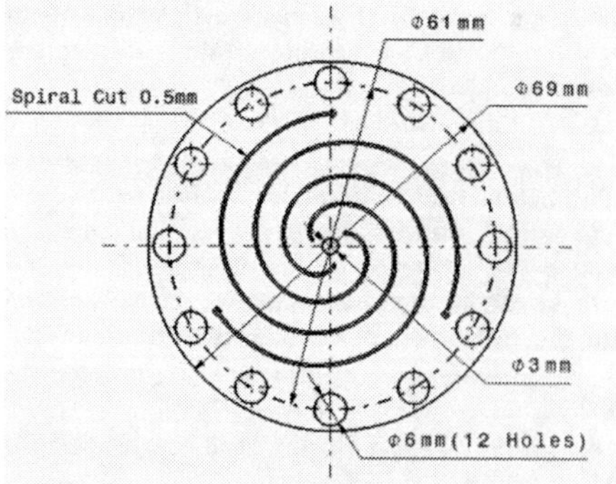

Fig. No. 2 Spiral flexure spring geometry

Figure 2 shows the spiral flexure spring geometry considered for the experimental investigation. Each flexure spring unit investigated had three spiral arms which would sustain the axial as well as radial loads. Each spiral arm traverses through an angle. The holes at periphery of the spring were provided to fix the disc rigidly for the experimentation purpose. To mitigate the misalignment in radial direction, the holes are slightly oversized with reference to bolt size for ease of mating components. The central hole was to allow shaft to get fitted. An abrupt change or discontinuity in spiral arm slit geometry leads to stress concentration. A material can fail potentially, via a propagating crack, when a concentrated stress exceeds the material's theoretical cohesive strength. Fatigue cracks always start at stress raisers, so removing such defects increases the fatigue strength. By drilling a hole at the end of the cut, the stress concentration may be reduced. The drilled hole, with its relatively large diameter, causes a smaller stress concentration than the sharp end of a cut.

The schematic illustration for an assembly of flexure springs is shown in Figure 3. The adequate numbers of flexure springs were stacked parallelly, in order to accomplish the intended axial and radial stiffnesses. During the experimental setup, the flexure springs were deflected through 5 mm stroke operating at the frequency of 50 Hz.

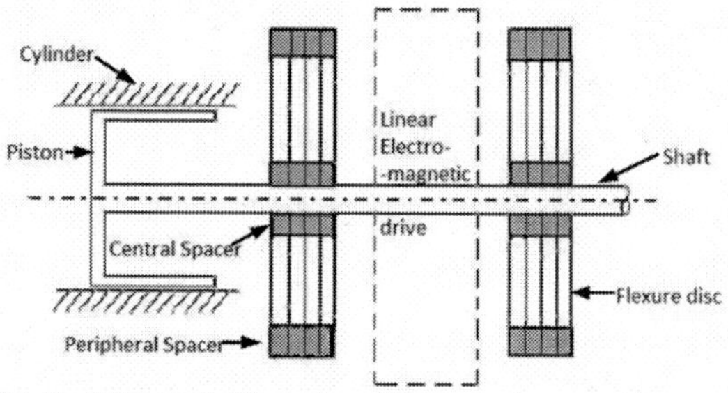

Fig. No. 3 Schematic Illustration of flexure spring assembly

Parametric considerations for FEA:

- Geometry of flexure spring
- Number of spiral arms
- Width of cut for the spiral/flexure arm
- Angle of traverse for spiral arms
- Disc thickness
- Effect of diameter of the spring
- Modification of the areas with higher stresses

Considering the above parameters individually, the exhaustive iterations through FEA were carried out. The geometry selected for the flexure spring was the spiral with helical arms. Then, the following materials were considered for the selection of the material for the flexure spring:

1. Stainless steel 301
2. Stainless steel 304
3. Cold drawn copper
4. Beryllium copper UNS C17000
5. Beryllium copper UNS C17200 (TH02)

So, it has been concluded to proceed with Beryllium copper UNS C17200 (TH02). It has been found that the material gives maximum stiffness.

For Finite Element Analysis, the following inputs were considered for further investigation:

- Thickness of flexure spring - 0.3 mm

- Material for flexure spring – Beryllium Copper UNS C117200 (TH02)
- Axial displacement – 5 mm from mean position

Then FEA was carried out to find out the no. of spiral arms keeping above inputs in consideration.

The observations are depicted using the following Fig.4

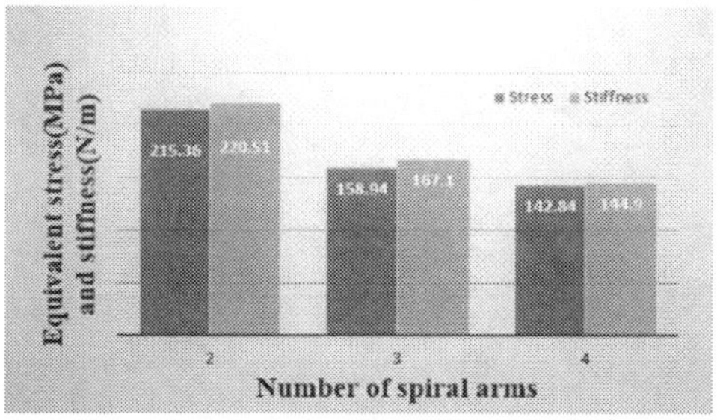

Fig. 4 No. of spiral arms vs eq. stress & stiffness

So, considering the best suitable results, three spiral geometry was finalized for the further investigation.

The finite element analysis for the selection of the width of spiral cut was done considering following options for the same:

1. 0.2 mm
2. 0.4 mm
3. 0.6 mm
4. 0.8 mm and
5. 1.0 mm

The observations recorded after doing analysis are illustrated in Fig 5

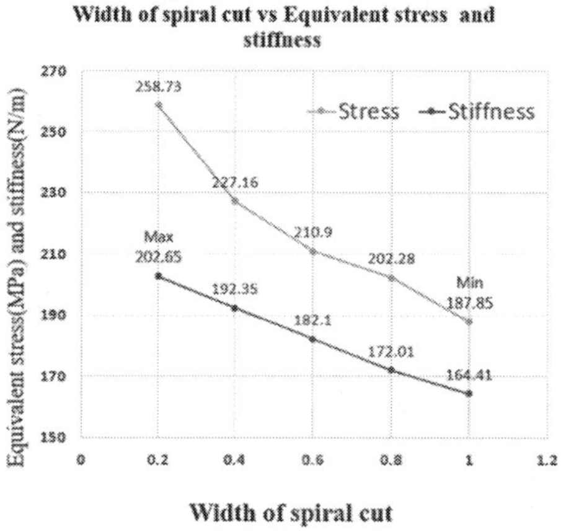

Fig. 5 Width of spiral cut vs Eq. Stress & Stiffness

From the results obtained, it has been concluded that as width of spiral cut goes on increasing, stiffness and equivalent stress both goes on decreasing gradually. For optimum design maximum stiffness and minimum stress is the prime criteria. If maximum stiffness is to be considered 0.2 mm spiral width is suitable and minimum stress is considered 1mm spiral width is suitable. Considering the design calculations, no. of flexure springs required would be more which is not acceptable.

The manufacturing for the flexure spring was considered by photochemical etching process and it would be difficult to get spiral cut less than 0.5 mm. So, taking the limitation in account, suitable width for spiral cut of 0.6mm has been taken.

For further analysis using FEA, the following inputs were considered for further investigation:

- Thickness of flexure spring - 0.3 mm
- Material for flexure spring – Beryllium Copper UNS C117200 (TH02)
- Axial displacement – 5 mm from mean position
- No. of spiral arms – 3
- Width of spiral cut – 0.6 mm

Then FEA was carried out to find out the spiral angle keeping above inputs in consideration. The following variations are considered:

1. 360°
2. 420°
3. 480°
4. 540°
5. 560° and
6. 570°

The observations are shown using the following Fig. 6

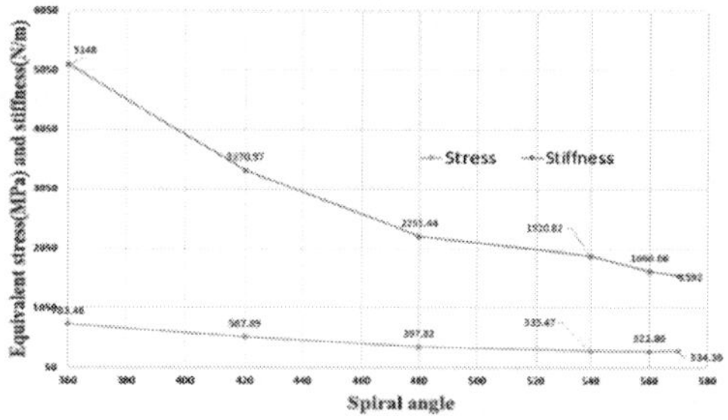

Fig. 6 Spiral Angle vs Eq. Stress & Stiffness

The results discussed in table 3 depicts that for 560° spiral angle minimum stress (322.86MPa) has been obtained and corresponding stiffness is 1666.66 N/m. So, spiral angle of 560° was selected for the further investigation.

The disc thickness as a variable parameter was considered later and the following variations were taken into account for optimum value:

1. 0.30 mm
2. 0.50 mm
3. 0.70 mm
4. 1.0 mm and
5. 1.25 mm

The following Fig. 7 shows the results for variation in disc thickness.

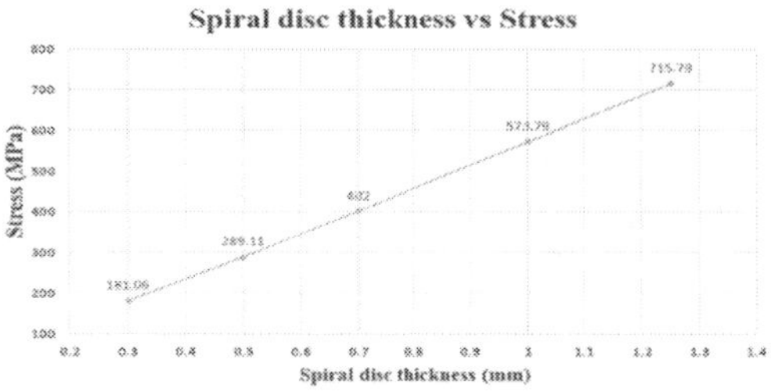

Fig. 7 Spiral disc thickness vs Stress

The stress induced in 1.00 mm and 1.25 mm thick material is more than its permissible limit, so 1.00 mm and 1.25 mm thickness are not suitable for flexure spring geometry. So, the simulation for deformation was carried out for 0.3 mm and 0.7 mm thick disc. It has been observed that, the stiffness value is more than 0.7 mm thick disc compared to 0.3 mm. And also, no. of disc units required for stacking was less in case of 0.7 mm. Hence, the value was considered for further evaluation.

5 Results and Discussion

The output from the parameter study carried out are as follows:

- Geometry of flexure spring – Spiral with helical arms
- Material selected - Beryllium copper UNS C17200 (TH02)
- Number of spiral arms - 3
- Width of cut for the spiral arm – 0.6 mm
- Angle of traverse for spiral arms - 560°
- Disc thickness – 0.7 mm

Table 2 Results and remarks

Parameter	Optimized value	Remarks
Flexure arm type	Spiral	Lowest stresses
Material	Heat treated Be-Cu (Yield = 1100 MPa)	High FOS
No. of spirals	3	Lower stress
Spiral Width of cut	0.6 mm	Manufacturing feasibility
Spiral angle	560^0	Low induced stresses
Disc thickness	0.7 mm	Max stiffness (FOS=2)
Disc Diameter	69 mm	For compact size

The above-mentioned parameters were evaluated against static stress and stiffness only. The flexure spring gets subjected to cyclic loading also. It becomes obligatory to verify whether these parameters are safe to fatigue loading. With reference to the above results, the value for stress was evaluated by comparing with S-N curve of the selected material.

The equivalent stress for 5 mm displacement and deformation was investigated, and the value was of 322.86 MPa. This stress later was used for finding out fatigue cycles by S-N curve to 10^8 cycles. The fig. 8 indicates the S-N curve for Be-Cu Alloy.

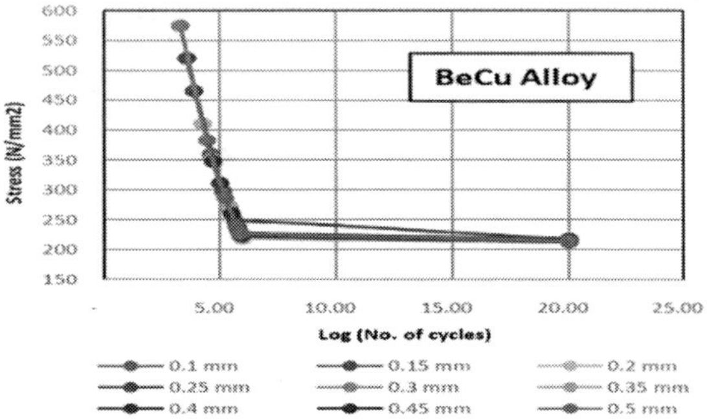

Fig. 8 S-N Curve of Be-Cu Alloy flexure spring

6 Result Validation using experimental set-up

The main objective of the parameter study was to optimize the design variables, geometric configuration, radial and axial stiffness, fatigue life, material, of a flexure spring. The endeavours have been taken in the research work to enhance the performance and improve upon the service life by achieving higher mechanical spring stiffness and lower operating stresses. As the flexure spring subjected to cyclic stress condition, the fatigue problem must be taken into consideration at the design stage in order to have flexure spring a long lifetime. Therefore, the validation of the parametric values achieved by FEA was critical and was carried out using three different set up namely:

1. Stiffness measurement test set-up.
2. Stress measurement test set-up.
3. Fatigue life test set-up.

The stiffness measurement was done using dead weights and dial gauge assembly. For the stress measurement, strain gauges were attached to the flexure spring surface. The surface was prepared with utmost care to ensure reliable bonding. And, the same had been integrated with multi-channel strain gauge data acquisition system for the better outputs in terms of strain value.

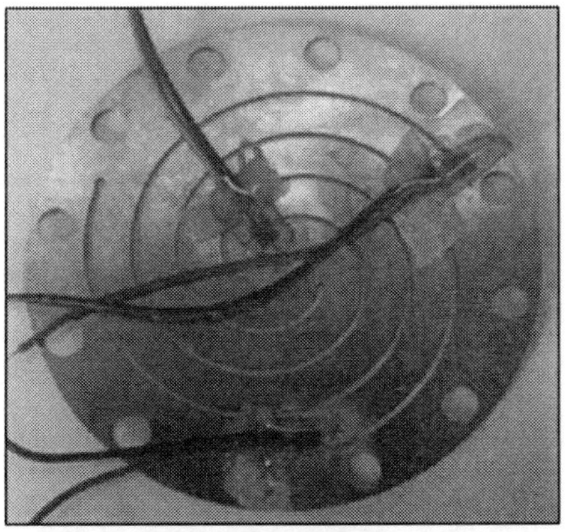

Fig. 9 Strain Gauge glued for readings

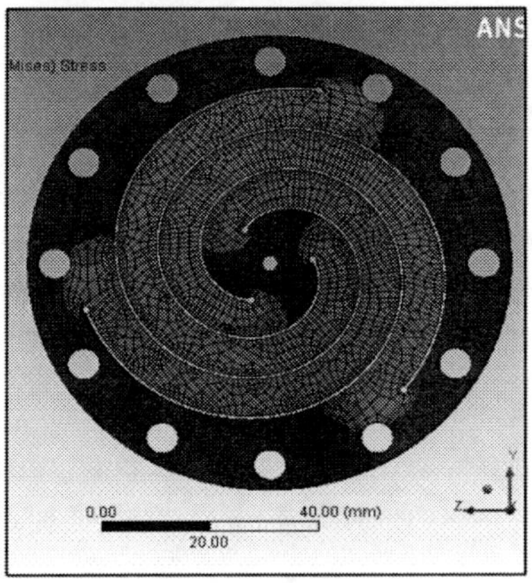

Fig. 10 Ansys image for the location fixing

The fatigue life setup had been fabricated considering the geometrical configurations for the flexure spring validation. The CAD model is shown in fig. 11 and actual setup in fig. 12

Fig. 11 CAD Model for fatigue life setup

Fig. 12 Actual fatigue life setup

The fatigue life predicted by the S-N curve was compared with the experimental number of cycles undergone by the component before failure.

Table 3 Result Validation

Stiffness Validation	
FEA results (N/m)	**Experimental results (N/m)**
1666	1729

Stress validation		
Fix point	**FEA results (MPa)**	**Experimental results (MPa)**
A	**139.68**	**127.55**
B	**168.39**	**130.38**
C	**39.76**	**9.67**

Fatigue Life Validation	
FEA results (N/m)	**Experimental results (N/m)**
10^8 cycles	10^8 cycles without failure

7 Conclusion

The design variables and geometric configuration parameters for a flexure spring were varied using FEA in order to determine their influence on axial and radial stiffness, as well as on maximum occurring stresses. The parameter study approach presented in the research work can be referred for optimization of flexure springs different specifications.

The summary for the research findings is:

- The Parametric study using FEA was carried out for design optimization.
- The stresses, stiffness, and fatigue life (using S-N curve) were predicted by FEA.
- The results were validated by experimental setup for stiffness, stress & fatigue life.
- The natural frequency for the flexure spring decreases with the spiral angle and no. of spiral arms.

- The parametric design approach for the flexure springs seems compromising in nature when stiffness ratio and the maximum stresses need to be considered.

8 Future Scope

- The experimental investigation and validation for stiffness measurement has been going on.
- The fatigue life testing is being done with the modular arrangement for the different diameters of the flexure discs.
- The future objective of the work will be mainly to achieve the maximum stroke and to study non-linearity.
- The optional material with better mechanical properties for the flexure disc will be tried out for improvement in service life.
- Design approach will be optimized for different applications.

9 Applications

The versatility of flexure springs being a spring in axial direction and being a bearing in radial direction gives a cutting-edge advantage as far as applications are concerned. The applications of flexure springs are extensive in nature and can be found across following industries like:

- Space Technology
- Automotive Sector
- Energy Sector
- Medical & Diagnosis
- Machine tools Manufacturing
- Micro-manufacturing, MEMS and NEMS
- Electronics & Telcom sector
- Optics and Image capturing technology
- Pharmaceutical industries

10 Acknowledgement

The research was funded by ISRO, SPPU ST Cell, Pune, India. Also, the research work would not have been possible without the inputs and resources provided by Mr. F.H. Kharadi.

References

1. Wolf A. et al., (1938), Vibration Detector US Patent N. 2,130,213.
2. Davey, C., "The Oxford University Miniature Cryogenic Refrigerator", International Conference on Advanced infrared Detectors and Systems, London, (1981), 39.
3. Wong, T.E., Pan R.B. and Johnson A.L.,"Novel linear flexure bearing", in: Proc 7th Int. Cryo Conf, (1992), 675-698.
4. Marquardt, E., Radebaugh, R. and Kittel, P., "Design equations and scaling laws for linear compressors with flexure springs", in: Proc 7th Int Cryo Conf, (1992), 783-804.
5. Wong T.E., Pan R.B., Marten H.D., Sve C., Galvan L. and Wall T.S. "Spiral flexural bearing" in: Proc 8th Cryo Conf, (1995), 305-311.
6. Gaunekar A. S., Goddenhenrich T., Heiden C., "Finite element analysis and testing of flexure bearing elements", in: Cryogenics 36, (1996), 359–364.
7. Lee C. C. and Pan R. B., "Flexure bearing analysis procedures and design charts", in: Cryocoolers 9, Plenum Press, New York, (1997), 413-420.
8. Al-Otaibi Z. S., Jack A. G., "Spiral flexure springs in single phase linear-resonant motors", in: Proceeding of: 42nd International Universities Power Engineering Conference (UPEC2007), 184-187.
9. Simcock C. J., "Investigation of materials for long life, high reliability flexure bearing spring for Stirling Cryocooler Applications", in: Cryocoolers 14, (2007), 335-343.
10. Amit Jomde, A. Anderson, Virendra Bhojwani, Fayaz Kharadi, Suhas Deshmukh, Parametric analysis of flexure bearing for linear compressor., Science Direct, Material Today: Proceedings 4 (2017) 2478-2486.
11. Yingxia Qi, Yinong Wu, Hua Zhang and Xi Chen, Simulation of deformation of diaphragm spring grooved with spiral slits by FE method, DETC2009-86371, ASME (2009) International Conference, IDETC/CIE 2009.
12. W.J.Zhou, L.Y.Wang, Z.H.Gan, R.Z. Wang, L.M.Qui and J.M.Pfotenhauer, The performance comparison of Oxford and Triangle flexure Bearings, Advances in Cryogenic Engineering, AIP Conf. Proc. 1434, 1149-1156 (2012)

Modelling and Performance Analysis of Electric Car-Sharing Systems Using Petri Nets

A. Hamroun[4], *K. Labadi*[*1,2,3], *and M. Lazri*[4]
[1]*ECAM-EPMI, 13 Bld de l'Hautil, 95092 Cergy-Pontoise, France*
[2]*LR2E Laboratoire de Recherche en Energétique et Eco-Innovation Industrielle*
[3]*Quartz-Lab (EA 7393) -* [4]*Université de Tizi-Ouzou, LAMPA, Algérie*

Abstract. Car sharing systems emerged as a new answer to mobility challenges in smart and sustainable cities. Despite their apparent success, design and exploitation of such systems raise crucial strategic and operational challenges. To help planners and decision makers, simulation, analysis and optimization models are unavoidable. Based on the formal modelling and analysis power of stochastic Petri nets, this paper proposes a discrete event simulation model for electric car sharing systems for performance and analysis purposes, taking into account their complex dynamic behaviour, organization and parameters including capacities of the stations, battery and energy availability, locations of charging stations and also their car maintenance activities, not negligible compared to the case of bike-sharing systems.

1. General introduction

With ecological and environmental transition, electric car-sharing systems emerged as new sustainable mobility in many cities around the world. They are introduced to face some crucial problems related to climate change, oil dependency, air pollution, and traffic congestion. Historically speaking, the original concept of car-sharing system was first realized by Carpooling, which was used as a rationing tactic in the United States during the Second World War. Later, the first professional car-sharing service is proposed by SEFAGE in Zurich in 1948. This concept have evolved dramatically due to the oil and energy crisis in the 1970s and today, the concept of car-sharing systems is developed and more attractive by using internet, mobile, and information technologies. Car-Sharing services are one of the solutions that currently spreading in big cities over the world such as Autolib', Bluely, I-Move in France, Bundesverband CarSharing, Share a Starcar, Stattauto München in Germany, Car2Go, Okanagan CarShare, Zipcar in Australia, Buffalo CarShare, City CarShare, eGo CarShare, Ithaca Carshare in United States, etc.). The concept is becoming a mainstream transportation solution with more than a million users in over 26 countries [12].

In its basic structure, a car-sharing network consists of a set of parking and charging stations and a fleet of electric vehicles to be used in a self-service mode by customers (users). For management of such systems, there are different possible organizations including commercial and business companies, public agencies, and cooperatives. In the field of transportation, electric car-sharing systems has emerged as an excellent mobility to cope with environmental issues affecting nowadays several dense urban areas and big cities in the world. Many schemes of the system have been implemented and can be classified into "one-way" and "two-way" systems, according to whether the users (i.e., customers) should return the rented car at a different or at the location they picked it up. "Two-way systems" require users to return the cars to their original pick up stations. Although relatively easy to manage, this operational mode is restrictive for some users. In order to make car-sharing service more flexible, the rented vehicles in "one-way" systems can be returned at any station destination.

2. Challenges and issues

Despite their apparent success over the world, car-sharing systems present several and crucial strategic and operational problems. So, to help professional organisation and managers of such "complex" urban transportation networks, modelling, performance evaluation methods, optimization models and simulation tools are requested. For a more comprehensive literature review, the reader can refer to [1, 5, 12].

The existing works in this research field addresses several issues including modelling, analysis and optimization of strategic, tactical and operational decisions. Strategic issues aim to optimize the number, size and allocation of stations, and the number of the electric-cars to be deployed [2, 4, 7]. Operational issues are notably related to the rebalancing operations to improve the level of service and to make the system more profitable [3, 9, 13]. About the literature, different

* Corresponding author: k.labadi@ecam-epmi.com

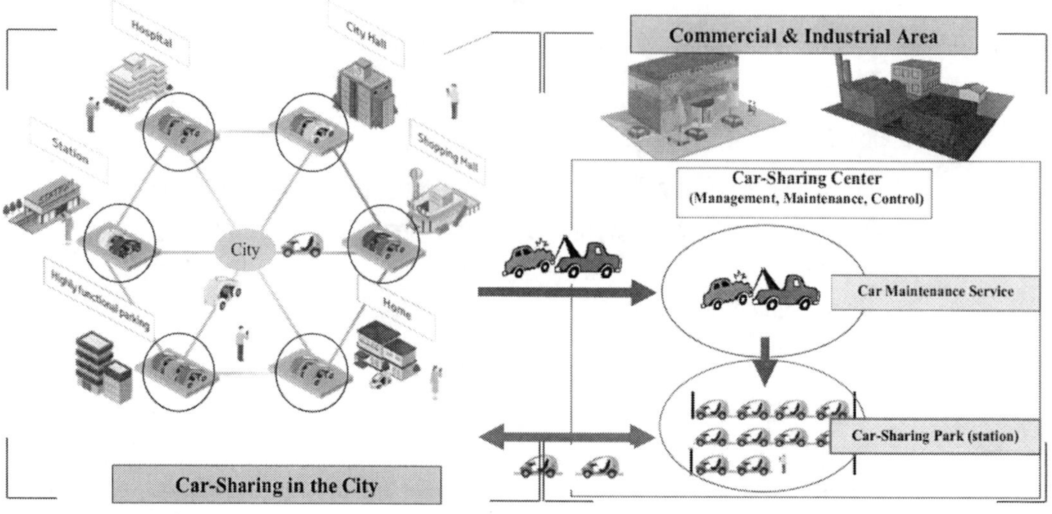

Figure 1. Car-Sharing design network

approaches and methodologies have been proposed, including mixed integer programming approaches, heuristic optimization approaches, data mining and statistical analysis models and simulation-optimization methodologies. However, as for bike sharing systems [6, 8], the recent and extensive literature on car-sharing systems is dominated by operational research (OR) approaches for solving some specific optimization issues such as routing, planning, scheduling, assignment and resource allocation problems.

For global performance evaluation and analysis of such complex systems, they may well be described through a discrete event dynamic model (SED). They can be characterized by a space of discrete states (e.g., number of available cars in a station) and dynamic changes of states according to some discrete events (e.g., return or departure of cars to (from) a station by user over time). Among the formalisms used to model complex dynamical systems, (stochastic) Petri Nets (PN) have arisen as a practical formalism for modelling DES, widely used thanks to their graphical and mathematical foundation, ready to be exploited for modelling, simulation, theoretical analysis, as well as performance evaluation [9, 11, 14, 16].

In this work, in line with our* previous original works on modelling, simulation and analysis of public bike sharing systems by using stochastic Petri nets (with marking dependent arc weights) [6, 8, 15], electric car sharing systems are viewed as discrete event systems (DES). Despite their similar concept, the exploitation and management of electric car sharing systems is more complex than bicycle sharing systems due to their specific operational constraints and activities including battery and energy availability; locations of charging stations, regulation of the stations by moving car by car, more road car incidents/ accidents and then car maintenance operations.

3. Electric Car-Sharing system (ECSS)

The basic concept of car sharing systems is simple: a fleet of cars can be shared between several users, which can drive a car when they need it. Even the idea is conceptually simple, in the last years several business and operational modes are adopted. Car-sharing systems can be classified in different configurations according to the rental conditions. In this work, we consider the case of "one-way" systems allowing people to park the cars anywhere in the covered area, whereas "round-trip" systems force the user to return the car to the location where it was picked-up. The type of vehicle (combustion, electric or hybrid cars) affects also the system's use. In our case, one-way Electric Car sharing Systems are considered, because of their increased flexibility towards the user compared to round-trip systems and their eco-friendly characteristics. On the other hand, the conception and management of such ecological transportation systems poses several additional challenges with respect to those based on traditional combustion vehicles, mainly related with the limited autonomy allowed by current battery technology.

As shown in Figure 1, we consider a realistic configuration of an Electric Car-Sharing system. The network system is structured around a city and its surrounding commercial and industrial area. It contains N stations (parks), noted by S = {S1, S2, ..., Sn} located around the city and one car-sharing center, located outside the city center, for management and maintenance operations of the system. Each station Si represents a private parking space with its limited (Ci) car parks (places) which are equipped by electric charging points. A car, when available, can be taken out from any station and returned to the same or any other station (one-way rental mode), provided that there is an available place in the station. After parking a vehicle at a given station, it becomes available for other users after a delay useful for

https://doi.org/10.1051/e3sconf/202017003001

battery charging and eventually for short maintenance operations locally performed by the service provider. Furthermore, car troubles and accidents that need to a "long" interventions and maintenance operations are not negligible unlike bike sharing systems. In this case, the concerned cars are moved to the maintenance park by the maintenance service. The system network contains a parking space that can be used by the maintenance service and also considered as another specific station for users. This parking space is strategically useful for linking the city to the commercial and industrial zone.

4. A Petri Net modelling of an ECSS

This work is developed in line with our previous works, based on the modelling and performance analysis power of (stochastic) Petri nets for discrete event systems including modelling, control and balancing of bicycle sharing systems [8, 6]. This section is dedicated to develop a realistic stochastic Petri net model for discrete event simulation and performance analysis of electric car sharing systems, taking into account their organization (Figure 1) and dynamic behaviour (e.g., station capacities, battery recharging, car maintenance, ...).

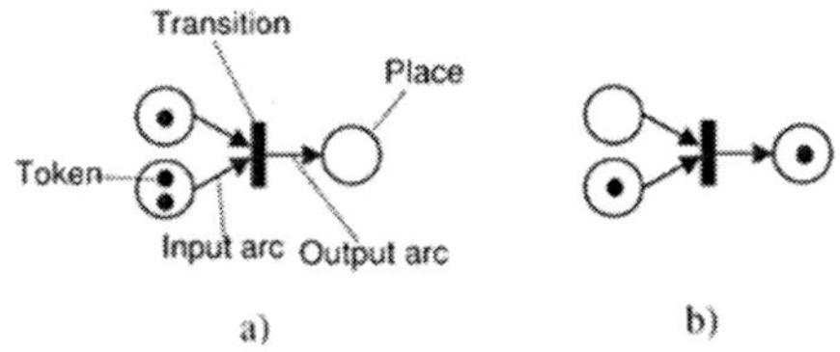

Figure 2. Illustrative example of a simple Petri net - (a) Before the transition firing - (b) After the transition firing

4.1. Basics of Petri Nets

This sub-section is dedicated for basic definitions and description of Petri nets [11]. A Petri net can be defined as is bipartite graph with places and transitions connected by directed arcs. Graphically, places are represented as circles and transitions by rectangles or bars (see Figure 2). The tokens, denoted by dots, contained in places can move from input places to output places by firing the transitions, over the time (firing delays associated to the transitions). At each firing of an enabled transition, the number of tokens removed/added is determined by the weights of the arcs connecting the transition with the corresponding places. A transition can be fired only if it is enabled. The enabling condition means all preconditions for the corresponding activity are fulfilled. In other words, there are enough tokens available in the input places of the transition. The signification of a Petri net model and the interpretation of its component depends on the system to be modelled. In the general case, the transitions are used to model "start" or "end" of events (actions, operations, processes and/or activities) and the occurrence of each event is represented by firing the corresponding transition (activity execution). The directed arcs between the places and the transitions are used to represent material, resource, information, or control flow directions. The ordinary Petri net formalism has been developed in different categories in order to extend its modelling power of diverse complex dynamical systems [14]. In this work, stochastic/ timed Petri nets are chosen to model ECSS.

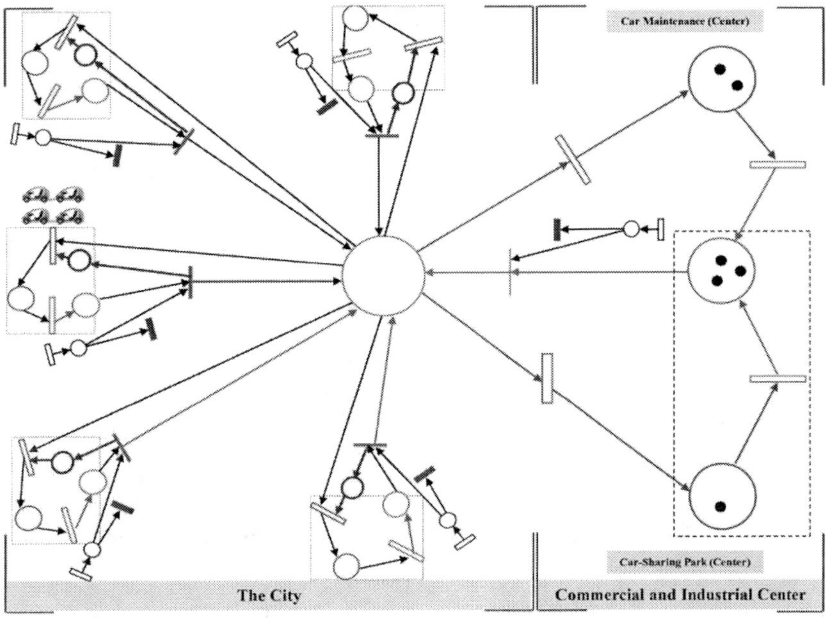

Figure 3. The Petri net model of the ECSS

 https://doi.org/10.1051/e3sconf/202017003001

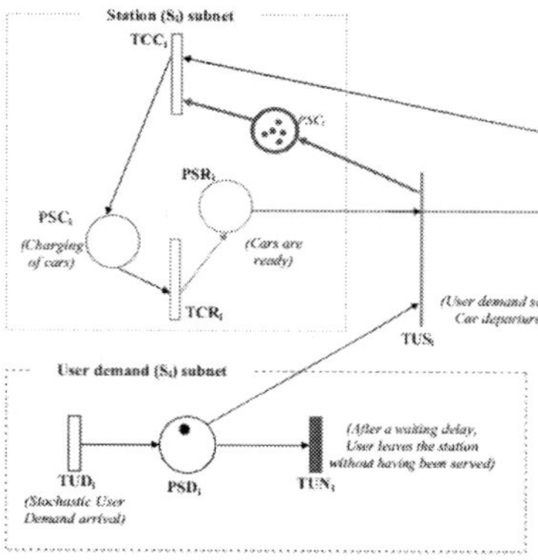

Figure 4. The "car station" and "user demand" subnets

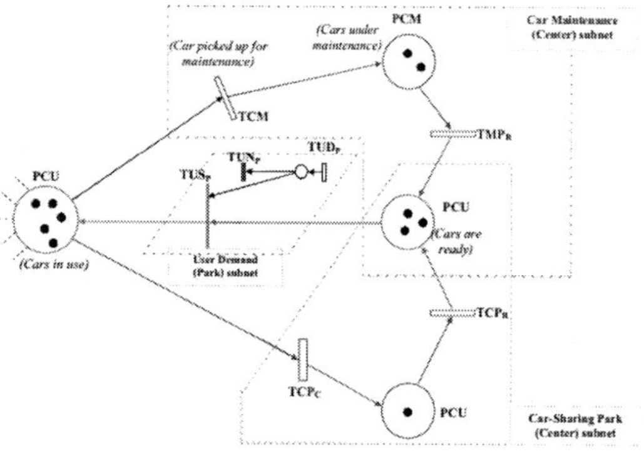

Figure 5. The "car station" and "user demand" subnets

Table 1. Interpretation of places of the model

Places	Meaning
PCU	Its marking represents cars in use by customers. In other words, it corresponds to the number rented cars, at a given moment.
PSC$_i$	Its marking represents rented cars parked at the station S_i but still under charging (or "short maintenance) operations and then not yet available for users.
PSR$_i$	Its marking represents cars parked at the station S_i, ready and then available for users.
PSC$_i$	Its marking represents the number of free places in the station. It is used to limit the capacity of the station S_i.
PSD$_i$	Its marking represents users (demands) waiting -during a given random delay- for available and ready cars in the station S_i.
PSM	Represents the maintenance service and its marking corresponds to cars waiting for "long" maintenance by the car maintenance service.
PCP	Represents the "station Park" used by the maintenance service and users. Its marking represents cars ready and available for users.

4.2. A Petri Net model of an ECSS

The Figure 3 and the Tables (1) and (2) allow readers, familiar with Petri nets, to quickly gain an understanding of the developed stochastic Petri Net model of the typical electric car sharing system represented in Figure 1. The model shows distinguishable parts representing different functions and operations according to the system description given in the previous section. The main function of each subnet is represented in a modular way by Figures 4 and 5. All components of the model are described in Tables 1 and 2.

Table 2. Interpretation of transitions of the model

Transitions	Meaning and events / Type
TUD$_i$	User demand arrival at station S_i (stochastic transition)
TUN$_i$	User demand "not satisfied" at station S_i. The user leaves the station without having been served. (Deterministic transition)
TUS$_i$	User demand "satisfied" and then "user departure" from station S_i. (immediate transition)
TCC$_i$	Car return to station S_i by user. Then, the car is in a charging situation (eventually for "short" maintenance) (stochastic transition)
TCR$_i$	Car is ready and available for other users, after its charging operation (stochastic transition)
TCM	Car picked up for "long" maintenance (e.g. due to an incident/ accident) by service provider (stochastic transition).
TMP$_R$	After maintenance, car is parked at the station Park. It is ready and available for users (stochastic transition)
TCP$_C$	Car return to station Park by user. This car will be in charging operation (stochastic transition).
TCP$_R$	The car is ready and available for other users, after its charging operation (stochastic transition).
TUD$_P$	User demand arrival at station park (stochastic transition)
TUN$_P$	User demand "not satisfied" at Park. The user leaves the station without having been served (stochastic transition)
TUS$_P$	User demand "satisfied" and then "user departure" from station park (immediate transition)

4.3. Performance evaluation and analysis

For the performance evaluation of any modelled system by using stochastic/ timed Petri nets, analytical or simulation techniques are possible. The analytic approach is applicable if the associated stochastic process has a finite number of states and its stochastic nature allows for analytical resolution under some assumptions. For complex and / or large systems, including car sharing systems, discrete event simulation techniques may be required. The simulation process leads to generate State/Time trajectories representing the dynamic behaviour of the modelled system. Based on its formal enabling and firing rules, the simulation process consists in the execution of the Petri net, dynamically calculating the remaining enabling times, the new markings (states) and the desired performance indices. As illustrated in Figure 3, each cycle corresponds to the firing of one transition (event) and the simulation time is increased accordingly to the nature of transition fired after its required execution delay and associated policy (restart or continue process). At the end of the simulation, after a fixed duration (Tsim), the state/ time trajectories are used to calculate different performance indices of the modelled system. Several performances can be formally expressed on the basis of the average marking of the places (e.g., average number of available cars in each station), the firing rates of the transitions (e.g., incident and maintenance rates), and the mean sojourn time of tokens in a given place (e.g., turnover / use of cars).

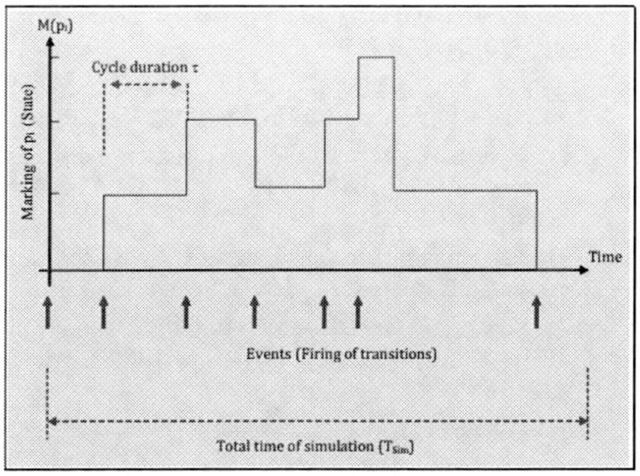

Figure 6. The State/Time Trajectory with Petri Nets.

Formally,

- Average marking of a place p_i:

$$\overline{M(P_i)} = \frac{1}{N_{Rep}} \sum_{k=1}^{N_{Rep}} \left(\frac{\sum (M(p_i)_k * \tau_k)}{T_{Sim}} \right) \quad (1)$$

- Effective firing rate of a transition t_j:

$$\overline{F(t_j)} = \frac{1}{N_{Rep}} \sum_{k=1}^{N_{Rep}} \left(\frac{NF(t_j)_k}{T_{Sim}} \right) \quad (2)$$

- Mean sojourn time of tokens in a place p_i:

$$\overline{S(P_i)} = \frac{1}{N_{Rep}} \sum_{k=1}^{N_{Rep}} \left(\frac{\sum (M(p_i)_k * \tau_k)}{NT_k} \right) \quad (3)$$

Where, as illustrated in Figure 6:

- T_{sim} represents the total simulation time
- N_{Rep} represents the number of replications
- k is the indice of the kth replication
- τ_k is the duration of the cycle
- $M(p_i)_k$ is the number of tokens at the beginning of the cycle
- $NF(t_j)_k$ is the number of firings of a transition t_j
- NT_k is the number of tokens that have passed through this place until the current cycle.

5. Real-life application of the model

To demonstrate the applicability and relevance of the developed model and approach, a real-life application on the case of YELOMOBILE in "La Rochelle" (France) is developed. By using data from: https://play.google.com/store/apps/details?id=com.yelomobile.larochelle&hl=en_US (e.g. station capacities, number of cars ...) and google-map (e.g, distances between stations, localisation, ...), the Petri net model is experimented and validated for several scenarios and parameters (demands, capacities, accident rate, ...).

To demonstrate the capability of the model to reproduce and analyze the system, some results about the dynamic behavior of the model are given graphically in Figure 8, where for each station, the average sojourn time and/ the average number of available cars are estimated by using the equations (3) and (1).

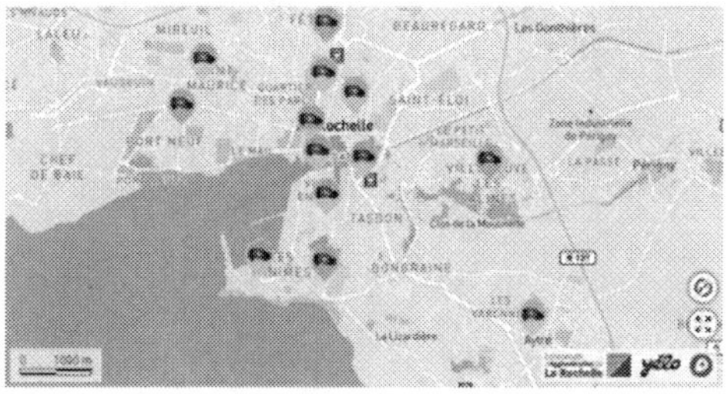

Figure 7. Yelomobile – La Rochelle (France) « Google map » https://play.google.com/store/apps/details?id=com.yelomobile.larochelle&hl=en_US

Table 3. The stations

Station	Designation	Capacity
S1	Médiathèque	6
S2	Minimes	6
S3	Piscine	5
S4	Préfecture	6
S5	Gare	9
S6	Pôle Techno forum	6
S7	Place Saint Maurice	6
S8	Marché de Portneuf	6
S9	Arsenal	12
S10	Verdun	9
S11	Aytré	6
S12	Villeneuve-Les-Salines	6
S13	EIGSI	6

6. Conclusion

Based on the formal modelling and analysis power of stochastic Petri nets, in this paper an original discrete event model is developed for electric car sharing systems (ECSS) for performance evaluation, analysis and simulation purposes. The model takes into account their

https://doi.org/10.1051/e3sconf/202017003001

complex dynamic behaviour, organization and parameters including capacities of the stations, battery and energy availability, locations of charging stations and also their car maintenance activities, not negligible compared to the case of bike-sharing systems. To demonstrate the applicability and relevance of the developed model and approach, a real-life application on the case of YELOMOBILE in "La Rochelle" (France) is presented. More results according to different scenarios are presented during the oral presentation of this work.

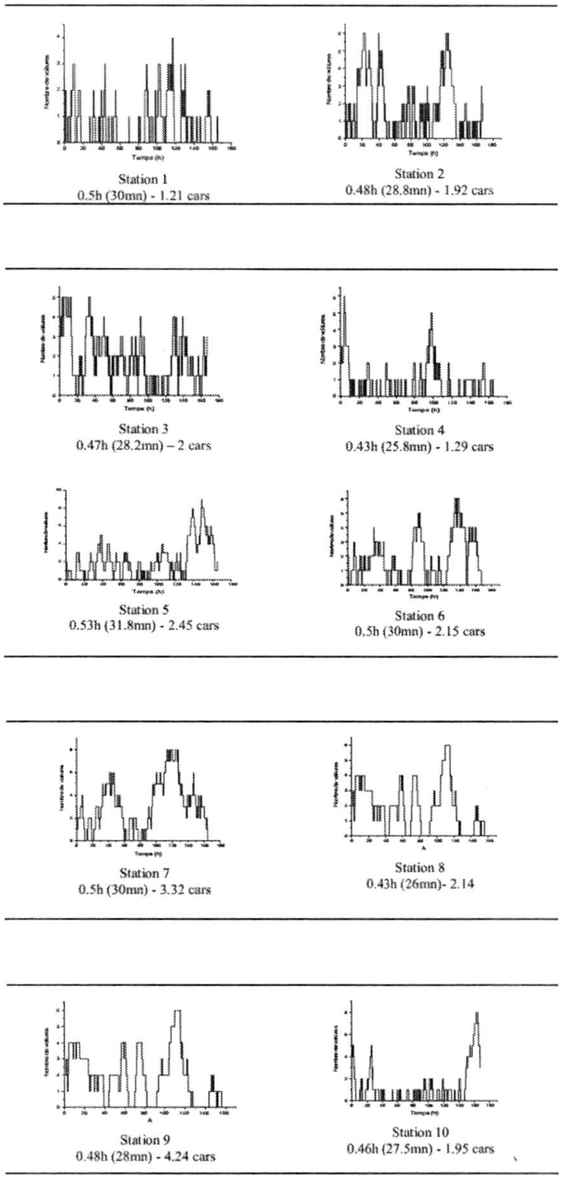

Figure 8. Discrete Event Simulation results

Reference

1. B., Boyacı, K.G., Zografos, N., Geroliminis, Eur. J. Oper. Res 240 (3), 718–733. (2015).
2. Burak Boyacı, Konstantinos G. Zografos, Nikolas Geroliminis, European Journal of Operational Research, Vol. 240, n° 3, 2015, pp. 718-733, (2015).
3. A.D., Febbraro, N. Sacco, M., Saeednia Transportation Research Record Journal of the Transportation Research Board, Vol 2319, n° 1, (2012).
4. Gonçalo Homem de Almeida Correia, António Pais Antunes, (2012), Transportation Research Part E: Logistics and Transportation Review, Vol. 48, n° 1, pp. 233-247, (2012).
5. Jorge, D., Correia, G.H., Eur. J. Transp. Infrastructure Res. Vol. 13, n°3, pp. 201–220, 2013.
6. A.A., Kadri, K., Labadi, I., Kacem, European Journal of Industrial Engineering (EJIE), Vol. 9, No. 5, pp. 638-663, 2015.
7. V. P., Kumar, and M. Bierlaire, Swiss Transport Research Conference, Ascona, Switzerland, (2012).
8. K. Labadi, T. Benarbia, J.B. Barbot, and A. Omari, IEEE TASE, Vol. 12, No. 4, pp. 1380-1395, Oct. (2015).
9. K. Labadi, H. Chen, L. Amodeo, Journal Européen des Systèmes Automatisés, JESA, Vol. 39, n° 7, pp. 863–886, (2005).
10. J., Lee, GL., In: Kim T., Ramos C., Abawajy J., Kang BH., Ślęzak D., Adeli H. (eds) Computer Applications for Modelling, Simulation, and Automobile. Communications in Computer and Information Science, Vol 341. Springer, Berlin, Heidelberg.
11. T., Murata, (1989) 'Petri nets: properties, analysis, and applications', in Proc. of the IEEE, Vol. 77, No. 4, pp.541–580, 1989.
12. S.A., Shaheen, , A.P., Cohen, International Journal Sustainable Trans. Vol. 7, n°1, pp. 5–34, 2013.
13. S.L., Smith, M., Pavone, M., Schwager, E., Frazzoli, and D., Rus, American Control Conference, Washington, DC, Jun. 2013, pp. 2362–2367, (2013)
14. R., Zurawski, and Zhou Mengchu. (1994), IEEE Transactions on Industrial Electronics, Vol. 41, No. 6, pp.567–583.
15. T. Benarbia, K. Labadi, A. Omari and J. P. Barbot, International Conference on Control, Decision and Information Technologies (CoDIT), pp. 112-117, 2013.
16. H. Chen, K. Labadi and L. Amodeo, 2006 International Conference on Service Systems and Service Management, Troyes, IEEE, pp. 575-582, 2006.

A Survey on the Revamped Future of Transport

Heena Tamboli[1,] Jyoti Malhotra[1], SambhajiSarode[1] and Shruti Deshpande[1]*
[1]*CSE, School of Engineering, MIT ADT University, Pune (MS), ,India.*

Abstract. Transport, a movement activity from home to work, studyor market place is one of the basic needs like food, shelter clothes. Nowadays, sharing and adaptive transport services are in demand due to increased traffic and rush to public transport; contributing to saving fuel, fuel cost, and above all time. Transport sharing consists of car sharing, auto-sharing and also bike sharing. . Where in, two or more individuals targeting the same location opts to share transport system. The transport sharing model, especially taxi services also adds various adaptive facilities a one-way and two-way station-based sharing system. These services help individuals to reserve their onlineseats with bus and car-sharing. This paper reviews various existing shared and effective ways of transportation. The observations include different route matching and applied machine learning techniques used by the models which are placed in working using Google maps, APIs, sensors, and Smartphone applications to get and track thedestination.In addition, the manuscript also presents the assessment taken on the communal sphere to acquire various transport facilities adopted by people and the different challenges faced by them.

1Introduction

There are 466 million working people in India and in the survey [1]it is found that 23 million and roughly 70 million people use public transport services like train and bus; moreover, individuals also prefer to either hire an auto, cab or travel by their private bike or car. Among the working category, some people are profiled with full-time jobs and some are associated with part-time jobs. People face many travel issues especially with public transports, leading to traffic congestion and sudden change in the scheduled time. Onthe other hand, few individuals don't preferthe hassles of owning a car like its maintenance, fuel, insurance, parking, and health issues. Such category-A individuals rely on publictransportservices like–bus, trains,auto,and cabs.

Category-B peopleprefer to take the journey on their private transport models with cars or bikes; taking up too much space on the roads. Considering these pros and cons of public and private transport models, traveling and traffic challenges are mostly faced in urban areas. To be specific, we have taken an example of Hinjewadi located in the Pune city – one of the largest cities in the Indian state of Maharashtra. Hinjewadi is well known IT hub for local professionals. Above 2.5 Lacs IT professionals travel daily to and from IT Park. Hinjewadi area is continuously facing serious traffic problems especially during rush hours where trafficisatitshighest.

Road traffic emissions produce greenhouse gases that contribute to global warming, causing serious pollution problems. Every Person riding in a private car increases the number of cars, stemming from escalating traffic and pollution. Individuals of Category-B can also willingly contribute in straight forward and convenient shared public transport services such as car or bike pool reducing the street space, fuel costs, and traffic congestion. The main purpose is to minimize congestion on the street by reducing vehicle flow by joining passengers of the same landing space.

Well-known taxi services namely Ola and Uber [2] are huge alternatives to public transportation filling a hole in our transportation ecosystem. Whereby, individuals can easily book the taxi by looking down into their apps and can easily track their ride with their relatives for security purpose. The major downsides of these services are- the expensive rates, ride cancellation, crimes, sexual assault, etc. Hence, shared transport is the most suitable solution for all. As sketched in figure 1, Car/bus/bike/auto are the different available transport sharing systems and it saves money, reduces traffic congestion, increases fuel efficiency, saves time, and above allincreasesmobility.

*Corresponding author: **heenatamboli1991@gmail.com**

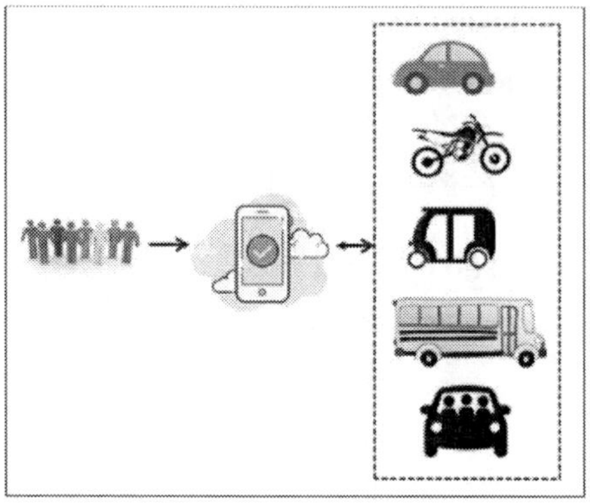

Fig. 1. Different ways of Shared Transport.

In this paper, we present the study of existing shared transport systems along with their upsides, downsides and the same is illustrated in section II followed by our observations covered in section III. At the end, conclusion and references are framed in section IV and V. We also indicate some comparative and algorithmic features with theexistingsolutions.

2 Literature review

This section reveals different solutions adopted by various authors on transportation systems. We place the transport reviewintotwo segments:
2.1. Questionnaire-based assessment of various public and private transport options preferred by individuals.

2.2. Existing solutions for economic transport.

2.1. Questionnaire-based assessment

A survey was taken by framing the following questions to understand various travel options grabbed by the individuals and challenges faced during the journey.

1. Their workplace location and how do they travel to their workplace?

2. What challenges do they face during travel?

3. Any suggestions to improvise the existing scenario?

4.Any preventive measures on working women's safety during transport?

The observations drawn from various responses helped us to understand existing options and the challenges. It was observed that nearly 41.6% ofindividuals reside inPune and among them, nearly 33.7% of people travel by public transport bus. The average percentage of individuals traveling by personal two-wheelers and four-wheelers are 32.6% and 12%. Moreover, only 6% of them prefer carpooling as shown in figure 2. Thus, we conclude that individuals should voluntarily contribute more to carpooling for social and economic benefit.

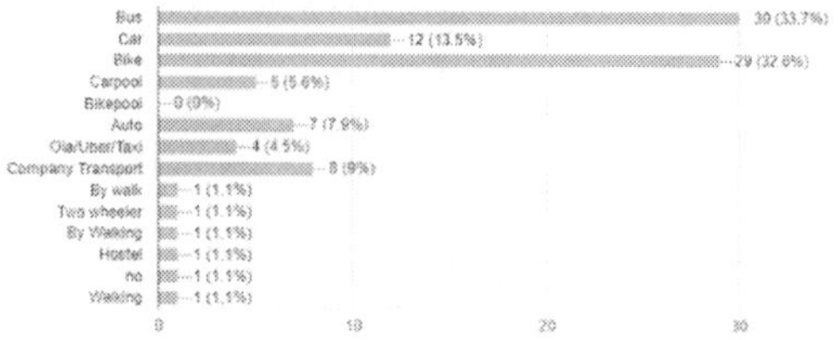

Fig 2: Work-place travel distribution

As shown in figure 3, out of various listed challenges including traffic, parking, time, road maintenance, unavailability of taxis and buses; it was also observed that nearly 76.4% individuals recorded traffic as one of the major challenges during their travel followed by time and parking contributing as 32% and 19%. Thus, we address that individuals should consider these components of parking problems, air pollution, traffic congestion, health complications, together with fuel and vehicle maintenance crisis and go for green ride i.e. shared public or private transport.

In addition to this, we also received some suggestions to overcome these challenges such as- Stringent rules for traffic, increase the police protection on roads for traffic regulations, date-wise parking, and many more. Majorly individuals focused on - use more public transport than private transport and car-pooling.

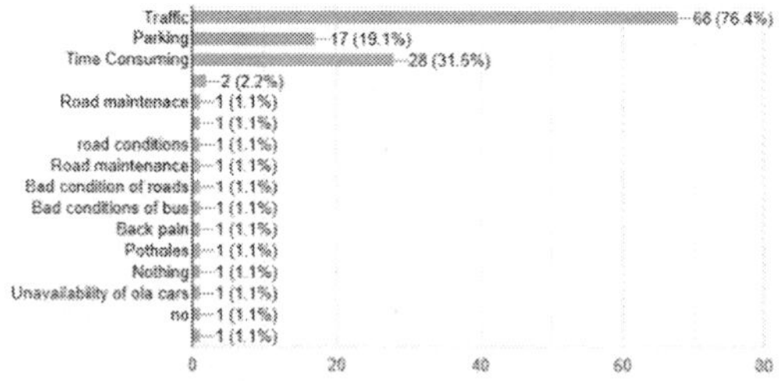

Fig3: Challenges faced during the travel

2.2. Existing solution for economic transport

This sub-section reveals the related study of diverse existing public and private transport systems along with their parameters, advantages, and disadvantages.

With respect to public transport such as bus, Palak Patel et. al. [3] addresses traffic congestion on roads and its reduction through an intelligent public transport system using the Internet of Things. The author highlights various application areas including - traveler's information, improved public transportation, and traffic management along with emergency solutions for an ambulance or higher priority vehicles. The research focuses mainly on sensor-based automation for -

Electronic fare collection, GPS based highway data collection to control traffic, vehicle data collection improving transport and tracking efficiency. Authors plotted the importance of IoT based sensor intelligence for public transport benefitting traffic management, seat occupancy, and security.

The author Kaijun Liu et. al. [4] has stated the importance of bus sharing over the car-sharing system. Car sharing has a drawback of low capacity and high cost. Whereas the bus shared service is comparatively economical. The paper optimizes the ride-matching service for its success rate with the help of a Contraction-Based algorithm scanning the count of visited and unvisited people. Divide and Conquer method is used for energy-efficient location-allocation in the system. It also provides decision support for government authorities for urban bus management and its planning. This is an effective approach for solving the taxi-hailing difficulty in urban areas.

The author Xiangjie Kong, Tao Tang et. al. [5] has proposed bus route planning. It is useful for reducing traffic congestions. The authors have focused on planning dynamic routes. It also includes the prediction of travel requirements. Bus Selection algorithm, along with station selection algorithms are defined for bus selection. This approach is used for shared subway shuttle bus services consisting of mobile data. Origin and Route location parameters of individuals are recorded for getting optimal and effective routes.

The authors of [6] have proposed another pool system based on bike-sharing. For the people traveling by public transports like buses and trains; the author suggested providing bicycle parking stands near public transit stations. This solution can help people to reach their workplaces within or before time. The only issue raised for this solution was unnecessary stands for unused bikes. This can be addressed by opting Steady State method, recording information such as bike allocation, it's return time and demand rates. Electric bicycles are more expensive but can be used for long distances. Bikeshare pool must be provided near the customer's home and workplace.

In this paper, the author Nicola Sacco[7] has proposed a one-way car-sharing system. The system is effective for users that may leave the vehicle anywhere in the network or at any station. The vehicle relocation task is challenging here. The system must decide when and how to distribute the vehicle. Heuristic Strategy used to minimize the waiting times of pending requests. In User-based relocation methodology, customers can drop a vehicle off in exchange for a fair discount. The user-based relocation allows increasing the profit of the sharing system.

In this paper, [8] the authors have proposed the reservation scheme to manage rental reservations of the car-sharing system. In an auction scheme, drivers bet for rental start time contributing to reduced rental fare. For combinatorial auctions, Vickrey-Clarke-Groves (VCG) mechanism is used. This mechanism guarantees a fair assignment for the drivers. This is for two-way station-based car-sharing systems. The potential benefits are higher in the system with shorter rentals from drivers.

With respect to carpooling author Fu-Shiung Hsieh et. al. [9] mentioned that carpooling is a collective transportation model based on shared use of private cars. Carpooling reduces the number of cars as it groups people targeting the same destination or multiple destinations in the journey. It helps in reducing pollution, users traveling cost and time. The author also discussed on daily carpooling and Long-term carpooling. In carpooling two or more passengers traveling together in the same direction so they get benefited from sharing the driving costs, parking fees, and economic advantages. This research uses Android applications in Java languages. The android platform makes it easy and handy to use sensors on the device. It uses sensors and Google Maps to determine locations and motion capabilities. Another car-pooling solution based on route-matching and machine learning is elaborated in the article published by authors of [10].

Table 1. Parametric summary of prior research

Approach	**Travel Cost**	**Passengr seating Capacity**	**Travel Time**	**Algorithmic strategy**
Bus Pooling [4]	Least	Range of 10-50 passenges	More	Divide and Conquer method
Bus shuttle service[5]	Least	Range of 10-20 passenges	More	Bus and Station Selection algorithm
Bike sharing[6]	Average	Two passenges	Relatively Less	Steady State method
One-way carsharing [7]	High	Range of 4-8 passengers	Less	Relocation algorithm
Carpooling [8]	Average or high	Range of 4-8 passengers	Less	First come first serve
Carpooling [9]	Average or high	Range of 4-8 passengers	Less	Matching Algorithm
Carpooling [10	Average or high	Range of 4-8 passenger	Less	K-means algorithm
Bike-sharing [11]	Average	Two passengers	Relatively less	Greedy algorithm

In yet another research the author Rahman, M. S., & Kiyomoto, S.[11] mentioned the importance of more economical Bike Sharing Systems (BSS); as it is one of the major components of intelligent transport systems in smart cities. To ensure a user-focused and robust service, Bike Sharing System should be a part of a multi-modal journey environment such that a user can plan her/his whole journey conveniently. Bike Sharing Systems should offer services to both bike users and Bike Sharing Service Operators. It proposes a secure bike-sharing system.

Authors of [11] have also proposed dual-stage algorithms. The first algorithm allows a user to reserve a bike and parking space according to her/his journey plan through her multimodal journey planning application. In the second algorithm, it shows how bike distribution management can be done in a secure way so that the system can distribute bikes to various stations according to the real-time demand. It provides a secure bike sharing system using encrypt decrypt methods. User authentication controls are designed to verify identity to process ticketing, reserving, parking services. They use Sensor-integrated smart bikes, and Smartphone apps and Google Maps for tracking locations. The review of prior research is summarized in table 1 along with the features including travel cost, passenger seating capacity, travel time and algorithmic strategies.
In this paper ,[15]the authors have propoesd public bike sharing system. In that bikes were provided for public use. User can find and take bike, ride it for his destination and leave for the next one. It provide the real-time bike rental information to the public and records the information of how bikes move each day inside the system. This infirmation is collected and used for reference to understand patterns of bike flow. They conduct statistical analysis or data mining to understand pattern of demands.

3 Observations

Post review, we believe and propose that an efficient transport pooling system can be designed using location-based APIs and/or ZigBee based wireless as mentioned in [12],[13],[14] infrastructure for better location detection, tracking, and communication. Hence, the importance of pooling can contribute to a convenient and secure ecosystem.

The Pooling system shall be equipped with online API services and offline intranet facility services to handle extreme cases. However, the Pooling system shall not necessary to be dependent on online API services considering the network connectivity issues in an emergency service. Having both the facility at the place is always better to handle real-time scenarios. It is essential especially for urban regions at great extent compared with rural regions. Presently, Urban regions are badly suffering from pollution, traffic jams, accidents, and safety. To overcome these issues, we need to design a system to manage the transport problem with the highest level of safety at an affordable rate to everyone. Presently every person has a mobile device, so utilizing it for location identification becomes easy and will not demand any additional hardware requirement. Women's safety is an important concern when we develop a pooling system.

4 Conclusion

In this paper, we have reviewed various transport sharing systems including car, bus, and bike-sharing. It allows the rider to customize the route and pick up and drop to the desired location. This study helped us to understand the process of matching the rides and sharing the transport for the same destination(s) using different clustering algorithms like K-means algorithm, greedy algorithm, relocation and matching algorithm we determine the transport sharing system. We summarize that implementation of car-sharing or pooling can be energy-efficient, traffic reducing, time-efficient and economic benefits leading to stress-free society and green ride.

References

https://www.indiaspend.com/only-4-75-million-join-indias-workforce-annually-not-12-million-as-claimed-70548/

https://www.olacabs.com/; https://www.uber.com/

Patel, P., Narmawala, Z., & Thakkar, In Emer. Res. in Computing, Info., Comm. and App., 231,(2019)

Liu, K., Zhang, J., & Yang, Q., IEEE Access, **7**, 74248, (2019).

Kong, X., Li, M., Tang, T., Tian, K., Moreira-Matias, L., & Xia, F. IEEE Trans. on Auto. Sci. and Engg., **15**, 1507, (2018).

Tang, G., Keshav, S., Golab, L., & Wu, K., IEEE Transactions on Intelligent Transportation Systems, **19**, 2279, (2018).

Di Febbraro, A., Sacco, N., & Saeednia, M., IEEE Transactions on Intelligent Transportation Systems, **20**, 628, (2018).

Roca-Riu, M., & Menendez, M., IEEE Access, **7**, 154627, (2019).

Hsieh, F. S., IEEE 31st International Conference on Advanced Information Networking and Applications, 972, (2017)

Samy, M., & Elkorany, A. M. IEEE/ACM International Conference on Advances in Social Networks Analysis and Mining, 702, (2018).

Rahman, M. S., & Kiyomoto, S., IEEE International Conferences on Big Data and Cloud Computing (BDCloud), Social Computing and Networking (SocialCom), Sustainable Computing and

Communications (SustainCom)(BDCloud-SocialCom-SustainCom), 437,(2016).

Sambhaji, S., & Bakal, J., Journal of Telecommunication, Electronic and Computer Engineering, **10**, 65, (2018).

Sambhaji, S., & Bakal, J., Journal of Telecommunication, Electronic and Computer Engineering , **9**, 37, (2017).

Sambhaji, S., & Bakal, J., Proceedings of the International Congress on Information and Communication Technology, 535, (2016).

Wang, I. L., & Wang, C. Second IIAI International Conference on Advanced Applied Informatics, 306, (2013).

Real-time Location Based Shared Smart Parking System

Pragati Kanchan[1,], Chhaya Mhaske[1], Reena Pagare[1], and NikhilKumar B. Shardoor[1]*
[1]Department of CSE, School of Engineering, MIT ADT University, Pune, India

Abstract. With the growth in population, traffic congestion and parking have become a serious problem. There is the explosive growth of the per capita amount of vehicles. In this paper, a Location-Based Shared Smart Parking System is proposed to solve the parking problem. This system is designed for both private and public parking areas. This system helps the user to find the nearest possible parking area and gives the information related to the availability of parking slots in that respective parking area. The main focus of this system is to reduce the time for finding the parking slot and also avoids unnecessary travelling through occupied parking slots in a parking area. The proposed system is designed using IoT technology, WSN(IR sensors), QR code and RFID technology. IR sensors are used to detect the presence of the car in the parking slot and QR code is used to authenticate the user in the public parking area, RFID technology is used to authenticate the user in private parking area. The system is expected to reduce traffic problems, fuel consumption which results in reduces carbon footprints in an atmosphere.

1 Introduction

With the increase in Indian population, and change in people's lifestyle, every family own a vehicle today. As India's population increases rapidly this has led to rapid increase in number of vehicle on the road. Also inconvenience in public transportation is one of the major reason for increase number of vehicle on road. Parking space in Indian cities has remained constant or reduce due to a growing population. In the current situation, there is no standard system for finding parking space in parking area. Existing systems are time-consuming and depends on human interaction. During festive season people visit the mall and after reaching to the parking area they face the problem related to limited parking area. Thus, there is a requirement for a shared smart parking system. This will help the user to find a suitable location for parking.

People who are searching for parking location causes traffic congestion. Mostly this issue is occurring in city areas. F. I. Shaikh et al. [11] , researchers have found that in one year, car cruising for parking burning 1.7 lakh litre of fuel and producing 730 tons of CO2, which increases air pollution. Smart Parking systems generally collects the information related to available parking slots in a particular geographic area and the process is real-time to help the user to place vehicles at available positions also, make use of shared parking resources. Shared parking help to optimize parking space in urban areas. Where private parking owners such as hotels, companies, hospitals, etc. and one can rent out their parking space at times they don't use it. In this paper, a real-time location based shared smart parking system based on IoT technology is proposed which enable the user to find the nearest suitable parking space based on their location via mobile application. QR code and RFID technology are used to increase safety and security.

Nowadays, searching for a parking slot in city areas becomes a serious issue. Most of the time people are wasting their time searching for a parking slot. Drivers keep searching for parking space leads to an increase in traffic flow. Searching for empty parking slot during peak hours not only time-consuming but also result in wastage of fuel. Searching for parking burn excess amounts of fuel-producing CO2 results in increased air pollution. Hence shared smart parking becomes a need of the day.

2 Related Work

T. N. Pham et al. [1] in this paper introduces a novel algorithm for the current cloud-based smart parking system to increase efficiency and develop a network architecture based on IoT technology. This system helps the user to find free parking space with the least cost. In this paper, authors have proposed and implemented a smart parking system based on IoT technology. This system uses WSN with RFID technology for monitoring car parks. In this paper system is design by considering only distance parameter from user location to parking location but percentage of available parking space is not consider.

S. R. Basavaraju et al. [2] in his work proposed a smart parking system. which helps the user to search the

*Corresponding author: **pragatikanchan04@gmail.com**

nearest parking area and availability of parking space. The aim of this system is to reduce the time for searching parking space which, leads to reduce fuel consumption and also reduces air pollution. This proposed system is based on IoT Technology which, is consists of a Raspberry Pi-based parking system and Pi camera. It uses a centralized server to maintain databased which having information about parking space available in the city.

B. M. K. Gandhi et al. [3] this study proposed a parking management system that helps a user to find the availability of parking space using IoT technology. This system consists of IR Sensors, which are placed at the centre of each parking slot to detect a car. Users can access or check the availability of slots and prices remotely using this system. Where all data is stored in the cloud at the centralized system. RFID tags technology is also used in this system to detect the car details incase of identifying or detecting car theft.

C. Shiyao et al. [4] in this study, the system is built using the ZigBee network to send urgent information to the PC through a coordinator and for making updation in databases. This paper presents design and implementation of an Smart parking system.

J. Xiao et al. [5] in this paper Markov M\M\C\C queue model and stochastic queuing model is used. This model are used to predict the future occupancy from previous occupancy data. This paper are used to estimation of model parameter and occupancy prediction parameter. Accuracy attribute is used in this system.

W. Alajali et al. [6] in his work GBRT and k-cross validation technology for car parking prediction in smart city is used. In this paper three dataset are used like a street car parking dataset, pedestrian data and car traffic data. Here accuracy attributes is used.

E. Simhon et al. [7] in this study RMSE (minimize root mean square error) is used. These prediction occupancy rate of all neighbour area and target occupancy rate. The prediction is based on level of occupancy rate of past data. This method is cost efficient for parking space. The main advantage of this paper is through mobile app find the parking slot.

Y. Zheng et al. [8] in his work the three algorithm like a regression tree, SVM and NN are used for real time prediction of car parking occupancy rate. The parameter are used like public safety, transport logistics and environment management. Here dataset are used for data collection of cities. This system are based on time interval.

S. Yang et al. [9] in this paper GCNN, RNN and LSTM technology are used for prediction of occupancy of parking. This system uses input parameter like a parking meter transaction, traffic speed and weather condition. The accuracy attribute are used by using mean absolute error formula.

J. Hu et al. [10] in his work introduces the block-chain based parking system for managing parking services and securely storing user information. The proposed system is based on the BlockChainOpenSource(BCOS) framework and smart contract technology for shared space. This system uses a consortium blockchain derived from ethereum technology. A smart contract is used for registration and transaction module for both the parking space owner and users. It allows users to search for a vacant parking space. The aim of this system is to reduced system maintenance and recovery cost by omitting a centralize party to control the entire system. This system is designed for paid parking resources. The vehicle user needs to pay the amount after leaves the parking lot to the owner's payment address through BCOS cash. It also provides a one-time pad mechanism and a group signature mechanism.

The study shows that there is a need to optimize management and labour costs for optimum use of resources for car-park owners. There is also less work found which covers the security aspect during parking. The time utilization can also be optimized by reducing the average waiting time of users for searching parking space. This will result in reduced air pollution and traffic congestion.

3 Problem Statement

A shared smart parking system is help the user to find the most suitable and nearest parking space in real-time based on their location. It enables the user to make online reservation of parking slot also allow parking administration to define and manage parking area.

4 Proposed System Architecture

The real-time location based share smart parking is designed for both the public and private shared parking.

4.1 Scenario I

This system is based on IoT technology. The system uses WSN(Infrared Sensors). Infrared sensors placed in parking slots used to identify the presence of a car at the parking slot in public parking area. It gives information about the parking slot is empty or not. After login into the system, the user will be able to access location-based information and make a reservation for parking slot through the mobile application. The user makes the request for an available parking slot. Information related to the parking slot gather by IR sensors is send to microcontroller and later stored on the centralized system i.e. on a cloud through web services. Shared smart parking system check the availability of parking slot based on information stored on a cloud. If the parking slot found empty it allocates that slot to the user who makes the reservation for that slot and sends QR code to user in response as shown in Fig. 1. The QR code technology is used to identify and authenticate the user against their parking reservations. After getting QR code user makes use of that code for making entry into parking. Whenever a new car enters into parking area system will updates parking space status in real-time. The Google Map API use the users geolocation and location of the parking spots to determine the nearest available parking[11].

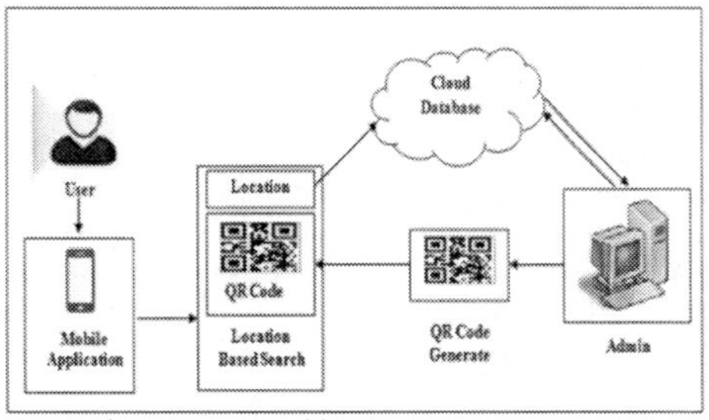

Fig. 1. Shared Smart Parking System.

4.2 SYSTEM FLOWCHART

Whenever the user wants to search for a parking, user log into the system. After, login user request for parking slot based on their location or given area. Then, the user's request is sent to the system, it checks the availability of given slots into the database. If a parking slot is available that slot is reserved for that user and then system generates a QR code and sends it to the user in response. If the parking slot is not available it sends a message to the user that no parking slot available. If a parking slot is available and the user gets acknowledgment. At the time user enters into parking QR code is verify and allow the user to park their vehicle. Otherwise, the user will not get entry into parking, as shown in Fig. 2.

4.3 Scenario II

Private or localized share smart parking system enables users to use the parking area of hotels, hospitals, commercial buildings, etc. at a time when there is a free parking slot available. This can be done by taking permission from owners of the parking area. Suppose, During festive season person visit mall and after reaching to parking area he experience that there is no slot is available for parking. In that situation, he\she can make uses of the private parking slot with the help of shared smart parking system. A real-time location-based shared smart parking system for private parking area is consists of IoT technology, Infrared Sensors, and RFID technology. The system is placed on the private parking area which continuously monitoring empty parking slots with the help of IR sensors and the same information is get reflected into the cloud sever. IR sensors are placed in the parking area which will continuously check the availability of slots in parking and send information to the microcontroller. Information gathered by an Infra-red sensor is later stored on the cloud database. If there is a vacant parking slot then Green LED is blown and if the parking slot is reserved then the Red LED is blown. The RFID technology is used for authentication of the user so that only authorized persons can make entry in the parking. To use this system user first need to register through the mobile application. Information filled by the user(like Id, Name, Vehicle number) is stored on a cloud database and RFID Tag.

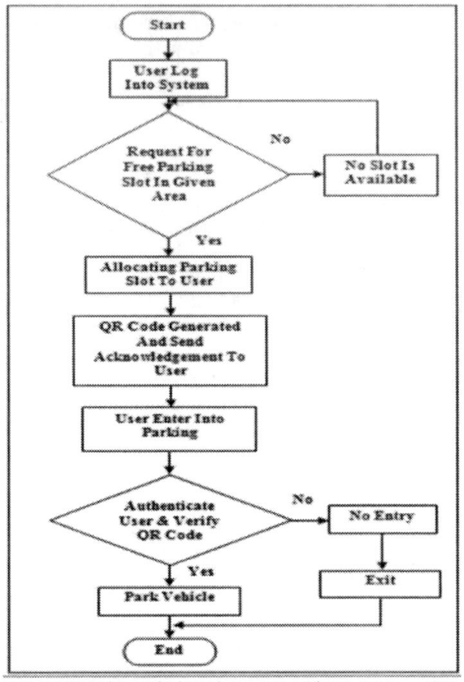

Fig. 2. Flowchart of Shared Smart Parking System

RFID Tag Id is given to registered users for identification and authentication purpose.

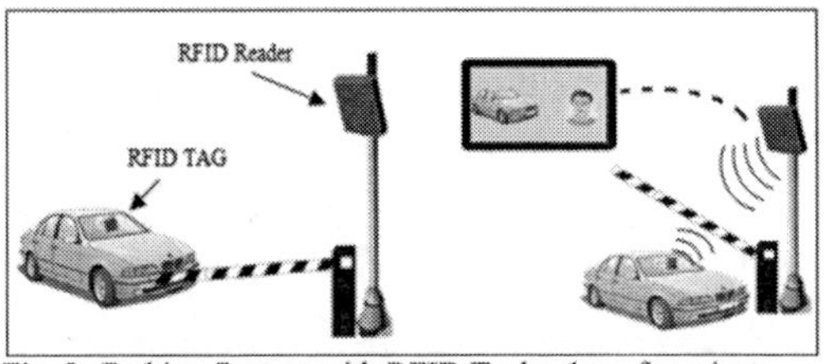

Fig. 3. Parking System with RFID Technology for private uses.

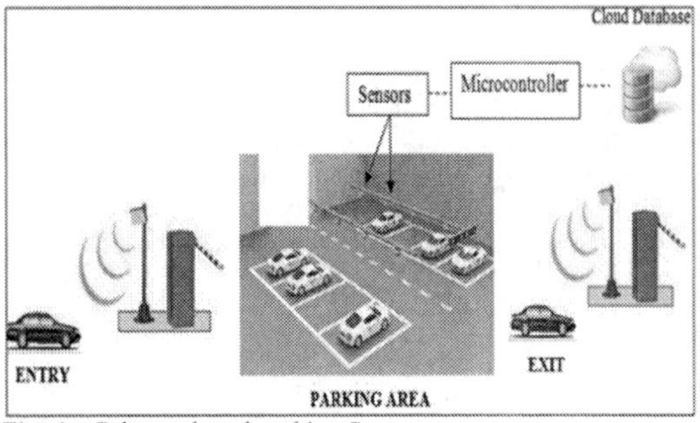

Fig. 4. Private shared parking System

The tags having electronically stored information. There are two types of RFID tag, Active Tag and Passive Tag. Passive Tags take energy from the nearest RFID Reader's interrogative radiowaves[12]. User first logs into the system, after login user can search for parking location based on their choice. Based on location entered by the user, the application will show related parking areas and availability of slots in that particular location. The user makes the reservation for an available slot through the mobile app. RFID Tag is placed on the car, when the user reached to that parking area at entry gate, RFID Reader reads the data from that tag through radiowaves, that data is checked with the cloud database. If user data is matched then allow the user to make entry into the parking otherwise if the information is not matched or if there is no RFID Tag on the car then that user will not get entry into the parking area. Similarly, when the user leaves a parking area user data is scan and status is updated to the database as shown in above figure Fig. 3 and Fig. 4. The vehicle user needs to pay the amount after leaving the parking slot to the owner's payment address.

5 Results And Discussion

Comparison of two scenarios shown in Fig. 5. and Fig. 6.

1) Parking System without Shared Smart Parking
2) Parking System with Shared smart Parking

With an existing parking system, people are wasting their time finding parking space, searching for parking space is not only increased traffic on road but also burn excessive amounts of fuel, producing CO2 results in increased air pollution. As shown in Fig. 5., the given graph depicts that if searching time increase, the traffic will also increase, which results in an increasing carbon footprint in the environment.

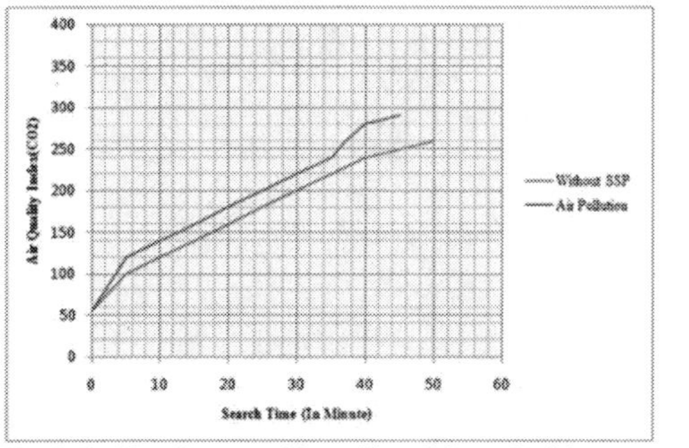

Fig. 5. Parking System without Shared Smart Parking

As shown in Fig. 6., the given graph depicts that with the help of a shared smart parking system, the searching time of the user for parking will be decreased, traffic will be reduced, and hence it will result in reducing air pollution.

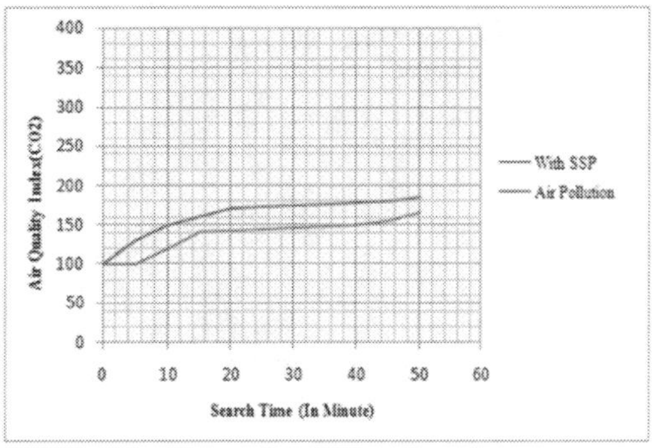

Fig. 6. Parking System with Shared Smart Parking

6 Conclusion

In this paper, a shared smart parking is proposed which facilitates the sharing of parking resources between the owners of parking slots and users. This system is not only helpful to the user to find the nearest suitable parking slots in minimum time but will also provide an effective solution to reduce air pollution and traffic problems.

References

1. T. N. Pham, M. F. Tsai, D. B. Nguyen, C. R. Dow, D. J. Deng, IEEE Access, *3*, 1581-1591, (2015)
2. S. R. Basavaraju, International Journal of Scientific and Research Publications, **5**, 12, 629-632, (2015)
3. B. M. K. Gandhi, M. K. Rao, Indian Journal of Science and Technology, **9**, 17, 1-6, (2016)
4. C. Shiyao, W. Ming, L. Chen, R. Na, IEEE, *2014 Sixth International Conference on Measuring Technology and Mechatronics Automation* ,741-744, (2014)
5. J. Xiao, Y. Lou, J. Frisby, TRANSPORT. RES. B-METH. , **112**, 19-39,(2018)
6. W. Alajali, S. Wen, W. Zhou, Springer, *International Conference on Security, Privacy and Anonymity in Computation, Communication and Storage* ,641-652,(2017)
7. E. Simhon, C. Liao, D. Starobinski, *2017 IEEE Conference on Computer Communications Workshops (INFOCOM WKSHPS)* , 641-646, IEEE, (2017)
8. Y. Zheng, S. Rajasegarar, C. Leckie, Sensor Networks and Information Processing, (2015)
9. S. Yang, W. Ma, X. Pi, S. Qian, Transp. Res. Part C Emerg. Technol. , **107**, 248-265, (2019)
10. J. Hu, D. He, Q. Zhao, K. K. R. Choo, IEEE CONSUM. ELECTR. M. , 8, 4, 45-49, (2019)
11. F. I. Shaikh, P. N. Jadhav, S. P. Bandarkar, O. P. Kulkarni, N. B. Shardoor, Int. J. Comput. Appl. , **140**, 12, (2016)

12. https://en.wikipedia.org/wiki/Radio-frequency_identification

Smart Transportation Based Car Pooling System

Celesty Gedam[1] *,Madhavi Sahare*[1,*] *,Rajneeshkaur Sachdeo*[2], *and Nilima Kulkarni*[2]
[1]School of Engineering, MIT ADT University, Pune, India.
[2]School of Engineering, MIT ADT University, Pune, India.

Abstract. Sharing of car is Carpooling, so that multiple number of people can travel in single car. the use of car pooling reduces single individuals Travelling cost , reduces the fuel cost and reduces the number of cars. Due to the growth in the population there is inadequate transportation through their own car . Rather than using different mode of Transportation. It results in increasing number of Traffic on roads also increases pollution and increases time to travel to there destination. So by Smart transportation using car pooling system the individual can travel and share there rides with different people of a same destination. In this paper we have carried outs survey. By reviewing various Literature papers on carpooling it aims to reduce the number of cars by sharing the rides.

1. Introduction

Due to increasing amount of peoples every single people travel to their Destination alone in their own car. Travelling alone leads to increases individual fuel cost. The carpooling system merges multiple new people in a car which leads to meet new people in a car, reduces air pollution and noise pollution. The carpooling system save economy of every people as they share their rides also share the cost with the other member in car. It will stop endless Spending money in travel.Rather than using public transportation such as local train , buses and metros. The people can conveniently travel to there destination in the comfortably by giving same cost. The carpooling is the best ideas to reduces the traffic jams as it reduces the car on road. Safety is an important aspect in every means, so travelling with the different people is also a prior thing not all the people are comfortable to travel with the unknown people. For example, women are not feeling safe to travel with unknown people. So the carpooling system also provide ride of only woman.

2.RelatedWork

Nusrat et.al.[2] in this paper introduced the Euclidean distance for measuring the nearest vehicle. This system is based on 3 tier architecture where it had client tier, middle tier and database tier. In this system the users location can be tracked by the GPS. And after finding their location the system suggest the nearest vehicle to the users. In this paper for implementing the system the searching algorithm had also been used. This system helps to reduce empty seats of cars.

Fu-shiung et.al.[4] the main intention of this paper is to find the match between passenger and driver. Depending on their trajectories. In this system for solving the car pooling problem heuristic techniques has been applied, this techniques is used for searching and it also used a matching algorithm for allocating passenger to drivers car depending on their trajectories. The system had developed based on mobile platform as well as Google API. Based on matching algorithm and various different techniques the system was used in reduction of distance.

Mena et.al.[5] In this paper Semantic features used for car pooling system is based on the user trips. In this proposed system the two main techniques has been used that is route matching and the other is applied matching algorithm the route matching techniques is used for to find the match between the passenger based on their point of interest. It also apply machine learning cluster algorithm likek-mean and the density based spatial clustering of application with noise (DBSCAN) is applied for optimizing the rank between different match trips.

Raza Hasan et.al.[1] In this paper, author hadproposed the system smart peer car pooling system(SPCPS) for minimizing the number of cars and increasing the parking slot areas by using the linear programming problem it also uses the Dijkstra algorithm for finding shortestpath. In this paper optimal result is obtained from the dijkstra algorithm. The author had proposed the SPCPS system which is beneficial consuming fuel and also helped to improve the socialism.

Mattia et.al.[7] in this paper, it uses two types of techniques to propose carpooling system, first is learning-to-rank which is basically used for ranking model of the individual. Secondly they used learning algorithm in their system. The author proposed GOTOGETHER system which was used for carpooling

*Corresponding author:**madhavi021296@gmail.com**

system which uses machine learning ranking system. The system used twitter and foursquare dataset for the evaluation. Those dataset are used for the evaluation for the pattern and ride requests.

Fu-Shiung et.al.[6]This paper author had formulated the carpooling problems and formulated thier solutions.Using meta heuristic technique to evaluate differential evolution problem. To evaluate integration programming problem with combination of binary decision variable, it is using Differential Evolution technique in this system. Initially carpooling optimisation problem is formulated in the form of integer programming problem and then based on it different variants of differential evolution algorithms had been developed.

Shih-Chia Huang et.al.[7], the author had formulated carpool service problem (CSP). The paper had used Ant colony optimisation (ACO) algorithm, for the carpool service system . The author used ACO algorithm for choosing optimal path between ant colony and food source over graph.

3. Methodology

Car pooling system provides the benefits of sharing rides with different peoples. By surveying different papers, it is found that it uses heuristic searching techniques for searching the cars. By using Euclidean distance techniques, it finds the nearest car that has to be assigned to passenger. It usually uses Global Positioning System(GPS) for finding the locations of cars. In study it is found that, Route matching techniques are used in order to match drivers and passengers locations by using Google maps API , and then based on their route, driver is allocated to passenger. In most of carpooling system, it is observed that the Dijkstra algorithm is used for detecting the shortest route of destination of passenger. Dijkstra algorithm is usually used for detecting the optimal minimum path. It is found that is checks for maximum possibilities and then chooses the minimum route for thejourney.

4. Car Pooling System

After studying the paper[1][2][3] the generalized car pooling system is highlighted in fig.1. The system shows that, the passenger has to do registration, if already registered than directly has to logged into the system. All the passengers details are stored into the database then. Then passenger needs to enter the pickup and drop locations. If the location of passenger and driver matches, then the driver accepts the ride from the passenger. Security is prior thing in carpooling system. So as soon as the passenger is assigned with driver, the passenger has to share his/her live location, drivers details, car's details, other passenger details(if present) to their contacts. Then driver will check for optimal path. Driver has to accept the request of same destination. Depending upon the availability of seats and drop location, driver has to accept the ride.

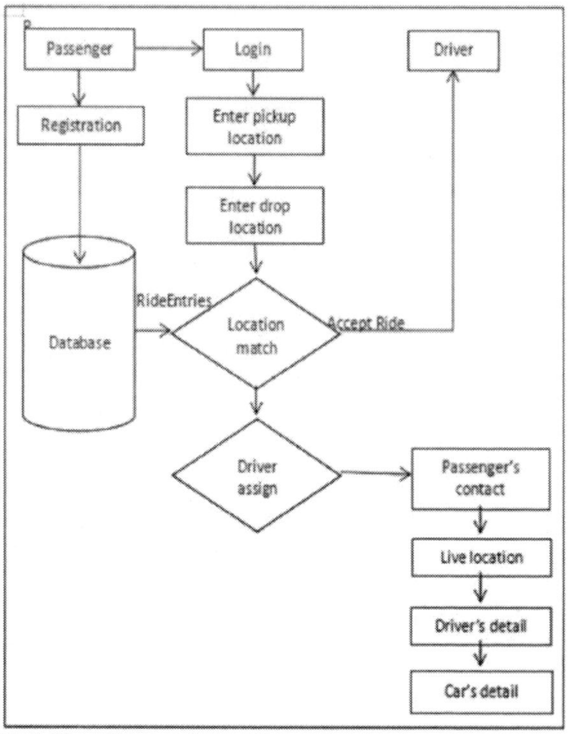

Fig.1. Flow diagram of carpooling system.

5. Literature Survey

Sr. no.	Author	Methods	Features	Observation
1.	Nusrat et. al.[2]	Searching techniques and Euclidean distance measuring	Euclidean distance ride sharing.	Eliminate space misuse
2.	Fu-Shiung et.al.[4]	Matching algorithm and heuristic searching method and google distance	Drivers cars based trajectories, google distance matrix API use to find two distance location.	Distance without carpooling is 48.5% and with carpooling is 22.6% and reduction of 49.59%.
3.	Mena at.al.[5]	Route matching method and machine learning algorithm	Use K-means as well as density-based spatial clustering cluster for the similar trips and most optimal trip to be joined is assumed.	Route matching, reducing the number of trips with 22.3% and machine learning technique 66% reduction

4.	Raza Hasan et.al.[1]	Dijkstra shortest path algorithm and linear programming problem	Dijkstra algorithm is used for finding shortest distance , linear programming used.	Trajectories between passengers and drivers are determined.
5.	Mattia et.al.[7]	Learning-to-ran k and the carpooling learning method	LR, ranking methods.	Validity and robustness.
6.	Fu-Shiung et.al.[6]	Differential Evolution(D.E)	Discrete differential evolution (D.E) algorithm.	Differential evolution problem is formulated.
7.	Shih-chia Huag et.al.[3]	Ant colony optimization (ACO)	Optimal path is chosen between the ant colony as well as food source over a graph problem.	ACPA with path based structure is achieved.

6.Conclusion

The papers are chosen from the literature is studied in this paper. It is observed that most of the paper uses matching algorithm and Dijkstra algorithm for finding shortest distance. Thus we have discussed Matching algorithm and Dijkstra algorithm in detail. In this paper we have studied that Carpooling system helps in Reducing pollution, Fuel expense, Parking and Tolls. It is observed that there is much scope for the research in the car pooling system for ex. in reducing the time to find the required ride, calculating the minimum distance etc. In this paper we have tried to give reader an idea about complete car pooling process.

References

1. R.Hasan, A.H. Bhatti, M.S. Hayat, H.M. Gebreyohannes, S.I. Ali, A.J. Sayed, *2016 3rd MEC International Conference on Big Data and Smart City (ICBDSC),* 1--6, (2016).
2. Farin, N. Jahan and Rimon, M. N. Ahsan Ali and Momen, Sifat and Uddin, M. Shorif and Mansoor, Nafees, *2016 International Workshop on Computational Intelligence (IWCI),* 204--208, (2016).
3. Kalogirou, Kostas and Dimokas, Nikos and Tsami, Maria and Kehagias, Dionysis, *2018 IEEE 20th International Conference on High Performance Computing and Communications; IEEE 16th International Conference on Smart City; IEEE 4th International Conference on Data Science and Systems (HPCC/SmartCity/DSS),* 1271--1278, (2018).
4. Hsieh, F. Shiung, *2017 IEEE 31st International Conference on Advanced Information Networking and Applications (AINA)*, 972--978, (2017).
5. Samy, Mena and Elkorany, Abeer Mohamed, *2018 IEEE/ACM International Conference on Advances in Social Networks Analysis and Mining (ASONAM)*, 702--708, (2018).
6. Hsieh, F. Shiung and Zhan, Fu-Min, *2018 IEEE 42nd Annual Computer Software and Applications Conference (COMPSAC)*, **1**, 577--582, (2018).
7. Campana, M. Giovanni and Delmastro, Franca and Bruno, Raffaele, *2016 IEEE 19th International Conference on Intelligent Transportation Systems (ITSC)*, 1856--1862, (2016).
8. Carrese, Stefano and Cipriani, Ernesto and Giacchetti, Tommaso and Sottile, Eleonora and Zamberlan, Leonardo, *2019 6th International Conference on Models and Technologies for Intelligent Transportation Systems (MT-ITS)*, 1--6, (2019).
9. Kulkarni and J. Amudha,*Journal of Intelligent \& Fuzzy Systems*, **34**, 1713–1722, (2018).
10. Zhang, Sijia and Jia, Shunping and Ma, Cunrui and Wang, Yuqiong, *2018 IEEE 3rd International Conference on Cloud Computing and Big Data Analysis (ICCCBDA*, 280--284, (2018).
11. Opasanon, Sathaporn , *2018 Smart City Symposium Prague (SCSP)*, 1--7, (2018).
12. Xi, Xiaoming and Yu, Zhilou and Zhan, Zhaolei and Yin, Yilong and Tian, Cuihuan,*IEEE Access*, **7**, 98061--98068, (2019).
13. Huang, Shih-Chia and Jiau, Ming-Kai and Lin, Chih-Hsiang, *IEEE Transactions on Intelligent Transportation Systems*,**16**, 352--364, (2014).
14. Lin, Chih-Hsiang and Jiau, Ming-Kai and Huang, Shih-Chia,*2012 6th International Conference on New Trends in Information Science, Service Science and Data Mining (ISSDM2012)*, 266--271, (2012).
15. Bresciani, Chiara and Colorni, Alberto and Costa, Francesca , Alessandro and Studer, Luca, *2018 International Conference of Electrical and Electronic Technologies for Automotive*, 1--4, (2018).
16. Elbery, Ahmed and ElNainay, Mustafa and Rakha, Hesham, *2016 IEEE 12th International Conference on Wireless and Mobile Computing, Networking and Communications (WiMob)*, 1--8,(2016).

Video summarization approaches in wireless capsule endoscopy: A review

Vrushali Raut[1,*], *Reena Gunjan*[2]
[1]MIT School of Engineering, Pune, Maharashtra (India)
[2]MIT School of Engineering, Pune, Maharashtra (India).

Abstract. Wireless capsule endoscopy (WCE) is medical examination process for gastrointestinal tract (GIT). This noninvasive multi advantageous procedure can be made more popular by overcoming the problem of prolonged analysis time. Video summarization is a concise and meaningful representation of a video. Along with automated detection and segmentation methods, summarized video will serve as an additional source for confirming the analysis of WCE video before the final diagnosis without any dropouts. Nowadays, Internet of Things (IoT) environments are predominant in healthcare sectors. Considering limited resources of smart phones and long duration of WCE process, it is impractical to send all the WCE data to health-care centers or gastroenterologists. This paper reviews video summarization types and the techniques used for WCE video summarization by various researchers. Feature set selection, clustering methods and key frame selection techniques play important role in performance of summarization technique.

1 Introduction

Initially capsule endoscopy was developed to see the inner area of small intestine, which is not easily reachable with the help of traditional endoscopy procedures. From 2001 onwards wireless capsule Endoscopy (WCE) was introduced for total gastrointestinal tract (GIT) inspection[1].

The WCE process involves swallowing of a camera loaded in a capsule which takes images of GIT during its movement through the entire tract. The images captured by the camera are then transmitted to the recording unit present on a belt worn by the patient[2]. These images are loaded in the work station (computer) in the form of video. The capsule is then expelled out from the body naturally after about seven to eight hours. During this total process, the number of images captured are significantly large (per patient 50,000 to 60,000) and with a diverse content, which makes the detection of gastro-intestinal (GI) abnormalities a challenging task. Inspecting a WCE video is laborious and time-consuming. An experienced doctor will generally take 60 to 90 minutes to evaluate the images of one patient. The workload of gastroenterologists can be reduced thereby increasing their efficiency by aiding the process with intelligent approaches in the diagnosis procedure. This way, the application of such approaches can contribute in speeding up the analysis of total WCE video consequently increasing clinical productivity and an overall cost reduction for healthcare systems is possible.

There is huge amount of redundancy in collected video. It is very difficult and time exhausting task for a medical practitioner to look for a peculiar abnormality from this large number of frames. Multiple image processing techniques like classification, detection and segmentation are proposed by researchers for analysis of WCE video. These techniques filter the video for specific images which may contains abnormality. However, in actual clinical practice, to avoid risk of missing something in the WCE examination, medical expert would always like to confirm the computer aided detection results. Acquiring summarized video can be a plausible solution for solving this problem. Video summarization process tries to abstract the vital occurrences, scenes, or items in a clip for providing easy interpretation of synopsis. In summarization process important steps are selection of features considering the application, different clustering methods and key frame selection. The later section gives information about different video summarization techniques based on multiple aspects and contribution by researchers in the field of WCE video summarization over past five years.

2 Video Summarization

Video Summarization is the technique with which abstract view of entire video can be created and presented in very less amount of time as compared to original video.

Video summarization is analogous to summary creation of a video. Important aspects of video summarization are as follows:

- The created summary should be as short as possible
- Summarized video should contain the most important scenes and events from the video

- There should not be any redundancy

If an application needs processing of a video then summarized video will reduce processing time as well as computation complexity.

2.1 Static and dynamic

These are the two broad categories of summarization techniques. The static type summarized video contains set of key frames from the base video without consideration of time and sequence. The dynamic type summarized video contains the most significant, small portions of audio and video. They are also called as video skims and are similar to trailers of movies.

2.2 Summarization with low/high/multiple features

In the applications where quick response is needed and the scenes of video are not complex low level features does the work e.g. color, texture, motion etc. But use of only low level features may lead to loss of complex scenes having significant information.

In the applications where contents of the image are of critical importance, high level features contribute more in the task of summarization. Some of the high level features that plays major role in proper understanding of the contents of the video are object recognition, face detection, gesture detection, event and emotion detection etc. Summarization process with high level features is time consuming and computationally expensive.

Multiple features investigate the content from various aspects and utilize them for video summarization. The evaluation of summarization technique with multiple features is necessary considering the area of application, practicability and its cost efficiency.

2.3 Shot boundary based video summarization

A prolonged video can be segmented as per the shot or scenes.

- A scene in a video is a group of temporally and semantically associated class of elements of that video which communicates a clear meaning.
- A shot can be seen as a series of activities picked up by a single camera which does not encounter any major deviations in the visual data.

The shot boundary is located at sudden changes in the frames. Located shot boundaries can be used for summarization process. The process generally involves selection of key frame from each segmented shot. These key frames can be stitched later to form a summarized video. The algorithms used for such techniques need proper threshold value for execution of the method efficiently. If the threshold value chosen is very small, then there will be large number of segments that ultimately results in elongated summarized video. If the threshold value is a large number then there are chances of missing some important information from the video. So selection of threshold value is a critical procedure.

2.4 Clustering based Video Summarization (static)

In clustering procedure initially random frames are considered as cluster centres. Application of clustering algorithm gives considered number of clusters which contains frames with nearly similar contents. The cluster centers then can be revised and selected as representative key frame. There are many clustering algorithms like k means[3], guarded zone clustering, spectral clustering[4], MST (Minimum Spanning Tree) clustering[5] etc. In this procedure the representative key frames are not in the sequential order. The applications in which the contents are of much important than there occurrence sequence can prefer this methodology. Time constraint clustering algorithm to maintain the sequence of the video is proposed in [6].

2.5 Non- Clustering based Video Summarization

In this type of summarization, the representative frames are selected by using thresholding techniques. Sequential selection of key frames by using thresholding technique is presented in [7]. The procedure has following steps, 1) Calculate histogram of frames 2) Calculate sum of absolute differences of histograms 3) Selection of threshold 4) Selection of key frames: The frame which have the sum of absolute difference greater than the threshold value. The efficiency depends on the threshold value.

3Summarization in wireless capsule endoscopy

In literature WCE video summarization is done in two ways as shown in fig 1. The video can be summarized by eliminating the redundant frames and second way to summarize video is to select only those frames which have salient contents.

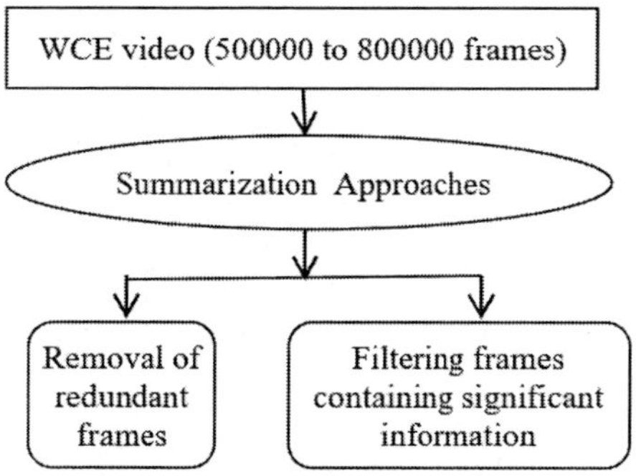

Fig. 1. Summarization approaches for WCE video.

Video summarization technique based on possibilistic clustering and feature weighting algorithm is proposed in [8] by Mohamed et al. The algorithm generates possibilistic membership which is used to detect the degree of similarity the WCE video frames. The frames are selected based on this membership function. The advantage of the algorithm is having an objective function from each cluster which reduces optimization of possibilistic membership values and optimal feature weights. This optimization is achieved by updating the membership values, the feature weights and the centroids repetitively. The experimental result indicates that the possibilistic weights can identify noisy frames and provides superior description of the data. This method does not require determining number of clusters. The unique part in this methodology is cluster dependent feature subset weights.

Emam et al. proposed summarization based on adaptive feature extraction techniques [9]. As per the results LBP performance is superior as compared to statistical features. For similarity detection cosine similarity distance from Matlab is used. 90 % similarity threshold is applied for all feature extraction techniques.

Chen et al. addressed the problem of low level features that they cannot capture semantic similarity between WCE frames and used Siamese Neural Network (SNN)for the extracting high level semantic features[10]. This methodology requires pairs of similar and dissimilar WCE frames with labels. SNN training is done with contrastive loss function for extracting features with high level semantics. Linear SVM is used for classification based on similarity judgment. This process avoids the problem of setting correct threshold manually.

Adaptive redundant images elimination technique was proposed by Chen et al. in [11]. Similarity of features between consecutive WCE frames and temporal correlation is used for summary creation. HSV color histogram model and Gray Level Co-occurrence Matrix are used for generating color and texture features. W-parametric mean value threshold (WMVT) is developed which is data driven. The low level features similarity of consecutive WCE frames are compared with the threshold and then the matching images are grouped into the same clip. Adaptive K-means clustering algorithm is used to generate representative frames and to exclude redundant frames by using gradient dependent characteristic in each clip.

Hamza et al. proposed secure video summarization technique to aid diagnosis of WCE video in IoT (Internet of things) environment [12]. Recently the health care services started using IoT techniques for transmitting data to hospitals or medical practitioner. Paper proposes framework for transmission of WCE data considering summarization and encryption methods. In summarization part feature computation is done by using concept of integral image which is computed based on COC model. The features considered are multi-scale contrast, MI (moments of inertia) and curvature map. The saliency scores are calculated for all these features. The scores are then normalized in the range 0 to 1. Then the representative frames are uprooted using aggregated attention curve. This curve is generated by combining normalized scores.

Mohammed et al. in [13] used multiple features for colon video summarization. Methodology involves grouping of video frames based on power spectral density. Key frames are then selected for multiple feature extraction. Multiple features such as handcrafted (color and texture) and deep were extracted. Pre-trained GoogleNet model is used for extracting deep features. Next step is sparse coding of these features using a learned dictionary. SVM classifier is trained by using sparse coded features. The output is categorized as frames with salient content and frames without any information.

Figure 2 gives pictorial representation of key frame extraction process from redundant frames in WCE video. Process summary is - First WCE video is divided in to frames with proper frame rate then different feature extraction methods are used for optimal feature set selection. Key frame can be extracted by using clustering algorithm or thresholding techniques. The hurdles to overcome in this field are availability of dataset and standard evaluation parameters to validate the result for summarization. The dataset for training purpose should be developed with the help of guidance from gastroenterologist for annotating the images. Standard evaluation measures will ease comparison of summarization techniques based on their performances.

Table 1. Comparative analysis of related work

Year	Author	Features	Dataset	Performance metrics
2014	Mohamed et al. [8]	RGB Color Histogram HSV Color Moments Edge histogram	4 CE videos from 4 different patients. 75% for training and 25% for testing	Classification rate Relative cluster evaluation measures
2015	Emam et al. [9]	Color Histogram, Color Moment HSV, Color Moment RGB LBP Features, Statistical Features	10671 images	Reduction rate
2015	Chen et al.[10]	HSV color histogram Texture Features	One dataset = 500 continuous WCE frames (4000 data sets)	F-measure Recall Compression ratio (CR) Precision
2016	Chen et al.[11]	CNN as a feature extractor	Similar frames: 7875 pairs Dissimilar frames: 7986 pairs	Compressed ratio (CR) F-measure Computation time
2017	Mohammed et al.[13]	Mean color, power spectral density (PSD), hue and opponent histogram, LBP histogram, deep features	500 images	Average Accuracy Sensitivity Specificity

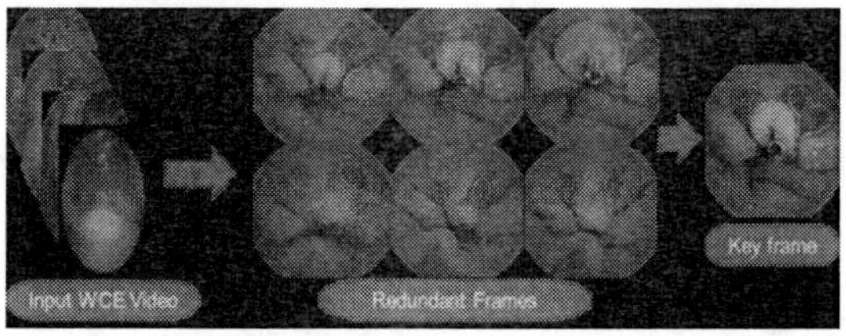

Fig. 2. Key frame extraction in WCE video

Source: Images adapted from [14]

4 Conclusion

This paper reviews types of video summarization techniques and particularly the methodologies used by various researchers to summarize WCE video in past five years. The aim of this paper is to survey various WCE video summarization techniques and provide a reference document which can be explored for further result improvement. Summarization techniques which are based on low level features are good for real time applications because of their speed and computationally inexpensiveness. The techniques which are based on high level features or user attention model are exclusively suitable for applications where the main constraint for summarization is accuracy rather than time. The various techniques have their own merits and demerits as per the application. The results are improved if semantic features are combined with visual descriptors. Video summarization field still demands more standard evaluation techniques to validate the results.

References

1. A. G. & P. S. Gavriel Iddan, Gavriel Meron, "Wireless capsule endoscopy," Nature, vol. **405**, no. May, pp. 417–418, (2000)
2. A. Wang et al., "TECHNOLOGY STATUS EVALUATION REPORT: *Wireless capsule endoscopy," Gastrointest. Endosc.*, vol. **63**, no. 4, pp. 539–45, (2013)
3. S. E. F. De Avila, A. P. B. Lopes, A. Da Luz, and A. De Albuquerque Araújo, "VSUMM: A mechanism designed to produce static video summaries and a novel evaluation method," *Pattern Recognit. Lett.*, vol. **32**, no. 1, pp. 56–68, (2011)
4. A. Ioannidis, V. Chasanis, and A. Likas, "Weighted multi-view key-frame extraction," *Pattern Recognit. Lett.*, vol. **72**, pp. 52–61, (2016)
5. R. Panda, S. K. Kuanar, and A. S. Chowdhury, "Scalable video summarization using skeleton graph and random walk," *Proc. - Int. Conf. Pattern Recognit.*, pp. 3481–3486, (2014)
6. J. L. Lai and Y. Yi, "Key frame extraction based on visual attention model," *J. Vis. Commun. Image Represent.*, vol. **23**, no. 1, pp. 114–125, (2012)
7. C. V. Sheena and N. K. Narayanan, "Key-frame Extraction by Analysis of Histograms of Video Frames Using Statistical Methods," *Procedia Comput. Sci.*, vol. **70**, pp. 36–40, (2015)
8. M. M. Ben Ismail, O. Bchir, and A. Z. Emam, "Endoscopy Video Summarization based on Multi-Modal Descriptors and Possibilistic Unsupervised Learning and Feature Subset Weighting," *Intell. Autom. Soft Comput.*, (2014)
9. A. Z. Emam, Y. A. Ali, and M. M. Ben Ismail, "Adaptive features extraction for Capsule Endoscopy (CE) video summarization," in *Proceedings - International Conference on Computer Vision and Image Analysis Applications, ICCVIA 2015*, (2015)
10. J. Chen, Y. Zou, and Y. Wang, "Wireless capsule endoscopy video summarization: A learning approach based on Siamese neural network and support vector machine," *Proc. - Int. Conf. Pattern Recognit.*, pp. 1303–1308, (2016)
11. J. Chen, Y. Wang, and Y. X. Zou, "An adaptive redundant image elimination for Wireless Capsule Endoscopy review based on temporal correlation and color-texture feature similarity," *in International Conference on Digital Signal Processing, DSP*, (2015)
12. R. Hamza, K. Muhammad, Z. Lv, and F. Titouna, "Secure video summarization framework for personalized wireless capsule endoscopy," *Pervasive Mob. Comput.*, vol. **41**, pp. 436–450, (2017)
13. A. Mohammed, S. Yildirim, M. Pedersen, O. Hovde, and F. Cheikh, "Sparse Coded Handcrafted and Deep Features for Colon Capsule Video Summarization," *Proc. - IEEE Symp. Comput. Med. Syst.*, vol. **2017-June**, pp. 728–733, (2017)
14. Koulaouzidis, D. K. Iakovidis, D. E. Yung, E. Rondonotti, U. Kopylov, J. N. Plevris, E. Toth, A. Eliakim, G. W. Johansson, W. Marlicz, and others, "KID Project: an internet-based digital video atlas of capsule endoscopy for research purposes," *Endoscopy International Open*, vol. **5**, no. 06, pp. E477–E483, (2017)

Doctors+

Vishwam Shukla[1*], *Sudeshna Ray*[1], *and Sukhpreet Kaur*[1]
[1]Dept. of CSE, School of Engineering, MIT ADT University, Pune, India

Abstract: "Doctors+". The primary goal of our project is to give up to 90% of paper-less hospitals. It expects to give minimal effort solid mechanization of existing frameworks. The framework gives incredible assurance of information at each degree of a client framework collaboration.

1 Introduction

It is essential for hospitals to screen their everyday exercises and the records of their patients, specialists, attendants, ward young men and different workers, who are running the hospital easily and effectively. However, following all exercises and their records on paper is exceptionally awkward and mistake inclined.

It is very inept and a time-consuming process that is seeing a steady increase in the number of hospitalized population and people. The chronicle and upkeep of every one of these records are exceptionally problematic, wasteful and mistake inclined. It isn't financially and conceivable to keep up these records on paper.

In this way, remembering crafted by the manual framework as the premise of our undertaking, we have built up a computerized variant of the manual framework assigned as "Doctors +".

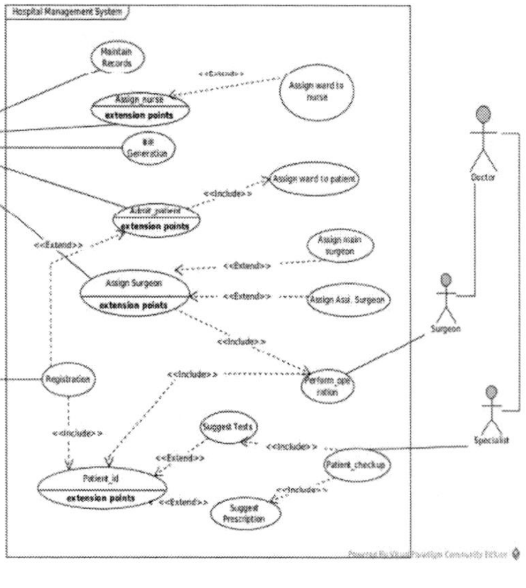

Fig.1. Use case diagram

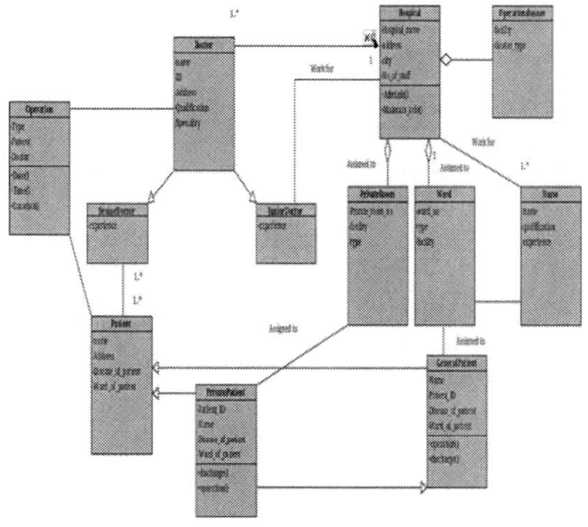

Fig.2. Class Diagram

2 Functionality

Doctors+ is a user-friendly android application that can able to manage patients. The application will start with a beautiful animated screen and there will be an option to continue where users can find their roles i.e. Doctors, Receptionist, Nurse, Pharmacist, Emergency department etc. And from their responses, the application will direct them to the login page where they can access their account. Several options will be given which includes sign in with email and password, sign in with phone number, Google/Facebook/Twitter/Microsoft. User needs to be registered with the system first.

Several options will be given to the reception department which includes adding patients to OPD, adding patients to IPD, Leave applications, Chats, IPD Requests, Profile etc. where reception department can add a patient to the respective department taking details like Name, Email, Phone Number, Age, Gender, Relation -

*Corresponding author: **shuklavishwam@gmail.com**

Status, Blood Group, Allergies, Weight, Blood Pressure, Pulse, Complaints, New or Old visit, referred by, Image etc. These details will be updated to the respective doctor on a real-time basis. Same procedure to be followed in the IPD section. Reception department will be able to view leave applications of doctors hence no appointments can be booked on those days. Reception department will be able to communicate with other hospital parties through the real-time chat option given in the application. Reception department will be able to the IPD requests from the doctor if in case a patient to be hospitalized. The receptionist will be able to update their information on the profile option in the application.

The portal for doctors will start will the list of appointments and on clicking/tapping on a particular appointment, doctors will be able to view the detailed information of that particular patient which includes details like Name, Email, Phone number, Image, Weight, Blood Pressure, Allergies Medical History etc. Several options will be given for a particular appointment which includes IPD Request in which doctors can notify the reception department to admit the patient, Connect to Pharmacy in which doctors can complete the appointment by suggesting some medicines to the patient and notifying the pharmacy department as well, Complete the appointment etc. Doctors will be given several options to view the admitted patients in IPD. Doctors will be notified if there are any emergencies in the hospital. Doctors will be able to communicate with other doctors and other hospital parties as well. Doctors will be able to update their information on the application as well. Doctors will be able to search a particular record of a patient in case of multiple records. Doctors will get a smart history portal for keeping and viewing information of patients. Several controls will be also given to the nurses where they can view assigned patients by a doctor and will be able to update the information such as Blood Pressure, Pulse, Recovery status etc. of a patient on a real-time basis. Nurses will be able to communicate with doctors and other hospital parties as well. A nurse will be able to update their information on the application.

Pharmacy department will be able to view the prescription of a patient so that that they can keep ready those medicines before the arrival of that particular patient. Pharmacy department will also communicate other hospital parties through a chat option. Emergency ward boys will be able to add emergencies in the hospital. They will be able to view the records of the emergencies as well.

The advantages of the applications are paperless(90%), cost-effective, time-effective, reduce power consumption, easily accessible to doctors, easy access to patient data, automated version of the existing system, excellent security of the data, better storage facilities, better communication between doctors and other parties as well, improved patient care, reduces the scope of error, improved efficiency.

Considering the above, Doctor application "Doctors+" can prove useful to all doctors, allowing them to give up keeping patient records on paper.

3 Modules

3.1 Category page of the application

The interface includes cross-communication of data between 6 categories including the Doctors, Receptionist, Nurse, Pharmacist, Emergency and Admin.

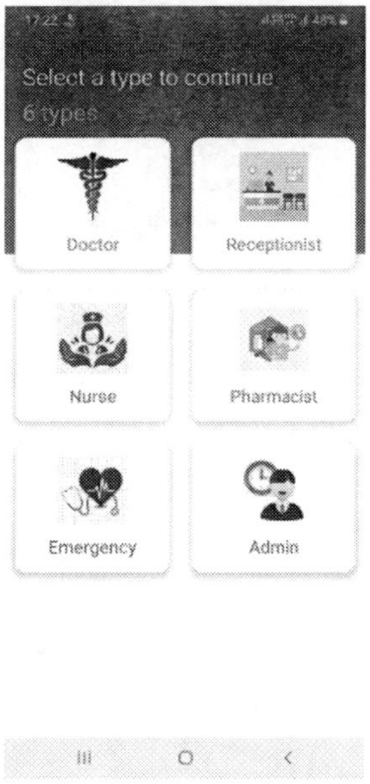

Fig.3. Category page

3.2 Login portal

The login portal is available for the use of all parties involved.

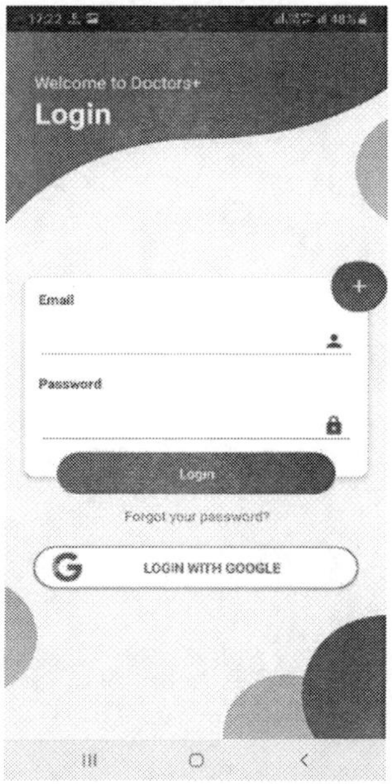

Fig.4. Login page

3.3 Reception dashboard

The Reception dashboard includes facilities of adding to the OPD, adding to the IPD, leaving applications, making IPD requests, accessing chats and profiles.

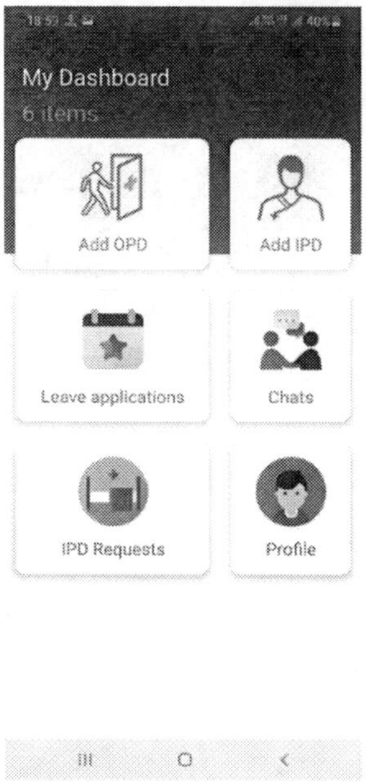

Fig.5. Reception dashboard

3.4 Adding patients to OPD

The Patient's referral information can be registered in the Out-Patient Department which can be accessed by various parties.

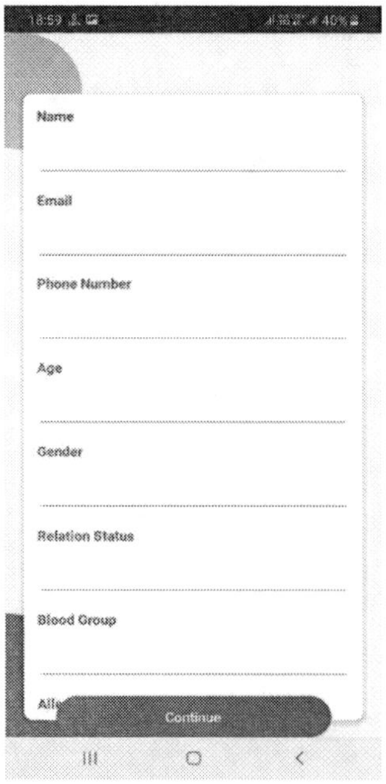

Fig.6. OPD admission

3.5 Adding patients to IPD

The Patient's referral information can be registered in the In-Patient Department which can be accessed by various parties.

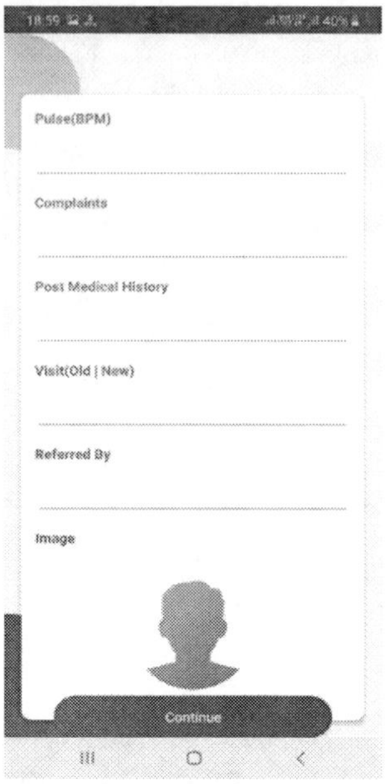

Fig.7. IPD admission

3.6 Home page for doctors

The appointments and schedules can be kept track of with the date and time details.

Fig.8. Home page for doctors

3.7 Detailed information on appointments

The detailed information regarding the medical issues and specifications of the patient can be accessed by the doctor while being updated in real-time.

Fig.9. Appointment details

3.8 Drawer for doctors

The drawer slider displays an array of features including cross consultancies, emergencies and other account settings.

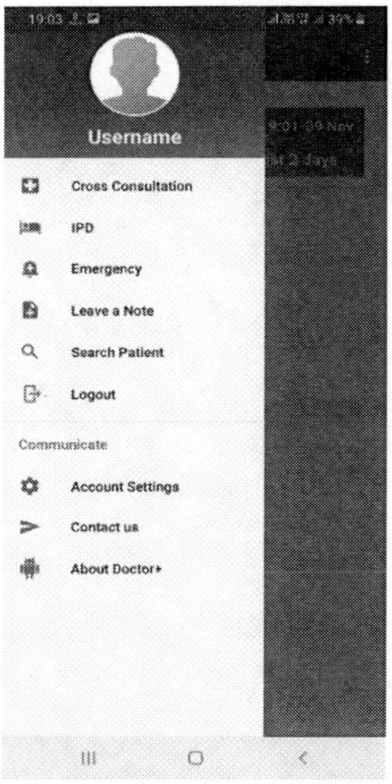

Fig.10. Drawers for doctors

3.9 Other controls given to the doctors

Appointments can be managed by giving several controls to the doctors which include Transfer to IPD, Connect to pharmacy, Complete appointment etc.

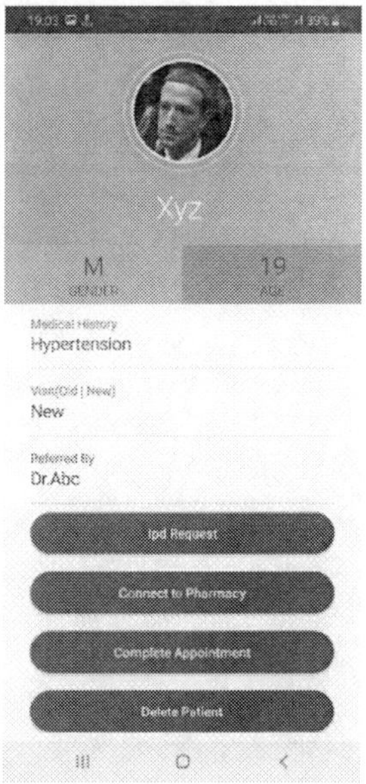

Fig.11. Options to doctors

3.10 Dashboard for nurses

Nurses will be able to track the assigned patient from the doctor and manage better.

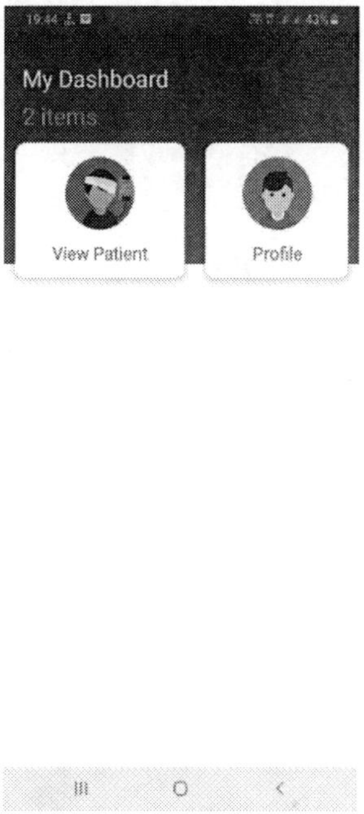

Fig.12. Nurse dashboard

3.11 Dashboard for pharmacists

The pharmacist can keep ready those assigned medicines by viewing prescription.

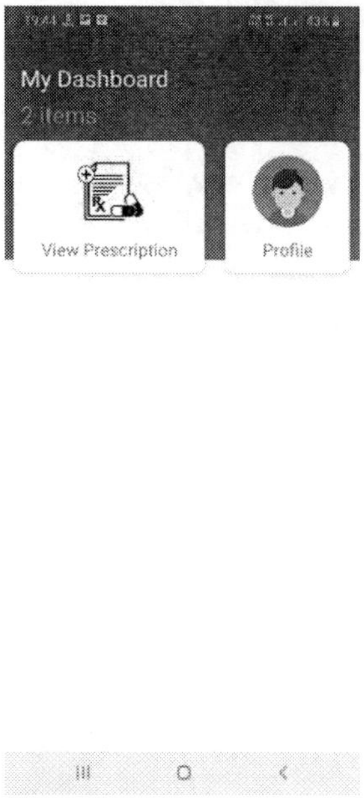

Fig.13. Pharmacy dashboard

3.12 Dashboard for emergency department

Emergency ward boys can add emergencies and can also view emergencies in the hospital.

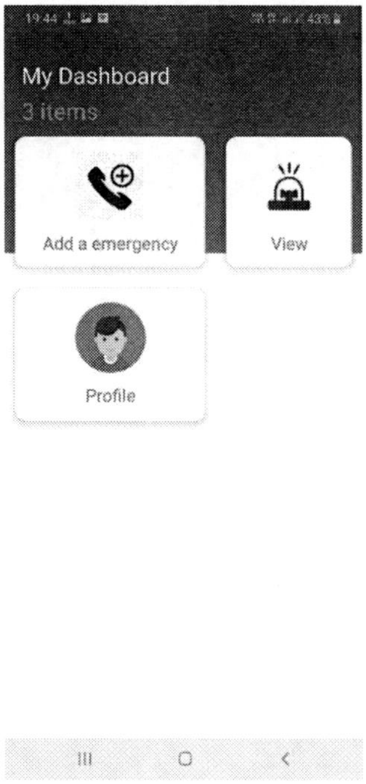

Fig.14. Emergency dashboard

4 Future enhancements

This will be an intelligent digital app helping doctors with not just appointments and patient details as a description but moving from 'descriptive' to 'predictive' to 'prescriptive' to 'cognitive'. It will help doctors to predict if there is any emergency in patients sitting in OPD. Predict what type of illness will be there in what season so that doctors can keep medicine stock ready in the hospital. Predict the date of discharge of an IPD patient Predict the issue and suggest medicines as per the complaints. Diet recommendation for a doctor to prescribe a diet based on real-time reports of patients. Waiting time for patients sitting in the OPD area through SMS and Email. Prescription, Reports (Pathology, X-ray, MRI, etc.) and medicine invoice in .pdf format to patients through SMS and Email. Messages to be secured with end to end encryption. AI-Powered Chatbots for patients etc.

5 Conclusion

Thus Doctors+ can be used to provide paper-less hospital up to 90% and can provide better healthcare to the patients.

References

1. D. Griffiths, D. Griffiths, *Head First Android Development,* **2** (2017)
2. G. D. Kunders, *Hospitals: Facilities Planning and Management* (2017)

Bleeding Detection in Gastrointestinal Images using Texture Classification and Local Binary Pattern Technique: A Review

Aparna Goyal[1,*] *and Reena Gunjan*[1]

[1]Electronics and Communications Engineering Department, MIT School of Engineering, MIT ADT University, Pune, Maharashtra, India.

Abstract. Texture analysis has proven to be a breakthrough in many applications of computer image analysis. It has been used for classification or segmentation of images which requires an effective description of image texture. Due to high discriminative power and simplicity of computation, the local binary pattern descriptors have been used for distinguishing different textures and in extracting texture and color in medical images. This paper discusses performance of various texture classification techniques using Contourlet Transform, Discrete Fourier Transform, Local Binary Patterns and Lacunarity analysis. The study reveals that the incorporation of efficient image segmentation, enhancement and texture classification using local binary pattern descriptor detects bleeding region in human intestines precisely.

1 Introduction

Bleeding in the intestine is an indication of several possible gastrointestinal (GI) diseases like ulcerative colitis (UC), tumors, and inflammatory diseases etc. Hence, detection of regions of bleeding becomes crucial in diagnosis of several diseases related to intestines. Among thousands of images received only few image frames might contain bleeding or very miniscule bleeding regions which may not be detected by manual inspection. Therefore, computerized automated methods have become the need of the hour for accurate, reliable and fast diagnosis. Among the techniques of analysis of medical images, texture analysis has played a crucial role. Texture is a tool to sub divide the images and then classify particular region of interest. It identifies the peculiarity of images and helps us in analyzing its features. Texture gives us information about structural arrangement of surfaces and their relationship to the surrounding environment. In several diseases, such as Pulmonary Fibrosis causing the thickening of tissues in the lungs, the textural changes are more visible in the X-ray images. Similar applications require more efficient texture analysis method helps to provide correct diagnosis. In gastroenterology, a distinct and specific texture pattern can be seen in ulcerous region in the internal mucous membrane of the digestive tract. The quality of the extracted features plays a crucial role in deciding the accuracy of bleeding detection method. In different medical image analysis systems, grey-scale images are used to calculate textural features. The Local Binary Pattern (LBP) operator is one of popular texture methods that has proven to perform well in texture classification applications.

2 Related work

The very first method of calculating LBP was introduced by Ojala[1] in 1996 which analyzed the images by their LBP histograms. Features based on histogram can identify the bleeding areas irrespective of their dimension and shape. Local Binary Pattern (LBP) is a comprehensible and an adequate texture operator. It remains uninterrupted if in gray scale occurs and provides good performance. Local Binary Pattern (LBP) is logical and consistent texture operator. It assigns labels to the pixels of an image. While labeling the pixels, its thresholds the pixels in the proximity of central pixel and assigns it a binary digit zero or one. The histogram of different intensity labels thus obtained were used for description of the texture. LBP Patterns distribution in an LBP histogram was used for analysis of texture image. Mekha Mathew, Varun P Gopi [2] had proposed a new method to discriminate bleeding regions and normal regions

*Corresponding author: aparnagoyal30@gmail.com

using Local Binary Pattern (LBP) and Contourlet transform. The images were resized to 256x256. In order to focus on intrinsic geometric structure, the two-dimensional contourlet transform was applied. The contourlet transform filters the noise and detects the smooth contour in different directions of images. This transform performs 2 level decomposition resulting in coarse sub band having images of low frequency and fine sub-bands having images with high frequency. RIULBP (Rotation Invariant Local Binary Pattern) were applied on these bands. Using the combination of Laplacian Pyramid (LP) and the Directional Filter Banks (DFB) a contourlet filter bank is obtained which is a double filter bank to analyses various texture feature.

Baopu Liand Max Q.-H. Meng[3] worked on the method that emphasized upon color texture feature using Chrominance moment by dividing the image into components as luminance and chrominance. The method used Tchebichef polynomials and illumination invariance of hue/ saturation/ intensity color space. The authors have used additionally uniform local binary pattern, to differentiate bleeding and nonbleeding regions in CE images.

Another method was developed by Konstantin Pogorelov and Shipra Suman [4] to detect bleeding in GI images using color feature and texture feature. The authors have first utilized color information to detect image frames with bleeding. To modify bleeding and non-bleeding images, detection algorithm makes use of classification methods based on machine-learning to execute segmentation at pixel-level for bleeding frames in WCE images. Adriana Florentina, Mihaela [5] proposed a method for reducing the overall frames number in medical imaging so to reduce shorter analysis time. The authors had used a software application for texture analysis based on local binary pattern (LBP) operator. This operator detects and removes non-informative frames and then identifies potential lesions. Using LBP operator, the texture of intestinal mucosa, food residues and air bubbles were differentiated, then the number of non-informative frames were detected with different LBP operators. The classification of feature vectors was done using K-Means clustering technique. Yuhui Quan [6] proposed a method based upon lacunarity analysis to distinguish spatial patterns of image structure. In this work, a robust statistical approach for texture description was developed, to analyze photometric changes and geometric changes. The method incorporates multi-scale local binary patterns of image. Aliahmadvand et.al. [7] had introduced LBP with discrete wavelet transform using multi resolution analysis. By utilizing frequency information i.e., the spectral features of textures in LBP based methods, the DWT(a spatial frequency filter) was used for extraction of the spatial frequency channels and then LBP features were extracted from these channels. The combined approach using LBP and DWT improved the classification rate of LBP for invariant texture classification.

3 Texture analysis

The analysis of the intensities of nearby pixels in an image without consideration of their color is distinguished by the texture feature. The orientation and shape of a surface and inspection of a surface are machine visions tasks which necessitate efficient analysis of texture. It defines the intrinsic properties of surfaces especially those that do not have a smoothly varying intensity. Several image properties such as smoothness, coarseness, depth, regularity etc. can be associated with texture. Texture can also be defined as a descriptor of local brightness variation from pixel to pixel in a small neighborhood through an image. In any digital image, the spatial arrangement of the pixel defines the texture. It is a neighborhood property of an image point. The dimensions of the observation neighborhood vary according to texture measures.

Different Texture Evaluation methods depends upon the technique used for extracting features of the texture in an image. The extraction of textural features is categorized as follows: Statistical methods, Structural methods and Transform-based methods as illustrated in fig.1.

As texture is a property of spatial arrangement thus the relationship between the gray level pixel is analyzed by the statistical analysis approach using histogram. These histograms illustrate statistical moments such as variance, mean, skewness and kurtosis of the intensity histogram of an image to analyze various texture features. One of the most accepted method for texture analysis is based upon two-dimensional grey level co-occurrence matrix.

Structural texture analysis techniques represent texture as the constitution of perspicuous texture elements. The image texture is defined by the properties of the texture elements. The structural approach describes the image symbolically.

Texture analysis method such as Transform Based approach, alter images into a new domain . The intensity variations in the image pixel provides the spatial frequency. The transform technique used to separate out textural trait from the image shows the efficiency of this method.

 https://doi.org/10.1051/e3sconf/202017003007

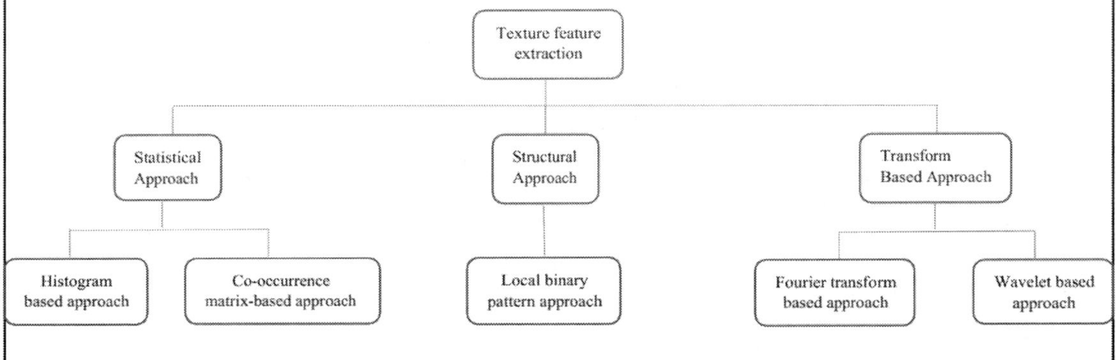

Fig.1. Texture features extraction methods

4 Detection of bleeding in GI images

Several methods have been developed incorporating statistical, structural and transform based approach for detection of bleeding in gastrointestinal images. A few methods have been outlined in brief using these approaches as indicated in the fig. 2. given below.

4.1 Contourlet Transform and Local Binary Pattern (LBP)

By using a double filter bank structure, the contourlet transform smooths the contour of images. In double filter bank, the point discontinuities are first detected by the Laplacian pyramid (LP), then converting point discontinuities into linear structures is done by directional filter bank (DFB). The properties of contourlet transform are multi resolution, localization, directionality, critical sampling and anisotropy. Multi scale and multidimensional decomposition are its basic functions.

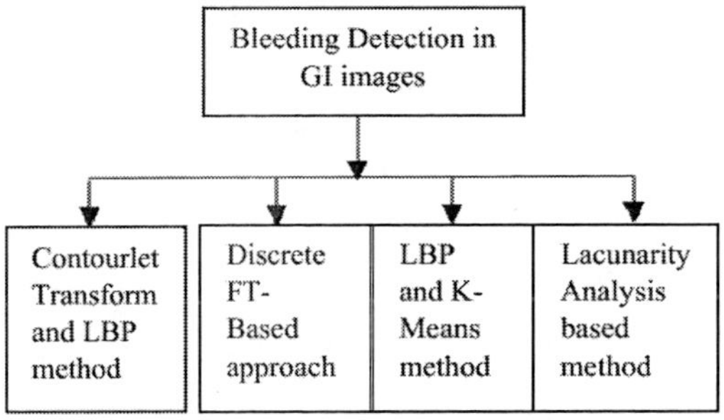

Fig. 2. Bleeding detection techniques

Local Binary Pattern (LBP), is a combination of structural and statistical approach. LBP has proven to be effective and efficient texture operator. In the image frame any monotonic transformation does not alter the corresponding LBP code and LBP is invariant against brightness

LBP divides an image into pixels. Each pixel present in a cell is compared to the neighboring pixel, which follows circular pattern that is clockwise or anticlockwise direction. If the centre pixel value is greater than the neighbor pixel, the value assigned is 0 or if not 1 is assigned.

As a result, a binary digit is obtained, which can also be named as binary string. Then histogram of the image is constructed. intensity values of pixels are shown by the histogram of an image. Finally, normalization of the histogram is done. The texture descriptor can be taken as different labels of the histogram. Thus, by the distribution of LBP patterns texture of an image can be typified by an LBP histogram.

4.2 Discrete Fourier Transform Based Approach

Transform methods represent an image in a space whose coordinate system defines the attribute of a texture. The application of Discrete Fourier Transform (equivalent to short time Fourier Transform) of small patches of the images is used in this feature extraction method. These images, as shown in the fig.3, are passed through frequency band filters. Inverse discrete Fourier transform is performed on resulting image and an image with frequency in selected band is obtained. Thus, rotation invariant texture descriptors are obtained in the form of local frequency descriptors.

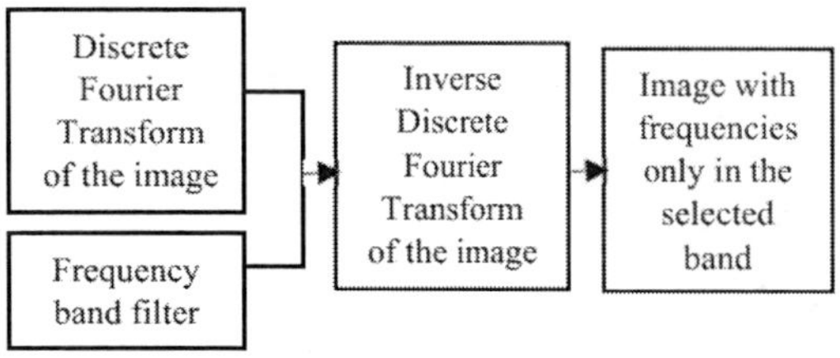

Fig. 3. Feature extraction method using DFT

4.3 LBP and K-Means Clustering Technique

Among various data analysis technique, clustering is used most extensively used to analyses the preliminary structure of the data set. K-Means clustering identifies subgroups in the given data set and divides the data based on the similarity in distinct cluster. This technique finds homogeneous subgroups within the data by finding Euclidean distance between the data points or correlation based distance. The determination of similarity depends upon the particular application for which K-Means clustering is used. Clustering analysis can be done basedon features where subgroups of samples based on features are found. K-Means subdivides clustered observations into a specific number of disjoint clusters. This technique minimizes the distance between the centroid of the cluster and the given observation. The procedure is repeated iteratively by appending an observation to the cluster and procedure is terminated when the lowest distance measure is observed.

4.4 Lacunarity Analysis for Texture Classification

Lacunarity suggest how the space in the data set is filled by the pattern. Lacunarity is more if there is a wide range of gap in pattern. For homogeneous texture lacunarity should be Low and transitionally invariant, as all gap size are of the same dimension, for heterogeneous texture lacunarity is considered high. It is a scale dependent estimation of the homogeneity of the texture pattern. The method is used to discriminate the different textures incorporating local binary pattern technique. Then multi-scale local binary pattern approach is applied on images to compute lacunarity-related parameters which characterizes the texture of the image. This distinguishes the spatial patterns in the texture of the image. The spatial distribution on multiple scales characterizes the features of local image structures.

5 Conclusion

A large collection of accumulated images during intestinal endoscopy emphasizes the need for efficient texture-based image classification methods for better diagnosis. This review was performed to study several texture feature extraction techniques. The contourlet-based texture feature combines uniform local binary pattern approach for texture classification. The method yields good result and thus increases better diagnosis. The Fourier transform based techniques contain information confined to the frequency domain only. It cannot be used for the space domain. Since frequency spectrum does not reflect enough on the spatial description of the texture. Therefore, the frequency response thus obtained from the spectrum may not be used for the spatial localization in the image texture directly. One of the constraints of the Fourier transform being computationally complex, local variations of textures cannot be described by Fourier Transform. Henceforth, other methods such as Gabor decomposition-based approaches or wavelet-based approaches are proposed for improved spatial localization in the texture. Capturing localized frequency and spatial information is possible by the wavelet transform. The wavelet transform requires less complex computation. In order to analyze the image texture at different scales, spatial resolution can be varied using wavelet transform. However, the wavelet transform is neither translation-invariant nor rotation invariant. K-Means clustering technique or the nearest neighbor technique is simple to implement. K- means classifiers are helpful in analyzing a small dataset. The Local Binary Pattern descriptors have been powerful tools for feature extraction. Texture features in Wireless Capsule Endoscopy image data set in GI tract can be extracted efficiently by incorporating proper image segmentation technique together with local binary pattern descriptors. It is observed that the methods based upon structural and statistical technique like Local Binary Pattern approach increase the performance of texture analysis. It is found that the method is simple to implement and requires less computational cost.

References

1. T. Ojala, M. Pietikainen, D. Harwood, Pattern Recognition, **29**, 55, (1996).
2. M. Mathew, V.P.Gopi, IEEE Sponsored 2nd Int. Conf. on Electro. and Comm. System, **13**,1223, (2016).
3. B. Li., Max Q. H. Meng, Comp. in Bio. and Medicines, 141, (2008).
4. P. Konstantin, S. Shipra, J. of Applied Clinical Medical Physics, **20**, 141, (2019).
5. A. Florentina, M. Ionescu, Applied Medical Informatics Original Res., **36**, 31, (2015).
6. Y. Quan, Y. Xu, Y. Sun, Y. Luo, IEEE Conf. on Computer Vision and Pattern Recog., 160, (2014).
7. A. Ahmadvand, R. Ahmadvand, M. T .Hajiali, M. Mosavi, 2nd Int. Conf. on Knowledge-Based Eng. and Innovation, 152, (2015)

Analysis and Examination of Heart distress from ECG signal using Artificial Intelligence.

Leena B. Chaudhari[1*] *and Dhananjay E. Upasani*[2]
[1] Bharati Vidyapeeth's College of Engineering, Lavale, Pune, India.
[2] MIT School of Engineering, MIT ADT University, Pune, India.

Abstract: Health data analysis is usually based on a comparison of derived health measures to predefined thresholds. Symptoms can be observed if a value is above or below a threshold. Early detection of signs of heart failure allows the prediction of strokes of heart failure and can therefore prevent these. So identifying "accurate" criteria is the most important task. The accuracy of an experiment depends strongly on the accuracy of the criteria used. Congestive heart failure (CHF) occurs when the heart can not pump sufficient blood for a stable physiological condition. CHF usually occurs when the coronary artery blockage causes the heart tissue to become acidic. The data used to analyze data such as Linear Regression, Missing Enrollment Data, Search Signal, Clinical Data Protection Programs, and Early Adaptive Alarm. The proposed system involves models including server and data warehouse processing, pre-processing, extraction classification characterization in this paper. Classification of heart defects and prediction of heart failure by using applied classifier for hybridization, guideline for treating patients as a gym, level of stress management. In this article, the program tracks the heart disease patients, predicts atrial fibrillation and ventricular fibrillation, and alerts patients when the critical condition occurs.

1 Introduction

According to the World Health Organization (World Health Organization, 2016), chronic diseases such as coronary heart disease, cancer, chronic obstructive pulmonary disease and type 2 diabetes mellitus are the world's leading cause of death, accounting for around 60% of all deaths. Chronic diseases are due mainly to heart failure, currently the leading cause of death in most western countries. The American Heart Heart Association (AHA) 2016 Heart Disease and Stroke Statistics update estimated that 15.5 million people in the United States suffer from cardiovascular disease, a prevalence that is growing with age for both men and women [1]. Throughout Europe, the predominance of individuals describing heart problems for both sexes was 9.2 per cent over the last year [2]. Cardiovascular diseases are the main cause of death in Italy in particular, accounting for 44 per cent of all deaths. Cardiovascular disease passages in India grew from 1.3 million in each of 1990 to 2.8 million in 2016, and most of the passages were caused by heart disease.

Social insurance is a pending problem with worldwide difficulties[3]. The acute socio-economic situation confronting a significant portion of developed countries is generating greater pressure to find more cost-effective solutions to health and social care delivery. In addition, healthcare systems in developing countries face serious difficulties in providing care and assistance, primarily due to a lack of staff and resources.Over the last few years a few proposals have emerged as a part of the new period of human services. Drug 2.0, Wellness 2.0/3.0, ePatient, and eDoctor are usually dispersed examples of such terms, among other commonly established words such as eHealth, telehealth, or telemedicine. Likewise, frameworks might be utilized in a pervasive way and give consistent observing abilities.One of the major disadvantages of the up-to-date portable health monitoring tools from a technical perspective is the processing and storage capacities. Checking gadgets consist regularly of many sensors which measure physiological sizes and convert them into discernible machine data [4]. While some of these devices contain dedicated resources for processing the information, they are very sparse and limited. Although some of these devices contain dedicated information processing tools, they are very sparse and minimal. On the other hand, mobile devices are the perfect way to collect data from health monitoring systems, provide local processing capabilities or even provide access to cloud services (high-performance computing, large capacity storage, and data analytics) [5] . Through this, not only the data collected from the patient could be taken into account for their health assessment and treatment, but also the records of hundreds, thousands or even millions of patients, which might share similar needs and conditions [6].

Atrial fibrillation (AF) has been depicted by doctors as the commonest heart arrhythmia in clinical everyday

*Corresponding author: leenapc23@gmail.com

practice, with an expected predominance of 1.5%–2% of the overall public in the created world [7]. In excess of 6,000,000 individuals in Europe and 3,000,000 individuals in the USA experience the ill effects of this arrhythmia [7]. It is additionally expected that its predominance will twofold in the following 50 years [8] . Today, three distinct sorts of AF are clinically stratified relying upon the scene length. The arrhythmia would then be able to be grouped into paroxysmal AF (PAF), persevering or long-standing tenacious AF and, at long last, perpetual AF . Indeed, the nearness of AF is related with a five-crease danger of stroke and a three-overlay frequency of congestive cardiovascular breakdown, accordingly inciting that AF patients have double the danger of death than sound individuals of a similar age [7] . Inside this specific situation, an early discovery of AF may help with decreasing that hazard by reestablishing typical heart mood or by improving the blood stream with antithrombotic treatment. This early finding may likewise include prominent advantages for social insurance benefits the world over, in light of the fact that the high hospitalization paces of AF, just as its extensive weight on wellbeing assets could be fundamentally restricted . Cardiovascular sign from both surface and intracardiac accounts have been generally utilized in AF thinks about planning to comprehend the atrial electrical conduct to improve adequacy of interventional removal treatment, however the nearness of ventricular movement (VA) on these sign has hampered the investigation with probability of mutilated outcomes. It is imperative to think about atrial action (AA) without the impact of the VA to improve comprehension of AF beginning and support [9].

Ventricular fibrillation is the quick, disarranged, and offbeat withdrawal of ventricular muscle. Hoffa and Ludwig recommended in 1850 that the inconsistency of these developments is on the grounds that the individual anatomical components lose their interrelationship and quit contracting as one. In 1887, MacWilliam depicted "a condition of fibrillar compression comprised by a quick progression of in facilitated peristaltic constrictions, and in 1914, utilized the term ventricular fibrillation to portray what he thought about the reason for unexpected passing. Superficially electrocardiogram, ventricular fibrillation is portrayed by the nonappearance of unmistakably characterized QRS edifices. It frequently happens in patients with auxiliary heart variations from the norm. Be that as it may, in certain patients, show coronary illness is missing, and the reason for the arrhythmia can't be distinguished or must be learned by broad assessment (e.g., hereditary investigation). Incredible consideration has been given to ventricular fibrillation since it speaks to the last basic pathway for death in many patients who experience out-of-clinic heart failure, and its pace of repeat is about 30% in the primary year in effectively revived patients. Since the particular reasons for ventricular fibrillation are examined in different sections, this section gives a progressively broad outline of the clinical issue and its administration. Ventricular fibrillation (VF) is a significant reason for dismalness and abrupt passing (SD), yet it is one of the most troublesome arrhythmias to viably treat. The implantable cardioverter defibrillator (ICD) offers delayed endurance by ensuring against death because of arrhythmia and is the best quality level treatment for essential and auxiliary anticipation for in danger patients. Be that as it may, gadget implantation doesn't influence the fundamental substrate. Moreover, up to 20% of patients with an ICD experience repetitive scenes of VF or electrical tempests, bringing about numerous gadget treatments, which fall apart the nature of life1 as well as increment related mortality. Antiarrhythmic medications, for example, amiodarone, β-blockers, and lidocaine have been utilized forever undermining arrhythmias, including VF, yet their adequacy has not yet been exhibited on an enormous scale. In addition, they are related with different symptoms, which frequently exceed the questionable prognostic advantage.

2 Objectives of the Study

1. To develop system for individuals tormented by chronic diseases that modify the gathering, sharing and exchange of physiological knowledge, like pressure level, heart and respiration rate and cardiogram signals.
2. To predict heart disease strokes by police investigation connected symptoms.
3. To give notice the specialist and therefore the patient once a heart disease stroke has been foreseen.
4. To work out mechanically and update thresholds in keeping with recorded historical knowledge.
5. To test rules and compare the value to predefined threshold of each derived health indicator.
6. Updating threshold values according to requirement.
7. To do the review and save the original files, which can be accessed and sent to the medical specialist whenever the user wants.

3 Approach of the study

3.1 Quantitative approach

- Constant Mathematical Model: The consistent work could be a work whose esteem is the same for each input esteem.
- Step Function Mathematical Model: The constant perform may be a perform whose price is that the same for each input price.
- Average and Confidence Level Mathematical Model: A simple method to update thresholds that comes back to adding the average of p past data values to the confidence level of this average.
- Moving Average (MA): MA is one of the widely known technical indicators used for time series analysis. During its development, many variations and implementations

have been proposed by researchers. Basic type of moving average is the Simple Moving Average (SMA).

3.2 Qualitative approach

- Allow physiological data, such as blood pressure, heart and respiration rate and ECG signals, to be recorded, transmitted, and exchanged.
- The audit report and the original files are saved, can be accessed and sent to the medical specialist whenever the user wants them to.

4 Architecture

In this design, the limit esteems are consequently processed by utilizing factual methodologies. The produced edges rely upon the patient state and his/her authentic estimations. The thought comprises of recognizing cardiovascular breakdown manifestations and thusly foreseeing basic circumstances (i.e.,strokes).

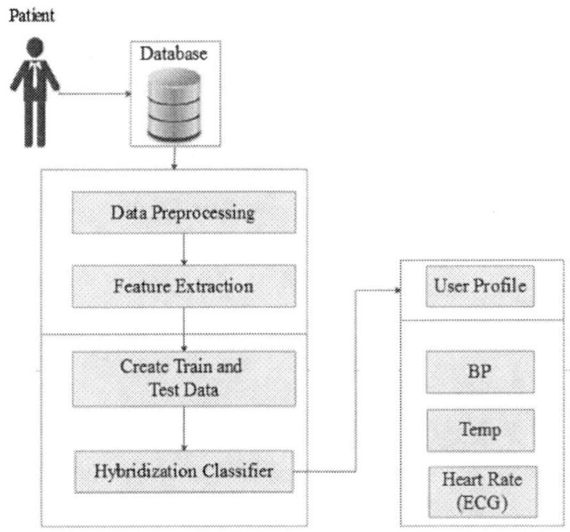

Fig. 1. Proposed System Architecture

1) Dataset: In this architectural study, the MIMIC II waveform database can be used. This is in line with the subgroup from the Psionet Benchmark, which is available online, and is collected from bed patient monitors in the adult and neonatal ICU (intensive care units).
2) First, the device stores data, such as CSV files, in a database. So, patient history is established.
3) Data is preprocessed in the preprocessing stage and the replacement values with threshold values or None are lackinga.
4) The machine performs the extraction process of the element once the preprocessing is completed. The values of the data function are determined using methods of main component analysis. The PCA's goal is to cut down on data and rate columns with a high impact point. This eliminated characteristics from this column after choosing heavily impacted columns such as weight, blood pressure, body temperature.
5) Then build the training and test files using extracted features and use implemented hybridization classifiers to perform classification. The system detects Ventricular Fibrillation and Atrial Fibrillation.
6) Finally, this program suggests care for patients such as Gym, the control of stress levels.

5 Scope

The program can be used to predict strokes that cause heart failure. By frequent monitoring along with support and suggestions it can be helpful for CHF patients. Patients with cardiovascular system disorders may measure their weight, blood pressure, activity, and other factors related to the health.

6 Advantages

1) Patients with heart disease are tracked by the machine.
2) Predicts ventricular and atrial fibrillation.
3) Warns patients when the situation is serious.
4) Automatically calculates and updates thresholds based on historical data collected.

7 Conclusion

Develop system for people suffering from chronic diseases. Rules are based on discussion with cardiologist and comparing extracted parameters with threshold. Feature values of data are calculated using the principal component analysis (PCA) techniques. Rules are based on discussion with cardiologist and comparing extracted parameters with threshold. This technique can be used to the prediction of Atrial Fibrillation and Ventricular Fibrillation.

References

1. Md Saiful Islam, Md Mahmudul Hasan, Xiaoyi Wang , Hayley D. Germack, and Md Noor-E-Alam, Healthcare 2018, 6, 54, (2018).
2. Afef Mdhaffar, Ismael Bouassida Rodriguez, Khalil Charfi, Leila Abid, and Bernd Freisleben, IEEE TRANSACTIONS ON NANOBIOSCIENCE, VOL. **16**, NO. 8, (2017)
3. Salvatore Naddeo, Laura Verde, Manolo Forastiere, Giuseppe De Pietro and Giovanna Sannino, SCITEPRESS Science and Technology Publications, (2017).
4. A. Dhillon, S. Majumdar, M. St Hilaire and A. El. Haraki, Conference: IEEE International Conference on Internet of Things, (2017).

5. Sebastian D. Goodfellow, Andrew Goodwin, Robert Greer, Peter C. Laussen, Mjaye Mazwi, Danny Eytan, Computing in Cardiology 2017; VOL **44**
6. Francesco Nocera, Tommaso Di Noia, Marina Mongiello, Eugenio Di Sciascio, SEMAPRO 2016 : The Tenth International Conference on Advances in Semantic Processing, (2016).
7. A. Javari and M. Jalili, Knowledge Information System, volume-**44**, no.3, (2015).
8. Y. Rao, N. Zhang, and H. Zou,Knowledge Information System, volume. **44**, No. 3, PP. 663–688, (2015).
9. R. Pathak and V. Vaidehi, in Proc. 3rd Int. Conf. Eco-Friendly Comput. Commun. Syst. (ICECCS), pp. 61-66,(Dec. 2014).

Development of Disease Self-Management (DSM) Regulatory, Non-communicable Diseases Negotiable Emerging eHealth driven Food Processing Technologies

Amol Dagadkhair[1]*, *Shradha Rodge*[1], *and Vasant Pawar*[1]
[1]MIT School of Food Technology, MIT Art, Design and Technology University, Pune, MH, India- 412 201

Abstract. In the modern food technology digital era, innovation and consumerism driven hypothesis generation plays important role to make the system summarily applicable under information and communication technologies (ICT) domain. eHealth conceptualization and promulgation based on scientific and technological information dissemination, recorded as a need base intervention to update food technology scenario. Nutrition transition leading to goodness and smartness has direct relationship with health of the consumer. Moreover, lifestyle transition generative non-communicable diseases (NCD's) is becoming a vibrant challenge to food technologists. The sedentary status and individual inactivity jointly invited physio-nutritional health problems like NCD's world over. Disease self-management (DSM) by eDiet requires hand in hand association of innovative food technology products and processes. eHealth appears to be one stop solution to prevent, cure and stimulate non-recurrence through regulatory dietary pattern. This may also be justify more efficiently with regulatory mechanism of DSM. Food processing being an art of material transition, leading to consumable security, requires to bring it under health claim base food processing scenario. Hence, concerted efforts of eHealth and novel food processing technologies are rightly awaiting to modify the DSM to ensure health security to consumers. The review base focus on tailoring of eHealth and novel food processing technologies is a techno-economical option to develop a regulatory mechanism for negotiation of NCD's in future generation.

1 Introduction

The complete well-being of not only physical and mental state but also social status and not just a free from diseases or infirmity is nothing but the health (WHO). The fulfillment in the most imperative reachable standard of living is one of the significant benefits of every person without race, religion, and political opinion, budgetary or social state of a person. Unequal development in various nations in the advancement of health and control of infections, particularly communicable diseases, is a typical peril. Governments have a commitment with respect to the quality of their society which can be fulfilled interestingly by the plan of acceptable well-being and social measures [19].

eHealth is a process of better utilization of Information and Communication Technologies (ICT) for the sake of health. It practices in alliance with worldwide, regional and national agencies to promote and strengthen the usage of ICT in health improvement. This unit is situated at the Department of Service Delivery and Safety in the Cluster of Health Systems and Innovation [20]. In the World Health Assembly, 2018, it has been recorded that the digital technologies are playing an important role in improving society health. The present delegates also have agreed on development and implementation of digital technologies to improve status of NCD's. [7].

The dreadful Noncommunicable diseases (NCDs), includes coronary heart diseases, cancer, stroke, diabetes and different lung diseases, which are reason behind death of about 70% of all deaths around the world. In the case, 75 % of whole NCD's deaths, and 82% of the 16 million individuals who died early, or before accomplishing the age 70 years, in specially low and middle income nations. The rising of NCDs has been driven by on a very basic level four noteworthy peril factors viz. constant use of tobacco, physical inactivity, over consumption of alcohol and deplorable weight control plans [21]. Disease Self-management is a practice which includes dynamic, collaborative, and regular process in which people deals to cope up a chronic diseases. Self-management refers to "the capability of the person, to interact with family, society, and healthcare professionals, to control symptoms, treatments, lifestyle changes, and psychosocial, cultural, and spiritual consequences of health conditions" [3]. Thus present review paper enlightens the various eHeath technologies, DSM and advanced food processing technologies can bring a possible solution on NCD's.

2 Emerging eHealth Technologies

Advanced innovations are always developing and finding new applications in health care services, even while the industries are battling with selection and 'digital transformation'. Every year new applications rise, yet the fundamental advancements driving them continue as before. In 2019, they surveyed many organizations around the globe with single question: "Please indicate the key technology which you believe will have the most profound impact on the healthcare industry during 2019". [14].

2.1 mHealth

mHealth is a one of eHealth fragment. Till today, nobody has defined mHealth. For the survey purpose, the Global Observatory for eHealth (GOe) has characterized mHealth or mobile health as public and medical wellbeing practice upheld by mobiles, personal digital assistants (PDAs), patient monitoring devices and different remote gadgets. It also includes utilization of SMS and voice services with 3G, 4G, Global Positioning System (GPS), Bluetooth technology and General Packet Radio Service (GPRS) [9].

The text messages on mobiles is one of common approach for DSM, either with or without any other suitable technology. Current researches on T2DM and T1DM reported that the text messages and demonstration are having an impressive elevation in diet and exercise, an increase in weekly exercise and introduction of fruits and vegetables in regular diet, helps in minimizing fat intake and regulating blood sugar level. Also, there was a continuous improvement in the desired positive behaviors. Additionally, with the quantity of days members practiced every week expanding from two days to very nearly four days, and the sugar level of members who consumed products of fruits and vegetables daily expanding ranged from 61 % to 79 %. The scores for diabetes self-care increased for healthy diet taking and practicing 30 minute exercise regularly. [2,10,18].

The European Heart Network (EHN) [4] has noted that an effective implementation of e/mHealth solutions could increase the prevention and remedies of CVD's. Mobile applications approached by people of different age groups, together with educated or health conscious individuals. Moreover, smartphone applications may help out the CVD's patients to understand the do's and don't after a cardiac attack, and they could find possible solution to it immediately.

2.2 Websites

Websites are the information delivering online systems through various media including audiovisual and written content. Moreover advancement in technology allows the web-based programs access by mobile devices. It has been observed that number of internet-based technologies are being used for the transfer of DSM instruction. Most of the studies are concentrated on change in glycemic level than in dietary pattern and physical activity. Furthermore it is recorded that educating people through internet based technologies found more efficient than typical care, with the capacity to attain additional patients with more convenience and efficacy [8].

2.3 Wearable Technologies

Wearable innovations involves pedometers, accelerometers, and shrewd watches are creating as novel ways to deal with assistance DSM by self-regulating physical movement and diet and outfitting the customer with brief updates, reminder and input. [11] An analysis of randomized selected eleven controlled trials associated with pedometer use, physical activity and glycemic control in person with T2DM. A consistent pedometer use resulted in an increased physical activity (increased by 1,800 steps/ day). Fascinatingly, it shows that it affected physical activity than without objective setting. Notwithstanding this ascent in physical activity, there was little or no change in HbA1c content, despite the fact that subgroup with objective setting remembered for their intervention.

2.4 Virtual Reality

The virtual reality is a computerized formation of the total virtual condition which recreates reality (Audio-visual) without sharing the entirety of its physical characteristics. In certain occurrences, the client can explore through and cooperate with a virtual condition. T2DM disease self-management training could be conveyed in an augmented experience condition or in a traditional in-person position. The study was conducted in association with multidisciplinary group (dietitian, dialectologist, and medical attendant) to convey care to the member in an underlying individual counsel, trailed by eight groups per week for several weeks. The eHealth mediation utilized food displays with exercise amenities in the virtual sphere. A face-to-face sessions were organized for the control group in a gathering passage. Furthermore it was observed that physical activity was lengthened by 18% in virtual, though it diminished by 25% with respect the face-to-face communication; in any case, there were not any contrasts between dietary plans [15].

2.5 Virtual Restaurants and Food Nutrition

The virtual restaurants like Ubereats, Zomato, Food Panda, Swiggy etc. should depict the information about foods they are processing on their webpage/application while purchasing the food item. The said initiative will make easy approach for consumers to purchase the required foods. India is country with young population (more than 50 % by 24 years and above) and increasing urbanization with intense speed [1]. The same things are responsible for increase in demand for convenience foods such as virtual restaurant. The virtual restaurants have to come up with an initiative to incorporate nutritional information about products on their websites/ applications to segregate the products by consumer as per their requirement. Hence the food items obtained through these virtual restaurants are examples of eDiet which will serve the consumers with their prime motto.

2.6 Short Term Online e-courses

The short term online courses are readily approachable on e-platforms like Coursera, Swayam, FSSAI video library etc. regarding food safety and quality (health claims) awareness. The courses are helping the consumers to know more about their disease, treatment, and no-recurrence. FSSAI's video Library is available to everyone with objective to create a sustainable culture and habit of eating safe and healthy foods by endorsing individual awareness as well as collective action. This won't just empower customer to settle on informed choices but also toughen the institutional system to promote, spread and support safe food and healthy diets [5].

2.7 Digital posters/panels

Displaying the digital posters/ panel regarding food quality and safety at eating places like Restaurants, Hotels, Mess, Public gatherings (Railway platform, Bus Depot, Airports etc.) to increases awareness about different processed foods. Example, one of novel initiative carried by FSSAI namely Eat Right India has to be spread through such digital posters/panels.

3 Novel food processing

3.1 Impact of Conventional Food Processing Technologies on Food, Nutrition and Health

Food processing is an art performed to covert raw foods into consumable foods from thousands of years. Foods processing not only makes foods ready to consume but also changes its physical, chemical and microbiological characteristics is such a way that it drastically affects its nutritional content and makes foods nutrient deficient. Most of water soluble nutrients (vitamins and minerals) are getting destroyed in thermal processing [12].

3.2 Novel Drying Techniques

The novel food drying techniques includes microwave, vacuum, freeze, ultrasound assisted drying, heat pump drying, high intensity electric field drying, and refractance window drying are making food processors more capable to minimize the nutrition losses while processing.

3.3 Nano-technology

The quick improvement in nanotechnology has been encouraging the changes of traditional food areas, especially smart packaging, active packaging, nanosensors,. Various new nanomaterials have been created to improve food food quality and safety. Moreover, it has recently studied that nanotechnology has made its approach towards pathogen detection, food packaging, disease treatments and bioactive compounds transfer to target sites [7].

3.4 Cold Plasma Technology

Initially cold plasma was utilized for cleansing the delicate things and now it is stretched out to food processing sector as an emerging technology. For a considerable length of time it has been seen as helpful for microbial inactivation while keeping up natural qualities of new products. In this method cold gases are used purify the surfaces of packaging or food products items. The procedure can possibly inactivate microorganism without raising temperature above 40^0C [13].

3.5 Artificial Intelligence and Robotics in Food Processing

AI is an advanced computational technique has wide applications in different manufacturing industries including food processing industries. In food processing it has numerous application from sorting and grading, cleaning, manufacturing, quality analysis, GC, HPLC, to food safety and quality evaluation of foods and packaging materials too [17].

Innovative technology in various areas have extend their application skyline of robotics to an incredible extent. Robots basically have the ability to transmute the food processing, handling, etc. practices. Therefore, robotics is emerging technology in deployment of robots in food sector. Subsequently, the features belonging to robots are kinematics, hygiene, dynamics, human-robot interaction, economic efficiency, operation and maintenance and safety and protection are of great significance. Now a day's food processing segment having potential where numerous research opportunities are open from various technology perspective. It is new potential territory where plentiful research openings in 'robo-food' by coordinating headways from different innovation spaces [6].

4 Conclusion

The intention behind this review article was to exemplify the various benefits and opportunities in the e-health technology domain and novel food processing technologies. An amalgamation of these two emerging technologies can satisfy the need of individual who is eclipsed by the different NCD's.

5 References

1. Anonymous, Report on Indian Food Services Industry: Engine for Economic Growth & Employment: A Roadmap for Unlocking Growth Opportunities, FICCI, New Delhi, India (2019).
2. E. Burner, M. Menchine, E. Taylor and A. Arora, J Diabet Sci Technol. **7(1),** 111-118. (2013)
3. S. G. Dena, J. Sarah, J. N. Scholarsh. **44(2),** 136-144 (2012).

4. European Heart Network. Cardiovascular patients and eHealth – a European Heart Network paper. Available at: http://www.ehnheart.org/component/downloads/downloads/1438 (2013).
5. FSSAI Video Library, https://fssai.gov.in/fssaivideolibrary/about (2019).
6. I. Jamshed, H. Zeashan, K. Azfar, Food Sci Technol. **37(2),** 159-165 (2016).
7. L. Rashidi, K. Darani. Crit Rev Food Sci Nutr. **51**, 723–730 (2011).
8. E. R. Megan, J. A. Elroy, L. W. Rebecca, W. Katie, K. Michelle, C. Robin, E. C. Clare. Diabet Metab Synd Ob, **9**, 381-390 (2016).
9. mHealth, WHO New horizons for health through mobile technologies, Global Observatory for eHealth series, World Health Organization, **3**, 6-7 (2019).
10. S. Nundy, A. Mishra, P. Hogan, S. Lee, M. Solomon, M. Peek, Diabet Edu. **40(6),** 806-819 (2014)
11. S. Qiu, X. Cai, X. Chen, B. Yang, Z. Sun, BMC Med. **12**, 36 (2014).
12. R. Devi, Sch J Agric Vet Sci. **2(4A),** 304-311 (2015).
13. R. Thirumdas, C. Sarangapani, U. Annapure, Food Biophy, **10**, 1-11 (2014).
14. R. Das. https:// www.forbes.com/sites/ reenitadas/2019/02/04/ the-top-five- digital- health-technologies- in- 2019/#24af9aab6c0f (2019).
15. M. Rosal, R. Heyden, R. Mejilla, M. Rizzo, C. Veerappa, M. Wiecha. JMIR Res Protoc. **1(2),** e24. (2012).
16. Seventy one World Health Assembly, https://www.who.int/about/governance/world-health-assembly/seventy-first (2018).
17. A. Sofu, D. Gıda, S. Cilt. Applications of Artificial Intelligence in Food Science and Technology Area (Turkish with English Abstract). agris.fao.org. (2007).
18. K. Waki, H. Fujita, Y. Uchimura, K. Omae, E. Aramaki, S. Kato, H. Lee, T. Kobayashi, K. Ohe, J Diabet Sci Technol. **8(2)**, 209-215 (2014)
19. WHO https://www.who.int/ about/ who-we-are/constitution (2019a).
20. WHO https://www.who.int/ehealth/en/ (2019b).
21. WHO https://www.who.int/ncds/en/ (2019c).

Municipal Solid Waste Management: A comparative study between Sydney (Australia) and Pune (India)

Daniella Sohkhlet[1*] *and Shekhar Nagargoje*[2]
[1] RICS School of Built Environment, Amity University, Mumbai, India
[2] RICS School of Built Environment, Amity University, Mumbai, India

Abstract. As rapid urbanization continues to take place, with a projected 68% of the world's population expected to live in cities by 2050 according to the UN, waste management in proximity to human settlements poses problematic. Although waste generation and failure of its management is seen as a huge problem in developing cities, wealthier cities generate more waste than less affluent cities, and thus waste management is a crucial issue in developed as well as developing cities for obvious ecological reasons and for sustainability. However, while collection of waste by high GDP cities is more frequent than collection in low GDP cities, it does not imply that the issue of waste management has been resolved. A significant portion of the world still dumps its waste in landfills. Furthermore, it is projected that by 2025, lower middle-income groups will produce the most amounts of waste, while high-income groups are expected to generate the second largest amount of waste. Accordingly, it is important to understand methods of collection and disposal of solid waste carried out by cites varying in GDP and HDI to combat the issue of waste management through landfills for the purpose of creating a sustainable future. This paper is a study of existing waste management methods implemented by the governing bodies of selected cities and progress towards future policies that these governing bodies have laid out. Comparisons of existing problems faced by each city, collection methods and coverage, recycle coverage and rates, unique waste management approaches, and plans to tackle waste are carried out. The analysis can thus serve as a reference for emerging cities lacking resources that can adapt strategies described, as well as for established cities that can easily adopt certain measures to safeguard their respective ecologies and take steps towards sustainability.

1 Introduction

Solid Waste Management within the framework of the city is a core element of its infrastructure. As the urban centre takes its place on the international stage, more sophisticated management methods in solid waste management is required to relieve the stresses of increasing relocation of the world's population to cities. Additionally, it is required for the smooth operations of social economic activities. However, this is not the only reason we must devote our efforts to understanding solid waste management practices. The environment bears the cost of poor practices. The world generated 1.6 billion tonnes of carbon dioxide equivalent greenhouse gas emissions from solid waste management in 2016, making up 5% of global emissions [1]. If renewable energy is to increase its foothold, waste management and resource recovery can not only generate large amounts of income and employment but can also spur the shift to clean energy. However, current solid waste management practices are poor and observed to be detrimental to the micro-level as well as macro-level functioning of the world's communities and the planet's ecosystem. Solid waste that has not been collected ends up in drains, clogging drainage systems and creating unsanitary conditions and breeding grounds for diseases. Burning of solid waste also adversely impacts the air as well as human health. Moreover, at least 2 billion people do not have access to regular waste collection, and only 5% of waste in squatter areas is collected [2]. When waste is collected, it is often the case that waste is openly dumped or dumped in landfills that are unmonitored and uncontrolled. Globally, 37% of waste is discarded in landfills, 8% of which are sanitary landfills with gas collection systems [1]. Another 33% of waste is disposed through open dumping and 11% is incinerated. Only 19% of the world's waste is recovered through recycling and composting [1].

Unsurprisingly, most low-income countries rely on open dumping while high-income and upper-middle-income countries have access to controlled landfills and other waste management infrastructure. However, landfills continue to be the most used form of waste disposal. 54% of landfill disposal is done by upper-middle-income countries [1].

For this reason, Sydney Australia and India on the basis of their differences are compared. Australia has a GDP per capita of 43,560 USD and HDI of 0.939, while India has a GDP per capita of 6,353 USD and HDI value of 0.640 [3].

2 Factors of Comparison

2.1 Gross Domestic Product (GDP)

GDP values taken into consideration in this paper are taken from UNDP 2018 statistical updates and are expressed as the sum of all incomes of a country's residents and businesses including taxes in a particular period. To understand GDP per capita, the GDP is divided by the total population in the same period. The values are expressed in international dollars using purchasing power parity (PPP) rates. [3]

2.2 Human Development Index (HDI)

The index accounts for three dimensions of development. The first dimension is measured by life expectancy at birth, while access to knowledge is measured by mean years of schooling and expected years of schooling, whereas a decent standard of living is measured by gross national income per capita. [3]

2.3 Municipal Solid Waste (MSW)

MSW is waste that consists of non-hazardous trash generated from "households, commercial and business establishments, industrial and agricultural premises, institutions such as schools and hospitals, public spaces such as parks and streets and construction sites" [2]. Municipal solid waste is further divided into *regularly collected* and *uncollected municipal solid waste,* where the former is collected by municipal authorities or licensed private contractors on a routine schedule and the latter remains scattered through the city due to a lack of collection services, particularly seen in informal settlements areas. This paper will be examining only *regularly collected* waste.

2.4 Waste Classification and Composition

Study of waste generation sources and composition and classification of waste by respective local municipal authorities. With Sydney, it is the City of Sydney and Pune is governed by Pune Municipal Corporation.

2.5 Waste Trends

Past and projected trends in waste generation as well as their sources.

2.6 Waste Collection

Defining comparison between the two cities and how their local governments differ in how they charge fees for waste management, how they collect and segregate waste, how often they collect the waste and the extent of the coverage.

2.7 Waste Disposal

Examination of current waste infrastructure in how waste is disposed/processed.

2.8 Successes and Strategies

Achievements and policies used in solid waste management as well as strategies for the future

2.9 Challenges and Opportunities

Existing challenges and whether these challenges can be turned into opportunities.

3 Sydney

Sydney covers 26.15 sqm of area and has a population of 210,000 with a transient population of 1.2 million per day, which includes workers, residents, visitors, and students [4]. The city is projected to grow by 70,000 residents and 120,000 workers by 2031. Sydney represents 22% and 7% of the NSW and national GDP respectively. [5]

The City of Sydney Council is made up of 10 village precincts and governs 33 suburbs and 23 localities. The council is responsible for the collection of waste and recycling in compliance with the Local Government Act 1993. The Environment Committee of the City of Sydney Council in particular oversees environmental issues within the government area. The City works in conjunction with and complies with regulations of the NSW state government. [6].

3.1 Waste Classification and Composition

While the City of Sydney (C.o.S) has categorized waste into three categories, explicitly, Residential and Public Space waste, Commercial and Industrial waste, and Construction and Demolition waste, the City only collects Domestic waste which it defines as "all waste generated by the ordinary use of residential premises and collected by the City or its agents". [7] Businesses are responsible for collecting their own waste, typically carried out by engaging a commercial waste contractor who operate under the Waste Contractors and Recyclers Association of NSW (WCRA) Code of Practice. The WCRA Code of Practice is subject to the Waste Policy in case of a conflict. [7]

Residential and Public Spaces waste contribute up to 4% of Sydney's total landfill waste. Waste generated by businesses consists of largely organic or biodegradable waste and is a major contributor to landfill waste. C&D waste generates more than 1.2 million tonnes per year out of which 400,000 tonnes are generated directly or indirectly through maintenance or contracts. [4]

While the New South Wales Environment Protection Authority (NSW EPA) guidelines of waste classification categorize solid waste into putrescible and non-putrescible, the C.o.S limits its scope to domestic waste from single dwelling or houses and multi-unit dwellings or apartments in the area of its jurisdiction. In general, NSW EPA describes putrescible solid waste as

https://doi.org/10.1051/e3sconf/202017004001

household waste, waste from litter bins, manure and night soil, disposable nappies, incontinence pads or sanitary napkins, food waste, animal waste (including dead animals and mixtures of animal parts), grit or screenings from sewage treatment systems without free liquids, and mixtures of the above. Non-putrescible wastes range from glass, plastic, rubber, plasterboard, ceramics, bricks, concrete, metal, paper or cardboard, to building and demolition waste and other inorganic material. According to Sydney's 2017 Waste Policy, domestic waste collected is largely made up of:

3.1.1 Garbage

Garbage includes food waste, nappies, meat trays, foam packaging, plastic wrappers and crockery.

3.1.2 Recycling

The City collects recyclables which includes paper, cardboard, aerosol cans, metal cans & tins, glass containers, plastic containers.

3.1.3 Garden Organics

The City collects grass clippings, flowers and weeds, leaves, tree and shrub prunings, sticks and branches. Building materials, rocks and soil, tree stumps and big branches are not collected.

3.1.4 Bulky Household Waste

Made up of furniture, mattresses, whitegoods or excess green waste.

3.1.5 Building and Renovation Waste

Includes asbestos, rubble, soil, timber, fencing, wooden pallets.

3.1.6 Hazardous Household Items

The City provides other methods for storage of hazardous household items (flammable liquids, gas bottles, oil, poisons or other chemicals, paints & solvents, batteries, energy-saving/fluorescent light globes, e-waste, clinical waste).

Table 1. Domestic Waste Content in Sydney. Source: *City of Sydney Interim Waste Strategy, 2011*

Domestic Waste Content	
Renovation Waste	4%
Hazardous	2%
Textile	3%
Glass	9%
Metal	2%
Plastics	12%
Other	7%
Food	37%
Greenwaste	2%
Paper and Cardboard	21%

Table 2. Commercial + Industrial Waste Content. Source: *City of Sydney Interim Waste Strategy 2011*

Commercial + Industrial Waste Content	
C&D material	8%
E-Waste	1%
Hazardous	14%
Textile	4%
Glass	2%
Metal	2%
Plastics	12%
Wood	13%
Other	13%
Food	14%
Greenwaste	3%
Paper and Cardboard	14%

3.2 Waste Trends

More than 5,500 tonnes of waste are produced by the metropolitan city every day amounting to 65,000 tonnes of domestic waste each year from more than 115,000 households contributing to 8% of the city's total greenhouse gas emissions. However, in 2016 -17, carbon emissions decreased by 7% due to waste diversion and advanced waste treatment practices.

Table 3. Waste Generated vs. 2030 forecast

Type of waste	Tonnes per year	KG per person per day	Of Total Waste	2030 forecast (tonnes)
Households	65,000	1	19%	102,000
Public Places	11,000			
Commercial & Industrial	700,000	2	52%	800,000
Constructio n & Demolition	1,300,000	4	29%	1,500,00 0

Table 4. Waste Diverted from Landfills

Source	Tonnes generated	Diverted from landfills
Buildings	400	35%
Construction works	4000	90%
Public spaces	11,000	18%

While 69% of domestic waste is recycled, 2000 tonnes and above each day goes to landfill without further means to re-use, recycle or recover for energy. [4]

3.3 Waste Collection

3.2.1 Charges

Residents are charged annually for provision of waste management services which account for administration, collection, processing, treatment, community education, and other activities associated with domestic waste services. The domestic waste charge is calculated on the size of garbage bin, amount of waste generated, and frequency of collection. A standard charge applies for single-dwellings based on the size of waste bin and collection frequency. For multi-unit dwellings, the charge is based on the number of units sharing the service, frequency of collections, size of the bins, and whether the waste is compacted. Schools within the local government area are provided recycling collection services free of cost up to a maximum of four 240 litre recycling bins per week. [7]

3.2.3 Segregation and Storage

Containers are provided ranging from 55 L carry bins to mobile bins and bulk capacity skip bins, which the City defines as, "a container designed and used for the storage, transport and disposal of waste" [7]. While garbage and recyclables are disposed in colour coded bins, garden organics must be bundled, boxed or bagged and then placed into green lid bins. If garden organics are too big to fit in the provided bins, bookings can be made for collection. Bulky household waste is collected by calling the City's Customer Service Centre or can independently be taken to a commercial waste facility to dispose of for a fee. The City instructs disposal of whitegoods to be placed separately to other items where they collect no more than one cubic metre of waste. Building and renovation waste such as asbestos, rubble, soil, timber, fencing, or wooden pallets must be disposed of at the appropriate waste facility or by hiring a waste contractor for collection who often provide residents with a skip bin which is subsequently collected. Hazardous household items and e-waste can be independently disposed of by calling the City's Customer Service Centre for details of disposal locations.

Table 5. Storage Methods in Sydney

Waste Type	Disposal Method	Standard Size
Garbage	Red lid bins	120 L
Recycling	Yellow lid bins	120 L
Garden Organics	Green lid bins	240 L

3.2.3 Frequency and Mode of Collection

Garbage and recyclable waste is collected once a week from single dwellings. Multi-unit dwellings or apartments without communal bins receive same collection as houses and have a variety of frequencies. While the City provides collection only for domestic waste, it has allotted time zones for commercial collection for waste contractors as well. 45 vehicles which include rear-lift compaction trucks for picking up domestic waste and recycling are maintained or contracted by the C.o.S. [7]

Table 6. Frequency of Collection in Sydney

Waste Type	Frequency
Garbage	Weekly
Recycling	Weekly
Garden Organics	Fortnightly / Booking
White goods & Bulky Household items	Weekly through booking
E-Waste & Hazardous waste	Drop-off events

Table 7. Domestic Waste Collection Time Zones in Sydney

Zones	Time	Days
Main and Arterial Roads Zone	5:30 AM – 10 PM	Every day
Residential Roads Zone	6:00 AM – 10 PM	Mon - Sat
	8:00 AM – 10 PM	Sundays
CBD Light Rail Collection Zone	1:00 AM – 5:30 AM	Every Day

3.2.4 Collection Coverage

The C.o.S covers its residential suburbs, commercial and CBD districts based on frequency, where commercial and domestic waste are collected individually and separately. Each category of waste has allotted routes classified by zones (main and arterial roads, residential roads, CBD light rail, and limited collection time zones for all other streets, etc.) with set routines and time slots that are made known to the public.

3.3 Waste Disposal

Australia's waste is managed by a combination of public, private and public-private players. The country's waste and resource recovery sector contribute 0.43% for the national GDP. Collection operations, landfilling, and operating waste infrastructure are also carried out by private players, who mostly operate in larger urban centres. (Fig 1). One of the large operators include Veolia, operating 250 km. away from Sydney. While smaller operators deal in focused activities such as composting, skip bin operation or operate regionally. [8]

While transfer stations, container systems, material recovery facilities and composting facilities are found to be operated by a combination of local governments and private entities who are contracted by local governments or private businesses, alternative waste treatment facilities (AWT), landfills and processing facilities for C&D and

Fig 1. Location of landfills closest to Sydney Area (a 50 km radius) Source: Google Earth Pro, 2020

C&I are exclusively run by private businesses, with the exception of landfills in rural areas being run by local governments. [8]

Landfill sites in the area are located away from the point of waste generation due to competing land priorities as seen in Fig. 4, which shows the location of some of the primary landfill sites used by Sydney and their proximity to the urban centre. These are Jack's Gully, Genesis Facility at Eastern Creek, Lucas Heights, and Belrose. In addition to these, the Woodlawn landfill at Tarago, operated by Veolia, 250 km. south-west of Sydney receives the largest amount of waste in Australia, more than 400,000 tonnes per year [9]. Waste from Sydney to Woodlawn is transported by rail.

In 2010, approved capacity of putrescible landfills was 40 MT and 4.7 MT in inert landfills. However, by 2022, Sydney's landfills would have to be replaced [9].

Current disposal pathways place recycling as the second stage of waste recovery and disposal. Recycling of garden organics, e-waste, household and whitegoods, and by-products of alternative waste treatment is carried out after collection. The third stage is treatment of only two types of waste - mixed waste and household hazard waste. The remaining get sent to landfill sites [10].

Out of the 65,000 tonnes and 11,000 tonnes of waste generated in 2016 by the city of Sydney area's residential and public spaces, 31% and 82% respectively goes to landfills and 69% is diverted [4]. Comparatively, in 2012, the City diverted 66% of 62,598 tonnes of collected municipal waste from landfills [10].

Although 69% of residential waste is presently diverted from landfills by reprocessing recyclables and composting food and garden waste, a large percentage of landfill waste is made up of other waste, particularly C&D waste. Recyclables also tend to enter landfills due to waste segregation from the source not being effectively managed. From a broader scale, 13.8 MT of MSW was generated nationally in 2016 – 17; 21.7 MT (40% of core waste generated) of Australia's waste was deposited in landfill in the year 2016 -17 [8], and almost all of the 140,000 tonnes of e-waste generated annually in Australia is sent to landfill.

3.4 Successes and Strategies

Although Sydney is one of the highest waste generating councils in Australia, it accounts for only 20% of domestic waste being sent to landfill [11]. Waste targets have been maintained and achieved, with the 66%

domestic waste diversion from landfill target rate being crossed in 2012. Recycling rates for public waste increased from 26% to 46%, while 100% of C&D waste was diverted for the year 2018-19, Several community and city-wide recycling, composting, upgradations and awareness initiatives have been increasing with guidance and support from C.o.S that foster innovation and improvement amongst local businesses, communities and residents, with a focus on the development of a circular economy approach. This includes an increase in recycling services, a 24/7 cleaning and waste response team, and consistent provision of advice to local businesses. These trends are only expected to grow as laid out in the council's as well as the state's strategic policies and master plans. [4, 5, 12].

NSW imposed a 'waste levy' in 2014 that charged EPA-licensed landfill operators in NSW and the Sydney metropolitan area for each tonne of waste being sent to the

landfill facility. The levy is an attempt to discourage landfill disposal and aid recycling and resource recovery. Together with the NSW government, initiatives such as Waste Less Recycle More (WLRM) funded by the waste levy has assisted in the development and upgradation of waste infrastructure and community recycling centres. [5, 13, 14].

The C.o.S as an independent entity has identified environmental sustainability targets in terms of waste management with a zero-waste target through various policies, benchmarks and guidelines as listed below.

Table 8. Policies and Guidelines in SWM by City of Sydney

Title	Time-frame	Type	Details
Environmental Action	2016 – 2021	Strategy and Action Plan	Set benchmarks higher than compliance requirements in six areas: low carbon, water sensitivity, climate resilient, zero waste, active and connected, green and cool.
Leave Nothing to Waste Strategy and Action Plan	2017 – 2030	Plan	Waste avoidance, reuse and better recycling
Sustainable Sydney	2030	Community Strategic Plan	As a commitment to the Sustainable Development Goals (12 SDGs).
Advanced Waste Treatment Plan	2013 - 2030	Master Plan	Integrating waste and energy solutions to divert from landfills and deal with changing nature of waste
Sustainability 2050	2050	Event	Engaged decision-makers, advocates, community leaders
Green Environmental Sustainability	Bi-annual	Progress Report	Benchmarks progress against sustainability targets

The City has identified Alternative Waste Treatment (AWT) as a solution to minimize waste being sent to landfill. C.o.S has reportedly ensured that domestic waste is not directly being sent to landfills due to initial processing at AWT facilities. In this effort, the City has upgraded 1,670 residential apartment building bin rooms to improve recycling and waste management. [10]

The model of contracting of private businesses and privatization of waste infrastructure has allowed for the creation of jobs and a requirement of higher-level expertise, and GDP growth. The waste and resource recovery sector provides a stable source of income within the commercial sector.

3.5 Challenges and Opportunities

Although the C.o.S has several targets and multiple action plans and strategies to move towards a zero-waste city, the divergence of local councils and state government has led to gaps in data and reporting. Limited information regarding waste processing plants and the amount of waste being processed is found because these operations are run by private entities, whether small or large. This creates obstacles in transparency that would be required to effectively manage waste. However, the national government has attempted to harmonize regulation by developing a National Waste Policy which has included efforts in reporting on the state of waste management of the country.

Due to the fact that operation costs of landfills are comparatively lower than waste recycling systems and due to their usually close proximity to the city, landfills have tended to be the go-to waste management method in NSW. The privatization of landfill ownership and the tradition of landfill use has led private owners to introduce larger and more revenue producing landfills [15]. 3% of Sydney's total greenhouse gas emission is contributed by landfill methane emission while other greenhouse gas emissions from landfill equate to 25,510 tons of carbon dioxide. While the 'zero waste to landfill' target by 2030 is progressing, the average capacity of 300,000 tonnes of Sydney's landfills provide cause for a laid-back approach where half of the waste still arrives in landfills. As the landfills increase in capacity, supplementary land would have to be acquired further away from the city, leading to larger carbon footprints. Additionally, these landfills have created barriers in the growth of the city.

Sydney is home to a steadily increasing population as well as worker and visitor groups. With 75% of residential units in the city being apartments, waste infrastructure will increasingly be burdened, and more effective management methods will be required.

With progressive complexity in the composition of products over the last few decades, the nature of waste is drastically changing. This includes superfluous packaging that is generated within the city. As consumption of products increases, the landfill method would prove to be a method that can be compared to dumping toxic waste into the sea if seen as hindsight by future generations.

A vital opportunity can be found in waste to energy systems that has not been fully implemented as yet, but has been identified by the Advanced Waste Treatment Plan 2013 -2030, which can reduce greenhouse gas emissions by 70% by 2030 in conjunction with the Renewable Energy and Trigeneration Energy Master Plan. [10]

4. Pune

With a population over 5 million in 2011 in the Pune Urban Agglomeration and over 4 million in Pune City, the Indian IT hub has been growing at a rapid rate. Contributing 26.5% to India's GDP, the city owes its increased standard of living not only to its IT and ITeS sectors, but also to its biotechnology, manufacturing and education sectors. The form of growth it has taken since 2001 has shown to be headed towards decreasing density at an annual rate of -4.8%. However, while urban sprawl is occurring, the city's overall population is projected to grow to a population of 6 million by 2026. 25% of properties are situated in slums, making up 40% of Pune's population. Pune is under the jurisdiction of P.M.C who has defined the city to occupy y 244 sq. km. P.M.C manages the city through 4 administrative Zones comprising of three or four Wards which is further subdivided into three to four Prabhags. At the lowest level, Prabhags are made up of Kothis all totalling 152 kothis, 76 prabhags, 14 wards and 4 zones. [16-18]

The management of solid waste in terms of collection, transportation and disposal, including reduction, reuse, recycling and resource recovery is the responsibility of the Pune Municipal Corporation (P.M.C) in compliance with the Bombay Provincial Municipal Act of 1949 and the Government of India Municipal Solid Waste Rules of 2000, among others. Specifically, the Department of Health headed by the Medical Officer of Health (MOH) who is assisted by the Deputy Health Officer of SWM oversees solid waste management for the city of Pune. [16, 18]

4.1 Waste Classification and Composition

The P.M.C has extensively detailed the classification of waste in the Public Health & Sanitation Bye-Laws 2014, listing and describing 25 types of waste. Municipal Solid Waste is defined as waste that consists of "commercial, residential, and other waste generated in (a municipal or notified area) either in solid or semi-solid form excluding industrial hazardous waste, and it also includes treated bio-medical waste" [19].

Classification of municipal solid waste as defined above depends on the source of waste generation, which can be categorized as Residential (individual households, government and private colonies/societies, bungalows, multi storied buildings, apartments, row houses, tenements, townships, slums and chawls), Commercial (ranging from hotels and eateries to flower markets, marriage halls, and construction sites) as well as Educational, Public and Private Gardens, Heritage Buildings, and Religious Places. [19] Composition of typical waste generated is shown in Fig. 2 and 3 below.

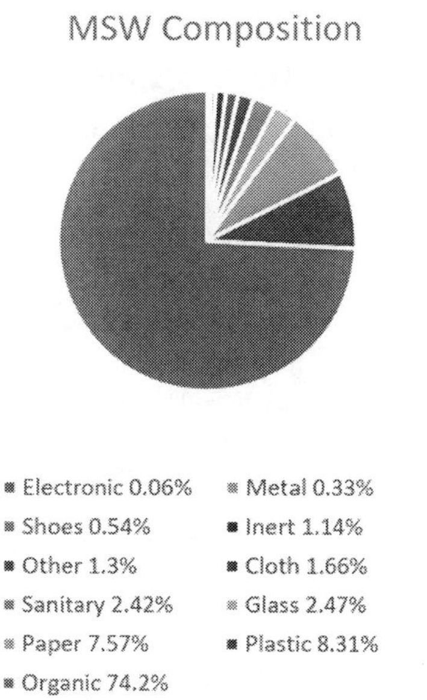

Fig. 2. Municipal Solid Waste Content in Pune. Source: *Pune Municipal Corporation, SWaCH Quarterly Report to P.M.C (Pune, 2018).*

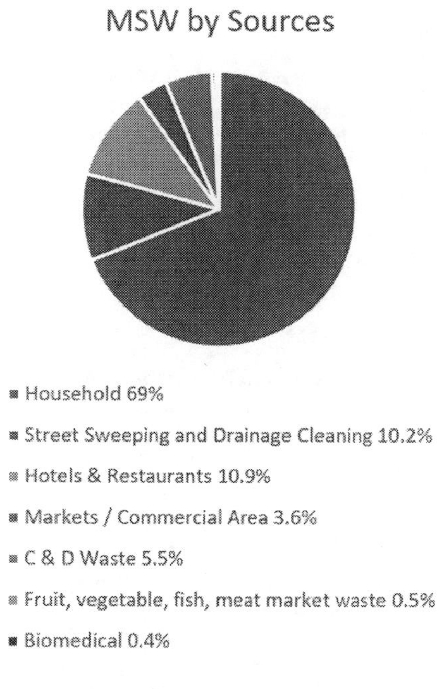

Fig. 3. MSW by Sources in Pune. Source: *Integrated Solid Waste Management, P.M.C, 2015*

Accordingly, the 2014 Bye-Laws describe municipal solid waste being made up of the following:

4.1.1 Bio-degradable Waste / Wet Waste

Consists of bio-degradable wastes such as tea leaves, eggshells, fruit and vegetable peels, meat and bones, garden and leaf litter, animal litter, soiled paper, house dust, etc.

4.1.2 Recyclable Waste / Dry Waste

Consists of paper, metal, plastic & rubber, glass and any other material which is recyclable in nature.

4.1.3 Construction and Demolition Waste (C&D waste)

Consists of waste generated by construction, refurbishment, repairs and demolition of houses, commercial buildings and other structures.

4.1.4 E-Waste

Consists of information technology and telecommunication equipment, consumer electronics, and related waste.

4.1.5 Commercial/Household Hazardous, Bio-medical & Sanitary Waste

Consists of any chemically hazardous components found in household cleaners, automotive products, personal care, and related items.

4.2 Waste Trends

Pune generated an average of 1650 MT of solid waste per day in 2016-17, with each person contributing approximately 364 grams each day across economic groups, translating to 0.07 tCO_2e or 4% of carbon emissions per capita. [20]

4.3 Waste Collection

4.3.1 Charges

P.M.C has outlined a uniform user fee for door to door collections that was enforced in 2017. Costs incurred for management and equipment procurement and maintenance are maintained by P.M.C. [16, 21]

4.3.2 Frequency and Mode of Collection

Waste is collected by various agents through primary, and secondary means. Municipal Solid Waste is largely collected through a door to door service making up 43% of the total collection, through the Solid Waste Collection and Handling (SWaCH), a social and civic cooperative involving informal workers. Secondly, accounting for 12% of primary collection, tipper trucks commonly known as "ghantagadi" translated as "hourly trucks" or "bell ringing vehicles" collect recyclable and non-biodegradable waste from gates of buildings according to programmed routes through prabhags. Inner city areas that are not accessible to trucks are provided with pushcart services. The waste is then transferred to a station within each ward. Waste from hotels are collected by a specific collection of trucks. 40% of primary collections are from community bins made by Dumper Placer trucks which also transfer the waste to the transfer station. Waste dumped along streets and roads are also collected by P.M.C to be taken to the stations, making up 5% of primary collections. [16, 18, 20, 22]

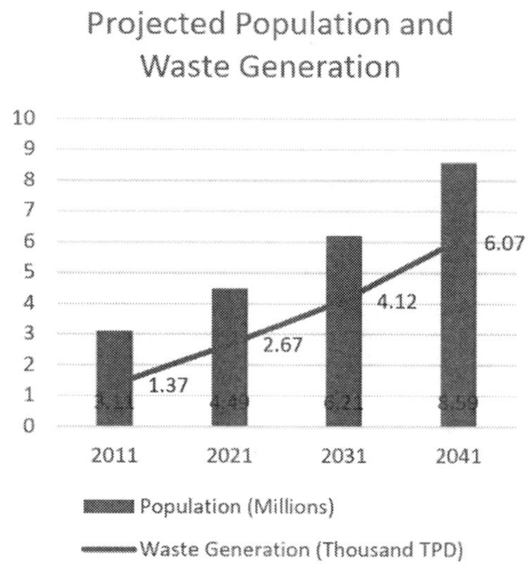

Fig. 4. Segregation of MSW at Source in Pune. Source: *Integrated Solid Waste Management, P.M.C, 2015*

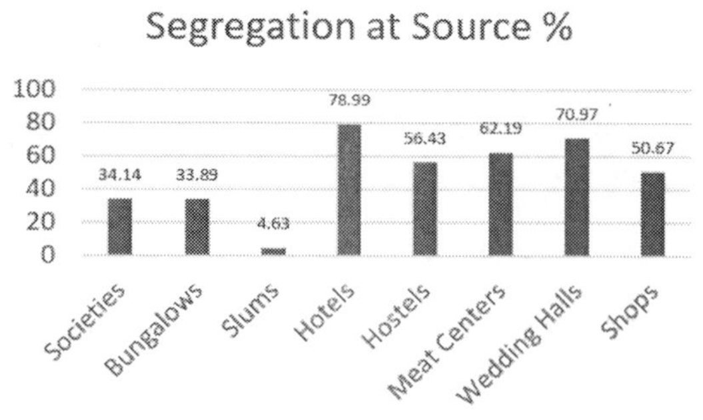

Fig. 5. Segregation of MSW at Source in Pune. Source: *Integrated Solid Waste Management, P.M.C, 2015*

Bulk Refuse Carriers (BRC) compactors then transport waste from transfer stations to dry or wet waste processing plants, respectively. [18]

C&D waste is currently being directly disposed in low lying areas along rivers. [23]

Table 9. Collection Power in Pune

Collection Mode	Strength in nos.	% of waste collected
SWaCH	2688	43
Community Bins	936	40
Tipper Trucks	160	12
Hotel Trucks	23	5
Tractors	10	
Dumper Placers	85	
BRC	59	
Other	88	

Table 10. Frequency of Collections in Pune

Collection Mode	Organic	Recyclable / Non - Biodegradable	Waste in TPD
Ghanta Trucks	Daily	Weekly	350
SWaCH	Daily	-	865
Community Bins and Containers	Daily	-	150
Private collectors	Daily	-	200

4.3.3 Collection Coverage

In 2010, P.M.C provided door to door services to 52.70% of households out of the 524,778 households, which implies that the remaining 47.3% disposed their waste in community bins. In 2017, this rose to 55%. Fig. 6 below shows the map of SWaCH services in Pune. In 2016 – 17, SWaCH covered 80.37% of properties across 41 prabhags in Pune, leaving 19.63% of households to be serviced and about 100 TPD uncollected. [16, 18, 20, 24, 25]

Table 11. Coverage of Households.

Department	Properties	% covered
SWaCH	564,421	47
Ghantagadi	316,415	27
Private	71,111	6

4.4 Waste Disposal

There are currently 7 transfer stations and 50 processing plants which include centralized (7) and decentralized biogas (25), decentralized biodigesters (14), mechanical composting facilities (4) which process 30% of waste.[20]

10% of waste collected is diverted by SWaCH. 60% of the remaining is sent to the Urali Devachi landfill, 25 km. away from the city. The landfill site is made up of 43 ha out of which 15 ha is at full capacity and has been permanently sealed off. Waste disposed at the landfill consists of both biodegradable and non-biodegradable matter. It has been estimated that the cost of collection, transportation and disposal of waste is 83.60 crore rupees (or 11.8 million USD) per annum. Unprocessed waste (36% of total waste) is directly sent to landfill. [20, 26]

Table 12. Processing Methods in Pune

Processing methods	% of total waste processing
Processed in situ	7.5
Diverted to farmers	7
Diverted to recycling	10
Sent to landfill	60

Table 13. Processing Capacity in TPD

Type	TPD
Landfill	500
Composting plants	300
Biogas plants	100
Community	350
Recycling	200

The plants are presently operating at 51% of installed capacity (Fig 6).

4.5 Successes and Strategies

The SWM model followed by Pune is commendable in terms of civil society participation through the engagement of informal workers in the form of SWaCH (Solid Waste Collection and Handling, also translated as "clean"). The cooperative initiative was launched by P.M.C and KKPKP, a union of self-employed waste pickers in 2007. This hybrid model not only empowers the urban poor but has also created a circular economy that has saved an estimated 900 million rupees (12.5 million USD) in labour, transportation and processing costs [27]. SWaCH diverted 52% of plastic waste from landfills through recycling, reducing annual greenhouse gas emissions [27].

P.M.C developed a solid waste management strategy plan for 2017 – 2025 [22], which details an action plan that aligns with G25 mission objectives through 4 pathways: the promotion of house-level and society level processing of wet waste, increasing processing efficiency of existing plants by resolving issues, commissioning new plants to meet gaps in processing capacity, and fast tracking the operationalising of commissioned plants. This includes moving towards a 100% door to door collection (DTDC) system through contracts and civic participation, strengthening existing infrastructure such as vehicles, community bins and route planning by setting a 90% source segregation target in 2020 – 2025. A noteworthy strategy outlined is the monitoring of operations through an integrated information technology system to track waste streams and provide information on indicators which would enable to allocate budgets accordingly. [22, 28]

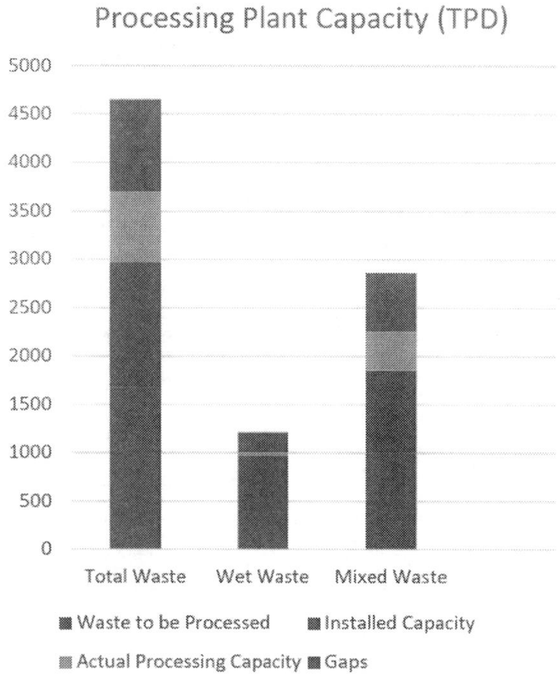

Fig. 6. Processing Plant Capacity (TPD). Source: *Solid Waste Management Strategy Plan 2017 - 2025, P.M.C 2017*

P.M.C has launched various initiatives such as Clean City Movement, Smart Garbage Management System which implements GPS and tracking technologies to monitor vehicles and to provide information to citizens through apps. Zero Garbage Ward (Katraj Model) which targets reduction in waste sent to landfills, Swachh awards and awareness programs have also been initiated.

Waste to energy models through the bio-conversion process to generate biogas has also been recognized as an important approach to managing waste sustainably, with 25 decentralized biomethanation plants installed and operating and various composting techniques underway [20]. Pune recently installed a plant with capacity 300 MTD capable of generating 11 MW [29]. When waste is processed is locally and decentralized, transportation requirements are reduced and will encourage community engagement. Incineration alternatives in the form of pyrolysis gasification as seen in the Rochem plant which processes 700 TPD has been installed [20]. Additionally, the prevention of leachate leakage through barriers has been adopted. P.M.C has stated that these scientific landfills have been in operation since 2010 [20].

4.6 Challenges and Opportunities

With a 50% segregation success rate, the subsequent chain process of solid waste management undergoes ineffectiveness due to inconsistent categorization of waste according to their properties. This includes physical spaces to facilitate segregation. SWaCH workers face issues in segregating between plastic types despite advancing plastic recycling in the country. This has resulted in a low-quality categorization between different plastic types for effective recovery and recycling [27]. Recycling options are also limited and a defined integrated system for reuse, recycling and resource recovery is lacking.

Although SWaCH makes up 43% of collection services, routines and scheduled times are not always followed and coverage of households and properties needs to be increased. The same issue applies to vehicular collection from secondary sources, which are often unreliable due to breakdowns and unpredictable in terms of timings and route movements [22]. Additionally, while it is important to fund civic infrastructure and services, the fee model may not be viable for the urban poor. P.M.C has also identified a deficiency of adequate staff and skill that lead to coordination between collection and transportation [22].

In terms of policy, P.M.C has faced difficulty in enforcing the SWM byelaws as mandated by the central government of India's Solid Waste Management Rules 2016 [22] as a step towards reducing waste sent to landfill.

Land requirements and acquisition for facilities and processing plants that are situated at an adequate distance from the core urban zone are also issues that would require either alternatives or a shift in prioritizing localized and decentralized methods of processing waste such as community composting.

Technological options that have been implanted such as waste incineration, and other mechanized and centralized processing options pose as threats to the SWM model of Pune due to the vulnerability of their size, financing, upgradation and skill requirements. This has been reflected in processing plants performances in the last few years, as seen in the gaps of processing capacity and actual processed waste.

Though the target to completely stop dumping at landfills within two years has been set and the implementation of scientific landfills has been in operation [22], the use of landfill sites for waste to energy conversion has not been identified. Urali Devachi Landfill for instance has the potential to generate 0.7 MW [30]. The capacity of current landfill infrastructure has been estimated to last only for an additional five years [22]. The gap between landfill operations and information also needs to be reduced by employing monitoring systems to check for failures and pollution levels [30].

As an IT driven economy, the nature of Pune's waste will pose problematic E-Waste generation is predicted to cross over 5000 MT per year. C&D waste also needs designated disposal methods since it is currently being directly disposed in low lying areas along rivers [20].

5 Discussion

While Pune belongs to a country that has a very low GDP per capita as compared to Sydney's, we see that some defining strategies are successful. This is particularly true in the case of the SWaCH workers. However, this model has grounds of comparison in Sydney, in that it has also been implemented on a larger scale through the contracting of private business owners and privatization of waste management infrastructure. In one case, the model is successful because it is based on the community level, whereas in the second, the model is also successful because of individual business agents, and the sector is contributing significantly to the country's GDP. It is important to remember, however, that because SWaCH workers belong to the informal sector, this model has larger potential in economies of scale as well as of scope, provided that the system set in place could be more streamlined and organized to increase efficiency in terms of timings and frequency. Furthermore, the framework in which this sector operates has raised awareness in inclusivity and has paved the way for a transformation in traditional modes of operation in India.

Secondly, the collection of waste is the most crucial aspect of solid waste management. Whether this is done by private or public agents, waste collection will reduce unsanitary conditions, reduce greenhouse gas emissions, prevent open dumping in public spaces, and reduce the chances of discarding waste in ecologically sensitive areas. In Sydney, the collection of domestic waste has been implemented systematically with 100% collection coverage. And although the collection of commercial and industrial as well as C&D waste is left to licensed contractors, it is observed that the City has maintained its eye on operations through tracking and monitoring systems. Pune's collection coverage can be improved. However, it is understandable that lower income groups may not have the capacity to be charged a fee, and the structural composition of the city is such that it may be somewhat difficult in collecting from all corners. As the cooperative increases in size, the collection coverage will also increase.

In terms of waste disposal and processing, it is important to learn from Sydney's City Council in how the dissemination of policies have led to an improvement in community recycling initiatives and sensitivity of waste management in city dwellers. Having several decentralized waste processing techniques could allow access to appropriate solid waste management even for corner areas of the city. This could range from community composting to materials recovery and larger waste processing infrastructure.

Following the initiative of SWaCH, a low-income country could join the challenge for a more sustainable future in not only how waste is collected, but also how it is recycled and processed through SWaCH handling of waste directly. Both high and low GDP per capita countries could also invest in alternative waste treatments and waste to energy models that could assist in the development of renewable energy. This has been observed in both cities.

References

1. L.Yao, S. Kaza, P. Bhada-Tata, F.V. Woerden, *What a Waste 2.0: A Global Snapshot of Solid Waste Management to 2050* (2018)
2. UN-Habitat, *SDG Indicator 11.16.1 Training Module: Solid Waste Management in Cities.* (United Nations Human Settlement Programme, Nairobi, 2018)
3. UNDP, *Human Development Indices and Indicators: 2018 Statistical Update.* (United Nations, New York, 2018)
4. City of Sydney, *Leave Nothing to Waste. Strategy and Action Plan 2017 – 2030* (2017)
5. City of Sydney, *Environment Action 2016 - 2021. Strategy and Action Plan* (2017)
6. City of Sydney, *Governance Guide* (2014)
7. City of Sydney, *Waste Policy, Sydney 2030* (2017)
8. J. Pickin, P. Randell, J. Trinh, B. Grant, *National Waste Report 2018* (2018)
9. R. Wainberg, J. Pickin, *Australian landfill capacities into the future* (2009)
10. City of Sydney, *Advanced Waste Treatment Master Plan 2013 - 2030* (2014)
11. Department of Environment, Climate Change and Water NSW, *Review of Waste Strategy and Policy in New South Wales* (Government of Australia, 2010)
12. C.o.S, *Green Environmental Sustainability Progress Report* (2019)
13. NSW EPA, *Waste Levy* (2019)
14. NSW EPA, *Waste Less, Recycle More* (2019)
15. C. Harvey, *Dial a dump: Sydney's $300m hole will become the biggest rubbish dump in the southern hemisphere* (Daily Telegraph, 2011)
16. P.Chikarmane, L. Narayan, *Solid Waste Generation and Collection in Pune* (2006)
17. Office of the Registrar General & Census Commissioner, *Census 2011* (Government of India, 2011)
18. P.M.C, *Pune City Sanitation Plan* (2012)
19. P.M.C, *Public-Health & Sanitation Bye-laws* (2014)
20. P.M.C, *Integrated Solid Waste Management* (2016)
21. P. Chikarmane, *Integrating Waste Pickers into Municipal Solid Waste Management in Pune, India* (2012)
22. P.M.C, *Solid Waste Management Strategy Plan 2017 - 2025* (2017)
23. P.M.C, *Revising/Updating the City Development Plan of Pune City - 2041* (2012)
24. P.M.C, *Total quantity of waste collected for 2017* (2017)
25. P.M.C, *Property wise waste collected by the solid waste department for the year 2017 (2017)*
26. M.T.T, H.H.N, *Existing Situation of Solid Waste Management in Pune City, India,* Res. J. Recent Sci, **1**, 348 (2012)
27. H. Moora, H. Barde, *Closing the Loop: Pune, India Case Study (2017)*

28. P.M.C, *Pune Towards Smart City* (2015)
29. Central Pollution Board, *Consolidated Annual Report for the year 2016 - 2017* (Government of India, 2017)
30. T. Paul, N. Soren, *An overview of municipal solid waste-to-energy application in Indian scenario*, Environ. Dev. Sustain. **22**, 575 (2018)

Investigation and Improvement of Content of Methane in Biogas Generated from Municipal Solid Waste.

Shyamsing Thakur[1*] , and *Rahul* Barjibhe[2]
[1] KBCNMU, Jalgaon, Maharashtra, 425001 India.
[2]SSGBCOET, Bhusawal, Maharashtra, 425203 India.

Abstract The methane yield and overall biogas generation reduce drastically in the winter and summer season. The Biogas plant operators reported better results with the co-digestion of the municipal solid waste (MSW) with cow dung in changing weather conditions. In this research work the quality and content of methane in biogas generated from biogas plant is improved by co-digestion of MSW, cow dung along with the urine with better carbon to nitrogen (C/N) Ration. We took number of experiment using different ratio of MSW and additives to improve biogas. Rigorous experimentations concluded that the co-digestion of the MSW, cowdung and urine in the proportion of (55:35:10) with equal amount water in a portable bio digester for anaerobic digestion results into better methane production with maintaining C/N ratio and reducing time duration for flammable biogas production.

1Introduction

Biogas is clean, concrete, convenient and capable source of energy generation from waste. Biogas has potential to replace conventional fuels e.g. fossil fuels, burnable oils, etc. which are hazardous to the environment .Biogas comprises of many gases constituents generated from the decomposition of organic waste in absence of oxygen. Biogas is generally produced in the container which is called as bio digester. Biodigester is closed container and mixed of organic waste took place to produce biogas. The bio digester is feed with certain organic waste which includes MSW, kitchen waste, agriculture waste and water at some extent. The contain of biogas consist of methane, CO_2, N_2 and H_2S. At one point some of these gases are harmful and poisonous. Biogas is usually non convention source of energy, eco-friendly, simple and low cost technology .The generation of biogas also produces organic fertilizer. This biogas also used for the generation of electricity, fuel and for healthy cooking.

The process of conversions of biogas from organic waste, agriculture waste, kitchen waste etc through anaerobic digestion is effective from decade. This biogas generation consists mainly of methane. This methane generation is highly effective for rural areas. Biogas also contains certain amount of gases such as CO_2, H_2S, O_2, N_2, etc. Thermal, Mechanical and Thermochemical methods of pretreatments are most efficient and effectively implemented as compared to other pretreatment methods. After the implementation of the most efficient thermal, mechanical and thermochemical methods resulted into better methane yield, drop in retention time and improved quality of the manure [1]. Dry MSW contains slow to biodegrade 'lignin' which took longer time to degrade and occupy top level of slurry in the biodigester resulting into scum formation and pretreatments are essential for smooth functioning [2].

Additives should have the nutrition values and enhance biodegradability. It plays vital role in biogas yield and improves metabolic conditions of microbes. The thermophilic digesters are cheaper than mesophilic digesters over a period of time with high initial investments. The Co-digestion means the mixing of multiple feedstocks had best results than single AD digester in terms of nutrition enrichment, less toxicity, simplified implementation, and higher biogas yield. Several pretreatment techniques are available optimization of the current technology is essential [3].

Major problem in biogas generation is larger retention time (30 to 50 days) and less productivity in summer and winter. Use of additive is beneficial but limited due to seasonal availabilities of additives. Implementation of implant pure culture of microbes should be look as future scope in order to taken into account human health and circular economy. Unwontedly research related to high total solid content cowdung was limited to lab scale only. Large scope to implement these on fields conditions of WWTP's [4].

The mixing facilitate co-digestion of more than one substrate and progressively the acculturate the C/N Ratio [7].Optimum mixing ratio of food waste to cowdung 3:1 reported highest methane production [9].In order to reduced the contain of this unwanted gases certain additives is use. Cow urine has important N component and traces of P, K, Ca and Mg [10]. According to literature survey, it indicates that the addition of substances such as urine could increase the efficiency of

*Corresponding author: **thakur.shyamsing23@gmail.com**

biogas generation and maintain better content of methane in biogas. Availability of animal urine and its nitrogen rich nature could be beneficial. The animal urine can use as additive with MSW. C/N ratio is moderately higher in urine due to its nitrogen content promising way to enhance the biogas production. Carbon (C) /Nitrogen (N) ratio of 20: 1 to 30:1 is generally maintained. Methane contents were enhanced by the mixing cowdung, food waste and organic portion of MSW when applied in the optimum condition with lesser retention time model of 30 days [6] surprisingly, the optimum combination were not evaluated in the study.

Half of the volatile solid of the cow dung waste is biodegrade and decompose into biogas . The animal urine used as additive is easily avilable from cattle farm. The discharge urine from animal is practically of no used and also polluted the environment. Thus use of urine the environmental pollution would be decreased at quick rate. So, considering the overall situtaion for the experimentation the use of urine in the MSW for anerobic digestion will increases the production of biogas at high rate, reduces the additive cost and so on thus thus experiment was conducted for the effect of MSW with urine for biogas production.

Table1. Biogas Constituents

Components	Content (Volume of constituents / Total Volume)
Methane (CH_4)	50% to 60%
Carbon Dioxide (CO_2)	35% to 40%
Water (H_2O)	2 to 7%
Hydrogen Sulphide (H_2S)	2%
Ammonia (NH_3)	0 to 0.05%
Nitrogen	Upto 0.2%
Oxygen	Upto 0.2%
Hydrogen	Upto 0.1%

2 Material and Methods

2.1 Preparations of Feeding Constituents

MSW (Municipal Solid Waste) was collected from different part of city for mixing in biodigester. For additive purposes, we have selected animal urine and cow dung [9] as it maintains C/N ratio after anaerobic digestion and animal urine easily available. Thus, we collected cattle urine from cattle farm nearby to our experimentation location. The properties of the MSW, cowdung and urine are shown in table 2.

Table2. Properties of feeded mixtures and types of the digesters

	MSW	COW DUNG	URINE
VFA (g/l)	8.9	0.2	150 µmole/l
VS (g/l)	169.2	12.8	-
COD (g/l)	13.8	158.3	-
TS (g/l)	222.4	21.2	-
PH (g/l)	4.6	7.2	5.8 – 6
C/N RATIO	25	6	17.54

*Traces of P, K, Ca and Mg found in PPM

2.2 Construction of biodigester

For experimentation purpose we have prepared a lab scale portatble biodigestor.

i. 20L container.
ii. Ballon for gas storage.
iii. PVC pipie for inlet and outlet.
iv. Connecting tube.
v. Flow control valve

The lab scale portable model was used for the further experimentations.

2.2 Experimentations

The experiments were conducted in three different biodigesters as per tests. The biodigester compramises of the 20L container, Ballon for gas storage, PVC pipie for inlet and outlet ,Connecting tube and flow control valves.The pressure is mainteained on the top of biodigester by weight. The Biodigster is covered to restrict extrageneous factors like sunlight etc.The flow control valves are used to regulate the flows.The digester temperature kept at aound the 35°c.

Test 1 : MSW with equal quantity of water.

Test 2 : MSW mixed with cowdung (3:1) and equal amount of water [9].

Test 3 : MSW, Cow dung and urine (55:30:10) with equal amount of water.

Biogas collected in the storage tube,observations were recorded a regular intervals.For formation of biogas at small scale it needs at least 7-8 days.The reading were taken upto 25 to 30 days of feeding when biodigeter attains the stabillity [5].The tests were rigoursly repeated to confirm the results.

2.4 Analytical Methods

The PH, TS, COD, VFA and TVS were determined by APHA 2005 standards. Gas chromatograph (Method-TCD) used to measure quality of the biogas.

3 Results and Discussion

The Test 2 and Test 3 concluded that the biogas was collected in the storage tube at faster rate as compared to ideal rate of biogas formation Test 1 of mono-digestion. The time required for formation of biogas (test 2 and test 3) was reduced as compared to standard time required for biogas formation without additives (test 1). The collected biogas was feed into gas chromatography (Method TCD) machine and the outputs were recorded. Biogas consists of methane, nitrogen, carbon dioxide, hydrogen sulphite.

3.1 Effect of co-digestion on the production of the Biogas

The Experimentations were done with different Co-digestion and mono-digestion the Bio gas generation were evaluated.

Test run 1 (MSW)
Test run 2 (MSW+ COWDUNG)
Test run 3 (MSW+COWDUNG+URINE)

Above tests gave peak biogas production and burnable biogas at 25 days, 23 days and 21 days respectively. The Methane yield and Quality is found out to be best in test run 3. Urine contains the nitrogen which is favorable for the bacteria growth. The codigestion with cowdung and urine enhances the concentration of soluble organic constituents which in turns reduces the time required for hydrolysis stage of Methanogenesis process and results into better biogas production.

3.2 Effect of co-digestion on Contents of Methane in produced biogas.

The Methane contents were experimentally tested on weekly basis by Gas Chromatographs (Method TCD). The Experimentations were done to evaluate methane contents in biogas generated from different compositions.

The Test run 1 (MSW), Test run 2 (MSW+ COWDUNG) and Test run 3 (MSW+COWDUNG+URINE) gave directly proportional biogas production along biodigestion time in weeks. The Methane yield and Quality is found out to be best in test run 3.

The co-digestion with Cowdung and urine breed the methanogenic bacteria and developed the better culture of bacteria which results into improved methane contents in generated biogas.

3.3 Effect of co-digestion on The pH

The pH is most drubbing factor for the growth and survival of microbes during AD methanogenesis. The drastic pH reduction inhibits the methane formation. The pH is a function of the Volatile Fatty Acid and carbon dioxide generations. Stable maximum pH obtained during co-digestion of the cow dung and MSW is 6.7 to 7.2 similar to the study of Kumar Awasthi [6]. In all the tests with different additives the digester reaches to the stable pH that means the digestion is neutral at the end of the experimentations .The codigestion helps to stabilize the PH which is essential for bacterial growth and hence the biogas production.

4 Conclusions

The Test run 1 (MSW), Test run 2 (MSW+ COWDUNG) and Test run 3 (MSW+COWDUNG+URINE) gave peak biogas production and burnable biogas at the 25 days, 23 days and 21 days Respectively. Methane yield is found out to be best in test run 3 due to the better C/N ratio maintained. Stable maximum pH obtained during co-digestion of the cow dung and MSW and mono digestion is 6.7 to 7.2.The Test 2 and Test 3 concluded that the biogas was collected in the storage tube at faster rate as compared to ideal rate of biogas formation Test 1 of mono-digestion.The retention time required for formation of burnable biogas (test 2 and test 3) was reduced as compared to standard time required for biogas formation without additives (test1). The maximum methane content in biogas obtained from the co-digestion of MSW, cowdung and urine was found out be around 68%. The upgradation of biogas purification and compressing it to CNG technology development at affordable cost is biggest challenge for future.

Acknowledgement

Special thanks to the Research and Development department of Mailhem Ikos Pvt. Ltd. for providing the access to the testing lab along with the guidance and mentoring during tenure of project.

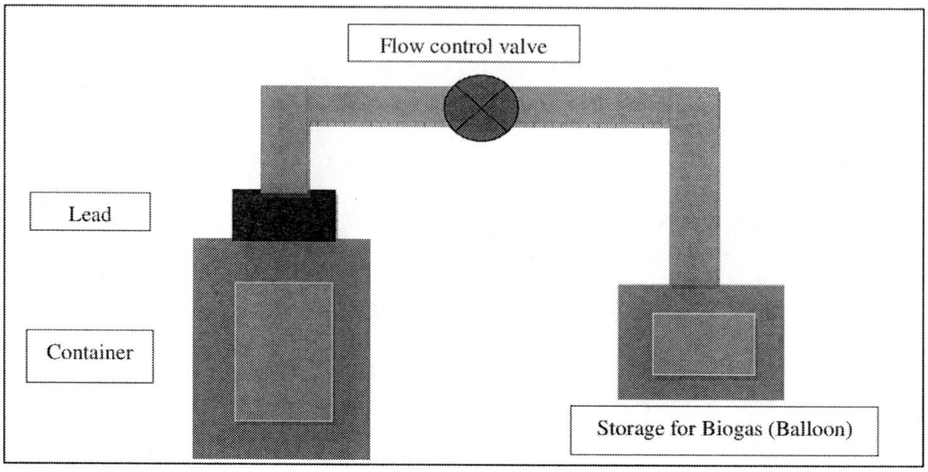

Fig.1. Construction of bio-digester

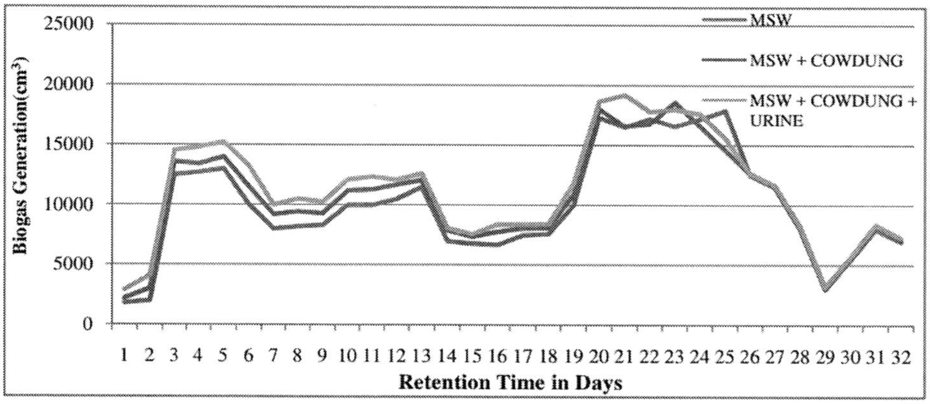

Fig.2. Biogas Generation (CM^3) Vs Retention Time (Days)

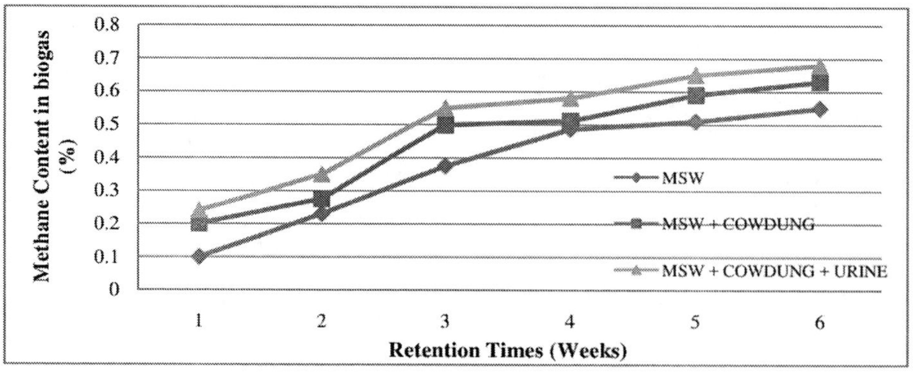

Fig.3. Methane Content (%) Vs Retention Time (Weeks)

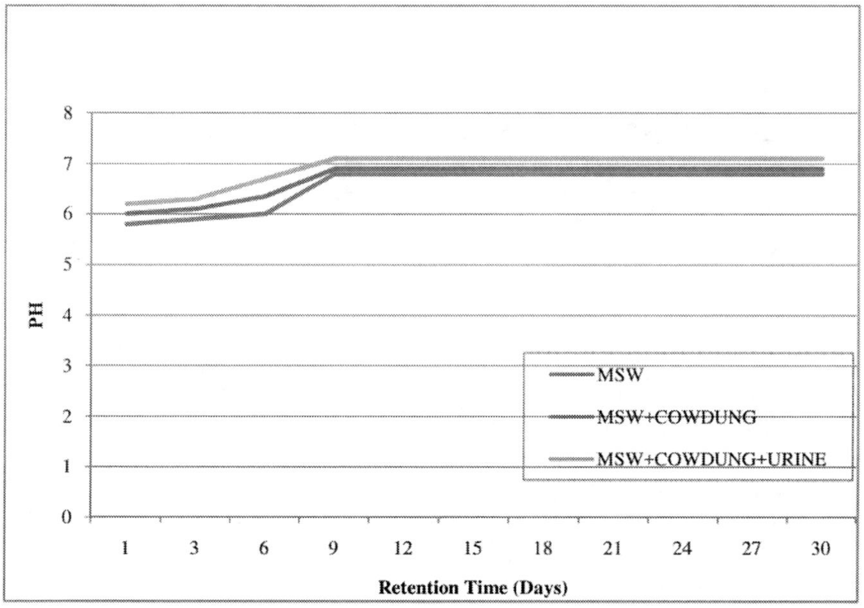

Fig.4. PH Variation Vs Retention time

References

1. S. Jain, I. T. Wolf, J. Lee, Y. W. Tong, Renewable and Sustainable Energy Reviews, **52**, 142-154 (2015).
2. M. Krishania, V.K. Vijay, R. Chandra, Energy, **57**, 359-367 (2013).
3. C. Mao, Y. Feng, X. Wang, G. Ren, Renewable and Sustainable Energy Reviews, **45**, 540-555 (2015).
4. Y., S., T.R Sreekrishnan, S. Kohli & V. Rana, Bioresource technology, **95**, 1-10 (2004).
5. A.K. Jha, Li, J., Ban, Q., Zhang, L., Zhao, B., Pak. J. Biol. Sci. **15**, 1111–1118 (2012).
6. S.K. Awasthi, R.Joshi, H. Dhar, S.Verma, M.K. Awasthi, S. Varjani, S.Sarsaiya, Z.Zhang, S.Kumar, Bioresource Technology, (2018).
7. R.Zhang, R.M .Mashad, K.Hartman, F.Wang, G.Liu, C.Choate, P.Gamble, Bioresour. Technol. **98**, 929–935 (2007).
8. J. K. C. L. Zhang, J. Lee, C.H. Wang, Y. Dai, Y.W.Tong, Scientific Reports, **7**, 1269 (2017),
9. R. Li , S. Chen & X. Li, Part A: Recovery, Utilization, and Environmental Effects, **31**, 1848-1856 (2009).
10. W. H. M. Saunders", New Zealand Jour. of Agri. Research, **25**, 61-68, (1982).
11. C. Okeh, C. O. Onwosi, F. John, C. Odibo, Renewable Energy, **62**,204-208, (2014).
12. M.O.L. Yusuf, N. L. Ify, Biomass and Bioenergy, **35**, 1345-1351, (2011).
13. S. Zhou, Y. Zhang, Y. Dong, Energy, **46**, 644-648 (2012).
14. Stromberg, Sten, M. Nistor, and J. Liu., Waste Management, 2014.
15. C. Ma, J. Liu, M.Ye, L. Zou, G. Qian, Y.Y.Li, Renewable and Sustainable Energy Reviews, 2018
16. J. Yang, D. Wang, Z. Luo, W. Zeng, Waste Management, 2019.
17. D. barik, S. Murugan ,Waste and Biomass Valorization, 2015.
18. S. Achinas, Y. Li, V. Achinas, G. Jan W. Euverink, Sustainable Environment Research, **28**, 240-246 (2018).

Development of shrimp-based chitosan film and assessment of its mechanical, barrier and antimicrobial properties

Jagruti Jankar[1,*], *Yogesh Nagargoje*[2], *Yogita Chavan* [1], *Jaydevi Jankar* [3], *and Akshay Kumar Sahoo*[2]
[1] MIT College of Food Technology, MIT ADT University, Pune, India
[2] Department of Technology , Shivaji University, Kolhapur, India
[3]Asian College of Arts and Science, Pune University, India.

Abstract. You Utilization of biodegradable films is a need of food packaging industries in order to reduce the hazards related to plastic use and to extend the food's shelf life. Various polysaccharides are in use for the purpose of making the films. In this research, chitosan based films were developed and its mechanical, barrier and antimicrobial properties were checked to fulfill the packaging requirements. Chitosan was extracted from shrimp waste and films were produced using 0.5-2 per cent concentrations. At ambient temperature, specific characteristics such as mechanical, barrier, and antimicrobial analysis were performed at an interval of two days. Among all, the films with 2 per cent chitosan showed best results in terms of tensile strength, thickness and percent elongation. Also, the films exhibited maximum resistance to water vapor permeability. The extracted chitosan at 2 per cent concentration had shown the maximum resistance against Staphylococcus aureus and Pseudomonas aeruginosa. From the current investigation it can be said that films with 2 per cent chitosan could be used as biodegradable food packing materials and can serve as material which would maintain a good city and future of world by minimizing plastic hazards..

1 Introduction

Plastic is used for various purpose in food sector such as food packaging in the form of plastic films, bags, trays, etc. been used for several years, but one should be aware of the toxic effects of plastics on human as well as on the environment. It results awfully, not only on nature but also on health of living beings. Soil pollution, water pollution, and air pollution are the worst effects of plastics. Moreover plastic is a vehicle for various diseases like cancer, headache, lung problems, heart disease, skin disorders, etc. These are caused by the harmful gases released after burning the plastics. Plastic packaging materials are typically made from synthetic products that are non-renewable and non-biodegradable. Because of the increasing demand for plastic use, particularly in the food packaging industry, plastics are largely manufactured in line with human population growth(13). The non-biodegradable feature of petroleum-based plastic products renders plastics improperly disposed of, leaving an enormous amount of unrecyclable and untreated plastic trash to pollute the earth. To compensate these severe problems, there is a need for such a technique which will not be causing any pollution and will protect the food quality and human health too. The use of films and edible coatings which degrade naturally can be a solution to these types of glitches. These films are obtained from different nature based polymers like polysaccharides and they can be developed from renewable sources (7). Significant progress has recently been made in the manufacturing of biodegradable plastics, primarily from renewable sources, for the creation of biodegradable products of comparable versatility to harmful plastics. It is expected that use of biodegradable plastic would lead to recycling and increasing the environmental impact associated with the processing oil based polymer (18). However, chitosan is a bio-based polymer derived from de-acetylation of chitin from crustacean shells (1) have been used to as packaging aid for different food commodities. Also, it is a polycationic polymer which is soluble in diluted acids owing to protonation of amino groups. The characteristics of this polysaccharide such as biocompatibility, biodegradability and non-toxicity allow its use in the food industries, pharmaceutical and cosmetics. Development of chitosan edible films and packaging materials has been broadly studied at laboratory scale (4). Due to its film forming capacity and good mechanical properties, chitosan is an excellent source for films, edible coatings and can form transparent films that can meet various packaging needs (19). The present research was intended to develop edible films from different chitosan concentration and to evaluate its versatile properties for further use. The present investigation entitled "Development of chitosan based biodegradable films" was carried out in laboratory of Department of Technology (Food technology), Shivaji University, Kolhapur during 2015-16. Shrimp waste was collected from the local market of Kolhapur. It was originated from sea water of Ratnagiri, Maharashtra. The species found was *Penaus Monodon.*

2 Materials and Methods

*Corresponding author: jankjagruti@gmail.com

The present investigation entitled "Development of chitosan based biodegradable films" was carried out in laboratory of Department of Technology (Food technology), Shivaji University, Kolhapur during 2015-16. Shrimp waste was collected from the local market of Kolhapur. It was originated from sea water of Ratnagiri, Maharashtra. The species found was *Penaus Monodon.*

2.1 Chitosan extraction from shrimp waste

Chitosan extraction for further study was carried out (14). Collected shrimp shells were washed with clean water to remove the unwanted parts on the surface, then it was subjected to drying and grinding into fine powder. The fine powder was soaked into 4% hydrochloric acid for 36 hrs at room temperature for demineralization. It was then deproteinized using 5 %sodium hydroxide at 90°C for 24 hrs. The process followed by deacetylation using 40-70% sodium hydroxide for 72 hrs to give higher degree of deacetylation. The complete process flowchart of extraction was given in fig.1.

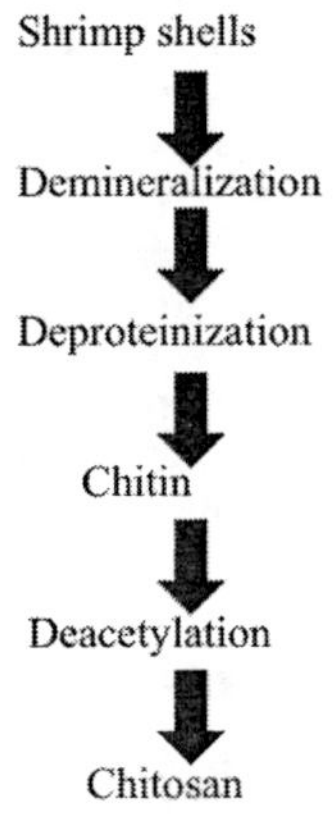

Fig.1 Extraction of chitosan

2.2 Preparation of Films

Films were prepared as reported by a specific method (5). Acetic acid was selected for making a film with 1% concentration. 1% acetic acid was used to dissolve different concentrations of chitosan viz. 0.5, 1, 1.5 and 2 per cent. Glycerol at the concentration of 0.5 % was added as a plasticizer and stirring was continued for 2-3 hrs until the complete homogenization of the mixture was obtained. Subsequently, after filtration the solutions were poured onto petri dishes and dried in a hot air oven at 80°C for 8 hrs. After drying, all films were kept at 25°C in a controlled environment chamber and 50 % relative humidity was maintained.

2.3 Mechanical and barrier testing of chitosan-based films

Physical and mechanical testing of chitosan-based films was carried out as earlier reported (5).

2.3.1 Thickness of the film

The strips of films with the dimension of 2 cm × 5 cm were taken to measure the thickness using micrometer (103–131, Mitutoyo, Japan). Random measurements at three different points were taken to calculate the mean value and further used for per cent elongation, puncture strength and tensile strength evaluation.

2.3.2 Tensile strength

Tensile strength was determined mechanically using a texture analyzer (TA.HD Plus, Stable Micro System). For the measurement of tensile strength and per cent elongation at break, standard method given by American society of testing and materials (ASTM) was followed.

2.3.3 Puncture strength and per cent elongation

Assessment of puncture strength and per cent elongation was carried out using the above mentioned texture analyzer. Strips of 2 cm were taken for analysis and cylindrical probe of 2mm diameter was passed through the strips at steady speed of 1mm/s.

2.3.4 Water Vapor Transmission Rate

The water vapor transmission rate (WVTR) of chitosan based films was analyzed by round cups with 100 ml volume capacity and mouth area of 120 cm^2. The cup was filled with 60 ml water and covered with chitosan film tightly using adhesive tape and it was put into desiccator by maintaining 90% relative humidity using sulphuric acid The loss of water from the cup was observed with time and WVTR was evaluated.

2.4 Antimicrobial Test

Antimicrobial analysis was done by following the protocol revious protocol (16). Agar well diffusion method measures the antimicrobial activity which includes swabbing of nutrient agar plates with broth culture of Staphylococcus aureus and Pseudomonas aeruginosa. In each of these plated wells of 10 mm diameter and 2 cm apart were made using sterile cork borer. The solution of chitosan stored as stock solution at the concentration of 0.5 -2 mg/ml in acetic acid were added into each well and kept at 4°C for 10 min for diffusion. The plates were transferred to incubator at

37°C for 24 hours. The diameter of zone of inhibition was measured after incubation with four replications and the mean value was used calculated.

2.5 Statistical Analysis

The complete randomized design was used to evaluate the proficiency of films. The SAAS software was used for statistical analysis.

3 Results and Discussions

Extraction of chitosan was performed in accordance with 2.1 procedure. Chitosan extraction was confirmed by dissolving 0.5-2 % chitosan in 1% acetic acid. Using casting techniques films were made from the 0.5-2 % chitosan and examined for different properties.

3.1Mechanical and barrier properties of chitosan based films

Table 1. Mechanical and barrier properties of different chitosan concentrations

Concentration of chitosan (g)	**Tensile strength (MPa)**	**Elongation (%)**	**Puncture Strength (N)**	**WVTR (g/m²/day)**	**Thickness (mm)**
0.5	26.24^{a}	13.89^{d}	1.06^{c}	113.62^{a}	0.13^{d}
1	28.08^{b}	17.98^{c}	1.13^{b} c	108.13^{b}	0.14^{c}
1.5	30.6[illegible]^{c}	22.42^{b}	1.34^{b}	101.2[illegible]^{c}	0.15^{b}
2	32.26^{d}	37.85^{a}	1.77^{a}	95.78^{d}	0.16^{a}
L.S.D.	0.4959	0.8266	0.2327	0.8225	0.384

Note: In the above table, the Average values of all properties were taken.

From the above table, it can be seen that chitosan films with 2 % concentration gave a stronger film as compared to other concentrations. It had the highest tensile strength i.e.32.26 MPa which showed that it increased with chitosan concentration. Obtained results were in line with (3) the researchers who found that the tensile strength of film increases with concentration of chitosan. Besides, same results were revealed (2) who produced the films by the composition of chitosan, fish gelatin and microbial transglutaminase (MTgase). They said that the concentration of chitosan increases, tensile strength of films also increases suggests that film thickness increases with chitosan concentration. Moreover, it can be depicted that films with 2% chitosan had good resistance to water vapor permeability. The results were resembled with the previous data (12) in which the chitosan films showed magnificent barrier properties to water vapor permeability. A film with 2 % chitosan showed elongation of 38.25 % and required maximum force of 1.606 N to puncture the film. It means that the % elongation and puncture strength of the film also increases as the concentration of chitosan in the film increases. The results are in agreement with (15) who developed starch –chitosan composite films and found that the elongation at break was increased as chitosan concentration increased. In addition, (21) it was concluded that, the film which was prepared from 2% chitosan concentration had higher thickness, % elongation and puncture strength, exhibited maximum resistance to water vapor permeability as well. The similar results were provided by (6) and (11) who intimated that high molecular weight chitosan films possess higher tensile strength and better barrier properties than low molecular weight chitosan films. Moreover, it can be depicted that films with 2% chitosan had good resistance to water vapor permeability. The results were resembled with the previous data reported by (12), the chitosan films showed magnificent barrier properties to water vapor permeability. A film with 2 % chitosan showed elongation of 38.25 % and required maximum force of 1.606 N to puncture the film. It means that the %elongation and puncture strength of the film also increases as the concentration of chitosan in the film increases.

3.2 Antimicrobial activity of chitosan

Experimental studies showed that tested concentration of extracted chitosan was antagonistic to gram positive bacteria Staphylococcus aureus and gram negative bacteria Pseudomonas aeruginosa. Four different concentrations (0.5, 1, 1.5, & 2%) of chitosan film were tested against the said bacteria and observed maximum inhibition at 2% concentration of chitosan. According to Goy et al.,(2016) 1% chitosan showed inhibitory effect against *S.aureus.* The reason behind the antimicrobial activity of chitosan was owing to charged group in the backbone of polymer and their ionic interactions with components present in cell was eliciting the discharge of intracellular electrolytes and bring about microbe's death. Antifungal and antibacterial properties of chitosan have been already reported ([8]), [9],[17]). The microbes showed maximum resistance at 2% chitosan at which it showed maximum zone of inhibition (Fig.2).

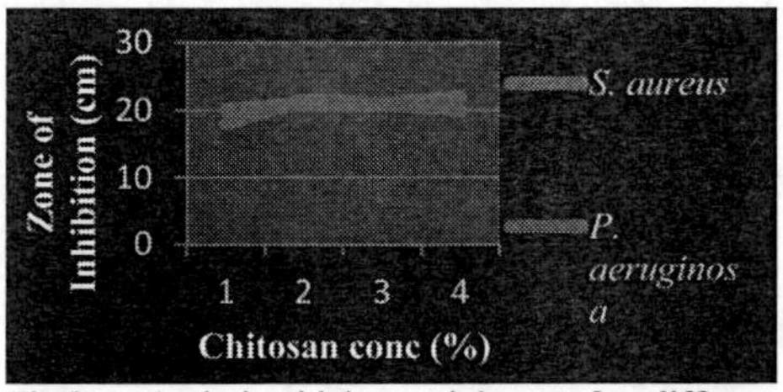

Fig.2. Antimicrobial activity of different concentrations of chitosan

4 Conclusion

Chitosan based films were developed by using different concentrations of chitosan (0.5-2 %). The extracted chitosan was examined for antimicrobial prperties and the films for mechanical and barrier properties. There was significant difference in the films made from each concentration and found the best results with 2 % chitosan. Extracted chitosan showed antagonistic property against Staphylococcus aureus and Pseudomonas aeruginosa. In addition the 2 % chitosan films exhibited maximum resistance to water vapor permeability. Therefore, the present investigation put forth an approach to use the chitosan films from shrimp waste as a packaging material in replacement of plastic with the view of avoiding the environmental pollution and extend the food's shelf life.

Reference

1. M. Aider LWT-Food Science and Technolgy 43, 837-842,(2010).
2. S .Alvarado, G. Sandoval, I. Palos, S. Tellez, Y. Aguirre-Loredo and G.Velazquez, Food Science and Technology, **35**(4), 690–695(2015)
3. J. Bonilla, A. Talon, Emma , L. Atares , M. Vargas and A. Chiralt A, Journal of Food Engineering. **11**, 271–278 (2013)
4. M. Cerqueira , B. Souza , J. Teixeira,A. Vicente, Food Hydrocolloids **27**,175-184(2012)
5. K. Chaturbhuj, S. Gupta, J. Bahadur, S. Mazumder , P. Variyar , A. Sharma Carbohydrate Polymers, **98**,1610-1617. (2013)
6. R.H. Chen, and H.D. Hwa, Carbohydrate Polymers, **29**, 353-358(1996)
7. Dai, H., Chang, P. R., Yu, J., and Ma, X.,Starch. **60**(12),676–684(2008)
8. R.K. Dhall, Critical Review in Food Science & Nutrition, **53**, 435– 450(2013)
9. M.Z. Elsabee, and E.S. Abdou, Materials Science and Engineering: C, **33**(4), 1819– 1841(2013)
10. R.C.Goy , S.T.B. Morais, and O.B.G. Assis, Revista Brasileira de Farmacognosia, **26**(1), 122– 127(2016)
11. K.T. Hwang , J.T. Kim, S.T. Jung, G.S. Cho and H.J. Park Journal of Applied Polymer Science, **89**(13), 3476–3484(2003)
12. I. Leceta, P. Guerrero, I. Ibarburu , M. Duenas and C.K. De la, Journal of Food Engineering, **116**, 889–899 (2013)
13. M.Z.I. Mollah, N. Akter, F.B. Quader , S. Sultana and R.A. Khan, Open Journals of Organic Polymer Materials, **6**,11–24 (2016)
14. Y.S. Puvvada, S. Vankayalapati and S. Sukhavasi, International Current Pharmaceutical Journal, **1**(9), 258-263(2012)
15. L. Ren, X. Yan , J. Zhou , J. Tong and X. Su, International Journal of Biological Macromolecules, 105, 1636–1643(2017)
16. N. T. Sena , B.J. Gomes, M. E. Vianna, V.B. Berber, A.A. Zaia and C. C. Ferraz, Int Endod J;**39**,878–85(2006)
17. R. A. Sheikh, M. A. Malik, S. A. AlThabaiti and M. A. Shiekh , Food Science and Technology Research, **19**(2), 139– 155(2013)
18. J. H. Song, R. J. Murphy, R. Narayan and G. B. Davies, Biol Sci,**364**(1526),2127–2139(2009)
19. P. C. Srinivasa Ph.D. Thesis, CFTRI, Mysore (2004)
20. T. N. Van, T. Hanh and P. Thien P The Open Biomaterials Journal, **4**, 1-5(2013)
21. Y. X. Xu, , K. M. Kim, M. A. Hanna and D. Nag, Industrial Crops and Products **21**(2), 185– 192(2005)

Basic Solutions to Waste Management Issues That Will Bring Clean Culture in India

Bharati Koot[1]
[1]305-Gulmohar Imperial, Dhayari, Pune - 411041, India

Abstract. The latest population of India is around 1.37 billion and India ranks number two by population in the world. With the increasing population, the obvious thing, the waste generated per day continues to increase. India generates around 65 million tonnes of waste every year and out of which most of the part is not collected, remains untreated at landfills or open areas. This is ruining the beauty of India. And this is already impacting largely on environment, climate, soil, water bodies, national health, and tourism. That also means the waste in India is heading to largely affect the national economy and the global environment. This paper attempts to go back to the basics of human nature and habits, needs and mechanisms for shifting them for the context of waste management and anti-littering to bring clean India forward and contribute to preventing the global environmental crisis.

1 Introduction

This section depicts the introduction to the topic of this paper.

Littering and poor waste segregation are a few of the key factors which make waste management difficult in India and everywhere. Littering has become the most convenient habit of people unknowingly since the past decades. Waste segregation rules have been in place since the year 2006, and after that also there are revisions to waste management rules, yet we lack in implementing it. India is a country of huge population and hence sometimes it lacks the infrastructure to establish the grounds for clean surroundings. But, in recent years many initiatives have been started by the government, NGOs, and individuals. These initiatives are a very welcome step towards the clean India dream. But the question still remains, are they enough? The impacts of poor waste management and littering habits on the global environment, national reputation and economy are threats to the world and the country. So, we really need to ask another important question to ourselves, are we going to the roots of the problem? Because the problem is so crucial from a local and global point of view, we must have concrete results soon across the nation and worldwide. And, hence when it comes to addressing city waste and litter solutions, we must go to the roots.

2 The Problems

This section briefly describes the problem due to increased waste generation and poor waste management.

The modern lifestyle, industrialization, urbanization, and the convenience brought by the magic called plastic has increased waste generation per person or per family. India generates around 65 million tonnes of waste every year and out of which most of the part is not collected, remains untreated at landfills or open areas. Landfills are the 3rd largest cause of greenhouse emissions in India [1]. They cause air, food, land, and water pollution. The toxic leachate generated at landfills and landfills catching up into fire is creating a severe impact on our life and environment. The lives of other animals are at risk. We are putting our next generations in danger. Plastic waste has become a major problem to the planet as oceans are full of plastic and micro-nano plastics are already entering the food chains. The lives of people working in waste management are at risk. Approximately 80% to 90% of waste handling workers die at the age of 45 to 55 due to side effects caused by handling the waste in an unsafe manner.

On one side we need to have proper waste handling and treatment mechanisms and on the contrary, there is a need for a major shift in the mindsets of people. The lack of infrastructure is not the only problem. The unintentional casual mindset towards waste and the environment is also a problem. Hence waste management and cleanliness should become a "Peoples Movement".

3 Recent Activities Addressing the Problem Area

This section briefly describes various (not all) initiatives taken by the government and non-government entities in India for addressing the waste-related problems.

The government and non-government bodies have understood the need for a change from national and global perspectives.

The initiatives like "Swachh Bharat Mission" [2], "Smart Cities Mission", "AMRUT (Atal Mission for Rejuvenation and Urban Transformation)" [3] are few to

*Corresponding author: bharatikoot.bbk@gmail.com

mention that government announced and is implementing across the nation.

Apart from these, many individuals, environment activists, waste warriors, NGOs, organizations are working to support government mission and national agenda about cleanliness and waste management.

These are bringing a positive change for sure in the society and country. Awareness under such programs is creating impact by spreading the word about sustainability and cleanliness. The cities like Indore, Bangalore, and Pune are ahead with the local community participation for awareness and cleanliness.

Yet, littering is a major hurdle in achieving big outcomes with such initiatives. We need the entire population to align with the national agenda to see the dream come true of clean India. The cleanliness in society creates a positive and optimistic environment around and that helps to boost creativity, wealth, happiness, and peace.

We need to put huge efforts into understanding the root causes and bringing solutions in the right direction with the right investment of money and resources.

As a citizen of India, everyone should understand how much the government needs to spend to clean up the country every day and still we are one of the unclean countries in the world.

We should encourage ourselves to take responsibility for own waste and let the government invest in other important things like education, national security, science, and technology.

We have many examples of global environmental activists, including the young generation, for example, Greta Thunberg. We have many countries that are the inspiration for bringing clean cities and clean habits like Germany, Canada, Singapore, Sweden, etc. We should learn from these inspirations.

What it takes at the heart of cleanliness is the love for our nation and our planet.

4 Root Causes

After describing the problem area, this section puts light on the root causes of the waste-related situation in India.

Any problem can be effectively addressed by reaching to the roots of it. A big reason behind waste management issues in India is that littering and dumping mixed garbage. The less sensitive and self-centric nature, poverty is another root cause.

The other factors are important in solving waste-related problems like infrastructure and mechanisms, but without having addressed the root causes, the efforts in these will be in vain.

We have given the message to the world, "Vasudhaiva Kutumbakam" [4] which means "The whole world is one family". We all must revise the meaning of this message for ourselves. If everyone understands this message, we will start caring for the world as we care for our limited family and little private homes. So, when we look at these root causes, we need a major shift on an individual level spreading across the nation. And to sustain that shift for our future generations, we should connect this shift to the education systems and ruling systems. Let's see in detail the root causes behind the waste problems in India.

4.1 Littering

Littering is an act of disposal of any waste item on undesired locations like roads, public places, etc. This act is impacting the environment in many ways. In general, littering is a normal act, the act as easy as the way we breathe. When we inhale and exhale, we hardly pay attention to it unless something serious is there, otherwise, it goes on naturally as long as we are alive. Littering is perceived in the same way. Littering seems to be second breathing sometimes. Most of the public do not know why they litter.

4.1.1 Unconscious Littering

Someone walks on the road, they get some advertising pamphlet on the way, they don't even read it, and it falls unknowingly on the road without their conscious attention. This is unconscious littering. These acts happen in public places like railway stations, parks, roads, bus stops, within vehicles, and while traveling. It hardly matters to understand where the litter goes in the end as the act of littering happens unknowingly. But the thing often missed to understand is that the littered items will remain on this planet itself unless attained. And otherwise, through obvious mediums, they end up in the soil, water bodies, oceans, and in the end our food chain.

4.1.2 Conscious Littering

Another way of littering is conscious littering. In this case, people know littering is bad, but somehow, someone else has already done it, and they don't find any good reason not to do it. They follow that first "someone". The psychology here is to assume that littering is a free act, it's not a crime, no one is going to ask penalty, no one is watching, and in fact, the place is already unclean. So, people again unintentionally follow the tribe called the "litter tribe". This type of tribe is very dangerous as they create mass litter piled on roadside and community areas. It only starts with the first person who chooses to litter. Others blindly follow without taking efforts to look at surrounding solutions. This not only makes the surrounding ugly but also creates serious health effects for animals and humans around.

4.1.3 Littering and Bystander Effect

Littering is a perfect example of the Bystander effect [5]. As per the source, it is defined as, "a social psychological claim that individuals are less likely to offer help to a victim when other people are present; the greater the number of bystanders, the less likely it is that one of them will help." [5]. It is like everyone is in the hope loop waiting for one to act and help.

When we look at the bystander effect for waste-related or littering habits, we can relate to what we see in society. The bystander observers continue staying in the hope loop that someone will act, someone, will clean up the space. It derives totally opposite behaviours in clean vs unclean place. At the place which is clean, it is less likely that people start littering. And opposite to that at the place which is unclean or becoming unclean, it is more likely that people will continue littering. And that is how the "litter tribe" grows. We as humans naturally tend to imitate the behaviours of others. But the point that is often missed here is that if one starts caring and acting for cleanliness, others will also start. The power of an individual can create big changes. The power of millions of individuals summed up together can create long-lasting changes. The sense of collective responsibility and collective ownership needs to be imposed in the public to deal with problematic social behaviours like littering.

And that means incorrect, misdirected social and individual behaviours are the root causes behind the littering and waste related crisis.

4.2 Mixed Waste

Mixed waste is very difficult to process. The major waste categories in India are degradable waste e.g. Food waste, paper waste, non-degradable waste e.g. plastic, and hazardous waste e.g. sanitary, medical waste. There are other types of waste which include metal, glass, fabric, e-waste, and industrial waste.

But, in most parts of India, the household waste collected daily turns out to be a mixed waste. The unsegregated waste is harmful to the people operating on it. The recycling and treatment of waste remain a big issue due to mixed waste. The regions where the waste collection doesn't happen, the mixed waste lands up in nature. Such mixed waste is then also littered in water bodies and drainages. They contaminate the water and surrounding air in large quantities.

So, mixed waste causes difficulties and need more resources to handle it.

4.3 Less Sensitive and Self-centric Human Nature

Another aspect that is widely assumed that cleaning is the duty of only a few people appointed by the government or related sectors. That means, "we pay taxes, so the government should take care of anything that is beyond my personal belonging and beyond my own doorstep" is a common unintentional attitude. This has also a psychological aspect. We want our house to be clean and neat. So, what happens in the neighbouring yard is naturally out of our scope. Also not following rules and policies is a major hurdle. There needs to be a strong awareness and strong force on following the rules and regulations. The force not necessarily has to come from respected sectors, we as common public also can ensure that every other person is following the rules. But we hardly do that. We focus on own life only.

And this means the increasing self-centric human nature is also a root cause.

4.4 Poverty

The poverty is another root cause. The reason is for obvious reasons, poor people buy items which they can afford and plastic is one of those as it is cheap. The products which are not durable go early into the waste. There is no blame on people who are economically backward. It is just that they don't find any option to make a living. Also, usually, they stay away from education due to poor financial situations and do not learn what's going on in the world and the environment. Many times, they are unaware of the government rules and policies about waste management. The lack of finances keeps them with a lower standard of living which means lower cleanliness and hygiene standards. The sustainable solutions which are currently available in the market are not affordable by most of the population.

So empowering economic growth and per capita income is something we need to address to solve waste-related issues.

5 Addressing the Root Causes

This section points out the solutions for addressing the above-discussed root causes so that we are able to solve the problem area defined earlier.

The following are the points to elaborate on in this section.

1. Public Knowledge and Awareness
2. Adopting Sustainable Lifestyle Habits
3. Upgrading Education System
4. Building more Humane Culture
5. Sustainable Solutions – Investing in Waste
6. Decentralized Waste Management
7. Economic Growth and Increasing Per Capita Income
8. Clean Strategies, Laws, and Policies

5.1 Public Knowledge and Awareness

Public knowledge and awareness on a national scale is a very important step in this path. This is because the casual nature of dealing with waste has made people unaware of its effects on the environment and themselves. Such programs can be arranged by anyone – the government, schools, colleges, corporates, community groups, NGOs, etc.

Knowledge and awareness programs should cover the following aspects.

1. Awareness of the current amount of waste generation and landfill sites, and their side effects.
2. Knowledge about global environmental effects due to waste issues that we have and the crisis situations that are nearing if we do not act.
3. Information about where we stand in the globe about cleanliness and waste management and what we can do to feel proud of being Indian.
4. Awareness of national vision about waste management and cleanliness
5. Awareness of global vision of environment and sustainability, for example, Sustainable Development Goals (SDG) [6] declared by the United Nations (UN).
6. Knowledge about the current policies, rules, and regulations regarding waste management and consequences if they are not followed.
7. Knowledge about waste management pyramid and how to minimize the use of landfills with the help of it.
8. Knowledge about waste segregation and waste processing at the source.

9. Knowledge about a sustainable lifestyle for example zero waste lifestyle.
10. Awareness about the national agenda for contributing to Sustainable Development Goals (SDG) [6].

5.2 Adopting Sustainable Lifestyle Habits

Personal habits play a major role in dealing with waste-related issues. They are difficult to change as they belong to an individual and become part of their lifestyles. But we can change them by practice if we determine to do so.

We do heavy plastic consumption due to the convenience that it offers. "Use and Throw" culture has grown up in recent years because of the need and comfort that comes along. The family structures have changed as situations demand the same. And all these things are causing more trash. However, we don't really need to go back to old styles, we only need to live the new lifestyle in a sustainable way.

So, when it comes to the habits in the area of waste, the following key things can be considered to create a huge environmentally positive impact.

1. Strictly prohibiting single-use plastic items.
2. Find alternatives for plastic made consumer products. We can talk to our grandparents to learn from there age how they lived with less plastic. And we can use today's creativity and technology to bring those alternatives.
3. Keeping track of our household waste will make us sensitive and help to find the motivation to reduce waste.
4. When we buy goods and products, we must think of the necessity of buying and its environmental effects after use or disposal.
5. Instead of "Use and Throw" products, we should go for sustainable products that last longer and reused many times. Those include items made from metal, glass, cloth, stone, paper, wood, etc.
6. We should donate excess items that we do not use but somebody else can use it.
7. Zero waste is a new lifestyle which makes living with zero-waste generation. By following environment-friendly products, the zero-waste lifestyle is achievable.
8. Littering must never be a choice. As we keep our home clean, we shall strive to keep our surroundings clean.
9. And above all educate the next generations regarding all these aspects.

5.3 Upgrading Education System

The education system needs significant changes. The knowledge and awareness that we talked about in section 5.1 should be part of the education syllabus. That will motivate kids to become more responsible citizens when they grow up. We can shape kids and teach good habits at an early age when their brain is ready to accept those things promptly. And it is very essential that their brain captures the right habits in the right age for the reason that they develop their personality for the rest of their life.

When we want to develop changes in the education system following things are the highlights.

1. The Education system should have practical exposure along with mere theoretical concepts.
2. The syllabus must include the environment, climate change, waste management as a compulsory topic with theory and practical.
3. Various educational sessions must be arranged for students for current scenarios of waste and environmental effects.
4. Various events must be arranged to encourage students to express creativity and innovation keeping environment and waste management as a central topic.
5. Good habits can be cultivated very effectively in schools by teachers. So, the teachers must lead by example and teach by example.
6. Education shall include effects not only on the environment but on the lives of waste pickers and their families.
7. Like industry visits, similar visits for students should be arranged which include visits to recycling units, landfill sites for site observation and study, visits to waste pickers families to learn their situation and issues.
8. As like inter-school sports competition, there should be inter-school cleanliness events to bring the motivation-based cleanliness culture.

5.4 Building More Humane Culture

Dealing with waste is a "Humane" thing. The self-centric nature of people these days is one of the reasons behind waste-related issues. The life of today is fast and technology-driven. People have own comforts and problems to deal with. And due to that, there are less chances of willingness to deal with social issues. The self-centric habits lead to acting less responsibly in society. The littering, spitting, the garbage piled roads are outcomes of this nature.

We care little about the families who work for cleanliness and waste processing. That is not because we do not want to care but because we are not exposed to their situations and worries. It is totally unfair that the waste produced by a larger population is handled by very little population and for less reward in return. This quote says that all - "We can't have landfills forever and we can't ask others to accept our trash. – Jaime Lerner". We must think more responsibly and humanely. And we can do the following for that.

1. Taking responsibility for own waste.
2. Strictly avoid littering under any circumstances.
3. Keeping public places and washrooms clean.
4. Being more sympathetic towards waste pickers and their families.
5. Providing safety kits like gloves, masks, shoes, sanitizers to the people who handle the waste.
6. Demanding a safe and healthy infrastructure for waste pickers.

7. Respecting the people who clean our surroundings, public places, and handle our daily waste.
8. Following the rules and regulations around waste management.
9. Educating the people who deal with the waste about new rules and regulations so that they can expect it from people.
10. Supporting for the education of children from waste pickers families.
11. Incentivizing those people for the great contribution they do for society and the environment.
12. Arranging recreational and rewarding events for the people and the families who handle the waste.

5.5 Sustainable Solutions – Investing in Waste

Along with awareness, education, and public knowledge, another aspect is bringing sustainable solutions so that people can use them. Investing in waste is a new need for today. That investment could be in education as we talked about it in section 5.3 or it could be in the industry and market. People will be more willing to contribute if they find alternatives in the market. Industrialists will be more willing to change if they understand the need and that they will not lose anything but just shift the production to sustainable goods. This will not have any effect on labour or existing business outcomes if they plan properly.

While we promote a sustainable lifestyle, we need to bring up sustainable alternatives in the market and that to at affordable prices. We can make surveys of current sources of waste and contents of waste to find out what solutions need to be brought up first on a large scale. That information can be also made available to the public for awareness. Plastic is definitely on the radar that must be minimized as far as possible. That means the study of municipal solid waste and current consumer options can help us invest in the right solutions.

This will generate new income streams and also replace existing non-sustainable income streams to sustainable income streams.

The government subsidization and incentivization must be available to businesses and NGOs who work for bringing sustainable solutions. This will encourage people to use modern technology in a sustainable way to bring solutions. This way the overall national creativity and innovation will increase.

In all, we can think of following in this regard.

1. Promoting sustainable goods and businesses, recycling projects, waste to energy projects.
2. Subsidize or incentivize businesses and NGOs working in sustainable solutions and environmental issues.
3. Investing in education as mentioned in section 5.3
4. Investing in public awareness and training.
5. Make sustainable goods available at affordable prices for consumers.
6. The key areas like supermarkets, malls, vegetable and fruit shops, milk vendors, restaurants, medicals should restrict using plastic bags and plastic items as far as possible and invest in sustainable alternatives.
7. Investing in recycling units.
8. Investing in segregation units.
9. Composting projects for all the religious places for the used flowers and leaves.
10. We should invest in infrastructure for proper and hassle-free waste management. This includes segregation bins at households, commercial places, public places, advanced-automated waste processing machines, modern waste transport vehicles preferably e-vehicles.
11. We should give the market for by-products of waste management, for example, compost or energy generated after waste processing.

5.6 Decentralized Waste Management

At present, the waste flows from households or industries to specific landfill sites that are away from the city area. This adds the costs of transport, middle agents like waste collectors, and adds up to city traffic and environmental pollution. We need to break down this system to a decentralized one.

The decentralized waste management implies the waste is processed within the same block or zone where it is collected. This avoids unnecessary transportation to longer distances. The solid waste management should be done in such a way that, wet waste has a composting facility in each block and other waste recycling and processing units are present in each block. Each block will have authority in charge of waste management and cleanliness. The local issues and concerns can be solved in a faster way with this system.

This way we are going to invest initially some extra money but the benefits are long terms. This will also create a competitive environment where each block will strive to be cleaner and zero waste. Part of this also is to encourage people to participate in waste processing at the source itself like at households, at townships, at residential zones, etc.

5.7 Economic Growth and Increasing Per Capita Income

Poverty or uneven distribution of wealth is another root cause behind waste-related issues. Poverty leads to a lack of education, knowledge, and unsustainable product usage on a large scale.

The equation that the rich become richer and poorer remains poor must be changed. For that, not only the government but the people from society also should come forward to uplift each other economically. Creating an environment where everyone has a basic living standard, employment to fulfill that living standard, education, and health facilities shall be created. Here, education is the key.

We should leave a self-centric nature and think beyond that. Few rich people, NGOs, the government are already doing a lot for this, however, that is not enough. If we all grow together then only, we will bring clean and wealthy India together.

The villages, slums, low-income farmers, low-income workers, and similar people and their children should be our focus to work on eliminating poverty. Various schemes announced by the government must be easily accessible to these people without any intervention that they can take benefit of those. The employment rate should be increased. The local and global investments in business and market should be promoted and supported. In our country the majority of the occupation is farming, we must introduce advanced farming methodologies to increase the income of farmers.

We must not let poverty being used as a tool for vote bank and leading to corruption. And that is why knowledgeable people and those who can spare their money and time for uplifting the society must come together to help to solve the problem of poverty along with the government.

5.8 Clean Strategies, Laws, and Policies

Policymakers from administration and government will play a key role in establishing the right policies and ensuring the implementation of those policies. They have broader access to the public and have a larger perspective for nation-building. On one side they can uplift key people and sectors for national building activities and on the other side, they can make policies that will lead to sustainable national development.

The policies, the rules, and the national agenda should be made accessible to the public. And provisions for not obeying the rules should be strictly imposed at the beginning of the implementation so that people act responsibly.

The following are key areas where strategic planning and policy rollout can be implemented.

1. A complete ban on single-use plastic and strict implementation of it.
2. A complete ban on plastic water bottles and establishing an alternative for purified water availability in public and required places.
3. Subsidization for businesses working for the environment.
4. Tax exemption for cleantech companies.
5. Green tax (tax dedicated for environmental impact reduction activities) provision from existing tax deductions or via new policy.
6. Continuous monitoring systems for chronic areas where people litter on a large scale so as to warn and collect fine otherwise until they stop littering.
7. Continuous monitoring systems at major water bodies and over bridges where garbage bags are typically thrown.
8. Installation of durable dustbins on large scale on roads, public places, public transport systems in an accessible manner.
9. Return and refund policies for promoting recycling of materials for example glass bottles.
10. Supporting zero waste and zero landfill projects at various blocks.
11. Strict laws and regulations for the industry by-products which are considered as waste.
12. Laws for corporates and businesses to handle their waste.
13. Laws for waste producers to take charge of waste processing and cost involved. This will indirectly promote the reduction of waste generation.
14. Decentralization of waste collection and processing.
15. Introducing legal committee for environmental protection and actions.
16. Penalty system for breaking the environmental and waste related laws.

6 Conclusion

We have more population who generates waste than who cleans up and processes it. We have more population who can follow the laws and regulations than who can generate the laws and regulations. That simply means the major actors in the process of waste management are, "we, the people". If we start understanding our own waste and taking responsibility for it, we can free the government to act and invest in more important things. It's our country, it's our planet and we owe something for it. And keeping it clean for all the creatures and future generations is our sole responsibility. We don't need rocket science for acting responsibly and following laws regarding the environment and waste management. We need the mindset to act on the root causes as described in this paper and become more humane. We need to talk not just about reducing taxes but also reducing trash. We need a national mindset for bringing a clean culture, and we can achieve that if we align the entire population to that national agenda of sustainability and cleanliness.

References

1. https://www.epw.in/engage/article/institutional-framework-implementing-solid-waste-management-india-macro-analysis
2. https://www.pmindia.gov.in/en/major_initiatives/swachh-bharat-abhiyan/
3. http://amrut.gov.in/content/
4. https://en.wikipedia.org/wiki/Vasudhaiva_Kutumbakam
5. https://en.wikipedia.org/wiki/Bystander_effect
6. https://sustainabledevelopment.un.org/?menu=1300

Cultural Infrastructure and the Planning of Future Cities

Sindhu S[1], *and Reshmi M K*[1*]
[1]BMS College of Architecture, Basavanagudi, Bangalore (India).

Abstract. Culture is an important aspect of human civilization. Preserving and giving value to the cultural heritage of a region can pave the way for local as well as regional development. This includes tangible, intangible and the natural heritage of cities. It is necessary to develop a cultural infrastructure plan along with other aspects such as transportation, built, green and grey in frastructure. Cultural infrastructure refers to places where culture is experienced, participated in or showcased in. This includes the existing cultural heritage of a place as well as the planning of spaces for cultural stimulation and involvement. With the advent of fourth industrial revolution smart cities are gradually becoming the way of life across the world. The Smart City uses Information and Communication Technology (ICT) and Internet of Things (IoT) to effectively manage transportation, water services, waste disposal, energy and other community services. In this scenario it becomes imperative to have strategic infrastructure planning. Indian Government has recently launched the Smart City Mission in India and several smart city projects are underway across the country. It becomes necessar y in this context that Indian cities with their rich tradition and cultural heritage do not lose their unique identity in this process of transformation into smart cities. Even as smart city projects stress the need for heritage preservation there is a lot of ambiguity in how they can be integrated and used to advance urban intelligence. The technologies of the smart city have considerable potential to be used for the management and enhancement its cultural heritage and can help in the creation of a cultural infrastructure plan. This paper will examine the significance of cultural infrastructure in future cities and how it can be integrated into the city planning process of Indian cities through the study of relevant case studies from around the world.
Keywords: Cultural infrastructure, Heritage, Smart cities, City planning

1 Introduction

As the world is getting increasingly urbanized there is a lot of pressure on the existing urban infrastructure and especially on the older heritage cities. As new cities are being planned and existing urban areas are getting renewed in the urban development process cultural infrastructure is usually given the least importance. The advent of newer technologies and emergence of smart cities have presented several opportunities in urban development. But as cities across the world are getting smarter the cultural aspect of cities are many times not given much consideration especially in the context of Indian cities.
The cultural heritage in cities can be categorised as
i)Tangible :- This includes movable heritage such as paintings, sculptures etc as well as mmovable heritage such as monuments and historical sites.

ii)Intangible :- This includes oral traditions, performing arts, festivals, customs and rituals.

iii)Natural :- This include natural features with cultural associations, geological formations including rocks, hills etc, sacred groves etc.

All these above categories can be found in varying ratios in all urban areas. They usually coexist along with modern infrastructure and give unique identities to the cities. They also form the crux around which the cultural life of the city revolves. They are aided in this by other structures including museums, art galleries, open spaces etc which together form the cultural infrastructure in the city.
Smart city concept is a comparatively recent trend where the ICT and IoT is being used in the urban development process. Vast amounts of data are collected and organized and made accessible to the population which enables sustainability and better management of cities. This can be used to enhance various aspects of cities including transportation, environment and overall well being of people. Smart city initiatives should be adapted to to the local context, building on existing assets of the city and the identity of place (Angelidou, 2016)

2 Cultural Infrastructure

Cultural heritage both old as well as new form important narratives in the city. But unfortunately these spaces are

*Corresponding author: reshmimk@gmail.com

rehearsal areas or any space where art , music or any other creative endeavours are generated. They can be full time or part time, single purpose or multipurpose , historical or contemporary structures or spaces. They can be either built for purpose or adaptively reused.
A cultural infrastructure plan helps to support and enhance these cultural spaces in a city. It becomes very important to recognize and incorporate this aspect into the planning process of cities by conserving and highlighting existing cultural assets in the city and at the same time planning and incorporating cultural spaces into future cities which will contribute to enhance creativity. The importance of a Cultural Infrastructure Plan is being recognized by many cities including London, Amsterdam and Toronto. Indian cities with their rich cultural heritage are yet to recognize the necessity of a Cultural Infrastructure Plan.

3. The Smart City

3.1 The concept of smart cities

The Smart City is the realization of Fourth Industrial Revolution in the spatial context. In Smart Cities 'People to People', 'People to Things' and "Things to Things' are connected in real time. Data collected from the citizens, devices and other assets are used to effectively manage the city. The Smart City uses Information and Communication Technology (ICT) and Internet of Things (IoT) to effectively manage transportation, water services, waste disposal, energy and other community services.

3.2 The Indian Scenario

India is the fastest growing economy and the second largest populated country in the world. The educated human resource that it posseses makes India one of the key players in shaping the fourth industrial revolution. As a part of the procedure towards e-governance India has been steadily enrolling its citizens into it's national database. Aadhar is the largest biometric database in the world.
The recent smart phone revolution and India's software capabilities enable it to maintain the position as an IT Superpower and 'info-nation'. India has been able to capitalize on the third industrial revolution and starting in the 1980's has managed to become a leading sourcing destination for the global IT industry. India also has the highest e-commerce market and the highest number of technological start ups.
Prime Minister Narendra Modi recently announced the creation of Centre for Fourth Industrial Revolution at the World Economic Forum meeting in order to take advantage of this situation and steer India as one of the frontrunners in this revolution. The centre which will be based in Delhi will be part of a network which includes USA, China and Japan and will be collaborating with the National Institute for Transforming India (NITI) . The centre will be preparing protocols and policies for emerging technologies mainly focusing on artificial intelligence, blockchain and drones.
India today is a country with the highest growth of internet access and has also managed to digitize its economic, social and political systems at an unprecedented scale. National Association of Software and Services Companies (NASSCOM) estimates that AI has the potential to contribute $957 billion to the Indian
Economy according to the Artificial Intelligence Primer 2018.Th main sectors which would benefit from AI would be agriculture, education, healthcare and infrastructure.

3.3 The Smart City Mission in India

'100 Smart Cities' mission was launched by the Government of India in 2016. According to the Smart City Mission , these cities will have:

i) adequate water supply,
ii) assured electricity supply,
iii) sanitation, including solid waste management,
iv) efficient urban mobility and public transport,
v) affordable housing
vi) IT connectivity and digitalization,
vii) good governance, especially e-Governance and citizen participation,
viii) sustainable environment,
ix) safety and security of citizens, particularly women, children and the elderly, and
x) health and education.

100 cities have been selected so far for this mission and work is underway in most of these cities. It is still a work in progress and will take few more years for the work to get completed.

4 Cultural Infrastructure In Smart Cities

Cultural Infrastructure plan in any city should recognize and address the local context. These should include smaller and larger hubs which nurture talent and also facilitate community connections and social cohesion.

4.1 Cultural Infrastructure in Smart City London

Fig. 1: Map of London (**Source :** London.gov.uk)

London is one of the oldest cities in the world. Situated on the banks of River Thames it has several historic layers.

4.1.1. Smart City Master Plan

London has set the goal to be the smartest city by the year 2020. Smart City London Plan has five key aims which include

i) User centered designs

ii) Open and secure city data

iii) Connectivity and Smarter streets

iv) Digital capability improvement of citizens

v) Smart City Collaboration

4.1.2. Smart City Master Plan

London has been a city which has been a frontrunner in developing a Cultural Infrastructure Plan in order to nurture creativity and has recently released Cultural Infrastructure Plan which is being planned alongside the smart city plan. The cultural infrastructure plan focuses on cultural spaces and how the Capital should best plan to meet its cultural needs

The seven prong action plan recently released by the Greater London Authority include the following points.

i) Understanding location of existing cultural infrastructure through a cultural infrastructure map.

ii) Planning and designing new cultural infrastructure

iii) Providing infrastructure for a world class city

iv) Increasing investment

v) Policies to enable culture and creative industries

vi) Providing training and networking

Cultural Infrastructure map helps to locate cultural infrastructure along with transport. Planning and demographic data. The user scan download information from this interactive map and also suggest missing infrastructure as well as edit information on existing infrastructure, check travel times etc.

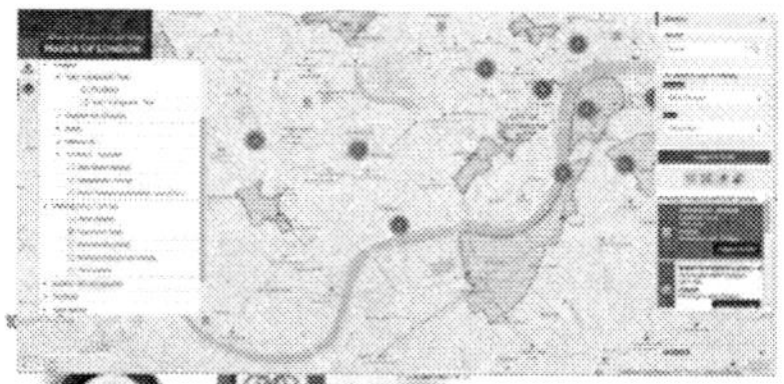

Fig. 2. Cultural Infrastructure Map, London (2019} **Source :** London.gov.uk

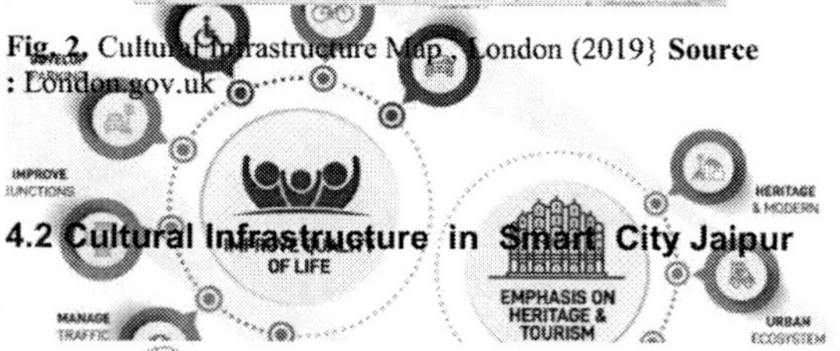

4.2 Cultural Infrastructure in Smart City Jaipur

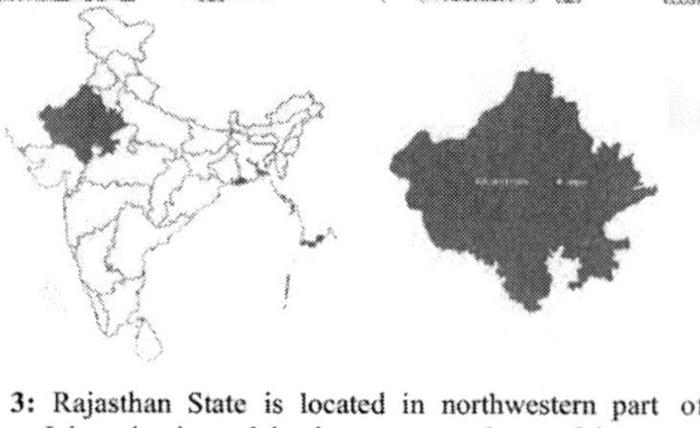

Fig. 3: Rajasthan State is located in northwestern part of India. Jaipur is situated in the east-central part of the state.

4.2.1 JAIPUR– History and Background

It was built by the Maharajah Sawai Jai Singh II. In 1727, Jaipur replaced Amber as the capital of Rajasthan due to the growing population and increasing scarcity of water. The suitability of the terrain, righteous defensive position, communication, and availability of water made Jaipur a suitable location for the new capital.

The new city capital comprised of huge fortification walls

along with seven strong gates. The town planning of Jaipur was very advanced for its time and the best in the Indian subcontinent in that period. When the Prince of Wales visited Jaipur in
1853 the whole city was painted pink to welcome him. Even today the buildings in the city are painted pink and give a distinctive appearance to the city.

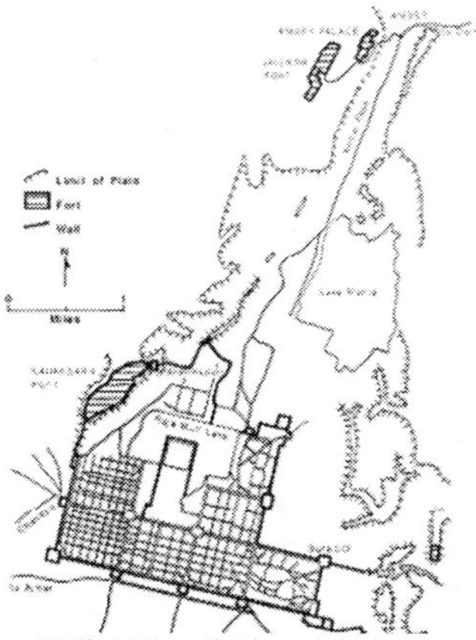

Fig. 4 – Walled City of Jaipur (1865 map of Jaipur) Source : Survey of India

The emergence of British government led to expansion of the city as they setup their agencies within fort resulted in walled city being moved out as the rapid growth could not be accommodated in the existing structure and the city couldn't cope with the force thrust on it. The expansion led to destruction of socio-economic structure, centripetal to centrifugal to suburbs, Dualism – old and new by the band of low status housing, cultural problem and haphazard growth.

And later the city has been expanding and evolving over a period of time yet retaining the cultural Identity of the city.

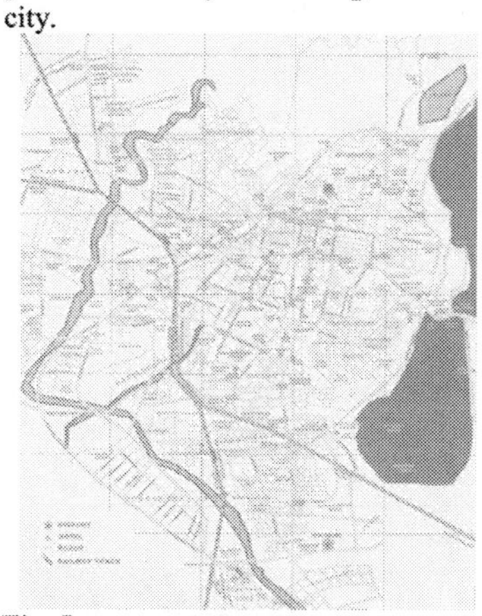

Fig. 5. Expanded City of Jaipur

4.2.2. Smart city initiatives in Jaipur

Jaipur Along with New Delhi and Agra which both lie approximately 250 km away, it forms the 'Golden Triangle' tourist circuit.
Among the list of the 109 smart cities of the country, Jaipur is one of those aspiring small cities which has been chosen by the Smart City Mission of Ministry of Urban Development (MoUD), Government of India launched in June 2015.

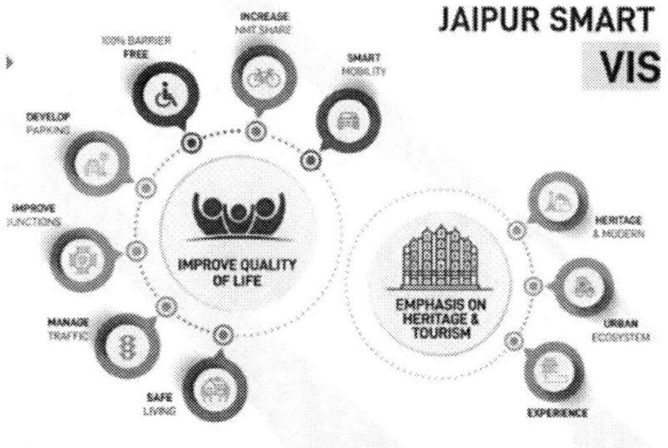

Fig. 6 : Jaipur Smart City – Vision Source : Jaipur Smart City LTD

4.2.3 Smart city: Cultural Infrastructure and treatment of heritage structure - blending modernity with Heritage

Smart city initiative for Heritage and Tourism serves as the purpose for protection of the cultural heritage, but also attracts tourism and promotes the economy of the place. Thus, important tourist destinations are also major urban centers and it is essential to protect the urban identity. Rajasthan is one of the major tourist destinations in the country because it has a large cultural resource.
The 18th century city of Jaipur city known for its architectural beauty and vibrant culture, is moving towards syncing its historic magnificence with modernity, as projects under the Centre's Smart City Mission gain momentum. The ancient walled city has been chosen for Area-Based Development under the Smart City Mission where a blend of heritage with modernity is taking shape through various projects such as :

Restoration and Adaptive reuse of heritage structures
Rejuvenation of Talkatora lake, Baoris
Heritage walks and Bazaar Street facade improvements, green roof tops
Integration with common mobility card for monument entry payments
Heritage App with QR Code for monument information
Training and rating of Tourist guides, Tourism Police
Promotion of traditional handicraft and textile workers
Night market and vending zones
Tourist interpretation centre

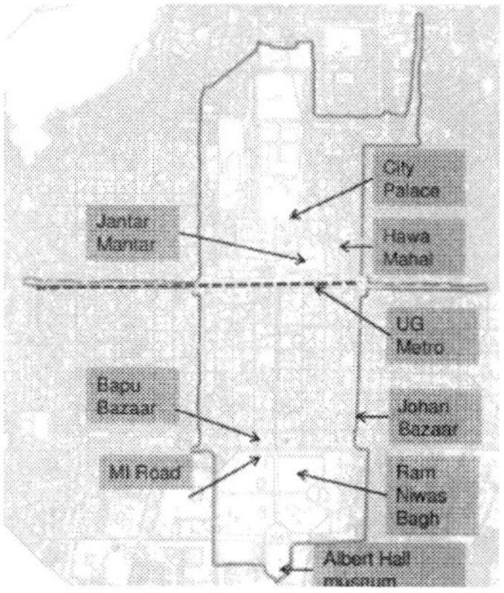

Fig. 7 : The chosen area for area based development is two chowkries of the walled city. Source : Jaipur smart city proposal

Conclusion:

Cultural infrastructure is an essential component of cities and should necessarily be a part of smart city initiatives. Heritage structures are important aspects of this infrastructure. There are several ways through which cultural heritage can be assimilated as part of smart city projects. Smart city applications can be crucial in the monitoring and management of heritage and this has been effectively incorporated in several cities in Europe and America. Indian smart cities are yet to incorporate this effectively into their planning process. As seen in the Jaipur case study there are initiatives being taken in a few cities, but most smart city projects are doing very little to fully utilize the potential of smart technologies and develop of the cultural infrastructure in those cities. The value of cultural infrastructure and heritage management need to be given more emphasis in the ongoing and upcoming smart city initiatives in India. Each city should develop a cultural infrastructure plan which should be implemented along with other developmental activities. The cultural aspect of vibrant Indian cities should not be ignored in the urban development process as it brings vitality and unique identity to these cities.

References

1.M.Angelidou , The International Archives of the Photogrammetry, Remote Sensing and Spatial Information Sciences, **XLII**, (2017)

2.Angelidou, M, Int. Jrnl. of Social Science Studies, 18, (2016)

3.Greater London Authority, *Cultural Infrastructural Plan* (2019)

4.Riganti, Int. Jo. for Arch. Res. ,16, (2017)

Holistic approach for cultural infrastructure with reference to future smart cities

Chaitanya Gokhale[1]

[1]MIT ADT University, Loni Kalbhor, Solapur Highway, Pune 412201, India.

Abstract - To take the initiative of smart cities to new heights is a responsibility of every citizen. Bona fide progress will be seen when people's approach towards their city will change through awareness and education about the prevalent systems. This paper focuses on improving the cultural aspects of a smart city, especially visually, and on improving governance with reference to the 21st century. A majority of the information deals with the must-haves of smart city, later moving on to more specific problems. In the following paper, evaluation, analysis, ideas, innovative solutions along with comparisons and examples from life have been suggested with a view of enhancing the pre-existent, and further developing the cultural infrastructure of a smart city. This is proposed in a way that will help old traditions and historical heritage keep pace with modernism and other urban developments in the age of internet without changing their essence. This, in turn, will lead to touristic developments. Attention has also been paid to heritage restoration and its methods. Special focus has been given to the aesthetics of proposed, self-designed signage systems of the city. It is hoped that the insights presented herewith encourage greater enthusiasm towards art and encourage appreciation towards the culture and heritage of a smart city. This will add value to the quality of life of its citizens. *Keywords: Smart City, Culture, Heritage, Aesthetics, Innovation, Governance*

1 Introduction

Let's begin with the most important question. What really makes a city "smart"? [1],The basic infrastructure needed for any city to make progress by using "smart solutions" is undoubtedly a major part of the answer. Some important areas where these smart solutions apply include the following but are not limited to:

- Clean, sustainable environment
- Assured electrical supply and safe water supply with 100% treatment of waste water
- Sanitation and solid waste management
- Efficient urban mobility and public transport
- Robust IT connectivity
- E-governance and citizen participation through public information and grievance redressal
- Renewable, efficient sources of energy and green building
- Safety and security of citizens
- Effective Visual Communication
- Aesthetic Pleasure

It is necessary to understand that the focus of such technological and infrastructural advancements is to ease human load. Not necessarily making the pace of life faster, but allowing citizens a peaceful space where they can have a rather decent quality of life, with the freedom to live their way, responsibly.

The art and culture of any city play a vital role in the quality of life of its people. [2,3],They have immediate as well as indirect effects on daily life apart from long-term effects. This paper talks about the different ways in which we can enhance the same, ranging from details like language to broader topics like city attractions. Of course, it is not an individual effort alone and a lot of government aid will be required in achieving idealistic scenarios. The citizens, however, must take centre stage regarding the smooth functioning and preservation of their art, culture, heritage and all the initiatives taken in that direction.

2 Attributes of a Smart city with respect to Art, culture and heritage

The next important question is -Why do we need to give a great amount of importance to art, culture and heritage? [4,5], It is necessary to understand that these three collectively result in the major takeaways for any citizen or visitor in the city. For example, a lot of Chinese cities render excellent souvenirs to people and the fact remains that their porcelain goods are known the world over. Another noteworthy aphorism is that the necessary environment needed to enrich a being in endless ways is provided due to the beauty and richness of the trio.

It doesn't come as a surprise that beautiful things make us happy. We may not be able to identify what beauty really is, but we manage to appreciate it when we see it. It is as though people have a basic common sensibility for beauty. For example, there have been

*Corresponding author: **chaigo@gmail.com**

https://doi.org/10.1051/e3sconf/202017005002

experiments where people were to distinguish between real and artificial abstract paintings. Some were original works by Mondrian and Pollock, painted based on strict principles like fractal patterns, whereas the artificial ones were not. A majority of the subjects selected the original works. A similar experiment was conducted on distinguishing originals from abstract artworks that were made either by children or animals. The artworks whose patterns were carefully planned rather than being unstructured were again selected by the participants. Smart cities must be planned carefully for similar results.

Beauty must be considered as one of the most important attributes of a smart city. [6,7], While shaping urban man-made environments and the objects in it, beautification has often been neglected by us in favour of functionality, cost or efficiency. We build concrete jungles where we don't even prefer living. Plain, standardised boxes beside each other. Humans tend to dislike arrangements that do not interest them much. Eye tracking software has shown that people keep focusing on details and ornaments of architecture as compared to swiftly brushing over blank walls. Such dull facades make us miserable. Experiments with skin sensors have shown that seeing them makes us feel bored and uncomfortable. Such boredom, in its turn is associated with increased pulse rates and stress levels. [8], Numerous other experiments have also concluded that aesthetically pleasant surroundings can improve our well-being, cognitive function and mindset. We react mentally and physically to everything in our immediate environment. This is not just visible, but it has been measured too. Beauty is one of those factors that regulates our well being to such an extent that creating objects of utility in a beautiful manner can make them much better! A 2017 study by a hospital employed observing and interviewing the patients as its research methodology to understand factors that help them recover. It was deduced that paintings or art in the areas that they occupied made them feel comfortable and happier about their stay in general. Pleasant surroundings led them to feel better physically! In a similar manner, beautiful music (especially classical music), that is a very integral part of culture, has measurable calming and soothing effects.

Aesthetics also affect us daily. Our happiness, in general, is directly associated with, and depends on aesthetics. Main factors affecting the happiness of adults have been studied (Leyden, Goldberg A., Michelbach, 2011), which reveal a rather unanticipated finding. Apart from factors such as good health and a harmonious family life, individual happiness is affected by how beautiful you find the city you live in. Beauty ranked above cleanliness and safety too! Humans have now become fine tuned to process visual input and assess the surroundings. We are understanding more about how beauty or aesthetics affect us. Beauty satisfies the innate need for meaning in life. It is a sense of feel-good that is an end in itself. It therefore seems like a practical opinion to give more importance to beauty in the new environments that we are creating.

As such, everybody has different ideas and expectations about their smart city. Thus no globally accepted definition can be formulated for it. But as discussed in the introduction, there definitely are certain common elements which contribute in making a smart city. There are major factors like infrastructure, governance, transportation, etc. All these things make a city livable, but they somewhere fail to make a city valuable. The difference here lies in experience and the ways in which a city manifests its rich experiences.

Value is both, an immediate and a long-term notion, one based on sustainable, affirmative experience. Cities, whether old or new, all have a certain story to tell. And, like any brand, it is this immersive experience or storytelling aspect that determines their success. Art, culture and heritage serve as the threads through which the plot of the story is built up and also serve as the characters through which the story unfolds its course. People treasure their rich past, be it their rich kingdoms and dynasties or their long-drawn struggle for freedom. And they also like to preserve and protect everything that reminds them of their heritage. Not only do people treasure their heritage, but they also learn a lot from it. For instance, some very important modern day American buildings like The Lincoln Memorial, The United States Capitol and the Supreme Court Building are actually based on classical Greek architecture and draw inspiration from some of the motifs and stone carving patterns that were used extensively in Mogul architecture are being emulated for various purposes till date. These examples show how much we depend and learn from art, heritage and cultures. The three intertwine from different parts of the world and bring people closer. Together, they make life interesting and break the everyday monotony.

Every city has its own culture and heritage. It reflects in every segment ranging from transport, buildings, shops, public areas, open landscapes to tourist destinations and monuments. Every person residing there represents the city. Public transport vehicles, signage, colours and typefaces play a vital role in visual aesthetics of the city. They are mainly connected with how people or visitors perceive the city, because art and culture have a lot to do with people's mindsets. The visual aspect is very important as it is the first impression of the city. It represents the culture or flavour of the city. The local food, monuments, signage systems communicate directly with the people of the city or those visiting it.

Major smart cities like London, New York have managed to transform their culture into visual communication very effectively. City of London reflects a blend of tradition and modern, whereas a city like New York reflects modern culture. In cities like London and Paris, the facades of many buildings have been maintained in their original state to give the city an even vintage appearance. This goes a long way in helping the city maintain and restore its heritage. On the other hand, New York is known for its sky-scrapers and graffiti culture. The walls of that city have been efficiently treated as the largest canvases that can be painted on. Another notable example is that of a combined effort by a group of artists, Pune Municipal Corporation and Pune Biennale Foundation, who painted India's largest mural at the Yerwada Central Prison in Pune City [10,11,12].

Fig. 1 India's largest mural, Yerawada Jail, Pune.

3 Art, Culture and Heritage in a Smart City – problem areas – weaknesses, Aesthetics Solutions - signage systems, Tourism solutions, Art narrative

Visual narration should be consistent everywhere in a smart city, especially at tourist destinations. The city must strive for cleanliness everywhere. For the tourists and visitors, whatever they see is what they get with any new city. In London, all the signage systems are uniform and maintain a systematic visual feel everywhere (Railway Stations, maps, city guides). As far as possible, a universally accepted language and a local language (if any) must be used for important signage, as this makes understanding easy for tourists and citizens alike. Mumbai city, for instance, uses English (considered by most as a universal language) as well as the Devanagari script for most of its important signage.

After understanding the flavour of the city, it is equally important to translate that into the appropriate visual language. For that, a selection of suitable touch points plays a very important role in communication. For any smart city, Airports, Railway Stations, Bus Stops, souvenir shops, restaurants and monuments can be treated as the tier-one touch points. These touch points contribute a lot when it comes to defining the visual language and style of that particular city. The visual communication should convey or reflect the core values of the city in relevant domains like historic, artistic, infrastructural or cultural.

Smart cities in different parts of the world have started including public art installations, pop arts and street displays. These social and personal endeavours leave individuals with an increased sense of pride, belonging and regard for their city, and, in turn, the public benefits from an increased sense of awareness and understanding. The charged atmosphere produced by such endeavours increases the enthusiasm of all. It leads to a feeling of overall well-being and a sense of being connected, which is not possible in a dry, dull environment where people tend to only focus on themselves or their work and are more or less, aloof. To execute such initiatives on city level, good governance, public awareness, right communication, ample opportunities with proper funding and economics are the key factors. If these are not available, then potentially brilliant ideas, very unfortunately, happen to fail due to lack of awareness and correct execution. Therefore, the whole process, from ideation to execution must be fairly well planned with proper government backing. The same applies to the processes of the various other segments of a smart city.

Once the city has developed all the appropriate infrastructure and is sound enough at its own level, then cultural exchange programmes can be undertaken by schools, colleges, universities, special government bodies, NGOs or other bodies that are successfully capable of organising them. They may be national or even international programmes. It is a humble suggestion that these programs can also be added to institutional curriculums at affordable rates. Such vital programmes lead to the participants mutually taking back a plethora of information and memories about each other's culture and a sense of association with the same. This goes a long way in promoting tourism.

3.1 Given below are some proposed visual design solutions for Pune City on similar lines:

3.1.1 City Logo

Every city has its own identity. Bengaluru city has its own City Logo. It is part English and part Kannada reflecting the true identity of the city. This mixed typography of two languages shows that the Silicon Valley of India is a city that is global as well as local in its culture and lifestyle.

Fig. 2 Bengaluru city logo

'Be' and 'U' in this logo have been emphasized in red. This is the idea of the logo. The creators of the logo say Bengaluru is a city that allows you to be you. [13]

Similarly, Pune city is known as IT hub as well as Cultural centre. So here is an Ambigram (Flip Script) Logo of Pune created by me.

Fig. 3. Pune city ambigram

3.1.2 Signage systems

Signage should carry the same flavour of the city. Here's my design suggestion for Info- signage plaque at Historical monument Shaniwar Wada.

Fig. 4. Shanivar Wada, Pune [14]

Fig. 5. Calligraphic Devanagari signage design for Shanivar Wada

3.1.3 Souvenirs

Souvenirs must be aesthetically pleasing and some may even follow a mood board (that suits the city's culture and the colours that represent it) for collective gifting options.

According to CNN, travellers spend a lot of time and money thinking about souvenirs. Looking at the global economy, over 400 million aggregated hours are spent in gift shops, raking in more than $2 billion PER YEAR! In this entire segment, India's contribution is very much negligible. [15]

3.1.4 City Bus

City tour buses must look their best and reflect the culture of the city they represent. They must be equipped with all the necessities like a knowledgeable tour guide, brochures, updated city maps, etc. for travel is a major area where tourists could face problems.

4 Conclusion

With the making of smart cities, we are witnessing the creation of a completely new era, and not just witnessing it, but creating it as well. In doing so, we must be very cautious about the direction in which we take things. Every little step we take will have a huge impact on the future. For instance, good thought will have to be put into the extent to which we must incorporate techniques like the Internet of Things (IoT) in order not to disrupt the balance between humans and technology or to make humans too dependent on technology. Moving forward in harmony with nature has always been and will always be the key to real growth.

We are fortunate enough to have the perfect chance to responsibly secure a better future for ourselves and for the generations to come. Extremely cautious planning will be required in every sector for the interlinked and overall development of a smart city and that is what will truly make it count.

References

1. What is smart city, (2017, May 09). Smart Cities Mission. Retrieved from http://smartcities.gov.in/content/innerpage/what-is-smart-city.php
2. G. Hoang, Application of Decision-Making Methods in Smart City Projects: A Systematic Literature Review. *Smart Cities,* **2**, 433 (2019)
3. Art in the Smart City, (n.d.). Retrieved from https://www.mtart.agency/press/read/art-in-the-smart-city/
4. S. Sharma, *Smart Cities Unbundled* , 35 (2018)
5. What is a smart city. (2015, May 02). *Times of India.* Retrieved from https://timesofindia.indiatimes.com/what-is-a-smart-city-and-how-it-will-work/listshow/47128930.cms?from=mdr
6. A. Firodiya, *Smart City,* 13 (2015)
7. Kurzgesagt, (2018, October 23), Why Beautiful Things Make us Happy , Retrieved from: https://www.youtube.com/watch?v=-O5kNPlUV7w
8. Philip C., (2018, July 04) Five Beautiful American Buildings Based on Classical Greek Architecture. Retrieved from https://usa.greekreporter.com/2018/07/04/five-beautiful-american-buildings-based-on-classical-greek-architecture/
9. L. Suzuki, (2018, November 29) Machine learning and AI – ensuring fairness in smart cities. Smart City Research – UCL – UK. Retrieved from http://www.smartcityresearch.com/
10. Mihir B, (2017, February 26) India's largest mural unfolds at Yerawada Jail. Times of India. Retrieved from: https://timesofindia.indiatimes.com/entertainment/marathi/movies/news/indias-largest-mural-in-pune/articleshow/57160142.cms
11. Payal M, (2017, November 15) India's Largest Mural Livens Up The Walls Of Maharashtra's Largest Jail. Retrieved from: https://homegrown.co.in/article/801972/indias-largest-mural-livens-up-the-walls-of-maharashtras-largest-jail

https://doi.org/10.1051/e3sconf/202017005002

12. Yerwada Mural (n.d.) JPEG, Pune Biennale. Retrieved from https://www.punebiennale.in/images/our-work/yerwada-mural/Song-of-the-City-1.jpg
13. Madur, (2018, January 02), Bengaluru Logo – Bangalore Gets Its Own Brand Identity. Retrieved from: https://www.karnataka.com/bangalore/bengaluru-logo/
14. Shaniwarwada, (2019), JPEG, Accessed on 20 November 2019, Retrieved from: https://in.bmscdn.com/nmcms/events/banner/desktop/media-desktop-shaniwarwada-2019-1-12-t-2-36-26.jpg
15. Alan B, (n.d.) The Grand Significance Of Cheap Souvenirs. Let's Roam. Retrieved from: https://www.letsroam.com/explorer/the-grand-significance-of-cheap-souvenirs/

 https://doi.org/10.1051/e3sconf/202017005003

People Centric Smart Cities

Sunil Rai[1], *Mukta Deshpande*[1*], *and Geeta Thakur*[2]
[1]MIT-ADT University, Pune, India 412201
[2]CGP, 1302, AH, SVR, A(W), Mumbai, India 400058

Abstract. Smart Cities Project is been undertaken in mission mode by the Government of India. The endeavor is to plan and build / rebuild pollution free, eco friendly living spaces with the efficient means of Transportation, Communication, Governance, Health Services and other civic requirements. There is also an emphasis on giving identity to these cities based on its main economic activity. The aspects of peaceful and happy living with regard to values and universal common good of mankind does not come out clearly in the definition and other mission statements. This paper endeavors to highlight the importance of the Culture and Spirituality in creating the cities that will promote Peace, Harmony, Happiness and the Virtue of co-existence amongst the dwellers in keeping with the rich traditions and value systems inherited from the glorious past. It is emphasized that the Smart City be given the identity which is representative of its character, culture and the unique value that it offers. The quest to build smart cities from the standpoint of enhanced economic activity and operational efficiency with the use of High-end integrated technologies for automation, Artificial Intelligence and Internet of Things the importance of "Culture" as bedrock of "Smartness" has been subdued. It is proposed that the richness of Culture, strength of Spirituality and importance of Identity based on values and character be emphasized in planning and building smart cities. A framework to assess the contribution of a People Centered Smart City in its role as "Culture Transmitter" in influencing the growth, sustainability and wellbeing of inhabitants is also presented.

1 Preamble

History bears testimony to the fact that cities that have a value system and intrinsic Culture that resonates with the belief system of the people over several generations, have been the influencers of growth and wellbeing of the society. Ancient civilizations were established and propagated with very fine examples of cities that have played the role of harbingers of Culture and beliefs: Athens, Rome, Mohenjo Daro, Harappa, Patliputra, Hastinapur, London to name a few.

2 The smart cities project by Government of India

The ambitious project-"100 Smart Cities Mission" was launched by Prime Minister of India on 25 June 2015.[1][2] A total of ₹98,000 crore (US$14 billion) has been approved by the Indian Cabinet for the development of 100 smart cities and the rejuvenation of 500 others. ₹48,000 crore (US$6.9 billion) for the Smart Cities mission and a total funding of ₹50,000 crore (US$7.2 billion) for the Atal Mission for Rejuvenation and Urban Transformation (AMRUT) have been approved by the Cabinet.[3][4]

2.1 Features of Smart Cities

Some typical features of comprehensive development in Smart Cities are described below. [5]
a) Promoting mixed land use in area based developments–planning for 'unplanned areas' containing a range of compatible activities and land uses close to one another in order to make land use more efficient.
b) Housing and inclusiveness - expand housing opportunities for all;
c) Creating walkable localities - reduce congestion, air pollution and resource depletion, boost local economy, promote interactions and ensure security.
d) Preserving and developing open spaces - parks, playgrounds, and recreational spaces in order to enhance the quality of life of citizens, reduce the urban heat effects in Areas and generally promote eco-balance;

*Corresponding author: mukta.deshpande@mituniversity.edu.in

e) Promoting a variety of transport options - Transit Oriented Development (TOD), public transport and last mile para-transport connectivity;
f) Making governance citizen-friendly and cost effective - increasingly rely on online services to bring about accountability and transparency, especially using mobiles to reduce cost of services and providing services without having to go to city offices;
g) Giving an identity to the city - based on its main economic activity, such as local cuisine, health, education, arts and craft, culture, sports goods, furniture, hosiery, textile, dairy, etc;
h) Applying smart solutions to infrastructure and services in area-based development in order to make them better.

2.2 The Smart City Project Implementation

The "Smart City Mission" project is being executed by various State Governments, local smart city project officers under the guidance of Government of India satisfactorily and on schedule says Ms Rupal Sharma, IAS, Project Director and Deputy commissioner, Smart city project office Pune, during the interview with the Authors.[6] She also informed that all the features as envisaged by the Smart City Mission Project, have been included in the master plans.
The interactions the authors had with several officials in the office, indicated inclusion of Technology aspects of most of the features regarding mobility, Governance, civic utilities, information etc. The other aspects of rejuvenating open spaces, theme parks, walk ways, waste disposal and other public amenities have also been catered for comprehensively in the execution plans. The information regarding promoting culture, spirituality and unique identity of the city was not evidently available either in the smart city offices or other literature published in public domain.

2.3 The Smart cities of Future

The smart Cities of Future for Sustainable development and wellbeing of the society need to be defined appropriately. The authors interviewed Honorable Professor Dr Vishwanath Karad, the founder of the MAEER's Trust, the MIT Group of Institutions, the UNESCO chair for World Peace, President of the MIT ADT University, Loni kalbhor, Pune and the proponent of the Philosophy of "The Union of Science and Spirituality" to achieve peace and well-being of mankind. Professor Karad questioned the limitation of the very definition of Smart Cities.[7]
Professor Karad along with his team of World Peace Mission which included the luminaries like Dr. Vijay Bhatkar, Padmashree (The noted Scientist "Param Super Computer" Fame) and Dr. R. A. Mashelkar, Padma Vibhushan (The noted Chemical Engineer and champion of Innovation, The fellow of the World Academy of Sciences), opine that the smart city is the one which fosters Peace and Harmonious living through Rich culture, Strong value systems and Practiced Spirituality aimed at the wellbeing of all with a spirit of equality and sustainability. The authors therefore propose the model of People centric Smart City with emphasis on aspects of "Fostering Spiritual consciousness Artifacts", "Culture transmitting spaces synchronized with Local culture", and "Identity of City as Transmitter of Culture and Uniqueness" In addition to the well understood and executed themes of "Energy & Environment Sensitivity" and "Technologically Smart Civic Life".

2.4 The Proposed Model of Smart cities of Future

The authors developed the Model proposed for building smart cities of future comprising of five components. The first Two components are addressed appropriately while planning and implementing the smart cities projects in Indian context in particular. However the other three components pertaining to Spirituality, Culture and Unique Identity of city need emphasis and attention.

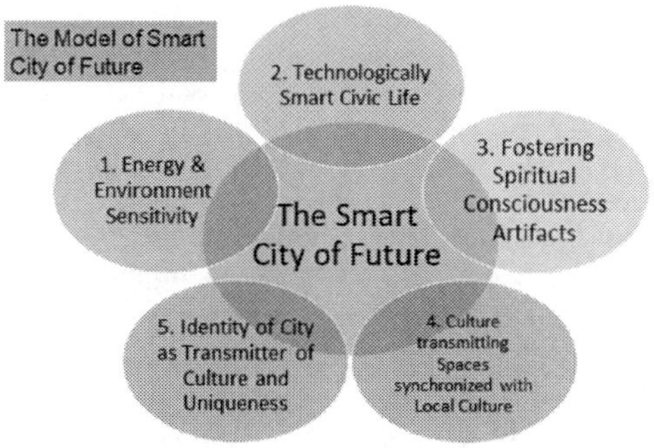

Fig. 1 Proposed model of people centric Smart City of the Future.

3 Spirituality for Peace and Sustainable Development

There is no one definition that can justifiably enumerate the meaning and sense of Spirituality.[8] The Cambridge dictionary defines spirituality as, "The quality that involves deep feelings and beliefs of a religious nature, rather than the physical parts of life." Collins dictionary says, "Spiritual means relating to people's thoughts and beliefs, rather than to their bodies and physical surroundings." The Oxford dictionary says, "Relating to or affecting the human spirit or soul as opposed to material or physical things."

3.1 The experts have put together the definition from practice perspective. [9]

According to Mario Beauregard and Denyse O'Leary, researchers and authors of The Spiritual Brain, "spirituality means any experience that is thought to bring the experiencer into contact with the divine (in other words, not just any experience that feels meaningful)."[10]
Nurses Ruth Beckmann Murray and Judith Proctor Zenter write that "the spiritual dimension tries to be in harmony with the universe, and strives for answers about the infinite, and comes into focus when the person faces emotional stress, physical illness, or death."

3.2 Spirituality for Peace

The 5th World parliament of Science, Religion and Philosophy organized by MIT World peace university from 2nd to 5th Oct 2019 resolved that the pathway to world peace is through the time tested philosophies of various nations of the world and can be achieved through the union of Science and Religion working in unionism for the benefit of mankind. [11]

3.3 Spirituality for Sustainable Development

The Sustainable development, Economical, Social and Ecological can be achieved through spirituality.[12] The empathy for other beings, sensitivity to the environment and the Spirit of common good promotes development that is equitable and environmentally friendly. Organizations that have the spirit of Sustainable spirit for all, have displayed continued good economic performance and have also lasted longer than their counterparts.[13] Societies and groups who follow spirituality as an intrinsic value are found to live harmoniously and enable their fellow beings to work together for the common good. [14]

3.4 Spirituality for Promoting Powerful Living

The above view point points at spirituality as a means of promoting the culture of powerful (self-esteem and sound health) and purposeful (for common, social and ecological good). This will enable scientific and technological developments to serve the needs and wants of mankind in a sustainable manner [15]. It is also logical that culture of powerful living empowered by spirituality and sustainable technologies will result in physiological and emotional wellbeing of the society and foster a spirit of creativity and innovation owing to collaborative and sensitive approach. While the literature and teaching of the society encourage sustainable development through culture and spirituality, the ecosystem in the city, which is aspired to be Smart has to be designed and implemented through appropriate artifacts and facilities, layout, systems and processes and non-formal monitoring mechanisms.

4 People Centric Cities

As per Nobel Laureates Amartya Sen, Sustainable Development can happen when cultural matters become an integral part of our life and enhancement of our living standards. [16] As per United Cities and Local Governments (UCLG) emphasize the role of cultural as a pillar of Sustainable development and driver of development and People Centered Cities taking Holistic and integrated approach to development needs with cognizance to Creativity, Heritage, Knowledge and Diversity.[17]

4.1 Developing People Centric Smart Cities

The People Centric Smart Cities are to be made sustainable through urban regeneration strategies to enhance the livability of cities by respecting their identities and spirit of powerful living ensuring social inclusion [18]. The principles of creativity and innovation including digital technologies should be used as resources for sustainable urban development and improve local livelihood. [19] The culture of spirituality can be used as a tool the shape the future Smart Cities to achieve the following.

- Plan people centric cities and make them live able
- Enhance the quality of urban environments
- Elevate holistic development of the mankind
- Inculcate culture of sustainable lifestyle
- Make the city inclusive, safe and resilient
- Promote intrinsic link between cultural diversity and biodiversity to ensure greater environmental sustainability
- Culture and spirituality as an asset to eradicate poverty by linking economic and social dimension.

5 Cities as Cultural Transmitters (Cultural Capitals)

5.1 Interrelation of Culture and Economic growth

The Cultural growth and Economic growth of cities manifest in regenerative manner[20]. Economic growth enables cities to make investments creation and maintenance of Culture Assets and indulge in Art, Performing Arts and Literary pursuits (Theaters, Operas, museums, Art galleries, Libraries, Science centers, Research centers etc.). Sound financial strength of the city results in creation of Sports facilities, Common civic Utilities and places of recreation (Indoor & Outdoor Sports arenas, Public baths, Parks & Gardens etc...). The abundance of wealth promotes the development of high quality and lasting architectural assets (Forts, Palaces, Clock towers, Fountains, etc).[21] The grandeur of rich Cultural assets and indulgence of city in Cultural, Sports,

Literary, Scientific discovery, Festivals and events creates wealth generation which in turn fuels urban development resulting in economic growth [22].

5.2 Cultural Transmitters (Cultural Capital)

The contribution of cities that are rich in cultural assets and pursuits and hence centers of economic growth, is notable in establishing and nurturing of rich cultures and civilizations.[23]. In the Pre Fordist era the cities that have been forerunners of Culture enabled centers of vibrant commercial activities stimulating economic growth and influencing progress of society and the nation [24]. Glaring examples of such cities include Athens, Amsterdam, Rome, Vienna, London, Paris, Madrid, Istanbul, Venice, Prague. Even in current times these cities enjoy the status of "Cultural Transmitters" or "Cultural Capital" of their nations and contribute immensely to growth and development [25]. These cities have earned their preferred status blessed by rich cultural heritage that has transcended from generations of hundreds of years.

In the Post Fordist Era modern cities have emerged that have built their own identity as Dominant Culture in bringing recreation and happiness to the inhabitants and visitors and in turn have also become centers of economic, scientific and industrial growth. The prominent modern "Cultural Transmitter" Cities include New York, Los Angeles, Shanghai, Hong Kong, Singapore, and Mumbai [26]. UNESCO has classified cities in three categories based on the degree of influence the "Cultural Transmitter" cities have in promoting Economic growth, Scientific & Industrial growth, Sports and Cultural Festivals and the wellbeing of inhabitants [27]. The three categories are Alpha (Profound Influence), Beta (Very High Influence) and Gamma (High Influence). The examples of Influencer City types are as below

a) *Alpha* - London, Paris, New York, Tokyo
b) *Beta* - Chicago, Frankfurt, Hong Kong, Los Angeles, Milan, Singapore
c) *Gama* - Beijing, Moscow, Delhi, Mumbai, Tehran

6 Top "Cultural Transmitter" Cities of the World - A Perspective

The cultural transmitters can attain their status by making them Livable for all by excelling economically through the Rich Cultural Assets and Distinctive City Life. "For Cities to thrive the need to be livable and citizens need to feel connected. They need to work hard to close the gap between rich and poor. And in the phase of significant growth they need to maintain their distinctive character." Says Justin Simons the Chairperson of WCCF.[28]

6.1 Characteristics of "World Cultural Transmitter" Cities (WCTC)

Cities and the people living there, all over the world, are greatly influenced by the prevalent culture. "The Leadership, Government, Business, Education and Health must understand the essential contribution Culture makes to Prosperity and Quality.[29] The world cities culture forum in its Annual Summit in Amsterdam reviewed top 25 "Cultural Transmitter" Cities. It was found that culture is a key ingredient of the success of "Cultural Transmitter" Cities owing to inclusion of culture aspects in urban planning and policy. The following observations were made in the report.

a) WCTC enjoy economic growth and are leading drivers of global economic recovery.
b) Culture has contributed to the economic growth as revenue drivers.
c) Economic success has shown negative consequences in certain cases that includes over stretched infrastructure, social inequality and risk of homogenization.
d) Economic Success is also driving out culture and creativity by making cities uncomfortable for many local artists and cultural producers.
e) There is a need to integrate new communities, manage diversity and improving public places to create the cities of the future.
f) For a city to be a "Cultural Transmitter" culture needs to be the golden thread of Urban policy.

6.2 Features of WCTC

The findings of the World Cities Culture Report 2015 are summarized by the authors placed at Exhibit 1.[30] The features are tabulated under following parameters for the purpose of assessing the impact of these parameters on economic performance and the number of tourists visiting the city.

a) Population in million
b) GDP (USD million)
c) GDP Contribution Per Person (USD million)
d) GDP Contribution Per SQM Area (USD million)
e) Rank as per GDP
f) Rank as per GDP Contribution per Person
g) Rank as per GDP Contribution per Sq. Km Area
h) Rank as per No of Tourists per year Rank
i) Area in Sq. Km
j) Number of Heritage Sites, Museum, Art Galleries, Libraries
k) Number of Theaters and Cinemas
l) No of Tourists per Year in million
m) Creative Industry Employee in percentage of population.
n) Number of Branded Festivals & Celebrations

6.3 Top ten Cultural Transmitter Cities based on Economics Performance

The top ten Cultural Transmitter Cities based on the GDP contributions are given in table 7.1 which is deduced from exhibit 1 (refer 7.2 above).

Table 1: Top ten "Cultural Transmitter cities" based on annual GDP (2015)

City	Population in m	GDP USD in m	Area in Sq Km	Rank as per no of Tourists per year
Vienna	1.77	7794200	415	14
New York	8.41	1406000	1215	6
Los Angeles	10.12	826826	10510	11
Tokyo	13.16	743826	2130	12
Paris	12.01	730819	12012	3
Shanghai	24.26	667448	6341	10
London	8.42	492584	1572	1
Singapore	5.47	425155	718.3	2
San Francisco	0.81	388272	121	16
Hongkong	7.27	381186	2755	13

The above analysis result indicates that there is a positive correlation between the GDP and the contributions made "per person" and" per square meter". This infers that the population and size of the cities contribute effectively towards GDPs in case of top Cultural Transmitter Cities of the world. It is also interesting that the top ten Cultural Transmitters Cities based on GDPs are not amongst top ten with respect to the number of tourists visiting the city. Vienna is top in GDP based on high per person GDP contribution whereas London and Singapore which are ranked one and two based on the number of tourists are seventh and eighth on GDP performance. This indicates that Cultural Transmitter Cities have influence on non tourist income that contributes to GDP. In other words, culture influences revenue generation even on non tourist revenues coming out of activities such as Science, Technology and Innovation.

6.4 Top ten Cultural Transmitter Cities based on number of tourists

The top ten Cultural Transmitter Cities based on the number of tourists together with the influence of Cultural Assets and related operations are given in table 7.2 below which is deduced from exhibit 1. (Refer 7.2 above)

Table 2: Top ten World Cultural Transmitter Cities Based on Number of Tourists per year. (Million)

City	GDP	Culture assets		Cultural activity	
Ranked based on no of tourists per year	In USD annually (2015)	Heritage Sites, Museum, Art Galleries, Libraries	Theatre and Cinemas	Festivals & Celebrations	Creative Industry Emp %
London	492584	215	1105		16.3
Singapore	425155	319	211		5.1
Paris	730819	1413	490		9.0
Dubai	337907	69	7	35	2.2
Istanbul	349000	38363	189	150	13.4
New York	1406000	142	1093		8.6
Seoul	335493	814	526		12.0
Rome	163005	200	292		2.7
Amsterdam	43401	350	0	350	9.2

Shanghai	667448	2471	214		22.7

The data in table 2 indicates that the larger number of tourists visiting a city are influenced by the size of the cultural assets of the city that includes Heritage Sites, Museum, Art Galleries, Libraries, Theatre and Cinemas. It is also observed five of the cities ranked on the basis of number of tourists also figure in the list of cities with top ten GDP performance. This highlights the contribution of cultural assets in generating revenues for the Cultural Transmitter Cities. The importance of deploying employees to harness the potential of cultural assets is indicated by the fact that there is positive significant correlation between Creative Industry emp % with number of cultural assets(P = 0.011). The numbers of tourists visiting the city are generally higher where larger number of employees is deployed by creative industry, as depicted in figure 7.1

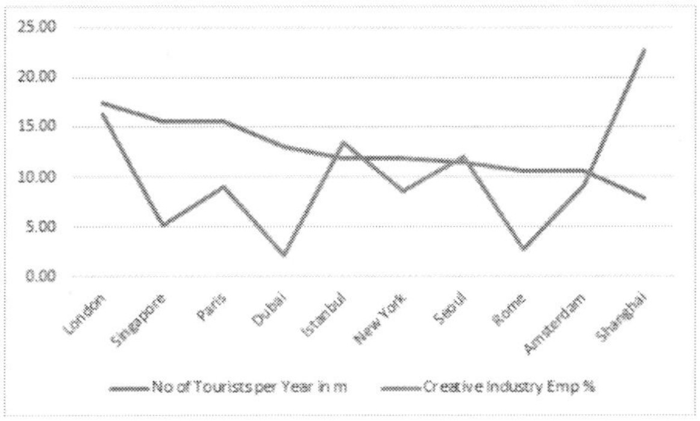

Fig 2: Influence of no of employees in creative industry on number of tourists

7 Framework for making44 Smart Cities People Centric

A framework to build/ rebuild Cities as "People Centric Smart Cities of the Future" is developed and presented below. The framework is structured in three sections:

a. Preparation - The activities that need to be undertaken before undertaking the project to develop Smart Cities.
b. Execution of the ProjectvAssessment Framework to measure the Levels achieved,
in terms of completeness of desired features of "People Centric City ".

7.1 Preparations

The salient steps required to make necessary planning and preparations [31] before launching on the project to develop "People Centric Smart Cities" are listed below.

7.1.1 Setting the objectives (Uniqueness & Identity)

The project team must ascertain as to what will be the ingredients that will give the city a unique identity. Few examples of the themes are listed below

a) Fostering spiritual consciousness
b) Holistic development to integrate creativity, heritage, knowledge and diversity
c) Enhance the livability of the cities
d) Innovation for sustainability of cities
e) City to be safe, resilient and inclusive
f) Fostering Environment Sensitive Powerful living Architecture
g) Unique Identity of City based on its Cultural/Technical/Economic/Environment ... prowess

7.1.2 Organization involving all Stakeholders

It is recommended that the project team for planning, execution and monitoring must have representation of Stakeholders from all walks of life. List is given below.

a) Main Administrative Control Office system
b) Citizens Forum
c) NGOs
d) Prominent Economic Groups
e) Prominent Industrial Groups
f) Institutions of Higher Education
g) Institution of Primary and secondary Education
h) Spiritual and Literary consortium
i) Religious Heads
j) Prominent achievers of various fields from the city

7.1.3 Identification of Constituting Elements of Culture and Spirituality (CECS)

The elements that would give the "People Centric City" its unique identity and richness depending on the History and Cultural Environment must be identified and relevant specifications/characteristics should be detailed. A list is given below.

a) Heritage Sites

b) Historical and Archaeological Centers
c) Museums
d) Art Galleries
e) Libraries
f) Theatre including Amphitheaters
g) Prominent artifacts of Significance
h) Places of Worship
i) Places of Spiritual Awakening and practice
j) Markets & "Haats"
k) Chaupaals & Common places of Meeting
l) "Ghaats" and Community bathing spaces
m) Artists and Artisan Colonies
n) Factories and Economic activity centers of yesteryears
o) Prominent Factories, Business offices and Economic Activity Centers identifying City
p) Residences or memoirs of Citizens with outstanding contributions
q) Animal parks and conservatories
r) Natural Activity Sites
s) Signages and Identity Displaying Artifacts
t) Sports arenas, grounds and sites representative of identity

7.2 Execution

The salient steps that could be undertaken for execution of the "People Centric Cities" are recommended below.

7.2.1 Layout impregnated with CECS

The layout of the smart city is to be designed in a manner that promotes interaction of people, exchange of knowledge, involvement in Cultural and Spiritual pursuits, promotes harmonious living in an inclusive manner such that all people are able to enjoy the fruits of the Cities wealth as per their economic standards. Themes to be considered are listed below.

a) Layout from mobility and operational perspective
b) Facilitate interactions
c) Aesthetics and beautification
d) Cultural infrastructure
e) Technologically smart civic life
f) The inclusive approach in city planning
g) Safe and resilient city planning
h) Sustainable (environmental, economic and social) city

7.2.2 Awareness and Promotional activities Schedule and System

The information regarding the unique artefacts that uniquely identify the city must be disseminated in all forms using all medium and practices available to promote them. Few of the methods are listed below.

a) Kiosks and Apps
b) Performing and visual arts
c) Cultural events/camps
d) Spiritual events/camps
e) Media and broadcasting
f) Literature

7.2.3 Signature Events and Specific Initiatives

Signature events like the "Tomatina Festival" of Spain, "Carnival" of Goa, Pushkar Mela of Rajasthan must be factored in the scheme of City Layout, Preparation of special spaces and appropriate municipal arrangements. A list of such activities is given below

a) Summit
b) Conference
c) Rally
d) Competitions
e) Seminars

7.3 Assessment Frameworks

A framework to assess the status and levels achieved by the "People Centric Smart Cities" in delivering the promise to the inhabitants and visitors in terms of providing powerful living, inclusive attitude and respect for sustainability. The assessment is recommended to be made under three categories to obtain a comprehensive index suggested by the three sets of indicators.

a) Economic Indicators
b) Social Indicators
c) Media Indicators

7.3.1 Economic Indicator

The Economic Indicator that the Smart City is delivering promise can be ascertained by

Net Revenue Earned (Culture) {NRE(C)}	=	(Sum of Income from all Cultural Contributor Elements [CCEn]) { Sigma CCEn}	-	Sum of (Direct Expenses [DECAn] on operating the Contributor Asset / Element + Amortized Investment [AICAn] on it) {Sigma (DECAn / AICAn) }

The cities can be ranked on the basis of Net Revenue Earned from Cultural / Scholastic pursuits

a) Annual Revenues earned from Cultural / Archaeological asset operations
b) Annual Income (Hospitality Expenses + Transportation + Purchases) from tourists
c) Income from the Events conducted in the city
d) Additional knowledge Industry income
e) Additional income through arts
f) Income from Sports events
g) Income from Spiritual events

highest attained anywhere as a guideline Indicator employing the following scheme:

7.3.2 Social Indicators

It is recommended to have multiple criteria based on Social recognition and contributions made by the city to ascertain the level of "People Centric Smartness".
A comparative index for each measure / theme could be worked out as a fraction of the number achieved by the particular city with the number achieved by the city with highest achievement for the theme which is to be converted on a 10 point scale.

Number achieved by city) {CNn}
Social Index [for the theme n] {SITn} 10/*
(Number achieved by the city with highest Achievement for the theme n) {HCN}
Social Index City = Average of SITn (for 1 to m) where m = Total no of Themes considered
The themes that could be considered for the purpose are given below.

a) Events and artifacts awarded on international forum
b) City's luminaries achieved international awards and recognitions
c) No of peace and sustainability events organized by citizens
d) Happiness index of people
e) Satisfaction index of new migrants/asylum seekers.

All such indices can be averaged to get a score out of 10

7.3.3 Media Indicators based

Media presence and media related achievement are good indicators of cities contribution as "Cultural Transmitter". The brand image and mentions in media can be worked out based on multiple themes and expressed as a percentage of the best in class figures expressed on a scale of 10.
All such indices can be averaged to get a score out of 10
(Numbers achieved by city) {MNn}=

*Media Index [for the theme n] {MITn} * 10/*
(Number achieved by the city with highest Achievement for the theme n) {MHN}

Media Index City = Average of MITn (for 1 to m) where m = Total no of Themes considered
The themes that could be considered for the purpose are given below.

a) Brand image of city, citizens and artifacts
b) Preferred destination status
c) Social media presence
d) No of information centers/kiosks
e) Films made in city
f) Media accolades for events for cultural and scientific events

Social indicators and Media indicators present a sense of overall quality of life or well-being of people in specific city. These indicators give direction to tourists, researchers, job seekers, students, migrants with respect to their decision of visiting or moving to particular city. All the people who are connected and involved in creating and developing smart city would be able to replicate and modify the model of best cities according to the specific unique culture of their own city. Awareness about the status of the city with respect to Economic indicators, Social indicators and Media indicators would develop positive competitive spirit among all stakeholders to put conscious efforts towards sustainable and systematic development of cities. This also would enable to involve dwellers of cities in the development of cities and make it people centric in true sense.

8 Conclusion

The influence of Culture and Scholastic pursuits on Sustainable development and Powerful living to all has been highlighted. The Cities that are designed as People Centric Smart Cities of the Future have displayed high economic growth as well as provide equitable and inclusive living experience. It is paramount that Smart cities must promote Spirituality and interaction with all citizens to foster harmonious and enlightened living. Framework to design (Prepare), develop (Execute) and assess (Measure) the status of City as Cultural transmitter has been suggested.

References:

1. Wikipedia, Smart Cities Mission: https://en.wikipedia.org/wiki/Smart_Cities_Mission.
2. Press Information Bureau, 25 June 2015 Prime Minister launches Smart Cities, AMRUT, Urban Housing Missions. (n.d.). Retrieved November 15, 2019, from https://pib.gov.in/newsite/PrintRelease.aspx?relid=122788

3. Prime Minister launches Smart Cities, AMRUT, Urban Housing Missions. (n.d.). Retrieved November 15, 2019, from https://pib.gov.in/newsite/PrintRelease.aspx?relid=122788
4. Business Standard, BS Reporter, 30 April 2015, archived from the original on 25 November 2015 Cabinet nod To Rs 1 lakh cr for urban renewal, 98 smart cities to take off, New Delhi:
5. Ref: Smart cities mission, ministry of housing and urban affairs, Government of India. http://smartcities.gov.in/content/innerpage/smart-city-features.php
6. Extracts from the interview with Ms Rupal Sharma, IAS, Project Director and Deputy Commissioner, Smart city project office Pune, during the interview with the Authors on 12th November 2019.
7. Extracts from the interview with Honorable Professor Dr Vishwanath Karad, the founder of the MAEER's Trust, the MIT Group of Institutions, the UNESCO chair for World Peace, President of the MIT ADT University, Loni kalbhor Pune on 02 Oct 2019.
8. Sasson, R (2019). Success Consciousness - skills for success, positivity and inner peace Retrieved November 15, 2019 ,from https://www.successconsciousness.com/blog/spirituality/what-is-spirituality-meaning-and-definition/
9. Amy R. Krentzman and Carlson, D., Leonard, B. (2013). Spirituality in healthcare. Online module, Center for Spirituality and Healing. University of Minnesota. Retrieved November 4, 2019 from http://www.csh.umn/Integrativehealingpractices .
10. Beauregard, M. & O'Leary, D. (2009) The spiritual brain. New York: Harper Collins.
11. 5th World parliament of Science, Religion and Philosophy organized by MIT World peace university from 2nd to 5th Oct2019. https://mitwpu.edu.in/news/world-parliament-of-religion-science-and-philosophy-2019/
12. Ulluwishewa, Rohana and Kumarsinghe S, (2014). Spirituality and Sustainable Development, Palgrave and Macmillan publication. USA
13. John Mackey, Raj Sisodia and Bill George (2014), Conscious Capitalism, Liberating the heroic spirits of business. Harvard Business Review press.
14. Rajvanshi, Anil. K; (2002). Spirituality can help in sustainable development, available online www.nariphaltan. org/suspri.htm.
15. A Duxbury, N. (2014). Cultural governance in sustainable cities. Kult-ur: Interdisciplinary journal on the culture of the city, 1(1), 165-182. Special issue: Governança de la ciutat i drets culturals (Governance of the City and Cultural Rights). Retrieved from http://www.e-revistes.uji.es/index.php/kult-ur/article/view/1252/1229
16. Sen, A. (2012). Cultural for Sustainable Development. Rio+20, 1–20.
17. The policy Statement of the 3rd world congress of UCLG and Global Taskforce of Local and Regional Governments, UNESCO Report on Sustainability implementation, 2015
18. UNESCO (2016). Culture Urban Future: Global Report on Culture for Sustainable Urban Development. Paris: UNESCO
19. UCLG (2018), Culture in the Sustainable Development Goals: A Guide for Local Action. Barcelona: UCLG.
20. Richard Florida June 7, 2018 How Culture Shapes Economic Development, CityLab.com
21. Hosagrahar, J.; Soule, J.; Fusco Girard, L.; and Potts, A. (2016), 'Cultural Heritage, the UN Sustainable Development Goals, and the New Urban Agenda', concept note, ICOMOS.
22. Yang, M., et,al (2016). Culture: urban future, global report on culture for sustainable urban development
23. A. Scott, "The cultural economy of cities", International Journal of Urban and Regional Research. 21, no2, pp.323 -39
24. J. Jones, "The Global City Misconceived" : The Myth of ‹‹Global Management›› in Transnational Service Firms››, Geoforum, vol. 33, 2002, pp. 335-350.of World Cities››, Cities, 16, no6, 1999, pp.445-458
25. UCLG (2018), Culture in the Sustainable Development Goals: A Guide for Local Action. Barcelona: UCLG.
26. Nadarajah, M., and A. T. Yamamoto. (Eds.) (2007). Urban Crisis: Culture and the Sustainability of Cities. Tokyo: United Nations University Press.
27. J.V. Beaverstock, R.G. Smith and P.J. Taylor, A Roster British Council (2018), Cultural Heritage for Inclusive Growth, British Council.
28. Justin Simons, Chairperson world cities culture Forum
29. Henry Matisse BOP Consulting January 2015, World Cities Culture Report 2015
30. Exhibit 1, Features of World Cultural Transmitters Cities deduced from World Cities Culture Report 2015 and summarized in tabular form by the authors and placed it enclosure.
31. UCLG (2015), Culture 21: Actions. Commitments on the role of culture in sustainable cities. Barcelona: UCLG.

Study of impact of increased number of devotees on indoor environmental quality of heritage temples of goddess Mahalaxmi at Kolhapur, Maharashtra, India.

Anjali Jadhav[1,*], *Sushma Kulkarni*[1], *and AshishRege*[2]
[1]SHIVAJI UNIVERSITY, Kolhâpur, 416004-Maharashtra (India)
[2]GOA COLLEGE OF ARCHITECTURE, Panaji, 4030001-Goa, India.

Abstract. India has a great cultural heritage of temples across the country. There is a wide variety of temples of variety of deities constructed from East to West and North to South of India. Most of the temples are built centuries ago and still are in good condition and the centre of cultural activities. Mahalaxmi temple precinct is a very old precinct in Kolhapur, built in 8th century by Chalukyan rulers. Every year lacs of devotees are visiting this temple of goddess Mahalaxmi from most of the part of India. The temple is constructed of basalt stone with beautiful intricate carving on it. This paper focus on indoor environmental quality of the temple which is deteriorated due to the impact of increased number of devotees. Some of the devotees face the problem of suffocation, fatigue, fainting, irritation, nasal congestion; as provision of vents are absent and natural ventilation is not available inside the shrine of goddess Mahalaxmi. Due to lack of air exchange and natural ventilation the indoor environmental quality is deteriorated. It is observed that the CO_2 level is increased, decreased in O_2 level, rise in temperature and humidity above the standard level causing discomfort.

Keywords. Heritage temple, Indoor environmental quality, comfort, devotees, natural ventilation, temperature, humidity, O_2, CO_2.

1 Introduction

Mahalaxmi temple is a very old temple .It is in Kolhapur near the river Panchaganga. It is situated to the north east side of the old palace in Bhavani Mandap. It is the good example of Chalukyan style of Hindu Architecture. The temple is at the centre of the temple complex and the shrines are grouped around the main temple. The plan of the temple is star shaped resembles Vesara type of Hindu Architecture. [1]

1.2 History

Mahalaxmi temple is in the heart of Kolhapur City. It is constructed in 8th century. The oldest part of the present shrine is the work of early Chalukyan style. (550 A.D. to 660 A.D.) Shilahara rules of Kolhapur (990 A.D. onwards) added beauty to the temple. The old structure is constructed from close fitting mortar less stones which are brought from the quarries nearby. The stones are black basalt stones which can take high polish and intricate carvings. [2]

1.2 Heritage listing

As per ASI it is a grade I heritage structure and have prime importance. The city is a popular pilgrimage city and Mahalaxmi temple is the centre of the city. The old Kolhapur city is situated around this temple.

2 Indoor Environmental Quality (IEQ)

IEQ has major impact on the health of the building occupants. All the human beings spend maximum time of the day i.e. 80% to 85% indoors. If the IEQ is good it gives very good result on the productivity and health condition of the occupants. The efficiency, freshness, mind alertness, attentiveness increases in good IEQ. [3]

2.1 Indoor Environmental Quality (IEQ) in Mahalaxmi Temple

Mahalaxmi Temple is constructed in 8th century. It is not designed for the heavy flow of devotees which is in lacs. Table no.1 showing the number of visitors every year. Building envelope of the temple is inadequate to house such large number of devotees. Due to that the IEQ is not at comfort.

3 Parameters of IEQ

*Corresponding author: 02anjalijadhav@gmail.com

Air quality, light quality sound quality, odour quality visual quality are the parameters of IEQ.

3.1 Air Quality

The parameters of air quality are a) Temperature b) Humidity c) O_2 d) CO_2 e) Air exchange f) Particulate matter. For comfort the temperature should be 23°C, and 1°C plus and minus is advisable. The range of humidity should be 30% to 50 % and the ideal level should be 45%.Minimum 19.5 % Oxygen is required for the good health of human being, 20.9% of Oxygen is needed for the better cognitive ability, balance and overall good health.0.03% is the ideal percentage of CO_2. High percentage of CO_2 cause rapid breathing and confusion. An air exchange is how many times the air enters and exits the room in an hour. Air exchange are calculated on the base of area and number of occupants. The source of particulate matter are pollen, smoke, toxic mould and spores, bacteria, pet dander, construction dust. Corse particles are of 10 to 2.5 micrometres. [4, 5, 6, 7]

3.2 Light Quality

Light is the major parameter of livelihood. It is measure in terms of lux. The light should be available in appropriate level for specific task. Daylighting has adequate visibility without causing discomfort, impact on health and productivity and it is very much related to psychology of human being. Improper lighting cause eyestrain, eye irritation, fatigue and headache. [3, 8, 9]

3.3 Sound Quality

An unwanted sound is noise and it affects the health, wellbeing of occupants. Loud noise can cause hearing loss, stress and impact on heart. Annoyance can be disturbing among individuals due to increased noise. As per research, 40 db is low sound and above 55db it is matter of concern. Meaning of acoustic comfort is having the correct level and quality of sound to use the space as planned. [3, 10]

3.3 Odour Quality

Odour plays a great role in the indoor environments. They are quite significant for the IEQ. Odour is distinctive smell especially an unpleasant one. Bad odour cause headache, nausea and temporary symptoms which bother people.

3.3 Visual comfort

The space should be clutter free, with good circulation and good visual, aesthetical quality.

4 Problem Identification

Every year in the festival season of Navratri two medical camps are arranged for the devotees to take care of the medical emergency. Most of them report shortness of breath, fainting, giddiness, irritation, nasal conjunction. Most of the children cry as they enter the shrine which is full of devotees and very crowdie.

Table 1. Visitor's data 2018-2019

Month	No. of visitors
Nov-18	713583
Dec-18	577617
Jan-20	776190
Feb-20	452059
March-20	733580
April-20	905725
May-20	584848
June-20	489003
July-20	316503
August-20	451930
Sep-20	611353
Oct-20	3175745

5 Objective

The objective of the paper is to study and analyse the impact of increased number of devotees on the indoor environmental quality of the heritage temple of goddess Mahalaxmi at Kolhapur.

1. To study and understand the heritage value of the temple and the indoor environmental quality of the temple.
2. To study and understand the number of devotees visiting the temple through the year and on the festival seasons.
3. To measure all the parameters of indoor environmental quality with the help of various instruments and analysis of the data.
4. To derive a broad conclusion on the basis of the data collected and the analysis.

6 Methodology

The work is carried out in Mahalaxmi Temple, in the month of September 2019 and October 2019. These months are very important as the main festival of Navratri is in these month and lacs of devotees visit temple during this period. To study the IEQ of the temple the instrumentation is carried out at 5 locations of the temple. Figure 1 shows the location of the instrumentation. The temperature and humidity is measured by Kestrel-4000 pocket weather tracker.CO_2, O_2, Ch4, are measured by air sensor Drager-X-am 5600. Light is measured by light meter LX-101.

R1-Garbhagruha

R2-darshan Mandapa

R3-Kasav Mandapa

R4-Ganapati Chowk

R5-Outside the temple

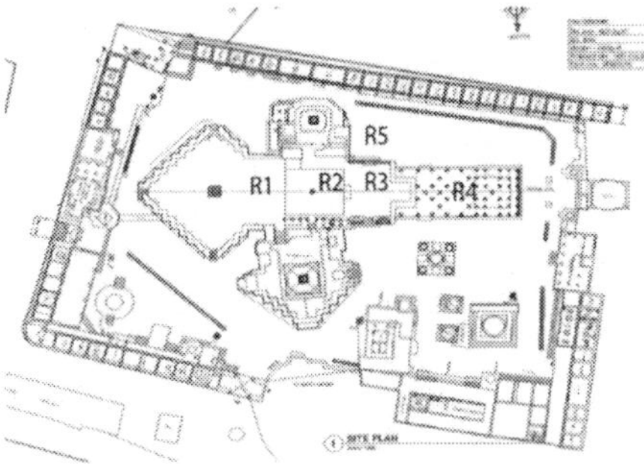

Fig. 1 Instrumentation in the temple.

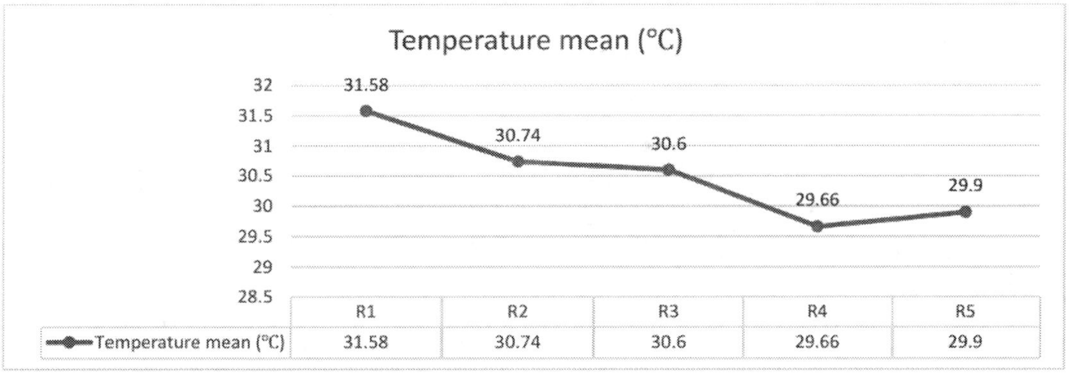

Fig. 2 Graph showing temperature at 5 locations of the temple

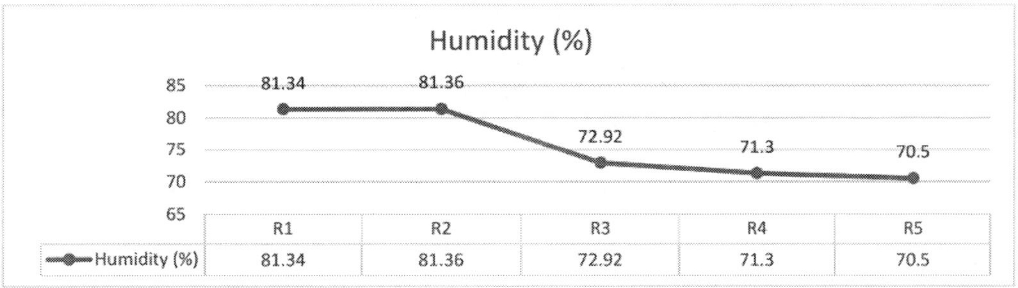

Fig. 3 Graph showing Humidity in the temple at 5 locations

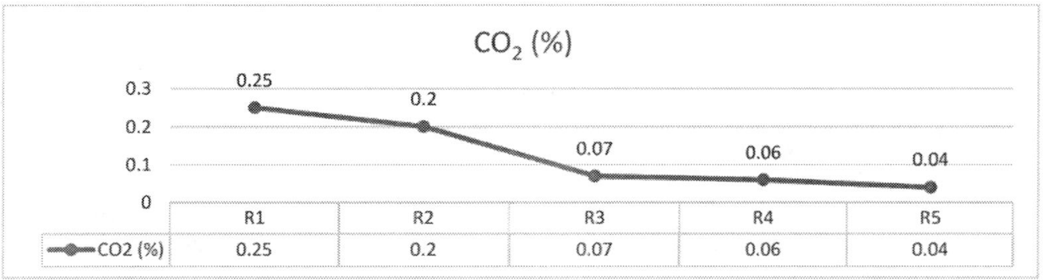

Fig. 4 Graph showing CO2 % in the air of the temple at 5 locations

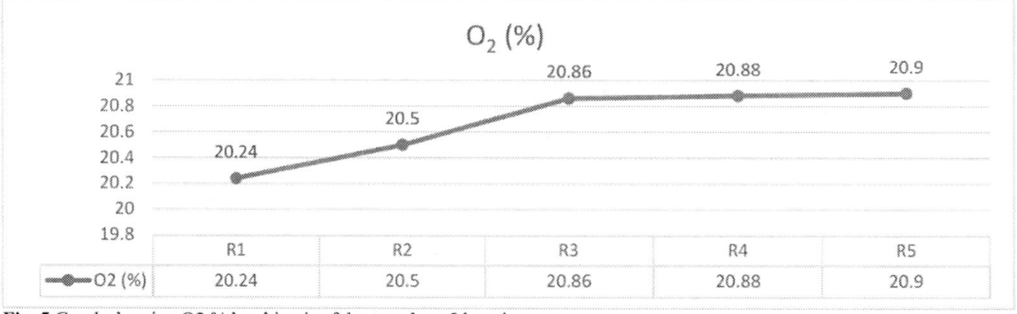

Fig. 5 Graph showing O2 % level in air of the temple at 5 locations

7 Data Analysis and Conclusion

From the analysis of the temperature graph (figure no. 2) it is observed that the temperature of the Garbha-gruha R1 is more than 1.7OC than the outside temperature R5. Devotees' density is maximum at location R2, they stay there at least 10 minutes. The area per person available is 0.35 sq. mt.; which is very suffocating. The humidity level is 81.34%; which more than the comfort standard, causing discomfort. (Figure no.3). The CO_2 level is 0.25%whcih is more than the standard level. (Figure no.4), O_2 level is also reduced by 0.66% (Figure no.5) The wind speed is 0m/s as there is no cross ventilation in the Darshan mandapa, and the vents above the shrine are closed. The air exchanges are not up to standards and very less per hour. The noise level in the shrine is 82 db. which is very chaotic and no silence is observed. Inside the temple there are various smell created by fresh flowers and decaying flowers, essence sticks, burning oil lamps and most objectionable and voluminous smell is of visitors sweat. Decreased O_2 level, increased CO_2 level, increased humidity and temperature, increased noise level causes discomfort. From the observations it is concluded that the IEQ in the temple is deteriorated and may cause an impact on the health of the devotees.

References

1) P. Brown, B. *Indian Architecture (Buddhist and Hindu Periods)*,62.
2) B. Maharaj, *Karveer Nivasini shree mahalaxmi* -, 20.
3) *ISHRAE Position Paper on Indoor Environmental Quality,* August (2015).
4) *ASHRAE STANDARD* **55** (2017)
5) L. Zagreus, C Huizenga, E Arens., D Lehrer. , I.A., J. 65(2004)
6) D.Kolokotsa, A.pouliezous, G.stavrakakis, C.Lazos. B. & E., J. **44** (2009)
7) S. Tanabe, M. Haned, N. Nishipana, Waseda Uni, Rehva J.-June (2007.)
8) B. Rechard Hugges, S. D., E.P., **105**, 2865-2870. (2017)
9) K. Fabbri, L. Tronchin, S.D, E. P. **78,** 2778-2783. (2015)
10) X. Leong, E. Essah, S.D., E. P., **132,** 87-92. (2017)

*Corresponding author: **02anjalijadhav@gmail.com**

Spiritual Revitalization of Indian Cities through Heritage Conservation

Sehgal Jagrati [1*], *Kumar Atul* [1], *Prakash Ila* [1]
[1]IIT Roorkee, Roorkee, Uttarakhand, India- 247667

Abstract. The aim and objective of heritage conservation with revitalization under Article 1, UNESCO Constitution is "To contribute to peace and security by promoting collaboration among the nations through education, science and culture in order to further universal respect for justice, for the rule of law and for the human rights and fundamental freedoms". Revitalization refers to bring back the essence of tangible and intangible aspects which has been diminishing on the urge of modernism. Through intervention of conservation techniques such as restoration, adaptation, preservation, reconstruction etc. the above aspect can be achieved efficiently. Conservation and revitalization of old Indian cities and its historic values have so far given low priority in urban development policies. There are 35 (27 cultural, 7 natural and 1 mixed) World Heritage Sites in India that are recognized by the UNESCO as of July 2016. Eagerness for the taste of development, the Indian cities are losing their architectural and historical assets. Thus, revitalization and conservation have to come in action to celebrate reputable past of our historic areas and to upgrade its way towards life. Various heritage sites are in dilapidated condition which are not being utilized or conserved in an effective manner Indian cities are going through a vibrant phase of urban renaissance; the study methodology of this research is based on continuous and organic approach of revitalization by amalgamating community and policy development followed by analysis of physical and spiritual attributes of the city. Research and study approach concludes with considerations for area revitalization approaches which maintain the typical urban tissue, essential qualities of the historic areas, life of the communities residing there and can also adapt the physical structures and activities to some of the present day requirements. In terms of intangibility, it is stressed to be seen as part of regaining the spiritual identity by suitable conservation approach, sustainability and revitalization will work effectively if they are unified into new concept of development.

1. Introduction

The aim and objective of heritage conservation with revitalization under Article 1, UNESCO Constitution is "To contribute to peace and security by promoting collaboration among the nations through education, science and culture in order to further universal respect for justice, for the rule of law and for the human rights and fundamental freedoms". Heritages are the root system of development; the higher one wants to grow the deeper should be the roots to promote the conservation of all historical monuments, culture, education as well as spiritual values of each and every city of the world and stressed on the complete sustainable development. In the era of industrialization, to maintain the world peace and security among the nations across the borders historical culture and education along with modernity plays a vital role. The term 'heritage' has evolved a lot in recent decades; Cultural heritage is not only about monuments and collections of objects, but it also comprise traditions and living expressions inherited from our ancestors and passed on to posterity, such as performing arts, social practices, civility, oral traditions, rituals, festivity, knowledge and practices regarding nature and the universe or the knowledge and skills to produce traditional crafts and thus with a fine line tangible and intangible heritages are separated. Heritage is being used in a broad sense which include both natural and cultural environment including beliefs, ideas and several ways of life and above all, the intimate link between human psyche and mystical nature. Heritage values are threatened by anthropogenic activity which lacks a commitment to spirituality and respect for symbolic identity. In this situation humanity is adulterated by thought pollution which is hazardous than most of the commonly accepted forms of environmental pollution which claims the world attention. Revitalization refers to bring back the essence of tangible and intangible aspects which has been diminishing on the urge of modernism. Through intervention of conservation techniques such as restoration, adaptation, preservation, reconstruction etc. the above

*Corresponding author: **jagratisehgal@gmail.com**

aspect can be achieved efficiently. An understanding of Heritage Conservation Cycle diagram (Simon Thurley, 2005) and its different factors help in manifestation of our past and sequentially part of our future.

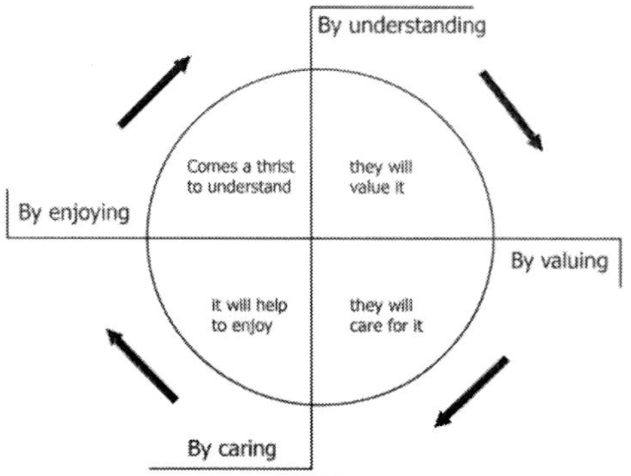

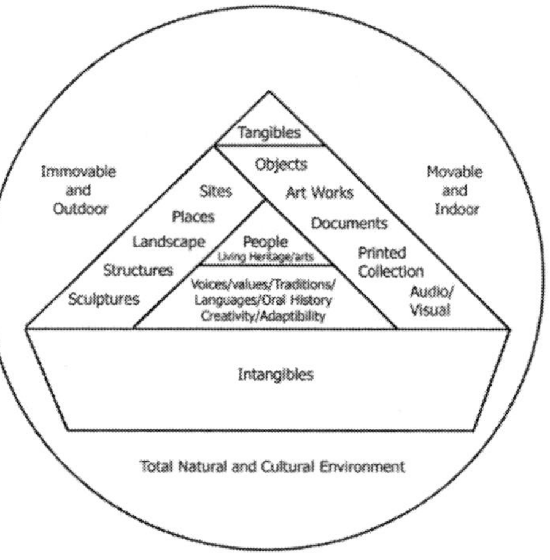

Figure 1 Holistic Representation of Cultural and Heritage Resources (source: Vol.03 2008 International Journal of Intangible Heritage pg-17)

- By understanding (cultural heritage) **people value it**
- By valuing it : **people want to care for it**
- By caring for it : **will help people enjoy it**
- From enjoying it : **comes a thirst to understand**
- By understanding it etc.

2. Indian perspective for heritage conservation

Indian cities are going through a vibrant phase of urban renaissance. Today, In India there are 475 urban agglomerations and over 31.16% of total population are living in 7935 towns out of one billion plus population *(Source: Census of India, 2011)*. By 2030, 40.76% of India's population is looking forward to inhabit in urban areas, according to the UN State of the World Population 2007 report. Eagerness for the taste of development, the Indian cities are losing their architectural and historical assets. Conservation and revitalization of old Indian cities and its historic values have so far given low priority in urban development policies. A mélange in Indian historic cities is because of the way in which development took place in correspondence to its natural and traditional neighborhood, local artisans, local crafts, traditions, its streets, water bodies, public spaces etc. along with all the outstanding monuments. There are 35 (27 cultural, 7 natural and 1 mixed) World Heritage Sites in India that are recognized by the UNESCO as of July 2016. In India, heritage conservation practice is categorized under five sections:

1. Development of Institutional Infrastructure for Urban Heritage;
2. Integration of Heritage in Urban Planning Framework;
3. **Revitalization of Urban Heritage through Urban Renewal;**
4. Community /Stakeholder Participation Approach; and
5. Generating Awareness of Heritage.

Despite of formulating rules and regulations for heritage conservation along with urban development, the historic sites are subjected to swift sink due to pressure of current urban scenario and various other components come up because of absence of relevant policies, legal frame-work, awareness, inadequate funding and technical skills as a result of which Indian heritage are in abysmal state. Thus, revitalization and conservation have to come in action to celebrate reputable past of our historic areas and to upgrade its way towards life. This can be achieved by addressing urban problem resulting from congestion, natural decay, space extension and disappearance of heritage etc., and assessing its history, custom, traditions, socioeconomic attribute and other characteristics involved in refurbishment and compatible use of recorded heritage.

Urban tissues of heritage sites can be intangibly revived with the concept of cultural heritage entrepreneurship so that the residing community can connect to their culture and pass on to their decedents.

3. Study Area

Jaipur, the capital city of Rajasthan, Indian, lies on the geographical coordinates of 26°55□ N and 75°49□, and is situated at an average elevation of 431 m. According to Master plan 2025, Jaipur city has occupied an area of 11,151 sq. km. Maharaja Swai Jai Singh II, has planned city as walled city in 1727. Maharaja Swai Jai Singh along with chief architect and planner, Vidhyadhar Bhattacharya, has laid foundations of city on very strong scientific principles. Since then this walled city, continues to act as central business district and holds congregation and trade and is responsible for originating 60% of city's commercial activity. Jaipur, being a section of Golden Triangle with Agra and Delhi attracts tourists both national and International. The city has a record of serving about 4900 tourists per day in 2015, including stay of 3- 4 days and also has marked 11.70% growth rate of foreign tourists in comparison to 20.27% of growth rate domestic tourist.

4. History

The design of old city has complex detailing with careful planning and clarity with combination of contemporary Mughal architecture and local architectural style to maintain high standard of habitat and living style. During1941 to 1946, Mirza Ismail the Prime Minister of the Kingdom has preserved the character and nature of the city and during this time major expansion was started outside the walled city. In 1821, during British rule many infrastructural work took place and many monumental buildings were constructed outside walled city and also attention was given to service layout in city which results in good network and planning of waterworks, gaslights and road network etc. After 19th century, urban transformations took place in the city which is influenced from British people and outcome of modern outlook of Sawai Ram Singh. Many attractive additions start taking place within the walled city, with buildings build on Indo- Saracenic Style, such as the Naya Mahal or Vidhan Sabha in edge with earlier Hawa Mahal.

Jaipur is known as the Pink City because of use of pink and its complementary colour scheme together with the use of red sand stone on building facades. Traditional features of Jaipur which includes domestic products like chuda, mojri, turban, ethnic dresses, local food, rajai etc. play a vital role in supporting the economical wealth of a number of families as it is the only source of income of considerable number of families. The ancient mahals, baulis, forts constructed thousands of years ago had a very scientific way of construction such as jharokhas for controlling the scorching heat of sun, bauli for various purposes such as self defence during enemy attacks, water supply etc., jalis for cross ventilation etc. all these structures are very beautiful example of how to adapt in local environmental conditions.

5. Methodology

Historical resources represent history embodying culture and tradition through architectural form, hence; need to be appraising as an asset on the contrary to liabilities. Qualities of cities such as historic, aesthetic, religious, social, ecological and traditional have to get acknowledgement and have to be utilized successfully in urban setup. The methodology adopted for documentation involved primary and secondary data collection and the approach is divided into three broad categories which are :

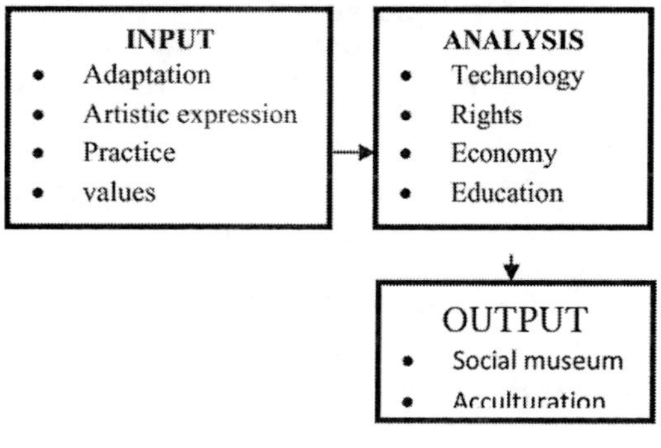

https://doi.org/10.1051/e3sconf/202017005005

The strategy for planning should therefore follow a middle path, maintaining the harmony of deeply implanted culture while, at the same time, taking account of future-oriented urban development. Planning through the approach of social museum will help in healing the lost "dharodhara" of the city by embracing the mode of conduct that holds everything together to gain the wholeness by acculturation to interrelate matter, life and mind. Heritage involves both natural and cultural environment and is being used in a wide range, including ideas, beliefs, and ways of life, and above all the intimate link between human psyche and mystical nature.

	Adaptation	Artistic expression	Practice	Values
Technology		√	√	
Rights	√			√
Economic	√	√	√	√
Education	√	√		√

Research is based on continuous and organic approach of revitalization by amalgamating community and policy development followed by analysis of physical and spiritual attributes of the city. Physical attributes includes TREE (**Technology, Rights, Economy, Education**) which are fundamental factors of development and spiritual attributes such as Adaptation, Artistic Expression, Practice and Values, these are the basic characteristics of heritage conservation. The connection between all these attributes are discussed in the above matrix and our study concentrates on how these spiritual attributes are subsets of physical attributes. Intangible heritage includes values, voice, traditions, oral histories, adaptation, artistic expressions, practice etc., since Jaipur has emerged out under the leadership of various rulers so it has amalgamation of various cultures, expressions and values, therefore focusing on four major and most vital aspects has been our main motto. These four aspects have been described with their importance in the developmental goals, as the physical attributes are incomplete without considering these factors as their subsets.

With respect to social museum and acculturation:

TECHNOLOGY: Adaptive technological advancement is an important part for conserving the societal values so that the liveliness of heritage sites can be restored. Artistic expressions and technology helps in changing and speeding up the cultural development. Practicing and mingling old with new i.e. vernacular with modernism.

RIGHTS: To maintain the intangible values the adaptiveness should come up with reduction of thought pollution which is more hazardous than the more commonly accepted forms of environmental pollution which claim world attention. Values threatened by human activities which lack a commitment to spiritual identity of the city.

ECONOMIC: In response to market opportunities there is a need of adaptation which is of more flexible and dynamic in nature with respect to cultural heritage enterprises. Since Jaipur is famous for its traditional handicrafts such as gems, jewelleries, textiles, wooden furniture's etc. which has a major contribution to countries GDP, labour force market development.

EDUCATION: Understanding the importance of our heritage will lead to development of community participation. Globalization of historical learning techniques, and merging old traditional techniques with the new one and marketing it globally will help in retaining the identity of the city at a larger level. Also the global market will help in encouraging the local artisans so that they may use their skills and try to restore the value of their culture. Values can be preserved by passing on our artistic talents to descendants; for this there is need to make them realize that this society and heritage is their own legacy and acculturation.

6. Result

Heritage is the mirror of human growth and development, and in this context it must be preserved. This ideology should be taken as an environmental ethic which has scarcity and deep attachment at its core. The paper draws attention towards the concept of social museum and acculturation which brings together spiritual and physical attributes of Jaipur. It is an initiative to restore the city heritage so that the society may itself behave as social museum, where museum refers to an institution where convivial values can be restored, and acculturation refers change in cultural values that can be accepted by community.

https://doi.org/10.1051/e3sconf/202017005005

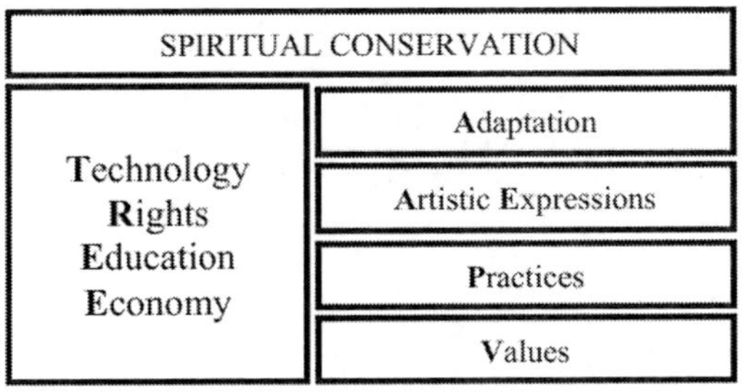

7. Conclusion

Our study concluded with the TREE analysis of Jaipur. This analysis focused on the regaining of spiritual identity of acculturation. Possible solutions are proposed under TREE aspects to develop social museum which should be emphasized and incorporated with values, creativity, language and adaptability. Heritage sites are continuously deteriorated due to natural and time factors and without correct revitalization strategy to sustain these qualities are very difficult. In order to be true to the existing local conditions in urban setup, a proper technique for museum is in need. Unidirectional aspects will only lead to conserving the tangible heritage but the time has come when there is need to conserve and revive the lost aura of Indian cities for which they have been known.

References

1. N. Aikawa, Mus. Inter., *An historical overview of the preparation of the UNESCO International Convention for the Safeguarding of the Intangible Heritage,* **56**, 137, (2004)
2. Appadurai, Universal Declaration on Cultural Diversity, *A Document for the World Summit on Sustainable Development*, (2002)
3. B. Feilden, Town Planning, *Architectural and Urban Conservation*, **56**, 2 (1985).
4. A. Galla, *UNESCO.* Hanoi, *Guidebook for the Participation of Young People in Heritage Conservation,* (2001)
5. J. Bianca Jackson, IEEE transactions on terahertz science and technology, *A Survey of Terahertz Applications in Cultural Heritage Conservation Science,* **1**, (2011)
6. S. L. Lee, Elsevier Science, *Urban conservation policy and the preservation of historical and cultural heritage*, **13**, 6, 399, (1996)
7. M.F. Jawaid, Elsevier, *City profile: Jaipur*, **68**, 63, (2017)
8. Hribara Mateja Smid, Social and Behavioral Sciences, *Sustainable heritage management: social, economic and other potentials of culture in local developmen,* **188**, 103, (2015)
9. E. Ruijgrok, Journal of Cultural Heritage , *The three economic values of cultural heritage: a case study in the Netherlands*, **7**, 206, (2006)
10. S. Burton, History with a Bottom Line, *Notes on Urban Conservation in Southeast Asia,* (1985/86)
11. Shahrul Yani, Social and Behavioral Sciences, *Heritage Conservation and Regeneration of Historic Areas in Malaysia,* **105,** 418, (2013)
12. R. P. Singh, Shaw, Brian and Jones, *Sacredscape and urban heritage in India: contestation and perspective,* 101, (1997)
13. Steinberg, F., elsevier science, habitat intl, *Conservation and Rehabilitation of Urban Heritage in Developing Countries*, **20**, 463, (1996)
14. R. Tagore, A Poet's School'. *Visva-Bharati Quarterly, October 1926*, 20, (1981)
15. Valentina Vasile, Social and Behavioral Sciences, *Innovative valuing of the cultural heritage assets. Economic implication on local employability, small entrepreneurship development and social inclusion,* **188**, 16, (2015)
16. G. Zaman, Cultural heritage entrepreneurship (CHE)- challenges and difficulties, **188**, 3, (2015)
17. R.K. Maikhuria, Forest Policy and Economics, **2**, (2001)

Energizing the Future with Memories of the Past: The Wadas of Pune City

Shilpa Nagapurkar[1], *Parag Narkhede*[2], *Vaseem Anjum Sheriff*[3]
[1]MIT ADTU's School of Architecture, Pune.
[2] BKPS College of Architecture, Pune.
[3] BMS College of Architecture, Bangalore.

Abstract: Pune, described as the Queen of the Deccan, [1] is located in the state of Maharashtra, India. It is a historic city associated with the Maratha Empire and seat of the Peshwa power. During the Colonial Period it was a British cantonment. Contemporary Pune city is considered as the cultural capital of Maharashtra and is also referred to as the Oxford of the East due to the presence of several well-known educational institutions. The old city of Pune is constituted by the seventeen Peths or localities. The wadas are a characteristic built-form that evolved during the Maratha Period. They were the residences not only of the Peshwas but also those connected with the administrative system of the times and are the manifestations of the culture of the period. They vary considerably in size and form. They have a characteristic spatial organization harmonizing form and space with distinct architectural features. They were once the seat of power, intrigue and grandeur. Now, they are the surviving witnesses of battle plans and palace intrigues at the height of glory of the Maratha Empire. After more than three hundred and fifty years the wadas themselves are waging a final battle for survival considering the apathy towards their woes and issues from both the civic body as well as their private owners. The objective of the paper is to explore the possibility of developing selected wadas as nodes in developing Pune city's culture infrastructure as well as heritage showcase. It seeks site specific solutions of 'Energizing the Future with the Memories of the Past' in Pune city.

1Introduction

For thousands of years, cities were having much simpler form although they rarely served single purposes. Instead, they supported a range of activities. Different land uses like Housing, commercial buildings, government offices and warehouses formed the built environment of the city. Urbanization took place at different chronological periods. The factors influencing urbanization were also different. Rapid urban growth is usually associated with time period of intense socio-economic or political change. Changes in the international trade modified the internal structure of national economics and the cities started getting flooded with migrants in the search of employment. Pressure on the existing housing stock increased with the swelling population, and people filtered in the central city through rented housing and hutments.[2] This scenario is continuously transforming the urban morphology of cities. The effect of these processes is seen through transforming social and spatial patterns of entire city.

The world is experiencing intense urbanization by the hand of extensive yet uneven processes of growth and expansion. More than half of humanity now lives in cities, and 80 per cent of the Earth's land surface has come to reflect the influence of city-based human activity. Dominating the urban world is a selective group of dynamic and highly specialized cities, as well as massively urbanized industrial regions.

The variation in influencing factors and historical circumstance gave rise to different urban fabrics in different parts of the world. The evolution of these urban forms or patterns is based upon main key factors that is social pattern and the built form. The characteristics of the social and the built form of the urban areas contribute to form an urban fabric for the area.

*Corresponding author: **shilpa.nagapurkar@gmail.com**

2Urban Fabric

Buildings, soft and hard landscaping, signage, lighting, roads and physical networks are the main physical characteristics of urban areas, which comprehensively form urban fabric of the cities. It is a reflection of the physical texture of the urban area. Changes in urban areas refers to long lasting spatial dynamics and is characterized by transformations of city objects that depends upon choices made upon by the residents and policy makers. Cities are with long lasting nature and thus characterized by the inertia of urban fabric and social structures. As cities go through continuous process of transformation in terms of physical and social structure urban fabric becomes very vital in nature. Heterogeneity of urban fabric is a key element of any developing urban area. [3] Throughout the process of this transformation cultural heritage of the area plays an important role. Within the changing urban pattern importance of this cultural heritage is getting weakened. It is a strong element of any urban fabric which needs to be strongly considered in terms of policy planning and regulatory framework as it represents origins of particular urban area, which is a main part of cultural infrastructure of the area.

Looking at Pune city, the city has not completely cut itself from its ancient character. Old city area and cantonment area still defines many of the areas and structures which represents moorings of the tradition and culture. The Pune city has also been famous for rich cultural heritage and long history of educational development. It paved strong base for achieving today's recognition 'Oxford of East'. Pune also has a good industrial base since independence, but after IT revolution, it

flourished as one of the ideal place in India as well as international as all-major software industries established here.
Along with the process of transformation Pune has witnessed transformations in urban fabric. Core city area of Pune is denser and represents old structures and architecture. It is developed in more organic pattern. [4] Instead new development is in more planned manner and with less density. These patterns kept changing along with chronological development of the city. The low rise high density which was the alternative to urban sprawl in the core continued due to few factors, thus assessing and understanding the phenomena of growth of traditional city by analysing the housing condition became more important.
The core area signifies the traditional Architecture which dates back to Peshwa era. It was the central trading area which attracted many artisans to settle within. Thus the Vernacular Architecture where the communities intending to build had close connections to housing, practical considerations and Architectural response was reflected in building complex. The type attempted to bring together the benefits of urban and sub-urban living.
The High density residential zone which was result of unplanned neighbourhood. The housing patterns and sizes shows variations in old core which can be the result of ecological footprints back then. Varieties of typologies were observed which interconnected through communal spaces highlighting the community biased development. The urban scenario of core city reflects the heterogeneous community with dense population which demands for housing causing congestion in the inner city. As a result of this the high income group which owned the wada's shifted to outer periphery of the core and rented out traditional housing.
The streetscape in the Peth areas still glorifies the old typologies, like wada, traditional houses, and chawls. It can be said that WADAs are most distinctive residential typology in the urban fabric of old Pune City. [5]

3Cultural Infrastructure

Along with the transformations in happening in the cities as a result of rapid urbanization, the cultural infrastructure in the urban areas is also witnessing a paradigm shift. Culture is recognized as an integral part of communities and a key element of creating great places for people to live, work, visit, play and do business. Cultural infrastructure is not solely hard infrastructure; people, technology, cultural collections and buildings form definition of cultural infrastructure. [6] Together, these components of cultural infrastructure enable delivery of heritage, arts and cultural experiences. Cultural heritage of any city plays key role in cultural infrastructure of the city. Any city which undergo through continuous process of transformation change its face chronologically. But origins of its character remains in olden architecture and built forms. Every city has a core area which represents original architectural character. Along with the urbanization and growth architecture, materials, built form keep changing over the period. Hence these old structures which represent character of the city plays an important role in cultural infrastructure of any city. This has never considered in any policy planning or regulatory framework as a component of development plans. Regulatory planning is essential to the development of existing and future cultural infrastructure.

5Image of the city

It is explained in the theory of Kevin lynch, image of the city; a public image of any given city is the overlap of many individual buildings. There are other influences on image ability, such as the social meaning of the area, its function, its history, or even its name. The contents of the city images so far studied, with reference to physical form can be classified into five key elements: paths, edges, districts, nodes and landmarks. These elements are integral part of any urban area which strongly represents the character of the city.
Paths are the channels along with movement happens. Edges are the boundaries between two phases, linear breaks in continuity basically lateral references rather than co-ordinate axes. Districts are sections of the city having two dimensional extent having some common, identifying character. Nodes and are points, the strategic spots of the city, which are intensive foci for an individual one. Landmarks are the points of reference person cannot enter into. These can be buildings, signs, stores, public arts etc. these are the physical elements with unique and special visual features that has point specific location and can be identified from the distance. These elements are integral part of any city's urban fabric.
Once the pride of the Maratha Empire, Pune exists as an intriguing blend of the past and the present. The city revealed in its image as Chhatrapati Shivaji's legacy before globalization hit on its door in the 1990s. Since then, the erstwhile capital of the Peshwas has transformed into a smart, vibrant metropolis that thrives with a modern vibe. However, Pune managed to preserve its bygone-era charm through its historical edifices and monuments. Exploring the city is a pleasant affair in itself because of this contrasting aspect.
City has many historical places which are nodes and landmarks representing the cultural and architectural face of the city. A walk through the streets of Pune will reveal to you a wealth of historic structures that offer a glimpse into the city's rich past. Dotted with ancient temples, palaces, forts, and Wadas. Which has its own importance in terms of cultural infrastructure of the city. Areas like cantonment area, old city clusters such as Kasaba Peth, Mandai precinct are the precincts representing old face of the city.
Together these old precincts, heritage buildings, wadas, mountains, forts, streets like MG rod, JM road form different layer of image of the Pune city. This image represents the architectural, cultural character and heritage of the city. Which needs to be integrated with the current face of the city to preserve the cultural infrastructure of the city.

4Wadas of Pune city

Wada symbolizes the tales of Pune's rich culture and heritage. They are historical symbol of Pune. They are famous for their typical architectural character and the traditional character of old city of the Pune. Wada's were used for both domestic as well as political purposes. These are basically two to three storied tall oldest structures in the city, generally square or rectangular in shape having courtyard at the centre. From the olden days till now, the designs of these houses have changed significantly but the main elements still preserve the essence. Wadas basically denote mansions with the courtyard having introvert design. [7] This type of planning was introduced by Marathas to safeguard their families from outsiders. The design is amalgamation of Mughal, Rajasthani, and Gujarati architecture combined with the local construction techniques. These houses are typically multiple storey buildings with internal open courtyards surrounded by rooms. These are designed to accommodate multiple or single families. The most important element of the wadas are the open courtyards known as 'Chowks'. The number of chowks vary based on the size of the house. The architecture is characterized by stone walls, wooden staircases and open chowks.
During the reign of the Peshwas the neighbourhood of Pune was well established and divided in wards known as Peths. The

major highlights of these old jewels are the woodworked railings of balcony, tinted glass windows and thick walls protecting the interiors from the changing climatic conditions.

If we see the planning of these wadas, the well-defined zoning of the internal spaces can be seen. The areas are divided in public, private, and semi-private zones. There are different elements that make these houses different and unique. Here are few important elements of these traditional houses. [8] Besides a central tank or a well, the central courtyard generally has a temple or a 'tulsi vrindavan'. These massive structures generally have very thick stone or mud walls, which are as thick as 6 to 8 feet. The wooden staircase is often set in the thick walls. Such Wadas are reminiscent of the society characterized by large joint families. They have generally number of bedrooms with a common drawing room, storeroom, kitchen and a toilet block. Formerly 'Wadas' used to be occupied by large joint families but with changing social set up they are occupied by many tenants and sub tenants.

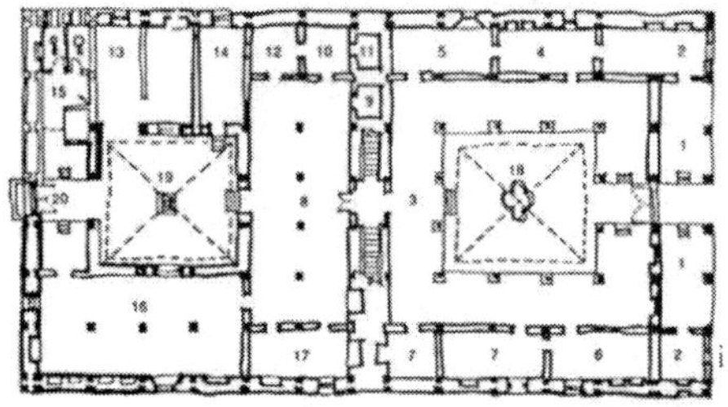

4	Kacheri	14	Balntinichi Kholi
5	Khalbatkhana	15	Nahani / Sandas
6	Baithak	16	Swayanpakghar
7	Dalan	17	Kothar
8	Majghar	18	Karanja
9	Balad	19	Tulsi Vrindawan
10	Devghar	20	Rear Entry

Fig. 1. Typical image of space distribution of wada plan

Fig. 2. Image showing wooden windows, doors and Typical railing pattern in wadas

Wooden Balconies with wooden doors and windows. Windows are wooden windows generally starting from the floor. Wooden Railings with wooden vertical posts are common feature of wada architecture.

Any house cannot be called a Wada without the courtyards or chowks. This is the main element in the design of wadas and its number can vary from 1 to 7 based on the size and capacity of the house. All the rooms are built facing the courtyard. This private open space is commonly used by the families residing there and can be converted to public space during festivals. This space is used to perform rituals, prayers, for gathering, gossip sessions and more.

Fig. 3. Image showing wooden balconies with wooden windows & railing

Apart from residential Wadas there were Wada structures built for administrative purpose also. [9] Throughout the second half of the century the Peshwas spent on improving civic amenities. Building activity was flourished during this period. A number of government wadas were built during this period. The Wadas become larger and lavish. Usually two storeyed, some Wadas became much higher this period.

The courtyard is surrounded by semi-covered passageway known as 'Sopas'. This space connects all the rooms and provides shelter from sun and climatic changes. This space is also used for recreational purposes where the families would put swings and play indoor games. [10]

The transitional space between outside and inside is referred as threshold. At the entrance of house, on the either side of the door, there are elevated platforms called Osari, which is supported by pillars. This provides space for the residents to connect with their neighbours. It also provides shelter from the harsh weather outside.

These wadas have huge wooden doors with smaller gates known as 'Dindi Darwaja'. For day to day purpose only the Dindi Darwaja is kept open. This was mainly for security purposes. The huge wooden gates have bolts called as 'Kadi Koyanda' in the local language.

https://doi.org/10.1051/e3sconf/202017005006

Fig. 4. Image showing ornamental wooden columns at external sopa of vishmbaug wada Pune

Fig.5. Image showing courtyards within the wadas

Fig. 6. Upper Floor of the wada which is mainly used for gatherings and programmes

Fig. 7. Internal Court of the Vishrmbag Wada showing different elements of Wada Architecture.

Here is list of some Heritage Structures which are still standing, reflecting the culture of their time and surviving as link connecting as past to future. [11]

- Shaniwar Wada
- Vishrambaug Wada
- Nana Wada
- Bhide Wada
- Kanya Shala
- Pawar Wada
- Mujumdar Wada
- Yelekar Wada
- Biniwale Wada
- Patwardhan Wada
- Deo wada
- Dagad wada

Apart from these there are some wadas still present in the core area of the city, many of the other old structures which are part of the cultural infrastructure on the city. [12]

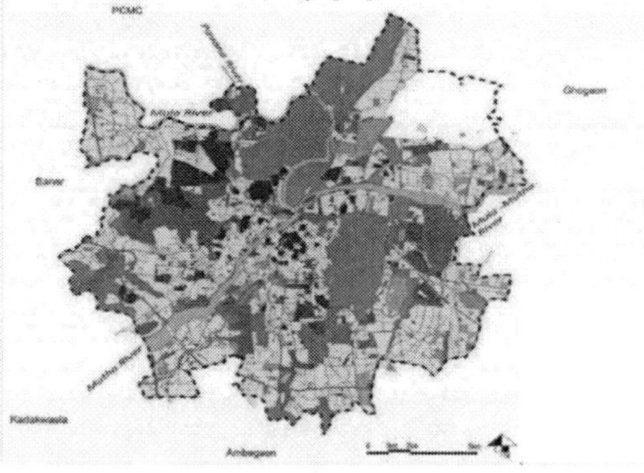

Fig. 8. Wada structures marked on development plan of the city

6Envisaging Pune's cultural infrastructure

Pune has been well known as the Oxford of the East, Queen of the Deccan, cultural capital of Maharashtra, Pensioner's City, Cycle City and Detroit of India. Pune has a long history and played a significant role in India's pre independence era.

Physical and social infrastructure of the city has always focused by regulatory bodies and policy makers. Still the heritage of historical Pune city is critically threatened. Since last 10 years around 50% of built heritage in Pune has disappeared and many more are in poor condition. The unique traditional values of old core area, the residential wadas are also vanishing. A big challenge is to preserve the city core area and must be solved gradually. How the structures can be preserved by doing additions to existing ones is important for preserving the old core of the city. The aim of the preservation has to be in totality of the urban fabric not only in the single building or structure. [13] Some of these wadas and ancient structures can be listed and featured as nodes and landmarks of the city which will be strong elements of city's cultural infrastructure. These structures are not only traditional architecture but the memories of the city which talks about the history, cultural and social structure of the ancient era. In the era of rapidly changing built form and urban fabric these nodes will stay intact and enrich the developing city with their presence. Some of them can be restored and revitalization of some wadas can be done in the form of museums which will showcase the particular era in terms of built form and socio-cultural aspects.

A network of these identified nodes and landmarks need to be established within the city and proper regulatory framework needs to be established to preserve and feature cultural infrastructure of the city. This will involve a coordinated approach to cultural network mapping along with research and consultation that enhances collective understanding of the diverse cultural and artistic circumstances of the community. Some of the residential wada structures like Belbag, Nana Wada can be identified as nodes and prominent built forms such as Vishrmbag wada, Ancient Temples can be identified as landmarks.

Apart from the wada structures there is a huge list of structures which can be part of Pune's cultural infrastructure. These small nodes weave a strong network of cultural infrastructure. A simple stroll across the narrow paths of Shaniwar peth, Kasba peth, Budhwar peth and the adjacent areas reveals a plethora of architectural marvels, some crumbling, some still standing tall. As you drift into the lanes at the cost of getting lost, you find more such interesting buildings that clearly shout out 'heritage.' Not only are there *wadas*, but also multi-storied buildings from the British era that adorn the narrow streets on both sides.

This network of nodes and landmarks is strongly associated with emotional cultural and use values. Those values are part of collective memory and create identity of the city. With regarding to the formation of urban space, the cultural infrastructure always plays a role of identity, for the cultural representation of a group in a given area and at a given time. It creates memories for the people living in it, for the particular urban area and for that particular era.

7Conclusion

Along with the process of urbanization, with continuous flow of migration city of Pune have become more heterogeneous and multicultural. Cultural infrastructure is very important factor needs to be taken into consideration while making regulatory framework for the planning of the city. [14] For the city of Pune, considering its rich cultural heritage along with the physical environment, non- physical elements of the city such as urban spaces need to be improved through creative means promoting the term 'cultural infrastructure' The city needs to establish a strong connection between cultural heritage, tourism and present urban fabric which will not only form cultural locality but also promote urban economic development. Culture is recognized as an integral part of communities and a key element of creating great places for people to live, work, visit, play and do business.

Cultural infrastructure planning needs to be integrated with development plan of the city. The wadas are the surviving built-forms which as architecture epitomize the culture of their times, through their conservation the heritage of the city can be showcased and a link can be established between the past and the future

The cultural heritage of Pune, old and new writes the story of the city. It reflects its character and authenticity in today's globalized world. [15] It will transfer the memories of the city through generations of residents of the city as well as through the conservation and showcase of the architecture of the past which can be perceived by the generations to come and add to the culture infrastructure of the city.

References

1. J.Didde, S. Gupta, *Pune Queen of Deccan,* (Intach Pune chapter 2003)
2. T. Y. Sing, S. Yoh, J-Stage, **3**, 1-20, (2016)
3. Architectural Conservation, Royal Institute of Art, Kungl, *Inclusive India – Re: Pune* (2011-12)
4. N.M. Mundhe, R.G. Jaybhaye, International Journal of Environment, Ecology, Family & Urban Studies,**7**, Issue 5, 33-50, (2017)
5. Mashal, *Housing Study for Pune Municipal Corporation* 61-65, (2009-2010)
6. Department of Culture & the Arts, Government of Western Australia, *Cultural Infrastructure Directions*, (2012)
7. R. Deshpande, R. Kotharkar, 10th International Space Syntax Symposium, 002, (2015)
8. S. S. Dhepe, C. Valsson, International Journal of Engineering Research & Application, **7**, Issue 3, Part 1, 12-22 (2017)
9. J. Chadchan, R. Shankar, International Journal of Society Systems Science, **6** , Issue 2, (2014)
10. www.nobroker.com/blog/architecture-of-traditional-houses-pune
11. www.heritage.intach.org
12. www.punemunicipalcorporation.org
13. Indian National Trust for Art & Cultural Heritage, National Institute of Urban Affairs, *Compendium of good practices, Urban Heritage in Indian Cities,* (2015)
14. www.intachpune.org
15. Confederation of Indian Industry, PWC India, *Transforming Urban India, Art and Culture play a pivotal role* , (2018)

https://doi.org/10.1051/e3sconf/202017006001

Redefining smartness in township with Internet of Things & Artificial Intelligence: Dholera city

Raghav Bang[1,*], *Manish Patel*[1], *Vasu Garg*[1], *Vishal Kasa*[1], *Jyoti Malhotra*[1], *and Sambhaji Sarode* [1]
[1]MIT School of Engineering, MIT ADT University, Pune (MS),India.

Abstract: Internet of Things (IoT) with Artificial Intelligence (AI) has the virtue to address the key challenges encountered by the excessive Urban population; contributing to water management, waste management, energy crisis, and many such affairs. The urban city has reached the level of water scarcity with no adequate water supply. The lack of interconnectivity within the city also leads to severe consequences, such as delayed responses to emergency situations along with irregular traffic and infrastructure management. "**Dholera**" the futuristic city attempt to solve these issues. Dholera is the biggest and India's first upcoming greenfield smart city solution developed under the Delhi Mumbai Industrial Corridor (DMIC) project in Gujarat, India. We have analyzed a few domains from this township project, to mention a few - Water Management, Waste Management, City Integrated Operation Centre (CIOC) and City portal. This paper spotlights on the novel ideas enhancing the smart city features and the working. Automating the city resources using futuristic technologies like big data analytics, Artificial Intelligence (AI) and the Internet of Things (IoT) would make the city well-functioning. In Dholera city, various sensors are mounted and interconnected to collect the data, monitor it, and communicate the values for dynamic action(s). Dholera has AI-based urban transportation, smart grids, renewable energy, solar power, waste and water management, along with urban farming, contributing to a reduction in carbon dioxide emissions and improving energy, water and managing traffic issues effectively. Smart cities are well classified as the growth bar contributing to the universal economy. This paper presents various models making the Dholera city a Fast Responsive, Sustainable, Intelligent and well-connected township.

1. Introduction

India is a country with a population of over 1.3 billion [1]. Out of which, over 31.6 percent of people live in urban areas as per the values [2] mentioned in the Fig.1.

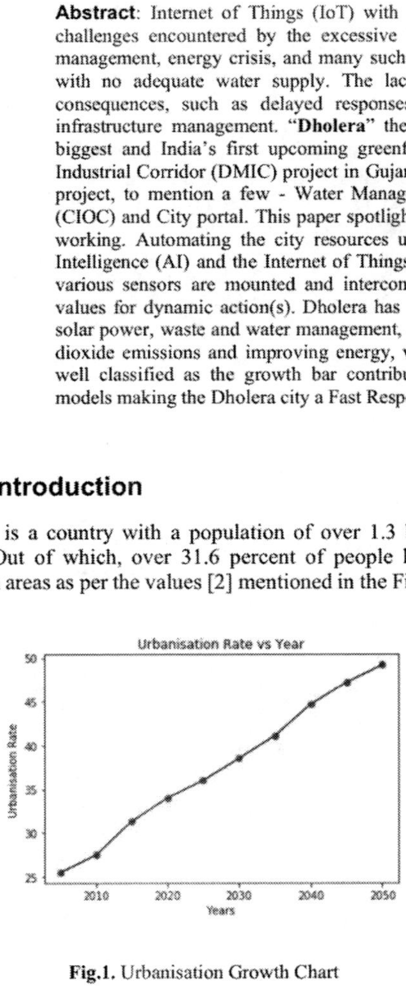

Fig.1. Urbanisation Growth Chart

No one can deny the fact that urbanisation helps in the economy and various other factors of the country but there is a major downside of this growth with sudden increase in the population rate leading to increase in the slum areas and decrease in the standard of living resulting in environmental damage

Despite of many issues that does not results to decrease the growth of urban area, we need to address them by developing the intelligent solutions. For instance, [3] on 28 September 2019, Patna has gone through a drastic rainfall, almost all areas were under knee-deep water and people were struggling to achieve their daily needs. Similarly on 29 August 2017, Mumbai has gone through the same situation. These problems occurred due to poor infrastructure and the lack of technology used in developing the drainage systems. Wastewater in cities is generally pumped into rivers and seas without any filtration leading to pollution, affecting aquatic life and directly damaging the ecosystem. These issues can be addressed or shedded through the smarter solution embarking the construction of smart cities. Smart cities make use of Information and Communication Technology to overcome these problems faced in ordinary cities and increase the standard of living. Using IoT along with Artificial Intelligence can contribute in decision making as IoT helps to collect data via sensors. Smart cities also have

* Corresponding author: **raghavbang2012@gmail.com**

proper infrastructure planning for industrial and residential areas.

Generally, a smart city is classified into two types Greenfield city and Brownfield city. Building a new city from scratch is called Greenfield city and converting existing cities into a smart city is called Brownfield city.

In the wake of this, the Government of India has launched the smart city mission with the aim to develop 100 smart cities and many Greenfield cities across the country making them citizen-friendly and sustainable. One of the cities in the mission is Dholera which is India's first Industrial Greenfield city. Dholera consists of a smart water management system that solves waterlogging and wastewater problems. It also has 4 ICT service centres, which will act as a central hub for all sensors. Supervisory control and data acquisition (SCADA), a computer system for collecting and analysing real-time data makes the city water supply robust. Smart city infrastructure elements includes adequate water supply with smart management and maintenance, assured electricity, proper waste management, efficient urban mobility, and urban transport. Dholera smart city project is divided into 3 phases. Each phase is of 10 years and the deadline for this project is 2042.

2. Literature Survey

This section summarizes the existing methodologies stepped towards smart cities.

Author [4] highlights the purpose of smart cities helping to drive the economic growth and increase the quality of life of people by harnessing the technology that leads to smart outcomes. Individuals are the most important stakeholders in the success of a smart city. It is not necessary that a successful smart city implementation in one country can be replicated in another. The reason is that the people who live in cities have different perspectives and priorities. One way to solve this is to involve people in the process, i.e. - By providing them the engagement platform to reach out to the officials. This way, the citizens will voluntarily suggest what they want in the city and officials can happily accept their suggestions, feedback, and complaints and function accordingly. In this transparent system, citizens will be satisfied with the efficiency of officials and the city administrators will have the opportunity to be proactive rather than reactive.

Palava[5] is an integrated live-work-learn-play city that is spread over 4500 acres. The city offers effortless connectivity due to its location that is in the centre of thane, Navi Mumbai, and Kalyan. The presence of over 60% open area and thousands of trees offer a better quality of life. In the city of Palava, The 911 service is used to provide emergency response teams. The Palava smart card enables secure access controls and enhanced security which is accompanied by 700+ CCTV cameras for monitoring along with video surveillance at all public places. Solar Energy is primarily used to augment power supply for common areas and ensures that the city uses an eco-friendly way to generate electricity. Also, Recycled water is used for all needs such as flushing and landscaping. A Net-zero SWM(Solid Waste Management) plants convert waste into manure for landscaping and biogas for electricity. The efficient functioning of public vehicles and the prevention of traffic congestion is also ensured in the city.

With the increasing population, the demand of the urban citizens for basic facilities such as water, energy, infrastructure, and clean environment is increasing correspondingly and the fact that urban resources and infrastructure are already used beyond capacity makes the matter even worse [6]. As a result, we are facing issues such as air pollution, waste management, poor water and electricity supply, aging infrastructure, and resource limitation. In order to overcome these problems cities need to alter their way of functioning and that is where the concept of smart cities comes into the picture. It is now time for India to transform its cities into smart cities to achieve the goal of smart urbanisation. Smart cities make use of Information and Communication Technology to reduce most of the challenges faced due to rapid urbanisation. They offer a suitable lifestyle to citizens by- a. Optimized resource usage. b. Connected and transparent public services. c. Better safety and security d. Improved quality of life. In January 2015, the secretary to the Ministry of Urban Development (MoUD) has stated the Indian government's mission to make 100 smart cities across the country.

Today the urbanisation is escalating rapidly[7] and there is no proper structured waste management system in urban cities. In this paper, the author proposed the solution of the "Sharp Garbage bin", which is an IoT recognizing the model. It will send an alert when the waste containing the compartment is completely filled. Sensors are used to detect the weight, height and abnormal behaviour of dustbin. Sensors data will be used to send an alert. The estimation of dustbin waste and sending the alert signal by SMS is actioned by proper utilisation of the data collected from the sensors. Strain Gauge driver, Smoke sensor, Node MCU, and Arduino Microcontroller are the key components of this system. With the help of this system, waste can be managed properly which would reduce the pollution, the smell of garbage and stop the flood of the waste compartment. The productive use of the Sharp Garbage bin can also help in reducing waste management costs and developing a Smart City.

The author [8] suggested the usage of real-time hydraulic sensor data for controlling and alarming in case of water leakage. This system can help in reducing water wastage and proper use of water in agriculture field. This system is specially developed to solve the water scarcity problem in a country like India. Various Wireless sensor technologies with ICT can be used to solve the problem of water scarcity. Leakage of water can be detected using GPR-Ground Penetrating Radar and other electromagnetic sensors. Non-Revenue Water (NRW) can be detected by using pressure sensors at each node. Crop sensors if used properly may help in saving 30% of water in the crop field. GPRS and RF module can be used to control the flow of water.

As well quoted in [9]; the smart city is a concept for the quality of life of people. Almost all resources are dependent on one another whereas some are completely dependent on energy. So, if there is no source of energy then those resources will be completely exhausted. It's the duty of government to prepare rules and administration to use technology to make proper use of all resources effectively. The smart city is made of three things: competence, cleverness, and convenience. A smart city should be strong interlinked and interdependent. This paper showcases all challenges, features, and policies related to smart cities.

In today's world, cities are the destination home for more than half of the world's population. People are facing environmental pressures and the need for infrastructure is increasing rapidly [10]. Smart technologies can help cities to solve challenges as cities will get smarter and will become more responsive. There are different aspects of technology that could eventually play an important role in urban development. Smart cities add different kinds of digital intelligence to existing urban systems. Digital intelligence can save lives, decrease the level of crime, and resolve the health issue. They can save time, reduce waste, etc. Smart cities also use data and technologies to make productive decisions and improve the quality of life.

In [11], the author has framed the concept of Smart city IoT is a which focuses on urban networking to deal with problems of new generation cities. Automated water supply and traffic management are two important aspects along with connectivity mesh networks to build a smart city. Traffic can be managed by doing big data analysis of regular traffic and using the proper algorithm by machine learning to manage it. Water supply can be managed by using PLC and monitoring the water resources. Automated water management could also be done using sensors and SCADA system.

3. Working system

3.1 Smart Infrastructure

The infrastructure of a city [12] comprises of housing, sanitation, water supply, sewage, electric power supply, transportation, gas line, Information and Communication Technology (ICT), waste management, street lights, and communication. Smart City Infrastructure brings smartness in the mobility, economy, living, governance, hospital, traffic management, and transportation. In the developing country, city needs immediate urban infrastructure to meet the rise in the population. To meet this increasing demand, smart infrastructure provides a better solution.

Generally in urban cities, all day to day utilities (gas line, power supply, commercial, ICT, etc.) are not underground, first the roads are made and then the utilities are implanted. Dholera, functions differently and smartly; it is the first smart city in India which has all utilities premounted underground. All those utilities are dependent on each other and none of these utilities clash with each other. Post implant; manholes are used when any underground utility needs servicing.

In water pipelines SCADA (Supervised software) sensors are used, which will send data to the ICT service centre. All issues related to communication technology will be solved by the ICT service centre. It is a place that will have access to all sensors' data, CCTV, etc. It will have an artificial intelligent system that will detect alarming situations such as - fire in the building, a vehicle breaking traffic signals, natural calamity and take the initial steps towards the solution of the problem. Also, it has prediction software that predicts the effect of any problem on other regions of Dholera. ICT building is like a command and control centre that integrates the safety and security of the city and makes Dholera a fully Internet-connected city.

The electrical connections and all other utilities required for the smart streets used in Dholera are also underground. They will work on solar energy and have provisions for installing CCTV cameras. If any street light goes off, the information will be sent to the ICT service centre via sensor alarm. Each street light as demonstrated in Fig.2 is connected to the streetlight gateway. 100 gateways are connected to one central software management server.

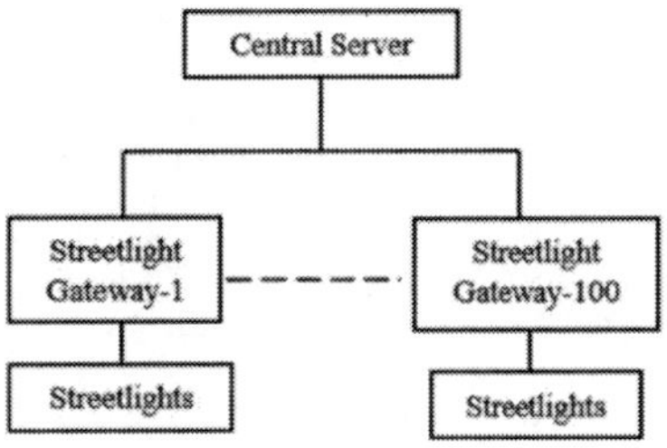

Fig. 2. Streetlight Architecture

Some specialties of Dholera are hospitals with helipads, affordable housing, electric mobility, strong public transport, greenery, and afforestation. In addition to these, it can generate 5000 MW electricity which is the amount of energy required by the Bihar state for a year. It has separate area for economically weaker sections (EWS) households, residential projects, along with commercial projects. Taking the connectivity under consideration - Dholera has 6 lane highway connectivity, seaport, international passenger airport, international cargo airport, metro connectivity within the city. For everyone's ease - all the government work will be done under a single roof i.e. Fig.3 ABCD (Administrative and Business centre for Dholera) building.

Fig.3. ABCD Building

3.2 Energy:

In cities, almost all resources are dependent on one another whereas some are completely dependent on energy. Most of the cities are already facing the energy and environmental challenges which puts 'saving energy and effective use of resources' as one of the main objectives of a smart city. Major cities in India have huge number of vehicles and there is no proper traffic management which causes an excessive amount of pollution which is continuously harming our environment. For example, Delhi the capital of India has unbreathable air due to heavy pollution which has made it India's most polluted city.

The replacement of Electric Vehicles with normal vehicles has a huge impact on protecting the environment. Electric Vehicles are cheaper to run as the cost of electricity is 1/3rd of the petrol. The efficiency of electric vehicles is nearly double because of the electric motor that is used in them in contrast to the combustion engine which is used in fuel vehicles which not only gives less efficiency but also emits harmful gases. Dholera is going to be India's first city with 100% Electric Vehicles [12].

Dholera has the charging slots for engine, car, and mobile. In addition, bus and walkway shelters are placed to make the mobility of people healthy.

An adequate supply of electricity is the most important need in urban cities. Most of the electricity is generated from coal (53%) and it is expected that coal will be completely consumed till 2050 and thus it's the need of the time to switch to the use of renewable sources of energy to generate electricity. Power generation is too costly in India so using solar power is the best alternative to make the environment healthy and liveable. Dholera [12] is in motion for establishing the world's largest solar park with 500 MV and it also has 200 MV wind parks. The Dholera solar park will be developed on a Public-Private Partnership (PPP) model and this will include pay out on works related to the fitting of solar PV panels, inverters and lying of DC cables. But as Dholera is near to sea, the proximity is high the cost is going to increase. The foundation of infrastructure is of very high quality. Generating the power from these alternatives will help tremendously as ultimately it will save energy resources and conserve our environment.

Wind Power and Hydropower are other ways to generate electricity with the use of renewable sources in Dholera. It also has an artificial river in which turbines are installed to be able to generate electricity.

3.3 Water waste management and water logging:

The proper planning of Wastewater management and waterlogging ensures that there can't be a flood like situation and all the excess water will be flown from the canal Fig.4 to the sea. Supervisory control and data acquisition (SCADA) sensors are everywhere even in the smaller pipes which will keep pinging to an ICT service centre. SCADA consist of water level indicator, pressure transmitter, flow transmitter, turbidity analyser, and chlorine analyser. For this management, Dholera has an artificial river (canal) that will be made on the basis of hydrology. The water treatment plant is constructed with the latest SCADA and automated technology to provide the intake and outflow of water which will be monitored by the ICT service centre. Different aspects like leakage of pipes all are monitored by sitting at one place by the use of SCADA.

Fig. 4. Artificial river (Canal)

To prevent the harmful effects of wastewater on flora and fauna of the sea, proper treatment of water will be done before dispersing the water in the sea and river.

A water treatment plant has been planned in Dholera. This water treatment plant will provide the basic requirements of potable water to residential, industrial and commercial activation areas. Also, there is recycled water distribution from the sewage treatment plants (STP) as recycled water will reduce the load of potable water demand. The rainwater directly flows into the canal while wastewater first goes through STP before coming to the canal. A barrier is being designed at the end of the canal, which will control the flow of outgoing water from canal to the sea. In Dholera, there are 9 Elevated service reservoirs (ESR) from which 2 are of potable water (ESR P1 and P2) and 7 are of recycling water (ESR R1...R7). Reservoir is an artificial lake which is used to store water.

Fig. 5. Drainage system in development

It also has a storm water drainage system as projected in the Fig.5 that is designed to sustain the rainwater and safely evacuate the rainwater to the ocean. Solid waste management is the key term for waste management; it is the process of collecting and treating solid waste. This system consists of some priorities of water distribution like network efficiency, water loss prevention, improvement in customer service. The water loss Prevention system is been used in Dholera. In it, the water Leakage management system is a central server, which has a database of all pipelines architecture. It receives live water flow data, water pressure data and meter reading. If any anomaly is found in data then it strikes a query to ICT building that water is being leaked at this particular node in the pipeline.

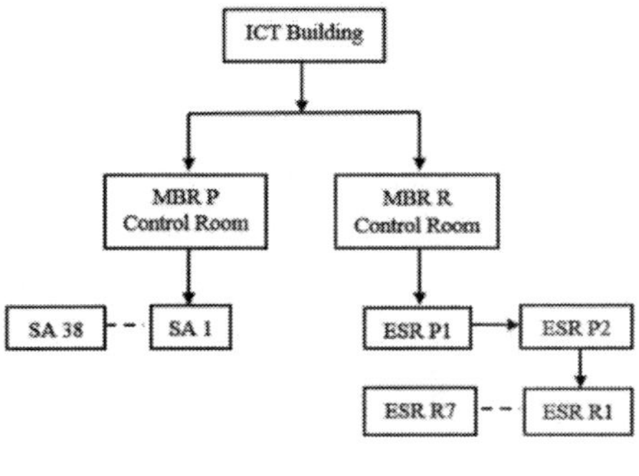

Fig. 6. Overall Instrument Connectivity

In Fig.6 there are a total of 38 service areas (SA), which are connected to the MBR-P (Mass Balancing Reservoir) control room. All ESR are connected to the MBR-R control room and all MBR control rooms are connected to each other. ICT Building controls this whole system.

4. Conclusion:

It is only the advancement in technology that has made a lot of human fantasies real. Of so many of its applications, one of them is in the making of cities smarter. One of the reports by the UN [13] suggests that until 2042, the world will need to create 10,000 new cities and as it seems the smarter option is to build smart cities instead. The idea of smart city is around the world for a while now; many countries in the world are already applying this idea such as Brazil, Taiwan, Denmark, US etc.

Many cities are racing towards the journey of smartness; but they are the brownfield cities. When it comes to Greenfield smart city, the count decreases strikingly. Brownfield Cities already have infrastructure built and making any changes to the existing infrastructure is neither simple nor practical. This constraint is nowhere to be found in case of the Greenfield smart cities as they are built from scratch. Greenfield smart cities could be considered as further advancement in the idea of brownfield smart cities.

Dholera SIR, also considered as Prime Minister Shri Narendra Modi's pet project is going to be the first Greenfield smart city of India which is being built on DMIC corridor. It has been selected as the Greenfield site for a number of reasons such as its large areas government land and strong transport links etc.

In the planning of the city, Dholera team has looked at the performance of the best cities in the world in each infrastructure discipline — such as water, waste-water and power. They took the key parameters and benchmarked Dholera against them, designing Dholera to match or improve on what they have achieved. The city of DHOLERA works on the concept of REAPE i.e.- Recreation, Education, Awareness, Public art and economic growth. In this paper, we have highlighted the need for the city and its importance to scale it as a big hit.

We have observed that when any ordinary city is transformed into a smart city, the involvement of technology in almost every aspect plays a huge role in generating the unique clusters of data. The technology used in Green fiel-d Smart city is embedded in it from the very beginning. Thus, it wouldn't be wrong when we say that the Dholera which is a Greenfield smart city when developed will result in huge data clusters containing really sensitive data hence a proper plan for risk management becomes the necessity considering the security concerns. Blockchain can play a significant role while addressing these security concerns.

The primary purpose of the paper is to review this future city of India. The majorly explored domains (but not limited to) are:

- City needs

- The key features of the city in the areas of smart infrastructure, Energy Saving, waste management and waterlogging.
- The working systems of the city.
- The role of technology in solving the problems of the city.
- The application of sensors and SCADA.

References:

1. https://www.statista.com
2. https://tradingeconomics.com
3. www.timesnow.com
4. https://www.ibm.com/blogs/internet-of-things/building-smart-city-solutions/
5. https://www.palava.in/
6. India: Surging to a smarter future-PWC.
7. Gopal Kirshna Shyam ; Sunilkumar S. Manvi ; Priyanka Bharti, 2017 2nd International Conference on Computing and Communications Technologies (ICCCT), *Smart waste management using Internet-of-Things (IoT)*,(2007).
8. Aditya Gupta, S. Mishra , Neeraj Dhanraj Bokde , K.D. Kulat, International Journal of Applied Engineering Research, *Need of smart water systems in India*, **11**,(2016).
9. Nallapaneni Manoj Kumar ; Sonali Goel ; Pradeep Kumar Mallick, 2018 Technologies for Smart-City Energy Security and Power (ICSESP) , *Smart cities in India: Features, policies, current status, and challenges*,(2018).
10. Jonathan Woetzel, Jaana Remes, Brodie Boland, Katrina Lv, Suveer Sinha, Gernot Strube, John Means, Jonathan Law, Andres Cadena, and Valerie von der Tann ,*Smart cities digital solutions for a more livable future*,(2018).
11. Himadri Nath Saha , Supratim Auddy ; Avimita Chatterjee ; Subrata Pal ; Susmit Sarkar ; Rocky Singh , Amrendra Kumar Singh , Priyanshu Sharan , Sohini Banerjee , Ritwik Sarkar, Ankita Maity. 2017 8th Annual Industrial Automation and Electromechanical Engineering Conference (IEMECON), *IoT solutions for smart cities*,(2017).
12. www.dholerasir.com
13. www.toi.com

The Patterns of Mobility and the Socio-Spatial Environments Created in and around the important multimodal nodes in a global city - A case of Navi Mumbai

Suhel M Makandar[1,*], *Saharsha A Naik*[2]
[1] Amity University, Bhatan, Panvel, Maharashtra, India
[2] Amity University, Bhatan, Panvel, Maharashtra, India.

Abstract. Globalization is seen as an important phenomenon shaping and reconfiguring the urbanism around the world. The urban fabric/society is directly impacted by the global network in which a city falls. For example an economic global city will have different issues than the political global city. Mumbai – Navi Mumbai is one of the cities in India which fulfils the most indicators to be a global city. The city had its own share of urban issues and the advent of globalization has led to origination of certain contemporary patterns of city development which address the past urban challenges as well as accommodate present global economy, yet it has been seen that these patterns rather address only the present conditions and the long term strategies with evolution of lifestyle and technology are uncertain. Transportation has evolved as an important aspect influencing the growth patterns of a city. This paper discusses the impacts of multimodal transport developments as a catalyst to growth of cities, its benefits and shortcomings.

1 Introduction:

Globalization along with global modern technologies has led urbanisation towards a completely new paradigm, yet in popular literature globalization is portrayed as a very recent phenomenon, but the fact is that we have already been living in a truly global world for ages [1].

The current trend of globalization is an offshoot of the tremendous power of information and communication technologies that decentralized the economy from one particular city to a number of cities which come in network of global loop [2]. It can be argued upon that, it is not only the important economic centres(cities), which form a network of the global cities; there are the networks of political global cities, religious global cities, global cities of importance for activistism. The urban fabric and the society are directly impacted by the global network in which a city falls. For example an economic global city will exhibit different issues then the political global city.

Many of the current global cities were a part of the influential cities in the past and when the current wave of globalization based on the ICT revolution swept the world, it created service based economy and this introduced a drastic change in the functioning of historic influential cities that were - say at one time manufacturing and industrial towns. Today the industries and the manufacturing units have largely shut down or relocated and the urbanism made way to cater the service based economy [3].

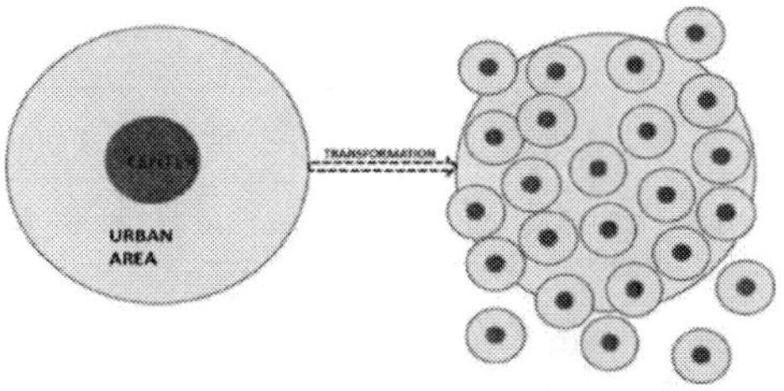

Figure 1 Decentralisation of urban Centre as envisaged due to Information and Communication Technology.

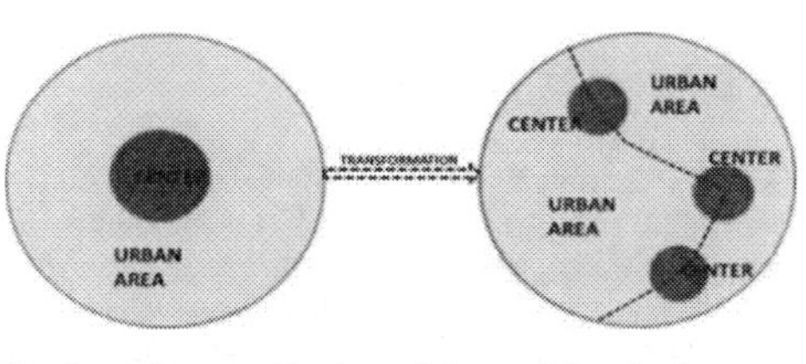

Figure 2 Recentralisation of the traditional urban center into a network of nodes which form a new centre

Bases upon the transition from manufacturing to service based urbanism, the changing fabric of the urban area was very pronounced in the urban centre. The electronic realm helped in decentralizing the centre and it was further envisioned that the electronic media and ICT would decentralize the urban centre to such an extent that the urban character would be lost 'as seen in Figure1. However this doesn't happen and there is a reverse action seen, the centre gets decentralized into

* Corresponding author: **smmakandar@mum.amity.edu**

various centres of special activities, eventually forming a network of nodes within the city 'as seen in figure 2. These networks of nodes connect, thrive and get accentuated through lines of transportation which forms a vital link. The hierarchy of transportation varies in these nodes from regional level to local ward level and hence they eventually start taking up a form of multimodal hub, weather the city designers make provision for it or not.

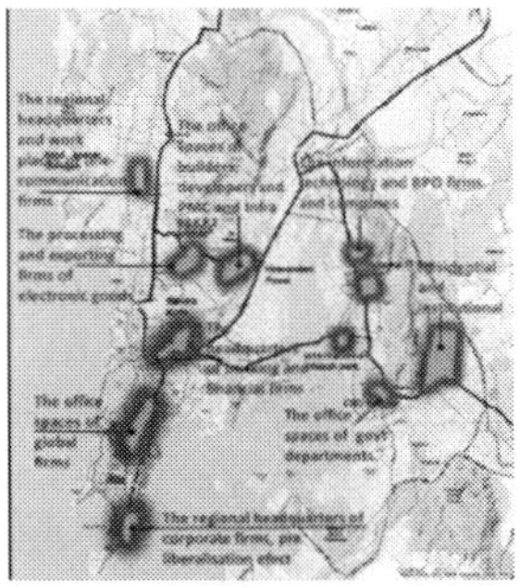

Figure 3 Mumbai - Navi Mumbai and its important activity nodes

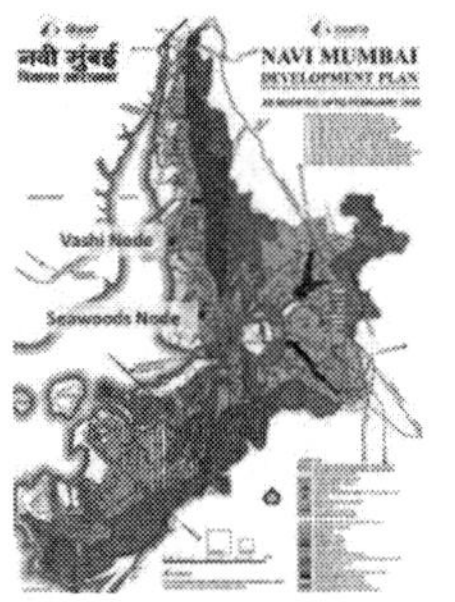

Figure 4 Navi Mumbai City development plan showing the two nodes of study area

In the further course of paper intentionally designed multi modal hub shall be investigated upon to ascertain the pattern and layers of mobility and the impact it creates on the socio-spatial aspect as well as urban design.

2 Method and Material:

Methodology: - The method adopted for research is through spatial map analysis. The attributes for evaluation are taken based upon the findings from review of various relevant research papers. The spatial integration and system of patterns created for the following elements are studied in this research paper:-

- Urban Centre & Node
- Patterns of connectivity
- Developments in and around the Multimodal node
- Typology of development of node
- Impact of Node on the surrounding urbanism
- User group interaction

Material:- The materials for study are primary visual survey, satellite imagery and mapping, research studies on impact of mobility patterns on the city and concept of 'urban Centre'.

3 Discussions:

Mumbai has always been an important destination on the network of global influential cities, the growth prospects and economic opportunities have led to the tremendous physical growth of cities into a system of sister cities. Today the Mumbai Metropolitan Region(MMR) comprises of multiple municipal areas such as Navi Mumbai, Panvel, Vasai Virar, Kalyan Dombivali, Mira Bhayandar, where all these cities together function as different organs of one large city i.e. Mumbai. All these organs function in synchronization as they are connected by common veins which are the Suburban railway network [4].

Out of all these cities Navi Mumbai is a planned city which was foreseen as an attempt to decongest the Island city of Mumbai in the early 1970's [5]. It was planned with consideration for the Local Suburban Rail Network as the backbone connecting the various nodes of Navi Mumbai to the Island city. Hence the Railway stations gained a pivotal role spatially as well as functionally in all the nodes. Today these railway stations have gained a central function as the urban nodes have got self-sustainable and depend lesser for day to day activities on the island city [6].

The railway stations achieving a central function was envisaged well in advance by the planners of Navi Mumbai and hence the railway stations were proposed to be developed as a multifunctional node having a combination of multi-layered transportation as well as economic activities [7]. The Nodes of Vashi and Seawoods are an example for development of multifunctional nodes having different activity juxtaposing with transportation in each case.

3.1. Vashi Node

This node was the first of the settlements to blossom outside the island city in Navi-Mumbai, the Vashi railway station is developed as a combination of a transportation node as well as an economic node in the form Railway Station at the Ground floor level and International InfoTech Park building on the floors above. This International InfoTech park has provision for office establishments thus resulting in bringing a certain amount of self-dependency with reference to employment in the newer nodes.[8] The Developments around the Vashi railway station comprises of commercial use catering to a wide usage from office spaces to retail and large scale shopping malls. Also there are large scale parking lots abutting the railway station.

Figure 5 Vashi Multimodal Railway station

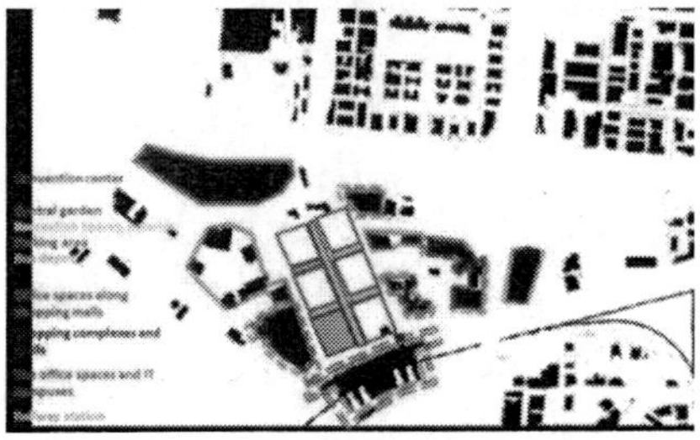

Figure 6 Figure-ground map Vashi railway station precinct

Certain challenges which are an outcome of the Multimodal Node typology in Vashi are as follows:-

3.1.1 Challenges in Vashi Node

- Commercial developments in the entire sector where Multimodal hub is located makes the entire sector deserted space post 9pm.
- Larger footprints of buildings and developments in the surrounding result in lack of human-built environment interactional experience.
- Lack of urban design elements like visual character, public engaging activities around the node
- As this multimodal hub attracts huge footfall the above issues are critical to the public comfort. [9]

3.2. Seawoods Node

Seawoods gathered its fame and recognition for the flagship NRI housing scheme implemented by City and Industrial Development Corporation (CIDCO). The Seawoods railway station is designed as a multimodal transport node where there is Railway station at the ground floor and Grand Central Complex on the floors above. The Grand central comprises of shopping mall, office spaces, recreational spaces and multilevel parking on floors above. The grand central complex now acts as a strong magnet for retail, entertainment and economic activities in Seawoods locality [10]. The developments in surrounding of Seawoods Railway station comprises of mixed use buildings which are a combination of upscale apartments as well as retail and commercial spaces.

Figure 7 Grand Central Complex(Seawoods Node)

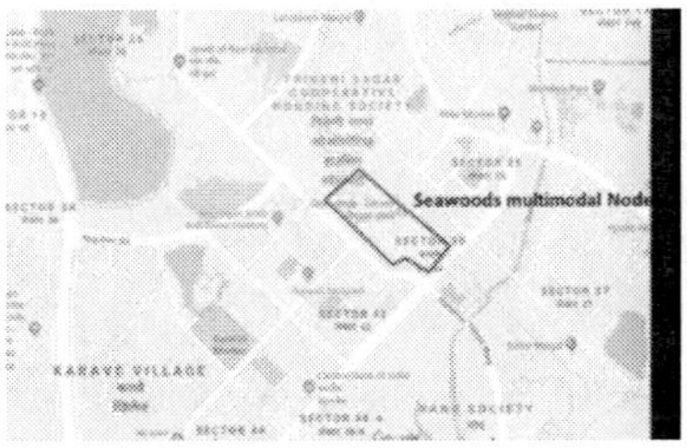

Figure 8 Seawoods Node Map

It can be seen from the example of Vashi and Seawoods nodes that the idea of a multimodal transport node has helped in the development of the city by garnering pace and self-sustenance (from island city) to the newly developing localities. This strategy is indeed helpful when viewed at a larger City development scale as a multi modal hub comprising of commercial and office spaces along the railway station helps in reducing an arm of modal split for people using public transport to reach these spaces. But when viewed at the individual node level, there are certain challenges which are an outcome of this typology of development.[11]

While the Multimodal hub becomes the primary node for transportation and economic/retail activity it attracts large traffic, especially during the peak hours. The public transport (referred to bus services here) in these newer areas are rather less in frequency and do not cater to larger mass volume routes. The commuters also rather do not prefer local buses due to the bad state of the buses, unpredictable waiting time, longer period of time taken to cover the same distance as compared to personal car, inconvenience in summer and monsoon seasons. [12] Hence in these newer areas personal vehicle usage has drastically increased especially for the Intra node commuting, thus it is evident that the Multimodal node becomes a magnet for vehicles especially during the peak hours. Certain challenges which are an outcome of the Multimodal Node typology in Seawoods node are as follows:-

3.2.1 Challenges in Seawoods Node

- Here the issues are related to the connectivity of surrounding localities and congestion around the Multimodal node.
- It is seen that there is a vibrant public realm and enclosure but the congestion leads to deterioration in public experience.

4 Alternative approaches:

In planned nodes, it is essential that there is a constant evaluation of the previous planned developments. The stereotype model should be constantly moulded in order to enhance the urban experience and quality of lifestyle. The following approach for newer developing nodes in Navi Mumbai can be adopted:

- The Railway stations which were envisioned as multimodal nodes now are acquiring the status of Major Node itself and thus in the newer areas there should be strategies for counterbalance through a network of micro-nodes.
- The technology of light mass transit like mono rail, PRT, Trams, are becoming affordable and sophisticated(requiring lesser road footprint) as technologies are advancing and thus becoming attractive for municipal corporations, the same shall be employed in the newer areas and recently developed areas.
- The option for connectivity shall be established taking into account the Inter-nodal (within city amongst suburbs) and Intra-nodal (within the suburb) transport and commutation patterns in order to understand the overlay and confluence points.
- By decongesting the Central node i.e. the multimodal hub from vehicles, the street space can be reclaimed for public realm, as the nodes attract large footfall it is essential to create a vibrant public realm using urban design elements like streetscape, furniture and breathing/halting spaces.
- The areas surrounding the Multimodal node shall have a mixed land use in order to avoid the chances of creating post working hour ghost spaces.
- The building blocks should have an ideal perimeter of 400m to 600m (reference to Barcelona blocks) and the street edge should be vibrant and visually permeable.
- In case of the metro stations/monorail projects being implemented the buildings surrounding the metro station shall have customised regulations in order to make most by establishing direct connectivity to the station and also enhance the visual realm.

All the above factors are to work in tandem and not a piece meal manner for a better socio spatial integration and enhanced urban experience in future.

5 Acknowledgements:

I am grateful to Prof. A. D. Shirodkar (Amity University Mumbai) for constant encouragement in taking up research work. I am thankful to Amity University Mumbai for their motivation to build up, compile and complete this research work. I am indebted to alma mater SPA Bhopal, SSSMS COA Sholapur, as well as Taranum Makandar.

6 References

1. Sassen, Saskia. Place and Production in the World Economy. *Cities in a World Economy*. New york : Sage Publishing, (1994)
2. *From Global Cities to Globalised Urbanism*. Brenner , Neil and Keil, Roger. (2014)
3. Bishop, Ryan, Phillips, John and Yeo, Wei Wei. Perpetuating Cities: Excepting Globalisation and the Southeast Asia Supplement. *Post Colonial Urbanism: Southeastern Asian Cities and Global Processes*. (2013)
4. Prajakta, R Rail network may reach Raigad district. Hindustan Times 22jan (2017)
5, NMMC history website. *https://www.nmmc.gov.in/history* (2019)
6. PWC, Revenue Maximizing Study Report for MRVC. (2014)
7. Livemint Report. *https://www.livemint.com/companies/news/l-t-realty-gets-strong-demand-for-500-flats-on-offer-in-navi-mumbai-project-1557905551843.html* (2019)
8. Amit C, Chattopadhyay RN. Satellite Towns in Neo-Metropolitan Development in India. Springer (2019).
9. Pranab JB. https://www.hindustantimes.com/cities/mess-cleared-from-vashi-railway-station-premises/story-a93z6t4bTiYf5NVinZXvsK.html (2019)
10. Moneycontrol. https://www.moneycontrol.com/news/business/real-estate/lt-realty-eyes-rs-3200-cr-revenues-from-navi-mumbai-project-3974001.html (2019)
11. Sandeep K. India Retailing, Driving the Change https://www.indiaretailing.com/2019/09/20/shopping-centre/driving-the-change-malls-use-technology-for-smart-parking-solutions/ (2019)
12. Karthik R. Navi Mumbai, Retrofitting a Planned City. The City Fix blog, (2010)

https://doi.org/10.1051/e3sconf/202017006003

Management of Various Safety Parameters in Tunnel Construction: A Critical Review

Priyanka Omar[1]*, *MVN. Siva Kumar*[1], *Subbarao Yarramsetty*[1]
[1]Department of Civil Engineering, NIT Warangal, Telangana State-506 004, India.

Abstract Population is growing rapidly and so are vehicles on roads. This leads to enormous need of tunnels and subways for easy locomotion and mobilization. From conventional tunnels, for sewage and water facilities to modern electrified tunnels, for mass transit and underwater highways, method of construction of a tunnel to its safety inspection and maintenance has taken a wide leap. Tunnel construction requires wide range of resources like human, technology, machinery, materials, energy, and finance. To handle all these resources simultaneously becomes complex and requires good management. Even with good management, accidents and hazards might occur. A critical review of various safety parameters while constructing a tunnel and its management is presented in this paper. This review paper discusses the history of safety of tunnels till present safety measures and techniques adopted around the world for different tunnels; various critical factors which affect the safety in tunnels; the extent of damage occurring due to these factors; numerous preventive measures which are adopted around the world to prevent loss of property and lives in tunnels; and advanced technology and software, which are being used in modern era to enhance safety in tunnel construction. The study from manual horse drawn-tramway to autonomous robotic system has been done. This paper also considers various tunnels around the world and gives a summary of factors for safety focused on for making these tunnels, with its adopted remedy.

1 Introduction

Present day tunnels differ completely from what they were in the past. From their usage to their construction practices, everything got evolved drastically. Trench digging was the most common conventional method used for constructing tunnels for waterways. In 1681, France used first power drove method of the form explosives and Nitro-glycerine dynamites, for drill and blasting. Due to lack of skill, knowledge and experience in construction fields, many others started focusing on human resource and machine management, which plays a crucial role in complex areas of tunnel constructions. With further increase in population and mobility, risks of accidents increased, and thereby power and safety management too started getting limelight, for better locomotion. And for making innovative tunnels, as a major source of attraction for tourists; cost, design and economy was also being focused. Tunnel construction deals with lots of initial challenges, as shown in fig.1. (a) For proper work order of all these conditions, skilled and unskilled human resources are needed, whose health and safety should be properly controlled on site. Many theories have been given for maintaining a good working environment. Many short time parameters like, vehicle speed, wind speed, wind direction, mix and age, play a crucial role than long time parameters like, monitoring and controlling, as latter ones are often hindered.

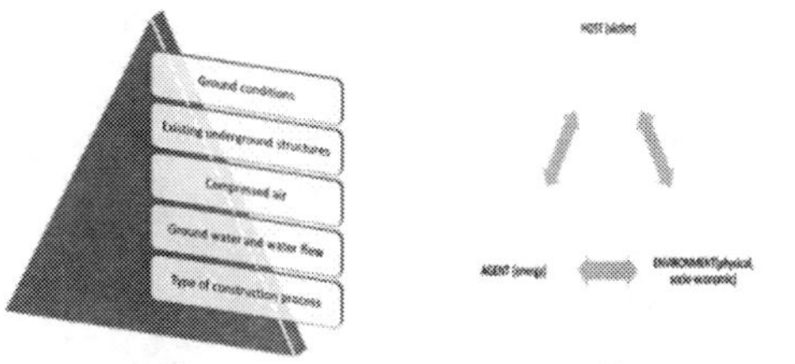

(a) (b)

Fig.1. (a) Challenges triangle, (b) Epidemiological triangle

Fig.1. (b) shows epidemiological triangle which is a model for controlling the spread of infectious diseases during the 19th century in Europe. By controlling one of the factors of this triangle, spread can be controlled. The same theory is also applicable in underground construction. Hence, it is essential to control factors like damaging of surface structures, settlements (before and after excavations), reduction in strength of concrete with aging, cracks due to vibration of moving vehicles, fire hazards due to freight carried by the trucks and many other factors, which are discussed further, in underground construction [1].

Various researches and publications have been made in the past years related to tunnel and its safety. Fig. 2 shows the number of publications versus year of publication graph. From the past five decades, around

*Corresponding author: **priyanka@student.nitw.ac.in**

 https://doi.org/10.1051/e3sconf/202017006003

forty thousand publications have been published on various internet sources, with maximum number of publications done in Springer, followed by ASCE (American Society of Civil Engineering) and Taylor and Francis.

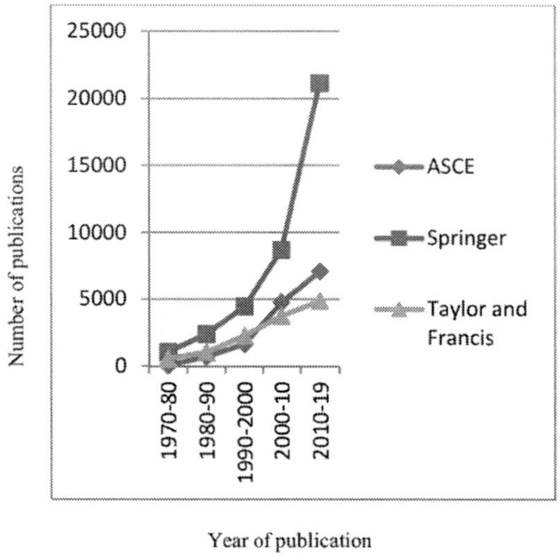

Fig.2. Graph showing past publications in tunnels

2 Progress of tunnel safety

Proper documentation of progress, resources used, construction methods, equipments, quality analysis, safety, site conditions, building information, data entry of manpower, quantity of materials and details, shall be done while constructing a tunnel, which can even reduce future disputes.

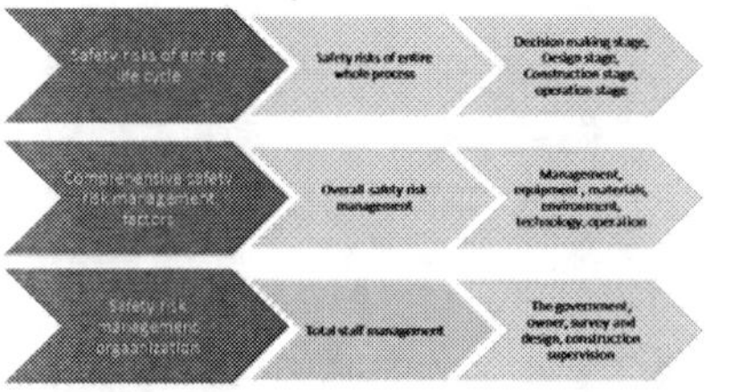

Fig.3. Methodology adopted by China for underground engineering

Fig.3. shows the procedure how China manages safety risk. Many companies now keep a track of all activities like approve payments, update schedules, field/shop adjustments, quality control using advanced computer-based software for our betterment and swiftness. Proper energy analysis has been done of cutter head drives, shield operations and various nonlinear parameters, which has become a challenge for tunnel engineers [2]. Various factors in the process of excavation like type of foundation, depth of tunnel section, size, geometric characteristics, makes it difficult to take risk management of each and every building [3]. For economic analysis, many designers have adopted dynamic soil parameters which were resulted from good ground investigation and soil testing. To reduce the impact of explosion for bomb type tunnel construction, fine protective layer over tunnel structure is proposed rather than constructing cost effective tunnels [4]. Earth pressure balanced shield machine (EPBS) is very efficient and economical for underground construction of metro tunnel in urban areas [5]. Displacement monitoring has become more important regulating factor based upon primary soil parameters. Electric heat tracing method is most widely used for frost damage tunnels and tunnels in cold regions [6]. Tunnel ventilation is one of the major factors which is often neglected. Natural and renewable wind has been used to produce power for proper ventilation of long tunnels at the tunnel construction site. Optimized energy consumption and feasible solutions have been given for maintaining proper ventilation inside the tunnel [7]. The main factor for initiating any construction is Communication [8].

3 Extent of damage in tunnels

Train tunnel fires are more devastating than road tunnel fires, due to wide mobilization of freights. Still the former ones are frequently used for easy locomotion, keeping the aim to reduce the risk of fatalities [9]. With growing population and vehicles, the safety of tunnels is a major concern. The major factors which contribute towards safety in an underground construction can be listed as:

1. Human health and accidental factors
2. Environmental factors
3. Technical factors

According to UNESCO [10], the level of tunnel safety can be classified into following groups:

1. Road users
2. Infrastructure
3. Operation
4. Vehicles

Drainage systems have been made adequate in design to avoid contact of water with the structure and added pressure on side walls and concrete slabs. This will reduce the maintenance costs of tunnels [11]. Shanghai has used grouting based treatment method for preventing the settlement in soft deposits. It has shown reduction in tunnel settlement rate and prevented further settlement [12]. Some of the important factors which can be used in advancing an underwater tunnel construction include: (1) Error-free forecasting of the topographical conditions and oozing characteristics of water lying ahead of the working portion; (2) High pressure water gushes shall be prevented [13].

The casualty factor as given by UNESCO is shown in Table 1, for the pre-calculation of risk and the extent of damage caused by several categories of transportation modes.

Table 1. Fatality factor adopted for calculation of risk

Mode of transportation	Casualties per 10^9 passenger- kms
Airline	0.06
Railway	0.1
Bus/coach occupant	0.19
Maritime vessel	0.27
Occupant	2.67
Automobile motorist	1.82
Automobile traveller	0.85
Fuelled two wheeler	37.8

There is a high degree of uncertainty in underground cavern groups as it has been observed that they commonly fail due to variation in construction over time. A key factor for determining 3D simulations is the uncertainty in construction and geological processes [14]. A number of hazards are being faced by human workforce during grouting operations. Grouting work consists of three major stages: 1) Involvement of powerful hydraulic and pneumatic drilling machines for drilling; 2) Usage of agitators and mixers of numerous capacities for mixing process; and 3) Usage of different capacity pumps for better output in pumping [15].

Floods are another factor which causes hindrance in tunnel construction. Detailed study has been conducted elicited on region wise flood risk assessment technique archives, which includes (a) statistical methods, (b) analysis using remote sensing (RS) and/or geographical information system (GIS),(c) multi-criteria analysis, and (d) scenario-based analysis. A perspective approach based on evaluation of overflowing risk of tunnel systems has been proposed. This approach has the following two characteristics: - from local to global, and from qualification to quantification. Overflow risk in tunnels is handled through the mix usage of GIS, satellite-based radio navigation systems like global position system (GPS) or our indigenously developed Indian Regional Navigation Satellite System (IRNSS): NavIC and build information modelling (BIM) has also been proposed [16]. Tunnel Dismantling Machine is as important as Tunnel Boring Machine [17].

Noise-induced hearing loss had been occurred among 92% of the tunnel constructing manpower [18]. There is lack of occupational medicine and industrial hygiene system with a low awareness of health and safety issues at all levels of management at many tunnel constructing sites. There should be a good functioning regulatory body, proper client supervision and sophisticated trade unions, which should have a high priority in construction project.

4 Critical conditions of focus

Many serious pollutants have both short and long term exposure, affecting the health of humans. For this, air quality standards have been set for minimizing health impact [19]. There should be rigorous distinction between 'core' and 'peripheral' staff, with a distinction between soft and hard human resource management as well [20].

There should be a strict distribution of the explosives into the rock and a specific drilling pattern should be followed for good blasting results. When designing the drilling pattern, several factors must be taken into account like rock drill and blast ability, type of explosives, restrictions in blast vibration and error-free requirements of the blasted wall etc [21]. In the process of localized tunnel construction, water table causes mud inrush; cost of construction, stability of the tunnel, schedule and safety, all gets affected by landslides and rock collapse [22]. The major step of construction that should be concentrated upon while controlling tunnel digging is the blasting. While constructing new tunnels, monitored value of blasting vibration of already made tunnels (maximum velocity) can be used in design parameters to lessen the impact of vibration [23].

Face stability is very critical point which should be kept in mind for the safety of underwater construction across fault zones [24]. Due to the complex nature of the tunnels, various important parameters had been noted for the sound operation of tunnel network. These parameters include the traffic, the behaviour of the people trapped there during evacuation, the latency of the tunnel personnel in traffic interruption or activating the mechanical ventilation, fire behaviour and the natural environmental conditions. While driving, drivers in road tunnels will gradually reduce their speed and increase their lateral position from the respective tunnel wall while driving [25]. Flame hindering on highway tunnel pavement made from asphalt shall be taken into account for constructing the tunnels [26].

Table 2. Parameters for tunnel design affected by various categories

Category	Parameter
Tunnel attributes	Gradient, Geometry, Ventilation
Vehicle fleet attributes	Age and mix, Emission Standards, Fuel quality, Engine type
Automobile operations	Stop and go traffic, Traffic density, Temperature, Inspection and maintenance

Table 2 gives a summary of parameters which should be taken care of while constructing the tunnel. Each of these parameters shall have an equivalent focus while designing the tunnels.

5 Preventive measures

Assumption of sequential accident process was made and major development took place since the Second World War for a number of safety analysis techniques [27]. Incorporation of risk-based analysis to safeguard and control had been proposed which can be classified into qualitative and quantitative risk analyses, to prevent heavy casualties and property losses resulting from safety violations [28]. Attempt had been made to formulate the basic event data as a fuzzy probability set and using them in subsequent computation [29]. A feasible method of adoption of the semi-quantitative analysis had been taken into effect. For underwater tunnels, methods which can be adopted are wire line engineering method, reverse method, double pipe reversing engineering method, pressure boring method, air boring method and grouting [13].

To release strain energy, it is relevant to induce spalling with a long period over roof and spandrel, whenever there is lower lateral pressure coefficient [30]. For preliminary check-up of underground conditions in the Himalayan region, a novel multiple-graph technique had been proposed. This technique is based on the estimation of three quantities, i.e. the ground condition, competency factor and rock mass strength [31].

Due to vulnerability in ground shaking, post-earthquake investigations had shown that tunnels are exposed to seismic risk [32]. Inclusion of mandated concepts and reactions shall be done for the driving licensing procedure of long tunnels and implementation of driving safety education should also be done for improving long tunnel safety. The designers should keep in mind the environmental influence while designing tunnels [33]. Fig 4 shows functioning inside a tunnel for proper alignment when engineers do not know the centre line during construction. People are very much concerned regarding the safety aspects of vehicular tunnels because of a severe past fire in Hong Kong, resulting from a truck carrying diesel outside a tunnel [34]. To prevent the loss of visual adaptation, proposal of an optical coupling of networked tunnels with translucent red structures shall be executed to weakly lit environments [35]. During tunnelling, radon level can be on the high side especially where poor ventilation is forecasted. Radon content in tunnel environment is found to be due to ventilation provision, groundwater ingress and topographical and geological factors. The situation had been improved a lot through the adoption of control legislation, education and more specifically, improvising of radon management plan in every tunnel drilling project [36].

Underground workplaces are a crucial element in modern urban planning. Proper research needs to be done on the health implications and the probable environmental vulnerability, to envisage future research paths including effects of artificial lighting particularly night time exposure to short wavelength (blue) light (light at night) and shift working, associated health effects, sleeplessness and circadian rhythm disruptions. Risks are manifold when exposed to working underground and may be linked to hypovitaminosis D, excessive noise, radar exposure and negative psychological effects.

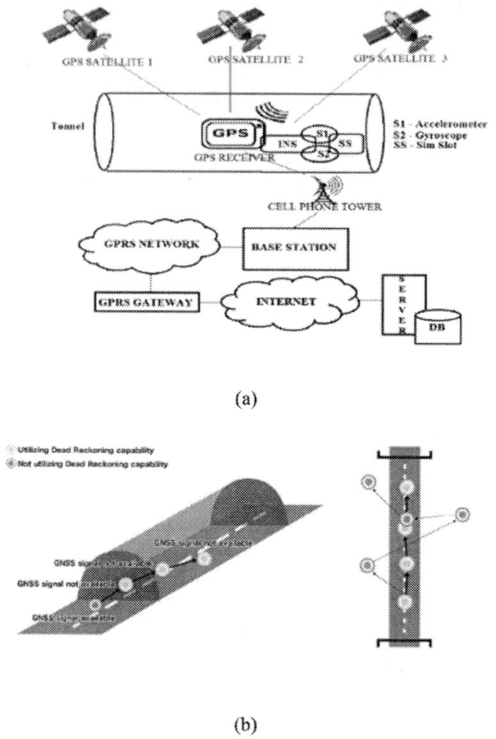

Fig.4. (a) GPS adoption for a tunnel construction, (b) ground tracking in tunnels where the GPS/GNSS signals are shielded and unavailable.

Research needs to be done on the health effects of underground environments so as to plan safety measures and remedial action accordingly [37].

6 Application of advanced technology to improve safety

Under well-controlled conditions, dynamometer-based emission factors are measured where proper monitoring of different emission-influencing parameters are done [19]. Serious efforts are involved in physical, analytical, or numerical models. Installation of durable and robust sensors enabled intelligent prediction and monitoring of real time behaviour of underground structures [25]. Introduction of a real-time online safety analysis approach which is based on four-dimensional technology, done during the construction of tunnel had been done. The 4D construction safety information models very well established, since it gives the complete picture of dynamic bond between the structural system and the construction process information, and thus are being widely used for rapid and automatic generation of 3D models. A structural analysis model with the construction schedule and material information can also be done.

There is a dearth of experience in implementation of construction management in underdeveloped nations. To improve the functioning of management, evaluation of human resource management processes shall be done in construction project, which should use inputs, outputs, utilities and tools and techniques recommended by the Guide to the Project Management Body of Knowledge (PMBOK, 2004) [38]. Numerous technologies had been adopted for the construction safety, including sensor-based technology, information communication technology, radio frequency identification, 3S (GIS/GPS/RS) technology, and virtual reality. With the proper usage of advanced seismic wave propagation modelling software SPECFEM2D, ample number of studies had been conducted on ground motion behaviour around excavations via analysis of velocity wave fields, PPV contours, and amplification factors under different k/D ratios. Both the loading conditions, ''quasi-static'' and ''dynamic interaction'', had been considered in the modelling analysis [39].

Extra safety measures should be provided during the remedial work. Installations like exhaust air blinds in the area prone to fire for full ventilation power and for emergency niches, a compressed air system shall be well equipped at the tunnel site [40]. The special light zones could help relieve drivers' vision tiredness to some extent and moreover develop certain optical stimulus that can intensify drivers' attention [41]. Building information models (BIMs) has the ability to support information accoutring over the entire lifecycle of infrastructure assets [42].

One of the main inferences that could be drawn out is that nine usual risk factors for mechanical tunnelling had been identified, i.e. excavation of tunnel, launching of TBM, segment assembly, special procedures and conditions, arrival of TBM, grouting, pilot excavation, mucking, and shaft construction [43]. Effects of TBM excavation rate can be analyzed by special laboratory testing which uses numerical models based on elasto-visco-plastic constitutive law. An important step of TBM tunnelling with segmental concrete lining has been taken towards hydroelectric power development and better transportation, especially in Himalayan regions of India. [44].

7 Discussions and Conclusion

Table 3 summarizes all the past tunnels with their adopted remedies.

Table 3. Tunnels constructed in past with adopted remedies

Tunnel name	Remedy
Tunnel beside Taipei Songshan airport	Adoption of dynamic soil parameters, obtained from good ground investigation and soil testing
Burata Railway Tunnel, Spain	Manual lowering of the water level
Shield tunnels under the Huangpu River of Shanghai	For joint sealing, preference of gasket sealing over paint sealing
Environmental effects of tunnel excavation in soft and shallow ground with EPBM: the case of Istanbul	Tunnels were excavated by using two Earth Pressure Balance Tunnel Boring Machines
Safety management in tunnel construction: Case study of Wuhan metro construction in China	ANSYS finite element software was employed to simulate a numerical model of the drilling process
Qingdao Metro subway tunnel	Grouting method was used to improve collapse-preventing treatment in the sand layer
Niayesh tunnel	Sequential excavation method (SEM) was selected for tunnel construction
MTR West Island Line (WIL) project and Overrun Tunnel (ORT	a customised tunnel dismantling machine (TDM) was innovated
Seikan Tunnel	Pilot boring and grouting methods used

After reviewing all the contents and publications on tunnel safety, following conclusions can be drawn out:

- China has shown an extent of development in tunnel engineering, which can help the world take a leap in tunnel construction and reducing the risks in its formation.
- Various factors like passengers, mode of transportation, type of tunnels and its usage, emissions caused by fuels inside the tunnels, etc. shall be considered while making studies for tunnel construction.
- Fire is one of the most hazardous factors which shall be taken care of. Proper fire prevention equipments and remedies shall be implemented at the tunnel site.
- Other factors like human resource management, technical factors, and environmental factors, shall be studied and analyzed well before hand, to avoid any serious damage or accidents.

A number of studies have proved that the tunnel can be used as a short-time shelter if a seismic secondary disaster such as landslide or avalanche occurs. It can also act as an important route for post-disaster rescue operations. Various diversified, innovative and continuous technologies have come to ensure tunnel safety during its construction. The extent of damage due to carriage of goods shall be taken into account and should be made safer to prevent fire hazards. Design safety, monitoring and safety information shall be kept updated by researchers and tunnel engineers. Strictly speaking, tunnel construction safety management is a crucial process of safety information collection, scheduling, transmission, storage, analysis, evaluation and response which should be done efficiently.

8 References

1. I. Ocak, Environ. Earth Sci. **59**, 347 (2009).
2. X. Li and H. Zhu, (2013).
3. J. Chen, J. Yang, H. Wang, B. Liu, and X. Yang, Geotech. Spec. Publ. 211 (2009).
4. M. W. Gui and M. C. Chien, Geotech. Geol. Eng. **24**, 227 (2006).
5. F. Meng, R. Chen, and X. Kang, Tunn. Undergr. Sp. Technol. **80**, 53 (2018).
6. Z. Hu, H. Ding, J. Lai, H. Wang, X. Wang, and S. He, Constr. Build. Mater. **185**, 670 (2018).
7. C. Guo, M. Wang, L. Yang, Z. Sun, Y. Zhang, and J. Xu, Renew. Sustain. Energy Rev. **53**, 1558 (2016).
8. J. K. Lemley, Tunn. Undergr. Sp. Technol. **10**, 9 (1995).
9. M. S. Tomar and S. Khurana, Tunn. Undergr. Sp. Technol. **85**, 149 (2019).
10. I. T. Committee, Group 1 (2001).
11. Osama Bedair, Pract. Period. Struct. Des. Constr. **16**, 73 (2011).
12. D. M. Zhang, Z. K. Huang, R. L. Wang, J. Y. Yan, and J. Zhang, Tunn. Undergr. Sp. Technol. (2018).
13. B. A. Kitamura and Y. Takeuchi, **109**, 25 (1983).
14. H. Salehi and R. Burgueño, Eng. Struct. **171**, 170 (2018).
15. K. R. Al-Labadi, H. M. Al-Humaidi, and F. H. Tan, J. Perform. Constr. Facil. **23**, 423 (2009).
16. H. M. Lyu, S. L. Shen, A. Zhou, and J. Yang, Tunn. Undergr. Sp. Technol. **84**, 31 (2019).
17. P. L. Ng, T. N. D. R. Barrett, G. Roux, S. Polycarpe, and M. Gonzalez, Procedia Eng. **189**, 560 (2017).
18. F. Chen, H. Peng, X. Ma, J. Liang, W. Hao, and X. Pan, Tunn. Undergr. Sp. Technol. **92**, 103034 (2019).
19. P. Taylor and Z. Hashisho, *Critical Reviews in Environmental Science and Technology Vehicular Emissions in Roadway Tunnels : A Critical Review Vehicular Emissions in Roadway Tunnels : A Critical Review* (n.d.).
20. C. Greed, Constr. Manag. Econ. **31**, 197 (2013).
21. A. Balasubramanian, (2017).
22. C. López-Fernández, D. A. Prieto, G. Fernández-Viejo, L. Pando, and E. C. Fernández, J. Geotech. Geoenvironmental Eng. **139**, 821 (2013).
23. D. Chen and S. Xu, Proc. 2nd Int. Conf. Transp. Eng. ICTE 2009 **2009**, 3748 (2009).
24. M. Karakus and R. J. Fowell, Tunn. Undergr. Sp. Technol. **18**, 513 (2003).
25. S. Bassan, IATSS Res. **40**, 35 (2016).
26. J. Qiu, T. Yang, X. Wang, L. Wang, and G. Zhang, Constr. Build. Mater. **195**, 468 (2019).
27. P. Swuste, C. van Gulijk, W. Zwaard, and Y. Oostendorp, Saf. Sci. (2014).
28. Q. Qian and P. Lin, J. Rock Mech. Geotech. Eng. **8**, 423 (2016).
29. S. K. Ayer, G. A. Macht, R. M. Leicht, and A. David, 349 (2014).
30. M. Tao, H. Zhao, X. Li, J. Ma, K. Du, and X. Xie, Geomech. Eng. **12**, 1 (2017).
31. D. Majumder, M. N. Viladkar, and M. Singh, Int. J. Rock Mech. Min. Sci. **100**, 278 (2017).
32. G. Andreotti and C. G. Lai, Tunn. Undergr. Sp. Technol. **91**, (2019).
33. T. I. Airport, T. County, T. County, L. District, and N. T. City, 78 (2014).
34. W. K. Chow, *Review on Fire Safety Management and Application to Hong Kong* (2001).
35. A. Peña-García, Tunn. Undergr. Sp. Technol. **91**, 103007 (2019).
36. K. Y. K. Li, A. T. Chan, and M. Asce, (n.d.).
37. N. Ei, E. Khaing, G. Abuduxike, P. Posadzki, U. Divakar, N. Visvalingam, N. Nazeha, G. Dunleavy, G. I. Christopoulos, C. Soh, K. Jarbrink, M. Soljak, and J. Car, **84**, 201 (2019).
38. K. El-dash, 65 (2007).
39. C. Li and X. Li, Int. J. Rock Mech. Min. Sci. **112**, 323 (2018).
40. A. Leitner, Tunn. Undergr. Sp. Technol. **16**, 217 (2001).
41. Y. Yan, X. Wang, L. Shi, and H. Liu, **9588**, (2017).
42. N. Vossebeld and T. Hartmann, 57 (2014).
43. W. Liu, T. Zhao, W. Zhou, and J. Tang, Saf. Sci. **105**, 98 (2018).
44. N. Swannell, M. Palmer, G. Barla, and M. Barla, Tunn. Undergr. Sp. Technol. Inc. Trenchless Technol. Res. **57**, 201 (2016).

 https://doi.org/10.1051/e3sconf/202017006004

Simple Design of Solar Based Watering System for Plants: Application in Multi-Story Buildings

Mrudula Hote[*], *Ila Dvivedi*
Delhi Public School, Roorkee, Uttarakhand (India)

Abstract. In this paper, the design of a solar-based watering system for plants in multi-story buildings is proposed. The main advantage of the design is that when the people are out of station or busy for particular days, the proposed solar-based system will water the plants for a particular time in a day. The proposed design is verified on a real-time hardware setup. The main advantage of the proposed design is that the mathematical approach is very simple and no need for any rocket science.

1 Introduction

East or west, going clean and green is always the best. Earth is called green due to the presence of plants. Plants are the backbone of all life forms on Earth and an essential resource for human well-being [1]."Go Green, India Green" this is what we are taught since our childhood this shows how much plants are important. From the air we breathe(by-product of photosynthesis) to the food we eat is directly or indirectly from plants. Plants help to maintain the water table thus regulate the water cycle. One-quarter of all prescribed medicines come directly from or are derivatives of the plants. Of course, aside from human usage, plants balance the ecosystem. Other species of fish and wildlife also depend for food and shelter on plants. Plants store carbon and have helped to reduce much pollution from excessive carbon dioxide produced from the burning of fossil fuels out of the atmosphere. Due to the increase in the number of Multi-storey buildings, there is a decrease in the number of trees and hence increase in pollution levels in various cities [2]. Further due to the increasing population, the ratio between the humans (animals): plants is not balanced. For this, people can plant plants in their apartments. This is useful when the continuous water is available for the plants. But, today's family structure is different than the older times, i.e., due to the rise of nuclear family structure seen in today's lifestyle creates another problem of watering the plants as the family has to go out for many reasons like official work, Leisure, away from the native place. During that time, it is a challenging task to serve water to plants, as nobody is present inside the house. Therefore, we need a mechanism by which we can save the water in the multi-story building. In view of this, in this paper, a solar-based watering system for plants in a multi-story building is proposed. In this proposed approach, using solar energy [3, 4], power is supplied to the motor gun which activates the tap of the tank by which water is supplied to the plants for some specific time. The specific time is decided based on the position of the solar tracker, which can move horizontally or in the vertical direction. The main advantage of the proposed approach is that there is no need for any advanced communication technology such as Wi-Fi, mobile technology etc.

2 Motivation

We live 1000s of kilometers away from our hometown and we go in summer and winter vacations over there but as we cannot carry our plants back with us as they stay in the balcony of our apartment of a multistory building. It is not possible to ask neighbours to water our plants as they have also their personal lives and sometimes they also go out for vacations and further asking your maid or any other person to water the plants is also point of security of home that made our mind for something that will give water to the plants when we are not at home. Moreover, the idea is to use solar energy because it is a renewable resource of energy, that will maintain the electricity bill and other advantages which we learned from various science books.

3 Literature survey

Various type of research work is carried on a solar-based watering system. The brief review of these works is as follows:

3.1 Jia Uddin et al., (2012) [5]

Authors of this paper, proposed a model of variable rate automatic microcontroller irrigation system. This is basically for farming purposes using sensors and mobile phones.

3.2 T A Binshad et al., (2016) [6]

[*]Corresponding author:**mrudulahote@gmail.com**

 https://doi.org/10.1051/e3sconf/202017006004

Authors of this paper proposed PV based pumping system for agricultural irrigation. In this approach, power electronic controllers were designed. They have been developed for the water pumping system. In this, the boost converter is used along with an inverter followed by an induction motor pump set but this is also again for agricultural purposes and this approach is complex.

3.3 K. Megalingam and Vamsy Gedela (2017) [7]

Authors of this paper have discussed the concept of a solar-based smart irrigation system. In this design, they have used a microcontroller-based circuit board, sensors for agricultural purposes and it is used in such a manner i.e., particularly the switch is ON only when the moisture content in the soil is less.

3.4 Abhishek Agrawal et al., (2018) [8]

In this paper, they have a microcontroller-based irrigation system in which the sensor detects the moisture of soil and provides a signal to the motor to turn either ON or OFF the irrigating pump for proper utilization of water by the farmers.

The existing work is based on microcontrollers etc. but our system can be used for farming as well as for multi-story buildings. There is no need for sensors and microcontrollers. The methodology is very simple. This is explained below.

4 Proposed set up

The block diagram of the proposed set up is shown in fig.1. In this figure, it is shown that the output of a solar panel is dc voltage. This voltage is possible only when the maximum solar energy from the Sunrays is directly in line with the solar panel. This solar panel is connected to the water tank with a motor gun that moves the water tap on the upper side through which water flows from the tank connected to a water pipe from where water reaches each plant by drip irrigation method.

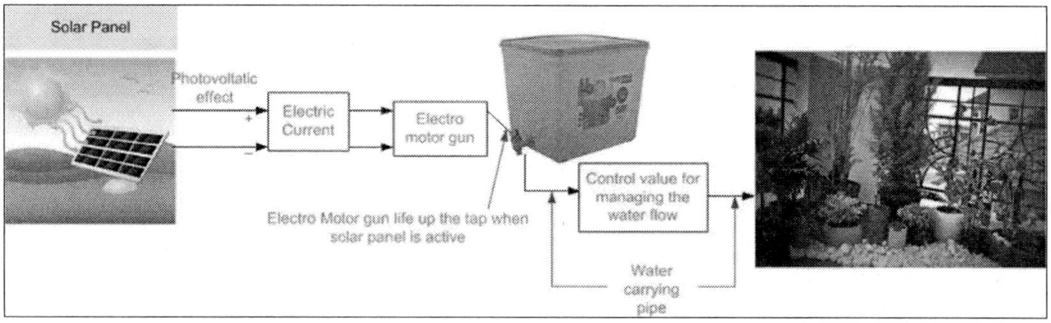

Fig.1. Proposed set up of solar water-based water system.

5 Proposed real-time setup of solar-based watering system

The proposed solar-based watering system is described below.

5.1 Solar tracker System [3, 4]

Solar panel voltage: 18 V, Angle set for solar tracker is: 30 Degree, Horizontal and vertical motions are possible using solar tracker system. Fig.2 shows the solar panel with the tracker system.

Fig.2. Solar tracker system

https://doi.org/10.1051/e3sconf/202017006004

5.2 Tank system

The various parts of the tank system are

- Water tank
- Motor gun (solenoid)
- Water tap
- Pipe
- Water flow controller

Fig.3. shows the tank system with a motor gun, water tap, pipe, and water flow controller.

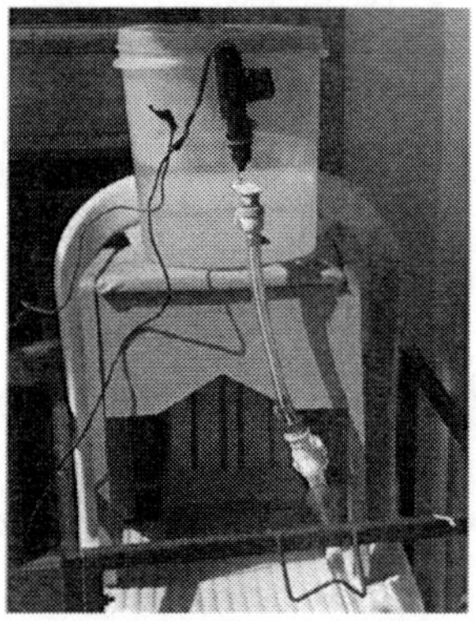

Fig.3. Tank system with motor gun

5.3 Electric switch

This switch is placed between a solar tracker system and the water tank system. This switch controls the opening of a water tap. When the switch is in off condition, even though there is solar energy, there will be no water flowing through the outlet. Fig.4 shows the electric switch, which activates the motor gun.

Fig.4. Electric switch which activates the motor gun.

5.4 Plants in the balcony (four plants)

There are four plants in the balcony. The pipe is placed near the plants as shown in fig.5. There are small holes made in the pipe by which water can drip easily.

Fig.5. Four plants with pipe in the balcony.

The complete set up is shown in fig.6.

Fig.6. Complete set up of solar panel with water tank system.

6. Proposed methodology

6.1 Proposed concept

The proposed methodology is based on the concept of time during which solar energy is available which activates the solenoid. Once the solenoid is activated, the water tap will move in the upward direction and hence water will flow to the plants through the pipe. There are small holes that have been created in a manner that the water would be supplied to the plants through the holes of the pipe. This water will be available for the plants. The most important part is the water flow controller by which we can adjust the flow of water. The mathematical concept and based on this, application to watering the plants in the balcony are given in the following subsections.

6.2 Mathematical formulation

Consider, there are 'n' number of plants in the balcony. Assume that 'y' ml of water is sufficient for each plant. The quantity of water needed (z, ml) is

$$z = ny \tag{1}$$

Suppose, we are out of station for 'd' days. Therefore, the total quantity of water requirement (T) is given by

$$T = nyd \tag{2}$$

Assume that say for 1 hour, approximately 'w' ml of water is dispersed. Therefore, for d days, we will require the following amount of water as given in eq. (3).

$$K = d \times w \tag{3}$$

The proposed design should be such that

$$K = T \tag{4}$$

6.3 Application

For experimental work, we have considered four plants in the balcony. It is assumed that 100-125 ml of water is sufficient for each plant. Therefore, for four plants, 125 x 4 = 500 ml of water is required daily. Suppose, we are out of station for four days. Therefore, total water requirement is 500 x 4 = 2000 ml = Approx.: 2 litres. Through experimentation, it is observed that for the position of solar tracker as mentioned earlier, water flow is continuous nearly for 1 hour for a fixed position of a solar panel in our balcony at home. After that, gradual intensity of sunrays reduces on the solar panel and the tap will automatically come in the off position. By trial and error procedure (experimental analysis), we have adjusted the position of water flow controller (additional tap) in such a way that in 1 hour, approximately, 0.5 litre (500ml) of water will be dispersed. For 4 days: 4 x 500 = 2000 ml =2 litre of water will be required in our case. If the number of plants is more, we can change the position of solar tracker and accordingly we can adjust the water flow controller tap and the size of the tank can be changed. This can be done using equations (2) and (3).

6.4 Proposed result

The actual values of voltage and current are measured using voltmeter and ammeter. From this, the total power produced by the solar panel is calculated. This is given as

The result of solar output:

Voltage (V): 2.315 V

Current (I): 0.28 A

Power (P) = $V \times I$ = 3.315 x 0.28 = 0.6482 W

7 Advantages and future scope of proposed work

The focus of this project is simplicity. The proposed system is automatic. No need for any advanced technology such as mobile, wifi... etc. The proposed system is cost-effective. This system can also be used for day to day watering of the plants. No rise in the electricity bill. The complete system is safe and no environmental issues.

We got inspiration for writing this paper after demonstrating a solar-based watering system for plants for multi-story buildings in Adi Shankara Asianet News Young Scientists Award-2018. We are thankful for the support from respected Mr. N. Singh, Principal, DPS Roorkee and parents for support and suggestions in writing this paper.

8 Conclusion

In this paper, a simple approach is presented to provide water to the plants which are in the balcony when no one is inside the house for a few days. The advantage of the proposed approach is that there is no need for any advanced communication technology such as wifi, mobile technology, etc. Here, only four plants have been selected. In the future, the proposed simple approach will be extended to a large number of plants. Further, the proposed idea can be extended for agricultural use also.

References

[1] https://www.workingmother.com/blogs/mommy-favorites/five-reasons-why-we-should-take-care-earth

[2] https://www.governancenow.com/views/interview/how-one-building-gets-pure-air-right-in-the-middle-of-delhi

[3] https://rgsenergy.com/how-solar-panels-work/the-basics-of-solar-power

[4] https://en.wikipedia.org/wiki/Solar_energy

[5] J. Uddin et. al, Automated Irrigation system using solar power, *7th International Conference on*

Electrical and Computer Engineering, Dhaka, Bangladesh (2012).

[6] T. A. Binshad, K. Vijayakumar, M. Kaleeswari, PV based water pumping system for agricultural irrigation, Front. Energy, **10**, 3, 319-328, (2016).

[7] R. K. Megalingam and V. V. Gedela, *Solar powered automated water pumping system for eco-friendly irrigation,* Proceedings of the International conférence on Inventive Computing and Informatics (ICICI), Coimbatore, India (2017).

[8] A. Agrawal et. al., *Microcontroller based Irrigation system solar powered using solar powered using moisture sensing technology,* 8th International Conference on Cloud Computing, Data Science & Engineering , Noida, India (2018).

E3S Web of Conferenced 170, 06005 (2020)
EVF'2019
https://doi.org/10.1051/e3sconf/202017006005

Investigation of phase equations for anaerobic digestion process of anaerobic bioreactor landfill

Priyanka R. Shinde[1], *Sandip. T. Mali*[2*]
[1]Department of Technology, Savitribai Phule Pune University, Pune, 411007, India
[2]Department of Civil Engineering, Pimpri Chinchwad College of Engineering, Nigdi, 411044, Pune, India

Abstract: The anaerobic digestion process consists of four phases hydrolysis, acidogenesis, acetogenesis and methanogenesis. In the present work anaerobic digestion process of waste from anaerobic bioreactor landfill is considered and leachate is analyzed chemically for 268 days and out of all 17 parameters, seven major chemical parameters are studied for two reactors R4 and R5 for different reactor conditions and their phase distribution is done by two methods standard and transitory method. These phase distribution also gives phase wise equation, for optimization of entire process each phase equation is essential. The accuracy of curve equations is defined by R2 value. These equations help in formation of mathematical model using MATLAB. This model is useful for defining phase values of parameters. Further quality and quantity of chemical parameters are also defined by these phase values.

Keywords: Anaerobic digestion, Hydrolysis, Acidogenesis, Acetogenesis, Methanogenesis, Anaerobic bioreactor landfill, leachate.

1 Introduction

Anaerobic digestion is the process occurs in the absence of oxygen by which organic waste is broken down to produce biogas. Anaerobic digestion is a series of processes produced by bacteria that transform organic materials into biogas through four different stages including hydrolysis, acidogenesis, acetogenesis and methanogenesis.[1] Organic processes are broken down step by step through these four stages towards the development of methane. In the first step of hydrolysis complex macromolecules and components are found in organic matter is converted into simple sugars, long chain fatty acids and amino acids, and conversion of these soluble micro molecules into volatile acetic acids, fatty acids, H_2 and CO_2 is occurs in the second stage called as acidogenesis. The transformation of volatile fatty acids to more acetic acid, CO_2 and H_2S gas is formed in third stage of acetogenesis. The final stage of methanogenesis has the potential to generate methane by using CO_2 and H_2S gas; otherwise acetic acid generated either in the second or third phases. Therefore, the Anaerobic Digestion process, if poorly controlled, will become inefficient and result in a reduced output of biogas, thus there is more room for further development of the production rate of biogas.

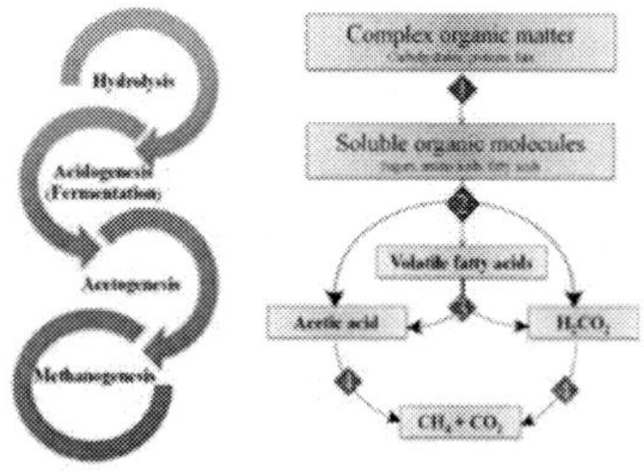

Figure 1:-Anaerobic digestion Process
(Source: B.P. 2017)

In the anaerobic digestion process there are total four stages Hydrolysis, Acetogenesis, Acidogenesis and methanogenesis. The phase distribution of each parameter from leachate chemical analysis is done on the basis of vesilind's standard values for phase distribution also the another method is on the basis of transitory changes in values of parameter leads to change in phase of anaerobic digestion.

*Corresponding author: **sandipmali05@gmail.com**

2 Materials and methods

2.1 Reactor configuration

Two groups of simulated Anaerobic Bioreactor Landfill were used in this study; first group for mixed MSWs with three Bioreactors (R1, R2, and R3) and the other group was consist of two Bioreactors (R4 and R5) of compostable MSWs. Various combinations of operating parameters have been aeration with introduction of aerobic microbial culture, anaerobic sludge, mixing of coarse gravel, soil cover as a intermediate layer and varying leachate circulation. For study purposes, we selected the reactor R5 [2].

2.2 Reactor conditions

Bioreactor R4, R5- Consist of Re-circulation of leachate simulations and bioreactor had aeration with addition of aerobic microbial culture, gravel mixing and introduction of sludge in re-circulated leachate intermediately.
The bioreactor was constructed of acrylic plastic with a thickness of 5 mm, an internal diameter of 0.3 m and a height of 1.2 m and was fitted with three ports. For leachate drainage and sampling there was first port was. For collecting gas samples and leachate recirculation other two ports are provided. A half-cut plastic ball was held at the center of the bottom above the leachate outlet to keep the leachate outlet from collapsing. The waste level in each reactor was 1 m and the freeboard was 0.1 m. 50 mm thick layer of coarse gravel was put at the bottom of the reactor for leachate drainage. The waste was packed in six layers with a thickness of 0.15 m and 10 mm thick layers of waste after each layer. 50 mm of soil layer was supplied at the top of the waste. The porous pipe was placed in the center of the waste mass to capture biogas. Every reactor was opened for sampling and temperature measurement at a distance of 0.30 m from the bottom of 50 mm in diameter. To prevent algal production of the reactors were wrapped in black cloth.
The reactors were kept in a closed room and run under atmospheric conditions. The temperature variability over the length of the experiment (270 days) ranged from 22.9 ° C to 39.4 ° C with a mean temperature of 29.8 ° C. Leachate was collected in a container (vol. 13 L) at the bottom of the reactor and transferred to an overhead container (a submersible pump made by Futura Industries India, model - KR 1900). The leachate was pumped into the reactor through the upper layer of the soil using a PVC pipe which is perforated. In the same day of charging, the bioreactor R4 was coated with an acrylic film. Waste material worked under anaerobic conditions for 270 days. After 14 days of aerobic treatment, the bioreactors R3 and R5 were sealed.

2.3 Analysis of Reactor

Leachate was collected and stored in 1 L air tight plastic bottle. The bottles were stored at -4°C in refrigerator. Chemical analysis of leachate was carried out after every 7 days for first 60 days (on every Monday analysis was started) and after that every 14 days up to 90 days and afterwards every 21 days to till the end of experimentation, as there was not too much variation in the characteristics of leachate. Leachate analysis was performed for 17 parameters (pH, electrical conductivity(EC), total solids (TS), total dissolved solids (TDS), volatile solids (VS), solids (FS), biochemical oxygen demand for 3 days (BOD_3), chemical oxygen demand (COD), total kejeldhal nitrogen (TKN), chloride, oxidation reduction potential (ORP), sulphate, sodium (Na), calcium (Ca), potassium (K), volatile fatty acids (VFA) and alkalinity. All the analysis was carried as per standard methods.

3 Results and discussion

3.1 Standard method of phase distribution

Based on vesilind's table the standard ranges of parameters are distributed as follows:

Landfill constituent concentration ranges as a function of the degree of stabilization:

Table 1: Anaerobic digestion Process phase values for parameters

Parameter	Transition	Acid formation	Methane fermentation	Maturation
COD (mgL^{-1})	480-18000	1500-71000	580-9760	31-900
Total volatile acids (mgL^{-1})	100-3000	3000-18800	250-4000	0
Ammonia (mgL^{-1})	120-125	2-1030	6-430	6-430
pH (mgL^{-1})	6.7	4.7-7.7	6.3-8.8	7.1-8.8
Conductivity (mgL^{-1})	2450-3310	1600-17100	2900-7700	1400-4500

(Source: Vesilind et al., 2003, pp. - 120)

3.2 Transitory phase distribution

In this method the phase distribution is done on the basis of transitory change in the parameter values and that forms the phases of anaerobic digestion.
After phase distribution of anaerobic digestion process comparison between both of these parameters is done as result of that we get the comparison of equation and R^2 value for each phase that comparison

https://doi.org/10.1051/e3sconf/202017006005

will decide the best suitable method for phase distribution.

Colour code for Phase Distribution:
Blue- Phase I
Red- Phase II
Yellow- Phase III
Violet- Phase IV
Green- Phase V

Results for R4 reactor Transitory phase distribution:

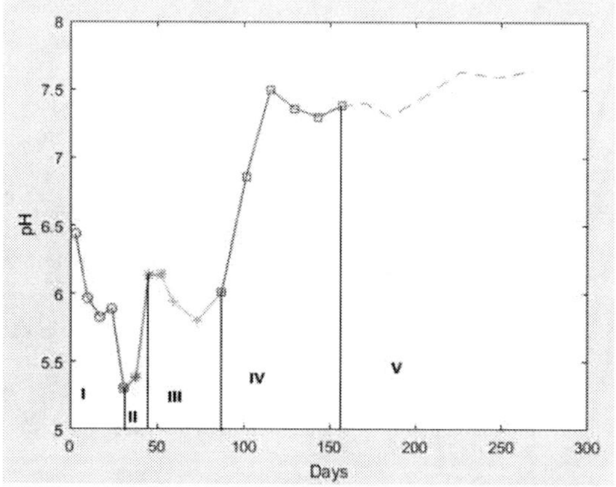

Figure 2:-pH Transitory phase distribution

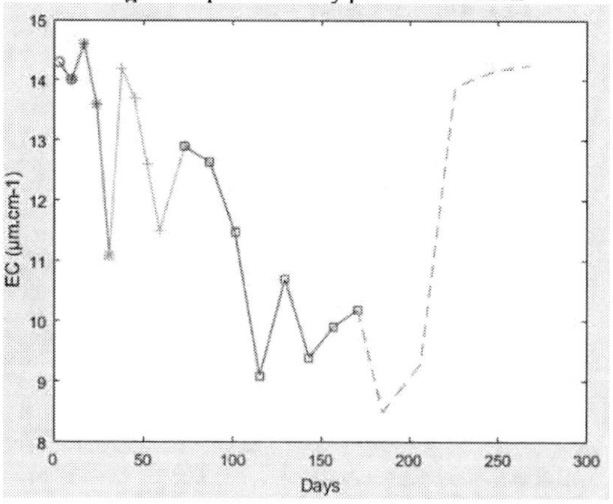

Figure 3:- EC Transitory phase distribution

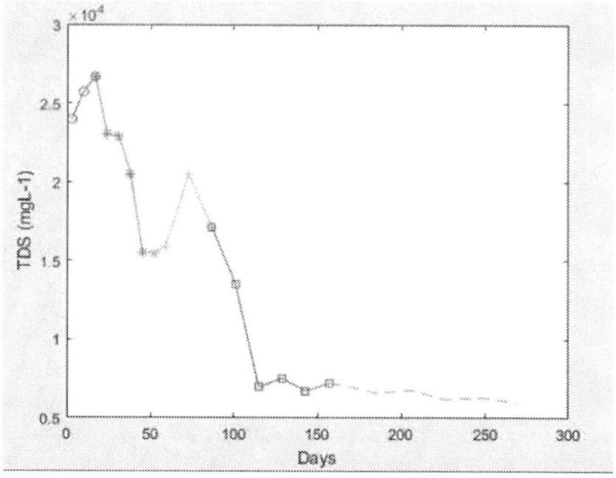

Figure 4:- TDS Transitory phase distribution

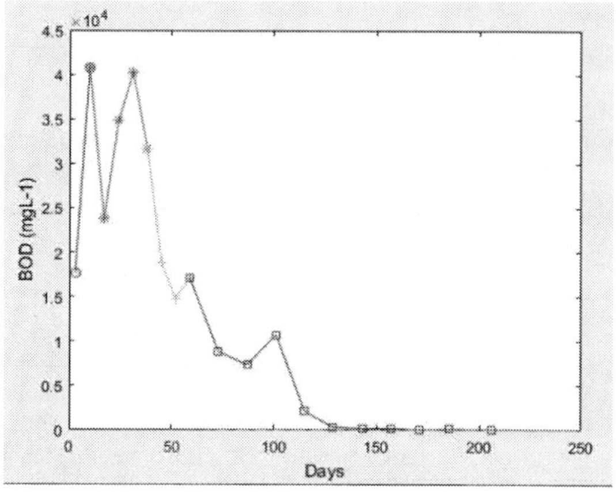

Figure 5:- BOD Transitory phase distribution

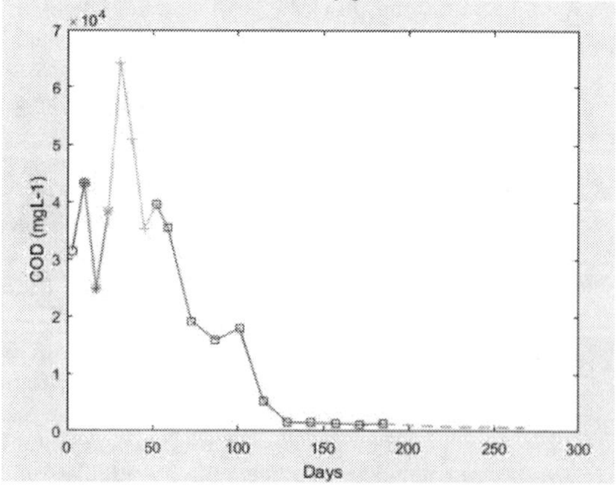

Figure 6:- COD Transitory phase distribution

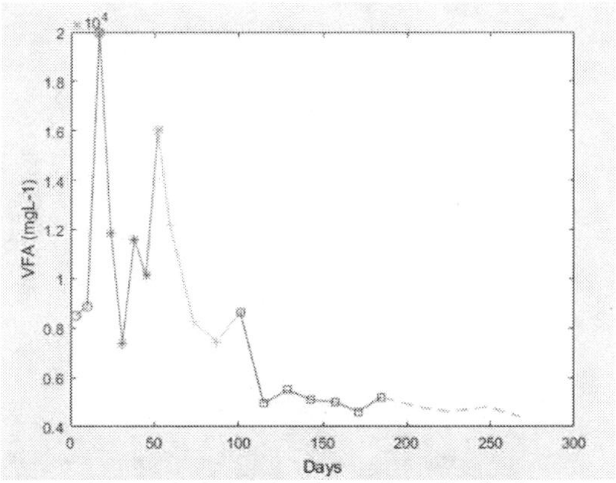

Figure 7:- VFA Transitory phase distribution

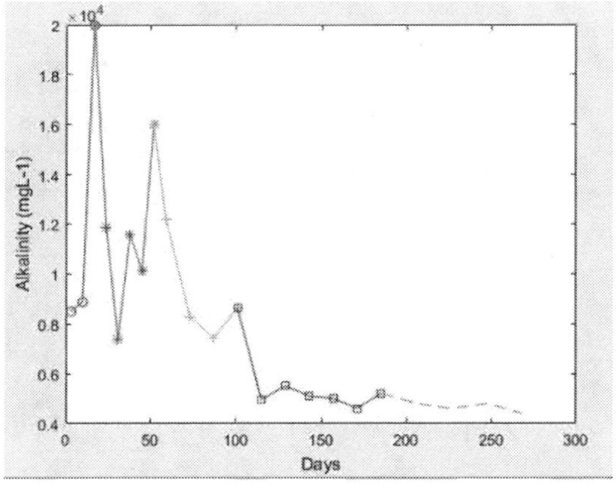

Figure 8:- Alkalinity Transitory phase distribution

Results for reactor R5 transitory phase distribution:

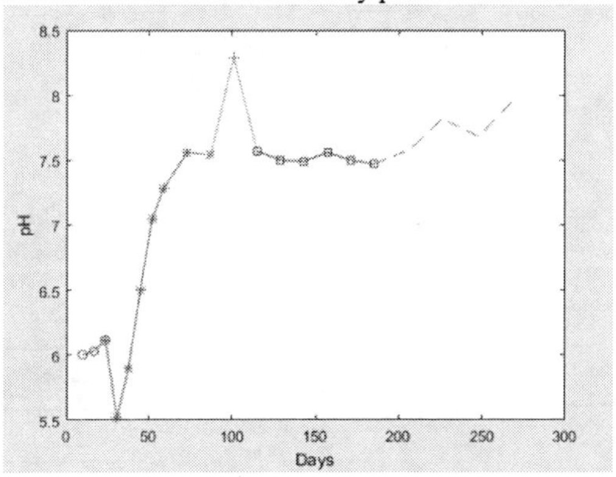

Figure 9:- pH transitory phase distribution

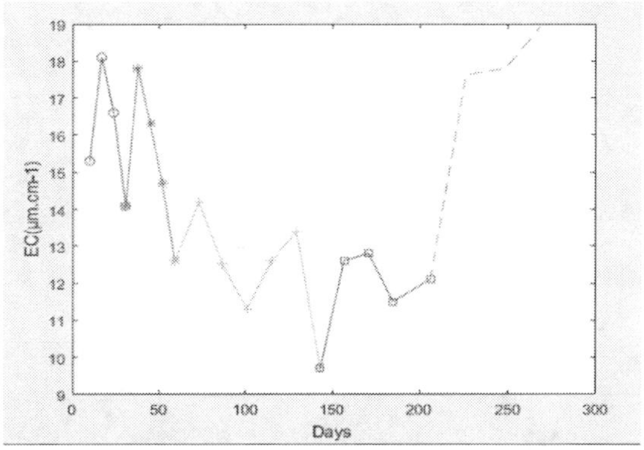

Figure 10:- EC transitory phase distribution

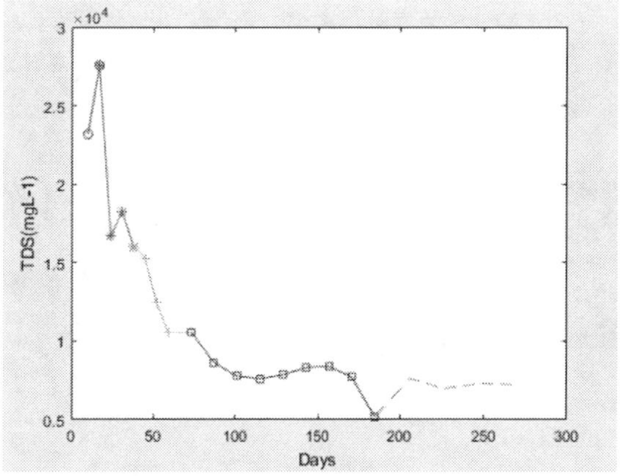

Figure 11:- TDS transitory phase distribution

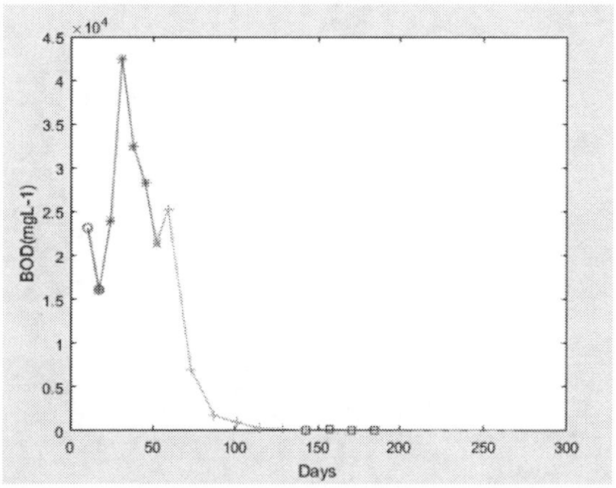

Figure 12:- BOD transitory phase distribution

https://doi.org/10.1051/e3sconf/202017006005

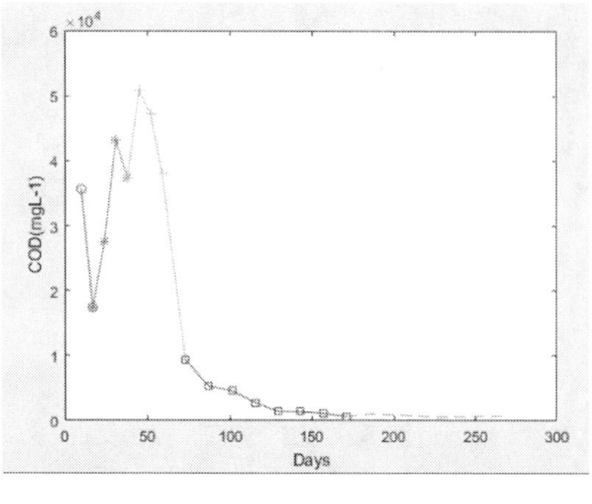

Figure 13:- COD transitory phase distribution

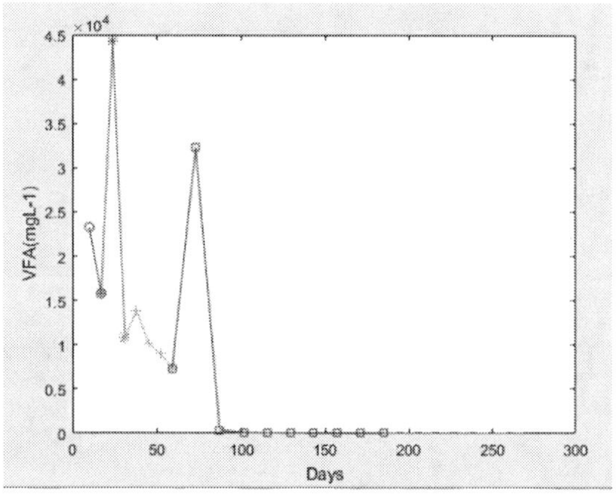

Figure 14:- VFA transitory phase distribution

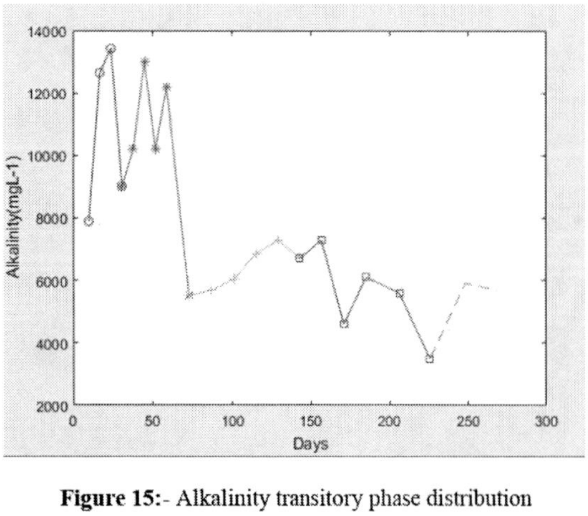

Figure 15:- Alkalinity transitory phase distribution

Coefficients in parameter Equations for R4 reactor by using transitory method:

Table 2: Equation for pH

pH				
	Standard values		Abrupt values	
Phases	Equation	r^2	Equation	r^2
I	$y = -5.8e\text{-}05x^2 - 0.031x + 6.448$	0.843	$y = -5.83E\text{-}05x^2 - 0.031x + 6.448$	0.843
II	$y = -0.000357x^2 + 0.076x + 3.205$	0.843	$y = -0.0004x^2 + 0.076x + 3.205$	0.843
III	$y = 0.00089x^2 - 0.127x + 10.36$	1	$y = 0.0009x^2 - 0.1275x + 10.3620$	1
IV	$y = -0.0006x^2 + 0.164x - 3.602$	0.9129	$y = -0.0006x^2 + 0.239x - 7.683$	0.974
V	$y = 4.05E\text{-}06x^2 + 0.0011x + 7.060$	0.765	$y = 4.05E\text{-}06x^2 + 0.001x + 7.060$	0.765
Overall	$y = -2E\text{-}05x^2 + 0.013x + 5.528$	0.77	$y = -2E\text{-}05x^2 + 0.013x + 5.528$	0.77

Table 3: Equation for EC

EC				
	Standard values		Abrupt values	
Phases	Equation	r^2	Equation	r^2
I	$y = -0.008x^2 + 0.200x + 13.50$	0.914	$y = -0.042x + 14.42$	1
II	$y = -0.021x^2 + 1.835x - 25.06$	0.92	$y = -0.015x^2 + 0.509x + 10.48$	0.999
III	$y = 0.0005x^2 - 0.0511x + 13.5520$	0.1845	$y = -0.001x^2 + 0.152x + 8.970$	0.058
IV	$y = 0.001x^2 - 0.318x + 31.55$	0.736	$y = 0.0007x^2 - 0.1991x + 24.1627$	0.7906
V	$y = 0.0004x^2 - 0.1054x + 16.8016$	0.7293	$y = 1E\text{-}04x^2 + 0.019x + 2.906$	0.731
Overall	$y = 0.0003x^2 - 0.0736x + 15.3551$	0.6407	$y = 0.0003x^2 - 0.0736x + 15.3551$	0.6407

Table 4: Equation for TDS

TDS				
	Standard values		Abrupt values	
Phases	Equation	r^2	Equation	r^2
I	$y = -12.30x^2 + 349.7x + 23292$	0.651	$y = -8.775x^2 + 371.2x + 22925$	1
II	$y = 11.93x^2 - 1380.x + 54607$	0.934	$y = -0.502x^2 - 295.7x + 31574$	0.935
III	$y = -10.75x^2 + 1572.x - 38011$	0.752	$y = -10.75x^2 + 1572.x - 38011$	0.752
IV	$y = 3.961x^2 - 1108.x + 83839$	0.943	$y = 3.961x^2 - 1108.x + 83839$	0.943
V	$y = 0.039x^2 - 26.65x + 10384$	0.88	$y = 0.039x^2 - 26.65x + 10384$	0.88
Overall	$y = 0.440x^2 - 194.2x + 27094$	0.913	$y = 0.440x^2 - 194.2x + 27094$	0.913

Table 5: Equation for BOD

BOD				
	Standard values		Abrupt values	
Phases	Equation	r^2	Equation	r^2
I	$y = -11.13x^2 + 939.0x + 19795$	0.371	$y = 3314.x + 7656$	1
II	$y = 23.07x^2 - 3187.x + 11751$	0.979	$y = 114.1x^2 - 4549.x + 73202$	0.697
III	$y = -0.388x^2 - 215.7x + 28512$	0.802	$y = 23.07x^2 - 3187.x + 11751$	0.979
IV	$y = 1.541x^2 - 516.9x + 42755$	0.709	$y = 1.12x^2 - 389.3x + 33590$	0.888
V	$y = 0.022x^2 - 10.17x + 1235$	0.593	$y = 0.024x^2 - 11.38x + 1410$	0.775
Overall	$y = 0.900x^2 - 368.9x + 35951$	0.796	$y = 0.900x^2 - 368.9x + 35951$	0.796

Table 6: Equation of COD

COD				
	Standard values		Abrupt values	
Phases	Equation	r^2	Equation	r^2
I	$y = 87.64x^2 - 2113.x + 42393$	0.695	$y = 1675x + 264$	1
II	$y = 89.74x^2 - 8733.x + 24982$	0.952	$y = 322.4x^2 - 11313x + 124$	1
III	$y = 15.79x^2 - 2925.x + 15026$	0.969	$y = -66.73x^2 + 4691.x - 297$	0.384
IV	$y = 3.598x^2 - 1134.x + 900$	0.834	$y = 3.257x^2 - 1041.x + 83513$	0.955
V	$y = 0.022x^2 - 15.55x + 3217$	0.874	$y = 0.114x^2 - 58.70x + 8211$	0.988
Overall	$y = 0.915x^2 - 433.5x + 50004$	0.761	$y = 0.915x^2 - 433.5x + 500$	0.761

Table 7: Equation for VFA

VFA				
	Standard values		Abrupt values	
Phases	Equation	r^2	Equation	r^2
I	$y = 6.246x^2 + 68.04x + 12155$	0.995	$y = 2.840x^2 + 150.4x + 11838$	0.998
II	$y = 63.19x^2 - 5970.x + 14628$	0.78	$y = -89.86x^2 + 5438.x - 61739$	0.999
III	$y = -45.94x^2 + 6511.x - 20704$	0.967	$y = -34.83x^2 + 4911.x - 15119$	0.939
IV	$y = 4.382x^2 - 1231.x + 85771$	0.942	$y = 5.397x^2 - 1456.x + 97898$	0.943
V	$y = 0.011x^2 - 4.946x + 552.9$	0.284	$y = 0.022x^2 - 9.907x + 1089$	0.674
Overall	$y = 0.250x^2 - 138.7x + 17930$	0.597	$y = 0.250x^2 - 138.7x + 17930$	0.597

Table 8: Equation for Alkalinity

Alkalinity				
	Standard values		Abrupt values	
Phases	Equation	r^2	Equation	r^2
I	$y = -42.15x^2 + 1444.x + 3099.$	0.57	$y = 109.8x^2 - 1375.x + 11660$	1
II	$y = 8.600x^2 - 366.5x + 11149$	0.78	$y = 29.88x^2 - 2148.x + 47103$	0.857
III	$y = 9.191x^2 - 1517.x + 69967$	0.998	$y = 7.452x^2 - 1282.x + 62261$	0.993
IV	$y = 0.603x^2 - 192.6x + 20286$	0.628	$y = 1.050x^2 - 329.9x + 30445$	0.728
V	$y = -0.045x^2 + 15.41x + 3606$	0.403	$y = 0.069x^2 - 39.26x + 10015$	0.754
Overall	$y = 0.130x^2 - 69.06x + 13300$	0.54	$y = 0.130x^2 - 69.06x + 13300$	0.54

Coefficients in parameter Equations for R5 reactor by using transitory method:

Table 9: Equation for pH

pH				
	Standard values		Abrupt values	
Phases	Equation	r^2	Equation	r^2
I	$y = -6E\text{-}05x^2 - 0.031x + 6.448$	0.843	$y = -6E\text{-}05x^2 - 0.031x + 6.448$	0.843
II	$y = -0.0004x^2 + 0.076x + 3.205$	0.843	$y = -0.0004x^2 + 0.076x + 3.205$	0.843
III	$y = 0.0009x^2 - 0.127x + 10.36$	1	$y = 0.0009x^2 - 0.127x + 10.36$	1
IV	$y = -0.0006x^2 + 0.164x - 3.602$	0.9129	$y = -0.0009x^2 + 0.239x - 7.683$	0.974
V	$y = 4E\text{-}06x^2 + 0.001x + 7.060$	0.765	$y = 4E\text{-}06x^2 + 0.001x + 7.060$	0.765
Overall	$y = -2E\text{-}05x^2 + 0.013x + 5.528$	0.77	$y = -2E\text{-}05x^2 + 0.013x + 5.528$	0.77

Table 10: Equation for EC

EC				
	Standard values		Abrupt values	
Phases	Equation	r^2	Equation	r^2
I	$y = -0.027x^2 + 1.035x + 7.810$	0.938	$y = -0.017x^2 + 1.447x - 13.84$	0.836
II	$y = -0.027x^2 + 2.248x - 29.36$	0.843	$y = -0.027x^2 + 1.035x + 7.810$	0.938
III	$y = -0.0004x^2 + 0.0128x + 14.3318$	0.264	$y = -0.0005x^2 + 0.0844x + 9.8089$	0.3658
IV	$y = 5E\text{-}05x^2 - 0.018x + 13.54$	0.02	$y = -0.001x^2 + 0.573x - 39.54$	0.563
V	$y = 0.000x^2 - 0.154x + 23.42$	0.812	$y = -0.002x^2 + 1.339x - 152.4$	0.92
Overall	$y = 0.0004x^2 - 0.1046x + 18.7111$	0.7018	$y = 0.0004x^2 - 0.1046x + 18.7111$	0.7018

Table 11: Equation for TDS

TDS				
	Standard values		Abrupt values	
Phases	Equation	r^2	Equation	r^2
I	$y = -14.59x^2 + 229.9x + 23736$	0.481	$y = 622.8x + 16971$	1
II	$y = -2.959x^2 - 11.81x + 21253$	0.963	$y = 43.98x^2 - 2892.x + 63226$	0.85
III	$y = 0.238x^2 - 126.1x + 17963$	0.855	$y = 5.376x^2 - 773.8x + 38182$	0.931
IV	$y = 0.661x^2 - 160x + 17397$	0.754	$y = -0.014x^2 - 22.77x + 11200$	0.536
V	$y = 0.338x^2 - 146.7x + 22618$	0.197	$y = -0.678x^2 + 326x - 31562$	0.703
Overall	$y = 0.542x^2 - 198.4x + 23681$	0.845	$y = 0.542x^2 - 198.4x + 23681$	0.845

Table 12: Equation for BOD

BOD				
	Standard values		Abrupt values	
Phases	Equation	r^2	Equation	r^2
I	$y = 131.4x^2 - 4450x + 54344$	0.997	$y = -1027.x + 33480$	1
II	$y = 15.18x^2 - 2224.x + 96373$	0.984	$y = -60.19x^2 + 4276.x - 39860$	0.782
III	$y = -1.982x^2 - 398.1x + 50298$	0.864	$y = 5.729x^2 - 1364.x + 79561$	0.901
IV	$y = 0.567x^2 - 160.1x + 11348$	0.988	$y = -0.245x^2 + 75.77x - 5704$	1
V	$y = 0.004x^2 - 2.299x + 386.6$	0.509	$y = -0.003x^2 + 1.141x + 2.420$	0.287
Overall	$y = 0.944x^2 - 378.6x + 35682$	0.734	$y = 0.944x^2 - 378.6x + 35682$	0.734

Table 13: Equation for COD

COD				
	Standard values		Abrupt values	
Phases	Equation	r^2	Equation	r^2
I	$y = 172.3x^2 - 6604.x + 83344$	0.927	$y = -2608.x + 61766$	1
II	$y = 10.93x^2 - 540.7x + 47532$	0.338	$y = -81.40x^2 + 5550.x - 54740$	0.906
III	$y = 34.66x^2 - 6105.x + 27321$	0.974	$y = -67.89x^2 + 6649.x - 11509$	0.97
IV	$y = 0.842x^2 - 270.2x + 22724$	0.958	$y = 1.132x^2 - 354.9x + 28624$	0.967
V	$y = 0.040x^2 - 20.27x + 3237$	0.431	$y = -0.014x^2 + 4.463x + 473.8$	0.155
Overall	$y = 1.053x^2 - 451.3x + 46808$	0.679	$y = 1.053x^2 - 451.3x + 46808$	0.679

Table 14: Equation for VFA

VFA				
	Standard values		Abrupt values	
Phases	Equation	r^2	Equation	r^2
I	$y = -20.99x^2 + 1607.x - 18305$	0.661	$y = -1065.x + 33882$ 1	
II	$y = -132.0x^2 + 5292.x - 21374$	0.263	$y = -630.6x^2 + 29920x - 31062$	1
III	$y = -75.95x^2 + 10537x - 33924$	0.648	$y = -9.870x^2 + 715.7x - 1017$	0.745
IV	$y = 0.130x^2 - 35.92x + 2473$	0.851	$y = 2.071x^2 - 632.6x + 47018$	0.372
V	$y = 0.001x^2 - 0.653x + 71.99$	0.562	$y = -0.002x^2 + 1.503x - 175.6$	0.755
Overall	$y = 0.771x^2 - 301.0x + 27465$	0.544	$y = 0.771x^2 - 301.0x + 27465$	0.544

Table 15: Equation for Alkalinity

Alkalinity				
	Standard values		Abrupt values	
Phases	Equation	r^2	Equation	r^2
I	$y = -46.93x^2 + 1983.x - 7329$	0.996	$y = -46.93x^2 + 1983.x - 7329$	0.996
II	$y = -20.40x^2 + 1785.x - 27092$	0.7	$y = -10.64x^2 + 1035.x - 13141$	0.817
III	$y = 2.045x^2 - 463.4x + 30039$	0.698	$y = -0.317x^2 + 92.11x + 280.7$	0.817
IV	$y = -0.408x^2 + 121.0x - 1814$	0.825	$y = -0.250x^2 + 57.98x + 3551$	0.602
V	$y = 0.512x^2 - 225.4x + 29526$	0.364	$y = -2.835x^2 + 1453.x - 18007$	1
Overall	$y = 0.139x^2 - 63.26x + 12411$	0.603	$y = 0.139x^2 - 63.26x + 12411$	0.603

3.3 Discussion

It was observed major seven parameters from leachate chemical analysis which is pH, EC, VFA, BOD, COD, TDS and alkalinity.
For all these seven parameters we have plotted graphs for standard and transitory phase distribution graphs. These graphs show the value of R2 which shows the perfect curve fit.
In standard and transitory phase distribution graph, nature of curve remains same but R2 Values are different because of different phase distribution methods. From the observation transitory method of phase distribution is best method. Hence by considering that method equations of each phase is obtained for reactor R4 and R5.

4 Conclusion

1. Obtained phase distribution for chemical parameters further phase equations are obtained.
2. Phase equations give phase values of parameters which defines quality and quantity of chemical parameter.

References

1. P. Bajpai, Basics of anaerobic digestion process, Anaerobic Technology, *Springer*, (2017).

2. S. T. Mali, K. C. Khare, A.H. Biradar, Enhancement of methane production and bio stabilization of municipal solid waste in anaerobic bioreactor landfill, *Bio- resource Technology, Elsevier,* **vol. 110**, pp. 10, 17, April (2012).

3. Vesilind PA, Worrell WA, Reinhart DR. *Solid Waste engineering, Singapore: Thomoson Asia Pvt. Ltd*, (2003).

Modular architecture shape design recursive algorithm

Raphaël Sasportas[1,*], *Feriel Mustapha*[1,2]
1Laboratory of Research on Eco-innovation of Industry and Energetic, ECAM-EPMI,13 Boulevard de l'Hautil, 95092 Cergy-Pontoise (France).
[2]Laboratoire of Mechanic and Materials of Civil Engineering, University of Cergy-Pontoise (France).

Abstract. The architecture and architectural shape of a building plays an important role in the interior housing design, many concepts have been developed to optimize this and to have a housing that reflects both visual and architectural comfort or even thermal one. The evolution of these houses however is not controlled, modular architecture has appeared because among its basic principles is being built from modules that can be assembled, then separately modified or even disassembled, the advantage of this typology of architecture is that it is fast in execution, light in installation and above all has very little impact on the environment in which it is part. These buildings are characterized by more dynamic geometries, which results in a very rich architectural language and an interesting sky line, however, to be able to project all the possibilities is a challenge in itself. In this article the discrete geometry is treated through the development of an algorithm that illustrates the process that allows to expose all possible cell combinations, the latter represents the basic module of modular architecture, the most attracting choices will be arranged to illustrate the formal and architectural optimization of a house. In this study the emphasis will be placed on the mathematical part in order to highlight the relationship between geometry and its influence on the final architectural rendering.

1 INTRODUCTION

Discrete geometry is a subgraph of square meshes composed of an infinity of several unit cells paired edge to edge, with adjacent cells forming the edges of the graph. These studies have a long history in the scientific literature, but they were first presented by Solomon Golomb, then by Martin Gardner in his American columns Scientific "Mathematical Games" [1], and by several research articles by David Klarner. They are now one of the most popular subjects in the field of mathematical recreation and have attracted the interest of mathematicians, physicists, biologists and computer scientists [2].
Work on the analogy between mathematics and the rest of the disciplines is now omnipresent, particularly between architecture and geometry, which are undeniably linked at the morphological and structural levels. Discrete geometry is applied in modular architecture, the latter is characterized, in the design phase, by the use of identical combined cells, by juxtaposition and/or superposition [3]. Among the adheres of modular architecture existing today, the architecture of containers, indeed this metal box is a standard product so considered as a basic cell, mainly used for logistics [4]. This archetype is inclined to be studied from a purely analytical point of view, and therefore the overview on the number of combination possibilities is difficult to visualize it without a numerical processing. These combinatorial morphologies have a relative impact, these impacts are mainly of an aesthetic nature, but also structural.

However, from a practical point of view these constructions face physical constraints that limit the choices in terms of purely morphological possibilities, because, for structural reasons, for example, the superposition of 15 containers cannot be achieved. From an architectural point of view, this type of construction reflects a strong ecological approach [5], insofar as the reuse of used containers is possible, therefore, during the design phase the architect decides on the number of cells, in this case containers, required for the projects, thus meeting the specifications from the surface point of view, and thus satisfy the criteria required by the project, the use of discrete geometry is first of all a temporal and morphological optimization, having the choice would considerably facilitate the design, while being in conformity with reality.
In this article, the main objective is to demonstrate the possibility of using discrete geometry in real conditions to serve modular architecture and architects who explore all possibilities. In this paper we showed that the concept of recursion and its application contributes greatly to the success of architectural projects of this type, more specifically we describe in the following section graphically and analytically an algorithmic approach to illustrate the possible combinatorial forms.

2 ALGORITHM DESCRIPTION

As already said, this section describes the basic recursive algorithms that generate all possible shapes with a predefined number of cells. The cells must be adjacent and every cell in the shape must be located above another cell or on the ground (see figure below)

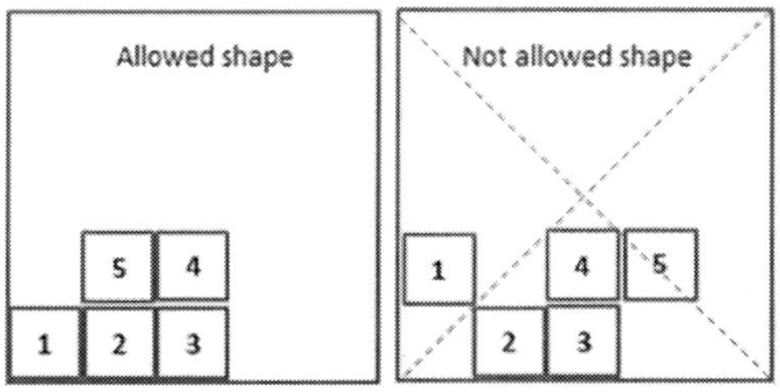

Figure 1: connexity constraints (examples of allowed/ not allowed shapes)

We first describe the basic recursive algorithm, outline its drawbacks and then we present a second algorithm that corrects the imperfections of the first.

2.1 Basic recursive algorithm

The recursive algorithm used to generate all possible shapes composed of $N + 1$ cells is shown below (**figure 2**). The algorithm has two input arguments. The first argument is a binary 2D matrix whose elements have 0 value except for a single element whose value is 1 and which represents the initial cell. The second argument is the number of cells to be added to the initial one. For each cell the algorithm checks its closest neighbours located in East, North and West directions. In the North direction (line 17 of the algorithm) whenever the location value is 0 the value is set to 1. In East direction, whenever the location value is 0 the value is set to 1 if and only if there is the ground or a cell in the South-East direction (constraint 1 line 11). In West direction, whenever the location value is 0 the value is set to 1 if and only if there is the ground or a cell in the South-West direction (constraint 2 line 23) (see **figure 2**). These two constraints allow to construct a subclass of 2D Polyminos [6] that correspond to physical construction: there is no suspended cell, each cell in the generated shape has a support cell or the ground. When a location value is set to 1, the same algorithm is executed with input arguments corresponding to the updated 2D matrix and number of cells minus one (lines 13, 19 and 25). The algorithm execution is stopped when the number of cells to be added is equal to 0. The output of the algorithm is a 3D array (**TabConfig**) in which each slice is a binary 2D matrix corresponding to a shape.

Algorithm 1.

```
1       function y=ShapeGenV1(M,N)
2  -    global nbcfg;
3  -    global TabConfig;
4  -    [nl,nc]=size(M);
5  -    if N~0
6  -        for i=1:nl
7  -            for j=1:nc
8  -                if M(i,j)==1
9                       %check East neighbor
10 -                    if M(i,j+1)==0
11 -                        if i-1 == 0 | M(j+1,i-1) == 1
12 -                            A=M;A(i,j+1)=1;
13 -                            ShapeGenV1(A,N-1);
14 -                        end
15 -                    end
16                      %check North neighbor
17 -                    if M(i+1,j)==0
18 -                            A=M;A(i+1,j)=1;
19 -                            ShapeGenV1(A,N-1);
20 -                    end
21                      %check West neighbor
22 -                    if M(i,j-1)==0
23 -                        if i-1 == 0 | M(i-1,j-1) == 1
24 -                            A=M;A(i,j-1)=1;
25 -                            ShapeGenV1(A,N-1);
26 -                        end
27 -                    end
28 -                end
29 -            end
30 -        end
31 -    else
32 -    nbcfg=nbcfg+1;
33 -    TabConfig(1:nl,1:nc,nbcfg)=M;
34 -    end
35 -    end
```

Figure 2. basic recursive proposed algorithm

2.2 Improved recursive algorithm

The major drawback of the basic recursive algorithm is that it generates, among all possible shapes, subsets of identical shapes. From an algorithmic point of view, due to its recursive processing, the identical shapes are in fact different shapes because the algorithm takes into account these two following points:

a) for a given spatial occupation the cells are differentiated as we can see in configuration B. (see **Figure3**)
b) spatial occupation of the shape as we can see in configuration D. and C. below. (see **Figure 3**).

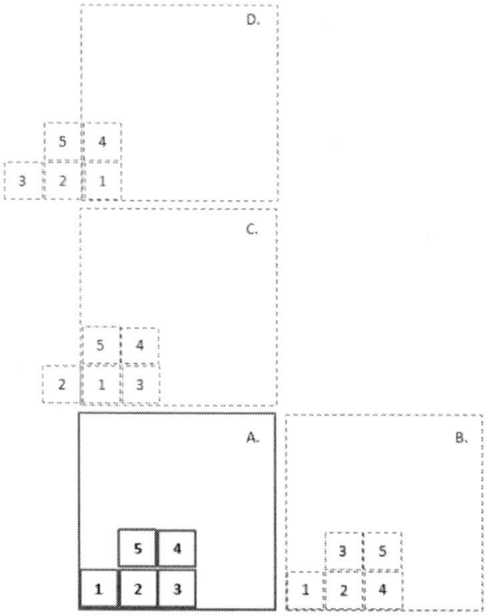

Figure 3. algorithm translation of the same shape

After the launch of the algorithm, the generation of shapes is efficient but we face the problem of N-uplet, which is the phenomenon of having duplicate, triplate, quadruplicate..., indeed we are witnessing the graphic generation of the same shapes, several times.

In order to reduce the number of identical shapes resulting from the remark in a. we added a condition in the algorithm (line 23) (see **Figure 4**) which avoids to generate shapes that are only translations of shapes already generated.
To eliminate the identical shapes resulting from the remark in b. we added lines of code starting at line 37 to line 45. This additional processing allows to verify that each generated shape is unique by comparing it with previously stored shapes. If there exists among the stored shapes a shape identical to that which has been generated then it is not memorized in the **TabConfig** array.

Algorithm 2.

```
1      function y=ShapeGenV2(M,N)
2  -   global nbcfg;
3  -   global TabConfig;
4  -   [nl,nc]=size(M);
5  -   if N~0
6  -       for i=1:nl
7  -           for j=1:nc
8  -               if M(i,j)==1
9                      %check East neighbor
10 -                   if M(i,j+1)==0
11 -                       if i-1 == 0 | M(j+1,i-1) == 1
12 -                           A=M;A(i,j+1)=1;
13 -                           ShapeGenV2(A,N-1);
14 -                       end
15 -                   end
16                     %check North neighbor
17 -                   if M(i+1,j)==0
18 -                           A=M;A(i+1,j)=1;
19 -                           ShapeGenV2(A,N-1);
20 -                   end
21                     %check West neighbor
22 -                   if M(i,j-1)==0
23 -                       if j-1 >= 10
24 -                       if i-1 == 0 | M(i-1,j-1) == 1
25 -                           A=M;A(i,j-1)=1;
26 -                           ShapeGenV2(A,N-1);
27 -                       end
28 -                       end
29 -                   end
30 -               end
31 -           end
32 -       end
33 -   else
34 -   if  nbcfg==0
35 -       nbcfg=nbcfg+1;
36 -       TabConfig(1:nl,1:nc,nbcfg)=M;
37 -   else
38 -       k=1;
39 -       same=0;
40 -       while(k < nbcfg+1)
41 -           if(TabConfig(1:nl,1:nc,k)==M)
42 -               same=1;
43 -           end
44 -           k=k+1;
45 -       end
46 -       if same==0
47 -          nbcfg=nbcfg+1;
48 -          TabConfig(1:nl,1:nc,nbcfg)=M;
49 -      end
50 -   end
51 -   end
```

Figure 4. improved recursive proposed algorithm

3 SIMULATION RESULTS

In this section we present results obtained by executing the second algorithm (see **Figure 4**) written in MATLAB. We have set the number of cells to 5 ($N = 4$) and in order to execute properly the algorithm we run the following script:

```
clear global;
global nbcfg;
global TabConfig;
nbcfg=0;
M=zeros(20,20);
M(1,10)=1;
tic;ShapeGenV2(M,4);toc;
```

Figure 5. script used to run the improved recursive algorithm.

The binary slices of **TabConfig**, converted in binary images are shown in table 1.

Table 1. graphic results with 5 cells

 $k = 3$	 $k = 4$	 $k = 3$	 $k = 3$
 $k = 2$	 $k = 2$	 $k = 2$	 $k = 4$
 $k = 3$	 $k = 3$	 $k = 4$	 $k = 5$
 $k = 4$	 $k = 3$	 $k = 2$	 $k = 1$

In the table below we have reported the results of the algorithm with different number of N. The execution time is measured on a CPU running at 1Ghz frequency (see **Table 2**).

Table 2. geometrical and temporal information's by cells number

N+1 (number of cells)	**Number of shapes**	**Execution time (s)**
3	4	0.015
4	8	0.016
5	16	0.078
6	32	0.141
7	64	0.936
8	128	8.174

3.1 Quantitative analysis

We can see from the table above that the number of shapes with $N + 1$ cells are equal to 2^N . To prove this result let us note by k the number of adjacent cells on the ground (see Table 1). For each value of k, k varying between 1 to $N + 1$, the number of shapes resulting from the positioning of $N + 1 - k$ cells on k cells are showed by the equation (1):

$$C^{k-1}_{(N+1-k)+k-1} = C^{k-1}_{N} = \frac{N!}{(N-k+1)!(k-1)!} \qquad (1)$$

The total number of shapes is then:

$$\sum_{k=1}^{N+1} C_N^{k-1} = \sum_{k=0}^{N} C_N^k 1^{N-k} 1^k = (1+1)^N = 2^N$$

(2)

4 CONCLUSIONS

Developing an algorithmic script based on recursive geometries, and allowing the generation of all morphologies is a considerable time saving, for architects and all other professionals working in the construction sector.
Indeed, thanks to this contribution, we can claim to have a global vision of all the possibilities from a purely geometric point of view, then the architect decides on the choice best suited to his specifications, both aesthetically, by choosing the volumetric composition that corresponds perfectly to urban language, and spatially by optimizing the interior layout that will form the spatial composition of the inhabitants. in addition thanks to this analytical work we are now also able to predict, from the beginning, the number of exact possibilities we can have, and those of course according to the conditions we have issued. That said, this optimization can indeed satisfy both constraints at the same time, and in this perspective, future work will focus on combining several criteria in an optimization, more precisely energy optimization, which today represents a major challenge and an international challenge.

References

[1] J. O. a. C. D. T. J.E. Goodman, Dis and Com Geo, Boc Rat: Pres, (2017).

[2] G. B. a. M.Moffie, «On the com of Jen alg for cou fix pol» *Jou of Disc Alg,* **5**. 348-355, (2007).

[3] W. Y. T. D. A. P.Mendis, «New adv cha and opp of mul mod bui – A sta-of-the-art rev» *Eng Str,* **183**. 883-893 (2019).

[4] «International Organization for Standardization,» [En ligne]. Available: https://www.iso.org/about-us.html.

[5] YASHADA, «Nat Sym on Nem Man: A Cha to Ind Agr in the Cha» *Nem Soc of Ind and Ind Cou of Agr Res* (2015).

[6] A. J. G. a. I. Jensen, Pol, Pol and Pol, Spr, Dordrecht, (2009).

Narratives of historial urban lakefront regenerations

Pallavi Dalal [1], *Sunita Kothari* [2]

[1] *Baliram Hiray College of Architecture,Kherwadi,Bandra,Mumbai,India*
[2] *Asmita College of Architecture,Mira Road,Mumbai,India.*

Abstract. Lakefront in the densely populated area is a place of relish, where people can enjoy and relax. As these kinds of areas are famous in urban context to provide ample visual and physical public access to water and land. The immediate alliance between settlement and water is inherent since decades, as many urban areas in the world located near to the waterfront. All these areas are directly or indirectly dependent on water for their day to day survival. Since last two decades development strategies are shifting from agriculture to industrial, these bring extensive changes in lakefront development. In precise, lakefront has become less significant for human life. This paper discusses the extensive phenomenon of lakefront development in Thane city, which is known as the "City of Lakes", because of manifold lakes found over there. At present most of the lakes present in city are at the verge of disappearance due to reckless development. Therefore it is necessary to emphasize at historical and social aspect of lakes taking into account the overall picture and presenting the opportunities to integrate and emphasize multidisciplinary approach with social and cultural life of people.

1 Introduction

Water is an element that is essential for both life and life threatening. The dual force within water represents the essence of Yin and Yang. Where good can't exist without evil (Toy 7) Human survival without water is unjustifiable and without the presence of water any human settlement can't exists. This implicit that water is integral factor of selection of site for human settlement. In fact micro-location of presence of water also shows positive aspects for human settlement in history.

For generating good aesthetic quality or representing the visual image of city or city's skyline water plays an important role. Which enhances the city's silhouettes, urban matrix and it provides scope for better future expansion. Water front provides strong base line for urban development and gives better opportunities for physical and visual connection of city and its people with water. City like Chicago or holy cities such as Banaras gives opportunities and dependence of water related structures to people. Water can even impact human activities physically, socially and even psychologically. As Cities water phenomenon defines cities growth shape and inheritance of human settlement.

In time human needs become more variants or complex with the change in social, economic and technological needs, it gives an number of options for collecting or disrupting water for human needs, This leads to exploitation of water sources. Furthermore, it defines the means of communication as well as micro endeavours in planning of some of the important infrastructural intervention in the cities development. For example, In Mumbai, Marine drive's Queens Necklace and Bandra's Bandstand areas defines water edge as well as building skyline.

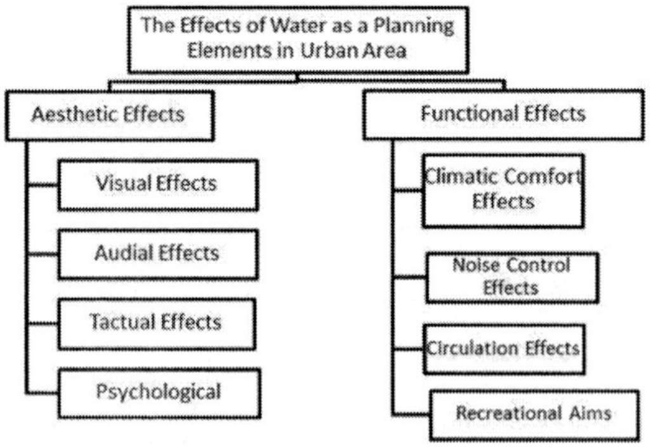

Fig.1. Adopted from Onen (2007), the effects of water as a planning element in Urban Area.

1.1 Pattern of Waterfront development-

In some of the recent decades, many waterfront cities or towns are facing an immense pressure of development from brownfield to large urban areas. As throughout the history the relationship between water and urban settlement is intact, it provides absolute living areas for human being. Such as in case of India, Banaras- Sacred

*Corresponding author: ***pallavidelite22@gmail.com***

place located on the western bank of the Ganga River, where many sacred activities are closely associated with the water. It shows the close relationship between human being and water.

In the contemporary era, many cities or towns in America, Europe and Asian countries are settled near the waterfront or on the bank of waterfront. In Mumbai Eastern waterfronts has historically been the hub of transportation, trade and commerce since early 19th centuries.

Therefore, major part of eastern edge is always inaccessible for common people. Which shows every urban waterfront has its own history, always tends to change as per the real estate impact, political scenario, physical and social needs of cities and people.

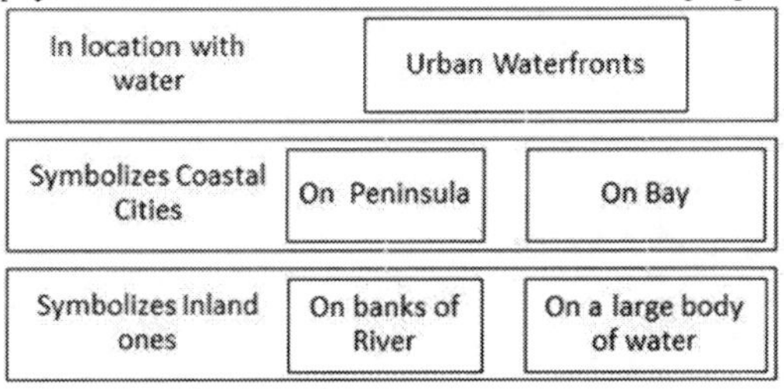

Fig 2: Water as a planning element in Urban Area

Cities which are located on peninsulas, bay, gulf, river, estuaries and deltas gets large area to deal with waterfront activities.

1.2 Typical pattern of waterfront development

Throughout history, waterfronts are the most ideal living area for human beings to be able to provide food, settling, reproduction, defence and learning etc. So, the many cities or towns Urban Waterfront Regenerations are established over water's edge from the history of civilization to until today.

In the contemporary era, the many cities or towns in America, Europe and Asian countries such as India, Singapore and China can be given as examples of waterfront settlement. In Mumbai, Eastern waterfronts has historically been the hub of transportation, trade and commerce since early 19th century .Therefore, major part of eastern edge were always inaccessible for common people. Thus, every urban waterfront has its own history which always tends to change as per the real estate impact, political scenario, physical and social needs of cities and people.

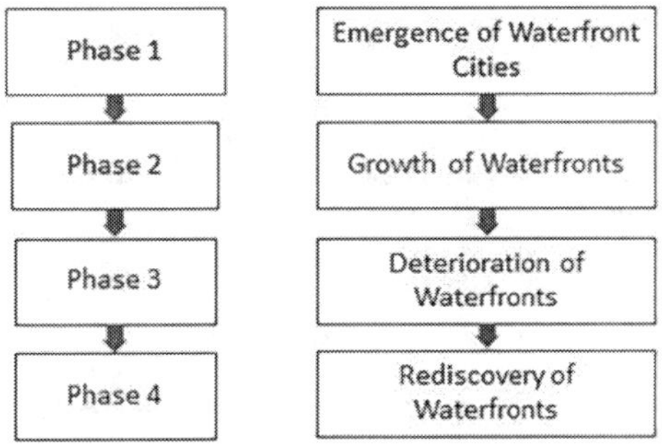

p

Fig.3.Adated from Wrenn et.al. 1983, Typical pattern of waterfront development phases.

2 Need of the Waterfront development

In last few decades because of the pace of urbanization and industrialization, many cities along the waterfronts are facing the issues of development. As areas around the waterfronts are in absolute degraded condition. Which even do not allow the access for recreation or open spaces around the water edge.

Pleasant water edge with minimum standards of recreational activities can provide boost or can attract, high investment and can motivate people to restrain which is being lost in the process of urbanization.

2.1 Waterfront development-a case of Chicago

The city Chicago- known for its skyline as well as its water edge development, is not as clean as it has seems today. In early 20th century it was the dump yard, the waste was directly discharged into either Lake Michigan or the river which flowed into the lake. As the solution to this situation they decided to reverse the flow of river and they build Chicago sanitary ship canal. Architect Daniel Burham proposed straightening of river channel, along with the 08 parks, the main branch runs through the heart of the city and it provides space for recreational activities, river walk and linkage with other places. · Strong emphasis on quality, functionality and sense of place.

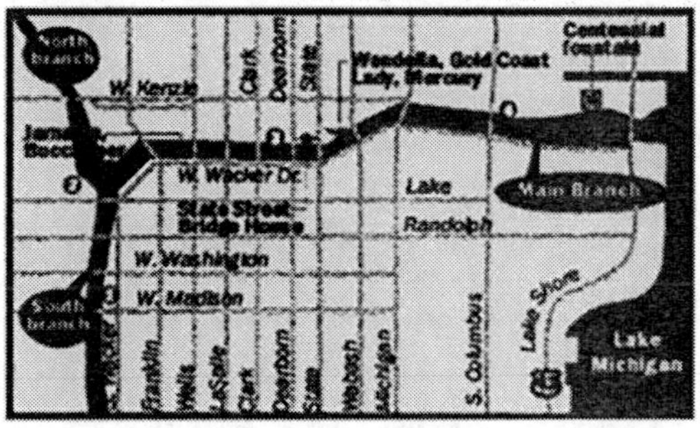

Fig.4.River walk plan showing various branches

*Corresponding author: *pallavidelite22@gmail.com*

Chicago Riverfront is examined and analysis under the urban design dimensions mentioned by Carmona, Heath, Oc and Tiesdell (2005)

Conclusion- It has been a successful place in its diversity for development and waterfront edge. It has provided a great example of waterfront development within in city.

3 Case of Thane city

The Thane is one of the Maharashtra's metropolitan city. As per national becemial census 2011 the population of the city is 11060148. This is the fastest growing city abutting to Mumbai. As the Mumbai has almost reaching to it carrying capacity of development, Thane is the one of the most favourable option for people as housing requirement. Which is already been looked up. The Thane city infrastructure and growth of the residential sector has given multi-facet to city's face. The geographical jurisdiction of the Thane city spreads over the area of 128.23 sq.mt. the city is called as "City of Lakes" as it comprise of 35 lakes in an area of 40 hector.The ancient Thane (Thana) is said to be derived from "Sthan".The capital of Silahara Kings of Kokan.Thane city has rich cultural heritage, mythological background and has been home to many events of historical importance.Thane city can be broadly categorized into 5 eras.

- Vedic period to 1300. : Comprising of Hindu or ancient period, Apparent, Shilhara and Bimba dynasties. 1300 1737: Muslim and Portuguese rule.
- 1737-1800: Maratha and Portuguese rule.
- 1800-1947: British Rule. 1947-till date: Independent India.
- The Shilharas ruled Thane from 810 to 1260.

During the 17th century place called Ghodbander was very famous for trading of horses. The Silahara emperors were disciples of Lord Shiva and they built Kopineshwar Temple during their rein. In all culture and religion water has always given a high spiritual importance and also utilized for recreational activities. Similarly, in ancient period these Thane lakes were integral part of such activities. However, over the period due to housing and infrastructure development pressure, these lakes have lost its originality and importance.

3.1 Need of conservation of lakefront

The principal images of Thane are triggered as a cultural centre, as well as for lakes. Earlier, Thane had more than 60 lakes, as eminent town planner Patric Geddes called it rightly as "City of Lakes". However today only 15 are remaining which are deteriorating or encroached upon and just occur incidentally while moving through the city. They are not maintained and are getting reclaimed slowly turning into city dump or wastewater collection ponds – perhaps waiting for an epidemic.

*Corresponding author: *pallavidelite22@gmail.com*

Table 1.showing elements of urban design and its implementation.

The Dimension of Urban Design	Elements	Comments
Morphological dimension	Land use Building structure Plot pattern Street pattern	Good connection with the rest of the Chicago City. It has good mixed land use.
Perceptual Dimension	Sense of place Meaning and symbolism' Image of the city	It form a small community within the site and has created a sense of place
Social dimension	People and space Public realm Safety Neighbourhood Accessibility, disability and exclusion	High quality of realm
Visual dimension	Aesthetic preference Building and architectural Amenity- footpath layout Landscaping	Good use of lighting and path layout
Visual dimension	Aesthetic preference Building and architectural Amenity- footpath layout Landscaping	Good use of lighting and path layout

Today it's difficult to call Thane a city of lakes. Is there a way to regain the ineffable quality of a city with lakes and how does one go about achieving it in the present situation is the question.

 https://doi.org/10.1051/e3sconf/202017006007

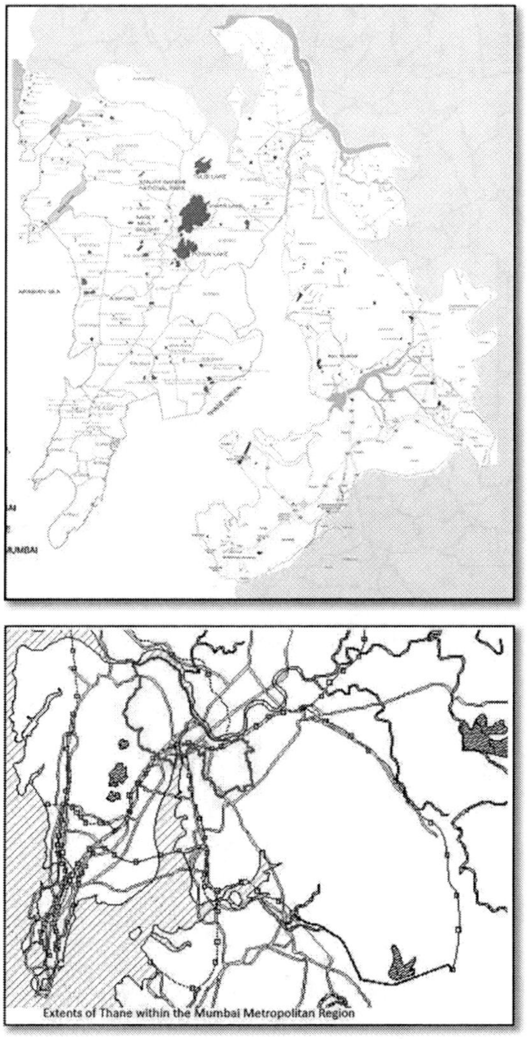

Fig.5. Map showing the location of water bodies in the city of Thane.

As population of Thane grows, this lakefront are becoming a dumping places.in last few decades population of thane is growing in tremendous way, as many other infrastructural development is becoming main part of urbanization in the society.as Thane is establishing good connectivity with other parts of Mumbai city, residential development as taken as boom.

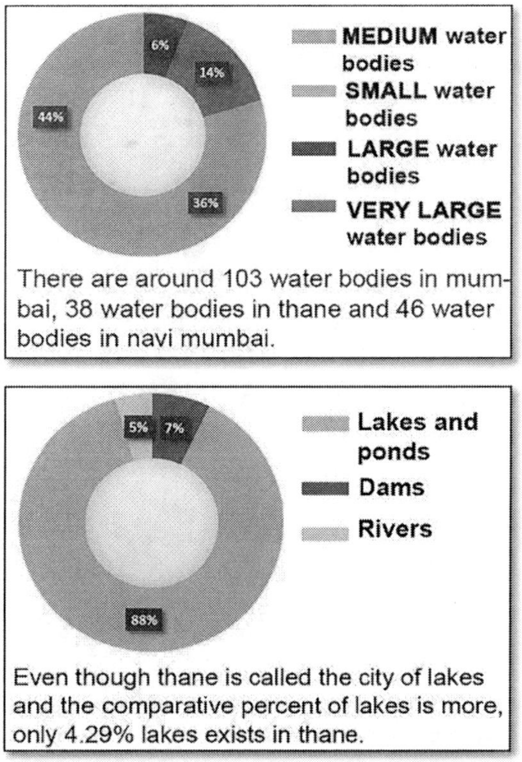

Fig.6. Pie chart showing the percentage of water bodies around the city.

There are around 103 water bodies in Mumbai, 34 water bodies in Thane and 46 water bodies in Navi Mumbai. In city of Thane, total number of lakes register by TMC(Thane Municipal Corporation)are 34,out of which only 9 lakes are conserves or having good public facilies.as urban growth increases the percentage of water bodies as well as open spaces is getting reduced.as per WHO norms of open spaces is required per person. But because of pace of urbanization and industrialization in the city of Thane per person open spaces percentage is reduced.

Downtown of Thane consisted of five major water bodies in close proximity to each other, which are having same peculiar charter and similar developmental problems.

By studying the downtown of Thane city and having done surveys for the same, we came to know that these spaces can be act as breathing spaces for the communities surrounding that. 2km radius, which are connected via roads or pedestrian movement within the settlement.

*Corresponding author: *pallavidelite22@gmail.com*

https://doi.org/10.1051/e3sconf/202017006007

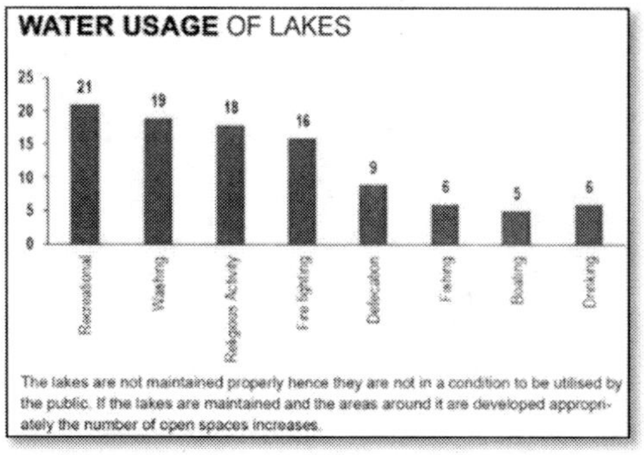

Fig.7. Percentage of water usage of lakes for different activities.

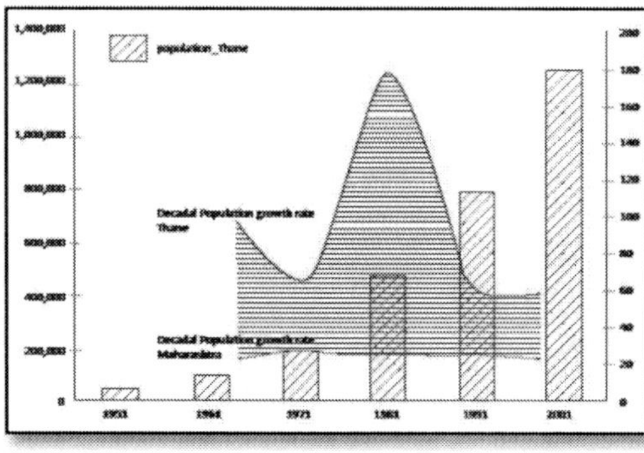

Fig.8. Graph showing population growth rate of Thane city.

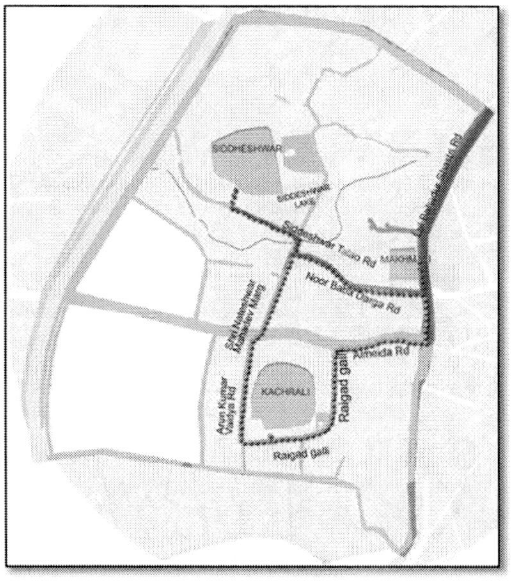

Fig.11. Map showing vicinity area of 3 lakes, which can reconnect to form network.

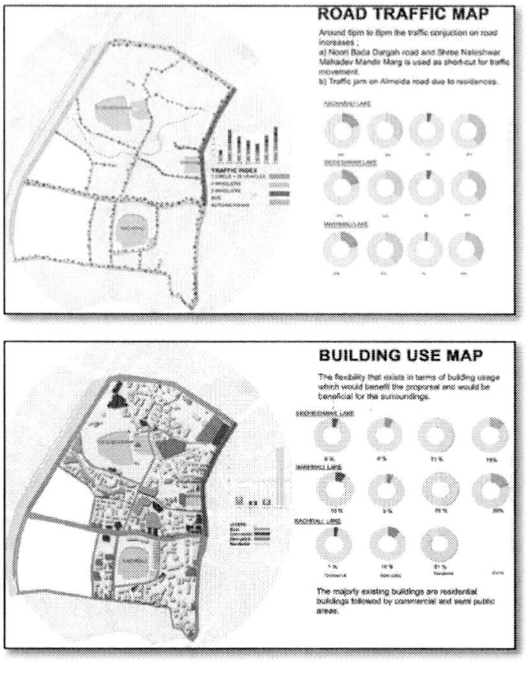

Fig.9&10. Map showing Road traffic map & Building use map. Out of five lakes, three lakes are within the periphery.

3.2 Survey findings-

- This route has been extensively used by passers-by as a short-cut from major roads, which increases the traffic movement in pick timings.
- The areas around this lake, has been encroached by slum dwellers resulting into unhealthy condition and reducing the water levels.
- The hawkers has been encroached on the footpaths resulting in to garbage littering.
- Illegal encroachment on roads and around the lakes making lakes smaller and decreasing the water level in the lakes.

As all of this survey findings point out towards the encroachment and negligence of the people surrounds the water body. This shows people are less interested in the areas which are not providing them any kind of activities or attractions.

4. Conclusion -

Rapidly growing city like 'Thane' the foremost challenge is to plan the city's growth in right direction and to improve the quality of water edges and its surrounding environment. Where lakeside land or surroundings can be utilize very carefully

*Corresponding author: *pallavidelite22@gmail.com*

https://doi.org/10.1051/e3sconf/202017006007

to improve the quality of life for close association between water people can be establish by using different activities as well as applying intervention and to create awareness among the people.

References

1) U. P. Timur, Urban Waterfront Regenerations, (2013)
2) D. M. Wrenn, Urban Waterfront Development, (1983)
3) G O. Moretti, M., TeMaLab Jrnl, of Mobility, **3**, 57, (2010)
4) Hussein, R.M.R., Int. Jnl. of Civil, Env. Struc. Engg., **8**, 488, (2014)
5) S. Barve, S. Sen., WITPRESS LTD., (2011)

*Corresponding author: ***pallavidelite22@gmail.com***

Case Study for Road Safety Audit of Aurangabad City

Prashant Awsarmal[1], S. L. Hake[2], Shubham Vaidya[1] P. K. Bhandari[2], M. P. Wagh[2]*
[1]*Department of Civil Engg., G.S. Mandal's Maharashtra Institute of Technology, Aurangabad, M.H., India*
[2]*Department of Civil Engg., Dr.V.V.P. College of Engineering, Ahmednagar, M.H., India*

Abstract: Efficient road network is a part-n-parcel of rapid industralization, urbanization and development of nation. While designing roads and highways, main emphasis is given on speed which will help to reduce time of journey and save fuel. But safety of drivers and passengers travelling along road is also important. In past, it was observed that while travelling, due to excess speed passengers safety was compromised. It will lead to accidents. It may cause severe injuries and loss of human life. Therefore it is important to check every aspect of vehicles as well as road during its design, construction and throughout the life of the road. Road safety audit is conducted to check performance of new road projects on grounds of offering maximum safety. Also checks are applied to study performance of existing roads to suggest repairs, rehabiliatation and maintenance work in order to improve condition of roads. During audit process, accident prone locations are identified. Past accident record from traffic department, Police department, hospitals etc are referred to understand damage that had occured. Even road geometry is investigated on technical basis. In present investigation, particular stretch of Beed Bypass Road passing through Aurangabad city in Maharashtra state, India was selected. On this road, accident sites where major accidents occurred in past were identified and investigated for different parameters. Based upon study, different causes of accidents and thereafter preventive methods were recommended during research work.

1 Introduction

1.1 Background

Planning and geometric design of road network has found place way back in history. In past, it was seen that during wars, roads were used for fast and strategic movement of army troops. Even for civilian use, roads are lifeline for development of nation. It is necessary to ensure that these roads are kept in working condition and should be safe for passengers to travel on roads. The process to ensure road safety is done through 'Road Safety Audit (RSA)'. It was first adopted in United Kingdom (UK) in 1980's. By next decade, in 1990 it was adopted in Australian continent. In America, RSA was adopted only after mid 1990's. It can be said that UK was pioneer in adapting RSA, thereafter successfully adapted in Australia and New Zealand. Austroads Guidelines were used to ensure road safety in Australia. Denmark was the country where RSA was made compulsory for all highway projects. By observing benefits achieved through RSA, World Bank has made RSA process compulsory in all road projects funded by them.
Road safety audit (RSA) includes investigating new as well as existing roads to study geometric design of roads [1]. Also find accident causes, effects and provisions made to avoid further accidents. It is believed that 'prevention is better than cure'. As per this proverb well designed road will ensure safety of people driving vehicles. RSA can be performed at any stage during lifecycle of road project [2, 3].

1.2 Objective

Since RSA is performed to ensure safety, it will also check that design of road components such as road junctions must be planned as per traffic safety guidelines. In present investigation, Beed Bypass Road passing through Aurangabad city was considered. Two road junctions, MIT chowk and Deolai chowk were selected for further investigation. The number of accidents which had occurred in past at these two black spots were studied in detail. Reasons for accident were investigated in scientific manner. Different remedial measures to avoid further accidents were also found and applied in model study to ensure its trustworthiness.

The objectives of the present study are:

1. To collect project documents of Beed Bypass road, particularly study road geometry of road. Also determine traffic intensity along this length of road.
2. Based upon data collected from traffic and police department, identify locations where maximum number of accidents occurred in past.
3. To identify defects and conflict points on the road network responsible for accidents.

*Corresponding author: **drsandeephake@gmail.com**

4. Mark locations of accident prone areas on maps.
5. To identify safety influencing parameter in reducing accidents rate.
6. To study the traffic flow pattern, type and number of vehicles running along selected stretch of road.
7. To study the causes of accidents and suggest corrective measures at potential location.
8. Create awareness amongst road users to follow traffic rules.

1.3. Road accident situation in India

Fig. 1 indicates the number of accidents have decreased by considering data of previos four years. By adapting safety measures and awareness amongst drivers helped to reduce accidents.

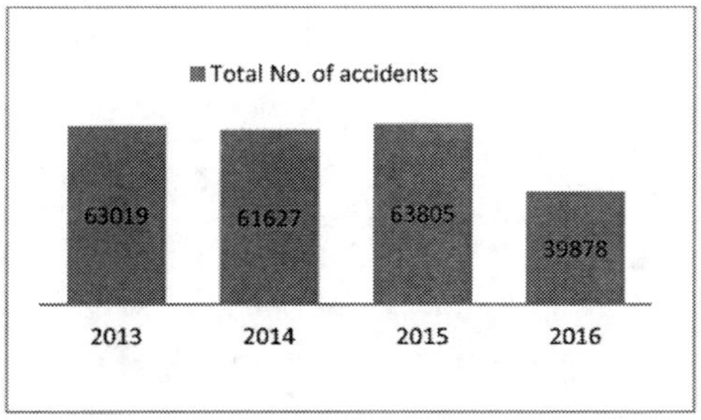

Fig. 1 Total Number of Road Accidents in India (2013-2016).

1.4. Problem statement

Fig. 2 shows the map of Beed bypass road passing through Aurangabad city, Maharshtra state, India. As per past data, more number of accidents occured along this road. Their was a need to investigate the reasons for accidents and thereby suggest measures to avoid accidents. By adapting Road safety audit, past accident data will be investigated and based upon analysis, factors affecting traffic pattern, accidents, drivers reaction will ne considered to propose safety measures required.

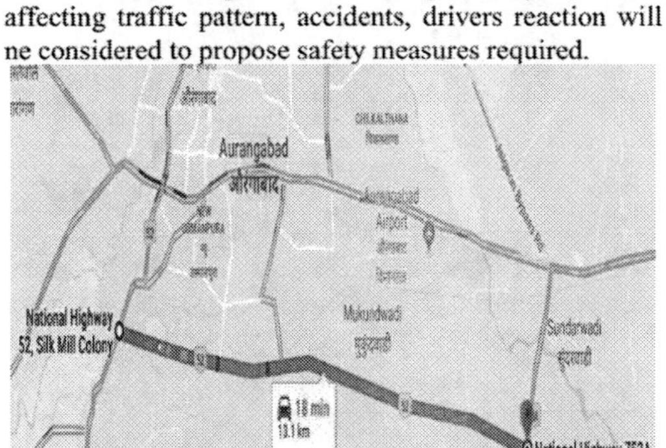

Fig. 2 Google map image of Beed Bypass Road

1.5 Scope of work

The scope of the study is to reduce accidents on road network, reduce severity of accidents also avoid costly remedial work. The road selected for the study is Beed Bypass road (13.150 km). It is stretched between two points, Mahanubhav Ashram (T- point of Paithan road) to Zalta Phata. These two road junctions MIT chowk and Deolai chowk are marked and further RSA investigation work carried at these two locations. The accident analysis will be done from previous year's data.

2literature Review

Investigation of past work has revealed several problems associated with geometry and vehicle safety. Although vehicles are designed by following all safety guidelines, but they are not able to keep people safe in event of accident. Thus, there was need to access causes of accidents and pin-point accident prone locations along the road to make retifications in road geometry.

K. Meshram and H.S. Goliya [4], performed accident analysis on NH-3 stretched between Indore to Dhamnod. Required information was collected for 3 years in-between 2009 to 2011. Both researchers made conclusion that, pinpoint location of accident was at Manipur region owing due to errors made while constructing road. Also accident rate is more in urban region especially in Indore city. It was about 35% of total accidents that have occurred in a year. Causes of accidents were also investigated and it was revealed that tendency of drivers to drive fast and more number of vehicles running on road was the major problem behind the accidents. Most of the accidents had occurred during night time between 6 to 8 pm. Since more number of buses plying on road during night time, carrying passengers from city back to village. Number of casualties during accidents was also recorded. And as per record one fatal and five deaths are reported per km stretch per year along this NH-3 road.

A. Theofilatos and G. Yannis [5] studied the impact of weather on roads. Effect of rains, heavy winds, storms, fog etc can reduce the speed of traffic and may cause an accident due to reduced visibility. Relationship of traffic parameters like traffic flow with number of accident was investigated. During rainy season, due to extreme weather and poor visibility chances of accidents are more. But it was observed that if drivers are following traffic rules then it will avoid accidents. As per actual data received from road survey, it was observed that other weather conditions like fog, wind storms, temperature had no direct effect on traffic flow.

A.B. Labana et al [6] performed road survey on NH-113 running from Dahod to Jhalod to find locations where maximum accidents occurred during span of 2010 to 2014. Based upon data received it was found that 509 accidents occurred during case study period. Investigation of accident spot showed that there was no median at centreline of road which leads to head collision of vehicles coming from opposite direction. 42.63% accidents occurred during summer season in month of March and April. 26.62% two wheelers and 27.5% cars were involved in accidents.

O. Gholap et al [7] has summarised reasons which were behind major accidents. As per author potholes cause health issues like back pain and even lead to fatal accidents. Information sign board and regulatory sign boards are important to give warning to drivers and make them to drive carefully with safety. All markings on roads must be painted and maintained from time to time. Thus, recommended need of regular road safety checking and maintaining audit report.

3. Study of Black Spot

3.1 Accident record

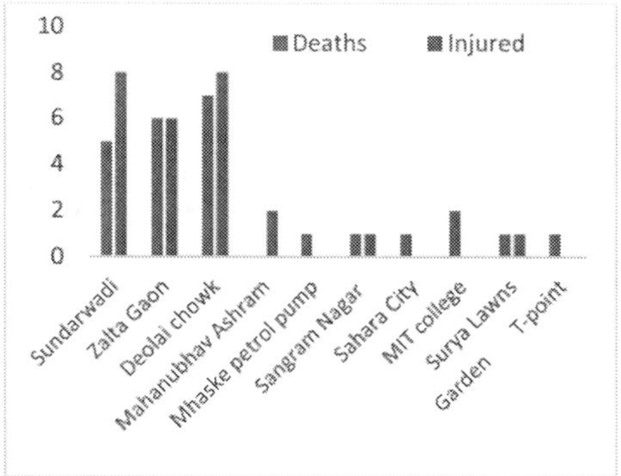

Fig. 3 Number of Injured & deaths at accident locations.

Fig. 3 showed different location along the stretch of Beed Bypass road where number of accidents occurred, were recorded. Corresponding injured and deaths also recorded. This data was made available from past records. It was observed that maximum accidents have occurred at Sundarwadi, Zalta gaon and Deolai chowk junctions. These locations also recorded maximum number of deaths and injured people. For further investigation Deolai chowk which recorded maximum accidents and MIT chowk which was near an educational institution, which may be prone to accident was considered for next investigation.

3.2 Observations

Fig. 4-10 shows photographs taken during investigation at MIT chowk and Deolai chowk. Different aspects of road geometry were studied from photographs and based upon actual investigation comments were made further.

Fig. 4 Sign board post at turning obstructs vehicles

Fig. 5 Open Gutter on road side

Fig. 6 Road Sign hidden behind plants

Fig. 7 Pot holes at MIT chowk

Fig. 8 Street Light not working at MIT chowk

Fig. 9 Uneven Service Road at MIT chowk

Fig.10 Insufficient Median Height

3.3 Major hindrances on Beed bypass road

1. Accidents occur on this road frequently.
2. Heavy traffic on road
3. People often drive from wrong side of road.
4. Drink and drive cases are seen more on this road.
5. Speed limit is frequently violated by drivers.
6. Rickshaw drivers and motorcyclists drive carelessly and try to overtake heavy vehicles without caution.
7. Road users do not follow signals properly.
8. There are obstructions to vision at intersections.
9. Unauthorized breaking of median at number of places.
10. Road markings are hidden behind trees and advertisement boards.
11. After closure of railway gate near Deolai chowk all traffic from Shivaji nagar road come on bypass road which makes the movement of fast-moving traffic difficult and dangerous.
12. People don't follow rules even if traffic police is available on the junction.
13. Road width is insufficient according to current traffic requirement.
14. The slope of the shoulder is not properly levelled at many places this leads to formation of cuts at the edges.
15. Absence of pedestrian markings.

3.4 Traffic volume study

Along the length of Bypass road, two black spots were identified as MIT chowk and Deolai chowk where maximum accidents were recorded in past. It was decided to perform traffic survey on these junctions. On 16th February and 31st August 2019, survey conducted on MIT chowk. On 14th March and 7th October 2019 conducted the survey on Deolai chowk. Manual readings were taken regarding counting number of vehicles crossing the roads. Vehicles were categorised as Heavy duty vehicles, Bus/ truck, passenger cars, Auto rickshaws and two wheelers. Vehicles type has direct impact on overall results and interpretation of data.

3.5 Factors affecting Traffic counting

Factors affecting traffic counting are as follows:

1. Weather conditions like rain, fog, storms, etc [8, 9].
2. Traffic composition includes assessing which category of vehicles are more in number.
3. Purpose of the traffic counting.
4. Method for vehicle counting which may be manual or automated counting machine
5. Location of the study which was already finalised as black spots which recorded maximum accidents.
6. Direction of flow which may be one way or dual [10].
7. Width of road lane meeting the junction.
8. Road type may be bituminous or concrete

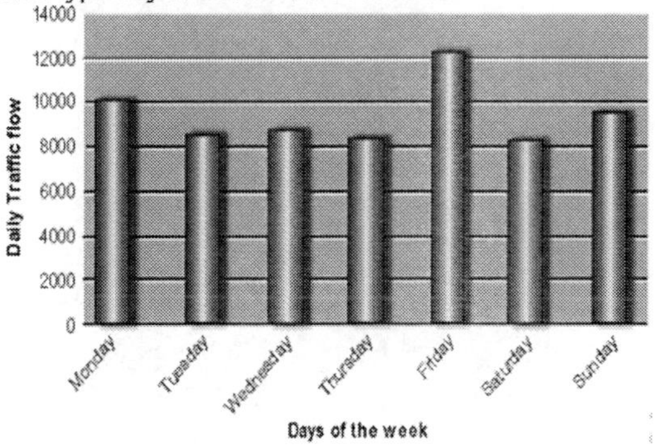

Fig. 11 Daily traffic flow Beed bypass road.

Daily traffic flow on Beed bypass road is represented as bar chart as shown in fig.11 in passenger car unit (PCU).

3.6 Identification of blackspots

After the study of accident data, traffic volume study and site observations as given above MIT chowk and Deolai chowk are identified as blackspots.

3.6.1 MIT chowk

MIT chowk is near an institutional area so there is presence of large number of students in day time. Also, at the junction, there is falling gradient which eventually make the vehicle reach the junction at high speed. Sufficient Set back distance (SBD) is not available while taking right turn from MIT college. All these factors lead to accidents. Fig.12 shows photograph of MIT chowk taken on investigation day.

Fig. 12 MIT chowk

3.6.2 Deolai chowk

Deolai chowk is one of the most crowded junctions in the Aurangabad city. Many accidents had occurred at this chowk. Small vehicles like two wheelers were involved in accidents. The road going towards Shivaji Nagar has a railway crossing and whenever this road is closed for the crossing of train then the traffic on this road gets diverted to the Deolai chowk making it more difficult for regular traffic at the chowk. Fig. 13 shows photograph of Deolai chowk taken on investigation day.

Fig.13 Deolai Chowk

4. Recommandations

4.1 Observations recorded

The road survey was done from Mahanubhav Ashram (T- point of Paithan road) to Zalta Phata. Following observations were made on investigation day.

1. Each and every corner of the road was surveyed.
2. Foot paths were checked.
3. Sign boards and their location were checked.
4. Pot holes where checked.
5. Lane marking was checked.
6. Position of speed breakers were checked and also their conditions.
7. People and vehicles travelling wrong side were photographed.
8. The total length of the road surveyed was 13.1km.
9. Almost footpaths were damaged and broken or at some places they were occupied by small shops.
10. Some sign boards were tilted, some were missing.
11. Total 39 pot holes were present
12. Lane marking was missing at some of the roads, at some places they were not at the center.
13. Almost all speed breakers have to be constructed properly.
14. More than 30% of the people as well as the vehicles were travelling wrong side.

4.2 Recommendations

Based upon observations made on investigation day and thereafter by studying the photographs are readings of vehicle counting following recommendations were made to improve the quality of road in order to avoid further accidents.

4.2.1 Immediate measures

1. Provide the speed breaker at MIT & Deolai chowk.
2. Hazard marker to be provided wherever they are missing.
3. Shrubs in shoulders to be removed. Shoulders to be dressed to proper slope.
4. Repairs of railing on culverts to be carried out.
5. Pavement marking is to be provided.
6. Median gap is to be closed.
7. Drainage needs to be improved.
8. Fallen shoulders to be restored.
9. Speed breaker shall be extended to full width of the pavement and shall be marked.
10. No. of electric poles close to the road need to be shifted. Shall be provided with hazard marking to highlight their position to the drivers.
11. Provide pedestrian way.
12. Provide street light at MIT junction.
13. Green time should be extended up to 45 – 50 seconds at MIT chowk of road Paithan to Zalta Phata & vice versa during 11am to 5pm.

14. Plantation at the junctions should be removed so that approaching vehicle can see the clear view of traffic.
15. Branches of trees projecting over the road at a height less than 7 m should be removed.
16. Pot holes and ruts should be repaired.
17. Traffic rules should be strictly followed.

4.2.2 Long term measures at Deolai chowk

Provide flyover on the given direction as shown by green arrow in fig. 14. It will have following advantages.

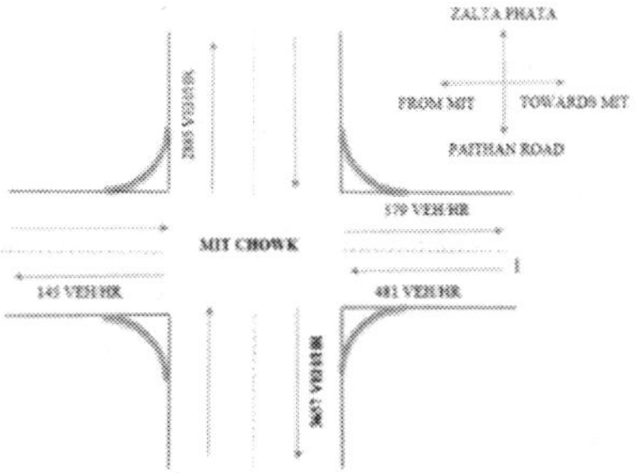

Fig. 14 Position of recommended flyover at MIT chowk.

5 Conclusion

Current research study regarding road safety audit highlighted issues in safety management. It examines the defects in the road safety in relation to vehicular traffic. It suggested the various recommendations which are easy to do and are economical. The audit is applied to the risks outside the framework of standards and codes.

The accident data, traffic volume and field observations showed risk involved while travelling on the road. Also, it gives the information about causes of accidents and vulnerable road users. Following observations where made while assessing traffic survey and accident data of Beed Bypass road, particularly at two junctions in Aurangabad city.

1. MIT chowk and Deolai chowk are identified as black spots.
2. It is observed that negligent driving, rules violation, over speeding of heavy vehicles and avoiding safety gears like seat belts and helmets are the prominent reasons for the accidents.
3. Further recommendations made at both accident spots must be adapted immediately to avoid further accidents. These recommendations may be subjected to revisions and alternations, but the immediate action will avoid further deaths and injuries.

References

1. A.S. Bagi, D.N. Kumar, *IOSR Journal of Mechanical and Civil Engineering,* 1, 6, pp. 1-8, (2012).
2. IRC:SP:84-2014, *Indian Road Congress,* (2014)
3. IRC:SP:88-2010, *Indian Road Congress,* (2010).
4. K. Meshram and H.S. Goliya, *International Journal of Application or Innovation in Engineering & Management,* 2, 7, (2013).
5. A. Theofilatos and G.Yannis, *Accident Analysis and Prevention,* 72, pp. 244-65, (2014).
6. A.B. Labana, V.A. Parikh, V.P. Parekh, *International Journal of Advance Engineering and Research Development,* 2, 5, (2015).
7. O. Gholap, N. Shinde, V. Shelke, N. Navale, and K. Deshmukh, *International Journal of Engineering Research & Technology,* 7, 4, (2018).
8. A. H. Ghods, F. Saccomanno and G. P. Guido, *Procedia in Social and Behavioral sciences,***53**, pp. 834–841, (2012).
9. M. Williamson and H. Zhou, *Procedia - Social and Behavioral Sciences,* 43, pp. 330-338, (2012).
10. R.R. Dinu, A. Veeraragavan, *Journal of safety Research*, **42**, pp.39-42, (2011).

Development of field-scale composite biorack constructed wetland – startup phase

Mitil Koli[1] and *Guru Munavalli*[2*]

[1]Department of Civil Engg., Walchand College of Engineering, Sangli, Maharashtra, India,
[2]Department of Civil Engg., Walchand College of Engineering, Sangli, Maharashtra, India

Abstract. Constructed Wetlands (CW) with Biorack (BR) technology have been studied extensively in recent years with positive outcomes compared to conventional CWs. A field scale application of the same has been worked upon in the Composite Biorack Constructed Wetland (CBCW) which is a secondary wastewater treatment facility of the Decentralized Wastewater Treatment System (DWTS) at Walchand College of Engineering, Sangli (M.S.), India. The CBCW is a multi-compartment system comprising of compartments with brickbats as supporting medium and BR Compartments (BRC) and has been in operation since May, 2018. The CBCW is vegetated with *Typha angustifolia* L. and *Canna indica* macrophytes. Being a unique CW, studies on difficulties aroused during the startup phase need to be addressed so as to arrive at proper troubleshooting techniques for future references. The paper addresses various issues, specifically growth of vegetation and their early wiltage in CBCW, for a period of 150 days. Using alternative modes of plantation, it is concluded that vegetation acclimatization, proper suspension and grip of the bulb and roots of saplings in racks, fluctuating inflow of wastewater are few reasons influencing development of vegetation. An improvement by about 12% is obtained in COD removal in the latter period of study implying establishing of vegetation in CBCW.

1 Introduction

Water scarcity has hit hard in most regions of developing countries like India. A huge iota of wastewater generated is disposed of untreated into surface water polluting the already scarce fresh water sources. With the emergence of new concepts in urban development such as Smart City in India, application of Decentralized Wastewater Treatment System (DWTS) can help curb this problem and treated wastewater can further be reused for non-potable purposes subsequently reducing load on freshwater sources. DWTS are gaining popularity due to their low CAPEX and OPEX as well as a potential source of non-potable water. Constructed Wetlands (CWs) being one such system has gained popularity in the recent years owing to its simplicity and robust treatment performance [2] CWs consist of aquatic macrophytes planted on a bed of porous media on which wastewater is fed either horizontally or vertically. The macrophytes have an extensive root system having a significant ability to transport oxygen from the atmosphere to the roots which then disperses a portion of this oxygen into the rhizosphere. The micro-organisms thriving on the substrate utilize this oxygen and stabilize the organic matter applied on the CW. Plant species such as *Phragmites* sp. show higher organic removal along with heavy metals and could sustain environmental stresses [7,11] Use of multispecies vegetation instead of singular species such as *Canna indica*, *Phragmites* and *Cyprus* provide habitats for development of a varied microbial population on their root matrix and thereby aid in enhanced removal of pollutants. Few other pollutant removal mechanisms in a CW include sedimentation, filtration, nitrification-denitrification and phytoremediation. CWs are basically classified according to three criteria: i) hydrology (surface or subsurface flow) ii) type of plant species (emergent, submerged or free-floating) and iii) flow path (horizontal or vertical) [10]. The disadvantages of single type of wetland can be overcome with improvement in pollutant removal efficiencies by combining two or more CWs in a series or using technologies of two CWs into one thereby forming a composite or hybrid CW. One of the major drawbacks of a CW is clogging of bed owing to high suspended solids. A more economical, compact and clogging free alternative to the conventional wetland system is Biorack (BR). A BR is a mesh of vertical PVC pipes arranged in a rectangular configuration within a tank. The pipes are provided with numerous perforations for liquid transportation. These BRs are involved in the treatment process by holding vegetation and providing abundant attachment site for microbial growth. BR system showed high rate of organic degradation of raw sewage effluent at lower HRT of 10h using *Phragmites* sp. [10] BRs were also found to be effective for wastewater with high total dissolved solids and heavy metal salts like cadmium, copper, nickel and zinc by using *Phragmites* sp. proving to be an ideal process for domestic wastewater treatment [11]. Vegetation such as *Thalia dealbata* planted in a BR was able to treat low concentration polluted river water due to higher nutrient removal rates of the plant species as well as high fine-root biomass [13]. [14] studied the performance of a BR fed with effluent through an anaerobic baffled reactor

* Corresponding author: guru.munavalli@walchandsangli.ac.in

and using *Typha* sp. and *Phragmites* sp. The integrated system showed low capital investment and high COD, COD and BOD_5 removal reaching 87%, 90% and 93% respectively at an HRT of 21h. BR was also efficient in Cu removal from Cu-contaminated freshwater [7]. BR planted with multispecies of vegetation viz. *Typha angustata* and *Canna indica* showed a significant increase in organic degradation and enhancement of dissolved oxygen as compared to conventional Biorack [9]

The present paper addresses the various difficulties encountered in the startup phase of a field-scale Composite Biorack Constructed Wetland (CBCW) as well as different alternatives applied inorder to overcome them and aid for a smoother performance. The outcomes would contribute to existing knowledge on starting up a CW.

2 Materials and methods

2.1. Source of wastewater

CBCW is a secondary wastewater treatment facility of the DWTS at Walchand College of Engineering (WCE), Sangli (M.S.), India. The college is supplied with two major sources of water, i.e., through municipal water supply; treated in campus water treatment plant for potable use and open wells within the college campus for non-potable facilities. Wastewater generated through different the residential sections in campus such as the boys' and ladies hostels and staff quarters mainly comprise of septic tank effluent and greywater. The wastewater is first collected in a storage sump and then supplied for primary treatment through screens and Anaerobic Baffled Reactor (ABR) after which it enters the CBCW. This anaerobically treated wastewater forms the basis for the current study.

2.2 CBCW details and operation

Technically, DWTS in WCE is a multi-level natural wastewater treatment plant of 42 m3/d capacity. Wastewater undergoes preliminary and primary treatment passing through screen chamber made up of coarse and fine screens for trapping large size floating matter and an underground chain of ABR for settlement of high suspended solids. The effluent from ABR enters CBCW for secondary treatment. CBCW is a brick masonry structure made up of twenty compartments comprising of an inlet chamber, alternate compartments having brickbat as supporting medium hereafter named as BSFW (Brickbat Subsurface Flow Wetland) and Biorack compartments (BRC), adsorption bed and an outlet chamber in a series with an effective wastewater depth of 0.64 m. BRC are made up of Asbestos Cement (AC) sheets arranged so as to create space for plantation of vegetation. An adsorption bed made up of coarse aggregates and charcoal forms the final compartment before outlet chamber. The primary treated wastewater enters CBCW and flows in an upflow and downflow mode through all the compartments and treated effluent is then discharged into the outlet chamber after which it is pumped to a Vertical Flow Constructed Wetland (VFCW) for final treatment. The treated wastewater is then sent for land application. Fig. 1 shows a schematic line plan of CBCW.

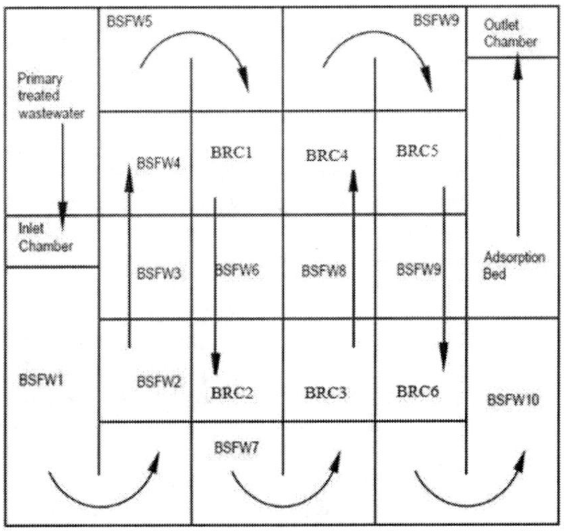

Fig. 1 Schematic line plan of CBCW

2.3 Vegetation details

Normally, locally available wetland plants called macrophytes are used in a CW due to their abundance and easy availability if required in larger quantity. The vegetation used in CBCW was *Typha angustifolia* L. and *Canna indica* which are wetland macrophytes found in the streams and marshy lands of Sangli city. *Typha angustifolia* L. belongs to cattail (Typhaceae) family which has narrow leaves of width about 0.01m and can grow upto 2m tall. Advantage of using *Typha angustifolia* L. is its effectiveness in removing pollutants such as carbonaceous organic matter, nitrogen and phosphorous. It also helps in enhancing dissolved oxygen level in wastewater due to oxygen leakage from root zone [5,6]. *Canna indica* is another widely found flowering plant with broad leaves and aesthetically pleasing flowers. The plant can grow upto 1m – 1.5m tall and has an extensive growth rate. It is also effectual in nitrogen, phosphorus, carbonaceous matter removal as well has specific ability to uptake some of the heavy metals. The root matrix of this plant is extensive and also helps in increasing the dissolved oxygen of wastewater [1,2,4,8,14]. Since both plants have an extensive root matrix, if planted together they can provide habitat for diverse group of microbes leading to higher pollutant removal rates. These plants were carefully dug up using simple hand tools such that their roots and bulbs remained intact. The roots were then washed thoroughly to remove clay and any foreign matter and then planted in CBCW using different approaches.

2.4 Monitoring of system

CBCW was monitored from August, 2018 to December, 2018 considering it as the startup phase for this CW. Initially, flow measurement was carried out daily for a week in the month of August, 2018 between 6:00AM to 6:00PM so as to determine the fluctuations in the flow rate as well as determining the times for peak and least flow. Flow was measured by fill and draw method and was found to range between 532 L/h and 7200 L/h (least and peak flow respectively).

Initially, around 80 young *Typha angustifolia* L. and 30 *Canna indica* saplings were planted in CBCW in late July, 2018. These saplings were gripped together and placed in compartments consisting brickbats, by creating space between them such that root tips reached wastewater. Similarly, they were placed in AC sheet BRC as shown in Fig.2 (a). Ideally, the plants should develop shoots after a few days. However, most of the vegetation was observed to be wilting out particularly in BRC. This problem was addressed in month of September, 2018 by typing up stems of two plants together with a thread and then kept on the edges of AC sheet with their roots suspended in wastewater shown in Fig. 2 (b). After few days of this method of plantation, it was noticed that although very few number of saplings had developed new shoots, the majority of plants had begun to dry out Thus, no significant plant growth was achieved. Vegetation brought thenceforth from month of October, 2018 was then washed thoroughly and kept in a barrel filled with the freshwater and then subjected to series of diluted wastewater, continued till the plants were subjected to the wastewater which they would be exposed in CBCW. Furthermore, these 3-4 saplings were gripped together into clean plastic water bottles assembled as a pipe cut on both sides and having perforations as shown in Fig. 3. These bottled plants were then inserted into the spaces between the AC sheets Biorack such that they remained suspended within those gaps. New shoots developed after a week of plantation in all the bottled plants with the parent plant wilting out. With daily observations it was found that the vegetation growth was significantly higher and new shoots developed within a week of plantation. It was noticed that around two to three new shoots were developed per *Typha angustifolia* L. plant and two new shoots per *Canna indica* using this approach of plantation. Harvesting of dried plants and replantation was done after every three weeks or once when large number of plants was wilted.

The effect of vegetation growth on removal of organic matter from CBCW was evaluated based on pH, COD and BOD_3. Wastewater samples were collected in clean 1L plastic cans using grab sampling procedure from the inlet and outlet chamber. Frequency of sampling was mostly on a daily basis. The samples were then brought in Environmental Engineering Laboratory at WCE for analysis as per standard procedures set by APHA [3].

Fig. 2 (a) Photographic view of vegetation inside AC sheet Biorack

Fig. 2 (b) Photographic view of vegetation inside AC sheet Biorack

Fig. 3 Photographic view of vegetation in discarded bottle as Biorack

3 Results and Discussions

3.1 Influent Characteristics

A thorough chemical analysis of influent revealed the significant characteristics as mentioned in the Table 1. The pH values indicate slight alkalinity in wastewater conforming to the fact that the influent to CBCW is from

ABR and supposedly has methanogenic activity taking place contributing to a slight high value of pH. The values of BOD_3 and COD indicate low strength of the influent along with BOD_3/COD greater than 0.5, which ascertains suitability of the influent into consideration to be suitable for biological treatment. CBCW being one such biological treatment unit can thus be an appropriate alternative for wastewater treatment.

Table 1. Characteristics of influent to CBCW

pH	COD	BOD_3 @ 27°C	TKN
-	mg/L	mg/L	mg/L
7.77 ± 0.16	196.19 ± 23.96	99.79 ± 13.05	63.88 ± 4.88

3.2 Effect on vegetation growth using different approaches of plantation

Vegetation being an important component for proper functioning of a CW, it is necessary to monitor its growth so as to maintain aerobic pockets in the CW bed required by aerobic microorganisms for survival and consequent degradation of organic matter. The vegetation in CBCW as mentioned in section 2.4 showed improper growth pattern. Initially, the vegetation brought was used as they were plucked from the streams directly into the spaces created in the mesh of AC sheets. There was a possibility that by directly inserting plants in the small BR spaces the bulb portion of plant could have got damaged after reaching the bottom of BR leading to its early death. The second approach of plantation wherein the stem/leaves of the plants were tied together and kept suspended onto the edge of BR too wasn't a success as the roots were either not fully exposed to wastewater or flooded with wastewater due to fluctuations in wastewater depth owing to diurnal flow variation as recorded from flow monitoring. Generally in subsurface wetlands, water level is the most critical aspect of plant survival during the first year after planting. A common mistake is to assume that because the plant being a wetland plant, it can tolerate deep water. Frequently, too much water creates more problems for wetlands plants during the first growing season than too little because the plants do not receive adequate oxygen at their roots. [15] This validates large wilting of vegetation and subsequent dieoff. Also, using a thread for tying formed an incision on the leaves which subsequently damaged the entire leaf and after a certain time period, the entire plant as a whole. Amongst these methods, the final method of first acclimatizing the plants (subjected to serial dilution of wastewater) and then gripping them into perforated pipes of plastic bottles was found to be most effective. Vegetation used in this study was brought from nearby marshy lands and for their best survival as well as growth, small stalks (1 - 2 inches) during the first growing season in the substrate should only be kept in saturated conditions, not flooded. As the plants become well-established (2 - 3 months), water levels can be raised. Plants should be allowed to become well established prior to wastewater being introduced into the system since the plants need an opportunity to overcome stress after plantation before other stresses are introduced. A gradual rather than sudden increase in the concentration of the wastewater applied reduces shock to the vegetation. Most wetland species are adapted to daily or seasonal fluctuations in water level but most wetland plants can tolerate neither extended periods of flooding nor drying of their roots. [15]. Also, using perforated plastic bottle pipes ensured continuous supply of wastewater to roots as well as proper grip and space for roots to spread.

3.3 Removal of organic matter in different plantation phases

The CBCW was analyzed for its performance from August, 2018 to December, 2018. The total sample size accounts to 110. As seen from Fig. 4, the removal of organic matter has an upward trend along with the time in months. It can be noted that COD removal in months of August and September, 2018 ranged from 17.85±5.51 % to 17.77±6.90 % for a hydraulic loading rate (HLR) ranging between 10.29 m^3/m^2.d and 0.76 m^3/m^2.d and organic loading rate (OLR) of 0.17 kgCOD/m^2.d and 2.89 kgCOD/m^2.d. The removal is lower which could be due to vegetation in CBCW still in their initial stage of growth along with improper growing pattern in these months. Also, early wilting of vegetation required removal of decayed plants from BRC. This could have created void of aerobic pockets at root zones of the plants. This may have led to non-growth of aerobic microorganisms which are mainly related to degradation of organic matter. However, the months of September and October, 2018 has seen a steady spike in COD removal increasing by about 4% and 12% respectively. A similar trend was observed for BOD_3 removal as well. This could be validated by the proper development of young shoots and root matrix of the bottled vegetation in BRC. The root matrix as well as the brickbat media provide abundant space for microorganisms to thrive and aid in further COD removal. The maximum COD removal was found to be 37.5%. There was a certain decrease in the removal efficiency in month of December, 2018. As the city of Sangli experiences cold winters in this month, the microbial activity is hampered due to low temperature which validates the drop in COD removal. Also, the lower COD removal can be validated by presence of residual BOD_3 entering the effluent due to decay of dead roots, shoots or leaves left within the wetland bed unchecked. The average effluent COD and BOD_3 observed was 130 mg/L and 60 mg/L respectively. An improvement in removal efficiency can be expected after proper and uniform growth of vegetation on the entire CBCW bed.

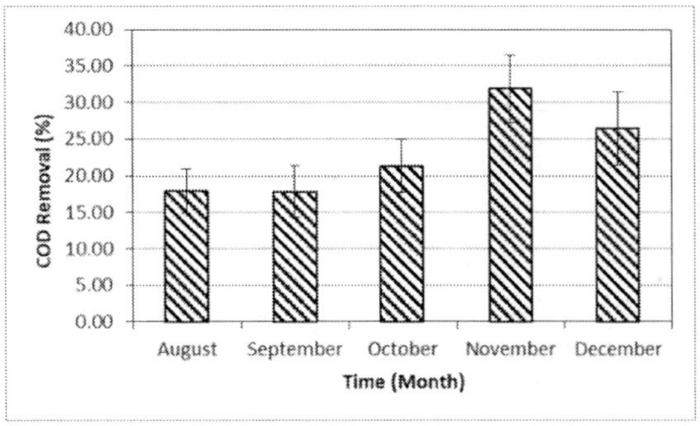

Fig. 4 Monthly variation in COD removal efficiency

4 Conclusions

The chemical characterization of wastewater reveals that biological treatment can be given to the anaerobically treated domestic wastewater and thus constructed wetland is a suitable alternative to conventional wastewater treatment systems. The problems faced while growing vegetation for adequate working of a Composite Biorack Constructed Wetland is highlighted in the present study by using different approaches. It is concluded that vegetation growth is majorly affected by improper suspension of vegetation into Biorack leading to damage to roots and bulbs of the plants as well as fluctuating flow of wastewater and wastewater level. It is also found that proper gripping of the vegetation in perforated plastic bottles and then suspending in BRs creates ample space for root development and growth of new shoots. Also, acclimatization of the new saplings in the present wastewater conditions plays an important role in vegetation growth. A maximum COD removal efficiency of 37.50% is achieved in CBCW with average effluent COD and BOD3 of 130 mg/L and 60 mg/L respectively. The values don't completely satisfy permissible land application wastewater standards set by CPCB, India and thus a further tertiary treatment is required before its disposal. From the removal trend, it can be deduced that vegetation growth affects COD removal significantly. Decay of roots, shoots and wilted leaves left unchecked within the BRs can contribute to drop in COD removal efficiency. Hence, care should be taken while harvesting to remove such unwanted plant debris which can affect the performance of the system.

The present study is a part of research project sanctioned by Department of Science and Technology DST) under Water Treatment Initiative (WTI) programme. The authors sincerely acknowledge and thank DST for providing financial support to undertake this study.

References

1. S. Abou-Elela and M. Hellal, Ecol Engg, **47**, 209–213, (2012)

2. S. Abou-Elela, G. Golinelli, A. El-Tabl and M. Hellal, Water Sci & Tech, Vol. **69 (1)**, 38-47, (2014)

3. APHA, *Standard Methods for the Examination of Water and Wastewater, American Water Works Association*, Water Environment Federation (2012)

4. L. Cui, Y. Ouyang, Q Lou., F Yang., Y. Chen, W. Zhu and S. Luo, Ecol Engg, **36**, 1083–1088 (2010)

5. D. Ghosh and B. Gopal, Ecol Engg, **36**, 1044-1051 (2010)

6. S. Jamshidi, A. Akbarzadeh, K. Woo and A. Valipour, J. of Env Health Sci & Engg, **12**, 1-12 (2014)

7. L. Marchand, F. Nsanganwimana, N. Oustriere, Z. Grebenshchykova, K. Lizama-Allende, M. Mench, Ecol Engg, **64**, 291-300 (2014)

8. Y. Patil and G. Munavalli, Ecol Engg, **95**, 492-500 (2016)

9. S. Sathe and G. Munavalli, J. of Water Process Engg, **28**, 240-249 (2019)

10. A. Valipour, V. Raman and V. Ghole, Ecol Engg, **35**, 1797–1803 (2009)

11. A. Valipour, V. Raman and V. Ghole, J. of Env Sci and Engg, **53(3)**, 281-288 (2011)

12. J. Vyazmal, *Water 2010'*, **2**, 530-549 (2010)

13. J. Wang, L. Zhang, S. Lu, X. Jin and S. Gan, J. of Env Sci. **24(6)**, 1006–1013 (2012)

14. A. Yadav, R. Abbassi, N. Kumar, S. Satya, T. Sreekrishnan and B. Mishra, Chem Engg J., **211**, 501–507 (2012)

15. *"Manual Constructed Wetlands Treating Municipal Wastewaters", United States Environmental Protection Agency, September 2000*

Feasibility Study of Adaptation of Biomimicry Approach in Green Building

Mahesh Bankar[1,*], *Vishakha Jogdand*[2]
[1]Maharashtra Institute of Technology, Aurangabad
[2]CSMSS'S Chhatrapati Shahu College of Engineering, Aurangbad.

Abstract. This paper deals with adaptation of Biomimicry strategies in building to make it environmentally sustainable. Biomimicry is the innovative approach which emulates the design principles of Flora and fauna as the genesis of the design. The approach is studied in context with the objectives of green building viz., Indoor air quality, Water efficiency, Energy efficiency and eco-friendly HVAC system. The study investigates the potential of emulating Biomimicry approach in green building technology. In this paper it is discussed that how the Biomimicry approach innovatively solves the purpose of green building, i.e. Resource efficiency and reduce environmental impact of the building.

1 Introduction

Out of total global emission, 26% is caused by buildings. Enormous amount of materials and resources viz., electricity, water, material, and energy are consumed by building sector. [1] At the same time, building sector has great potential of significant reduction of emission and environmental impact. The need of Green building raises here. Green building is the practice to build structure which is environmentally friendly and resource efficient throughout its life cycle. i.e., from designing to construction and from renovation to demolition according to U.S Environmental Protection Agency. The practice broadens to comfort, durability and well-being of occupant. Material Efficiency, Water efficiency, Energy efficiency, Enhancing Indoor Environment Quality are some of the core objectives of Green building.[2] In Indian context, The IGBC (Indian Green Building Council) Green Interior Rating System, Version 1.0, 2015 provides the design and construction specifications of green building. [3]

Green building combines different approaches and technological approaches to satisfy its objectives. One of the advanced approaches is Biomimicry. It is the study and emulation of the design principle of flora and fauna i.e., natural constituents and processes to find the solution of human design problems. Thus Biomimicry approach utilize the natural elements and processes as "Model, Mentor and Measure" and adapt the techniques in all levels of design.[4] There are two categories in which Biomimicry approach is divided. First category is solving the design problem by identifying the principle with which another organism or an ecosystem solves it. It is conceptualized as "Design looking to Biology". Second is to solve the human design problem by identifying a particular behaviour, principle, function or characteristic in ecosystem or in an organism. It is conceptualized as "Biology influencing design". [5] The three levels of Biomimicry approach are Organism level, Behaviour level and ecosystem level. Organism level refers to adaption of working principle of a particular organism such as plant and animal in part or whole in design process. Behaviour level refers to adaption of simple natural element's behaviour (air, water, etc.) in design. Ecosystem level includes adaptation of the working baseline of whole ecosystem in designing or solving a particular problem.

This study explores the potential of Biomimicry concept in order to achieve the objectives of green building. In Indoor air quality improvements, the selective plants naturally remove the Volatile Organic Carbons from the air and improve the air quality. In water efficiency natural principle i.e., increase in the volume of water by mixing the air is used and a faucet aerator is designed which saves the water from a building maintaining the same water pressure. In energy efficiency, the new kind of Light Emitting Diodes (LED) lights inspired from lantern of a firefly potentially improves the illumination efficiency of light and subsequently reducing the demand of the number of indoor lights; reducing the power consumption of the building for indoor lighting. And an eco-friendly HVAC system inspired from the air flow mechanism of termite mount not only provides the adequate ventilation to achieve the occupant's comfort level but also solve the purpose of energy efficiency too. It also finds the compatibility of Biomimcry approach in Green building in IGBC's rating system.

*Corresponding author: msbankar365@gmail.com

2Incorporation of Biomimcry in Green Building

The Biomimicry in green building is used at various levels. To evaluate the potential of Biomimicry approach adaptation in Green Building, the building is identified. The identified building is Guest house of Maharaja Sayajirao University of Baroda, Vadodara, Gujarat, India. Feasibility of Biomimicry is studied for following objectives,

2.1 Biomimicry in Indoor Air Quality (IAQ)

Benzene, Trichloroethylene, and formaldehyde are major VOCs emitted from the indoor products like waxed papers, furniture waxes, cleaning products, varnishes, and adhesives, rubber, paints plastics grocery bags, printing inks, disinfectants, paints, lacquers, and cosmetics, , facial tissues, and paper towels, gasoline, oils. Plants remove the VOCs. The leaves and roots of plants absorb the pathogens and VOCs. The microbes in the plants break them down and destroy them, by converting the air pollutants into new plant tissues. [6, 7] Hence incorporation of the indoor plants can be done according to the credit no. 6 of IGBC Green Interior Rating system, Version 1.0, 2015 which suggests providing one indoor plant for every 100 sq.ft.[3] The five best plants suggested in the IGBC Green Interior Rating system, Version 1.0, 2015's annexure 3, Chemical vapour removal rating, requirement of light, difficulty to grow and removal of all three major VOCs [6, 7] are Janet Craig, English Ivy, Peace Lily, Bamboo Palm, Dragon tree. (See Fig. 1) Incorporation of plants in indoor environment explores second type of Biomimicry i.e., adopting specific aspect of flora and fauna. It is also an organism level of the biomimicry. Hence it has great significance in improving indoor air quality.

To check the feasibility of Biomimicry in Indoor Air Quality improvement, first the total carpet area is calculated in order to find numbers of plants to be provided in each building spaces according to compliance standard given in IGBC GIRS - Indoor Environment Credit no. 6. It states that for every 100 sq. Ft of carpet area at least one plant should be provided. Then Indoor plants are selected on the basis of requirement of light, step number one, Chemical vapours removal rating, and removal of all three major VOCs and difficulty to grow. [5, 6] The plants are provided in different section of the building. Aesthetic liberty is considered for provision of plants. Then total removal of the three major VOCs; Trichloroethylene, Benzene and Formaldehyde by the total surface area of the provided plants for 24hr of exposure is evaluated on the basis of report of NASA [7]. Finally gain in points of IGBC GIRS - Indoor Environment credit 6 is evaluated in order to prove the acceptability of the Biomimetic approach in Indoor Environment.

2.2 Biomimicry and water efficiency

According to U.S Environmental Protection Agency, "water efficiency is smart use of the water resources through water saving technologies. Using water efficiently will help ensure reliable water supplies today and for future generation". A simple device is designed from inspiring from the simple natural principle; Volume of water increases when air is mixed in it. This is the second type (Biology influencing design) and Behavioural level Biomimicry. The air is drawn into the water stream breaking the stream into the tiny droplets mixed with air. The mixture of water and air passes through the screen and further mix the water and air which evenly spread out the resulting stream. Because there is less space of water to flow through, the water flow is reduced but water pressure is maintained and users don't notice the amount of water coming out of the aerator. This phenomenon saves the water. A structure of typical faucet aerator is given in fig 1.

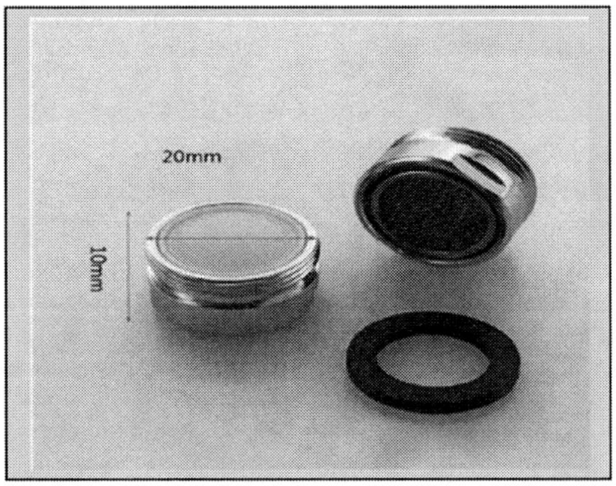

Fig.1. Faucet aerator

The Water efficiency credit criteria 1 of IGBC's Green Interior Rating system, Version 1.0, 2015 suggests to use the water fixtures with minimum 10% and maximum 40% less water flow rate than the standard baseline flow rate of the building according to uniform plumbing code, India-2008. For Water tap, baseline flow rate is 8 Liters per Minute. [8] In the market the faucet taps with flow rate of 6 to 4 Liters per Minute are available.
To study the feasibility of faucet aerator in identified building, total water consumption of the building is evaluated on the basis of number of occupants. The baseline flow rate of the water fixtures of the building is as per Uniform Plumbing Code- India, 2008. Then potential water saving is evaluated by replacing the current water fixture with the Biomimetic water fixture. Credit gained by only faucet aerator is evaluated on the basis of IGBC Green Interior Rating system's WE Cr 1.

Then the water saving potential is evaluated using additional water efficient water fixtures for checking the feasibility of Biomimetic technique in combination with modern water efficient technique. The total saving in water with the use of both the water efficient techniques is evaluated. Cost associated with the Biomimetic water fixture and modern water efficient water fixtures are evaluated. Finally, total IGBC credit gain is evaluated.

2.3 Biomimicry and Energy efficiency

Highly efficient Bioluminescent lighting system of fireflies inspired researchers to create an antireflection lens for Light Emitting Diode (LED). A firefly lantern is an organ in firefly to deliver an optical signal. A lantern organ is enclosed around the abdominal consist of 1) Cuticle, 2) Photogenic layer 3) Dorsal layer. (See Fig. 2).

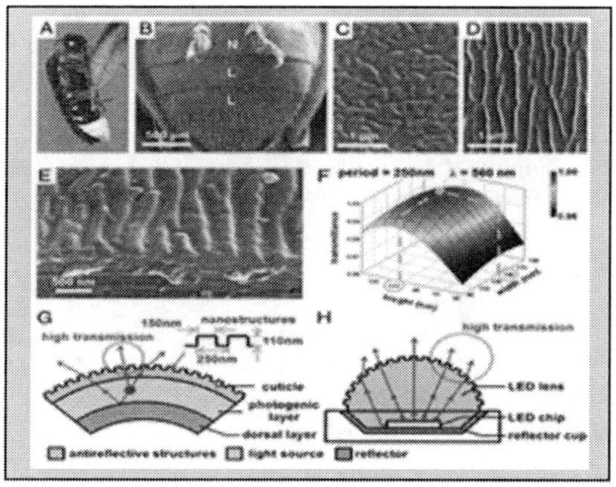

Fig.2. Highly ordered cuticular nanostructures of a firefly light organ inspiring the design of New LED.

In figure 3 observe, A: The optical image of a firefly in male, B: The SEM image of abdominal segment of a firefly in male including Normal N, and lantern (L), cuticles, C: amorphous nanostructure on the normal cuticle, D: Highly ordered nanostructure on the lantern cuticle, E: A transverse cross section of the light organ, F: Transmission of bioluminescent light through nanostructured surface on the lantern cuticle calculated by FDTD method, G: A firefly lantern, H: A high power LED package.

Bioluminescent light in all directions is emitted by the photogenic layer which passes through cuticle after reflecting off the dorsal layer. Highly ordered circular nanostructures surrounding veterans photogenic layer reduce the index mismatch between cuticle surface andair to have high extraction of bioluminescent light in all directions. The different space patterns created by each nanostructure cause the light to form pattern in different directions which reduce the loss of amount of light due to internal reflection. [9] The LED light have same configuration. It consists of reflector cup, LED chip and LED lenses. The light is emitting through LED chips and finally extracted either by reflecting from reflection cup or LED lens. Thus, new nanostructure LED lenses are discovered which reduce the index mismatch between air and lens material. These LED lights have 55% more emitting efficiency than conventional LED lights.

Hence for a particular building after finding the adequate illumination requirement to maintain visual comfort, these new kinds of LED lights can be provided. As the LED lights gives more light emitting efficiency the requirement of number of LED lights is reduced. This phenomenon can automatically reduce the consumption of the electricity contributing the objective of energy efficiency. This is the first category of Biomimicry and organism level of Biomimicry as it is directly adopting principle of the firefly.

To analyse the potential of this phenomenon, first evaluation of Lumen required as per Illuminating Engineering Society code of Lighting is done for each section of building. Then identification of Carpet area is done for evaluation of required illumination in each room of the building. Evaluation of standard LED light fixture is done assuming the standard LED light of 1100 lumen for No of lights, Energy consumption and cost associated. Evaluation of Biomimetic LED light Fixture is done assuming 55% more illumination than standard LED light fixture for No of lights, Energy consumption, Cost associated. Then comparison of Standard LED lights and Biomimetic LED lights is done on the basis of No of lights, Energy consumption and cost. Finally, total Gain in IGBC Credits is evaluated to understand the suitability of this Biomimetic approach.

2.3 Biomimicry and Energy efficiency

Another prime objective of the green building is maintaining the optimum comfort of indoor occupants. To achieve the objectives various conventional HVAC technologies are used. These technologies are costly in installation, operation and along with have significant environmental impact. However, Biomimicry provide the new approach to design the HVAC system of a building.

Termites in Zimbabwe build huge mounds for their shelter. However, the optimum temperature of 30°C is maintained by constantly opening and closing a series of heating and cooling vents throughout the mount over the course of the day. Air is sucked in the lower part of the termite enclosing with muddy walls and vented up through the channel to the peak of termite mound. Observe fig 3.

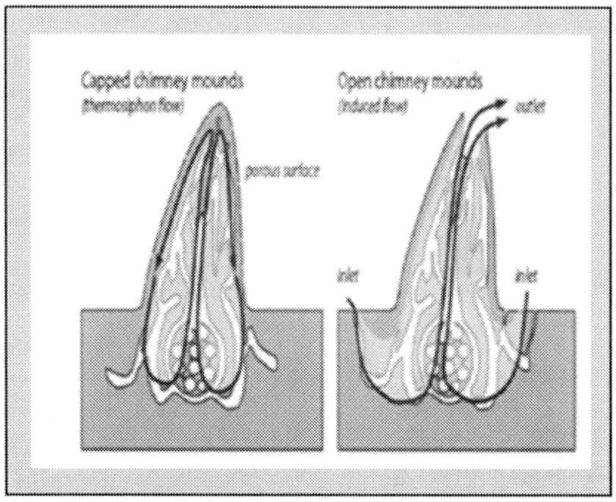

Fig.3. Air flow mechanism in termite mound

The air is sucked in the bottom section of mount, circulated in the mound and vented from the chimney. [10] The Eastgate Centre, Harare, Zimbabwe designed the ventilation system works same as that of the termite mount. Observe Fig. 4.

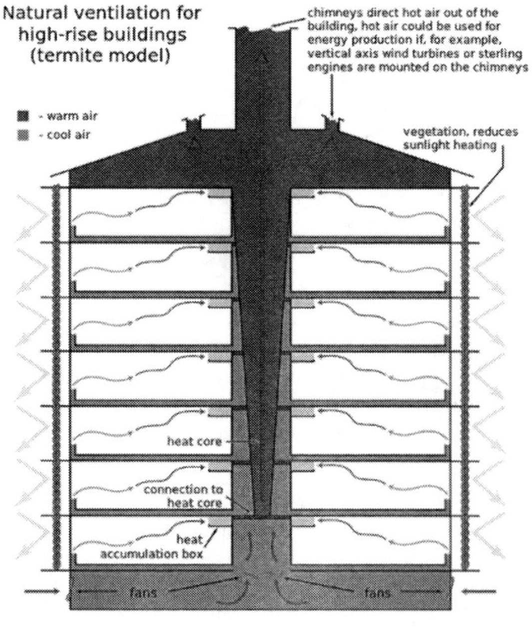

Fig.4. Air flow mechanism of Eastgate Centre, Harare, Zimbabwe

Air is continuously drawn from the open space by fans on first floor as shown in Fig. 5. It is then pushed up through vertical supply of the section of the duct located in central spine of each building. The fresh air replaces dead air that rises through and exit through exhaust ports. The exhaust ports are in the ceiling and through it flushed out of building through chimneys. As compared to same size conventional building the Eastgate centre uses less than 10% the energy. The owners also saved 3.5 million dollars alone because of this system. [10] Hence it is observed that the innovative HVAC system not only improves the air circulation of the building but also saves cost.

To check the potential of this phenomenon for identified building, first cooling load calculations is done in order to design the Biomimetic HVAC system and Conventional HVAC system. Then Biomimetic HVAC system designed on the basis of calculated cooling load. The system is evaporative cooling system supported by ducting. At the same time conventional HVAC system is designed for same identified building using same cooling load for comparison purpose. After that comparison of Conventional and Biomimetic HVAC system is done on the basis of Capital cost, Operational cost and Energy consumption. And finally, IGBC credit gain is evaluated for Biomimetic HVAC system.

3ResultsAnd Discussions

From the study and evaluation, it is observed that Biomimicry approach in the Green Buildings plays a significant role. Following results are obtained from the study in identified aspects of green building.

3.1 Indoor air quality enhancement

It is observed that indoor plants have potential to remove the hazardous VOCs considerably. For the identified building, considering the carpet area 74 plants are required. However, 170 plants are provided considering aesthetic and Biophilic aspect. The plants selected and number of each plant provided are Bamboo palm (59), Janet Craig (22), English Ivy (62), Peace lily (10), and dragon tree (17). The total plants have potential to remove the major VOCs considerably and can enhance indoor air quality. For 24 hrs. of exposure, all provided plants having total surface area of 12,14,525 sq.cm have potential of removing 25,82,566 ppm of Trichloroethylene (TCE). Similarly, with total surface area of 12,36,534 sq.cm they have potential of removing 44,38,459 ppm of Benzene while with total surface area of 14,49,182 sq.cm potentially 31,58,169 ppm of Formaldehyde can be removed. The total cost for provision of plants is Rs. 57,930. The provision of plants gains 2 points from IGBC's GIRS- Indoor Environment, Credit No.6. Also adequate ventilation provided by the Biomimicry principal inspired HVAC system helps to remove the indoor pollutants more effectively.

3.2 Water Efficiency

The water fixture of identified building is observed and replaced with Biomimicry inspired tap i.e., faucet aerator. For adequate water supply, the total daily maximum water demand of the occupant as per IS 1172:1993(2007) of identified building is 135 lit/capita/day i.e., 15862 lit/day in total. However, for adequate water supply, baseline flow rate of the tap as per Uniform Plumbing Code- India, 2008 is 8 LPM and the estimated daily use is 4 minutes. Thus, by replacing the regular tap with biomimetic faucet aerator, the flow rate is reduced by 2 LPM i.e.6 LPM. Hence the water saving from the saved flow rate is 8 litters/ capita/ day. The total water saving from required water is 752 lit/day. Also 11% flow rate is saved from baseline flow rate. Due to this, 1 point in IGBC Green Interior Rating System's Water Efficiency Credit 1, is gained. The total cost of the faucet aerators is Rs. 33,800. However to ensure more water saving and credit points to gain, the Water closets and the urinals are evaluated for replacements with water efficient closets and urinals. It is observed that the efficient water closets and waterless urinals save 6 LPF flow rate than total baseline flow rate. This is 33 % of the total baseline flow rate. Also, both the fixtures save 752 litre of water of total requirement per day. The total cost of the additional water fixture is Rs. 46,500. Hence the biomimetic fixture and the additional efficient water fixture together, save 1052 litre of water per day. The total water fixture also saves 44% of the baseline flow helps gaining 13 points in IGBC's Green Interior Rating system; Water Efficiency Credit 1.

3.3Energy Efficiency

Assumed conventional LED bulb of 15 Watts emits 1100 lumen of light. For the same wattage, assumption for biomimetic LED bulb emits 1700 lumen of light. Hence for identified building, to achieve the visual comfort of occupant. Total 167 conventional LED bulbs are required. The total cost of the conventional bulb is Rs. 33,400. The total electricity cost is 2626.2 Rs/ month. For same criteria of illumination, 112 Biomimetic LED lights are required. The total cost of the Biomimetic LED lights is 29,130 Rs which is considered as same as 1700 lumen LED light's market price. The total electricity cost is 1753 Rs/month. Hence, for the same illumination requirement less number of biomimetic LED bulbs is required, the capital cost is lower than conventional LED bulbs and the electricity consumption and cost is also less as compared to the conventional LED bulbs. It helps to gain 2 points in IGBC's Green Interior Rating System; Energy Efficiency Credit 2.

3.4 HVAC system

For identified building, new HVAC system is designed which is inspired from the biomimetic principle of the termite mount. The total cooling load to design HVAC system is 15 TR. The total Design air to be supplied in the spaces is 5036 cfm. HVAC system in fusion of evaporative cooling system and ducting system is designed. For same cooling load, the conventional but environmentally friendly 5 star split air conditioners are also identified. The total capital cost of the Biomimetic HVAC system considering all the components is Rs. 379,150 while the capital cost of the total conventional Split AC system is Rs. 531850. The electricity consumption cost of the split AC system for 4 months of summer and 24 hrs. operation is Rs. 26,35,77 while for same criteria electricity consumption of the Biomimetic HVAC system is Rs. 68080. Hence biomimetic system saves the capital cost and operational cost. The Biomimetic HVAC system gains 1 point of Energy Efficiency Credit 1, 8 points in Energy Efficiency Credit 2; Fulfils Mandatory requirement number 2 of Indoor Environment Credit, 2 points of Indoor Environment Credit 1 from IGBC Green Interior Rating System. Hence the Biomimetic HVAC system helps to gain 11 points in total of IGBC Green Interior Rating System.

Hence from overall study it is observed that Biomimicry helps to get the points in credit system of IGBC Green Interior Rating system by satisfying the criteria for it. Table 1 gives details of credits gain by Biomimicry

Table 1. Credit Gain by Biomimetic Green Building

Title	**Credit**	**Credit Name**	**Available Points**	**Achieved Points**
Water Efficiency	WC Credit 1	Water Conservation	12	12
Energy Efficiency	EE Credit 1	Eco-friendly Refrigerants & Halons	1	1
Indoor Environment	EE Credit 2	Energy Efficient Interiors	10	10
	IE Credit 1	Enhanced Fresh Air Ventilation	2	2
Total			25	25

At certification level as Biomimicry is achieving 26 points, the identified building have potential to get recognition of Best practices as the criteria for it is to get points between 22 – 31.

3.5 Cost analysis

To achieve the identified green building objectives, total cost required for conventional techniques is 128.55 Rs/ Sq.ft while total cost required for Biomimetic system is 100.78 Rs. It is observed that Biomimetic techniques save 21.62 % cost as compared to conventional techniques in order to achieve the Green Building objectives viz., Indoor Air Quality, Water efficiency, Energy efficiency and Environmental friendly HVAC system.

4Conclusion

It can be concluded that Biomimicry approach increase the Indoor Air quality of the building by using indoor plants by removing 10179194 ppm VOCs. Biomimetic water fixture saves 6.63 % of total daily water demand. Also, Biomimetic LED light fixtures save 33.26 % of total monthly power demand for illumination. Biomimicry inspired HVAC system provides desired comfort, saves the capital cost by 28% and operational cost by 74.17% as compared to the conventional AC system to provide the indoor comfort. Provision of Biomimicry techniques along with the other efficient green building technologies has potential to gain 25 points of IGBC Green Interior Rating system as well as to amplify the building's resource efficiency and reduce its environmental impact. Along with advantages mentioned above, the cost of applying the Biomimicry technique for conventional building is resulting in saving 21.62 % cost than conventional techniques to achieve the Green Building objectives.

References

1. U. Diana, IPCC, **4**, 3, (2007)
2. D. Emily, N. Kriti, N. Zeenat, S. Andrew , D. Varsha and K. Vijaya Lakshmi , "Green Building- Case study" , ODI,**1**, 11,(2013)
3. IGBC, IGBC Green Int. Rating Syst., IGBC,**1.0**,51, (2015)
4. B. Janine M, Harper Perennial,**1**, 2-7, (2002).
5. Z.Maibrit, SB07,**7**, 3, (2013)
6. K. Kent, J. Andrew, G. John and M. James, furtherance of Cooperative Extension work, CTAHR,**39**, 4, (2007)
7. B. Wolverton, J. Anne, B. Keith, NASA, ALCA, (1989)
8. Indian Plumbing Association, Uniform plumbing code, IPC, **1**, (2008)
9. E. Nour ,PES, **34**, 375 -386, (2016)

Exploring societal participation in making cities smart

Shraddha Mahore Manjrekar[1,*] *and Manali Deshmukh*[2,]
1 Brick School of Architecture, Pune (India)
2 Brick School of Architecture, Pune (India)

Abstract. The word 'Future' has been synonymous to 'Development' for most of the people. Architects of the Nation strive for a well-developed state, where it reaches towards enhanced living conditions than the present. In this regard of devèlopment, structure of Indian government is well defined. The policies designed at central level envision the total development. These policies make the local bodies function on the same lines. They make an impression in common people's life that there are some positive changes in the cities. These policies also mark an identity, e.g., after implementation of city Development Plans under JNNURM (Jawaharlal Nehru, National Urban Renewal Mission), the buses bought for BRT (Bus Rapid Transport) had become identity of this policy. The meaning of Smart Cities in India also has been interpreted by people in many ways. For cities to become smarter it needs a holistic approach in developing a sustainable urban eco system and thus the roles of various stakeholders become very important in executing this mission. Not only the investors, financial organizations, planners, architects and people in construction industry but the citizens too have role in making cities smart. People of all age, gender, ability and disability are responsible in making a city smart. Education and awareness is the key for making them aware of their participation in the process of development. This paper also mentions the role of educational institutes to make people understand the essence of development.

Key words: *Smart, Futuristic, Sustainable, Public participation, Education*

1. Introduction

The objective of the Smart Cities Mission is to promote cities that provide core infrastructure and give a decent quality of life to its citizens, a clean and sustainable environment and application of 'Smart' Solutions. The local governance and various stakeholders are involved in this process. In line with the objectives of the mission, the project typology too has a common pattern in these cities. The categories of the works are, riverfront development projects, open spaces development, housing, App based projects for systemization of the administration, infrastructure and transportation etc. Dashboard of official website of the smart cities mission states that there are 100 winning proposal from various cities and estimation of 99,630,069 people getting affected by this mission [1]. The impact is also much visible in these respective cities. For example, Coimbatore city is working on desilting, bund strengthening, removal of weeds, turfing and lining, drains development, inter-linkages boating and water sports facilities, other amusement and recreational facilities, theme parks, medicinal gardens etc. Kochi city is working on upgradation of infrastructure in housing through credit linked schemes under the same mission. Pune city is working on intelligent government service for Bus System Intelligent Traffic Management System (ITMS), where the Smart Public is going to be equipped with Real time tracking of buses using GPS, maintenance using vehicle health monitoring system, Passenger Information System (PIS) with LED screens in buses and bus-stops and Mobile app to track -bus location and ETS real time, Ride quality monitoring using smart sensors, in -bus wifi and entertainment systems, etc. Digital India has reached till the limit where even a small vendor does digital transactions. Technologies have made the tasks easier. However, it cannot be said that the urban life has become easier with technologies. The complications, difficulties and the side effects of this technological up gradation have also been realized by people. Question that arises here is about total smartness, and its real perception in people.

*Corresponding author: **iitrshraddha@gmail.com**

Smartness is not only about the superficial aesthetic quality or speed of works, but it is also about up gradation of civilization done together by people, places, things and systems. For example, the beautification of streets is a cosmetic surgery, as its going to add only into aesthetic part of it. ITMS, PIS, public screen may bring impression of technological up gradation but the convenient last mile connectivity of the mass rapid transport, and good quality of roads, pedestrian safety on these roads etc., will bring real smartness in the system.

This paper is based on a qualitative research, an attempt to identify and analyse of various parameters defining smartness with examples. It points out the gaps, between the policies and public participation. As a next step to this, the areas have been identified where people and education system can contribute to make their lives smarter. The role of education and educational institutes in conceiving the real essence of smartness has been elaborated in the further part of the paper.

2. Defining smartness in urban context

Smartness is both about the tangible and intangible aspects. Aesthetics of the city are important for cities to look smart, at the same time the functionality, efficiency and nearness to human values automatically represent smartness. A city can be smart with its people, places and systems. The various parameters defining smartness in the urban areas are:

- Efficient systems in place
- Intellectual smartness (people and system)
- Empathy for environment and people
- Value system for places and practices
- Futuristic, Sustainable and healthy
- Vibrant, Dynamic and Culturally alive

2.1. Efficient systems in place

The systems are of two types the administrative systems and infrastructural system. The urban local bodies and their functioning both have major role. Though there are similar systems of administration, however few cities have adopted some policies and set goals that have taken them till a level. Management of system with upgraded technologies and people's participation too have equal role in these systems.

For example, Indore city has been winner of 'Swachha Bharat Award' since three consecutive years (2016-19). The major reason for its success is healthy competition between the wards. Every ward competes with each other in order to maintain their areas clean and beautiful. The efficient system of collection and transportation of garbage, processing and disposal of waste, clean sanitation (public toilets, proper drainage, etc.), people's participation and making people aware of cleanliness have worked well for Indore [2] (The Indian Express n.d.). City schools too have oriented the students to keep their surroundings cleaner.

Fig. 1 A street in Indore city with its prototype waste collection system (India today 2019)

2.2. Intellectual smartness (people and system)

City is made with people. The knowledge, education and intellect of people to understand the surroundings is a must in executing the smart city mission. To develop these qualities, city's system should have infrastructure like educational institutes, libraries, laboratories, art schools, auditoriums, concert halls etc. For example, city of Oxford is known for its University, but other than that its identity is made also by theatres, concerts and other events. East oxford is the area with students' population and related infrastructure, press building, etc. The Plain, Cowley Road in East Oxford is the preferred place for trendy, both old and young, with a cinema, music venues, tattooists, and many good ethnic shops. West Oxford is full of a diverse landscape of waterways and islands (Oxford city n.d.)[3]. Figure 2 is a picture of a street in Oxford, showing prestigious institutional building, bicycle stand and people using bicycles.

People in oxford generally prefer walking and bicycling. They indulge in academics, arts, theatre, etc. Most of their activities are eco- friendly and add into character of the city. Hence other than buildings, streets and infrastructure, it is the people who have made an identity of the city.

https://doi.org/10.1051/e3sconf/202017006011

Fig. 2 A view of Oxford city (Source: UKtourcentre.com)

Fig. 3 Cowley Road, Oxford city (Source: UKtourcentre.com)

Ladakh, India is another example of smart citizens. Known by the name and works of Sonam Wangchuk, the area got its 'Education and cultural movement' for young kids in year 1988. Sonam Wangchuk[a] has come up with the uniquely systematic, collaborative & community-driven reform of learning systems in remote Ladakh region. This system has helped to improve life opportunities for youths (Examrace n.d.)[4]. Education with strong application base has changed the picture of the area. Pune based activist Sarang Gosavi has also taken help of science and education to change people's lives in the same area. At one hand the area got good leadership by well qualified social reformer, and at the other hand it had been smart participation of local people to introduce better living conditions.

2.3. Empathy for environment and people

The developed state of civilization is sensitivity for humanity and environment. People should have value for fellow citizens and environment. Knowing about surroundings, owning, and taking care of it, makes area sustainable. Very good example here is of Central Park, New York City. New York City felt the need of a breathing place for people in year 1853. After city's joint call, the New York State Legislature enacted into law the setting aside of more than 750 acres of land central to Manhattan Island to create America's first major landscaped public park [5] called Central Park (July 21, 1853). Since then there were several ups and downs in operation and maintenance of park, but the park sustained with people's efforts. (Central Park Conservancy n.d.) Today also with private public partnership the park is in good condition and visited by many people every day. The separate zones in this park are maintained by different groups of people. This zone management system has brought accountability, pride of workmanship, and clear and measurable results to the Conservancy and Parks Department staff.

Central park is an example of being sensitive about environment, however having sense of ownership to the surroundings has been culture of ancient India too. Sacred groves were actually the reserved forests and people regarded the trees and forests as almighty. Ironically, today in Indian cities, there are concrete walls around reserved forests to safeguard them from people. In the same lines, enforcement of 'Tree act' in Maharashtra state is a consequence of anti-environmental incidences in the past. At present there is Tree act, however there is a long way to sensitize people about total environment and ecology.

People are also part of environment; they have different abilities and disabilities, and city's ecosystem need to have respect for them and their convenience. Kautilya's Arthashstra has mention of city to have arrangements for looking after the aged, the children and informal people. Our built environment needs to be redesigned to meet the needs of physically challenged people. The vulnerability of the younger generation and senior citizens is more in present Indian cities, as they are not safe in terms of commuting on road because of heavy and unmanaged traffic and inappropriate road designs. Lessons need to be learnt from the cities that offer better living conditions to all. For example, in Singapore city, a six-year-old can alone commute in metro to go to school. Also, a senior citizen can commute anywhere in the city without any manual help. City's roads, public transport system, digitization of services support their easy life. At the same time, the people share these amenities, respect each other and give warm gestures while using these public amenities, and infrastructure.

[a] Works of Sonam Wangchuk have been acknowledged by a Bollywood movie, "Three Idiots", in which Amir Khan has enacted Sonam in the character of Phunsukh Vangdu.

 https://doi.org/10.1051/e3sconf/202017006011

Fig. 4 Marina Bays, Singapore, an urban space having favourable commuting facilities for all

2.4.Value system for places and practices

Some places in the urban context have environmental, ecological or historical importance. Smartness of citizens includes having knowledge about these places, appreciating and valuing these. For example, Rome is a city of monumental buildings, monuments, sculptures and also of the ruins. The city administration and people have regarded and preserved the ruins equally like other historical undamaged buildings. The buildings and cityscape reflect the Christian culture and also history of war and victory over the enemies. A number of buildings in Rome have been built in memories of victories. After Second World War, the city grew in different fashion and had developed a vision to attract the world to appreciate the city through tourism. In order to make city ready for hosting international tourists, the new buildings have been built. However, it is noticeable to see that these newer buildings too have taken a distinct place in the harmony of entire city. Every new building has been built with due regard to its surrounding areas and the existing architecture. Figure 5 represents a picture from a Roman street where people have maintained the same façade to meet the street character.

Not only the construction but day to day activities too matter a lot in valuing the heritage component of the city. In Indian context there are old temples, residences, market places, open spaces, which need respect from daily commuters. The very first attempt is needed to make people aware about the context of the area. The heritage of the area is first respected by the locals than the tourists.

Indore city, India has demonstrated a good example of giving respect to historical monument, Holkar's Rajwada, by matching the façade character of the surrounding buildings with it. The uses of the surrounding buildings are different but they together make a market place with cultural identity.

Fig. 5 A Street in Rome

2.5.Futuristic, Sustainable and Healthy

The historical places with good construction are still part of city and are lively, with their timeless character. The design of the spaces needs integrity that would be appreciated by future generations too. The resources are depleting, and the present time has complete liability to leave resources for the future. The design of buildings and urban spaces should be functional, aesthetically acceptable and at the same time energy and resources efficient. Climate is changing and dependency on mechanical means of comfort is increasing, however the building design has capacity to reduce the need of mechanical means of comfort. There are a number of Green buildings rating systems (LEED, GRIHA, Eco-Housing, etc.) that give recognition to the sustainability of buildings; however, sustainability feature can be adopted as an accepted norm for being futuristic. Regarding urban spaces, too much dependency on mechanical means (lifts, motorized vehicles, etc.) has affected lifestyle and in turn it has degraded the public health. The cities need to be healthy. A simple example is pedestrian friendly roads with clean, and shaded walkways with lush green vegetation encourage people to walk, reduce fuel usage and keep the environment pollution free. Also, when people walk, they interact and the social life on streets is also seen. Pedestrian-priority street design is need of the hour, which is an urban public space with the highest priority for pedestrians and in some cases cyclists [6]. It has equivalent terminologies such as; a car-free street, a pedestrian zone, a pedestrian mall and a walk-able street (Basil Kamel 2017)

City of Bogota the capital of Columbia has set an example in world by reducing motorized vehicles on streets. The initiative that changed city in many positive ways is called Cyclovia that began in year 1974. Though implemented much later in year 1982, it is a success story. They started with a car free day event. Later when walking had become popular, people have accepted changed cityscape with promoted systems of walking and bicycling. Cyclovia brought a lot of environmental, health, economic and social benefits. Also, the social interaction between individual was significant, that resulted in increased sense of place and equity among the different classes of public realm. [7] Also a lot of economic benefits are obtained due to program investments and medical cost savings (Yassin 2019).

2.6. Vibrant, dynamic and culturally alive

Vibrancy in the urban environment can be brought both by spaces and people. Space design has potential of bringing vibrancy in the environment. The urban fabric is formed with vistas between buildings, sky and land. The landscape naturally changes with season. The vegetation in the area attracts birds and butterflies. Urban spaces integrated with landscape design have vibrant character. Cities must have pause points, where people stop and experience this vibrancy of the urban environment. Urban design may include water fountains, changing building facades, light and sound show elements to make the urban environment more vibrant.

Indian culture has many festivals that make streets lively. Keeping culture alive adds value to vibrancy of the area. Festivals change cityscape by decorations, festive markets, processions, and dance and music. Figure 6 represents a procession during Ganesh festival in Pune city, India. During the festival people and activities make the city colorful and vibrant. Any area full with people of different age, gender and communities make city vibrant.

Law garden is a park in Ahmedabad, is another good example of vibrancy. The street gets converted into market space in the evening time, and hence it has different character at day time and in the evenings. In the evenings, its souvenir shops, visitors and shopkeepers wearing traditional Gujarati costumes add a lot of vibrancy in the space. In the same city, Manek Chowk is a good example of space dynamics. This dynamism happens with people and their activities. The area is fully crowded with eateries and people since the evening to late nights. Early morning people get cattle in this area to feed the fodder. After this all the spaces are cleaned by municipal workers and by 10:00 am the shopkeepers open their jewellery shops. A same space offers environment for different activities and is dynamic in nature. The changes are so different in nature that a new person cannot imagine the next use of the space.

Fig. 6 Procession of Ganeshotsava, Pune

3. Roles of various stakeholders in making cities smart

India is a huge country and is known for its diverse topography and culture. Still the nation is united in many aspects. Indian government has remained successful in implementation of central policies with its efficient hierarchical system. The Niti Ayog, formerly 'Planning Commission' of India, has been working on a number of plans that have set language of translation at various stages. The administrative system of government is comprised of Central, State, Regional, District, and then respective local authorities. These function and title of these authorities depend on population, area and population density of the area (municipal council, municipal corporation, or Gram Panchayat). The census of the country done in every ten years is a detailed and systematic survey that conveys status of various socio-economic, cultural and environmental status of the country. The central plans like five-year plans, ten year plans are result of the thoughtful process of National development based on these surveys.

For any mission of the central government there is healthy competition between the local authorities and at their level they work in the line with mission. For example JNNURM was a nationally accepted mission and local authorities were the most active role players in the same. Smart cities mission is also on the same lines.

However, for any governance to be successful in all regards, people's participation is must. Indian constitution has very well defined the roles and rights of the citizens. However, it has always remained a challenge to convey the responsibilities of the citizens. The perception of responsibility as a citizen is limited to giving vote to select the government, attending national festivals and paying tax on time. The other duties towards nation and environment are not so well defined, and hence peoples' contribution in making area smarter does not remain united. Every individual think about self-development but not for the area development. A general awareness about smart cities is also not evident, as people have noticed only the superficial component of the mission. As the previous part of paper conveyed the real essence of smartness, the next part of the paper highlights that how educational institutes play role in making area smarter with public participation.

4. Contribution of Education and educational institutes to make cities smart

Implementation of government policies aspiring to provide an enhanced infrastructure thus plays a vital role in the comprehensive development of a city. It can happen by involving organisations at a ward-level as a responsible instrument and will help in sensitizing the local people in respecting their neighbourhood and improving the urban eco-system. These organisations include educational institutes like secondary schools, undergraduate colleges, universities, and also non- profit organisations, government and semi government institutes in that ward. However, as the students constitute the majority of the population of any neighbourhood, it is best suited to involve the educational institutes in propagating the smart city mission and work towards the long-term goal of developing a sustainable urban eco -system.

Winner of the Magasasy award of year 2018, Engineer turned into Educationist Sonam Wangchuk says that the Education needs to be linked with life and should take into count the ground situation in different parts of India [8]. Various projects with an aim to respect their neighbourhood and improve the living environment can be included as part of their curriculum which addresses the responsibility of an individual to societal concerns. These points can be imparted in educational system.

4.1. Traffic and transportation

The areas near the periphery of the schools have a number of student and parents' commuters. In order to control the use of single occupancy cars, the commuting to school can act like a key. Institutes may formulate policies that govern commuting to schools/institutes by use of public transport, bicycles or walking. The defined area around school can make an identity by demonstrating reduced use of private cars.

4.2. Sensitization about heritage sites, and environment

Knowledge of the area and valuing it can be imparted best in schools. The academic lessons conducted in the premises of these structures and projects that trigger their thinking. It can also help in relating it to the literature taught, thus generating a sense of ownership and pride about the location. Pune, India has been a place for organizations like Balbhavan where cultural sessions are conducted for social interaction between different age groups (elderly and toddlers). Such organization can play a vital role in Smart cities by involving people in awareness sessions. This can remain instrumental in up keeping and regarding the heritage of the area.

Schools and institutions can promote activities to make people aware about the bio-diversity, natural resources and their role to respect and preserve it. The field trips and outdoor teaching can include these topics. At the same technical and professional education institutes can orient their pedagogy for area development through technological support. As an application of advancement in technology in the advanced sustainable infrastructure that needs to be incorporated. Projects dealing with smart solutions with emphasis given to energy and resource management can help in sustainable infrastructure and maintaining a better living environment. The development of technologies developed- robotics, sensory equipments to support this mission can encourage.

4.3. Ownership of public spaces

Urban spaces include parks and amenity areas. These spaces have major role in health and happiness quotient of the area. Areas could involve activities related to toddlers and elderly at the grass root level, e.g. Nana-

Nani Park concept of Maharashtra, India. This is a concept of parks that has an intention of playful and caring interaction between elderly and toddlers. It gives a platform of learning and sharing of cultural intellect as well as newer versions of systems. This could help instil a sense of ownership and pride in respecting the culture along with the basic need of state of play for both the elderly and young. International Play Association (IPA) has wisely identified 'play' as right of every individual and an activity that improvises health and intellect. Giving priority to play can happen only when the urban areas and spaces support the safe and playful activities. IPA is approaching schools and neighbourhoods for this mission. Such missions can be clubbed with smart cities mission to activate institutes' participation for area development.

4.4. Festivals and Cultural events

Culture and festivals bring vibrancy in urban environment. Institutes can be actively involved during festivals in maintaining the balance of the cultural events with urban systems. The traffic management, waste management are common concerns during festive times. The voluntary roles of the institutes in this may include, supporting the assigned government officials in traffic management, spreading awareness for eco-friendly practices, etc. Educators and students together can play major role to educate the people for eco-friendly celebration of all the festivals. These activities can be helpful in making students smart, active and realistic. For institutes and smart cities mission it can be a win-win situation in the process of educating and making cities smart.

5. Findings and Discussions

Educational institutes play a vital role in strengthening the gap between the strategic planning in making smart cities. The study opens up the thought process for the institutes on how they can contribute to the area in which they are located. The various parameters of defining smartness in this paper convey that smartness is also about being aware of responsibilities as an individual in urban system. Reiterating on democratic approach- 'city of the people', citizens need to actively participate and share the responsibility to make an area smart. Educational institute have strength to unite the people and their efforts in multiple ways. A win-win situation between the governance and educators is in educating the younger generation about the land, ecology, environment, health, heritage, culture and festivals and their roles in developing a sustainable model in implementing the smart city mission.

References

1. NIUA, *Ministry of Housing and Urban Development, Smartnet.* https://smartnet.niua.org/smart-cities-network (accessed October 17, 2019).

2. The Indian Express, *https://indianexpress.com/article/india/indore-indias-cleanest-city-swachh-bharat-mission-5662774/* (accessed October 17, 2019).

3. Oxford city, *Oxford visitors Information. https://oxfordcity.co.uk/about-oxford/east-oxford/* (accessed OCtober 18, 2019).

4. *Examrace, https://www.examrace.com/Current-Affairs/NEWS-Mumbai-Psychiatrist-Bharat-Vatwani-Ladakhi-Education-Reformer-Sonam-Wangchuk-Win-Magsaysay.htm* (accessed November 2018).

5. Central Park Conservancy, *https://staging.centralparknyc.org/?utm_expid=.7uMp4LtfRrmxBFWTxEhVoA.1&utm_referrer=.*

6. Kamel B., Wahba S., Kandil A., Fadda N. *,"Reclaiming Streets as Public Spaces for People: Promoting Pedestrianization Schemes in Al-Shawarbi Commercial Street - Downtown Cairo." SSRN Elecronic Journal,* (January 2017).

7. Yassin, Hend H, *"Livable city: An approach to pedestrianization." Alexandria Engineering Journal,*

8. India times, *'Here's What India Needs To Do To Improve Its Education System By Sonam Wangchuk'*(Aug 2018)

 https://doi.org/10.1051/e3sconf/202017006011

9. India today, *Swachha Bharat. June 2019. https://www.indiatoday.in/mail-today/story/swachh-bharat-delhi-narendra-modi-mcd-june-5-world-environment-day-980517-2017-06-02 (accessed October 18, 2019).*

Salutogenic design with healthcare approaches to indoor environment for office premises

Alexey Danilov [1], *Andrei Benuzh* [2,*]
[1] Institute of Interdisciplinary Medicine, 18/19 3-rd Rybinskaya street, Moscow (Russia), 107113
[2] Moscow State University of Civil Engineering, 26, Yaroslavskoye Shosse, Moscow (Russia), 129337

Abstract. It is statistically proven that a person on average spends about 90% of the time indoors. Therefore, it is not surprising that more and more scientific studies demonstrate that the design, construction and operation of buildings directly affect health. Air, microclimate, water, light, decoration materials - have a powerful effect on both the physical and mental state of people. No company striving for economic growth can ignore the problems arising from the incorrect organization of the workspace, since the internal environment of the premises determines the health of employees, and ultimately the productivity of employees and the company's profit.

1 Introduction

According to the World Health Organization, office workers are increasingly suffering from "Sick Building Syndrome" (SBS), which is manifested by increased fatigue, headache, allergic reactions, asthma, and other disorders. In Russia, according to a survey of 2011, 60% of respondents consider their office unsafe and not environmentally friendly (employees of 140 domestic companies took part in the survey).

1.1 The trend towards a healthy built environment

Today a new trend is emerging, combining the concept of green design with the idea of preserving and strengthening people's health [1, 2]. When developing architectural projects, more and more architects, designers, and builders strive to take into account not only environmental factors (reduced energy consumption, etc.), but also what impact the design will have on the well-being, mood, and performance of the building's inhabitants [3, 4]. To solve these problems, interdisciplinary teams are created, in which, in addition to architects, engineers, designers, medical experts, psychologists, ergonomics specialists, and others are involved. [5] Advanced companies that equip their offices in this way receive additional profit in connection with increased productivity. For example, Bank of America, thanks to the salutogenic approach, increased labor productivity, which allowed the company to save $ 15 million a year. In Russia, there are regulatory, technical and sanitary-epidemiological documents (GOST, SanPiN (Sanitary Regulations and Norms)), following which ensures safety for the inhabitants of the building. However, these standards do not take into account many factors that could contribute to better health and increased productivity [6]. Tools are needed to evaluate the health of office workers, the impact of the workspace on the condition of workers and their productivity. A health assessment will help you understand what measures you need to take to improve your health, increase your ability to work, and increase productivity. Such tools, recommendations and standards are being developed today at the Moscow State National Research University of Civil Engineering and the Institute of Interdisciplinary Medicine [7]. This article presents recommendations for designing an office environment based on a scientific evidence base and research to achieve a synergistic effect [8]: improve the health of office employees, increase their productivity, and also reduce the environmental burden.

2 Salutogenic design

The design of the premises affects the general well-being, determines the emotional state of employees, the level of stress and performance. Salutogenic design involves such an organization of space that evokes a sense of well-being, pleases, inspires work and creativity. To ensure this effect, knowledge and a certain skill of architects are required. The sensations and functioning of a person in space are better when the height of the ceilings is combined with the width and depth of the room; the color palette and wall patterns are in harmony with the floor covering and the shade of the ceilings. Carefully thought-out navigation improves the mental state (it should be intuitive, easy to navigate in space). Integration into the design of the nature theme has a beneficial effect on well-being. Biophilic office design increases productivity and creativity by 15%, creates comfort, and ultimately reduces staff turnover, which significantly saves company expenses. Moreover, employees, whose workplaces look out onto nature or trees, take on average 57 hours of sick leave per year, and people without a beautiful view of the workplace take 68 hours a year. And even indoor plants increase work efficiency and satisfaction with their work [9]. (Fig.1) The layout is of particular importance, since visual and acoustic noise during operation significantly affects labor productivity. German scientists conducted a

* Corresponding author: ABenuzh@gmail.com

https://doi.org/10.1051/e3sconf/202017006012

study, revealing that during the day at work, employees are distracted by visual interference an average of 600 times, which is about 40 hours a year or 5 workdays. This is especially relevant in the case of open space [10]. The absence of partitions or the use of low structures creates a strong noise background, visually distracts and amplifies computer radiation. It tires people and causes stress. With intense background noise, labor productivity decreases by 66% [11]. At the same time, 99% of employees say that they are most distracted by phone calls from colleagues and outside conversations. The use of special perforated acoustic panels, reflective screens, directing sound waves to the ceiling or into the aisles, reduces the impact of acoustic noise on a person. It should also include areas where employees can socialize in an informal setting. This increases the sense of cohesion by 18%, reduces stress by 6%, and reduces the "turnover" of staff from 40 to 12% (Fig.2). According to research, this increases labor productivity, and allows the company to save up to $ 15 million per year. Such free communication zones should be calm and comfortable so that employees can concentrate in silence, or collectively discuss work issues, talk in a relaxed atmosphere, present their ideas to colleagues, or simply relax.

Fig. 1. Reception of the Institute of Interdisciplinary Medicine

Fig. 2. Meeting room of the Institute of Interdisciplinary Medicine.

2.1. Daylight

A major factor affecting health and performance is access to daylight. Lack of daylight can cause sleep disturbance, stress, obesity, fatigue, and anxiety [12]. Daylight, a pleasant view from the window, help to improve mental activity by 10–25% and increase labor productivity by 18% (according to Russian data) and 15% - according to foreign ones. Studies show that thanks to bright classrooms with a beautiful view from the window, students receive higher grades and study 20–26% faster, and 20–25% of office workers report a decrease in health problems if their workplace is located next to the window.

However, it is important that the daylight does not cause discomfort, does not overload the visual system. To do this, it is necessary to diffuse lighting using special cover-ups on windows that redirect daylight - this is a simple and at the same time effective way to improve daylight due to the projection of natural light into the interior of the room. You can also install blinds that direct light to the ceiling, reduce glare and reduce the discomfort of direct sunlight.

Computer screens should be positioned at an angle of 20 degrees from the vertical plane of the nearest window so that the screen does not reflect, and employees do not strain their eyes.

2.2 Artificial light

Skillful use of daylight reduces energy consumption for artificial light by 40–80%. However, it is important to pay special attention to artificial light, since in the vast majority of cases, electric lighting, unlike natural lighting, provides one level of illumination and color spectrum. This negatively affects the hormonal state of a person, on circadian rhythms - the sleep-wake cycle, which determines the work of the body, affects all life support processes. So, ordinary fluorescent lamps consume alternating current, which causes an invisible flicker - 100-120 flashes per second, which negatively affects vision, causes rapid fatigue and can lead to sleep disorders. According to medical statistics, about 10-15% of adults complain of poor sleep, and 8-12% constantly take sleeping pills.

Today, energy-saving lamps are popular, which allow you to save 50% of electricity (compared with fluorescent lamps). However, the use of these lamps raises many questions because of the mercury content, which belongs to the 1st class of health hazard: negatively affects immunity, affects enzyme systems, causes a violation of hormonal regulation, diseases of the central nervous system, digestive organs, eyes, kidneys, and skin.

To maintain health, it is better to choose lamps with high color rendering index (CRI) and high R9 values — this index affects the transmission of human skin tone, this lamp will display warmer tones, which is more pleasant for the eyes.

In recent years, it is recommended to use customizable LED luminaires, which also consume 50% less energy compared to fluorescent lamps with the same light output, can work together with a lighting control system and reduce the amount of $CO2$ emissions into the air. Such lamps are able to change color temperature during the day, simulate changing natural light, and also have a built-in system that automatically reduces or increases electric lighting depending on the access of daylight.

In addition, it is important to choose finishing materials for ceilings, walls and floors that reflect and evenly distribute lighting.

Employees can also protect their eyes from computer screens using glasses that block harmful blue light. Or install computer programs that automatically adjust the light intensity and color rendering of the screen depending on the time of day.

2.3 Air quality

Studies show that the air in homes and offices is often more polluted than on the street, because it stagnates in the room and comes in with impurities from the outside. Emissions from furniture, household chemicals, finishing materials are added to it. In 2011, exposure to selected chemicals in the environment and in the work environment resulted in 4.9 million deaths worldwide. It is predicted that in the period up to 2050 the market for chemicals will grow by 3% annually, which is likely to affect the increase in mortality.
Substances such as lead, pesticides, fluorides, benzene, arsenic, formaldehyde are especially dangerous. Moreover, many of them, for example, formaldehyde, according to the Research Institute of Human Ecology and Environmental Hygiene, are almost always present in residential and office premises. Concentrations of hazardous substances (styrene, benzene, ethyl benzene, etc.) in rooms can be 10-15 times higher than safe levels. Studies show that volatile organic compounds (VOCs) and carbon dioxide directly affect cognitive functions, including strategic thinking, creativity, and the ability to quickly and correctly respond in critical situations. In addition, polluted air can lead to various diseases, headaches, office syndrome, allergies, bronchial asthma, etc.
And, on the contrary, high air quality increases the efficiency of employees and work efficiency by 5-10%, and reduces the number of absenteeism by 1/3. US experts have calculated how much this will bring the company in cash, and found that in an office with 100 employees, average annual productivity growth could be about $ 107,000 per year. And all investments in air purification, installation of air conditioners, reconstruction of premises - will pay off in 12 months due to increased productivity.
One of the easiest ways to ensure high quality indoor air is a good ventilation system and regular room ventilation, which can reduce the level of carbon dioxide and pollutants by several times.

2.4 Environmental materials and finishes

Studies on the effects of the environment and building materials on human health and labor productivity have been carried out for more than 40 years, and a number of experiments show that cognitive abilities are 61% higher when working in rooms that meet green standards (VOCs - Volatile Organic Compounds - are reduced approximately up to 50 µg / m3 and 20 cubic feet of water per person) and 101% higher in rooms that comply with the WELL standard [13].
The environmental friendliness of materials is:

- lack of emission of harmful chemical compounds;
- durability;
- non-toxic to humans and animals;
- use of renewable resources;
- safety for the environment after the expiration of the useful life;
- low energy consumption during production and transportation.

Careful selection of materials is crucial, since indoor air quality is largely dependent on this. When choosing finishing materials, it is important to pay attention to certification, which would confirm the compliance of materials with quality standards.

2.5 Healthy food in the office

Lack of adequate nutrition in the office leads to the development of chronic diseases, reduces concentration, increases stress, negatively affecting work efficiency. Conversely, a healthy, regular diet increases the efficiency and productivity of employees.
Studies show that eating habits, although persistent, can be changed in the addictive centers of the brain [14]. To do this, it is necessary to provide employees with the opportunity to eat healthy foods:

- To organize a dining room, creating a healthy diet - for the well-being of employees and increase efficiency, as well as the prevention of cognitive impairment.
- To establish a healthy food delivery system in the office.
- To inform employees about the importance of functional nutrition.
- To provide the possibility of snacks, the ability to heat food, store homemade food.

In addition, in the office it is important to provide employees with access to clean drinking water. Water or its deficiency (dehydration) have a direct effect on memory, cognitive functions and performance. Weak levels of dehydration can cause impaired concentration and short-term memory in adults. Moderate levels - impair memory, arithmetic abilities and psychomotor skills [15].

2.6 Office fitness

One of the key factors contributing to the increase in the prevalence of diabetes mellitus, obesity and cardiovascular disease among the urban population today is low physical activity.
Practical solutions of physical in office:

- Office rental near the fitness center.
- Organization of a sports zone with exercise machines in the office itself. The hall should have good ventilation and air purification. There should be a shower.
- If the office does not have a sports zone, it is possible to create conditions for employees to walk, for which, for example, large stairs or long passages between rooms are suitable.
- Creation of places for safe storage of bicycles, or installation of a bicycle parking. However, this solution is only suitable for offices located far from highways and near green forested areas, since cycling or running along busy roads does more harm to the body than good. During physical activity and when running, in particular, huge volumes of air are required to maintain the gas exchange function of the body. The larger the volume, the higher the chance of harmful particles entering the lungs. The constant inhalation of car exhaust can provoke the development of lung diseases, bronchial asthma and lead to intoxication.

Cycling or jogging is recommended no closer than 2 km from the main traffic flow.

- Providing employees with fitness bracelets or installing fitness applications on smartphones that measure heart rate, the number of steps taken per day, calories burned, allow you to track diet, weight and health and can stimulate a healthy lifestyle.

3 Stress management room

People - the main asset of the company, and employee illness, fatigue, office syndrome - negatively affect both business processes, profit levels, and the budget, because the employer is forced to pay for bulletins, etc. Therefore, investing in the health and well-being of employees through the organization of a healthy environment will help to increase employee productivity, achieve goals and increase company revenue.

Taking as a basis the biofeedback technology (BFB technology - psychological techniques with which a person learns to understand and control the reactions of his body: blood pressure, muscle tension, palpitations and nerve impulses. Such technologies are very effective in counteracting stress and are becoming more and more popular all over the world). Experts at the Institute of Interdisciplinary Medicine developed a stress management methodology for office workers called "Manage stress", which includes a repertoire of programs and training modules that allows you to quickly recover after an overload or optimize your psychophysiological state before a responsible meeting or complex activity requiring increased mobilization of cognitive and emotional resources in. Employees of such companies can take such trainings in their office in a specially equipped room - stress management room (Fig. 3).

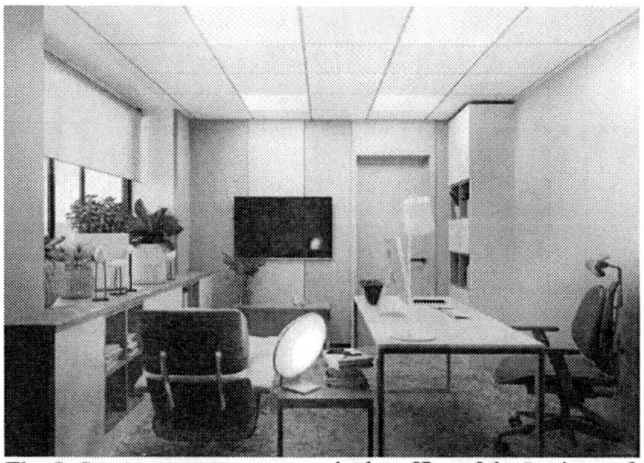

Fig. 3. Stress management room in the office of the Institute of Interdisciplinary Medicine.

Stress management contributes to the normalization of hormone levels, the proper course of metabolic processes, which improves mental activity and gives a feeling of mental well-being. The acquired skills of stress management will help to avoid overwork and protect against psychosomatic diseases, activate attention, memory, creative resources; increase the ability to solve complex intellectual problems; improve your mood and appearance.

This study was performed with the financial support of the RF Ministry of Education and Science, President Grant #NSh-3492.2018.8.

References

1. Telichenko V.; Benuzh A.; Mochalov I. Landscape Architecture and green spaces in Russia // MATEC Web of Conferences, 2017, № 117, **00164**
2. Telichenko V.I.; Benuzh A.A.; Mochalov I.V.; Using of green roof technology on buildings for creating a comfortable urban environment. Realty: economics, management. 2017, № **1** 30-33
3. V. Telichenko, A. Benuzh, G. Eamesa, E. Orenburova, N. Shushunova, Procedia Engineering, Volume **153**, pp. 726-730 (2016)
4. A. Benuzh, S. Fedorov, E. Orenburova, MATEC Web of Conferences 251, **03052** (2018)
5. V. I. Telichenko, A. A. Benuzh, V. V. Fateeva. IOP Conf. Series: Materials Science and Engineering, **456** (2018)
6. V. Telichenko, V. Roitman, A. Benuzh. Complex safety in construction: tutorial // UNIVERSITY. (2015). 144 p
7. Valery Telichenko, Gavin Dunn, Andrey Benuzh. The process of adaptation "green" standards BREEAM international in Russia and role of participants. MATEC Web of Conferences 193, **05001** (2018)
8. M. Slesarev. Formation of systems of ecological safety of construction: monograph / M.Y. Slesarev; M-in Education and Science of the Russian Federation, FGBOU VPO "MSSU". - Moscow: MGSU. (2012). 352 p
9. Arumi Ikei, Misako Komatsu,Chorong Song, Eri Himoro, Yoshifumi Miyazaki. Journal Physiol Anthropol. 2014; **33**(1): 6. 2014 Mar
10. Laro W. Office Kaizen: Transforming office operations into a strategic advantage. / Laro W. - Minsk: Grevtsov Publisher, 2009. - 35 p
11. Banbury SP. and Berry DC. Disruption of office-related tasks by speech and office noise. British Journal of Psychology **89:3**, (1998), pp 499–517
12. Robbins, C.L.; "Daylighting Design and Analysis", Van Nostand Reinhold Company (1986) Rosen, L. N., et al.; "Prevalence of seasonal affective disorder at four latitudes", Psychiatry Research, vol. **31**, no. 2, p. 131-144 (1990)
13. Joseph G. Allen, Piers MacNaughton. Health Perspect; DOI:10.1289/ehp.1510037. October 26, 2015
14. T. Deckersbach, S. K. Das, L. E. Urban, T. Salinardi, P. Batra, A. M. Rodman, A. R. Arulpragasam, D. D. Dougherty, S. B. Roberts. Pilot randomized trial demon- strating reversal of obesity-related abnormalities in reward system responsiv-ity to food cues with a behavioral intervention. Nutrition & Diabetes, 2014; **4** (9): e129 DOI: 10.1038/nutd.2014.26
15. Cian C., Barraud P.A., Melin B., Raphel C. Effects of fluid ingestion on cognitive function after heat stress or exercise-induced dehydration. Int J Psychophysiol. 2001; **42**:243–251

Transforming 'traditional' cities into 'smart' cities

*Laxmi Nagaraj**

BMSCA, Bangalore, India

Abstract. This paper aims to discuss the challenges of transforming 'Traditional' cities to 'Smart Cities' and the tools that can be used to transform 'Traditional' cities to 'Smart' cities in the Indian Context. In this context, this paper discusses the expectations and goals of the Smart City India Mission for the 100 Smart cities, the existing scenario of the 'Traditional' cities, the current status of the Smart cities in India and concludes that 'Traditional' cities can become 'Smart' by developing a base line scenario and developing a 'Road Map' to become 'Smart'. The 'Road Map' must consist of the following four stages: Assessment, Vision, Project Plan and Metrics.

1 Introduction [1]

A 'Smart' City is defined as a developed urban area that creates sustainable economic development and high quality of life by excelling or becoming "smart" in multiple key areas; economy, mobility, environment, people, living, and government. Recent interest in smart cities is motivated by major challenges, including climate change, economic restructuring, and the move to online retail and entertainment, ageing populations, and pressures on public finances. Basically, a "Smart City" uses information technology to make more efficient use of physical infrastructure, support a strong and healthy economic, social and cultural development and engage effectively with citizens in governance and decision making

1.1 Smart Cities in the Indian Context [2]

32% of the total population in India is urban and it contributes to over sixty percent of India's GDP. It is projected that urban India will contribute nearly 75% of the national GDP in the next 15 years.

India is at a point of transition where the pace of urbanization will speed up. The relatively low base presents an opportunity to plan the urbanization strategy in the right direction by taking advantage of the latest developments in technology especially in ICT (Information and Communication Technology). It is in this context that the Government of India has launched the Smart City Mission in 2014 and has decided on developing a hundred (100) "Smart Cities" in the country.

The Ministry of Urban Development, in its concept note, has visualized that a smart city "will have to provide a very high quality of life,

i.e., good quality but affordable housing, cost efficient physical, social and institutional infrastructure (water, sanitation, 24/7 electricity), clean air, quality education, cost effective health care, security, entertainment, high speed connectivity and efficient mobility; it must also attract investments, experts and professionals.

The 100 smart cities are to be developed in population ranges of up to one million, one to four million and over four million.

The 100 smart cities mission intends to promote adoption of smart solutions for efficient use of available assets, resources and infrastructure with the objective of enhancing the quality of urban life and providing a clean and sustainable environment,

Special emphasis will be given to participation of citizens in prioritizing and planning urban interventions.

The India Smart City Mission has developed several guidelines and components for developing Smart cities.

The major components are:
1. Area based Approach – including retrofitting, redevelopment, pan-city and greenfield development initiatives
2. Citizen engagement for visioning - to involve all sections of society who are affected by a decision in the decision-making process and come up with proposals that are citizen-driven.
3. Financing proposals with revenue models - to attract Public Private Partnerships (PPPs) for infrastructure development and other projects by setting up Special

* Corresponding author: laxminl0@hotmail.com

 https://doi.org/10.1051/e3sconf/202017006013

Purpose Vehicles (SPVs) that would be responsible for the PPPs.

2 Current Urban Challenges in India[2]

India is facing many of the global urban challenges of large urban populations and overcrowding in the cities.

Some of these challenges are:
- Managing size and rapid growth
- Providing Urban Services
- Reducing Poverty
- Housing for all
- Making Cities Healthy
- Making Cities Environmentally Friendly
- Solving Traffic and Transportation Problems
- Supporting Social Development

2.1 'Traditional' Cities – Current Scenario

Economy - No strategy for attracting investment, no specific industrial or economic zones and no identity developed for the city.

The long-term dependence on Central and Local funds have made the cities complacent and they have not yet developed a working culture of developing strategies and policies at the local level for attracting investments. Therefore, they have not yet felt the need to develop a strong identity for the city that will help attract investments.

Environment - No vision for energy efficient buildings and systems, concept of green buildings not introduced, no policies for renewable energy, no strong policies for sustainability, environmental preservation and resource management.

Several of the Smart Cities Mission strategies require the cities to strongly embrace environmental stewardship, resource conservation and environmental preservation. Until now, the cities have not been proactive in these areas. In order to be proactive in these areas, strong implementation environmental policies must be in place and the staff must be capable of implementing and enforcing these policies.

Government - Departments are scattered with no data sharing /collaborations, limited and scattered online access to citizens, citizens cannot make optimum use of city services

At present, the Government Departments are scattered and work in isolation. There is no data sharing and collaboration between the departments. This is because a lot of the data is still in paper form and have not been transferred to digital data. With digital data, data sharing and collaborations are relatively easy and services provided by the various departments are more effective. A work culture of a multi-disciplinary approach to finding solutions is not yet implemented.

Governance/Public Participation – Online services for services has not been adopted to its fullest capacity and citizens have never participated or have been involved in decision making for citizen centric solutions.

E-services and citizen participation in decision making from conceptual stage is not yet implemented.

Living - Inadequate social infrastructure, parks and open spaces, health facilities and services are scattered, crime prevention is inadequate.

City services have not kept pace with the rapid population growth in the cities and therefore, social infrastructure and health services are inadequate for the population they serve.

Mobility - Existing transportation systems are inefficient and not cost effective

Transportation services have not kept pace with the rapid population growth in the cities and therefore they are inadequate for the population they serve.

People - Education systems not available for all levels of education, affordable housing and services for the poor are inadequate.

City services have not kept pace with the rapid population growth in the cities and therefore, social infrastructure and health services are inadequate for the population they serve.

The Smart City India Mission strategies and components have been developed to address and overcome the above drawbacks and help cities upgrade themselves.

2.2 Current Status of Smart Cities in India [3]

The 100 selected Smart Cities, with the help of consultants, have developed several proposals based on the Smart City requirements. However, the development of Smart Cities in India has been slow, and the funding has not been fully utilized according to the time-lines for utilizing the funds. Given the current scenario of the "Traditional" Indian cities, the cities are not fully capable of implementing the Smart cities proposals and components.

This current scenario is creating several roadblocks to achieving the goal of 100 Smart Cities by 2020.

Some of the road blocks are:
- Inaction
- A lack of understanding
- A dearth of technical training & skills
- Non existent security framework for data.

The reason for this inaction is that the Local and State Government system has been used to a relationship of getting grants from the Central Government that did not require strict accountability measures with respect to the "quality" of delivery.

Now, accounting systems must be improved and quality control measures must be included for accountability.

Now, suddenly, the Local governments must come up with sustainable solutions and be accountable for them. This requires a change in mindset. Even if the mindset is changed, capacity building is the biggest hurdle. Training a large work force to upgrade their skills to be able to understand and implement the new components is a major task that cannot be accomplished in a short period of time.

In addition, the legacy of the old technologies complicates the ability to make valuable solutions. Sudden upgradation of old technologies is complicated and hinders progress.

Finance is another obstacle with the focus on the PPP model to fund smart cities. Most of the city staff are not trained to implement PPP financial models.

Many cities are also facing resistance in the execution of projects. Citizens have opposed user charges for services provided under the mission.

The citizens have not been made fully aware of the benefits of the projects with effective citizen participation resulting in opposition during various stages of the projects.

The Smart Cities Mission strategies and components can only be fully utilized to bring the much-needed value addition if the cities develop a "base line" scenario.

The "base line" scenario is a scenario where the "Traditional" cities have upgraded themselves and improved in all of the above components to be "ready" to develop, implement and enforce the Smart City components and strategies.

The base line scenario must be used to develop well-planned approaches to effectively utilize the Smart City strategies and components.

Therefore, the 'Traditional' Cities must develop a Road Map to develop a Base Line scenario and transform themselves into 'Smart' Cities.

3 Road Map [4]

For a 'Traditional' city to develop a Base Line scenario and transform itself into a 'Smart' city, innovations and progress must be made for the following:

1. Organization – An established and efficient organization structure to lead to a smarter city. This could include organizational changes and/or restructuring to address and utilize the new strategies and components.

This would require a major organization changes with the addition of new departments or the reorganization of old departments to handle the new requirements.

2. Technology – Improvements to evolve the right way. This could include making investments in appropriate technologies and training staff to fully utilize the capabilities of the technologies

This would require major investments in appropriate and new technologies and elimination and phasing out of out-dated technologies to implement the new requirements and provide services efficiently.

3. Social aspect – To make the population evolve the right way. This could include educating the citizens regarding capabilities of online services, educating the citizens about sustainability, waste segregation, conservation of resources and the benefits of citizen participation in Governance.

A major program to fully involve citizens in Governance must be developed and implemented either entirely by the Government or with the help of NGOs.

4. Capacity Building – To develop sustainable solutions and be accountable for them. This could include highly specialized training for staff for all aspects of Governance to make them understand and implement Global best practices locally.

This would require specialized training for all levels of staff to help them understand and implement the new components. This training will also require a major investment.

The Road Map must consist of the following four stages:

1. Assessment

A 'Traditional' city can assess the current scenario with respect to the key areas of a 'Smart' City including Financing. This assessment will help the City assess their needs for improving their capabilities to effectively utilize the Smart City strategies.

This would require creating an up to date inventory of existing infrastructure, services, population, housing, services, green areas, resources etc. to understand what is lacking, what is working and what is redundant. This will essentially be a "snap shot" of the current scenario and existing conditions in the City.

2. Vision

Although the selected Smart Cities have developed proposals according to the requirements of the Smart City Mission, the implementation of the proposals has been very slow. Developing a 'base line' scenario and timeline for accomplishing what can be achieved in next year, 5 years, 10 years will help.

The Smart Cities guidelines also prescribe various deadlines for implementing the proposals. Release of funds is after each phase of the proposals is implemented. The cities have not been able to fully implement the proposals to seek a release in funding.

Therefore, developing a "Base Line" scenario and developing phases of implementation and accomplishment will help. The implementation targets must be set for the immediate next year, 5 years and 10 years. At each phase, the targets must be reviewed to see what has been achieved, what has not been achieved, why certain things have not been achieved, what corrective measures must be taken to achieve these targets etc.

3. Project Plan

A project plan must be developed on how the above vision will be achieved within the targeted timelines. Milestones must be developed for the next one year (immediate), next five or ten years (short term) and next 20 years (long term).

For the immediate milestones, improvement of existing mechanisms, data collection and assessment of requirements and other targets and goals that are easily accessible (low hanging fruit) must be accomplished.

For the short-term milestones, certain major frameworks, policies and implementation must be completed.

For the long-term milestones, all the tasks must be completed and implemented.

In addition to the above, a detailed funding plan listing how the funding will be obtained for each milestone must be developed.

4. Metrics

Performance Metrics must be utilized for each milestone phase to determine if the tasks were completed within the projected timelines, within the budgeted amounts and whether the targets fulfill the smart city vision and goals. The metrics should also assess tasks that could not be completed, the reason they were not completed and how these affect the overall vision and goals.

By effectively incorporating the Smart City strategies and components, the 'Traditional' cities can become 'Smart' and meet the performance targets for Economy, Environment, Government, Governance/Public Participation, Living, Mobility and People.

By becoming 'Smart', Indian cities the Indian cities can effectively accomplish the following:

- Tackle Climate Change
- Use Local Resources
- Construct Green and Eco-friendly Buildings
- Grow the Cities Green Footprint
- Make Dynamic Downtowns
- Enhance Social Infrastructure
- Provide Housing for various income groups
- Provide multimodal connectivity
- Plan for Disasters
- Promote Economic Development

4 Conclusion

Although the selected Smart cities in India have developed proposals for becoming 'Smart' according to the Smart City India Mission strategies and components, implementation of the proposals has been very slow.

Given the current scenario of the "Traditional" Indian cities, the cities are not fully capable of implementing the Smart cities proposals and components.

This current scenario is creating several roadblocks to achieving the goal of 100 Smart Cities by 2020.

Some of the roadblocks are:

- inaction
- a lack of understanding
- a dearth of technical training & skills
- non existent security framework for data.

The Smart Cities strategies and components can only be fully utilized to bring the much-needed value addition if the cities develop a "base line" scenario and develop well-planned approaches to effectively utilize the Smart City strategies and components.

The base line scenario must be used to develop well-planned approaches to effectively utilize the Smart City strategies and components.

Therefore, the 'Traditional' Cities must develop a Road Map to develop a Base Line scenario and transform themselves into 'Smart' Cities.

For a 'Traditional' city to develop a Base Line scenario and transform itself into a 'Smart' city, innovations and progress must be made for the following:

Organization, Technology, Social Aspect and Capacity Building

The Road Map must consist of the following four stages:

Assessment, Vision, Project Plan and Metrics.

By effectively incorporating the Smart City strategies and components, the 'Traditional' cities can become 'Smart' and meet the performance targets for Economy, Environment, Government, Governance/Public Participation, Living, Mobility and People.

References

1. L. Nagaraj, *What is a Smart City?* SCIF (2015)

2. *Smart City Mission Statement & Guidelines*, MOHUA, GOI (2015)
3. *Demystifying the current state of India's ambitious Smart City Mission,* digit.in (2019)
4. *Smart cities - background paper,* BIS, GOV.UK (2013)

Study on hybrid biorack constructed wetland

Guru Munavalli[1*], *Pratap Sonavane*[2], *Sagar Wandare*[3], *Kusum Biradar*[4], *and Nikhil Aswale*[5]
[1-5] Department of Civil Engg., Walchand College of Engineering, Sangli, Maharashtra, India

Abstract. Decentralized Wastewater Treatment Systems (DWTS) based on natural mechanisms of purification are appropriate for Indian conditions. Biorack constructed wetland are recent developments in the domestic wastewater treatment. In the present study the concept of biorack wetland is extended to Hybrid Biorack Constructed Wetland (HBCW) in which bioracks are supplemented by brickbat as support medium for wetland vegetation. Three separate HBCW reactors were developed with *Typha angustifolia, Canna indica* and dual-species (*Typha angustifolia and Canna indica)*. Batch studies were carried out on these reactors for secondary treatment of domestic wastewater. COD removal was determined for feed wastewater strength of 100 to 200 mg COD/L by these developed reactors. The growth of vegetation was also monitored. *Typha angustifolia and Canna indica* were found to be an appropriate combination as dual-species. The COD removal by dual-species is 5 to 10% greater than mono-species under similar conditions of feed and support medium. HBCW is efficient to an extent of 65 to 75% for COD removal. This is preliminary study conducted on HBCW. Further studies on pilot/field scale will provide better insight to assess its suitability as an alternative biological treatment.

1 Introduction

Domestic wastewater is composed of greywater and blackwater. The generation of this type of wastewater is increasing with time. The disposal without treatment is problematic as it leads to contamination of groundwater and surface water. The treatment facilities wherever available are either centralized or decentralized. The centralized treatment facilities are normally provided and operated by municipal authorities. Such facilities have pre-requisite of collection/sewerage system. The fund required to build and operate these systems is large. It is a fact that the larger systems have more operational difficulties and thereby necessitating more mechanization. Further, the application of technologies based on natural mechanisms of purification may not be feasible for large scale application. Decentralized Wastewater Treatment System (DWTS) may be an alternative to centralized treatment systems for on-site treatment of domestic wastewater. The proven benefits of DWTS include cost effective, incorporation of natural mechanisms of purification in treatment system, reduce load on municipal treatment systems, better control over operation and maintenance, and scope for recycle and reuse. Further, individual households, isolated domestic establishments, small housing townships, educational organization, and commercial establishments can be provided with DWTS. The treatment in DWTS may be aerobic, anaerobic, or combination of these. A typical DWTS may have either anaerobic treatment in the form of baffled reactor/filter followed by primary and secondary Constructed Wetland (CW) or aerobic moving bed bioreactor. In Indian context, first option is more feasible provided space is available due to the use of natural mechanisms in treatment line. However, there are studies conducted on CW and possible modifications to conventionally developed CW. Biorack CW (BRCW) are one of such modifications. Conventional Bio-Rack Wetland System (CBRWS) is one of the latest configuration in wetland technologies that was developed by [10] to overcome the drawbacks of CW. CBRWS is the grid arrangement of racks having perforations for liquid transport in which plants are placed in the absence of any other supporting media. This is a robust biological practice that can be applied to an efficient and reliable elimination of pollutants, even under unfavorable influent conditions stressed by high total dissolved solids and heavy metal [7, 10, 11]. The reported studies on CBRWS include [6, 10, 11] for domestic wastewater treatment and [13] for treatment of low polluted river water. All these studies have demonstrated the potential use of CBRWS in wastewater treatment. Further, [9] modified CBRWS by using non- perforated bioracks and inducing plug-flow pattern into the system and demonstrated

* Corresponding author: guru.munvalli@walchandsangli.ac.in

enhancement in the performance of BRCW for the removal of Organic Carbon. The primary component of BRCW is system of pipes and cost of the system is governed by this component. [9] tried to minimize the cost of BRCW by using discarded plastic package drinking water (1 L) bottles. The development of root matrix of wetland plants plays a significant role in the performance of BRCW. Vertical spread of root matrix is more encouraged in BRCW. It is hypothysed that the combination of vertical and lateral spread of root matrix enhances the treatment potential of BRCW. In the present study, the following contributions to existing knowledge on BRCW are addressed;

1. The brickbats are used as support media for vegetation and thus replacing the bioracks partially. This modification reduces the cost of system and provides scope of vertical and lateral development of root matrix.
2. The multi-species vegetation is used in bioracks and brickbat medium. This enables proper root matrix in the system. The developed system is referred as Hybrid Biorack Constructed Wetland (HBCW) in the present study.

2 Materials and methods

2.1. Source of wastewater

The settled domestic wastewater generated from campus of Walchand College of Engineering, Sangli (Maharashtra, India) was collected from primary settling tank and used as influent to HBCW. The contributing sources to this wastewater from college campus included food establishments (mess and canteen), greywater and septic tank effluent from residences, and hostels.

2.2 Experimental HBCW

Fig. 1 shows schematic sketch and photographic view of HBCW. The developed systems include HBCW with *Canna indica* (Reactor 1), HBCW with *Typha angustifolia* (Reactor 2) and HBCW (Multispecies with *Canna indica and Typha angustifolia*). Each reactor of HBCW was made up of seven 80 mm diameter 0.70 m deep corrugated plastic pipe (as bioracks) surrounded by brickbats medium in 80 L plastic tank of 0.70 m diameter. The wetland plants were vegetated in the respective reactors as specified above. *Canna indica* and *Typha angustifolia* were collected from a nearby stream and acclimatized before initial plantation in HBCW. The size of brickbats used was 20 mm to 40 mm and measured porosity was 45% of brickbat medium. The locally available *Canna indica* and *Typha angustifolia* were chosen as these were well proven as wetland vegetation [1, 2, 4, 5, 8, 14] surviving, growing and multiplying in saturated conditions and capable of releasing the dissolved oxygen through roots. Initially, 10 plants each were vegetated in reactors 1 and 2, whereas reactor 3 has 5 plants each of *Canna indica* and *Typha angustifolia*. The average initial depth of roots of each plant was 50 mm. The lateral spread was also measured and was 30 mm.

2.3 Operation of HBCW and Experimental Procedure

The systems were operated for performance evaluation study after development of roots. Then, all systems were subjected to batch mode of operation for 30 d. Hydraulic Retention Time of 24 h was used in the study for all systems for a week and systems were monitored by sampling and analyzing influent and effluent for every cycle of applied HRT. The experiments were conducted on all three systems to assess their performance for the removal of Chemical Oxygen Demand (COD). The samples were analyzed for COD following procedure of APHA [3].

Fig. 1 Photographic view of HBCW a) reactor 1; b) reactor 2; and c) reactor 3

3 Results and discussions

3.1 Wastewater Characteristics

The characteristics of influent wastewater to HBCW were analyzed to check its suitability for the biological treatment. The characteristics are summarized in Table 1. It can be seen that the BOD_3/COD ratio of wastewater is in the range of 0.52 to 0.63 indicating amenability of settled wastewater for biological treatment.

3.2 COD Removal by HBCW

Fig. 2 shows the influent COD, effluent COD and removal of COD by HBCW system planted with *Typha angustifolia*, *Canna indica* and Dual-species (*Typha and Canna* together). The comparative performance is shown for the reactors fed with COD of influent settled wastewater in the range 100 to 250 mg/L. The results show that the removal efficiency is higher in HBCW reactor vegetated with Dual-species.

Table 1. Characteristics of influent wastewater

Parameter	pH	TSS (mg/L)	COD (mg/L)	BOD_3 27°C (mg/L)
Settled wastewater (Influent)	7.05 ± 0.35	138 ± 33	102 ± 50	65 ± 14

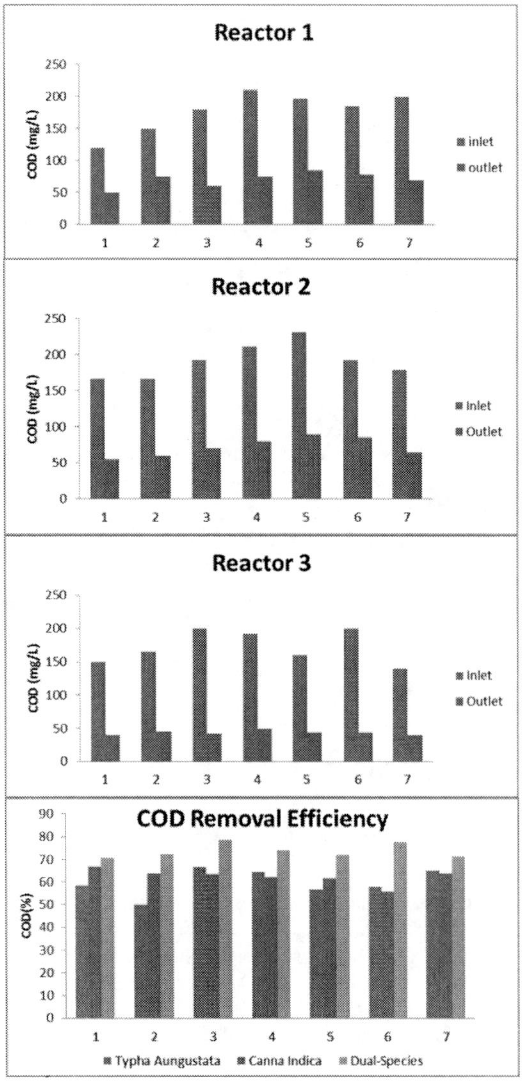

Fig. 2 COD removal by HBCW systems

The comparison of the relative performance of *Canna indica and Typha angustifolia* indicates both are effective. The average percent removal for both reactors 1 and 2 is 55% to 65%. The removal efficiency in reactor 3 is significantly higher that of reactors 1 and 2 the same range of influent COD. These results are suggestive of the mixed type of vegetation enhances the efficiency as all the three reactors have brick bats as the support media. The combination of shallow and deep roots developed into root matrix enhances the removal. The reactor 3 with dual-species consistently has shown better removal (65 to 75%). In order to validate the effect dual-species these results were correlated with growth of vegetation and discussed in the following section.

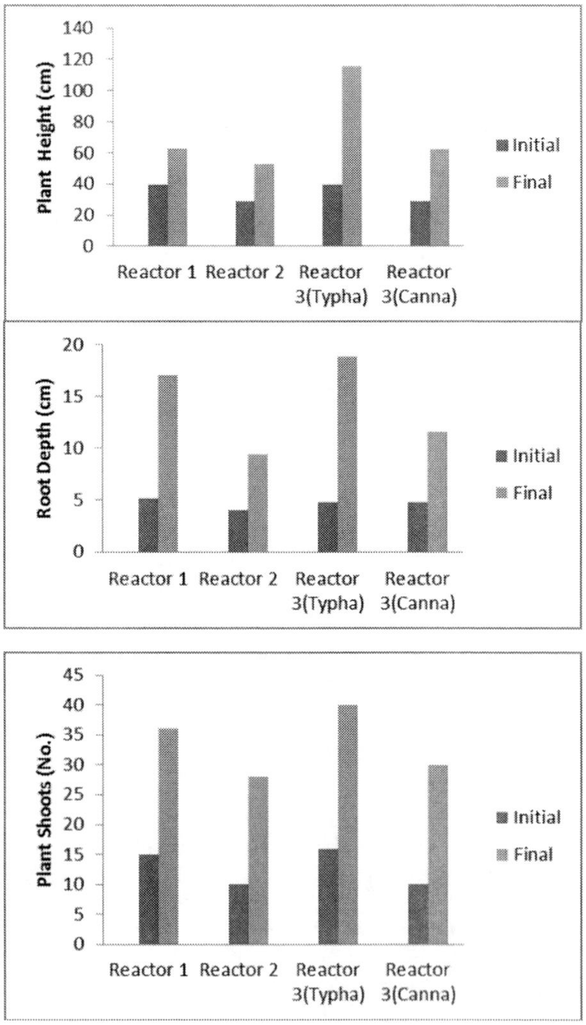

Fig. 3 Quantitative measurement of vegetation growth

3.3 Vegetation growth in HBCW systems

The growth of wetland vegetation in all the three reactors was monitored for a period of 30 days. The height of plants, number of shoots, depth of roots and spread of roots laterally were observed and recorded as a part of monitoring. The photographic view of vegetation growth in sample bioracks taken from respective reactors is shown in Fig. 4. It can be seen from the figure that bioracks support the wetland plants individually and combined/as a dual species. The roots in bioracks spread both vertically and laterally. The lateral spread of roots is evident as roots emerging through the perforations of biorack are clearly visible. The individual root matrix is also well developed.

Fig. 3 shows the quantitative measurement of vegetation growth in terms of height, root depth, root spread and number of shoots. The growth of both these plants in terms of the stated parameters is significant considering the shorter period of monitoring/study. The vegetations were well grown in brickbat medium as well as in bioracks. The average root depth of *Typha angustifolia* (in reactor 1) and *Canna indica* (in reactor 2) was observed to be 175 mm and 100 mm respectively in reactors 1 and 2 after 30 days. The root growth for both plants was 2 to 4 times the initial depth. Interestingly, the corresponding the average root depth of these plants was 200 mm and 125 mm respectively in reactor 3. This indicates that there is no interference in the relative growth of these plants. The combination of shorter and longer roots provided by dual- species has resulted in dense matrix of roots. The matrix was so dense that root system in bioracks of reactor 3 could not be pulled out. This extra grip provided by the developed roots in dual-species bioracks might be the reason for extra root depth of plants. The effect of development of dense matrix of roots is evident in enhancement of COD removal efficiency in reactor 3 compared to other two reactors (Fig. 2). The growth rate of plants with reference height was also significant for both the plants. The growth of *Typha angustifolia* is relatively more than *Canna indica* as mono-species and in dual-species. Its growth is more in dual-species (reactor 3). The lateral spread is to an extent of 50% more than the corresponding initial values. The spread of *Typha angustifolia* in reactor 3 is higher. All these results demonstrate that i) *Typha angustifolia* and *Canna indica* are suitable species for domestic wastewater treatment; ii) the combination of these two species is appropriate; and iii) dual-species HBCW is an alternative to CBRWS.

4 Conclusions

The present was carried out to assess the potential of HBCW in treating domestic wastewater. COD removal was primarily studied on three reactors planted with mono-species and dual-species in bioracks and brickbat medium. The relative growth of individual plants is higher in reactor with dual-species. The use of dual-species (*Typha angustifolia* and *Canna indica*) was found to be more effective than mono-species systems in COD removal by 5 to 10%. The overall COD removal that can be achieved by dual-species HBCW is 65 to 75% for HRT of 24 h. HBCW has a potential to be an alternative to existing biological treatment for domestic wastewater.

The present study is a part of research project sanctioned by Department of Science and Technology DST) under Water Treatment Initiative (WTI) programme. The authors sincerely acknowledge and thank DST for providing financial support to undertake this study.

Fig. 4 Photographic view of vegetation growth in bioracks of reactor 1, 2, and 3

References

1. S. Abou-Elela and M. Hellal, Ecol Engg, **47**, 209– 213, (2012)

2. S. Abou-Elela, G. Golinelli, A. El-Tabl and M. Hellal, Water Sci & Tech, Vol. **69 (1)**, 38-47, (2014)

3. APHA, *Standard Methods for the Examination of Water and Wastewater, American Water Works Association*, Water Environment Federation (2012)

4. L. Cui, Y. Ouyang, Q Lou., F Yang., Y. Chen, W. Zhu and S. Luo, Ecol Engg, **36**, 1083–1088 (2010)

5. D. Ghosh and B. Gopal, Ecol Engg, **36**, 1044-1051 (2010)

https://doi.org/10.1051/e3sconf/202017006014

6. S. Jamshidi, A. Akbarzadeh, K. Woo and A. Valipour, J. of Env Health Sci & Engg, **12,** 1-12 (2014)

7. L. Marchand, F. Nsanganwimana, N. Oustriere, Z. Grebenshchykova, K. Lizama-Allende, M. Mench, Ecol Engg, **64**, 291-300 (2014)

8. Y. Patil and G. Munavalli, Ecol Engg, **95**, 492-500 (2016)

9. S. Sathe and G. Munavalli, J. of Water Process Engg, **28**, 240-249 (2019)

10. A. Valipour, V. Raman and V. Ghole, Ecol Engg, **35**, 1797–1803 (2009)

11. A. Valipour, V. Raman and V. Ghole, J. of Env Sci and Engg, **53(3),** 281-288 (2011)

12. J. Vyazmal, Water, **2**, 530-549 (2010)

13. J. Wang, L. Zhang, S. Lu, X. Jin and S. Gan, J. of Env Sci. **24 (6),** 1006–1013 (2012)

14. A. Yadav, R. Abbassi, N. Kumar, S. Satya, T. Sreekrishnan and B. Mishra, Chem Engg J., **211**, 501–507 (2012)

Performance Evaluation of Baffled and Floating Hybrid Constructed Wetland Operated Under Field Conditions

Shrutee Dasare[1], *and Guru Munavalli*[1*]
[1]Department of Civil Engineering, Walchand College of Engineering, Sangli, Maharashtra, India

Abstract. Baffled and Floating Hybrid Constructed Wetland (BFHCW) consisting of constructed wetland with brickbat medium and Floating Treatment Wetland (FTW) was developed and operated under actual field flow conditions. BFHCW was vegetated with dual-species (*Typha angustifolia* L. and *Canna indica)*. The performance of continuously operated system was evaluated for varied Hydraulic Loading Rate (HLR) and Organic Loading Rate (OLR) for a longer period. The system was also assessed for controlled flow rate resulting more uniform OLR and HLR. The results showed that COD removal efficiency was enhanced by 5 to 15% with controlled flow than uncontrolled highly fluctuating flow conditions. HLR has more impact on COD removal than OLR. BFHCW is effective to an extent of 25 to 40% for COD removal. BFHCW is found to be cost effective and reasonably efficient treatment for pretreated domestic wastewater.

1 Introduction

The volume of domestic wastewater generated has increased with population, urbanization, improved living conditions, and economic development. There are two systems for the treatment of domestic wastewater viz. centralized and decentralized. It is disadvantageous to implement centralized systems in developing countries due to financial constraints, requirement of technical expertise for operation and management, and necessity of large sewerage network. Further, wastewater generated from satellite areas or city outskirts require spending even more cost to transfer wastewater to centralized treatment facilities due to the absence of sewer systems. Decentralized Wastewater Treatment System (DWTS) means to treat, reuse or dispose of wastewater at or near its source of generation. DWTS is a cost effective and an appropriate option for small clusters/ household/ housing colony/ institution/ commercial complex. Constructed Wetland (CW) is one of the suitable and affordable options as secondary treatment in DWTS. CWs have emerged as an economical and feasible option for the treatment of domestic wastewater, as they can be built with locally available materials, simple construction, and utilization of natural processes, lower operation and maintenance cost. The aspects related to environmental sustainability and decentralization are present in these treatment technologies [2]. Constructed wetlands (CWs) are proven to be efficient for treating urban and domestic wastewater [1]. The problems associated with CW such as longer start-up time and high risk of clogging of media resulting waterlogging can hamper the treatment process. The choice of an appropriate coarser media and combining CW with another treatment system helps to overcome these problems. In this context, CW system with brickbat as a support media for CW vegetation and Floating Treatment Wetland (FTW) arranged in a typical hybrid sequence can be devised. [4] reported that FTW seem to be effective for the treatment for wastewater and reviewed applicability and design of FTW. A system based on this concept referred as Baffled Floating Hybrid Constructed Wetland (BFHCW) was implemented for the treatment of domestic wastewater in the campus of Walchand College of Engg., Sangli (WCE). BFHCW was a secondary treatment in DWTS of WCE which was fed with wastewater pre-treated in anaerobic baffled reactor.

In the present study, BFHCW was assessed for its performance to remove organic content when operated with actual flow conditions. The primary objective of the work was to study the system behavior under varied Hydraulic Loading Rate (HLR) and Organic Loading Rate (OLR).

2 Materials and methods

2.1. Source of wastewater

Wastewater is generated from different eating establishments and hostels in WCE campus. It is comprised of washings, septic tank effluent and greywater. The wastewater is pre-treated with bar rack

* Corresponding author: guru.munvalli@walchandsangli.ac.in

screen and Anaerobic Baffled Reactor (ABR) as part of DWTS. This anaerobically treated wastewater was fed to BFHCW.

2.2 BFHCW details and operation

DWTS in WCE is an integrated multi-stage natural wastewater treatment system treating 45 KLD. Wastewater is treated in preliminary and primary treatment consisting of screen chamber and ABR respectively. The effluent from ABR enters BFHCW for secondary treatment. Fig. 1 (a) and (b) shows a schematic sketch and photographic view of BFHCW. BFHCW is made up of 24 compartments with an inlet chamber, alternate compartments having CW with brickbat as supporting medium [referred as Brickbat Subsurface Wetland (BSFW)], FTW, adsorption bed and an outlet chamber. The effective wastewater depth in the system was 0.80 m. Each BSFW was made up of brickbats of size 50 to 75 mm with bed porosity of 50 to 60% and vegetated with dual-species. In FTW different floating mats were tried and that included Thermocol sheet, plastic bottles, and perforated basket. These are shown in Fig. 2 (a) and (b). Charcoal and aggregate were used to provided proper anchorage to vegetation. Dual-species (*Canna indica* and *Typha angustifolia* L). were used in both BSFW and FTW. These were chosen as they were locally available and found to be effective to remove pollutants such as carbonaceous organic matter, nitrogen and phosphorous. Initially, plants were acclimatized by growing them gradually increasing strength of wastewater and then vegetated in BFHCW. An adsorption bed made up of coarse aggregates and charcoal forms the final compartment before outlet chamber. The primary treated wastewater enters BFHCW and flows in an upflow and downflow mode through all the compartments and treated effluent is then discharged into the outlet chamber.

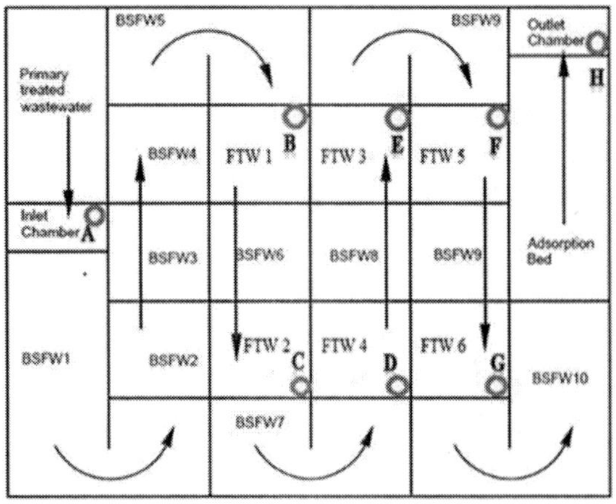

Fig. 1 (a) Schematic line plan of BFHCW

2.3 Sampling strategy and analysis

Composite sampling method was adopted. The sampling locations are shown in **Error! Reference source not found.**(labeled in blue colored circle). Normally samples were collected after every alternate day. Timings for sampling were decided by considering peak flow (9 to 10.30 am), average flow (3 to 4 pm) and minimum flow (before 6.20 am). The parameters such as pH, EC, COD, and BOD_3 were analyzed to evaluate performance evaluation of the system. The analysis was carried out by referring to APHA [2].

Fig. 2 (b) Photographic view of BFHCW

Fig. 2 (a) Photographic view showing configuration of FTW with bio-rack

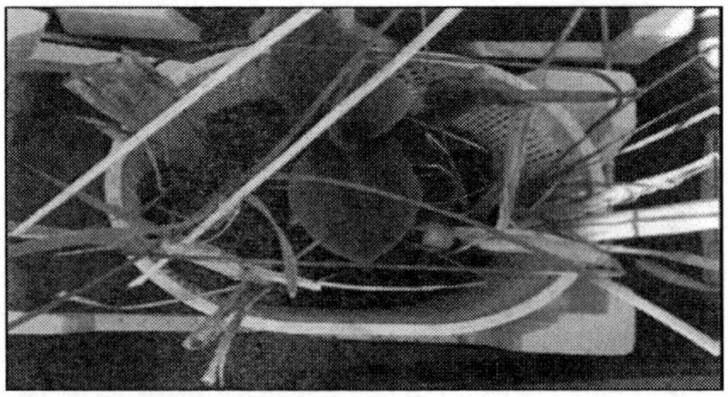

Fig. 2 (b) Photographic view showing configuration of FTW with basket configuration

3 Results and Discussions

3.1 Influent Characteristics

Pre-treated wastewater from ABR was taken as an influent to BFHCW system. The summary of influent wastewater characteristics during the study period is given in **Error! Reference source not found.**. pH values show that wastewater is slightly alkaline in nature. The ratio of BOD_3 to COD is greater than 0.5 indicating its amenability for biological treatment. Thus the ABR effluent is an appropriate feed to BFHCW.

Table 1. Characteristics of influent to BFHCW

pH	COD	BOD_3 @ 27°C	TKN
-	mg/L	mg/L	mg/L
7.35 ± 0.17	268.90 ±45.77	144 ±7.65	45.54 ±10.43

3.2 Effect of varied HLR on COD removal

Error! Reference source not found. shows the effect of varied HLR on COD removal under uncontrolled flow conditions. The results show that the COD removal varies from minimum of 6.2% to maximum of 47.72 %. The corresponding HLR for these removals are 4.8 m^3/m^2d and 1.98m^3/m^2d respectively. The COD removal varies inversely with HLR. The observed high and low HLR correspond flow rate of 4000 L/h (peak flow) and 1980 L/h (low flow). Hydraulic Retention Time (HRT) was also determined using tracer (Fluoride as tracer) study and found to be 49 min whereas theoretical HRT was 1.30 h during peak flow. The respective low flow HRT values were 1.66 h and 2.60 h. Normally, HRT of FTW systems were considerably larger (> 12 h). In the present BHFCW system, HRT values significantly lower than those normally used. The system performance is said to be significantly better owing to the conditions to which it is subjected.

3.3 Effect of varied OLR on COD removal

Error! Reference source not found. shows the effect of varied OLR on COD removal. The system was subjected to OLR variation from 0.57 to 1.2 kg $COD/m^2.d$. The variation is induced by varying flow rate and influent COD. The COD removal was observed to be in the range 6.2 to 34.18%. The effect of OLR is similar to that of HLR as discussed in the previous section. The COD removal efficiency is not very high but significant considering the high variability of influent COD, lesser HRT, and fluctuating flow rates.

3.4 Performance evaluation of system based on controlled HLR conditions

Fig. 6 shows the effect of controlled HLR on COD removal. The trend of COD removal with HLR variation is same as that of uncontrolled conditions. The comparison of the performance of two studies viz. controlled and uncontrolled conditions was done. The system was operated at HLR of 1.71, 3.43 and 4.29 $m^3/m^2.d$ and OLRs considered were 0.36, 0.74 and 1.44 $m^3/m^2/d$. **Error! Reference source not found.** 7 (a) and (b) indicate system performance for COD removal under controlled and uncontrolled conditions for the above specified HLR and OLR values. COD removal rate was observed to be higher by 5 to 15% for controlled HLR condition than that of uncontrolled condition. Flow normalization and avoidance of sudden change in organic loads are the reasons for relatively better performance under controlled conditions. The results also indicate that HLR is more governing factor than OLR.

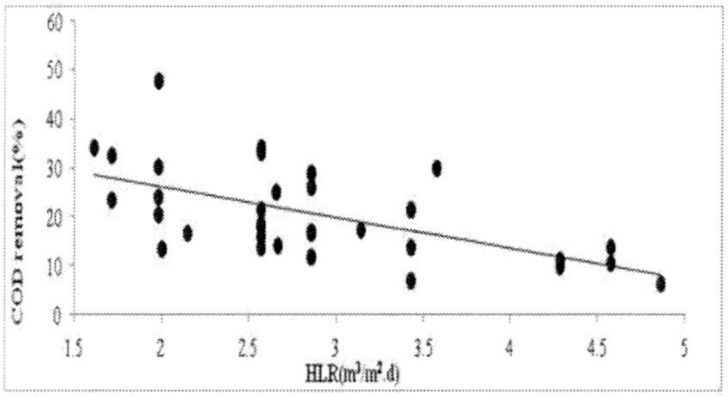

Fig. 3 Effect of varied HLR on COD removal

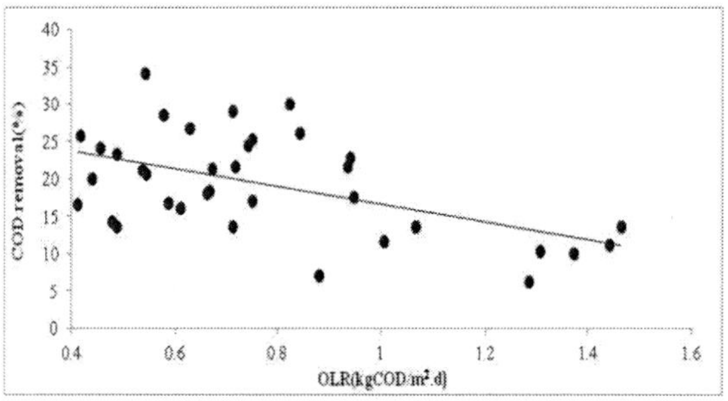

Fig. 4 Effect of varied OLR on COD removal

3.5 Vegetation development

The growth of vegetation was monitored during the study in terms of height of plant and root depth. The average initial and final root depth and height of plant are recorded. Plants grew well in BSFW compartments in terms of root depth and plant height. It was observed the pores in the brick-bats media facilitated spaces for deeper root penetration. Vegetation in FTW showed relatively lesser root growth because of no provision of media to support the root development. Compared to *Canna indica, Typha angustifolia* L. had shown high growth rate due to its dense and fibrous root matrix providing more surface area for attachment.

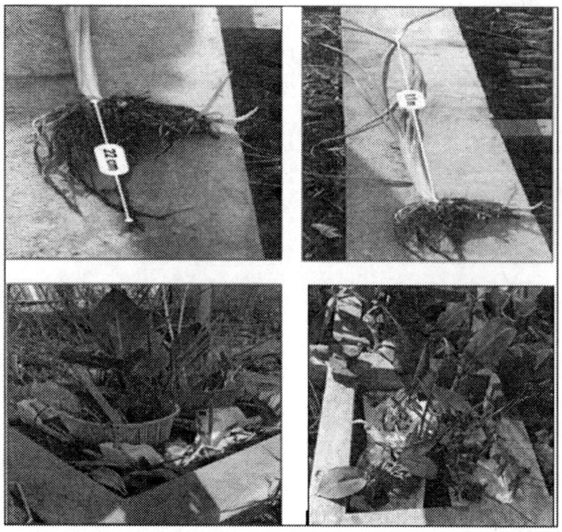

Fig. 5 Photographic view of vegetation with root and height measurement and development of vegetation in FTW at the end of study

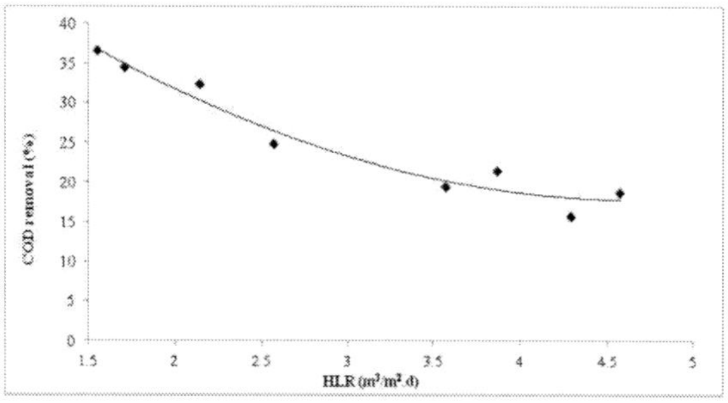

Fig. 6 Effect of controlled HLR on COD removal

4 Conclusions

The study was conducted on BFHCW for treatment of an anaerobically pretreated domestic wastewater. The system was operated under high diurnal variation of HLR and OLR. The performance was also assessed when operated under controlled flow conditions leading to uniform HLR and OLR. COD removal rate is increased by 5 to 15 % under controlled conditions. HLR is a dominating parameter compared to OLR and COD removal is significantly affected by varying HLR. BFHCW supported dual-species vegetation. BFHCW is found to be effective for COD removal to an extent of 25 to 40%.

The present study is a part of research project sanctioned by Department of Science and Technology DST) under Water Treatment Initiative (WTI) programme. The authors sincerely acknowledge and thank DST for providing financial support to undertake this study.

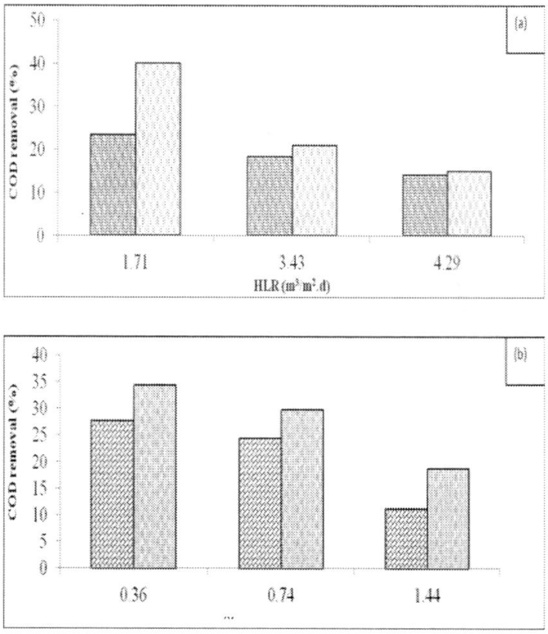

Fig. 7 Comparison of COD removal under controlled and uncontrolled conditions for a) varied HLR (b) varied OLR

References

1. J. Álvarez, C. Ávila, P. Otter,, R. Kilian, D. Istenič, M. Rolletschek, P. Molle, N. Khalil, I. Ameršek, Mishra, V. K., Jorgensen, C., Garfi, A., P. Carvalho, H. Brix and C. Arias, Water Sci & Tech. **76 (6)**, 1474–1489 (2017)

2. APHA, *Standard Methods for the Examination of Water and Wastewater, American Water Works Association*, Water Environment Federation (2012)

3. G. Dotro, G. Langergraber, P. Molle, J. Nivala, J. Puigagut, O. Stein and M. Von Sperling, IWA Publishing, London, UK, **7**, (2017)

4. N. Pavlineri, N. Skoulikidis, V. Tsihrintzis, Chem Engg J. **308**, 1120–1132 (2017)

Cost-Effective Housing Technology with Carbon Neutral Blocks

Narsing Rao Sudom[1*], *and Kishore Ravande*[2]
[1]Government Polytechnic Masabtank, Hyderabad, Telangana, India.
[2]MIT-ADT University, Pune, Maharastra, India.

Abstract. Conventional materials are considered to be major contributors of Green House Gas (GHG) emissions. New cost effective technologies for housing is, therefore, a need of the hour. To make mass housing cost effective in shortest possible time, each of the housing components has to be optimized for cost and production time. Disposal of fly ash is posing a great threat for its disposal problem. There is a dire necessity for alternate materials replacing conventional costly material with cost effective and environmental friendly materials making them energy efficient. This paper presents the technology and advantages of construction of cost effective and energy efficient construction using the developed Carbon Neutral Soil-Cement Fly Ash (SCF) blocks. Construction methods adopted were Arch foundation, Rat-Trap bond using SCF blocks for masonry, precast SCF block Lintel elements and partially precast SCF Slab Panels for flooring/roofing. Overall Cost to the extent of 42% can be saved using this cost effective construction technology over the conventional method. Energy of the buildingcan be drastically brought down from. 36.39 GJ (4.2T of CO^2) for conventional construction to 9.51 GJ (1.33T of CO^2) for cost effective construction. Energy to the extent of 74% can be saved using this Technology.

1 Introduction

Housing shortage is the major problem faced by the developing countries. Migration of rural people to urban areas has increased enormously and compelled them to live in unauthorized accommodation and slums. Nearly 38 million rural housing units are required for additional population during next decade [1]. Development of soil-cement-fly ash (SCF) blocks and all relevant building components using SCF blocks is an effective solution to meet the challenge of providing cost effective mass housing. Soil is the most abundant and cheaply available low enrgy material traditionally used for construction of houses. It is also known fact, that traditional bricks are manufactured from the locally available suitable soil followed by baking (burning) to develop the desired strength. While on one hand burning of bricks causes air pollution, on the other hand, millions of tones of unutilized fly ash produced by industries pose great threat to the environment. Thus, there is a need to overcome these environmental related challenges while meeting the demand for mass housing.

2 Need for the Present Work

Adequate Shelter is a basic human need, yet most of the urban population in developing countries still lives in spontaneous settlements as they cannot afford the high cost of building materials. There is lagging in Mass housing programmes due to high costs of conventional building materials. They are also high energy intensive and are becoming scarce giving rise to threats to mass housing. Disposal of waste materials of the industries is becoming a major problem. Hence the need of the hour is to replace conventional construction materials by alternate energy efficient materials with emphasis on alternate technologies. To meet the unprecedented demand for mass housing there is a huge requirement of building blocks. Burnt clay bricks pose threat to environment due to emission of carbon dioxide due to burning of bricks. It is estimated that 24% of green house gas (GHG) emissions [1] is contributed by the construction industry in India. Also, millions of tons of unutilized fly ash produced by industries pose great threat to environment. Thus, the need of the hour is to overcome these environmental related challenges by developing bricks/building blocks which are produced by using local soil, cement, fly ash and water curing without any baking or burning. The present paper disseminates developed SCF blocks which are in turn utilized in casting of cost-effective building components with an emphasis on optimizing construction time.

*Corresponding author: snarsingrao2006@gmail.com

3 Carbon Neutral Soil-Cement-Fly Ash (SCF) Blocks

New cost effective and energy efficient technologies for housing are essential to cope up with the ever increasing demand of housing and reduction in carbon dioxide emissions. Cement being the main component of construction activity, construction industry is considered to be the major contributor of Green House Gas emissions. Hence a block had been developed to overcome the above mentioned drawbacks and resulted in the development of Soil-Cement-Fly ash (SCF) blocks which are cost effective and energy efficient. This sustained research work has resulted in to a product with strength and durability, besides making it carbon neutral.

3.1 Method of Manufacture of SCF Blocks

The process of making these blocks are as under: Any locally available soil, other than black cotton soil, can be used for the production of masonry blocks with suitable modification in grading of soil to derive maximum density. Ideally, the particle size distribution in a soil sample should match the Fuller curve[2]. Hence, the soil proposed to be used in production of blocks was made up for the deficiencies in particle sizes so as to bring it close to the Fuller curve requirement. 50 to 85% of the locally available soil is reconstituted by blending and mixed with 1 to 5% cement and 5 to 50% of fly ash. Blending of different soils with fly ash was done to achieve the grading of blended soil which closely matches the Fuller's curve of minimum voids. Optimum fly ash content to be added in the mix was determined from experimental work.

The method of blending was computerized and adopted to arrive at the proportions of natural soils/manufactured fine aggregate and fly ash to be blended. Required quantity of SCF mixture was filled in the mould in three layers and the whole assembly of the mould was placed on the machine and was operated to effect the vibro-compaction to get maximum density by a vibro-compacting block making machine. The mould assembly was demounted and the SCF blocks were demoulded. Standard SCF block specimens of size 215x100x75mm were prepared and cured. The SCF blocks were then subjected to compressive strength test and durability test. The strength and Durability Characteristics of SCF Blocks are given in Table 1. There is significant increase in the ratio of wet to dry strength of SCF blocks and reduction in loss of their weight on wetting and drying test with increase in fly ash content, thus reflecting the improved durability of blocks.

This may be attributed to addition of fly ash which fills in micro pores and proper grading of blended soil resulting in increased density followed by strength and durability[3,4].

There was extensive studies carried out on energy values of SCF blocks. SCF blocks so developed are carbon neutral (viz. the energy value is zero) with addition of fly ash and cement, whereas the energy content of burnt clay brick is 4.35 MJ per brick.

Table 1. Strength and Durability Characteristics of SCF Blocks

Properties	Mean	Standard deviation	Coefficient of variation
Wet compressive strength (MPa)	5.22	0.68	13%
Water absorption (MPa)	11.92	0.75	6.27%
Durability on alternate wetting and drying (% reduction in weight)	0.35	0.028	8.07%

The cost of manufacture of each SCF block is worked out at Rs.4-00 and details are shown in Table 2. SCF bricks are 63% cheaper than conventional burnt clay bricks at prevailing market rate.

Table 2. Cost of Manufacture of SCF Brick

Description of Item	Amount (Rs.)	Description of Item	Amount (Rs.)
Soil	0.43	Total	2.27
Cement	0.85	Misc @ 28%	0.64
Fly ash	0.22	Profit @ 20%	0.45
Labour	0.67	Transport	0.60
Power	0.10	Final cost of each brick	3.96
Rounded 4.00*			

*63% cheaper than conventional burnt clay brick

3.1.1 Embodied energy of SCF blocks

Embodied energy content of SCF blocks dependent upon cement content and fly ash addition. The embodied energy due to addition of cement is 862.5MJ. In addition 21 MJ of energy is required for operating block making machine having 1 H.P. Transport of the soil to the casting yard consumes 67.5 MJ. Fly ash addition plays a dual role viz., becomes cementitious in the presence of excess lime and has energy saving value of 6.17 MJ per kg. Fly ash to the extent of five percent replaces cement and remaining five percent is taken as otherwise land filled. The energy values of fly ash replacing cement and land filling are -5.966 MJ/kg and 0.617 MJ/kg respectively as specified by EPA [5]. Considering the energy values due to energy intensive cement and energy savings associated with fly ash addition, the net energy value of the block is -0.036 MJ or -0.011 MJ/kg. Since there are no emissions of GHG with SCF block, the block is regarded as Carbon Neutral[6].

3.1.2 Abatement of CO2

If SCF blocks are replaced by burnt clay bricks to the extent 100 billion bricks, the CO_2 emission reductions would be 43.12 million Tons. In addition, the reduction in GHG emissions would be 7.12 million Tons due to better surface of wall requiring no plastering and thus resulting in further saving of cement.

4 Cost-Effectiveness for Sub- Structure

4.1 Arch Foundation

This foundation was constructed for the research work taking into consideration economy and energy efficiency (low carbon material). Arch foundations are adopted for soils of low bearing capacity with semi-circular or segmental curves so that the ends of arches rest on side abutments and they in turn supported on deeper hard stratum. Since these foundations are cheaper than conventional construction they are best suited for cost-effective housing in hard stratum at shallow depths.

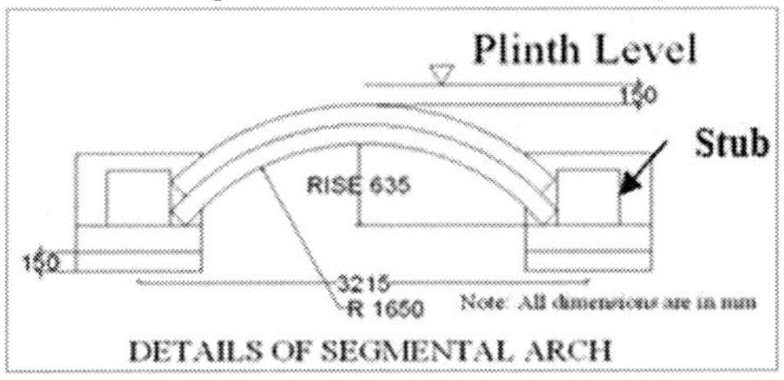

Fig. 1. Details of Arch Foundation

4.1.1 Arch Foundation Construction

Stubs are compression members to receive thrust from arches. After excavating a trench of size 1m x 1m x 1 m, P.C.C. of proportion (1:5:10) was laid at the bottom to a thickness of 150 mm and compacted well to make a levelled surface. First footing consisting of CRS in CM(1:6) and constructed to a thickness of 200 mm. The CRS stubs of size 500 mm x 500 mm were constructed to a height of 500 mm above first footing. Segmental arches using cement concrete blocks in CM (1:4) were constructed with a span of 2300 mm and a rise of 635 mm. Bottom profile of the arch was made with CRS masonry and top surface was made smooth to receive cement concrete blocks. To prevent moisture from entering in to masonry, R.C.C damp proof course of 75 mm thick was cast. Details of arch foundation are shown in Fig 1 and Fig 2 .

Fig. 2. Arch Foundation Construction in progress

4.2 Soil Cement Fly Ash (SCF) Mix Foundation

In case the SCF mix which is used for in situ work such as in foundation, there are no energy values associated with transportation and also no casting work takes place. On incorporating above two values, the net energy value of SCF mix would be -0.039 MJ/kg. Since there are no emissions of GHG with SCF mix, the mix also is regarded as Carbon Neutral.

4.2.1 SCF Mix Foundation Construction

This foundation is an alternative to cost-effective arch foundation. This type of foundation is similar to spread foundation, except that the excavated trench is filled and rammed with SCF mix. The soil which is excavated from the trench foundation is sieved and then measured at the same time on the side of the trench. Blending is resorted to as discussed in section 3.1.The mix should be calculated for 1 bag of cement per mix. A team of workers mix and ram manually or mechanically. Usually the top level of the foundation is at the level of the original ground. The section of the foundation should normally be square. Two-floor building section of foundation should be 75 cm x 75 cm. This mix is not only Carbon Neutral but also cost-effective.

5 Cost-Effective Housing Technology for Super Structure

5.1 Rat-trap Bond Masonry

In this bond of brick work, bricks are laid on edge with alternate rowlocks (brick on edge showing its breadth and height)and shiners (brick on edge showing its length and height) leaving 65 mm gap (215-2x75=65 mm) in between bricks. In this bond each alternate course begins with two rowlocks followed by a shiner.

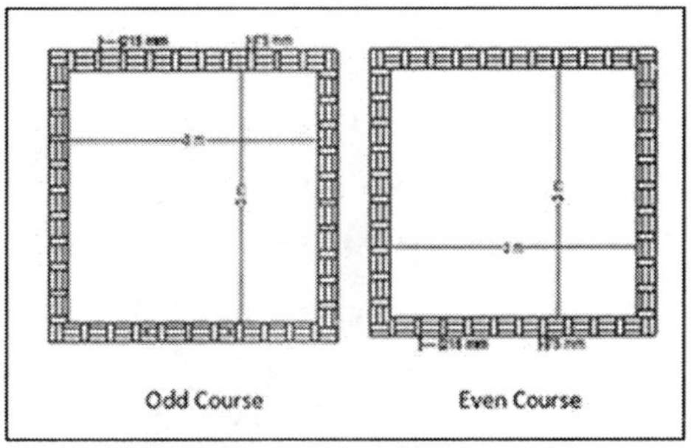

Fig 3 Plans of Odd & Even Courses in Rat-Trap Bond

The intermediate course begins with a shiner succeeded by a rowlock as shown in Fig. 3. Overall wall thickness of 215 can be maintained with around 21% cavities. Economy to the extent of 21% in bricks and 25% in cement mortar can be achieved (lesser number of joints).

This bond is simple to implement and can be employed for fast track construction and has thermal and acoustical comfort due to internal cavity. Plastering can be avoided since SCF blocks have sharp and straight edges that enhance beauty of the building. Details of construction of rat-trap bond masonry is shown Fig. 4.

5.2 Provision of Vertical Reinforcement in Rat-trap Bond

5.2.1 Quetta bond

This bond is adopted when it is necessary to provide vertical reinforcement in walls for areas subjected to earth tremors. Minimum thickness of wall in this case is one and half brick. With the adoption of this bond, quarter brick by half brick pockets, which are continuous through full height of the wall, are formed along the length of the wall. In these pockets steel rods are placed, and pockets are filled up with cement concrete as the work proceeds.

Fig. 4. Rattrap bond brick masonry

5.2.2 Vertical reinforcement bars

Bureau of Indian Standards (BIE) suggests provision of vertical reinforcement at corners of walls, junctions of walls, jambs of windows and jambs of doors. Code specifies use of Quetta bond for vertical reinforcement in walls for areas subjected to earth tremors. Since Quetta bond is employed for one and half thick brick wall , for one brick thick wall the Rattrap bond is best suited to provide vertical reinforcement in cavities (65 mm x 65 mm) of walls for areas subjected to earthquake.

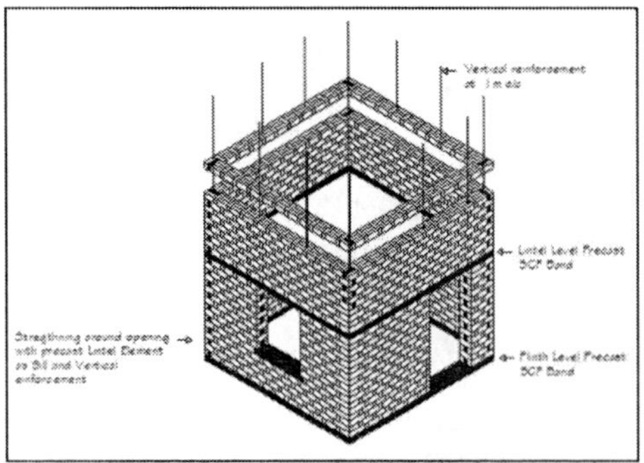

Fig. 5. Earthquake Provisions in SCF block Rattrap bond Masonry and Horizontal bands (DPC/Lintel/Sill)

5.2.3 Embodied Energy of Conventional Masonry Vis-À-Vis SCF Block Masonry

The embodied energy of burnt clay brick masonry is 2573.2 MJ/m^3 whereas it is 193.1 MJ/ m^3 with SCF block masonry. Embodied Energy of SCF block masonry drastically reduced due to Carbon Neutral block and is only 8% of the burnt clay brick masonry [7].

5.3 Precast SCF Lintel

As soon as the SCF block masonry construction reaches lintel level, precast lintels will be placed over the door and window openings and masonry work is further continued above the lintel level. Construction features of SCF lintels are similar to that of partially precast SCF Slab Panels. Fig 6 shows the details of lintel over door opening. This structural element can be used as Plinth band (as DPC), window Sill level element which becomes integral part of vertical reinforcement of openings and lintel level band for earthquake resistant design as shown in Fig. 5. This element is designed in parts of 1.0 m span and joined together.

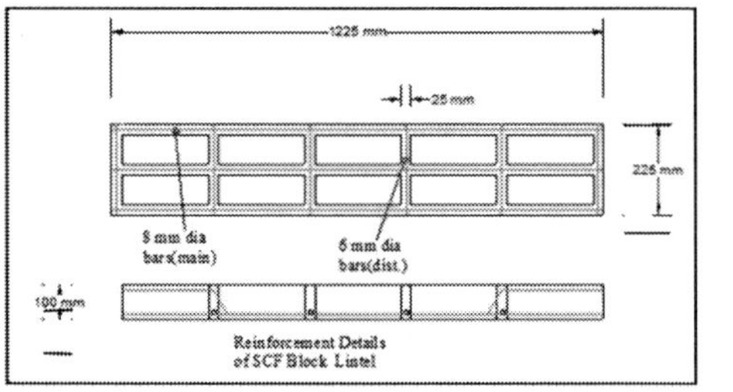

Fig. 6. Details Of Reinforced S.C.F Block Lintel of span 1225 mm

5.4 Partially Precast SCF Slab Panels for Floor/Roof

For casting purpose a wooden mould was prepared using wooden pieces of cross section dimensions of 75 mm x 50 mm to prepare the mould of internal dimensions of 1805 mm x 350 mm x 75 mm.

The wooden mould with an opening at the base was kept on a leveled ground having thin layer of sand. The bricks were wetted and arranged in the mould with a 50 mm gap between them. Steel reinforcement cage (8 mm diameter main reinforcement and 6 mm diameter distribution reinforcement) was inserted in the gap of assembly of bricks with a provision of 15 mm cover. M20 grade concrete was then poured in the gaps and finished flush with the bricks. After demoulding, the slab panel was cured for 14 days in water and 14 days in air. The required number of slab panels were made ready. Figure 9 shows arrangement of blocks and reinforcement in a slab panel. The required number of slab panels of dimensions 1805 mm x 350 mm x 75 mm were produced for construction of a assembled slab system over a room of size 3 m x 3 m and tested it for deflection recovery test (prototype testing).

Fig. 7. R.C Joist Supported over Bearing Walls

5.4.1 Partially precast RC Beam

A partially precast beam was cast to reduce span of SCF slab panel. A mild steel mould of 150 mm x 150 mm cross sectional dimensions and length equal to 3m (room width) was used for moulding a partially precast reinforced cement concrete joist. After placing the mould on a levelled surface, the reinforcement cage was placed in the mould with a provision of 25 mm cover. Steel reinforcement provided for the partially precast joist was designed as per the procedure laid down in IS 14142 – 1994[8]. The concrete was poured in the mould and compacted properly. After demoulding, the joist was cured for 14 days in water and for 14 days in air prior to placing it over SCF block masonry wall for assembly of slab panels as shown in fig.7.

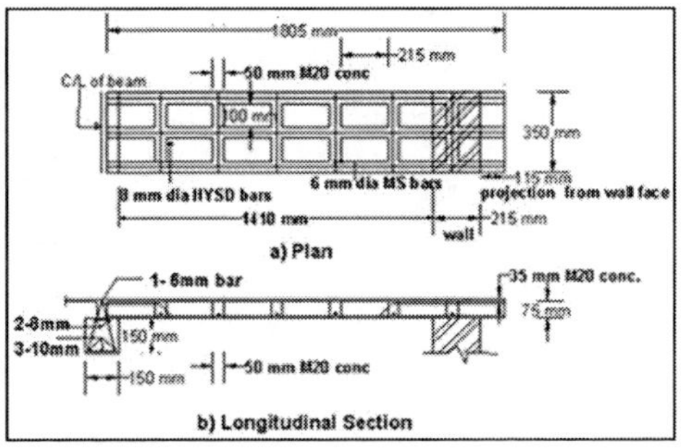

Fig. 8. Details of Reinforced S.C.F Block Slab Panel of length 1805 mm

5.4.2 Arrangement of SCF slab panels over the RC Joist and wall

The partially precast joist (beam) was placed over SCF block wall with a 15 mm thick cement-coarse sand (1:4) mortar beneath it. The alignment/ placement of joist was checked and then propped at one third span locations. The brick panels with overall size of 1805 mm x 350 mm x 75 mm , shown in Fig. 8, were placed over the joists/wall, side by side after laying a 6 mm thick layer of cement coarse sand (1:4) mortar over the joists/walls.

Arrangement of brick panels over masonry wall and precast joist (beam) is shown in Fig. 9. The gaps between the panels were filled up with rich mortar. Shrinkage and temperature reinforcement of 10 g x 10 g – 100 mm x 100 mm welded wire mesh was placed over the panels. A 35 mm thick screed concrete layer was poured on the top of the assembled slab panels. The roof was finished with a floating coat of 1:3 cement fine sand mortar of 6 mm thickness just after laying the in-situ concrete and cured for two weeks. The prefabricated SCF brick slab panel was designed as simply supported between a joist and a wall as per IS: 14142- 1994[8]. Extensive deflection recovery test was carried out by K. Ravande and N. Sudom [9] as specified in clause 17.6.3 of IS: 456-2000 [10] on the entire assembly of SCF slab panels and joist provided over a room of dimensions 3mx3m.

5.4.3 Embodied Energy of Conventional RCC Slab and SCF Slab Panels

The embodied energy of conventional RCC (Reinforced Cement Concrete) slab is worked out to be 272.32MJ/m3 whereas the embodied energy with SCF block slab panel [7] is 2539.47MJ/ m3. The energy content of SCF slab is 59% of the energy content of the RCC slab.

Fig. 9. Details of Reinforced S.C.F Block Slab panel supported on RC joist and wall

6 Cost Analysis

Cost comparison of the Cost-effective housing using CNB vis-a-vis the conventional construction is given in Table 3 and is evident that there is a saving of 42% by adopting CNB with partial prefabrication. It is evident from table 2 that the cost of each SCF Block worked out at Rs.4-00. For rate analysis, Telangana State Common Standard Schedule of rates of 2019-20 were considered. Area allowance of 40% in GHMC Limit, Hyderabad, Telangana was also taken into consideration. The plinth area rate of cost-effective construction using SCF blocks works out to be Rs.1268/sft or Rs. 13640/m^2 compared to conventional cost of Rs.2189/sft or Rs. 23554/. Saving to extent of 42% is achived by adopting cost effective construction with a plinth area of 126.6 sq.ft or 11.77 m^2. If benefitiary directly under takes construction of the residence, the plinth area rate works out to be Rs.600-00 per sft or Rs.6456-00 per m^2 without any agency.

7 Embodied Energy of Conventional Load Bearing and SCF Load Bearing Construction

The embodied energy of conventional brick load bearing with RCC slab is worked out to be 36.39 GJ for a plinth area of 16.6 m^2 whereas the embodied energy with SCF block load bearing using SCF slab panels is 9.51 GJ. Energy to the extent of 74% can be saved using SCF Block Technology in the construction of a house. Carbon dioxide emissions can be brought down from 4.2 T for a conventional construction to 1.33T for cost effective house using SCF Block Housing Technology [7].

7.1 Saving in Cement Compared to Conventional Construction

In mass housing programme with a plinth area of 16.6 m^2, the saving in cement with SCF blocks construction would be around 1150 kg compared to conventional construction. At national level considering shortage of housing in rural and urban areas at 20 million housing units the saving in cement would be 23 million Tons. The benefit arising out of saving in cement is Rs.115 billion ($1.65 billion). The saving in cost due to GHG emission reduction in terms of Carbon Credits is Rs.32 Billion ($ 460 million).

8 Time of Construction

With partial prefabrication and using precast units for DPC, Sill and Lintel Level the time of construction can be drastically reduced. With proper scheduling, SCF Block residence can be completed within 30 days.

9 Climate Change Mitigation and Sustain -able Development

GHG emission is the reason for global warming. Burnt clay bricks pose threat to environment due to emission of GHG due to burning of bricks. While construction industry cannot do away with the bricks or building blocks for meeting the housing demand, the concern for the environment, particularly embodied energy associated with the bricks and its reduction is the need of the hour. The alternate construction technologies play a vital role not only in cost reduction but also in the reduction of GHG emissions[7] and help in protection of environment which lead to sustainable development. The construction of houses using SCF block and housing elements would mitigate climate change considerably.

10 Conclusions

1. Carbon Neutral SCF blocks are alternative to energy intensive burnt clay bricks and are cost effective.
2. Construction using Carbon Neutral SCF blocks has in-built advantages to perform better against earthquake forces by providing SCF Plinth band, Sill, Lintel level band and vertical reinforcement in the masonry.
3. Construction of experimental prototype room disseminates the construction Technology and brings awareness among the end users and develop confidence to utilize the material and Technology.
4. Plastering can be completely avoided taking durability of the block into consideration. For smooth and elegancy of inside surfaces, luppum finishing to walls and plaster of paris to ceiling may be recommended instead of plastering since bricks are true and square.
5. The Construction Technology developed using Soil Cement Fly Ash blocks has resulted in significant Reduction of cost by 42% over conventional construction thus paving way for implementing mass housing projects for the benefit of poor homeless population.

References

1. Dev. Alt ,News Lr. N. Del. (2014) https://www.devalt.org/newsletter/may14/of_3.htm
2. G.Rothfuchs, Betonstrasse, **14**, 12, (1939)
3. N. Del, B.I.E, Specifications for Soil Based Blocks Used in General Building Construction, IS 1725, (1982)
4. K.Ravande, N. Sudom, P. REDECON, India, 139, (2007)
5. Anne Choate, Report of U.S. Environmental Protection Agency, (2005)
6. N.Sudom, K. Ravande, P. ISEC-6, Zurich, 21, (2011).
7. N.Sudom, K. Ravande, J. Adv. Mat. **R 935**, 197, (2014)
8. B I E, Design and Construction of Floors and Roofs with Prefabricated Brick Panel, N. Del. IS 14142, (1994)
9. K. Ravande, N. Sudom, P. SEWC,India, 197, (2007)
10. B.I.E, Code of Practice for Plain and Reinforced Concrete, N. Del. IS 456, (2000)

Table 3 Cost Comparison of Cost-effective Construction with Conventional Construction

S No.	Description of Work	Conventional Method			Cost Effective Method			Savings	Remarks
		Quantity	Rate	Amount	Quantity	Rate	Amount		
1	Earth work Excavation in all soils	13.16 m^3	323.00/m^3	4250.68	4.00 m^3	323.00/m^3	1292.00	2958.68	Conventional -Strip footing Cost Effective-Stub footings
2	PCC bed in (1:5:10) prop	1.31 m^3	5203.00/m^3	6815.93	0.40 m^3	5203.00/m^3	2081.20	4734.73	same as item (1)
3	a) CRS Masonry in CM (1:6)	8.37 m^3	5642.00/m^3	47223.54	1.95 m^3	5642.00/m^3	11001.90	26790.24	same as item (1)
	b) Arch base using CRS in CM(1:4)				0.55 m^3	9169.00/m^3	5042.95		Arch foundation using concrete blocks over CRS arch base
	c) Arch work with Cement blocks in CM(1:4)				0.55 m^3	7979.00/m^3	4388.45		Arch foundation using concrete blocks over CRS arch base
4	Filling in basement with moorum	3.60 m^3	667.00/m^3	2401.20	3.60 m^3	667.00/m^3	2401.20	nil	
5	Damp Proof Course	0.30 m^3	14432.00/m^3	4329.60	0.30 m^3	6549.00/m^3	1964.70	2364.90	Conventional -R.C.C Cost Eff-SCF R C Panels
6	Brick masonry in CM(1:6)in Rat-Trap bond using Soil-Cement flyash bricks	8.91 m^3	7217.00/m^3	64303.47	8.91m^3	5054.00/m^3	45031.14	19272.33	Conventional -Flemish bond with burnt clay bricks Cost Eff-SCF bricks using Rat-trap bond
7	Precast Reinforced S-C-F Lintels with M20 CC in joints	0.14 m^3	14574.00/m^3	2040.36	0.14 m^3	7261.00/m^3	1016.54	1023.82	Conventional -R.C.C Cost Eff-SCF R C Lintels
8	a) Precast RC SCF slab Panel	11.97 m^2	1182.00/m^2	14148.54	0.91 m^3	7261.00/m^3	6607.51	3345.71	Conventional -R.C.C Cost Eff-SCF R C Panels
	b)Screed concrete 35 mm thick				0.32 m^3	6848.00/m^3	2191.36		Screed concrete of 35 mm thick with welded wire mesh reinforcement
	c) welded wire mesh reinforcement				11.97 m^2	1000.00/10 m^2	1197.00		
	d) Partially precast RCC beam				0.08 m^3	10087.00/m^3	806.96		Additional central beam for cost effective construction
9	Steel Reinforcement Fe 250/415								Steel reinforcement required for Cost-effective construction is less since the slab panel effective lenth is 1.5 m.
	a) PRECAST SALB PANEL	79.01 kg	68205.00/T	5388.88	35.91 kg	68205.00/T	2449.24	2461.52	
	b) PRECAST lintel	9.76 kg	68205.00/T	665.68	9.76 kg	68205.00/T	665.68		
	c) partially precast beam				7.01 kg	68205.00/T	478.12		
10	Flooring with Shahbad stone rough slabs set in CM(1:8)	9.00 m^2	804.00/m^2	7236.00	9.00 m^2	804.00/m^2	7236.00	nil	
11	Doors	1.78 m^2	3000.00/m^2	5346.00	1.78 m^2	3000.00/m^2	5346.00	nil	
12	Windows	2.16 m^2	2500.00/m^2	5400.00	2.16 m^2	2500.00/m^2	5400.00	nil	
13	a) 20 mm thick plastering in CM(1:6) & CM(1:4) to walls	80.28 m^2	569.00/m^2	45679.32				41755.32	Conventional-Wall Plastering Cost-eff-Flush Pointing
	b) Flush Pointing in CM(1:4) for brick work				36.00 m^2	109.00/m^2	3924.00		
14	Plastering to ceiling 12 mm in CM(1:5)&CM(1:3)	9.00 m^2	554.00/m^2	4986.00				4986.00	Conventional-ceiling Plastering Cost-eff-No plastering
15	Water Proofing over RCC slab	11.97 m^2	577.00/m^2	6906.69				6906.69	For Cost effective construction water proofing is nor necessary as the screed layer with welded wire mesh offers resistance to wheathering
			TOTAL	227122		TOTAL	110522	116600.00	
Provision is made for Electrical, water supply and sanitary, painting and miscellaneous @22% on cost of conventional construction									

Future city - challenges and opportunities for water-sensitive sustainable cities, in India

Vandana Pusalkar[1*] *Vimala Swamy*[2] *and Anand Shivapur*[3]

1. Visvesvaraya Technological University, Belgavi, Karnataka, India
2. Reva University, Banglore, Karnataka, India.
3. Visvesvaraya Technological University, Belagavi, Karnataka, India

Abstract- Rapid urbanization, depleted resources, imbalance in urban hydrologic cycle, tremendous pressure on ecological functions, unhygienic and unhealthy environment is the scenario in almost all Indian cities. Water resources in the urban areas are fast depleting while the demand for the same is fast growing. Now more than ever before, water resources face threats from climate change, population growth, aging infrastructure, declining revenues, and a variety of other localized challenges. The challenge is to find ecologically sustainable, socially justified, economically viable, culturally transferable, and technological and managerial innovations which are appropriate to meet the needs of the future generations i.e. sustainable urban development. What is sustainable development? Detail study of sustainable development goals especially related to water, various recent trends of city development at major parts of world, from water perspective like Water- sensitive city, Water-wise city. Water-centric sustainable community can prove to be an ideal model of city development that will be truly sustainable with full focus on ecological balance so that social and economical sustainability also can be achieved.
Keywords: - Sustainable development, water- sensitive city, water- wise city, water centric sustainable community.

1 Introduction

Cities have multiplied and expanded rapidly worldwide over the past two centuries. Urban centers having world's economic activities, utilizes maximum natural resources and generate most waste and pollution (1). Since ancient times, cities started as walled villages or settlements on the bank of rivers, streams. Without water, life in the cities could not be sustained. Water also provides for cleaning and hygiene, transportation, irrigation of crops and gardens, defense and transportation (2). But, what is the urban water scenario all over the world? Especially in developing countries, like India. In India almost all rivers flowing through the cities are getting polluted and converted to drainage carrying corridors i.e. nallahs (polluted water stream).

As stated by almost all professionals like urban planners, urban designers, landscape designers, environment engineers and scientists that urban water infrastructure, its planning and management needs to be changed drastically (2). The conceptual models employed in most cities have not changed much since Roman times. They comprise rapid-conveyance piped water to carry away human and industrial wastes for disposal. Managing water quality, both for potable and disposal purposes, is generally accomplished by removing contaminants at the beginning or end of the pipe. Historical paradigm of municipal sewerage and drainage system was adopted for mainly two purposes, 1) Public health and hygiene and 2) Prevent flooding (3). One of the most pervasive challenges is the alteration of the urban water cycle (3)

1.1 Methodology

This review paper is to study, the consequences of urbanization, and other economic development activities on flowing water bodies in Indian cities. How this urban water scenario can be transformed to make them sustainable cities? In first part of the paper, an exhaustive background study regarding number of challenges and reasons behind this unsustainable situation due to polluted urban rivers is explored, and in second part of the paper, case studies have been done to study various approaches worldwide to restore the degraded urban rivers, as the problem is complex and multifaceted.

2 Challenges in water sector

*Corrosponding Author :- vandana.pusalkar@gmail.com

Unprecedented urbanization and industrialization imposing environmental degradation on such a vast scale, that results into polluted water bodies inside cities putting extreme pressures on urban local bodies to make available and manage safe and clean water supply to growing population (4). The challenge is to supply quality water as per C.P.C.B. (Central Pollution Control Board, India) standards to each and every individual as per their needs by keeping in mind ecological sustainability as a main concern (4).

2.1 Water: International, national scenario

The world water council says, " There is a water crisis today. But the crisis is not about having too little water to satisfy our needs. It is a crisis of poor water management that affects billions of people and the environment suffer badly". (5)

Out of total available water resources on Earth, 97.5% water is in oceans, 2.5% fresh water (1.5% water trapped in glaciers, 0.9 % ground water and 0.1% surface water) (6). Extensive use of water resources for various purposes like domestic, agricultural, industrial, etc. leads to water scarcity. (7). It is persistent due to deterioration of flowing (Rivers/Streams) or stagnant (lakes) water bodies as various pollutants mixed with fresh water. All professionals think that, water reuse is the only solution to reduce this water scarcity. Degraded quality of water as a result of impact of socio-economical systems is the main concern of urban local bodies at city level. Water scarcity is a quality problem rather than quantity (5). In Table:-1 Institutional challenges in water sector are catagorised with respect to urban water governance in India.

Table 1:- Institutional challenges and opportunities in water sector **(8)**

Sr.No	Institutional challenges and opportunities	
1	Planning and co-ordination	Planning is not co-ordinated. No inter-departmental collaborative innitiatives right from central level to local level. No long-term solutions and lacunas in implimentation of it in flexible manner.
2	Finance management	Lack of innovative financially viable solutions, conventional cost intensive centralised approach.Inefficient to manage the situation.
3	Acts, bye-laws, rules and regulations	key drivers how organizations works to strategize, plan, budget, and implement projects
4	Community participation	Ensure community participation and development of active citizen,s forum.
5	Culture/Knowledge/Capacity	Cultural changes needed with appropriate knowledge of urban water management is challenging and results can be unpredectable.

2.2 How climate change affects cities?

On one side, cities are playing main role in economic developments, but in due course of time on the other side human well being in cities is very neglected part. Now climate change will have direct or gradual impact on both these roles. The direct impact may be in the form of sudden shocks like, storms, typhoons, heat waves and floods or gradual impact such as sea level rise, temperature increase and whether changes, etc (9).

2.2.1 Resilience and Urban Ecosystem services

The capacity of water system to sustain and adopt to the adverse effects and retain its basic identity, function and performance is called resilience. (10). Same can be true to the context of cities. The capacity of cities to sustain and adopt to the adverse effects of environmental degradation, ecological crisis, water crisis and climate change, global warming, etc. and still try to retain its basic identity, function and performance is called resilient city. (11).

"Urban ecosystem services are generated in a diverse set of habitats, including: green spaces, such as parks, urban forests, cemeteries, vacant lots, gardens and yards, campus areas, landfills; and blue spaces, including streams, lakes, ponds, artificial swales, and storm water retention ponds" (12) .

2.2.2 Urban water and human health and cities (urban) health

Since historical times nobody denies the relationship of cities health and water resource quality. (13). Urban water management and Water health management plays a major role in protecting the health of urban populations worldwide (14). In 1984 the Healthy Cities movement originated in Toronto, Canada. To protect health of citizens, water health plays major role which is accepted and agreed by WHO (World Health Organization) and hence, promoted healthy cities movement worldwide since 1986.

3 Discussions

Impact posed by economic development activities on water systems are getting more and more complex today that creating chaotic condition in cities. As a result,

 https://doi.org/10.1051/e3sconf/202017006017

whatever actions taken against nature at local level has directly or indirectly adverse effects on the regional and global level. (15). To assess sustainable development goals related to water, needs knowledge of urban Hydrological Cycle and how its imbalance have repercussions both at local and global level (15).

3.1 Sustainable Development goals, 2015

Efforts to meet human water needs at local scales may cause negative environmental impact and stress on the water system at regional and global scales. Hence, assessing Sustainable Development Goals (SDGs) targets requires broad and in-depth knowledge of the global to local dynamics of water availability and use (16).

Superimposing the different elements of global change, the question thus arises of how sustainable human development can be ensured while safeguarding earth's vital life-support system on which the welfare of current and future generation depends? As stated in the first sentence of the World Water Development Report 2015, "water is at the core of sustainable development and the latter is strongly connected to the availability and access to sufficient quantity and quality of water for the preservation of healthy ecosystems and is critical for socio-economic and human development" (17). Sustainable Development Goals (SDG), SDG61and SDG112, are fully empasised on sustainable urban water management for resilient cities (18). To achieve this we need to harness the power of collaboration: a new paradigm for water management infrastructures and policies, enabled by new governance that spurs broader collaboration, motivates stakeholder engagement and active citizenship involvement as shown in table 2.

3.2 A sustainable water vision (Opportunities)

A sustainable urban water vision, as part of a resilient city vision, enables people to work together at different scales and across disciplines. It enables the political will needed to invest in long term measures. It provides consistency beyond political cycles. The International Water Association published final draft "Principles for water-wise cities". In this document, they determined to play a decisive role in making it possible, inspiring a shared vision of sustainable urban water as shown in table 3.

3.3 Water Sensitive Eco-cities

In Year 1896, Ebenezer Howard launched the garden city movement in his book 'Garden Cities of Tomorrow' in which he describes how cities can be planned with green open spaces to line with nature without much impacting it. (19). Cities thus have increasingly large ''ecological footprints'' Its definition was coined by Register (1987) as "a sustainable city, or eco-city is a city designed with consideration of environmental impact, inhabited by people dedicated to minimization of required inputs of energy, water and food, and waste output of heat, air pollution - CO_2, methane, and water pollution". Global warming, climate change and other environmental externalities of haphzard growth made compulsory to all countries to promote eco-cities and resilient cities (19). To develop sustainable water-centric eco-cities, water management different strategies at different level can be adopted as per table no. 4

Table 2:-The four level of action (8)

Level A. Regenerative Water Services for all:	Level B. Water Sensitive Urban Design:	Level C.Watershed management	Level D. Water-Wise Communities:
Rejuvenate water bodies and their ecosystems to protect the water sources.	Plan and implement urban design ensuring rejuvenating water services. cleaner waterways, benefiting ecosystems and people	Secure the water resource plan for drought mitigation strategies by sharing water for agriculture, industry and energy sectors	Citizens involved in the progressive urban water vision. opportunities (resource recovery, future resources, well-being).
Reduction in water and energy useage.	Design urban spaces to reduce flood risks.	Protect the quality of the water resource	Policy makers enable the implementation
Reuse and use diverse sources of water	to enable regenerative water services.	drinking water quality with minimal treatment	Trans-disciplinary planning
Recover energy from water Recycle nutrients or organic matter from water	Urban design strategies minimise their impact on water pollution:	Plan for extreme events, including flooding by managing flow regimes in rivers	Experts from various Professions
Increase modularity	Enhance liveability with visible water	Investing flood warning systems.	Water wise policies and financing
Use a systems approach integrated with other services	improving social and urban amenities by building green infrastructure	by maintaining adequate vegetation in the basin to minimise flash floods	Leaders provide the vision and a sound governance structure

3.4 Smart cities: A vision for development from water perspective, India

"Cities in the past were built on riverbanks, now they are built along highways. But in the future, they will be built based on availability of optical fiber networks and next-generation infrastructure." Mr. Narendra Modi

The 100 number of Smart Cities program declared by Govt. of India is to make cities smart in all respect. In Smart Cities, every citizen ensures quality of life along with all basic amenities and facilities. Economic social and environmental sustainability is the main agenda of development (20). The concept of smart city varies from one place to another. It depends on the level of development, willingness to change and reform, resources and aspirations of the city residents. The smart city should provide a clean and sustainable development considering the aspects of social, economic and environmental sustainability.

A smart city should be able to provide a better Quality of Life to its citizens, which includes safety and security for women, children and the elderly, inclusiveness, entertainment, access to public Utilities, cost efficient healthcare, quality education, transparency, accountability and opportunities for participation in governance (20).

3.5 Indian water scenario

In 2000, International Water Management Institute published, one survey of projected water scarcity in 2025 worldwide, which says more than 85% India, will be water scarce as the population is more than 125 million today (21). 4% water available for 17% world's population. Unavailability of water, overuse and exploitation of water resources create trouble to people as water is life, without which survival is impossible. In Indian context the crisis of water is viewed as crisis of availability of water.

In urban India, Urban water management is a significant challenge in front of govt. authorities, as due to unavailability of water and sanitation infrastructure, waste water is often laid in open drains leading to unhygienic condition and environmental pollution of rivers and streams. Public officials believe that environmental problems are solved by rules and regulations. Storm water rules and regulations in India, are borrowed from USA/ EU and not tailored to fit the local environment.

3.6 The current water paradigm in India

The current paradigm of treating all forms of rainwater, storm water, ground water and wastewater separately leading to polluted water bodies in most of the cities. Mithi river at Mumbai, Musi river at Hyderabad, Ganga-Yamuna at North India, Panchganga, Krishna and Jayanti at Kolhapur and so many other rivers in India, most of them are polluted and water is not safe for drinking purpose as per C.P.C.B. (Central Pollution Control Board, India) (3). This The recent fifth paradigm of sustainable urban water management (all forms of water like rainwater, storm water, ground water, wastewater) supports the importance of river ecosystems and agrees rivers are the lifelines of cities. In this paradigm micro level green infrastructure linked to macro level water shed management to balance urban hydrological cycle. To deliver sustainable urban water five major aspects like, resilient vision, urban water governance, knowledge capacities, planning tools and implementation tools as explained in Table No. 3. (3).

Table 3 Five building blocks to deliver sustainable urban water (8)

Vision	Governance	Knowledge/ Capacities	Planning Tools	Implementation Tools
Resilient city vision enables people to work together for greater benefit of urban community.	Governance and institutions provide the framework for urban stakeholders to work together	Urban stakeholders have knowledge and capacities at different scales within the city.	Asset management, master plans decision support systems to initiate action.	Regulations based on quality assurance, transparency, accountability
A shared vision implementation of new policies	Intégration water in the city services and design.	Through a sustainable water "vision"	identifying benefits and co- benefits of projects,	Financing tools to adopt more efficient solutions
A sustainable urban water vision	incentives for urban stakeholders	sustainable urban water successes,	defining levels of service	circular economy mechanisms,
It provides consistency beyond political cycles.	Maximizing the benefits of water to cities.	The ability to pool resources, share tools	ensuring ownership by stakeholders	Integrated services with investment cycle,

It is also very often economically viable, even under prevailing economic models, provided that the multiple services and all their associated benefits for the large number of beneficiaries in cities are properly quantified and recognized. Such information is essential to include in decision-making processes related to land use and management in urban landscapes, and to help guide urban and landscape planners, architects, restoration practitioners, and public policy makers, as well as private and institutional stakeholders.

4 Results and recommendations

Always there are tensions between preservation and management, nature and culture, tradition and technology, theory and practice. Prolonged and immediate effect of population explosion and unprecedented urbanization decision makers always faces new challenges to provide all basic infrastructure services and to connect them at city level basic planning, management, implementation and monitoring of all water systems. The challenges may resemble same wherever you go, throughout India, irrespective of different cultures, politics, institutional frameworks etc.

As rivers are lifelines of the cities, new paradigm of sustainable water centric eco-cities are based entirely on role of watercondition inside the cities. The urban eco-region concept is depending upon sustainability of urban water cycle. The wastewater (gray water, black water) management and solid waste management and the prime most parts of city functioning and cannot be neglected for River's ecological health (2). While thinking of a development paradigm form cities resilience point of view, it is important to fill the gaps of knowledge fields like urban design, urban planning, sociology and ecology. Urban nexus of water – food – energy is always depending upon what kind of human and nature relationship persistent at city level. Recent rapid expansion of urban areas on one side putting tremendous pressure on environment and eco systems, but on the other side providing fundamental challenges and opportunities to design more water sensitive eco-cities (2), as per table 4.

Table 4 :- Sustainable water centric eco-cities (2)

Water	The focus of sustainable cities movement
Micro scale	Green buildings.
Subdivision	Eco-blocks
Macro scale	Eco-cities and ecologically engineered urban water sheds.
Urban-eco	Region concept, Sustainable urban water cycle.
Waste water	Used water provides energy in the form of electricity, biogas, hydrogen, fertilizer, raw materials for reuse and heat. Used water , Reclaimed water, Resource recovery
hybrid	Partially decentralized or fully decentralized water/storm water/used water system

5 Conclusion

Geographer Matthew Gandy writes – "Water is not simply material element in the production of cities but is also a critical dimension in the social production of space."

The calls for "Water centric sustainable communities," "Cities of the Future," "Sustainable blue-green urbanism" may sound today more like futuristic dreams than a potential reality (2). But the future is always an extension of history and change is forced, and guided, by new incremental discoveries and stresses. To solve multi-aspect problem, multi-disciplinary rather trans- disciplinary research is absolute necessary (2). The time has come to look for and implement new concepts for urban planning and landscape design. In the "Cities of the Future" context, one has to look to, and learn from, the past about the importance of water in cities (2). Resilient cities able to manage all the consequences and ill effects of rapid urbanization and industrialization also bring many new opportunities in front of future urban planners. Water sensitive sustainable cities based on low impact development strategies also promotes social interaction and green economy along with ecological restoration of rivers.

This research was supported by Visvesvaraya Technological University, Jnana Sangama, and Belagavi 590018 for Grant of financial assistance.

References

1. J. Anthony. "The urban environment and health in a world." *Bulletin of the World Health Organization*. Vol. **78 (9)** (2000).
2. V. Novotny. *WATER CENTRIC*. United States of America : JOHN WILEY & SONS, INC. ISBN 978-0-470-47608-6 (cloth); ISBN 978-0-470-64282-5 (ebk); ISBN 978-0-470-64283-2 (ebk);(2010)
3. V. Novotny. *A New Paradigm of Sustainable Urban Drainage and Water*. Oxford University : Oxford Roundtable Workshop on Sustainability, (2008).
4. *Paris statement*. United Nations, (2007).
5. *Metadata on Suggested Indicators for Global Monitoring*. UN-Water(2016).
6. *Consolidated Technical Input from UN Agencies*. (2015).

7. E. Rietveld. *Improving health in cities through systems.* Manchester, UK : Environmental Health (2016).
8. *Principles for urban stakeholders to develop a shared vision and act towards sustainable urban water in resilient and liveable cities.* The IWA.
9. D. Satterthwaite, S. Huq, M. Pelling, A. Reid, Romero-. *Adapting to climate change in.* London (UK) : Human Settlements Discussion Paper Series –, (2007).
10. C. Folke. *Resilience: the emergence of a perspective for social-ecological.*Global Environmental Change, Vols. **16 (3)**, pp. 253–267. (2006)
11. G. Barnett, X. Bai. *Urban resilience research.* Stockholm : Stockholm University, (2007).
12. T. Elmqvist, H. Setala , S. Handel, S. Ploeg, B. Elsevier. *Benefits of restoring ecosystem services in urban areas.*, Current Opinion in Environmental Sustainability, Vol. **14**, pp. 101–108, 22nd May (2015).
13. D. Koutsoyiannis, N. Zarkadoulas, A. Angelakis, G Tchobanoglous. *Urban* : Water, Vol. **134**, pp. 45–54 (2008).
14. D . Grey, C . Sadoff. *Sink or Swim? Water security for growth and* Water Policy. Vol.**9**, pp 545 (2007).
15. C. Vörösmarty, J. Hoekstra, S. Bunn , D . Conway, and J. Gupta, *Fresh water goes global.* Vol. Science **349**, pp. 478–479.
16. *Achieving Sustainable Development.* A. Bhaduri, J . Bogardi, A . Siddiqi, Front. Environ.Sci, Vol. **4**, pp. 64. (2016),
17. *Theunitednationsworldwaterdevelopmentreport:.* S.l. : UNESCO-WWAP, (2015).
18. *Principles for urban stakeholders to develop a shared vision and act towards sustainable urban water in resilient and liveable cities.* THE IWA.
19. S. Pickett, M. Cadenasso. *"Linking ecological and built components of urban mosaics: an open cycle of ecological design."*. Journal of Ecology, , Vol. **96(1)**, pp. Pp. 8-12 (2008).
20.V.Upadhyaya , *Smart Cities: A Vision for Development of Indian Cities,* Imperial Journal of Interdisciplinary Research (IJIR) , Vol. **2.** ISSN: 2454-1362. (2016)
21. *Water scarcity.* S.l. : International water management institute, (2000).
22. D. Schaffer, D. Vollmer. *Pathways to urban sustainability,* Washington D.C : National Academies Press .

Effect of Glass Fibers on Self Compacting Concrete

Sandeep L. Hake[1], *S. S. Shinde*[1], *Piyush K. Bhandari*[1], *P. R. Awasarmal*[2], *B. D. Kanawade*[1]
[1] *Department of Civil Engineering, DVVP College of Engineering, Ahmednagar, Maharashtra (India).*
[2] *Dept. of Civil Engineering, G.S. Mandal's Maharashtra Institute of Technology, Aurangabad, Maharashtra (India).*

Abstract. Self Compacting Concrete (SCC) is a specially developed concrete for concreting under extreme condition of inaccessibility from heights. It is capable to flow under influence of its own weight. It could be used when encountered with dense reinforcement and complex structural design. Problem of segregation as well as bleeding is eliminated and vibration is not required for compaction. As concrete is strong in compression and weak in tension. Hence to make it strong in tension, discontinuous Anti-Crack high dispersion glass fibers are added. SCC mix prepared with addition of discontinuous glass fibers is called as Glass Fiber reinforced Self Compacting Concrete (GFRSCC). In this paper an experimental study has been carried out to check the effect of Anti-Crack high dispersion glass fibers on the compressive strength, split tensile strength and flexural strength of SCC. The result show that, as compared to the Normal SCC, the compressive strength of GFRSCC increases by 2.80% and 12.42%, the split tensile strength of GFRSCC increases by 4.47% and 25.12% and the flexural strength of SCC increases by 6.57% and 14.34% when the Cem-FIL Anti-Crack HD glass fibers were added as 0.25% and 0.50% respectively by the weight of total cementitious material contents. The addition of 0.25% Cem-FIL Anti-Crack HD glass fibers to SCC has not much affect on the workability of Normal SCC. Whereas, addition of 0.50% Cem-FIL Anti-Crack HD glass fibers reduces the workability of SCC.

1 Introduction

Today Self Compacting concrete (SCC) has proven to be a big milestone in the construction industry because of its advantages. It can pass easily through congested reinforcement and the gap between formwork and steel reinforcement, also capable to fill the formwork without any voids. It is common experience that, dense reinforcement causes problems during pouring and compaction of concrete. Therefore, concrete must be capable enough to pass through the dense reinforcement arrangement without blocking or segregating. Also poor placement and the lack of compaction can lead to the inclusion of voids and loss of long term durability of concrete structures. To avoid this concrete capable enough to flow by itself and filling the spaces should be used. And this need has been solved by introduction of SCC. It is superior over normal vibrated concrete as it eliminates requirement of vibration for compaction of concrete, thus the reduction in noise at construction site is achieved. It also gives better surface finish. Due to high workability, it is easier to place SCC, thereby reducing the number of workers on site. It increases construction speed leading to economy. Since SCC requires more amount of binder and to achieve economy, cement is replaced in percentage by cement replacement materials like fly ash, GGBS, silica fume, etc [1, 2, 3].

In 1980's, Japan encountered shortage of construction workers. To overcome this problem, Okamura from Japan proposed necessity of SSC as an effective solution for labour intensive concreting work. SSC was superior to normal cement concrete in terms of properties in fresh state as filling ability, ease in passing and segregation resistance. To prepare SSC mix, cement paste content has to be increase, coarse aggregate content and water-powder ratio must be reduced. To compensate these changes suitable admixture, preferably Superplasticizer is added [2]. Prepared SSC mixture is checked as per EFNARC guidelines for fresh concrete properties. Tests like Slump flow test, T_{50} slump flow, L-box test, V-funnel test and V- Funnel at $T_{5minutes}$ are

* Corresponding author: **drsandiphake@gmail.com**

performed to comment on workability of SSC mix [3].

Concrete is considered strong in compression by assessing its high compressive strength. But at the same time it is weak in tension and shows brittle nature. Hence fibre reinforced composite (FRC) were incorporated to solve this issue within concrete [4]. Different types of fibers like steel, glass, asbestos, polypropylene etc are already used in concrete to get fiber reinforced concrete. Addition of fiber helps to arrest the development of cracks and improve modulus of elasticity of composite material. Since load carrying capacity is improved, tensile strength also found to have increased. Fibers helped to improve toughness, tensile and flexural strength also reduce cracks and shrinkage in concrete.

Present investigation was carried to study outcomes of glass fiber addition in normal SSC mix. Discontinuous Cem-FIL Anti-Crack High Dispersion glass fibers were added in percentage by weight of cementitious material to normal SCC mix. Past investigation showed glass fibers addition increases ductility of mix [5, 6]. Also reduction in crack length and total cracked area observed due to early age shrinkage in composite mix. By performing workability test, performance of fresh prepared GFRSCC is investigated. Properties of hardened concrete upon fiber addition must provide glimpse about performance of GFRSCC mix in long run. Plot of load versus deflection values during flexural strength test will provide behaviour of beam specimen cast from normal SCC and GFRSCC under applied point load. Aim is to investigate the effect of glass fibers addition within normal SSC mix.

2Literature Review

Concept and inception of SCC was done in 1986 by Okamura in Japan. Since then number of researchers has contributed in enhancing properties of SCC. Noticeable amongst them were studied and their findings are considering for improving workability as well as strength properties. Addition of fibers also contributed in increasing the compressive, split tensile strength and flexural strength of SCC.

Nan Su et al. [5] developed mix design method which was found to be more economical and easy to use as compared to conventional Japanese mix design method. The Packing Factor (PF) parameter was adopted which has influenced workability, durability and strength of SCC mix. Author adopted PF values in gradual decreasing order from 1.18, 1.16, 1.14 and 1.12. On reducing PF value in SCC mix compressive strength was found to be enhanced from 27.5 to 48 MPa. Requirement of cement binder was reduced upto 424 kg/m^3 as compared to previous requirement of 500 kg/m^3. Even workability and durability was improved by using optimum content of Packing Factor.

J. K. Su et al. [7] suggested SCC mix by varying sand ratio (S/A) in increasing order from 0.3, 0.4, 0.45, 0.475, 0.5, 0.525, and 0.55. It showed increase in workability upon increase in S/A ratio. Also rheological properties were enhanced. Optimum S/A ratio recommended were upto 47.5 %. Investigation also revealed that S/A ratio has negligible effect on elastic modulus of SCC.

T. Suresh Babu et al. [8] prepared glass fibre reinforced self compacting concrete (GFRSCC) by adding 600 gm/m^3 of Cem-FIL Anti-Crack High Dispersion glass fibers along with suitable admixtures. SSC mix was prepared as per Nan-Su method mix design steps and Okamura's guidelines. By adding fibers, prepared GFRSCC showed increase in compressive strength, split tensile strength and flexural strength by 2.0 to 5.5%, 3.0 to 7.0 % and 11.0 to 20.0 % respectively. Ductility, Energy absorption capacity, Secant modulus of modified SCC was also improved.

M Chandrasekhar et al. [9] prepared M50 grade GFRSCC by adding glass fibers within normal SCC. By implementing different percentage of confinement in the form of hoop steel, showed increase in compressive strength from 7.25% to 44.30%. Also values of Young's modulus increased owing due to confinement.

3 Materials and Experimental work

3.1 Materials

The ingredients used for making the SCC and GFRSCC are as follows.

3.1.1 Cement: The Ordinary Portland Cement (OPC) of 53 grade (Ambuja Cement) having specific gravity of 3.15 was used in study. It was tested to find physical properties in accordance with Indian Standard specifications.

3.1.2 Fly Ash: Locally available fly ash confirming to IS: 3812 (Part-1) -2013 [10].

3.1.3 GGBS: GGBS confirming to IS: 12089-1987 having the surface area about 350 to 450 m^2/ kg was used [11].

3.1.4 Fine Aggregate: Natural River Sand, clear from all sorts of organic impurities was used. The fine aggregate passing through 4.75 mm sieve and having specific gravity of 2.70 and confirming zone II were used.

3.1.5 Coarse Aggregate: Crushed basalt angular aggregates passing from 12.5 mm sieve and confirming to IS: 383-2016 and having specific gravity of 2.74 was used [12].

3.1.6 Water: Potable water is generally considered satisfactory for mixing concrete. Water free from acids, oils, alkalis, vegetables or other organic impurities is preferable for investigation. Ordinary tap water was used for preparing concrete mix.

3.1.7 Superplasticizer: The Superplasticizer Conplast SP 430 confirming to IS: 9103-1999 having specific gravity of 1.1 to 1.2 was used in this study [13].

3.1.8 Viscosity Modifying Agent: VMA-Mastermatrix 2 (Glenium Stream 2) having specific gravity of 1.19 was used in this study.

3.1.9 Glass Fibers: Cem-FIL Anti-Crack High Dispersion glass fibers were used for this experimental study. Table 1 provides information related to physical properties of fibers used in the investigation

Later activator solution added at gradual rate during continuous mixing to get desired mix.

Table 1. Properties of Cem-FIL Anti-Crack High Dispersion Glass Fibers

Sr. No.	Property	Typical Value
1	Number of Fibers	214 millions/kg
2	Aspect Ratio	857: 1
3	Specific Surface Area	150 m^2/kg
4	Tensile Strength	1700 MPa
5	Modulus of Elasticity	72 GPa
6	Corrosion Resistance	Very High
7	Specific Gravity	2.68
8	Density	26 kN/m^3
9	Filament Diameters	14 microns
10	Filament Length	12 mm

3.2. Mix Design Procedure for Normal SCC and GFRSCC

As per literature study, in previous investigation SCC mix was prepared without performing mix design [14]. Whereas, in present study, ACI 237-2015 guidelines and IS-10262:2019, sec. 4, both were used to prepare mix design [15, 16]. The proportion of materials required for M30 grade normal SCC mix is given in table 2.

Table 2. Ingredients of SCC for M30 Grade.

Sr. No	Ingredients of SCC	Proportion by Weight of Cement	Quantity in (kg/m^3)
1	Cement	1	325.68
2	Fly Ash	0.3571	116.315
3	GGBS	0.0714	23.263
4	Fine Aggregate	3.1066	1011.77
5	Coarse Aggregate	2.8421	794.96
6	Water	0.5428	176.8
7	Superplasticizer	0.0357	11.63
8	VMA	0.0049	1.62

3.3 Mix Design Procedure and fiber dosage

For making GFRSCC, the Cem-FIL Anti-Crack High Dispersion glass fibers were added in two different percentages as 0.25% and 0.50% by the weight of total cementitious materials content. Prepared modified SCC mix used to study the fresh and hardened properties of SCC. Workability of fresh prepared 0.25% GFRSCC was assessed using tests like, Slump flow by Abram's cone, T_{50cm} slump flow, L-box test, V-funnel test and V-funnel at $T_{5\ minutes}$ test. The flow tests values were satisfying as per EFNARC guidelines. Since mix design data for the 0.25% GFRSCC was similar to Normal SCC mix design data, the design was adopted for further assessment. In case of 0.50% GFRSCC, it was observed that workability is affected in some extent due to addition of 0.50% glass fibers. In order to achieve workability requirements, the Superplasticizer dose was increased from 2.5% to 2.55% keeping the VMA dose constant. The 0.50% GFRSCC was prepared using new mix design data and again workability tests were carried out. This new design data was then adopted for further assessment. Table 3 denotes the quantity of ingredients used for GFRSCC for different percentage combinations.

Table 3. Mix Design Data for Different GFRSCC Mixes of M30 grade

Sr. No	GFRSCC Contents	Unit	Quantity	
			For 0.25% GFRSCC	For 0.50% GFRSCC
1	Cement	kg/m^3	325.68	325.68
2	Fly Ash	kg/m^3	116.315	116.315
3	GGBS	kg/m^3	23.263	23.263
4	Fine Aggregate	kg/m^3	1011.77	1011.77
5	Coarse Aggregate	kg/m^3	794.96	794.96
6	Water	$lit./m^3$	176.8	176.8
7	Super-plasticizer	kg/m^3	11.63	11.86
8	VMA	kg/m^3	1.62	1.62
9	Cem-FIL Anti-Crack HD Glass Fibers	kg/m^3	1.163	2.326

4 Results and Discussions

The workability tests were carried on the Normal SCC and GFRSCC mixes. The cubes, cylinders and beams were cast for Normal SCC and GFRSCC mixes and were tested for the compressive strength, split tensile strength and flexural strength respectively.

4.1. Fresh Properties of Normal SCC and GFRSCC Mixes.

The workability test were performed on the fresh Normal SCC and GFRSCC. The test results for flow values are given in Table 4. It was evident from flow test that there is decrease in workability with addition of glass fibers. In case of 0.50% GFRSCC mix, workability was improved by adjusting the dosage of Superplasticizer. Even on assessing T_{50cm} time, shows more time is required for spreading of fresh concrete. Thus, fiber addition reduces workability and makes concrete more stiff. Flow value in L-box test showed decrease in workability on addition of fibers. V-funnel test was performed on normal as well as GFRSCC. For GFRSCC, with increase in percentages of fibers as 0.25% and 50%, more time was required for concrete to flow out of funnel. $T_{5minutes}$ measured in seconds showed same nature for GFRSCC mix.

Table 4. Fresh Properties of Normal SCC and GFRSCC Mixes

Test	Mix Designation		
	Normal SCC	0.25% GFRSCC	0.50% GFRSCC
1)Slump Flow Test			
a) Horizontal Flow in mm	689	681	676
b)T_{50cm} time in sec	3	4	4
2) L-Box Test	0.86	0.85	0.84
3) V-Funnel Test			
a)Time for complete discharge in sec	8	9	9
b)T_5 time in sec	9	10	11

4.2 Hardened Properties of SCC and GFRSCC

The hardened concrete properties of Normal SCC and GFRSCC mixes such as compressive strength, split tensile strength and flexural strength were studied in this work. The effect of addition of Cem-FIL Anti-Crack HD glass fibers on the hardened properties of Normal SCC and GFRSCC were investigated.

4.2.1 Compressive Strength

The compressive strength test was performed on the standard cubes made of the Normal SCC and GFRSCC mixes under CTM [17]. Ultimate load upon failure was recorded and corresponding compressive strength was calculated. Fig. 1 shows compressive strength values of Normal SCC and glass fibers added GFRSCC mixes. It indicates increase in strength upon addition of Cem-FIL Anti-Crack High Dispersion glass fibers in normal SCC. For 0.25% addition, the compressive strength was increased by 2.8% whereas, 0.50% addition showed 12.42% increase in strength as compared to normal SCC.

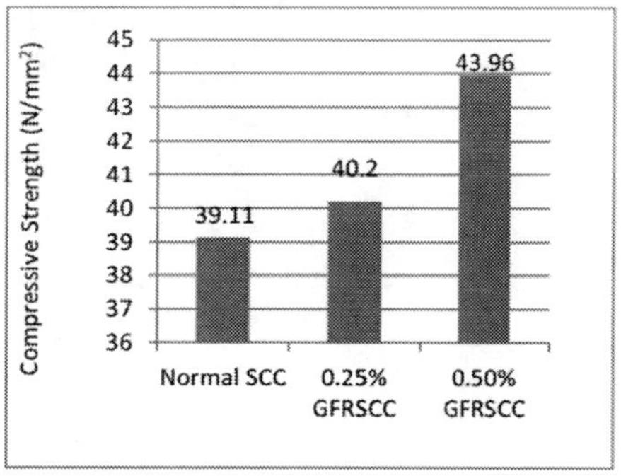

Fig. 1 Compressive Strength Vs % of Cem-FIL Anti-Crack HD Glass Fibers

4.2.2 Split Tensile Strength

The split tensile strength test was performed on the cylindrical specimen cast from Normal SCC and 0.25%, 0.50% fiber added GFRSCC [18]. It was observed that upon addition of 0.25% and 0.50% Cem-FIL Anti-Crack High Dispersion glass fibers to SCC, theres is increase in split tensile strength by 4.47% and 12.25% respectively. Fig. 2 shows tensile strength values of fiber added GFRSCC mix being more than normal SSC mix.

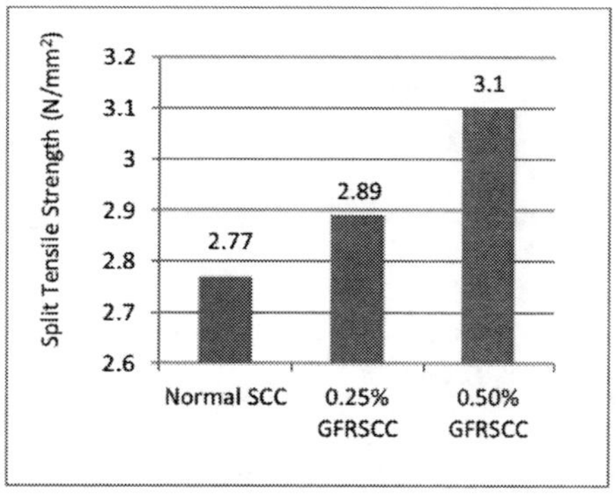

Fig. 2 Split Tensile Strength Vs % of Cem-FIL Anti-Crack HD Glass Fibers

4.2.3 Flexural Strength

Standard beam specimens were cast to perform flexural strength test for normal SCC and GFRSCC mix. Fig. 3 indicates increase in flexural strength upon addition of Cem-FIL Anti-Crack High Dispersion glass fibers. For 0.25% fibers addition, strength increased by 6.57% and for 0.50% fibers addition, showed maximum increase in flexural strength by 14.34%. Thus, glass fibers addition has made enhancement in hardened properties of SCC mix.

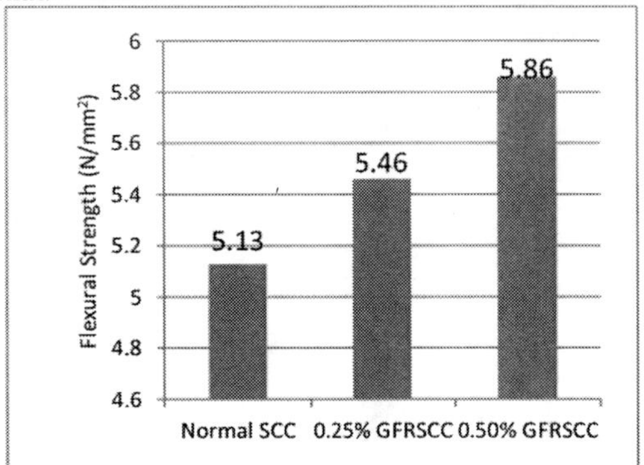

Fig. 3 Flexural Strength Vs % of Cem-FIL Anti-Crack HD Glass Fibers

4.2.4 Load Vs Deflection values

During flexural strength test, the ultimate load corresponding to failure of beam was recorded for normal SCC and GFRSCC beams. The deflection values under applied load were recorded with help of dial gauge. Fig. 4 gives plot of load Vs deflection for normal SCC as well as GFRSCC mix consisting of 0.25% and 0.50% fibres addition.

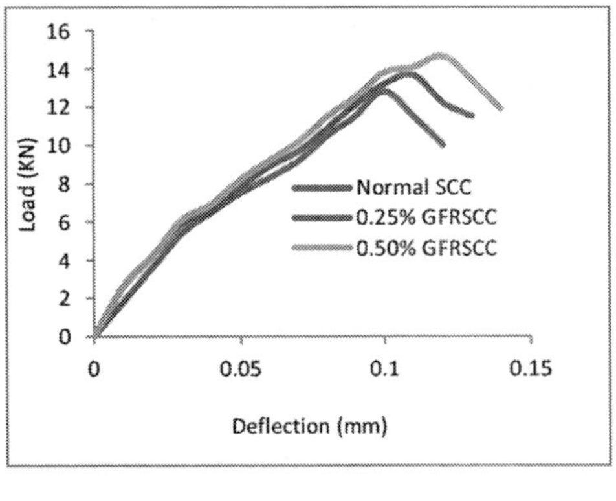

Fig. 4 Load Vs Deflection for Normal SCC Beams

In case of normal SCC mix ultimate load value was found to be 12.824 KN which denoted failure of beam. Thereafter applied load value has decreased and wide cracks were observed at beam bottom. After the addition of Cem-FIL Anti-Crack High Dispersion glass fibers, the load carrying capacity of beams increases compared to Normal SCC beams. Propagation of cracks was observed in these beams which mean the beams shows ductile failure. Also the load carrying capacity of 0.50% GFRSCC beams was more than 0.25% GFRSCC beams and Normal SCC beams. Propagation of cracks observed in these beams denotes same ductile failure pattern.

5 Conclusion

From past literature investigation and based upon experimental work, normal SCC and Cem-FIL Anti-Crack High dispersion glass fibers added as 0.25% and 0.50% by weight of total cementitious material contents in GFRSCC mix were prepared. Both mix were investigated for workability test as well as for hardened properties. Following conclusions are drawn based upon results obtained in above investigation:

1. Workability of SCC mix depends upon flow value of fresh concrete mix. Upon addition of glass fiber as 0.25% and 0.50 % by weight, the workability was found to decrease due to presence of fibers, making concrete mix stiff and difficult to flow. Extra dosage of Superplasticizer is required to make concrete flow through mould while testing.
2. The compressive strength of GFRSCC increases by 2.80% and 12.42% due to addition of glass fiber as 0.25% and 0.50 % by weight, as compared to the compressive strength of Normal SCC. This indicates presence of fibers increase toughness of concrete. Hence preferable for SCC mix.
3. The split tensile strength of GFRSCC increased by 4.47% and 25.12% due to addition of glass fiber as 0.25% and 0.50 % by weight, as compared to the split tensile strength of Normal SCC. Fiber addition has reduced the crack propagation in concrete making it more durable and sustainable for tensile stresses.
4. The flexural strength of SCC increases by 6.57% and 14.34% due to addition of glass fiber as 0.25% and 0.50 % by weight, as compared to the flexural strength of Normal SCC.
5. Load-deflection curve plotted based upon flexural strength test deflection values indicates, concrete failure was gradual under applied load. Fibers addition helps to make concrete ductile and gives prior warning before failure of member.
6. For better and homogeneous mixing of Cem-FIL Anti-Crack High Dispersion glass fibers in the SCC, fibers were added in the wet mix after the addition

of all ingredients of normal SCC along with Superplasticizer.

References

1. H. Okamura, *Concrete International*, **1**, 4, pp. 378-383, (1998).
2. T.R. Naik and S. Singh, *ACI Materials Journal*, **94**, 5, pp. 355-360, (1997).
3. H.G. Russell, *Concrete International*, **19**, 8, pp. 62-63, (1997)
4. S. Prashanth, K.M. Subbaya, K. Nithin and S. Sachhidananda, *Journal of Material Sciences & Engineering*, **6**, 3, (2017).
5. N. Su, K-C Hsu and HW Chai, *Cement and Concrete Research*, **31**, pp. 1799-1807, (2001).
6. J.K. Su, S.W. Cho, C.C. Yang and R. Huang, *Journal of Marine Science and Technology*, **10**,1, pp. 8-13, (2002)
7. T.S. Babu, M.V.S. Rao and D.R. Seshu, *Asian Journal of Civil Engineering*, **9**, 5, pp. 457-472, (2008).
8. M. Chandrasekhar, M.V.S. Rao and M. Janardhana, *Proceedings of International Conference on Advances in Materials and Techniques for Infrastructure Development, India*, pp. 28-30, (2011).
9. S. L. Hake, R. M Damgir, *Global Journal Of Civil Engineering, Knowvel Publication*, **01**, 1-5, (2019).
10. S. L. Hake, S. A. Urphate, P. R. Awasarmal, R. M Damgir, *Global Journal Of Civil Engineering*, **01**, 12-15, (2019).
11. Sandeep L. Hake P. R. Awasarmal R. M. Damgir *Global Journal Of Material Science & Engineering*, **01**, 13-17 , (2019).
12. Sandeep L. Hake, Dr R. M. Damgir, Dr S.V. Patankar, *Advances in Civil Engineering Hindawi Publication*, (2018).
13. The European Guidelines for Self-Compacting Concrete, EFNARC, (2005).
14. ACI Committee 237: *American Concrete Institute, Denver, CO*, (2015).
15. IS-10262:2019, Concrete mix proportioning-Guidelines, *Bureau of Indian Standards, New Delhi*, (2019).
16. IS-456:2000, Plain and Reinforced Concrete-Code of Practice, *Bureau of Indian Standards, New Delhi*, (2000).
17. IS-9103: 1999, Concrete Admixtures-Specification, *Bureau of Indian Standards, New Delhi*, (1999)

AUTHOR INDEX

AUTHOR INDEX

AUTHOR INDEX